Antimutagenesis and Anticarcinogenesis Mechanisms

BASIC LIFE SCIENCES

Alexander Hollaender, General Editor

Council for Research Planning in Biological Sciences, Inc., Washington, D.C.

Recent volumes in the series:

Volume 29 SISTER CHROMATID EXCHANGES: 25 Years of Experimental Research
Part A: The Nature of SCEs
Part B: Genetic Toxicology and Human Studies
Edited by Raymond R. Tice and Alexander Hollaender

Volume 30 PLASMIDS IN BACTERIA
Edited by Donald R. Helinski, Stanley N. Cohen, Don B. Clewell, David A. Jackson, and Alexander Hollaender

Volume 31 GENETIC CONSEQUENCES OF NUCLEOTIDE POOL IMBALANCE
Edited by Frederick J. de Serres

Volume 32 TISSUE CULTURE IN FORESTRY AND AGRICULTURE
Edited by Randolph R. Henke, Karen W. Hughes, Milton J. Constantin, and Alexander Hollaender

Volume 33 ASSESSMENT OF RISK FROM LOW-LEVEL EXPOSURE TO RADIATION AND CHEMICALS: A Critical Overview
Edited by Avril D. Woodhead, Claire J. Shellabarger, Virginia Pond, and Alexander Hollaender

Volume 34 BASIC AND APPLIED MUTAGENESIS: With Special Reference to Agricultural Chemicals in Developing Countries
Edited by Amir Muhammed and R. C. von Borstel

Volume 35 MOLECULAR BIOLOGY OF AGING
Edited by Avril D. Woodhead, Anthony D. Blackett, and Alexander Hollaender

Volume 36 ANEUPLOIDY: Etiology and Mechanisms
Edited by Vicki L. Dellarco, Peter E. Voytek, and Alexander Hollaender

Volume 37 GENETIC ENGINEERING OF ANIMALS: An Agricultural Perspective
Edited by J. Warren Evans and Alexander Hollaender

Volume 38 MECHANISMS OF DNA DAMAGE AND REPAIR: Implications for Carcinogenesis and Risk Assessment
Edited by Michael G. Simic, Lawrence Grossman, and Arthur C. Upton

Volume 39 ANTIMUTAGENESIS AND ANTICARCINOGENESIS MECHANISMS
Edited by Delbert M. Shankel, Philip E. Hartman, Tsuneo Kada, and Alexander Hollaender

---

A Continuation Order Plan is available for this series. A continuation order will bring delivery of each new volume immediately upon publication. Volumes are billed only upon actual shipment. For further information please contact the publisher.

Antimutagenesis and Anticarcinogenesis Mechanisms

Edited by
Delbert M. Shankel
University of Kansas
Lawrence, Kansas

Philip E. Hartman
Johns Hopkins University
Baltimore, Maryland

Tsuneo Kada
National Institute of Genetics
Mishima, Japan

and
Alexander Hollaender
Council for Research Planning in Biological Sciences, Inc.
Washington, D.C.

Technical Editors
Claire M. Wilson
and
Gregory Kuny
Council for Research Planning in Biological Sciences, Inc.
Washington, D.C.

PLENUM PRESS • NEW YORK AND LONDON

Library of Congress Cataloging in Publication Data

International Conference on Mechanisms of Antimutagenesis and Anticarcinogenesis (1985: University of Kansas)
Antimutagenesis and anticarcinogenesis mechanisms.
(Basic Life Sciences; v. 39)
"Proceedings of the International Conference on Mechanisms of Antimutagenesis and Anticarcinogenesis, held October 6–10, 1985, at the University of Kansas, Lawerence, Kansas"—T.p. verso.
Includes bibliographies and index.
1. Carcinogens—Antagonists—Congresses. 2. Mutagens—Antagonists—Congresses. 3. Carcinogenicity—Congresses. I. Shankel, Delbert M. II. Title. III. Series. [DNLM: 1. Carcinogens—antagonists & inhibitors—congresses. 2. Mutagens—antagonists & inhibitors—congresses. W3 BA255 v. 39 / QZ 202 I5677 1985]
RC268.6.I6 1985 616.99′4071 86-15070
ISBN 0-306-42375-8

The opinions expressed herein reflect the views of the authors, and mention of any trade names or commercial products does not necessarily constitute endorsement by the funding sources.

Proceedings of the International Conference on Mechanisms of Antimutagenesis and Anticarcinogenesis, held October 6–10, 1985, at the University of Kansas, Lawrence, Kansas

A Division of Plenum Publishing Corporation
233 Spring Street, New York, N.Y. 10013

Printed in the United States of America

DEDICATION

The editors gratefully and respectfully dedicate this Volume to Dr. Charlotte Auerbach and Dr. Takashi Sugimura.

"Lotte" Auerbach was born in Germany, the grand-daughter and daughter of distinguished scientists. In 1933, she left Germany and completed her Ph.D. in the Institute of Animal Genetics at the University of Edinburgh where she remained until her official "retirement" in 1969. In the early 1940s, along with Dr. Robson, she studied the first known drastic chemical mutagens--the sulfur- and nitrogen-mustards. She is widely regarded as the "discoverer" of chemical mutagenesis for these studies. (See also the almost simultaneous reports on chemical mutagenesis by Oelkers and by Rapoport.) She is the author of numerous articles and books, has served as the editor of many significant journals, and has been a constant source of ideas and stimulation for innumerable colleagues working in the field of mutagenesis.

Takashi Sugimura was born in Tokyo and received his doctoral degrees (M.D. and Doctor of Medical Science) from the University of Tokyo. He has served in numerous positions at the National Cancer Center in Japan where he now serves as President. He, too, is the author of numerous scientific articles; has held significant visiting lectureships in Japan, Europe, and the U.S.; and has served on many editorial boards and as editor for significant journals and books in his areas of scientific expertise. He is widely known for his numerous studies on mechanisms of carcinogenesis, and for his work on nutrition and cancer which has helped to open up the fields of antimutagenesis and anticarcinogenesis.

For their pioneering scientific contributions, for the stimulation and encouragement provided by Drs. Auerbach and Sugimura, and for the leadership they have demonstrated in expanding our scientific horizons, we dedicate this Volume to these superb colleagues and friends.

ACKNOWLEDGEMENTS

For a conference of this nature to be successful requires the combined efforts of many organizations and individuals. Listed below, you will find the membership of the Scientific Organizing Committee and the Local Arrangements Committee; their contributions are obviously important. Also listed are those federal, state, and private organizations or companies that have provided the essential funding for the conference. Without their support, the conference would not have been possible; we are deeply grateful for that support. Less obvious, but also critical, has been the splendid secretarial and logistic support provided by the secretarial staffs in the Office of the Chancellor, the Department of Microbiology, and the Division of Continuing Education at the University of Kansas. The cheerful dedication of these groups is thankfully acknowledged.

More specifically, we are indebted to Dr. Tsuneo Kada for the suggestion that this conference should be held; to each of the major speakers for excellent presentations in areas of their special expertise and for the preparation of manuscripts; to the session chairs for presiding and preparing introductory comments on the major subject areas of the conference; to Marilyn Long and her staff in the Kansas University Division of Continuing Education for careful and thoughtful coordination of logistical details; to Claire Wilson and her production staff in the Council for Research Planning in Biological Sciences in Washington, D.C. and the staff at Plenum Press who have joined to publish these Proceedings rapidly and efficiently; and to John Naughtin and the staff in the Kansas University Office of University Relations for the design of the conference logo and other special assistance.

Major funding for the conference was provided by the following federal agencies and other organizations:

U.S. Environmental Protection Agency (Grant No. D007258-01-0 and Order No. 5D4863NAEX)

Argonne Universities Association Trust Fund

Association for the Propagation of Genetics, Iden-fukyu-kai, Japan

National Cancer Institute (Grant No. 1 R13 CA39971)

National Science Foundation (Grant No. DMB 8419850)

The University of Kansas:

- Offices of the Chancellor and Executive Vice-Chancellors
- Office of Academic Affairs
- Medical Center
- Division of Continuing Education

The Monsanto Company
The Upjohn Company

Additional support was provided by Marion Laboratories, Inc., Eli Lilly and Co., Revlon Health Care Group, the Lawrence Chamber of Commerce, the Lawrence Holiday Inn-Holidome, the Lawrence Automobile Dealers' Association, Allen Press (Lawrence), University State Bank (Lawrence), and K.C. Biological.

Members of the Scientific Organizing Committee were Alexander Hollaender, Colin H. Clarke, Barry Glickman, Philip Hartman, Tsuneo Kada, Thomas Matney, Earle Nestmann, H.F. Stich, Graham C. Walker, and Delbert M. Shankel. David Brusick assisted members of the committee in the development of the final program.

Members of the Local Arrangements Committee were Judy Billings, C.C. Cheng, Kurt Ebner, Morris Faiman, Takeru Higuchi, Paul Kitos, Marilyn Long, Carol Shankel, George Stewart, and Delbert M. Shankel.

To all of the above, and to all of those who participated, we express our gratitude.

Delbert M. Shankel
Philip E. Hartman
Tsuneo Kada
Alexander Hollaender

CONTENTS

NATURAL ENVIRONMENTAL ANTIMUTAGENS

MECHANISM OF ACTION OF ANTIMUTAGENS AND ANTICARCINOGENS

AVOIDANCE OF ERRORS IN DNA

THE FIXATION OF ERRORS

INTRODUCTION

Delbert M. Shankel

Departments of Microbiology and Biochemistry
The University of Kansas
Lawrence, Kansas 66045

Welcome to the "International Conference on Mechanisms of Antimutagenesis and Anticarcinogenesis." We are delighted that so many of you have chosen to attend this first meeting on this important topic.

The significance of genetic changes in cells has been recognized for many years. The seminal observations of Henri in 1914 (UV), Muller in 1927 (X-rays), and Auerbach in 1946 (chemical agents) established the fact that physical and chemical agents which may be present in our environment are capable of producing profound changes in heredity. It is now well-established, of course, that such changes can result in the development of cancer, produce hereditary birth defects, alter microorganisms to cause drug resistance, or other harmful (or even beneficial) changes; it is likely that the processes of mutagenesis and the intricate balance between mutagenesis and antimutagenesis are involved in aging, evolution, and other fundamental life processes. Consequently, we hope and believe that assembling this group of scientists to share current fundamental and applied research in these areas will lead to a better understanding of these processes and to long-term benefits for society.

As stated clearly by Garfield (4), "Almost every aspect of modern living exposes us to health risks. The air we breathe, the food we eat, the water we drink, the drugs we take, and the places where we work may be contaminated by toxic substances or additives. This is in addition to natural carcinogens such as sunshine, and habits such as smoking, or the way our food is cooked ... Their effects may include cancer and organ damage ... Other unfortunate effects are suffered by future generations--genetic mutations, birth defects, inherited diseases and so on."

In fact, in the past decade there has been a tremendous rebirth of interest in mutagenesis, particularly as it is related to carcinogenesis, environmental toxicology, and other health-related problems. There has not been, however, a concomitant development of interest in <u>antimutagenesis</u>--those cellular treatments or conditions that <u>reduce</u> the frequency of mutations.

The original published observations on antimutagenesis were made by Novick and Szilard in 1951 using their elegantly designed chemostat experiments. Growth of interest in this field proceeded slowly, however, and in 1975 it was possible to cover two decades of research on antimutagenesis in microbial systems in a single review article. A variety of chemical compounds had been reported to act as antimutagens against several types of spontaneous or induced mutations in a range of microorganisms. Such compounds included spermine, adenosine, guanosine, caffeine, quinacrine, acridines, phenothiazine tranquilizers, dibenzocycloheptene antidepressants, warfarin, methionine, histidine, and manganous ions. In addition, states of genic repression had been shown to affect mutagenesis, and some classes of repair-deficient mutants, e.g., uvr, lex, and rec, showed reduced or no mutagenesis by some mutagens. These aspects of antimutagenesis and antimutagens were reviewed thoroughly (3). In concluding that review, Clarke and Shankel stated, "At a pragmatic level it would be most useful if antimutagens should be found which strongly counteract the mutagenic activities of at least some environmental mutagens ... It is tempting to speculate regarding whether they might also have some practical use in cancer chemotherapy or prevention" (3).

Since that review was completed, a number of other "antimutagens" have been identified, including human placental extract (9), mammalian placental extracts from a variety of species (10), human saliva and its components (12), some naturally occurring plant phenols such as ellagic acid (17), human or mouse plasma (2), factors present in vegetables, and compounds as chemically diverse as cobaltous chloride (7,8,11) and cinnamaldehyde (13). In addition to these known <u>antimutagens</u>, certain kinds of compounds have now been described as "desmutagens," i.e., compounds that actually destroy the mutagenic agent. These include compounds such as ascorbic acid and cysteine (14) and extracts from cabbages and other plants (6). In short, there has been a tremendous expansion in the identification of antimutagenic and desmutagenic factors in recent years. In spite of this expansion, however, our knowledge of the <u>mechanisms</u> by which these factors affect the frequency of genetic changes is still extremely limited.

Sugimura (15) has discussed the occurrence of mutagens, carcinogens, and "tumor promoters" in our daily food. He pointed out the occurrence of a number of mutagens in foods. Compounds such as AF-2 (a nitrofuran derivative), nitrosamines, 4-nitroquinoline 1-oxide (4NQO), aflatoxin B, and many other compounds are either mutagenic or are "promutagens" which are readily converted to mutagens by metabolic processes. He also suggested that there are now some substances "known to suppress various steps of carcinogenic process," and cited as examples the inhibition of formation of nitrosamines by vitamin C, the inhibition by naturally occurring flavonoids of metabolic activation of procarcinogens such as benzo(α)pyrene, and the inactivation by vitamin E and sulfhydryl compounds (antioxidants) of reactants "derived from procarcinogens/promutagens."

In a thoughtful review entitled "Rethinking the Environmental Causation of Human Cancer," Higginson (5) discussed a number of the hypotheses on the causation of human cancer. Cancer caused by well-defined exogenous factors, cancer of <u>probable</u> environmental origin, and the roles of ambient environmental pollution and life-style (including diet) are all evaluated. Among his conclusions is the following: "It is becoming increasingly clear that combinations of modulating endogenous and exogenous factors may explain the influence of lifestyle based on identifiable biochemical mechanisms. This may eventually allow the possibility of tumor prevention by interference with other stages of carcinogenesis through chemoprevention, use of cancer inhibitors and dietary changes." Clearly, in addition to its inherent scientific and intellectual interest, the study of antimutagenesis and its mechanisms can be related to the study of mechanisms of anticarcinogenesis.

It is now widely recognized that naturally occurring substances, particularly those present in the food consumed by man, are one of the most promising sources of inhibitors of carcinogenesis and other deleterious effects of environmental agents. Epidemiologic studies also indicate that, since marked differences in cancer rates are observed among population groups in different geographic areas and between immigrants and people in their native countries, natural products and dietary practices are logically among the most promising areas for exploration. In his recent review, Ames (1) emphasized the role of oxygen radicals and lipid peroxidation in carcinogenesis and pointed out that many of the plant-derived anticarcinogenic and antimutagenic substances could operate by antioxidant mechanisms. There are natural cellular mechanisms by which host cells protect themselves from overenthusiastic oxidative reaction to cellular insult. These include the action of superoxide dismutase, glutathione peroxidase, DT-diaphorase, glutathione transferases, and so on. Dietary factors playing a supplementary protective role include vitamin E, β-carotene, selenium, glutathione, ascorbic acid, uric acid, and a variety of miscellaneous substances less well-known. In another recent review by Wattenberg (16) entitled "Inhibition of Neoplasia by Minor Dietary Constituents," a concise chart was provided in which the metabolism of carcinogens to ultimate carcinogenic species that react with target sites, as well as points at which protective mechanisms can intervene, were depicted as shown in Fig. 1.

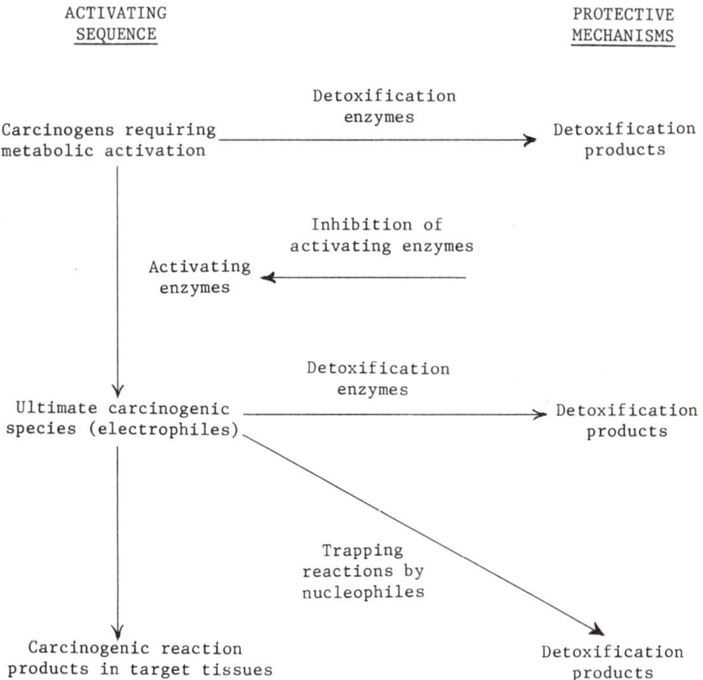

Fig. 1. Steps involved in metabolic activation of carcinogens requiring activation and some mechanisms for blocking activation.

As our knowledge of the interlocking roles of mutational changes in genes and the development of cancers has increased, it has become increasingly clear that antimutagenesis and desmutagenesis could have significant roles in lowering cancer incidence. It is also clear, of course, that increased knowledge of the mechanisms of antimutagenesis will be potentially useful in addressing other problems related to the rapid increase in numbers and types of chemicals in our environment, and will be inherently interesting for its contribution to our fundamental knowledge of genetic processes. In addition, an understanding of the mechanisms by which antimutagens and known and potential anticarcinogens act should enable the scientific community to develop improved means for dealing with all of the consequences, both harmful and beneficial, of the large and growing number of mutagens.

The major objectives of this conference, then, are: (a) to bring together scientists working at the molecular/mechanistic level in the fields of antimutagenesis and anticarcinogenesis in order to enhance knowledge in these fields; (b) to rapidly disseminate a thorough, current summary of a promising research area directed at a preventive rather than a curative stage of genetically related illnesses; and (c) to establish productive directions for future research in this field of growing interest and importance.

REFERENCES

1. Ames, B.N. (1983) Dietary carcinogens and anticarcinogens: Oxygen radicals and degenerative diseases. Science 221:1256-1264.
2. Aukerman, S.L., R.B. Brundett, J. Hilton, and P.E. Hartman (1983) Effect of plasma and carboxylesterase on the stability, mutagenicity and DNA cross-linking activity of some direct-acting N-nitroso compounds. Cancer Res. 43:175-181.
3. Clarke, C.H., and D.M. Shankel (1975) Antimutagenesis in microbial systems. Bacteriol. Rev. 39:33-56.
4. Garfield, E. (1982) Risk analysis, Part 2. How we evaluate the health risks of toxic substances in the environment. Current Contents 35: 5-11.
5. Higginson, J. (1981) Rethinking the environmental causation of human cancer. Food Cosmet. Toxicol. 19:539-548.
6. Inoue, T., K. Morita, and T. Kada (1981) Purification and properties of a plant desmutagenic factor for the mutagenic principle of tryptophan pyrolysate. Agric. Biol. Chem. 45(2):343-353.
7. Inoue, T., Y. Ohta, Y. Sadaie, and T. Kada (1981) Effect of cobaltous chloride on spontaneous mutation induction in a Bacillus subtilis mutator strain. Mutat. Res. 91:41-45.
8. Kada, T., and N. Kanematsu (1978) Reduction of N-methyl-N^{1}-nitro-N-nitrosoguanidine induced mutations by cobalt chloride in Escherichia coli. Proc. Japan Acad. 54:234-237.
9. Kada, T., and H. Mochizuki (1981) Antimutagenic activities of human placental extract on ultraviolet-light and gamma-ray-induced mutations in Escherichia coli. Radiat. Res. 22:297-302.
10. Mochizuki, H., and T. Kada (1982) Antimutagenic action of mammalian placental extracts on mutations induced in Escherichia coli by UV-radiation, X-rays and N-methyl-N^{1}-nitro-N-nitrosoguanidine. Mutat. Res. 95:457-474.
11. Mochizuki, H., and T. Kada (1982) Antimutagenic action of cobaltous chloride on Trp-P-1 induced mutations in Salmonella typhimurium TA98 and TA1538. Mutat. Res. 95:145-157.

12. Nishioka, H., and K. Nishi (1981) Effect of human saliva and its components on mutagenicity of carcinogens. Science Engr. Revs. Doshisha Univ. 22:124-133.
13. Ohta, T., K. Watanabe, M. Moriya, and Y. Shirasu (1983) Antimutagenic effects of cinnamaldehyde on chemical mutagenesis in *Escherichia coli*. Mutat. Res. 107:219-227.
14. Osawa, T., H. Ishibashi, M. Namiki, and T. Kada (1980) Desmutagenic actions of ascorbic acid and cysteine on a new pyrrole mutagen formed by the reaction between food additives: Sorbic acid and sodium nitrite. Biochem. Biophys. Res. Commun. 95:835-841.
15. Sugimura, T. (1982) Mutagens, carcinogens, and tumor promoters in our daily food. Cancer 49:1970-1984.
16. Wattenberg, L.W. (1983) Inhibition of neoplasia by minor dietary constituents. Cancer Res. 43(Suppl.):2448-2459.
17. Wood, A.W., M.-T. Huang, R.L. Chang, H.L. Newmark, R.E. Lehr, H. Yagi, J.M. Sayer, D.M. Jerina, and A.H. Conney (1982) Inhibition of the mutagenicity of bay-region diol epoxides of polycyclic aromatic hydrocarbons by naturally-occurring plant phenols: Exceptional activity of ellagic acid. Proc. Natl. Acad. Sci., USA 79:5513-5517.

CARCINOGENS AND ANTICARCINOGENS

Bruce N. Ames

Department of Biochemistry
University of California
Berkeley, California 94720

INTRODUCTION

There is the potential for major progress in the prevention of cancer in the next 5 or 10 years. In the short term, progress in the prevention of cancer depends on a better understanding of some of the risk factors and antirisk factors for the major human cancers. In the longer term, the spectacular progress in molecular biology is in understanding some of the fundamental aspects of the nature of cancer which should contribute substantially to further intervention possibilities.

The major types of cancer in humans should, in principle, be largely preventable either by avoidance of tobacco or by modification of infective, dietary, nutritional, or hormonal factors (2,3,54). Evidence for this view comes from epidemiological comparisons of the age-specific onset rates of various types of cancer in different peoples of the world and from studies on migrants. For example, colon, breast, and stomach cancers are major human cancers with widely different rates in different human populations, e.g., Japanese vs Americans. The argument that these are preventable comes from studies on migrants, suggesting that children of Japanese immigrants to the United States become more like Americans in their cancer rates. Thus the enormously high Japanese stomach cancer rate could presumably be lowered to the very low American stomach cancer rate if we understood the causative agent(s) in the Japanese diet [or the protective agent(s) in the American diet]. Similarly, Americans could presumably lower their colon and breast cancer rates if we could identify those dietary or other factors that are chiefly responsible for the difference between the very high American rates and the very low Japanese rates.

Human cancer appears to be mostly multicausal, with both risk factors and antirisk factors involved. Evidence suggests that these factors will be understood by a combined epidemiological and experimental model approach. I suspect that it will be unusual for human cancer to be caused by large-dose exposure to single chemicals, acting as both initiator and promoter, as is the case in a standard animal cancer test.

Some of the antirisk factors that epidemiologists are turning up may have to do with the normal protective factors that have been devised by evolution to protect us against the many natural carcinogens (2). I have discussed some of these factors and why it appears that most of the DNA-damaging carcinogens are likely to be "natural" rather than man-made (2). Antirisk factors (such as calcium or various antioxidants) may sometimes be as important as risk factors and in some cases may be more easily manipulated (2).

Several reviews have discussed the epidemiological evidence on dietary causes of cancer and dietary protective factors, such as selenium and carotenoids, in cancer and heart disease (54,126,140,159,160). There is still a lack of definitive evidence about the dietary components that are critical for humans and about their mechanisms of action. Laboratory studies of natural foodstuffs and cooked food are beginning to uncover an extraordinary variety of mutagens and possible carcinogens and anticarcinogens. In this chapter I discuss dietary mutagens and carcinogens and anticarcinogens that seem of importance and speculate on relevant biochemical mechanisms, particularly the role of oxygen radicals and their inhibitors in the fat-cancer relationship, promotion, anticarcinogenesis, and aging. This may provide some clues as to dietary carcinogens and anticarcinogens for epidemiologists to investigate as possible causes of human cancer.

NATURE'S PESTICIDES

Plants in nature synthesize toxic chemicals in large amounts, apparently as a primary defense against the hordes of bacterial, fungal, and insect and other animal predators (2). Plants in the human diet are no exception. The variety of these toxic chemicals is so great that organic chemists have been characterizing them for over 100 years, and new plant chemicals are still being discovered (2,3). However, toxicological studies have been completed for only a very small percentage of them. Recent widespread use of short-term tests for detecting mutagens (2) and the increased number of animal cancer tests on plant substances (102) have contributed to the identification of many natural mutagens, teratogens, and carcinogens in the human diet (2,3). Sixteen examples were discussed in my review in *Science* (2), and some others in the exchange of letters about the review (3).

The examples illustrate that the human dietary intake of "nature's pesticides" is likely to be several grams per day--probably at least 10,000 times higher than the dietary intake of man-made pesticides (2).

Levels of plant toxins that confer insect and fungal resistance are being increased or decreased by plant breeders (2). There are health costs for the use of these natural pesticides, just as there are for man-made pesticides, and these must be balanced against the costs of producing food. However, little information is available about the toxicology of most of the natural plant toxins in our diet, despite the large doses to which we are exposed. Many, if not most, of these plant toxins may be "new" to humans in the sense that the human diet has changed drastically with historic times. By comparison, our knowledge of the toxicological effects of new man-made pesticides is extensive, and general exposure is exceedingly low (2,3).

Plants also contain a variety of anticarcinogens, which are discussed on page 9.

COOKED FOOD AS A SOURCE OF INGESTED BURNT AND BROWNED MATERIAL

Work of Sugimura and others has indicated that the burnt and browned material from heating protein during cooking is highly mutagenic (24,136, 182,183). Eight chemicals isolated on the basis of their mutagenicity from heated protein or pyrolyzed amino acids were found to be carcinogenic when fed to rodents (182,183). In addition, the browning reaction products from the caramelization of sugars or the reaction of amino acids and sugars during cooking (for instance, the brown material on bread crusts and toasted bread) contain a large variety of DNA-damaging agents and presumptive carcinogens (106,155,176,177,179,202,203). The amount of burnt and browned material in the human diet may be several grams per day. By comparison, about 500 mg of burnt material is inhaled each day by a smoker using 2 packs of cigarettes (at 20 mg of tar per cigarette) a day. Smokers have more easily detectable levels of mutagens in their urine than nonsmokers (224), but so do people who have consumed a meal of fried pork or bacon (12).

In the evaluation of risk from burnt material it may be useful (in addition to carrying out epidemiological studies) to compare the activity of cigarette tar to that of the burnt material from cooked food (or polluted air) in short-term tests and animal carcinogenicity tests involving relevant routes of exposure. Route of exposure and composition of the burnt material are critical variables. The risk from inhaled cigarette smoke can be one reference standard: an average life-shortening effect of about 8 years for a 2-pack-a-day smoker. The amount of burnt material inhaled from severely polluted city air, on the other hand, is relatively small; it would be necessary to breathe smoggy Los Angeles air for 1 to 2 weeks to equal the soluble organic matter of the particulates or the mutagenicity from 1 cigarette (20 mg of tar) (2). Epidemiological studies have not shown significant risks from city air pollution alone (37,54,171,205). Air in the houses of smokers is considerably more polluted than city air outside (31).

Coffee, which contains a considerable amount of burnt material, including the mutagenic pyrolysis product methylglyoxal (at about 500 to 1,000 μg/cup), is mutagenic (104,182,183). Methylglyoxal has recently been shown to be a carcinogen (124). However, 1 cup of coffee also contains about 250 mg of the natural mutagen chlorogenic acid (11,74,175) [which is also an antinitrosating agent (174,178)], highly toxic atractylosides (137), the glutathione transferase inducers kahweal palmitate and cafestol palmitate (110), and about 100 mg of caffeine [which inhibits a DNA-repair system and can increase tumor yield (8) and cause birth defects at high levels in several experimental species (59)]. There is preliminary, but not conclusive, epidemiological evidence that heavy coffee drinking is associated with cancer of the ovary, bladder, pancreas, and large bowel (44, 80,115,118,143,194,212).

Cooking also accelerates the rancidity reaction of cooking oils and fat in meat (166,167), thus increasing consumption of mutagens and carcinogens.

ALCOHOL

Alcohol has long been associated with cancer of the mouth, esophagus, pharynx, larynx, and, to a lesser extent, liver (54,65,90,200,201,228), and it appears to be an important human teratogen, causing a variety of physical and mental defects in babies of mothers who drink (1,154). Alcohol drinking causes sperm abnormalities in mice (7) and is a synergist for chromosome damage in humans (173). Alcohol metabolism generates acetaldehyde, which is a mutagen and teratogen (20,34) as well as a carcinogen (63, 96), and also radicals that produce lipid hydroperoxides (172,181,206,220) and other mutagens and carcinogens (Ref. 45; see below). In some epidemiological studies on alcohol (65,90,200,201,228), it has been suggested that dietary green vegetables are a modifying factor in the reduction of cancer risk.

MOLD CARCINOGENS

A variety of mold carcinogens and mutagens are present in mold-contaminated food such as corn, grain, nuts, peanut butter, bread, cheese, fruit, and apple juice (87,187). Some of these, such as sterigmatocystin and aflatoxin, are among the most potent carcinogens and mutagens known (87,187). Dietary glutathione has been reported to counteract aflatoxin carcinogenicity.

NITRITE, NITRATE, AND NITROSAMINES

A number of human cancers, such as stomach and esophageal cancer, may be related to nitrosamines and other nitroso compounds formed from nitrate and nitrite in the diet (81,83,116). Beets, celery, lettuce, spinach, radishes, and rhubarb all contain about 200 mg of nitrate per 100-g portion (125). Anticarcinogens in the diet may be important in this context as well (174,178).

FAT AND CANCER: POSSIBLE OXIDATIVE MECHANISMS

Epidemiological studies of cancer in humans suggest, but do not prove, that high fat intake is associated with colon and breast cancer (54,64,105, 126). A number of animal studies have shown that high dietary fat is a promoter and a presumptive carcinogen (64,105,126,215). Colon and breast cancer and lung cancer (which is almost entirely due to cigarette smoking) account for about half of all U.S. cancer deaths. In addition to the cyclopropenoid fatty acids already discussed (2), two other plausible mechanisms involving oxidative processes could account for the relation (43) between high fat and both cancer and heart disease.

Rancid Fat

Fat accounts for over 40% of the calories in the U.S. diet (64,105), and the amount of ingested oxidized fat may be appreciable (166,167). Unsaturated fatty acids and cholesterol in fat are easily oxidized, particularly during cooking (166,167). The lipid peroxidation chain reaction (rancidity) yields a variety (48,147,167) of mutagens, promoters, and carcinogens such as fatty acid hydroperoxides (45), cholesterol hydroperoxide (23), endoperoxides, cholesterol and fatty acid epoxides (23,25,95,142), enals and other aldehydes (20,34,63,111), and alkoxy and hydroperoxy radicals (111,147). Thus the colon and digestive tract are exposed to a variety of fat-derived carcinogens. Human breast fluid can contain enormous levels (up to 780 μM) (142) of cholesterol epoxide (an oxidation product of cholesterol), which could originate from either ingested oxidized fat or from oxidative processes in body lipids. Rodent feeding studies with oxidized fat (9) have not yielded definitive results.

Peroxisomes

Peroxisomes oxidize an appreciable percentage of dietary fatty acids, and removal of each 2-carbon unit generates one molecule of hydrogen peroxide (a mutagen, promoter, and carcinogen) (88,97,144,150,151,198). Some hydrogen peroxide escapes the catalase in the peroxisome (39,87,99,150, 151), thus contributing to the supply of oxygen radicals, which also come from other metabolic sources (39,56,77,78,147). Hydroperoxides generate oxygen radicals in the presence of iron-containing compounds in the cell (147). Oxygen radicals, in turn, can damage DNA and can start the rancidity chain reaction which leads to the production of the mutagens and carcinogens listed above (147). Drugs such as clofibrate, which cause lowering of serum lipids and proliferation of peroxisomes in rodents, result in age pigment (lipofuscin) accumulation (a sign of lipid peroxidation in tissues) and liver tumors in animals (150,151). Some fatty acids, such as $C_{22:1}$ and certain trans fatty acids, appear to cause peroxisomal proliferation because they are poorly oxidized in mitochondria and are preferentially oxidized in the peroxisomes, although they may be selective for heart or liver (29,127,130,134,190).

There has been controversy about the role of trans fatty acids in cancer and heart disease, and recent evidence suggests that trans fatty acids might not be a risk factor for atherosclerosis in experimental animals (57, 93,157). Americans consume about 12 g of trans fatty acids a day (57,93, 157), and a similar amount of unnatural cis isomers [which need further study (223)], mainly from hydrogenated vegetable fats. Dietary $C_{22:1}$ fatty acids are also obtained from rapeseed oil and fish oils (29,127,130,134, 190). Thus, oxidation of certain fatty acids might generate grams of hydrogen peroxide per day within the peroxisome (29,127,130, 134,190).

Another source of fat toxicity could be perturbations in the mitochondrial or peroxisomal membranes caused by abnormal fatty acids, yielding an increased flux of superoxide and hydrogen peroxide. Mitochondrial structure is altered when rats are fed some of the abnormal fatty acids from partially hydrogenated fish oil (41). Dietary $C_{22:1}$ fatty acids and clofibrate also induce ornithine decarboxylase (29,127,130,134,190), a common attribute of promoters.

The recent papers from W.R. Bruce's laboratory (21,129,208) suggest an interesting alternate explanation for fat toxicity. They show that free fatty acids and bile acids cause damage to the protective epithelium in the colon, thus acting as promoters. They also show that dietary Ca^{++} forms a nontoxic soap with fatty acids and bile acids. Thus, they postulate that increasing dietary calcium intake may be an alternative to lowering fat intake in the prevention of colon cancer (21,129,208). High fiber also seems to be an antirisk factor for colon cancer in epidemiological studies.

A recent National Academy of Sciences committee report suggests that a reduction of fat consumption in the American diet would be prudent (126), although other scientists argue that, until we know more about the mechanism of the fat-cancer relationship and about which types of fat are dangerous, it is premature to recommend dietary changes (44).

ANTICARCINOGENS

We have many defense mechanisms to protect ourselves against mutagens and carcinogens, including continuous shedding of the surface layer of our skin, stomach, cornea, intestines, and colon (82). Understanding these mechanisms should be a major goal of cancer, heart, and aging research. Among the most important defenses may be those against oxygen radicals and lipid peroxidation if, as discussed here, these agents are major contributors to DNA damage (193). Major sources of endogenous oxygen radicals are from hydrogen peroxide (39) and superoxide (69,133,147) generated as side products of metabolism, and the oxygen radical burst from phagocytosis after viral or bacterial infection or the inflammatory reaction (73,186,214). A variety of environmental agents could also contribute to the oxygen radical load, as discussed here and in recent reviews (147,197). Many enzymes protect cells from oxidative damage; examples are superoxide dismutase (69,133), glutathione peroxidase (66), DT-diaphorase (113), and the glutathione transferases (209). In addition, a variety of small molecules in our diet are required for antioxidative mechanisms and appear to be anticarcinogens; some of these are discussed below.

Vitamin E

Tocopherol or vitamin E is the major radical trap in lipid membranes (147) and has been used clinically in a variety of oxidation-related diseases (19). Vitamin E ameliorates both the cardiac damage and carcinogenicity of the quinones adriamycin and daunomycin, which are mutagenic, carcinogenic, cause cardiac damage, and appear to be toxic because of free radical generation (207). Protective effects of tocopherols against radiation-induced DNA damage and mutation and dimethylhydrazine-induced carcinogenesis have also been observed (16,42). Vitamin E markedly increases the endurance of rats during heavy exercise, which causes extensive oxygen radical damage to tissues (47).

Beta-Carotene

β-Carotene is another antioxidant in the diet that could be important in protecting body fat and lipid membranes against oxidation. Carotenoids are free-radical traps and remarkably efficient quenchers of singlet oxygen (26,67,135). Singlet oxygen is a very reactive form of oxygen which is mutagenic and particularly effective at causing lipid peroxidation (26,67, 135). It can be generated by pigment-mediated transfer of the energy of light to oxygen, or by lipid peroxidation, although the latter is somewhat controversial. β-Carotene and similar polyprenes are present in carrots and in all foods that contain chlorophyll, and they appear to be the plants' main defense against singlet oxygen generated as a by-product from the interaction of light and chlorophyll (108,168,199). Carotenoids have been shown to be anticarcinogens in rats and mice (119,152). Carotenoids (in green and yellow vegetables) may be anticarcinogens in humans (54,65, 90,109,140,165,200,201,228). Their protective effects in smokers might be related to the high level of oxidants in both cigarette smoke and tar (35, 148,149). Carotenoids have been used medically in the treatment for some genetic diseases, such as porphyrias, where a marked photosensitivity is presumably due to singlet oxygen formation (120).

Selenium

Another important dietary anticarcinogen is selenium. Dietary selenium (usually selenite) significantly inhibits the induction of skin, liver, colon, and mammary tumors in experimental animals by a number of different carcinogens, as well as the induction of mammary tumors by viruses (22,72, 98,122,191,221). It also inhibits transformation of mouse mammary cells (40). Low selenium concentrations may be a risk factor in human cancer (159,217). A particular type of heart disease in young people in the Keshan area of China has been traced to a selenium deficiency, and low selenium has been associated with cardiovascular death in Finland (158). Selenium is in the active site of glutathione peroxidase, an enzyme essential for destroying lipid hydroperoxides and endogenous hydrogen peroxide and thus helping to prevent oxygen-radical-induced lipid peroxidation (66), although not all of the effects of selenium may be accounted for by this enzyme (22,72,98,122,191,221). Several heavy-metal toxins, such as Cd^{++} (a known carcinogen) and Hg^{++}, lower glutathione peroxidase activity by interacting with selenium (66). Selenite (and vitamin E) has been shown to counter the oxidative toxicity of mercuric salts (107,225).

Glutathione

Glutathione is present in food and is one of the major antioxidants and antimutagens in the soluble fraction of cells. The glutathione transferases (some of which have peroxidase activity) are major defenses against oxidative and alkylating carcinogens (209). The concentration of glutathione may be influenced by dietary sulfur amino acids (185,218). N-Acetylcysteine, a source of cysteine, raises glutathione concentrations and reduces the oxidative cardiotoxicity of adriamycin and the skin reaction to radiation (32). Glutathione levels are raised even more efficiently by L-2-oxothiazolidine-4-carboxylate, which is an effective antagonist of acetaminophen-caused liver damage (218). Acetaminophen is thought to be toxic through radical and quinone oxidizing metabolites (86). Dietary glutathione may be an effective anticarcinogen against aflatoxin (131).

Dietary Ascorbic Acid

Ascorbic acid in the diet is also important as an antioxidant. It was shown to be anticarcinogenic in rodents treated with ultraviolet radiation, benzo(α)pyrene, and nitrite (in forming nitroso carcinogens) (55,81,83,101, 116,125), and it may be inversely associated with human uterine cervical dysplasia (although this is not proof of a cause-effect relationship) (210). Ascorbic acid may have been supplemented and perhaps partially replaced in humans by uric acid during primate evolution (6).

Uric Acid

Uric acid is a strong antioxidant present in high concentrations in the blood of humans (6). The concentration of uric acid in the blood can be increased by dietary purines; however, too much causes gout. Uric acid is also present in high concentrations in human saliva (6) and may play a role in defense there as well, in conjunction with lactoperoxidase.

Edible Plants

Dietary plants and a variety of substances in them, such as phenols, have been reported to inhibit (cabbage) or to enhance (beets) carcinogenesis (27,110,211) or mutagenesis (106,155,174,176-179,222) in experimental animals. Some of these substances appear to inhibit by inducing cytochrome P-450 and other metabolic enzymes (Ref. 27; see also Ref. 110), although on balance it is not completely clear whether it is generally helpful or harmful for humans to ingest these inducing substances.

The hypothesis that as much as 80 percent of cancer could be due to environmental factors was based on geographic differences in cancer rates and studies of migrants (85). These differences in cancer rates were thought to be mainly due to life style factors, such as smoking and dietary carcinogens and promoters (85), but they also may be due in good part (see also Ref. 54) to less than optimum amounts of anticarcinogens and protective factors in the diet.

The optimum levels of dietary antioxidants, which may vary among individuals, remain to be determined; however, at least for selenium (22,72,98, 122,191,221), it is important to emphasize the possibility of deleterious side effects at high doses.

ENDOGENOUS MUTAGENS AS CONTRIBUTORS TO SPONTANEOUS MUTATION

Interest in environmental mutagens first focused on man-made chemicals and pollution. In recent years much interest has shifted to mutagens arising from the cooking of food and natural environmental mutagens such as plant toxins and mold toxins (2). Here we discuss endogenous mutagens arising out of our normal metabolism as contributors to spontaneous mutations, perhaps the major source of DNA damage in animals.

We have recently reviewed the subject of spontaneous DNA damage (162). Types of DNA damage caused by endogenous processes include: base loss, base deamination, and strand breaks caused by spontaneous hydrolysis (114); base oxidation caused by cellular reactive oxygen species formed via oxidative metabolism; base methylation; and base mispairing during replication caused by base tautomerization, base ionization, or base rotation. DNA methylation may be caused by the endogenous alkylating agent S-adenosylmethionine, which needs to be in the nucleus in order to make the normal methylated base 5-methyl-cytosine and may contribute appreciably to the background level of abnormal DNA methylation adducts (13). A specific DNA glycosylase repair enzyme recognizes and repairs the methylated bases, N-7-methylguanine and N-3-methyladenine, while an inducible system repairs O-6-methylguanine (114), attesting to their importance. A variety of aldehydes arising out of normal metabolism, such as formaldehyde, acetaldehyde, methylglyoxal (124), and enals arising from normal cellular lipid peroxidation (117), are mutagens, and the first three have been shown to be carcinogens. Numerous endogenous processes can lead to the formation of cellular oxidizing agents, and we discuss below why oxidizing agents may be the most interesting of the endogenous mutagens.

CANCER, METABOLIC RATE, AND LIFE SPAN

Cumulative cancer risk increases with approximately the fourth power of age, both in short-lived species such as rats and mice (about 30% of the rodents have cancer by the end of their 2- to 3-year life span) and in long-lived species such as humans (about 30% of people have cancer by the end of their 85-year life span) (52,138,146,156). Thus, the marked increase in life span that has occurred in 60 million years of primate evolution has been accompanied by a marked decrease in age-specific cancer rates, i.e., in contrast to rodents, most humans do not have cancer by the age of 3 (52,138) (Fig. 1). One important factor in longevity appears to be basal metabolic rate (46,77,169,184,192), which is much lower in man than in rodents and could markedly affect the level of endogenous mutagens

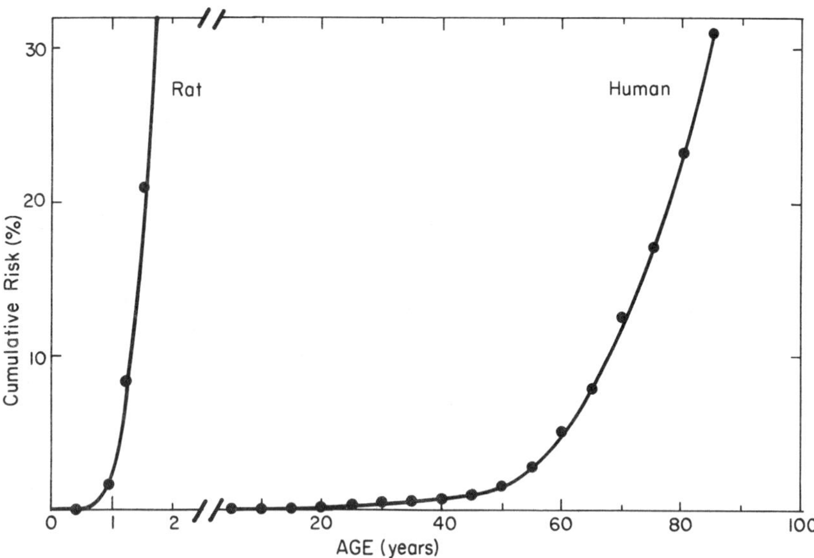

Fig. 1. Cumulative net risk of death from cancer for rats and humans. Rat data is for male Charles River COBS-CD outbred rats (156). Human data is from life tables for the year 1964 for men from 11 developed countries (146). (Reprinted with permission from Raven Press, New York, 1985.)

produced from normal metabolism. Oxidative DNA damage could be one major contributor to aging (2,77,193) related to metabolic rate.

OXIDATIVE DNA DAMAGE AS A MAJOR CAUSE OF ENDOGENOUS MUTATION

(i) The reactive oxygen species superoxide, hydrogen peroxide, and hydroxyl radical are generated in vivo as a consequence of normal metabolism as well as being the active agents of DNA damage due to radiation (2, 10,132,147). The oxidation of certain cellular components by these oxygen species could contribute both to aging and to age-dependent diseases such as cancer (2,38). Hydroxyl radicals have been shown to produce base damage and strand breaks in DNA (132,147) (Fig. 2).

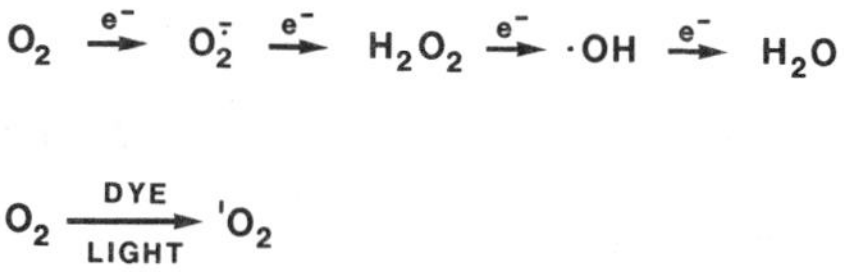

$$O_2 \xrightarrow{e^-} O_2^{\cdot -} \xrightarrow{e^-} H_2O_2 \xrightarrow{e^-} \cdot OH \xrightarrow{e^-} H_2O$$

$$O_2 \xrightarrow[\text{LIGHT}]{\text{DYE}} {}^1O_2$$

Fig. 2. The formation of superoxide, hydrogen peroxide, and hydroxyl radicals by successive additions of electrons to oxygen. Cytochrome oxidase adds 4 electrons fairly efficiently during energy generation in mitochondria, but some of these toxic intermediates leak out. Singlet oxygen is generated from oxygen by the absorption of energy from a dye activated by light. (Reprinted with permission from Raven Press, New York, 1985.)

(ii) Oxidants also initiate the process of lipid peroxidation (132, 147). The latter results in the formation of lipid hydroperoxides which, in the presence of cellular iron-containing compounds, can also break down to yield oxygen radicals (147). Unsaturated fatty acids and cholesterol in fat are easily oxidized, particularly during cooking (166,167). The lipid peroxidation chain reaction (rancidity) yields a variety (48,147,166,167) of mutagens, promoters, and carcinogens such as fatty acid hydroperoxides (45), cholesterol hydroperoxide (23), endoperoxides [cholesterol and fatty acid epoxides (23,25,95,142)], enals and other aldehydes (20,34,63,111), and alkoxy and hydroperoxy radicals (111,147). Thus the colon and digestive tract are exposed to a variety of fat-derived carcinogens. Human breast fluid can contain enormous levels (up to 780 μM) (142) of cholesterol epoxide [a mutagenic and carcinogenic (163) oxidative product of cholesterol], which could originate from either ingested oxidized fat or oxidative processes in body lipids.

(iii) Peroxisomes oxidize an appreciable percentage of dietary fatty acids, and removal of each 2-carbon unit generates one molecule of hydrogen peroxide (a mutagen, promoter, and carcinogen) (88,97,144,150,151,170,198). Some hydrogen peroxide escapes the catalase in the peroxisome (39,92,99, 150,151), thus contributing to the supply of oxygen radicals, which also come from other metabolic sources (39,56,77,147). Oxygen radicals, in turn, can damage DNA and can also start the rancidity chain reaction that leads to the production of the mutagens and carcinogens listed above (147).

(iv) Phagocytes, in the process of scavenging bacteria and dead cells, generate the carcinogen and mutagen hydrogen peroxide and the mutagens superoxide and hydroxyl radicals. Phagocytes also generate OCl^- and chloramine, oxidizing and chlorinating agents.

(v) Light, and also lipid peroxidation, can generate singlet oxygen, a very energetic form of O_2 and a potent mutagen and inducer of lipid and protein oxidation.

Numerous defense mechanisms within the cell have evolved to limit the levels of reactive oxygen species and the damage they induce (2,10,91,132, 147). Among the cellular defenses are the enzymes superoxide dismutase, catalase, and glutathione peroxidase (2,10,132,147), as well as the antioxidants β-carotene (2,10,132,147), the tocopherols (145), and uric acid (6). However, low levels of reactive oxygen species can escape these cellular defenses (39) and produce damage to DNA, protein, and unsaturated fats.

THYMINE GLYCOL AND 5-HYDROXYMETHYLURACIL

Two products which are formed in DNA in vitro as a consequence of chemical oxidation or ionizing radiation (an oxidative mutagen) are thymine glycol (5,6-dihydroxydihydrothymine) and hydroxymethyluracil (17,68,94, 188), both oxidation products of thymine (Fig. 3). We have recently described a new specific DNA repair enzyme, a DNA glycosylase from mouse cells, which repairs 5-hydroxymethyluracil (91). We also showed that a specific DNA glycosylase repair enzyme for thymine glycol in mouse cells is a different enzyme. Since these DNA glycosylases are specific repair enzymes, rather than general repair enzymes such as those for nucleotide excision repair, they point to the importance of oxidative DNA damage for cells.

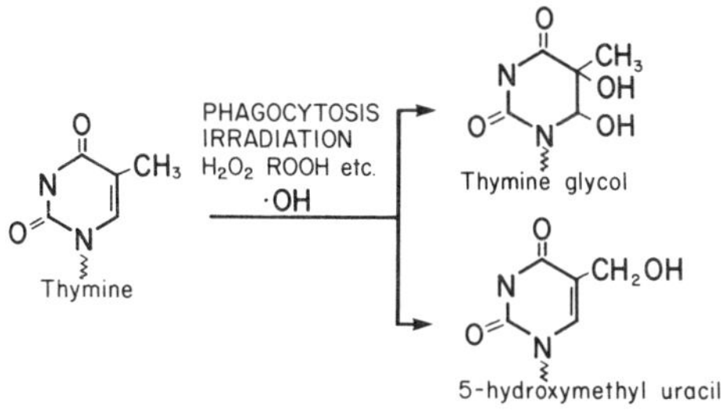

Fig. 3. Thymine residues in DNA can be oxidized by a variety of agents to yield thymine glycol or hydroxymethyluracil. (Reprinted with permission from Raven Press, New York, 1985.)

OXIDIZED DNA BASES IN URINE

In order to examine individual humans (and rodents) for DNA damage, we have looked for damaged DNA bases and deoxynucleosides excised from DNA by repair enzymes and excreted in urine (Fig. 4). We have developed a high-pressure liquid chromatography (HPLC) assay for thymine glycol, thymidine glycol, and hydroxymethyluracil in urine (Ref. 36; and unpubl. data). Our results indicate that humans excrete a total of about 100 nmol/day of these three compounds (Tab. 1). Preliminary evidence suggests that most of this total is derived from repair of oxidized DNA, rather than from alternative sources such as the diet or bacterial flora. This total may therefore represent an average of about 1,000 oxidized thymine residues per day per each of the body's 6×10^{13} cells.

Since these products are only three of a considerable number of oxidative DNA damage products described by radiobiologists (33,103), there are likely to be several thousands of oxidative DNA hits per cell per day in man. Rats, which have a higher specific metabolic rate and a shorter life span, excrete about 15 times more thymine glycol plus thymidine glycol (and about 10 times more hydroxymethyluracil) per kg of body weight than do humans (Ref. 36; and unpubl. data). Preliminary results with mouse urine indicate that mice have even higher levels of thymine glycol and thymidine glycol than rats (R. Adelman, R. Saul, and B.N. Ames, unpubl. data). These noninvasive assays of DNA oxidation products may allow the direct testing of current theories which relate oxidative metabolism to the processes of aging and cancer in man.

URINARY THYMINE GLYCOL, THYMIDINE GLYCOL, AND HYDROXYMETHYLURACIL APPEAR TO COME FROM REPAIR OF TISSUE DNA

We propose that urinary thymidine glycol and most thymine glycol and hydroxymethyluracil are derived from the repair of oxidatively damaged DNA (Ref. 36; and unpubl. data). This hypothesis is supported by the following 9 points of evidence.

(i) Both in vitro and in vivo thymine residues in DNA are converted to thymine glycol and hydroxymethyluracil by chemical oxidations and ionizing radiation (15,17,68,94,121,188,189).

(ii) Repair enzymes exist that are specific for the excision of these lesions from the DNA (28,49,75,76,84,91,114,121,128). The existence of

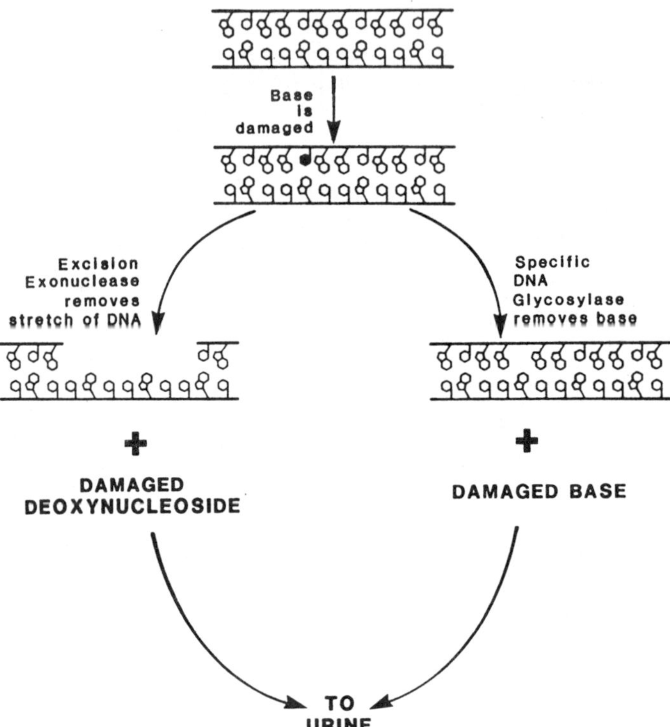

Fig. 4. Repair of a damaged DNA base by excision repair or a specific glycosylase. If these products are removed from the cell, are not further metabolized, and are fairly hydrophylic, they may be recovered in the urine. (Reprinted with permission from Alan R. Liss, Inc., New York.)

specific enzymes in itself suggests that their substrates, the thymine glycol-residue, and hydroxymethyluracil-residue exist in nature, and that these residues in DNA are biologically undesirable to the cell.

(iii) Our pharmacokinetic studies in the rat showed that once excised from DNA in the tissues, thymine glycol, thymidine glycol, and hydroxymethyluracil would each be eliminated within 24 hr from the blood to the urine without further metabolism (Ref. 36; R. Saul et al., unpubl. data).

(iv) Urinary thymidine glycol in rats is not of dietary origin since it is not absorbed from the gut, though thymine glycol and hydroxymethyluracil can be (Ref. 36; and unpubl. data).

(v) We have fed to rats (along with their usual chow) radioactive DNA containing high levels of tritiated thymine glycol-residues and hydroxymethyluracil-residues. The oxidized residues were recovered as the free deoxyribosides in the feces and none could be found in the urine, showing that oxidized dietary DNA would not contribute to the urinary levels of these products.

(vi) Most thymine glycol and hydroxymethyluracil are not of dietary origin, since these bases continue to be excreted when rats are fed a nucleic acid-free diet (Ref. 36; and unpubl. data).

(vii) The cis-thymidine glycol has two different possible stereochemical forms, cis-(5R,6S) and cis-(5S,6R), which are formed in similar amounts when made by chemical synthesis or radiolysis of thymidine. In contrast, urinary cis-thymidine glycol is primarily in one form, suggesting that it is derived from DNA and is stereo-selected either by differential synthesis in DNA because of steric hindrances or by differential repair (K. Cundy, R.L. Saul, and B.N. Ames, unpubl. data).

(viii) Thymine glycol, thymidine glycol, and hydroxymethyluracil are not of microbial origin, as seen in our germ-free rat studies (Ref. 36; and unpubl. data).

(ix) In the cell, the thymine moiety resides almost exclusively in DNA, so that these thymine damage products are more likely to be indicators of damage to DNA rather than damage to RNA, free thymine, or thymine mononucleotides.

TESTING RELATIONSHIPS BETWEEN DIETARY FACTORS AND OXIDATIVE DNA DAMAGE IN MAN

The level of thymine glycol, but not thymidine glycol, varies considerably from day to day in certain individuals but not in others (36). It remains to be determined whether the variability in human urinary thymine glycol is caused by dietary thymine glycol, dietary influences on DNA damage, or factors independent of diet. It is of considerable interest in this regard that calorie restriction in rats and mice causes a marked increase in life span (213,226). Calorie restriction also results in markedly fewer tumors in both rats and mice (153). We are currently exploring the effect of calorie restriction in rats, mice, and humans on thymine glycol, thymidine glycol, and hydroxymethyluracil excretion. This approach may also prove to be useful for testing current theories about proposed "anticarcinogens" such as selenium, β-carotene, vitamin C, and vitamin E (2). We hope that measurement of the oxidative DNA damage products thymine glycol, thymidine glycol, and hydroxymethyluracil will enable us to optimize the intake levels of anticarcinogens in individuals.

Tab. 1. Levels of DNA damage products, thymine glycol (Tg), thymidine glycol (dTg), hydroxymethyluracil (HMU), and hydroxymethyldeoxyuridine (HMdUrd), in human and rat urine. Data from Cathcart et al. (Ref. 36; and unpubl. data). (Reprinted with permission from Alan R. Liss, Inc., New York.)

	Levels (molecules/cell/day)		
Compound	Human	Rat	Rat/human
Tg	270	3,800	14
dTg	70	1,200	17
HMU	620	6,000	10
HMdUrd	trace	trace	--

OXIDIZING AGENTS CAUSE MANY TYPES OF DNA DAMAGE IN ADDITION TO THYMINE OXIDATION

Though the molecular mechanisms leading to aging and cancer are not yet known, it may be pertinent that oxidative damage seems to cause almost all known types of DNA abnormalities. Radiation is known to cause single-strand and double-strand breaks as well as a wide variety of different oxidized bases leading to deletions, chromosome abnormalities, and point mutations. Oxidizing agents, such as hydrogen peroxide or organic peroxides, generate the same oxidizing species as radiation, and have also been shown to cause deletions as well as point mutations (112).

We suggest that oxidizing agents also could cause loss of 5-methylcytosine, which may be important in relation to dedifferentiation and aging (51,53). By analogy, with the formation of 5-hydroxymethyluracil, 5-hydroxymethylcytosine could be formed readily by oxidative damage. We suggest that this would lead to lack of methylation after DNA replication because the DNA maintenance methylase would not recognize the opposite strand's hydrophylic 5-hydroxymethylcytosine as 5-methylcytosine. Since 5-methylcytosine may be important in formation of Z-DNA in turning off genes in differentiation, the 5-hydroxymethylcytosine might prevent this and cause dedifferentiation.

RISK ASSESSMENT BASED ON ANIMAL CANCER TESTS

Extrapolation of results from rodent cancer tests done at high doses to humans exposed to low doses is a method widely used by regulatory agencies in an attempt to prevent future cancer. This is a controversial extrapolation, based more on a desire for prudence than on scientific knowledge, and is viewed with great unease by eminent epidemiologists (89,139, 161). There is relatively little argument about regulating high dose exposure to occupational or medicinal drug carcinogens. There is a great deal of controversy over whether regulatory agencies are preventing cancer by extrapolating risks from rodents to humans at low levels of sometimes a few hypothetical cases of cancer in the country, with the assumption that alternate technologies with unexamined risks will be better (89,216).

One of the difficulties in current regulation is that there is almost total preoccupation with man-made chemicals, both in cancer testing and in regulatory effort, and little concern with doing the same potential risk calculations on common natural carcinogens. Yet, natural carcinogens are not rare--they are very common. As detailed above, there are large numbers of carcinogens in every meal, all perfectly natural and traditional (2,3), and no human diet can be entirely free of carcinogens and mutagens. Nature is not benign. Thus, the problem regulators face is not merely to identify carcinogens, but also how to sensibly assess relative potential risks and set priorities, and in particular how to identify the *important* carcinogenic potential risks among the vast number of trivial risks.

In addition, it should not be overlooked that major, or even most, causes of human cancer may not be traditional chemical carcinogens in animals (139), but may be due to agents causing cell death and consequent cell proliferation (e.g., viruses such as hepatitis B or HPV16, or chronic irritations), to changes in hormonal status which cause cell proliferation, to deficiencies in normal protective factors against endogenous carcinogens (e.g., selenium or β-carotene; Ref. 2 and 3), to lack of other anticarcinogens (e.g., fiber or calcium; Ref. 21), or to dietary imbalances such as excess fat or salt consumption. Nevertheless, animal bioassay and in vitro studies are also providing clues to epidemiologists as to which carcinogens and mutagens might be contributing to human cancer.

To identify a substance, whether natural or man-made, as a mutagen or a carcinogen in laboratory studies is just the first qualitative step. Beyond this, it is necessary to quantitate the approximate magnitude of the potential hazard and to consider the potential hazards of alternative courses of action. The quantitation of risk poses a major challenge (89). Carcinogens differ in their potency at high doses in rodents by over 10-million-fold (70), and the levels of particular carcinogens to which humans are exposed can vary more than 1-billion-fold.

We have recently made an attempt to achieve some orientation in the world of comparative carcinogenic potential hazards as indicated by experiments on rodents (B.N. Ames and R. Magaw, ms. in prep.). We have calculated the relative potential hazard for amounts of various common carcinogens that a human might be exposed to in a day. These potential hazards, though not risks to humans, are adjusted for the potency of each carcinogen so that they can be compared to each other. Since carcinogens differ enormously in potency, a comparison of risks of various carcinogens ingested by humans must adjust for the potency as well as the dose. We do this by expressing each human exposure (dose in mg/kg) as a fraction of the potency [tolerated dose $(TD)'_{50}$; in mg/kg] of each carcinogen in rodents. We call this the Exposure/Potency Index (EPI). The TD_{50} is approximately the daily dose (in mg/kg) to give 50% of the animals cancer when fed for a lifetime (141). The TD_{50} values of the various carcinogens in rodents are taken from our recently completed exhaustive database (3,200 experiments on 840 chemicals), which calculated the potency of carcinogens in experimental animals (70,71,141).

Our conclusions are that the amounts of carcinogens ingested from synthetic chemicals (except for cases of occupational exposure or drugs) are extremely tiny relative to the natural background and that it is extremely unlikely that pollution from pesticide residues, or other such sources, is contributing in any appreciable way to human cancer.

UNCERTAINTY IN RELYING ON ANIMAL CANCER TESTS FOR HUMAN PREDICTIONS

Species Variation

If a chemical is a carcinogen in a rat but not in a mouse (or vice versa), the risk to the other species may be nonexistent. Of 179 chemicals in our database that are positive in rats or mice and tested in both species, 42% were positive in the mouse and negative in the rat or vice versa (L.S. Gold, R. Magaw, and B.N. Ames, ms. in prep.), despite the fact that both species are very closely related with similar life spans. This shows the uncertainty of whether a substance carcinogenic to one (or even both) of these species is of any carcinogenic risk to humans. Conversely, important human carcinogens may be negative in rats and mice, as was for a long time the case for tobacco and alcohol, the two largest reliably known causes of neoplastic death in the United States.

Quantitative Uncertainties

In addition to these difficulties with qualitative correspondence, there are major problems with quantitative extrapolations to humans (54,89,

139,161). This quantitative uncertainty would decrease with more knowledge about species affected and understanding of pharmacokinetics and mode of action of the carcinogen, but it is rare to have the desired information. Quantitative extrapolation of risk from rodents to humans, particularly at low doses, is guesswork, which at present we have no way of validating. It is guesswork because of uncertainties in 5 major areas that prevent quantitation: (i) the uncertainty of which step in the multistage carcinogenic process is limiting, e.g., promotion, initiation, or progression (89); (ii) the multicausal nature of most human cancer (54,89); (iii) extrapolation from a short-lived rodent to a long-lived man; (iv) differences in anticarcinogens; and (v) differences in metabolism.

Can one extrapolate by scaling down linearly from a high-dose animal experiment to a low-dose level? This has been argued ad nauseam, but the simple fact is that it cannot be known because we cannot measure effects at low doses. The best fit in the massive ED01 megamouse study on one chemical appears to be J or hockey-stick shaped (30). We would like to bring up one other argument which bears on the question. Wilson, Crouch, and Zeise (219,227) have pointed out the remarkable finding that one can predict fairly precisely the potency of carcinogens in high-dose animal cancer experiments from the toxicity (the LD_{50}) of the chemical, though one cannot, of course, predict whether the substance is a carcinogen. We have now found the same result in a more detailed analysis of our carcinogenic potency database and have shown a close relation of maximum tolerated dose (MTD) to TD_{50}.

Our evidence suggests that this relation has a biological rather than just a statistical basis, and we have postulated an explanation (18). This is that a just sublethal level of a carcinogen is causing cell death, which allows neighboring cells to proliferate, which is an essential aspect of the carcinogenic process (50,60,79,100,123,180,195,196). Cell toxicity may be mutagenic in itself by oxygen radicals released from phagocytosis and aberrant methylation (14). Cell killing is a threshold process (61), and this raises serious questions about a linear extrapolation to low doses if a key step of carcinogenesis at high doses in rodents is cell proliferation (18).

Thus, if many animal cancer tests done at the MTD are really measuring cell killing and consequent cell proliferation as the limiting step in the carcinogenesis process, one might predict that the dose-response curves close to the MTD would generally be nonlinear. This, in fact, appears to be the case in those experiments that could be analyzed properly by a life-table analysis (70,71,141). A detailed analysis (D. Hoel, pers. comm.) shows that when a proper life-table analysis is done, the quadratic (or cubic) model gives a much better fit to the data than a linear model. Human cancer, except in rare cases, is not from high (just subtoxic) doses of a single chemical, but is likely to be from several risk factors often combined with lack of antirisk factors (164), e.g., aflatoxin (a potent mutagen) combined with an agent causing cell proliferation (hepatitis B virus; Ref. 204), or alcoholic cirrhosis.

Another uncertainty about animal cancer tests is the percentage of chemicals tested that will be scored as carcinogens. Currently about 50% of the chemicals tested are positive. Our database (tests meeting the criteria for calculating potency), for example, has 306 chemicals tested in both rats and mice: 59% are positive in at least one species (70,71,141). It is always assumed that the reason for this high proportion of positive results is that more suspicious chemicals are being tested. We do not really know, however, if the percentage of positives would be low among less suspicious chemicals. If cell proliferation is really the true limiting step in rodent carcinogenesis, as discussed above, the "carcinogens" may be chemicals toxic to particular susceptible tissues at the MTD, which could include a high percentage of all the chemicals in the world. Thus, even this fundamental assumption in our thinking can be questioned.

It is not possible to convert rodent numbers done at an MTD to human risks at low doses with any scientific credibility. When the Environmental Protection Agency (EPA) does it, for example, saying ethylene dibromide (EDB) residues in grain (EPI = 0.4) would cause 3 cases of cancer in 1,000 people and consequently bans the main fumigant in the country, it is based on a succession of worst case assumptions (several unique to EDB) (58). Our calculations (Ames and Magaw, ms. in prep.) on a daily peanut butter sandwich (formaldehyde in bread, aflatoxin in peanut butter) indicate 700 times the potential hazard of daily EDB residues, or more cancer than there is in the country for this single item by an EPA-type calculation--an absurd result. The EDB residues were about equal in risk to the chloroform (produced from chlorination) in the average glass of tap water.

Thus, we will have to rely on epidemiology for identifying the present major causes of human cancer (epidemiologists are doing this at a good rate considering the difficulties); on animal tests for preventing future cancer due to high doses in occupational or medicinal drug exposure, for giving clues (along with short-term tests) to epidemiologists, and for setting priorities; and on cancer research to bring us out of this morass of ignorance. Finally, we do have a threshold for carcinogens: a threshold of attention to important risks if we want to decrease risk. We could start with the 400,000 premature deaths each year from tobacco.

ACKNOWLEDGEMENTS

This work was supported by Department of Energy Contract DE-AT03-76EV70156, by a contract from the National Foundation for Cancer Research, and by National Institute of Environmental Health Sciences Center Grant ES01896. This chapter has been adapted in part from Ref. 2, 4, and 5, and from B.N. Ames and R. Magaw (ms. in prep.).

REFERENCES

1. Abel, E.L. (1982) Consumption of alcohol during pregnancy: A review of effects on growth and development of offspring. Human Biol. 54: 421-453.
2. Ames, B.N. (1983) Dietary carcinogens and anticarcinogens (oxygen radicals and degenerative diseases). Science 221:1256-1264.
3. Ames, B.N. (1984) Cancer and diet. Science 224:668-670; 757-760.
4. Ames, B.N. (1986) Overview: Food constituents as a source of mutagens, carcinogens, and anticarcinogens. In Genetic Toxicology of the Diet, Alan R. Liss, Inc., New York, pp. 3-32.
5. Ames, B.N., and R.L. Saul (1986) Oxidative DNA damage as related to cancer and aging. In Genetic Toxicology of Environmental Chemicals, Alan R. Liss, Inc., New York (in press).
6. Ames, B.N., R. Cathcart, E. Schwiers, and P. Hochstein (1981) Uric acid provides an antioxidant defense in humans against oxidant- and radical-caused aging and cancer: A hypothesis. Proc. Natl. Acad. Sci., USA 78:6858-6862.

7. Anderson, Jr., R.A., B.R. Willis, C. Oswald, and L.J.D. Zaneveld (1983) Ethanol-induced male infertility: Impairment of spermatozoa. J. Pharmacol. Exp. Ther. 225:479-486.
8. Armuth, V., and I. Berenblum (1981) The effect of caffeine on two-stage skin carcinogenesis and on complete systemic carcinogenesis. Carcinogenesis 2:977-979.
9. Artman, N.R. (1969) The chemical and biological properties of heated and oxidized fats. In Advances in Lipid Research, Vol. 7, R. Paoletti and D. Kritchevsky, eds. Academic Press, New York, pp. 245-330.
10. Autor, A., ed. (1982) Pathology of Oxygen, Academic Press, New York.
11. Aver'yanov, A.A. (1981) Generation of superoxide anion radicals and hydrogen peroxide in autoxidation of caffeic acid. Biokhimiya 46: 256-261.
12. Baker, A., A. Arlauskas, A. Bonin, and D. Angus (1982) Detection of mutagenic activity in human urine following fried pork or bacon meals. Cancer Lett. 16:81-89.
13. Barrows, L.R., and P.N. Magee (1982) Nonenzymatic methylation of DNA by S-adenosylmethionine in vitro. Carcinogenesis 3:349-351.
14. Barrows, L.R., and R.C. Shank (1981) Aberrant methylation of liver DNA in rats during hepatotoxicity. Toxicol. Appl. Pharmacol. 60:334-345.
15. Baudisch, O., and D. Davidson (1925) The mechanism of oxidation of thymine: 4,5-Dehydroxyhydrothymine (thymine glycol). J. Biol. Chem. 64:233-239.
16. Beckman, C., R.M. Roy, and A. Sproule (1982) Modification of radiation-induced sex-linked recessive lethal mutation frequency by tocopherol. Mutat. Res. 105:73-77.
17. Beer, M., S. Stern, D. Carmalt, and K.H. Mohlhenrich (1966) Determination of base sequence in nucleic acids with the electron microscope. V. The thymine-specific reactions of osmium tetroxide with deoxyribonucleic acid and its components. Biochemistry 5:2283-2288.
18. Bernstein, L., L.S. Gold, B.N. Ames, M.C. Pike, and D.G. Hoel (1985) Some tautologous aspects of the comparison of carcinogenic potency in rats and mice. Fund. Appl. Toxicol. 5:79-86.
19. Bieri, J.G., L. Corash, and V.S. Hubbard (1983) Medical uses of vitamin E. New Engl. J. Med. 308:1063-1071.
20. Bird, R.P., H.H. Draper, and P.K. Basrur (1982) Effect of malonaldehyde and acetaldehyde on cultured mammalian cells. Production of micronuclei and chromosomal aberrations. Mutat. Res. 101:237-246.
21. Bird, R.P., A. Medline, R. Furrer, and W.R. Bruce (1985) Toxicity of orally administered fat to the colonic epithelium of mice. Carcinogenesis 6:1063-1066.
22. Birt, D.F., T.A. Lawson, A.D. Julius, C.E. Runice, and S. Salmasi (1982) Inhibition by dietary selenium of colon cancer induced in the rat by bis(2-oxopropyl)nitrosamine. Cancer Res. 42:4455-4459.
23. Bischoff, F. (1969) Carcinogenic effects of steroids. In Advances in Lipid Research, Vol. 7, R. Paoletti and D. Kritchevsky, eds. Academic Press, New York, pp. 165-244.
24. Bjeldanes, L.F., M.M. Morris, J.S. Felton, S. Healy, D. Stuermer, P. Berry, H. Timourian, and F.T. Hatch (1982) Mutagens from the cooking of food. II. Survey by Ames/Salmonella test of mutagen formation in the major protein-rich foods of the American diet. Food Chem. Toxicol. 20:357-363.
25. Black, H.S., and D.R. Douglas (1972) Model system for the evaluation of the role of cholesterol α oxide in ultraviolet carcinogenesis. Cancer Res. 32:2630.

26. Bors, W., C. Michel, and M. Saran (1981) Generation and reactivities of various types of oxygen radicals. Bull. Europ. Physiopath. Resp. 17(suppl.):13-18.
27. Boyd, J.N., J.G. Babish, and G.S. Stoewsand (1982) Modification by beet and cabbage diets of aflatoxin B_1-induced rat plasma α-foetoprotein elevation, hepatic tumorigenesis, and mutagenicity of urine. Food Chem. Toxicol. 20:47-52.
28. Breimer, L.H., and T. Lindahl (1984) DNA glycosylase activities for thymine residues damaged by ring saturation, fragmentation, or ring contraction are functions of endonuclease III in Escherichia coli. J. Biol. Chem. 259:5543-5548.
29. Bremer, J., and K.R. Norum (1982) Metabolism of very long-chain monosaturated fatty acids (22:1) and the adaptation to their presence in the diet. J. Lipid Res. 23:243-256.
30. Brown, K.G., and D.G. Hoel (1983) Multistage prediction of cancer in serially dosed animals with application to the ED_{01} study. Fund. Appl. Toxicol. 3:470-477.
31. Brunekreef, B., and J.S.M. Boleij (1982) Long-term average suspended particulate concentrations in smokers' homes. Int. Arch. Occup. Env. Health 50:299-302.
32. CME (1983) Symposium on "N-Acetylcysteine (NAC): A Significant Chemoprotective Adjunct." Sem. Oncol. 10(Suppl. 1):1-230.
33. Cadet, J., and R. Teoule (1971) Peroxides produced from thymine by γ irradiation in aerated solution. Biochim. Biophys. Acta 238:8-26.
34. Campbell, M.A., and A.G. Fantel (1983) Teratogenicity of acetaldehyde in vitro: Relevance to the fetal alcohol syndrome. Life Sci. 32: 2641-2647.
35. Carmella, S.G., E.J. LaVoie, and S.S. Hecht (1982) Quantitative analysis of catechol and 4-methylcatechol in human urine. Food Chem. Toxicol. 20:587-590.
36. Cathcart, R., E. Schwiers, R.L. Saul, and B.N. Ames (1984) Thymine glycol and thymidine glycol in human and rat urine: A possible assay for oxidative DNA damage. Proc. Natl. Acad. Sci., USA 81:5633-5637.
37. Cederlof, R., R. Doll, B. Fowler, L. Friberg, N. Nelson, and V. Vouk (1978) Air pollution and cancer: Risk assessment methodology and epidemiological evidence. Env. Health Perspect. 22:1-12.
38. Cerutti, P.A. (1985) Prooxidant states and tumor promotion. Science 227:375-380.
39. Chance, B., H. Sies, and A. Boveris (1979) Hydroperoxide metabolism in mammalian organs. Physiolog. Revs. 59:527-605.
40. Chatterjee, M., and M.R. Banerjee (1982) Selenium mediated dose-inhibition of 7,12-dimethylbenz[a]anthracene-induced transformation of mammary cells in organ culture. Cancer Lett. 17:187-195.
41. Christiansen, E.N., T. Flatmark, and H. Kryvi (1981) Effects of marine oil diet on peroxisomes and mitochondria of rat liver. A combined biochemical and morphometric study. Eur. J. Cell Biol. 26:11-20.
42. Cook, M.G., and P. McNamara (1980) Effect of dietary vitamin E on dimethylhydrazine-induced colonic tumors in mice. Cancer Res. 40:1329-1331.
43. Correa, P., J.P. Strong, W.D. Johnson, P. Pizzolato, and W. Haenszel (1982) Atherosclerosis and polyps of the colon. Quantification of precursors of coronary heart disease and colon cancer. J. Chron. Dis. 35:313-320.
44. Council for Agricultural Science and Technology (1982) Diet, Nutrition, and Cancer: A Critique, Special Publication No. 13, Council for Agricultural Science and Technology, Ames, Iowa.

45. Cutler, M.G., and R. Schneider (1974) Tumours and hormonal changes produced in rats by subcutaneous injections of linoleic acid hydroperoxide. Food Cosmet. Toxicol. 12:451-459.
46. Cutler, R.G. (1984) Antioxidants, aging, and longevity. In Free Radicals in Biology, Vol. VI, W.A. Pryor, ed. Academic Press, New York, pp. 371-428.
47. Davies, K.J.A., A.T. Quintanilha, G.A. Brooks, and L. Packer (1982) Free radicals and tissue damage produced by exercise. Biochem. Biophys. Res. Commun. 107:1198-1205.
48. Demopoulos, H.B., D.D. Pietronigro, E.S. Flamm, and M.L. Seligman (1980) The possible role of free radical reactions in carcinogenesis. In Cancer and the Environment, H.B. Demopoulos and M.A. Melman, eds. Pathotox Publ., Park Forest South, Illinois, pp. 273-303.
49. Demple, B., and S. Linn (1980) DNA N-glycosylases and DNA repair. Nature 287:203-208.
50. Denda, A., S. Inui, M. Sunagawa, S. Takahashi, and Y. Konishi (1978) Enhancing effect of partial pancreatectomy and ethionine-induced pancreatic regeneration on the tumorigenesis of azaserine in rats. Gann 69:633-639.
51. Denda, A., P.M. Rao, S. Rajalakshmi, and D.S.R. Sarma (1985) 5-Azacytidine potentiates initiation induced by carcinogens in rat liver. Carcinogenesis 6:145-146.
52. Dix, D., P. Cohen, and J. Flannery (1980) Role of aging in cancer incidence. J. Theor. Biol. 83:163-173.
53. Doerfler, W. (1984) DNA methylation: Site-specific methylations cause gene inactivation. Angew. Chem. Int. Ed. Engl. 23:919-931.
54. Doll, R., and R. Peto (1981) The causes of cancer: Quantitative estimates of avoidable risks of cancer in the United States today. J. Natl. Cancer Inst. 66:1191-1308.
55. Dunham, W.B., E. Zuckerkandl, R. Reynolds, R. Willoughby, R. Marcuson, R. Barth, and L. Pauling (1982) Effects of intake of L-ascorbic acid on the incidence of dermal neoplasms induced in mice by ultraviolet light. Proc. Natl. Acad. Sci., USA 79:7532-7536.
56. Emerit, I., M. Keck, A. Levy, J. Feingold, and A.M. Michelson (1982) Activated oxygen species at the origin of chromosome breakage and sister-chromatid exchanges. Mutat. Res. 103:165-172.
57. Enig, M.G., R.J. Munn, and M. Keeney (1978) Dietary fat and cancer trends--A critique. Fed. Proc. 37:2215-2220.
58. Environmental Protection Agency (1983) Amitraz (BAAM): Position Document 4. Environmental Protection Agency, Special Pesticide Review Division, Arlington, Virginia.
59. Fabro, S., ed. (1982) Caffeine in pregnancy--State of the science. Reprod. Toxicol. 1:2-3.
60. Farber, E. (1984) Cellular biochemistry of the stepwise development of cancer with chemicals: G.H.A. Clowes Memorial Lecture. Cancer Res. 44:5463-5474.
61. Farber, E., S. Parker, and M. Gruenstein (1976) The resistance of putative premalignant liver-cell populations, hyperplastic nodules, to the acute cytotoxic effects of some heptocarcinogens. Cancer Res. 36: 3879-3887.
62. Feron, V.J., A. Kruysse, and R.A. Woutersen (1982) Respiratory tract tumours in hamsters exposed to acetaldehyde vapour alone or simultaneously to benzo(a)pyrene or diethylnitrosamine. Eur. J. Canc. Clin. Oncol. 18:13-31.
63. Ferrali, M., R. Fulceri, A. Benedetti, and M. Comporti (1980) Effects of carbonyl compounds (4-hydroxyalkenals) originating from the peroxidation of liver microsomal lipids on various microsomal enzyme

activities of the liver. Res. Commun. Chem. Pathol. Pharmacol. 30: 99-112.

64. Fink, D.J., and D. Kritchevsky, eds. (1981) Workshop on fat and cancer. Cancer Res. 41:3677-3679.

65. Flanders, W.D., and K.J. Rothman (1982) Interaction of alcohol and tobacco in laryngeal cancer. Am. J. Epidemiol. 115:371-379.

66. Flohe, L. (1982) Glutathione peroxidase brought into focus. In Free Radicals in Biology, Vol. V, W.A. Pryor, ed. Academic Press, New York, pp. 223-254.

67. Foote, C.S. (1982) Light, oxygen, and toxicity. In Pathology of Oxygen, A. Autor, ed. Academic Press, New York, pp. 21-44.

68. Frenkel, K., M.S. Goldstein, and G.W. Teebor (1981) Identification of the cis-thymine glycol moiety in chemically oxidized and γ-irradiated deoxyribonucleic acid by high-pressure liquid chromatography analysis. Biochemistry 20:7566-7571.

69. Fridovich, I. (1982) Superoxide dismutase in biology and medicine. In Pathology of Oxygen, A. Autor, ed. Academic Press, New York, pp. 1-19.

70. Gold, L.S., C.B. Sawyer, R. Magaw, G.M. Backman, M. de Veciana, R. Levinson, N.K. Hooper, W.R. Havender, L. Bernstein, R. Peto, M.C. Pike, and B.N. Ames (1984) A carcinogenic potency database of the standardized results of animal bioassays. Env. Health Perspect. 58: 9-319.

71. Gold, L.S., M. de Veciana, G. Backman, R. Magaw, P. Lopipero, M. Smith, M. Blumenthal, R. Levinson, J. Gerson, and B.N. Ames (1985) Chronological supplement to the carcinogenic potency database: Standardized results of animal bioassays published through December 1982. Env. Health Perspect. (in press).

72. Griffin, A.C. (1982) The chemopreventive role of selenium in carcinogenesis. In Molecular Interrelations of Nutrition and Cancer, M.S. Arnott, J. Vaneys, and Y.M. Wang, eds. Raven Press, New York, pp. 401-408.

73. Halliwell, B. (1982) Production of superoxide, hydrogen peroxide and hydroxyl radicals by phagocytic cells: A cause of chronic inflammatory disease? Cell Biol. Intl. Reports 6:529-542.

74. Hanham, A.F., B.P. Dunn, and H.F. Stich (1983) Clastogenic activity of caffeic acid and its relationship to hydrogen peroxide generated during autoxidation. Mutat. Res. 116:333-339.

75. Hariharan, P.V., and P.A. Cerutti (1974) Excision of damaged thymine residues from gamma-irradiated poly (dA-dT) by crude extracts of Escherichia coli. Proc. Natl. Acad. Sci., USA 71:3534-3536.

76. Hariharan, P.V., J.F. Remsen, and P.A. Cerutti (1974) Excision repair of γ-ray damaged thymine in bacterial and mammalian systems. In Molecular Mechanisms for Repair of DNA, Part A, P.C. Hanawalt and R.C. Setlow, eds. Plenum Press, New York, pp. 51-59.

77. Harman, D. (1981) The aging process. Proc. Natl. Acad. Sci., USA 78: 7124-7128.

78. Harman, D. (1982) The free-radical theory of aging. In Free Radicals in Biology, Vol. V, W.A. Pryor, ed. Academic Press, New York, pp. 255-275.

79. Harris, C.C., and T. Sun (1984) Multifactorial etiology of human liver cancer. Carcinogenesis 5:697-701.

80. Hartge, P., L.P. Lesher, L. McGowan, and R. Hoover (1982) Coffee and ovarian cancer. Int. J. Cancer 30:531-532.

81. Hartman, P.E. (1982) Nitrates and nitrites: Ingestion, pharmacodynamics, and toxicology. In Chemical Mutagens, Vol. 7, F.J. de Serres and A. Hollaender, eds. Plenum Press, New York, pp. 211-294.

82. Hartman, P.E. (1983) Mutagens: Some possible health impacts beyond carcinogenesis. Environ. Mutag. 5:139-152.
83. Hartman, P.E. (1983) Review: Putative mutagens and carcinogens in foods. I. Nitrate/nitrite ingestion and gastric cancer mortality. Environ. Mutag. 5:111-121.
84. Helland, D., I.F. Nes, and K. Kleppe (1982) Mammalian DNA-repair endonuclease acts only on supercoiled DNA. FEBS Lett. 142:121-124.
85. Higginson, J. (1979) Cancer and environment: Higginson speaks out. Science 205:1363-1366.
86. Hinson, J.A., L.R. Pohl, T.J. Monks, and J.R. Gillette (1981) Acetaminophen-induced hepatotoxicity. Life Sci. 29:107-116.
87. Hirono, I. (1981) Natural carcinogenic products of plant origin. CRC Crit. Rev. Toxicol. 8:235-277.
88. Hirota, N., and T. Yokoyama (1981) Enhancing effect of hydrogen peroxide upon duodenal and upper jejunal carcinogenesis in rats. Gann 72:811-812.
89. Hoel, D.G., R.A. Merrill, and F.P. Perera, eds. (1985) Risk Quantitation and Regulatory Policy, Banbury Report 19, Cold Spring Harbor Laboratory, Cold Spring Harbor, New York.
90. Hoey, J., C. Montvernay, and R. Lambert (1981) Wine and tobacco: Risk factors for gastric cancer in France. Am. J. Epidemiol. 113:668-674.
91. Hollstein, M.C., P. Brooks, S. Linn, and B.N. Ames (1984) Hydroxymethyluracil DNA glycosylase in mammalian cells. Proc. Natl. Acad. Sci., USA 81:4003-4007.
92. Horie, S., H. Ishii, and T. Suga (1981) Changes in peroxisomal fatty acid oxidation in the diabetic rat liver. J. Biochem. 90:1691.
93. Hunter, J.E. (1982) Trans fatty acids in tumor development. Letter to the Editor (reply to A.B. Awad). J. Natl. Cancer Inst. 69:319-321.
94. Iida, S., and H. Hayatsu (1971) The permanganate oxidation of deoxyribonucleic acid. Biochim. Biophys. Acta 240:370-375.
95. Imai, H., N.T. Werthessen, V. Subramanyam, P.W. LeQuesne, A.H. Soloway, and M. Kanisawa (1980) Angiotoxicity of oxygenated sterols and possible precursors. Science 207:651-653.
96. International Agency for Research on Cancer (1985) Allyl Compounds, Aldehydes, Epoxides, and Peroxides, IARC Monographs on the Evaluation of the Carcinogenic Risk of Chemicals to Humans, Vol. 36, International Agency for Research on Cancer, Lyon, France.
97. Ito, A., M. Naito, Y. Naito, and H. Watanabe (1982) Induction and characterization of gastro-duodenal lesions in mice given continuous oral administration of hydrogen peroxide. Gann 73:315-322.
98. Jacobs, M.M. (1983) Selenium inhibition of 1,2-dimethylhydrazine-induced colon carcinogenesis. Cancer Res. 43:1646-1649.
99. Jones, D.P., L. Eklow, H. Thor, and S. Orrenius (1981) Metabolism of hydrogen peroxide in isolated hepatocytes: Relative contributions of catalase and glutathione peroxidase in decomposition of endogenously generated H_2O_2. Arch. Biochem. Biophys. 210:505-516.
100. Jones, T.D. (1984) A unifying concept for carcinogenic risk assessments: Comparison with radiation-induced leukemia in mice and men. Health Physics 4:533-558.
101. Kallistratos, G., and E. Fasske (1980) Inhibition of benzo(a)pyrene carcinogenesis in rats with vitamin C. J. Cancer Res. Clin. Oncol. 97:91-96.
102. Kapadia, G.J., ed. (1982) Oncology Overview on Naturally Occurring Dietary Carcinogens of Plant Origin, International Cancer Research Data Bank Program, National Cancer Institute, Bethesda. Maryland.

103. Kasai, H., and S. Nishimura (1984) Hydroxylation of deoxyguanosine at the C-8 position by polyphenols and aminophenols in the presence of hydrogen peroxide and ferric ion. Gann 75:565-566.
104. Kasai, H., K. Kumeno, Z. Yamaizumi, S. Nishimura, M. Nagao, Y. Fujita, T. Sugimura, H. Nukaya, and T. Kosuge (1982) Mutagenicity of methylglyoxal in coffee. Gann 73:681-683.
105. Kinlen, L.J. (1983) Fat and cancer. Brit. Med. J. 286:1081-1082.
106. Kitamura, K., C.-I. Wei, and T. Shibamoto (1981) Mutagenicity of maillard browning products obtained from a starch-glycine model system. J. Agric. Food 24:378-380.
107. Kling, L.J., and J.H. Soares, Jr. (1981) The effect of vitamin E and dietary linoleic acid on mercury toxicity. Nutr. Reports Intl. 24: 39-46.
108. Krinsky, N.I., and S.M. Deneke (1982) Interaction of oxygen and oxy-radicals with carotenoids. J. Natl. Cancer Inst. 69:205-209.
109. Kvale, G., E. Bjelke, and J.J. Gart (1983) Dietary habits and lung cancer risk. Int. J. Cancer 31:397-405.
110. Lam, L.K.T., V.L. Sparnins, and L.W. Wattenberg (1982) Isolation and identification of kahweol palmitate and cafestol palmitate as active constituents of green coffee beans that enhance glutathione S-transferase activity in the mouse. Cancer Res. 42:1193-1198.
111. Levin, D.E., M. Hollstein, M.F. Christman, E. Schwiers, and B.N. Ames (1982) A new Salmonella tester strain (TA102), with A:T base pairs at the site of mutation, detects oxidative mutagens. Proc. Natl. Acad. Sci., USA 79:7445-7449.
112. Levin, D.E., L.J. Marnett, and B.N. Ames (1984) Spontaneous and mutagen-induced deletions: Mechanistic studies in Salmonella tester strain TA102. Proc. Natl. Acad. Sci., USA 81:4457-4461.
113. Lind, C., P. Hochstein, and L. Ernster (1982) DT-Diaphorase as a quinone reductase: A cellular control device against semiquinone and superoxide radical formation. Arch. Biochem. Biophys. 216:178-185.
114. Lindahl, T. (1982) DNA-repair enzymes, review. Ann. Rev. Biochem. 51:61-87.
115. MacMahon, B. (1982) Risk factors for cancer of the pancreas. Cancer 50:2676-2680.
116. Magee, P.N., ed. (1982) Nitrosamines and Human Cancer, Banbury Report 12, Cold Spring Harbor Laboratory, Cold Spring Harbor, New York.
117. Marnett, L.J., H. Hurd, M. Hollstein, D.E. Levin, H. Esterbauer, and B.N. Ames (1985) Naturally occurring carbonyl compounds are mutagens in Salmonella tester strain TA104. Mutat. Res. 148:25-34.
118. Marrett, L.D., S.D. Walter, and J.W. Meigs (1983) Coffee drinking and bladder cancer in Connecticut. Am. J. Epidemiol. 117:113-127.
119. Mathews-Roth, M.M. (1982) Antitumor activity of β-carotene, canthaxanthin and phytoene. Oncology 39:33-37.
120. Mathews-Roth, M.M. (1982) Photosensitization by porphyrins and prevention of photosensitization by carotenoids. J. Natl. Cancer Inst. 69:279-285.
121. Mattern, M.R., P.V. Hariharan, and P.A. Cerutti (1975) Selective excision of gamma ray damaged thymine from the DNA of cultured mammalian cells. Biochim. Biophys. Acta 395:48-55.
122. Medina, D., H.W. Lane, and C.M. Tracey (1983) Selenium and mouse mammary tumorigenesis: An investigation of possible mechanisms. Cancer Res. 43(Suppl.):2460s-2464s.
123. Mirsalis, J.C., C.K. Tyson, E.N. Loh, K.L. Steinmetz, J.P. Bakke, C.M. Hamilton, D.K. Spak, and J.W. Spalding (1985) Induction of hepatic cell proliferation and unscheduled DNA synthesis in mouse hepatocytes following in vivo treatment. Carcinogenesis (in press).

124. Nagao, M., Y. Fujita, and T. Sugimura (1985) Methylglyoxal in beverages and foods--Its mutagenicity and carcinogenicity. IARC Workshop (in press).
125. National Academy of Sciences, Committee on Nitrite and Alternative Curing Agents on Food, Assembly of Life Sciences (1981) The Health Effects of Nitrate, Nitrite, and N-Nitroso Compounds, National Academy Press, Washington, D.C.
126. National Research Council (1982) Diet, Nutrition and Cancer, National Academy Press, Washington, D.C.
127. Neat, C.E., M.S. Thomassen, and H. Osmundsen (1981) Effects of high-fat diets on hepatic fatty acid oxidation in the rat. Biochem. J. 196:149-159.
128. Nes, I.F. (1980) Purification and properties of a mouse-cell DNA-repair endonuclease, which recognizes lesions in DNA induced by ultraviolet light, depurination, γ-rays, and OsO_4 treatment. Eur. J. Biochem. 112:161-168.
129. Newmark, H.L., M.J. Wargovich, and W.R. Bruce (1984) Colon cancer and dietary fat, phosphate, and calcium: A hypothesis. J. Natl. Cancer Inst. 72:1323-1325.
130. Norseth, J., and M.S. Thomassen (1983) Stimulation of microperoxisomal β-oxidation in rat-heart by high-fat diets. Biochim. Biophys. Acta 751:312-320.
131. Novi, A.M. (1980) Regression of aflatoxin B_1-induced hepatocellular carcinomas by reduced glutathione. Science 212:541-542.
132. Nygaard, O.F., and M.G. Simic, eds. (1983) Radioprotectors and Anticarcinogens, Academic Press, New York.
133. Oberley, L.W., T.D. Oberley, and G.R. Buettner (1980) Cell differentiation, aging and cancer: The possible roles of superoxide and superoxide dismutases. Med. Hypoth. 6:249-268.
134. Osmundsen, H. (1982) Factors which can influence β-oxidation by peroxisomes isolated from livers of clofibrate treated rats. Some properties of peroxisomal fractions isolated in a self-generated Percoll gradient by vertical rotor centrifugation. Int. J. Biochem. 14:905-914.
135. Packer, J.E., J.S. Mahood, V.O. Mora-Arellano, T.F. Slater, R.L. Willson, and B.S. Wolfenden (1981) Free radicals and singlet oxygen scavengers: Reaction of a peroxy-radical with β-carotene, diphenyl furan and 1,4-diazobicyclo(2,2,2)-octane. Biochem. Biophys. Res. Commun. 98:901-906.
136. Pariza, M.W., L.J. Loretz, J.M. Storkson, and N.C. Holland (1983) Mutagens and modulator of mutagenesis in fried ground beef. Cancer Res. 43(Suppl.):2444s-2446s.
137. Pegel, K.H. (1981) Coffee's link to cancer. Chem. Eng. News Vol. 59 (July 20), p. 4.
138. Peto, R. (1979) Detection of risk of cancer to man. Proc. R. Soc. London Ser. B 205:111-120.
139. Peto, R. (1985) Epidemiological reservations about risk assessment. In Assessment of Risk from Low-Level Exposure to Radiation and Chemicals, A.D. Woodhead, C.J. Shellabarger, and V. Pond, eds. Plenum Press, New York, pp. 3-16.
140. Peto, R., R. Doll, J.D. Buckley, and M.B. Sporn (1981) Can dietary beta-carotene materially reduce human cancer rates? Nature 290:201-208.
141. Peto, R., M.C. Pike, L. Bernstein, L.S. Gold, and B.N. Ames (1984) The TD_{50}: A proposed general convention for the numerical description of the carcinogenic potency of chemicals in chronic-exposure animal experiments. Env. Health Perspect. 58:1-8.

142. Petrakis, N.L., L.D. Gruenke, and J.C. Craig (1981) Cholesterol and cholesterol epoxides in nipple aspirates of human breast fluid. Cancer Res. 41:2563-2565.
143. Phillips, R.L., and D.A. Snowden (1983) Association of meat and coffee use with cancers of the large bowel, breast, and prostate among Seventh-Day Adventists: Preliminary results. Cancer Res. 43 (Suppl.):2403s-2408s.
144. Plaine, H.L. (1955) The effect of oxygen and of hydrogen peroxide on the action of a specific gene and on tumor induction in Drosophila melanogaster. Genetics 40:268-280.
145. Porter, R., and J. Whelan (1983) Biology of Vitamin E, CIBA Symposium 101, Pitman Press, Bath, United Kingdom.
146. Preston, S.H., N. Keyfitz, and R. Schoen (1972) Causes of Death, Life Tables for National Populations, Seminar Press, New York.
147. Pryor, W.A., ed. (1976-1982) Free Radicals in Biology, Vols. I-V, Academic Press, New York.
148. Pryor, W.A., B.J. Hales, P.I. Premovic, and D.F. Church (1983) The radicals in cigarette tar: Their nature and suggested physiological implications. Science 220:425-427.
149. Pryor, W.A., M. Tamura, M.M. Dooley, P.I. Premovic, and D.F. Church (1983) Reactive oxy-radicals from cigarette smoke and their physiological effects. In Oxy-Radicals and Their Scavenger Systems: Cellular and Medical Aspects, Vol. 2, G. Cohen and R. Greenwald, eds. Elsevier Press, Amsterdam, pp. 185-192.
150. Reddy, J.K., and N.D. Lalwani (1983) Carcinogenesis by hepatic peroxisome proliferators: Evaluation of the risk of hypolipidemic drugs and industrial plasticizers to humans. CRC Crit. Rev. Toxicol. 12: 1-58.
151. Reddy, J.K., J.R. Warren, M.K. Reddy, and N.D. Lalwani (1982) Hepatic and renal effects of peroxisome proliferators: Biological implications. Ann. N.Y. Acad. Sci. 386:81-110.
152. Rettura, G., C. Dattagupta, P. Listowsky, S.M. Levenson, and E. Seifter (1983) Dimethylbenz(a)anthracene (DMBA) induced tumors: Prevention by supplemental β-carotene (BC). Fed. Proc. 42:786.
153. Roe, F.J.C. (1981) Are nutritionists worried about the epidemic of tumors in laboratory animals? Proc. Nutr. Soc. 40:57-65.
154. Rosett, H.L., L. Weiner, A. Lee, B. Zuckerman, E. Dooling, and E. Oppenheimer (1983) Patterns of alcohol consumption and fetal development. Obstet. Gynecol. 61:539-546.
155. Rosin, M.P., H.F. Stich, W.D. Powrie, and C.H. Wu (1982) Induction of mitotic gene conversion by browning reaction products and its modulation by naturally occurring agents. Mutat. Res. 101:189-197.
156. Ross, M.H., E.D. Lustbader, and G. Bras (1982) Dietary practices of early life and spontaneous tumors of the rat. Nutr. Cancer 3:150-167.
157. Ruttenberg, H., L.M. Davidson, N.A. Little, D.M. Klurfeld, and D. Kritchevsky (1983) Influence of trans unsaturated fats on experimental atherosclerosis in rabbits. J. Nutr. 113:835-844.
158. Salonen, J.T., G. Alfthan, J.K. Huttunen, and P. Puska (1982) Association between cardiovascular death and myocardial infarction and serum selenium in a matched-pair longitudinal study. Lancet 2:175-179.
159. Salonen, J.T., R. Salonen, R. Lappeteläinen, P.H. Mäenpää, G. Althan, and P. Puska (1985) Risk of cancer in relation to serum concentrations of selenium and vitamins A and E: Matched case-control analysis of prospective data. Brit. Med. J. 290:417.
160. Samet, J.M., B.J. Skipper, C.G. Humble, and D.R. Pathak (1985) Lung cancer risk and vitamin A consumption in New Mexico. Am. Rev. Resp. Diseases 131:198-202.

161. Samuels, S.W., and R.H. Adamson (1985) Quantitative risk assessment: Report of the Subcommittee on Environmental Carcinogenesis, National Cancer Advisory Board. J. Natl. Cancer Inst. 74:945-951.
162. Saul, R.L., and B.N. Ames (1985) Background levels of DNA damage in the population. In Mechanisms of DNA Damage and Repair, M. Simic, L. Grossman, and A. Upton, eds. Plenum Press, New York (in press).
163. Sevanian, A., and A.R. Peterson (1984) Cholesterol epoxide is a direct-acting mutagen. Proc. Natl. Acad. Sci., USA 81:4198-4202.
164. This Volume.
165. Shekelle, R.B., S. Liu, W.J. Raynor, Jr., M. Lepper, C. Maliza, A.H. Rossof, O. Paul, A. Macmillan-Shryock, and J. Stamler (1981) Dietary vitamin A and risk of cancer in the Western Electric study. Lancet 2:1185-1190.
166. Shorland, F.B., J.O. Igene, A.M. Pearson, J.W. Thomas, R.K. McGuffey, and A.E. Aldridge (1981) Effects of dietary fat and vitamin E on the lipid composition and stability of veal during frozen storage. J. Agric. Food Chem. 29:863-871.
167. Simic, M.G., and M. Karel, eds. (1980) Autoxidation in Food and Biological Systems, Plenum Press, New York.
168. Simpson, K.L., and C.O. Chichester (1981) Metabolism and nutritional significance of carotenoids. Ann. Rev. Nutr. 1:351-374.
169. Sohal, R.S. (1981) Metabolic rate, aging, and lipofuscin accumulation. In Age Pigments, R.S. Sohal, ed. Elsevier/North-Holland Biomedical Press, New York, pp. 303-316.
170. Speit, G., W. Vogel, and M. Wolf (1982) Characterization of sister chromatid exchange induction by hydrogen peroxide. Env. Mutag. 4: 135-142.
171. Speizer, F.E. (1983) Assessment of the epidemiological data relating lung cancer to air pollution. Env. Health Perspect. 47:33-42.
172. Stege, T.E. (1982) Acetaldehyde-induced lipid peroxidation in isolated hepatocytes. Res. Commun. Chem. Pathol. Pharmacol. 36:287-297.
173. Stich, H.F., and M.P. Rosin (1983) Quantitating the synergistic effect of smoking and alcohol consumption with the micronucleus test on human buccal mucosa cells. Int. J. Cancer 31:305-308.
174. Stich, H.F., and M.P. Rosin (1984) Naturally occurring phenolics as antimutagenic and anticarcinogenic agents. In Nutritional and Toxicological Aspects of Food Safety, M. Friedman, ed. Plenum Press, New York, pp. 1-29.
175. Stich, H.F., M.P. Rosin, C.H. Wu, and W.D. Powrie (1981) A comparative genotoxicity study of chlorogenic acid (3-O-caffeoylquinic acid). Mutat. Res. 90:201-212.
176. Stich, H.F., M.P. Rosin, C.H. Wu, and W.D. Powrie (1981) The action of transition metals on the genotoxicity of simple phenols, phenolic acids, and cinnamic acids. Cancer Lett. 14:251-260.
177. Stich, H.F., W. Stich, M.P. Rosin, and W.D. Powrie (1981) Clastogenic activity of caramel and caramelized sugars. Mutat. Res. 91:129-136.
178. Stich, H.F., P.K.L. Chan, and M.P. Rosin (1982) Inhibitory effects of phenolics, teas and saliva on the formation of mutagenic nitrosation products of salted fish. Int. J. Cancer 30:719-724.
179. Stich, H.F., M.P. Rosin, C.H. Wu, and W.D. Powrie (1982) The use of mutagenicity testing to evaluate food products. In Mutagenicity: New Horizons in Genetic Toxicology, J.A. Heddle, ed. Academic Press, New York, pp. 117-142.
180. Stott, W.T., R.H. Reitz, A.M. Schumann, and P.G. Watanabe (1981) Genetic and nongenetic events in neoplasia. Food Cosmet. Toxicol. 19: 567-576.

181. Suematsu, T., T. Matsumura, N. Sato, T. Miyamoto, T. Ooka, T. Kamada, and H. Abe (1981) Lipid peroxidation in alcoholic liver disease in humans. Alcoholism: Clin. Exp. Res. 5:427-430.
182. Sugimura, T., and M. Nagao (1982) The use of mutagenicity to evaluate carcinogenic hazards in our daily lives. In Mutagenicity: New Horizons in Genetic Toxicology, J.A. Heddle, ed. Academic Press, New York, pp. 73-88.
183. Sugimura, T., and S. Sato (1983) Mutagens-carcinogens in foods. Cancer Res. 43(Suppl.):2415s-2421s.
184. Sullivan, J.L. (1982) Superoxide-dismutase, longevity and specific metabolic-rate. Gerontology 28:242-244.
185. Tateishi, N., T. Higashi, A. Naruse, K. Hikita, and Y. Sakamoto (1981) Relative contributions of sulfur atoms of dietary cysteine and methionine to rat liver glutathione and proteins. J. Biochem. 90: 1603-1610.
186. Tauber, A.I. (1982) The human neutrophil oxygen armory. TIBS 7:411-414.
187. Tazima, Y. (1982) Mutagenic and carcinogenic mycotoxins. In Environmental Mutagenesis, Carcinogenesis and Plant Biology, Vol. 1, E.J. Klekowski, Jr., ed. Praeger, New York, pp. 68-95.
188. Teebor, G.W., K. Frenkel, and M.S. Goldstein (1984) Ionizing radiation and tritium transmutation both cause formation of 5-hydroxymethyl-2'-deoxyuridine in cellular DNA. Proc. Natl. Acad. Sci., USA 81: 318-321.
189. Teoule, R., C. Bert, and A. Bonicel (1977) Thymine fragment damaged retained in DNA polynucleotide chain after gamma-irradiation in aerated solutions. 2. Radiat. Res. 72:190-200.
190. Thomassen, M.S., E.N. Christiansen, and K.R. Norum (1982) Characterization of the stimulatory effect of high-fat diets on peroxisomal β-oxidation in rat liver. Biochem. J. 206:195-202.
191. Thompson, H.J., L.D. Meeker, P.J. Becci, and S. Kokoska (1982) Effect of short-term feeding of sodium selenite on 7,12-dimethylbenz(a)anthracene-induced mammary carcinogenesis in the rat. Cancer Res. 42: 4954-4958.
192. Tolmasoff, J.M., T. Ono, and R.G. Cutler (1980) Superoxide dismutase: Correlation with life-span and specific metabolic rate in primate species. Proc. Natl. Acad. Sci., USA 77:2777-2781.
193. Totter, J.R. (1980) Spontaneous cancer and its possible relationship to oxygen metabolism. Proc. Natl. Acad. Sci., USA 77:1763-1767.
194. Trichopoulos, D., M. Papapostolou, and A. Polychronopoulou (1981) Coffee and ovarian cancer. Int. J. Cancer 28:691-693.
195. Trosko, J.E., and C.C. Chang (1985) Implications for risk assessment of genotoxic and non-geneotoxic mechanisms in carcinogenesis. In Methods for Estimating Risk of Chemical Injury: Human and Non-Human Biota and Ecosystems, V.B. Vouk, G.C. Butler, D.G. Hoel, and D.B. Peakall, eds. John Wiley, New York, pp. 181-200.
196. Trosko, J.E., and C.C. Chang (1985) Role of tumor promotion in affecting the multi-hit nature of carcinogenesis. In Assessment of Risk from Low-Level Exposure to Radiation and Chemicals. A Critical Overview, A.D. Woodhead, C.J. Shellabarger, V. Pond, and A. Hollaender, eds. Plenum Press, New York, Basic Life Sciences, Vol. 33, pp. 261-284.
197. Trush, M.A., E.G. Mimnaugh, and T.E. Gram (1982) Activation of pharmacologic agents to radical intermediates: Implications for the role of free radicals in drug action and toxicity. Biochem. Pharmacol. 31:3335-3346.
198. Tsuda, H. (1981) Chromosomal aberrations induced by hydrogen peroxide in cultured mammalian cells. Japan. J. Genet. 56:1-8.

199. Turner, J.A., and J.N. Prebble (1980) Protection of cell viability and respiratory quinone levels by carotenoid in *Micrococcus lysodeikticus* (*M. luteus*). *J. Gen. Microbiol.* 119:133-144.
200. Tuyns, A. (1982) Alcohol. In *Cancer Epidemiology and Prevention*, D. Schottenfeld and J.F. Fraumeni, Jr., eds. W.B. Saunders Company, Philadelphia, pp. 293-303.
201. Tuyns, A.J., G. Pequignot, M. Gignoux, and A. Valla (1982) Cancers of the digestive tract, alcohol and tobacco. *Int. J. Cancer* 30:9-11.
202. Van der Hoeven, J.C.M., W.J. Lagerwey, C.A.J.M. Meeuwissen, P.C.M. Hauwert, A.G.J. Voragen, and J.H. Koeman (1982) Mutagens in food products of plant origin. In *Mutagens in Our Environment*, M. Sorsa and H. Vainio, eds. Alan R. Liss, Inc., New York, pp. 327-338.
203. Van der Hoeven, J.C., W.J. Lagerweij, I.M. Bruggeman, F.G. Voragen, and J.H. Koeman (1983) Mutagenicity of extracts of some vegetables commonly consumed in the Netherlands. *J. Agric. Food Chem.* 17:1020-1026.
204. Van Rensburg, W.J., P. Cook-Mozaffari, D.J. Van Schalkwyk, J.J. Van Der Watt, T.J. Vincent, and I.F. Purchase (1985) Hepatocellular carcinoma and dietary aflatoxin in Mozambique and Transkei. *Brit. J. Cancer* 51:713-726.
205. Vena, J.E. (1982) Air pollution as a risk factor in lung cancer. *Am. J. Epidemiology* 116:42-56.
206. Videla, L.A., V. Fernandez, A. de Marinis, N. Fernandez, and A. Valenzuela (1982) Liver lipoperoxidative pressure and glutathione status following acetaldehyde and aliphatic alcohols pretreatments in the rat. *Biochem. Biophys. Res. Commun.* 104:965-970.
207. Wang, Y.M., S.K. Howell, J.C. Kimball, C.C. Tsai, J. Sato, and C.A. Gleiser (1982) Alpha-tocopherol as a potential modifier of daunomycin carcinogenicity in Sprague-Dawley rats. In *Molecular Interrelations of Nutrition and Cancer*, M.S. Arnott, J. van Eys, and Y.-M. Wang, eds. Raven Press, New York, pp. 369-379.
208. Wargovich, M.J., V.W.S. Eng, and H.L. Newmark (1984) Calcium inhibits the damaging and compensatory proliferative effects of fatty acids on mouse colon epithelium. *Cancer Lett.* 23:253-258.
209. Warholm, M., C. Guthenberg, B. Mannervik, and C. von Bahr (1981) Purification of a new glutathione *S*-transferase (transferase μ) from human liver having high activity with benzo(α)pyrene-4,5-oxide. *Biochem. Biophys. Res. Commun.* 98:512-519.
210. Wassertheil-Smoller, S., S.L. Romney, J. Wylie-Rosett, S. Slagle, G. Miller, D. Lucido, C. Duttagupta, and P.R. Palan (1981) Dietary vitamin C and uterine cervical dysplasia. *Am. J. Epidemiol.* 114:714-724.
211. Wattenberg, L.W. (1983) Inhibition of neoplasia by minor dietary constituents. *Cancer Res.* 43(Suppl.):2448s-2453s.
212. Weinberg, D.M., R.K. Ross, T.M. Mack, A. Paganini-Hill, and B.E. Henderson (1983) Bladder cancer etiology. *Cancer* 51:675-680.
213. Weindruch, R. (1984) Dietary restriction and the aging process. In *Free Radicals in Molecular Biology, Aging, and Disease*, D. Armstrong, R.S. Sohal, R.G. Cutler, and T.F. Slater, eds. Raven Press, New York, pp. 181-202.
214. Weitberg, A.B., S.A. Weitzman, M. Destrempes, S.A. Latt, and T.P. Stossel (1983) Stimulated human phagocytes produce cytogenetic changes in cultured mammalian cells. *New Engl. J. Med.* 308:26-30.
215. Welsch, C.W., and C.F. Aylsworth (1983) Enhancement of murine mammary tumorigenesis by feeding high levels of dietary fat: A hormonal mechanism? *J. Natl. Cancer Inst.* 70:215-221.
216. Whipple, C. (1985) Redistributing risk. *Regulation* 9:37-44.

217. Willett, W.C., J.S. Morris, S. Pressel, J.O. Taylor, B.F. Polk, M.J. Stampfer, B. Rosner, K. Schneider, and C.G. Hames (1983) Prediagnostic serum selenium and risk of cancer. Lancet 2:130-134.

218. Williamson, J.M., B. Boettcher, and A. Meister (1982) Intracellular cysteine delivery system that protects against toxicity by promoting glutathione synthesis. Proc. Natl. Acad. Sci., USA 79:6246-6249.

219. Wilson, R., E. Crouch, and L. Zeise (1985) Uncertainty in risk assessment. In Risk Quantitation and Regulatory Policy, Banbury Report 19, Cold Spring Harbor Laboratory, Cold Spring Harbor, New York, pp. 133-147.

220. Winston, G.W., and A.I. Cederbaum (1982) A correlation between hydroxyl radical generation and ethanol oxidation by liver, lung and kidney microsomes. Biochem. Pharmacol. 31:2031-2037.

221. Witting, C., U. Witting, and V. Krieg (1982) The tumor-protective effect of selenium in an experimental model. J. Cancer Res. Clin. Oncol. 104:109-113.

222. Wood, A.W., M.-T. Huang, R.L. Chang, H.L. Newmark, R.E. Lehr, H. Yagi, J.M. Sayer, D.M. Jerina, and A.H. Conney (1982) Inhibition of the mutagenicity of bay-region diol epoxides of polycyclic aromatic hydrocarbons by naturally occurring plant phenols: Exceptional activity of ellagic acid. Proc. Natl. Acad. Sci., USA 79:5513-5517.

223. Wood, R. (1979) Incorporation of dietary cis and trans octadecenoate isomers in the lipid classes of various rat tissues. Lipids 14:975-982.

224. Yamasaki, E., and B.N. Ames (1977) Concentration of mutagens from urine by adsorption with the nonpolar resin XAD-2: Cigarette smokers have mutagenic urine. Proc. Natl. Acad. Sci., USA 74:3555-3559.

225. Yonaha, M., E. Itoh, Y. Ohbayashi, and M. Uchiyama (1980) Induction of lipid peroxidation in rats by mercuric chloride. Res. Commun. Chem. Pathol. Pharmacol. 28:105-112.

226. Yu, B.P., E.J. Masoro, I. Murata, H.A. Bertrand, and F.T. Lynd (1982) Lifespan study of SPF Fischer-344 male rats fed ad libitum or restricted diets--Longevity, growth, lean body mass and disease. J. Gerontol. 37:130-141.

227. Zeise, L., R. Wilson, and E. Crouch (1984) Use of acute toxicity to estimate carcinogenic risk. Risk Analysis 4:187-199.

228. Ziegler, R.G., L.E. Morris, W.J. Blot, L.M. Pottern, R. Hoover, and J.F. Fraumeni, Jr. (1981) Esophageal cancer among black men in Washington, D.C. II. Role of nutrition. J. Natl. Cancer Inst. 67:1199-1206.

ACTIVATION AND INACTIVATION OF MUTAGENS AND CARCINOGENS--PART I

THE RELATION OF ACTIVATION AND INACTIVATION TO ANTIMUTAGENIC PROCESSES

R.C. von Borstel

Department of Genetics
University of Alberta
Edmonton, Alberta, Canada T6G 2E9

INTRODUCTION

Antimutagenesis is the lowering of spontaneous mutation rates and induced mutation frequencies by exogenous conditions. The field has been divided into two parts by Kada (5): desmutagenesis and antimutagenesis per se. Desmutagenesis refers to inactivation of mutagens and carcinogens by whatever means, whether by antioxidants, by cytochrome P-450-associated activity, or by the active destruction or absorption of the toxic substances before they can enter the target cell. Antimutagenesis then becomes either the enhancement of the nonmutagenic repair of DNA, the inactivation of mutagenic repair, the inactivation of the SOS repair processes, or some interference with the expression of mutations.

Even though the functions of antimutagenic agents may be quite different, the endpoint of analysis is usually the same, that is, a reduction of mutation rates and frequencies (19). The processes that can lead to a reduction of a mutational yield are listed in Tab. 1. This list is merely a simple summary of the major processes leading to a reduction of mutations and cancer. Each item can be divided and subdivided.

DESMUTAGENESIS

The desmutagenic mechanisms of mutagen absorption and mutagen inactivation are simple to comprehend conceptually. Electrophilic compounds actively do their best to be neutralized; only when DNA is the compound that does the neutralizing is the cell subject to a mutational event, and then only when nonmutagenic DNA repair mechanisms do not repair all of the damage.

The desmutagenic mechanisms typified by the polysubstrate monooxygenases, the cytochrome P-450 species, are large in number (7,16), and are now being shown to be controlled by a large number of genes within gene families. There are at least four gene families of cytochrome P-450 terminal oxidase in the rat genome, as defined by chromosomal location and by the substances that induce them (3). The promoter regions of these genes

Tab. 1. Processes that lead to reduced mutational yield.

Detoxification (desmutagenesis)	Antimutagenesis
Extracellular	Induction of nonmutagenic repair.
Absorption by inert materials.	Inhibition of mutagenic repair.
Inactivation by antioxidants.	Inactivation of SOS repair.
Inactivation by other agents.	Inhibition of mitosis (given time, nonmutagenic DNA repair prevails over mutagenic DNA repair).
Intracellular	
Induction of polysubstrate monooxygenase (cytochrome P-450).	Interference with processes of phenotypic expression.
Inactivation of polysubstrate monooxygenase (cytochrome P-450).	
Inactivation by other agents.	

respond to different toxic inducers, although some toxic substances may induce the activity of more than one family of cytochrome P-450 genes. There is more that we need to know, such as the extent of the activity latent in the soluble form (cytochrome P-420) of cytochrome P-450 (21), as well as the possible association of the terminal oxidases with the other enzymes of a larger, detoxifying enzyme network.

In Tab. 1 it can be seen that inactivation of cytochrome P-450 by exogenous substances, e.g., polyaromatic hydrocarbons, is desmutagenic (15). If a promutagen cannot be turned into a mutagen, mutational yield will be nonexistent. The other side of the coin is that if induction of a terminal oxidase results in detoxification of a mutagen into a nonmutagenic form, mutational yield will be reduced.

The role of free radicals in mutagenesis and carcinogenesis has been hypothesized ever since it was found that ionizing radiation induces free radicals in water. Now it is known that free radicals are liberated under a variety of conditions (1,8,11). Free radicals will act upon DNA unless they are inactivated by enzymes such as catalase or superoxide dismutase, or unless they are trapped by natural antioxidants such as cysteine or glutathione.

The inactivation of free radicals by radical traps has a long history. The enhancement of cell survival and effective dose reduction of mutational processes by radical traps, such as cysteine, β-aminoethylisothiouronium, glutathione, and cysteamine, was a popular field of study during the 1950s. It was during these years that we performed an experiment using cysteamine on *Escherichia coli*. This experiment was never published, but a brief discussion of it now will serve to demonstrate that a "radical trap" may in fact work by doing something "radically" different.

Szilard (17) developed the notion that the aging process was caused by an accumulation of spontaneous mutations. We attempted to decrease the spontaneous mutation rate in the histidine-requiring mutant h^- of *E. coli*

(13) by adding increasing amounts of cysteamine to the culture. Conceivably, if Szilard were right, and the spontaneous mutation rate in *E. coli* were cut in half by cysteamine, a comparable dose reduction in people might double their life span.

What occurred was not expected. At the lowest concentrations of cysteamine that would reduce the effects of X-radiation (then believed to be due to radical trapping), the cells would no longer divide. In other words, the cysteamine blocked mitosis, permitting nonmutagenic DNA repair to function. In routine radical trapping experiments, the cells are washed after irradiation, so that after a brief pause, long enough for nonmutagenic repair to take place, the cells again begin to grow. The lesson is that experiments on radical trapping per se can lead to the misinterpretation of data, unless cell division rates are controlled carefully.

ANTIMUTAGENESIS

We now know that spontaneous mutation rates can be reduced by at least 90% in yeast by blocking the mutagenic repair systems (12). So far, this has been accomplished by obtaining antimutator mutants. Chemical inhibitors of mutagenic repair in yeast are conceivable, but these have not yet been found, except for the report on the antimutagenic activity of selenium (4). Of course, inhibition of mutagenic repair can take place at extremely high doses of a mutagenic agent when the mutagenic repair system becomes saturated with lesions in the DNA (18,20), but this is a different kind of chemical inhibition from the type we would like to obtain with routine antimutagens. Kada and his collaborators (5,9) have found a number of compounds that reduce SOS inducibility, but it must be kept in mind that this is different from chemical inhibition of mutagenic DNA repair systems which are already functional.

It is known that the spontaneous mutation rate is correlated with temperature, and thus with growth rate, in yeast (R.C. von Borstel and C.M. Steinberg, unpubl. data), but not in *E. coli* (6,14), which indicates that there are marked species differences in the processes that control the spontaneous mutation rate. On the other hand, it is known that, when cells are held in suspension in buffer after irradiation or chemical treatment, the mutation frequency declines as a function of time held before plating, both in *E. coli* (cf. Ref. 2) and yeast (10).

In both these organisms, nonmutagenic repair processes function when cell division is inhibited. It is usually assumed that this liquid-holding recovery is the normal excision repair of DNA damage which functions, at least residually, during stationary phase. We believe this because mutants lacking excision-repair processes do not exhibit liquid-holding recovery either in yeast or in *E. coli*. An inhibition of mitosis by a toxic chemical might subject the cells to liquid-holding recovery, and this would be the equivalent of inducing a nonmutagenic repair process. Thus, it is worth noting that, for experiments on antimutagenesis, it would be best to use mutant strains of bacteria, yeast, or mammalian cells that do not exhibit liquid-holding recovery, but are still excision-repair proficient. Such a mutant does exist in yeast (R.C. von Borstel and E. Moustacchi, unpubl. data). These kinds of liquid-holding recovery mutants should be sought in mammalian cells and *E. coli* in order to control properly the experiments on antimutagenesis carried out with these systems.

ACKNOWLEDGEMENTS

This work was carried out with the assistance of grants from the Natural Sciences and Engineering Research Council. The author is grateful to Dr. U.G.G. Hennig for reviewing the paper.

REFERENCES

1. Ames, B.N. (1986) Carcinogens and anticarcinogens (this Volume).
2. Drake, J.W. (1970) The Molecular Basis of Mutation, Holden-Day, San Francisco.
3. Gonzalez, F.J., D.W. Nebert, J.P. Hardwick, and C.B. Kasper (1985) Complete cDNA and protein sequence of a pregnenalone 16α-carbonitrile-induced cytochrome P-450. J. Biol. Chem. 260:7435-7441.
4. Hastings, P.J., and A. Galloway (1984) Do endogenous oxygen radicals damage DNA in yeast? Genetics 107:s45.
5. Kada, T. (1982) Mechanisms and genetic implications of environmental antimutagens. In Environmental Mutagens and Carcinogens, T. Sugimura, S. Kondo, and H. Takebe, eds. University of Tokyo Press, Tokyo, and Alan R. Liss, Inc., New York, pp. 355-359.
6. Kondo, S. (1973) Evidence that mutations are induced by errors in repair and replication. Genetics 73(Suppl.):109-122.
7. Lau, P.E., and H.W. Strobel (1982) Multiple forms of cytochrome P-450 in liver microsomes from β-naphthoflavone-pretreated rats. J. Biol. Chem. 257:5257-5262.
8. Nagao, M., K. Wakabayashi, Y. Suwa, and T. Kobayashi (1986) Inactivation and activation of mutagens by catalase, peroxidase, and superoxide dismutase (this Volume).
9. Ohta, T., K. Watanabe, M. Moriya, Y. Shirasu, and T. Kada (1983) Analysis of the antimutagenic effect of cinnamaldehyde on chemically induced mutagenesis in Escherichia coli. Mol. Gen. Genet. 192:309-315.
10. Patrick, M.H., R.H. Haynes, and R.B. Uretz (1964) Dark recovery phenomena in yeast. I. Comparative effect with various inactivating agents. Radiat. Res. 21:144-164.
11. Pryor, W.A. (1986) Cancer and free radicals (this Volume).
12. Quah, S.-K., P.J. Hastings, and R.C. von Borstel (1980) Spontaneous mutations in yeast. Genetics 96:819-839.
13. Ryan, F.J. (1955) Spontaneous mutation in non-dividing bacteria. Genetics 40:726-738.
14. Ryan, F.J., and K. Kiritani (1959) Effect of temperature on natural mutation in Escherichia coli. J. Gen. Microbiol. 20:644-653.
15. Shahin, M.M., and F. Fournier (1978) Suppression of mutation induction and failure to detect mutagenic activity with Athabasca tar sand fractions. Mutat. Res. 58:29-34.
16. Strobel, H.W. (1986) Cytochromes P-450 and the activation and inactivation of mutagens and carcinogens (this Volume).
17. Szilard, L. (1959) On the nature of the aging process. Proc. Natl. Acad. Sci., USA 45:30-45.
18. von Borstel, R.C. (1982) Thresholds and negative slopes in dose-mutation curves. In Environmental Mutagens and Carcinogens, T. Sugimura, S. Kondo, and H. Takebe, eds. University of Tokyo Press, Tokyo, and Alan R. Liss, Inc., New York, pp. 737-741.
19. von Borstel, R.C., and P.J. Hastings (1980) DNA repair and mutagen interaction in Saccharomyces--Theoretical considerations. In DNA Repair and Mutagenesis in Eukaryotes, W.M. Generoso, M.D. Shelby, and F.J. de Serres, eds. Plenum Press, New York, pp. 159-167.

20. von Borstel, R.C., and S. Igali (1975) Mutagenicity testing of antischistosomal thioxanthenones and indazoles on yeast. J. Toxicol. Environ. Health 1:281-291.
21. von Borstel, R.C., D.F. O'Connell, R.D. Mehta, and U.G.G. Hennig (1985) Modulation in cytochrome P-420 and P-450 content in *Saccharomyces cerevisiae* according to physiological conditions and genetic background. Mutat. Res. 150:217-224.

CANCER AND FREE RADICALS

William A. Pryor

Biodynamics Institute
Louisiana State University
Baton Rouge, Louisiana 70803

ABSTRACT

It is now clear that free radical intermediates often are involved in the activation of many types of procarcinogens and promutagens to their active forms as well as in the binding of these activated species to DNA. In this chapter, a general introduction to free radical chemistry is presented, with some discussion of radical lifetimes and reactivities. Potential biological targets of radical attack include lipids, proteins, and nucleic acids, and the reactions of all three of these target molecules with radicals are discussed. Finally, the evidence linking free radical reactions with chemical carcinogenesis is reviewed. A mechanistic scheme that divides the mechanisms for activating procarcinogens into 5 types is suggested; of these, 3 types of mechanisms involve free radicals, either in the activation of the carcinogen or in its binding to DNA or both. It also is suggested that a "reverse binding" can occur in which radicals produced on the DNA backbone attack and bond to <u>unactivated</u> substrates, rather than activated substrates (such as radicals) attacking unactivated DNA. It is known that systems that produce superoxide can lead to the production of hydroxyl radicals and that these HO· radicals form radical sites on DNA; thus, reverse binding could occur when any species that can add to a free radical is in the vicinity of the radical-damaged DNA.

INTRODUCTION

Some years ago, the evidence linking free radical reactions and cancer was tenuous and unconvincing to most scientists. In recent years, however, results from a number of fields of research have conclusively shown that free radical intermediates are very often involved in the enzymatic activation of chemical carcinogens. In this chapter, I wish to outline briefly some of the chemical characteristics of free radicals, including their lifetimes, discuss the types of biological target molecules that are subject to attack by free radicals, and then review the evidence linking free radical reactions to the development of cancer.

LIPID PEROXIDATION

Early studies concentrated on lipids as the primary targets for attack by free radicals since the polyunsaturated fatty acids (PUFAs) in lipids were known to undergo free radical autoxidation very rapidly (52,60). In recent years it has been recognized that PUFAs are not the sole target for free radical attack; both proteins and nucleic acids also are attacked by free radicals in processes that have important biological consequences. Nevertheless, it is important to understand lipid autoxidation since the peroxidation of lipids in cellular membranes can produce cell lysis. For example, free radical generating systems (e.g., hydrogen peroxide/ferrous iron) produce free radicals that can lyse red blood cells (RBCs); in fact, this radical-initiated RBC hemolysis is the basis of the clinical test for vitamin E deficiency. The study of the oxidation of lipids has become even more important in the past decade, since the identification of the hydroperoxide intermediates produced in the arachidonate cascade has demonstrated that lipid peroxidation occurs in mammals (41,42).

Scheme I (Fig. 1) shows the equations that govern the peroxidation of a lipid, LH. I have recently described this system in considerable detail and have derived these equations (60). The autoxidation of PUFAs can be used to measure the effectiveness of antioxidants (65). In the system we use, an initiator that is unstable at 37°C decomposes to form free radicals as shown in equation 1. The initiation sequence, shown in equations 1-3, leads to the formation of L•, the lipid radical. In our test system, we use linoleic acid as the PUFA substrate, so L• represents the conjugated pentadienyl radical. The propagation sequence then converts the L• radical into the conjugated diene lipid hydroperoxide, LOOH. In the absence of an added free radical scavenger such as vitamin E, termination occurs by equation 6, 7, or 8. The original studies of autoxidation were performed by chemists interested in the destructive reactions that organic materials undergo in the presence of one atmosphere of air; under these conditions, equation 8 is the only termination that occurs. In vivo, however, in organs other than the lung and skin, oxygen tensions that are from 1% to 10%

Initiation

$$\text{Initiator} \longrightarrow 2\text{R}\cdot \qquad (1)$$
$$\text{R}\cdot + \text{O}_2 \longrightarrow \text{ROO}\cdot \qquad (2)$$
$$\text{ROO}\cdot + \text{LH} \longrightarrow \text{ROOH} + \text{L}\cdot \qquad (3)$$

Propagation

$$\text{L}\cdot + \text{O}_2 \longrightarrow \text{LOO}\cdot \qquad (4)$$
$$\text{LOO}\cdot + \text{LH} \xrightarrow{k_p} \text{LOOH} + \text{L}\cdot \qquad (5)$$

Termination

$$2\text{L}\cdot \longrightarrow \text{Non-radical products (NRP)} \qquad (6)$$
$$\text{L}\cdot + \text{LOO}\cdot \longrightarrow \text{NRP} \qquad (7)$$
$$2\text{LOO}\cdot \longrightarrow \text{NRP} \qquad (8)$$

Termination by inhibitors

$$\text{LOO}\cdot + \text{InH} \xrightarrow{k_{InH}} \text{LOOH} + \text{In}\cdot \qquad (9)$$
$$\text{LOO}\cdot + \text{In}\cdot \longrightarrow \text{NRP} \qquad (10)$$

Fig. 1. The equations that describe the autoxidation of polyunsaturated fatty acids (PUFAs).

of those obtained under an atmosphere of air are commonly observed; under these conditions, equations 6 and 7 can be expected to contribute to the termination process. The involvement of reactions 6 and 7 changes the kinetics of the autoxidation quite substantially. To date, however, there is very little study of biological oxidations under low oxygen tensions that would model those observed in most internal organs and tissue.

In our system, using linoleic acid in sodium dodecyl sulfate (SDS) micelles as the autoxidizable PUFAs, we obtain a steady-state concentration of LOO• radicals of 2×10^{-7} M (60). The concentration of the L• radicals is much lower (4×10^{-12} M) since L• radicals are converted to LOO• very rapidly, as shown in equation 4. Under these conditions, where antioxidants are not present and linoleic acid is the only material available for LOO• to attack, the kinetic chain length is calculated to 26; that is, each L• free radical introduced into the system by equation 3 results in the conversion of 26 linoleic acid molecules to the hydroperoxide. In a RBC, this probably would be sufficient damage to result in lysis of the cell.

The effect of adding α-tocopherol is dramatic. The termination reaction shifts from being equation 8 to equations 9 and 10, and α-tocopherol becomes the predominant species that traps peroxyl radicals (LOO•) and terminates chains. When just 0.1 mole % vitamin E is added to the linoleic acid, the LOO• concentration is reduced from 2×10^{-7} M to 2×10^{-9} M. Furthermore, the kinetic chain length is now reduced to 0.3; that is, it takes <u>three</u> free radicals in order to cause the destruction of just one linoleic acid. In fact, the LOO• radicals attack vitamin E five times faster than they attack linoleic acid, even though linoleic acid is present in much higher concentration. Vitamin E is the most efficient antioxidant that occurs in vivo (13) as well as in model test systems such as the one I have described.

THE LIFETIMES OF FREE RADICALS

I have described a method for calculating the halflife for free radicals under conditions that model those that might occur in the cell (60, 61). Table 1 shows these calculated halflives for the hydroxyl, alkoxyl, and peroxyl radicals, and also the lifetime of the semiquinone free radical that is present in cigarette tar. As can be seen, the variation in these lifetimes is enormous. The hydroxyl radical is so reactive that it reacts with a biological molecule before diffusing more than 2 or 3 molecular

Tab. 1. Calculated halflives of some activated oxygen species of interest in free radical biology.

Species	Name	Halflife
HO•	Hydroxyl radical	10^{-9} sec
RO•	Alkoxyl radical	10^{-3} sec
ROO•	Peroxyl radical	10 sec
QH•	Tar semiquinone radical	days

Note: See discussion in Ref. 60 and 61.

diameters from the site at which it is produced. In contrast, the peroxyl radical has a lifetime of several seconds and can diffuse considerable distances in the cell. The cigarette tar free radical, as well as other semiquinone radicals of biological interest, has a very appreciable lifetime and a considerable stability (69). Thus, while it is true that small carbon- and oxygen-centered free radicals are highly reactive and short-lived, it is not true that all free radicals have the same lifetime. Furthermore, large complex radicals can have lifetimes that are substantial.

RADICAL ATTACK ON PROTEINS

In recent years it has become clear that proteins and enzymes are attacked by free radicals and inactivated (1,46). Recently, a group of workers (26,55) has shown that some enzymes become catalytically less efficient as animals age and that mixed-function oxidase (MFO) reactions account for the loss of efficiency that is observed. Free radical activity in the cell over the life span of the organism could lead to the inactivation of enzymes and cause enzymes of lower fidelity to accumulate during aging (1, 43).

Certain types of functional groups in proteins cause them to be particularly susceptible to free radical attack. One functionality that has drawn attention recently is the sulfide-type sulfur present in methionine, which can be oxidized to methionine sulfoxide by a range of oxidants including singlet oxygen, peroxyl radicals, and superoxide-generating systems present in phagocytes. The biological systems in which methionine sulfoxide appears to accumulate with age include the eye lens, among others (10). The gradual oxidative inactivation of proteins by oxidative processes may be quite common (50), and this may contribute to aging (1,3).

My research group has been studying the oxidation of alpha-1-proteinase inhibitor (a1PI) by the free radicals in cigarette smoke (19,62-64). The a1PI has a methionine at the active site, and when this amino acid is oxidized to the sulfoxide, the protein is made inactive. Since a1PI is the principal antiprotease protecting human lung tissue against proteolytic activity, the inactivation of a1PI is thought to be related to the development of emphysema in smokers.

ATTACK OF FREE RADICALS ON NUCLEIC ACIDS

Several systems have been studied in which radicals attack DNA (21a, 22,32,75). For example, doxorubicin, one of the most effective clinical anticancer drugs now available, generally is accepted to act by reducing oxygen to superoxide, which then dismutates to form H_2O_2 (54). The hydrogen peroxide then undergoes an iron-catalyzed decomposition, leading to the production of the extremely reactive hydroxyl radical (21,31). These reactions are illustrated below, where Q is the quinone drug:

$$Q \xrightarrow{e} Q^{\overline{\cdot}}$$

$$Q^{\overline{\cdot}} + O_2 \longrightarrow Q + O_2^{\overline{\cdot}}$$

$$2O_2^{\overline{\cdot}} + 2H^+ \longrightarrow H_2O_2 + O_2$$

$$H_2O_2 + Fe^{2+} \longrightarrow HO\cdot + HO^- + Fe^{3+}$$

$$Fe^{3+} + O_2^{\cdot -} \longrightarrow Fe^{2+} + O_2$$

Redox cycling of this type has recently been reviewed (8). Quinones can undergo this type of redox cycling, but they also can be reduced in a 2-electron process by DT diaphorase, for example (18). Certain types of semiquinone radicals also can themselves bind to DNA (4a,36).

Ames et al. have shown that thymine glycol and hydroxymethyluracil can be detected in rat and human urine (2,5,15). These derivatives of thymine are known to result from attack on DNA by hydroxyl radicals (33). In addition to attacking the DNA bases, radicals also can attack sugar molecules in the DNA backbone, nicking the chain and producing thiobarbituric acid-reactive materials (TBARMs) (29,31,32).

The tumor promoter tetradecanoyl phorbol acetate (TPA) initiates a cascade of chemical changes that leads to DNA strand scission in vitro, and it is thought that oxy-radicals are involved in these processes (6,17). In addition, the classical studies of the genetic effects of high energy radiation, where free radicals are known to be largely responsible, clearly demonstrate that radicals can attack nuclear DNA (33,56).

FREE RADICALS AND CANCER

Implication of Radicals in Chemically Induced Tumorigenesis

In recent years it has become clear that free radicals are involved in many of the biological processes that occur when chemicals transform cells to produce tumors. In particular, the following areas of research implicate radicals in chemically induced tumorigenesis:

Promotion involves radicals. In the usual mouse skin test, tumorigenesis can be divided into 2 stages, initiation and promotion; in addition, promotion appears to involve at least 2 stages (70,78). While the mechanism of promotion remains elusive, the involvement of radicals in this extremely important process is indicated by many lines of evidence (17,37,38, 40,73). For example, the structure-activity relationships in the phorbol esters show that those compounds that cause the greatest production of superoxide are the strongest promoters. Also, many radical-producing compounds, such as benzoyl peroxide, lauroyl peroxide, cumene hydroperoxide, etc., are themselves promoters (71).

Thus, it is clear that superoxide and radicals derived from it play a critical role in promotion. In general, if promotion does not occur, the development of cancer is less likely. This being true, the long-term administration of antioxidants might protect an organism that is exposed to a "complete" carcinogenic system containing both initiators and promoters (such as smog or tobacco smoke).

Antioxidants protect against carcinogenesis. It has been known for many years that a number of types of antioxidants protect a variety of experimental animals against the effects of many types of chemical carcinogens (74,76,77). The protection in many cases is impressive; nevertheless, it should not be inferred either that all radical scavenger drugs are anticarcinogens or that those antioxidants that do show antitumorigenic properties necessarily do so because of their antioxidant activity (72). As the chapters in this Volume have made clear, many phenolic compounds, including a number of plant phenols, have striking antitumorigenicity. However, these compounds appear to protect against cellular changes initiated by radical-generating compounds (such as H_2O_2) and also against ionic alkylating agents such as ethyl methanesulfonate. Phenolic compounds can act either as nucleophiles or as radical scavengers and thus can react both with alkylating agents and with free radicals. For this reason, even when an antioxidant protects against carcinogenicity, it cannot be inferred that radical reactions are necessarily involved. At present, the hypothesis that certain classes of compounds are antioxidants and also anticarcinogens provides a useful working hypothesis for the design of new antitumor drugs. However, it must not be assumed that antioxidants necessarily are anticarcinogenic because they protect against radical reactions.

Prostaglandin synthetase causes xenobiotic oxidation. The prostaglandin synthetase (PGS) system of enzymes contains a peroxidase component that can oxidize xenobiotics, as Marnett originally showed (51). This peroxidase can use either endogenous hydroperoxides [such as prostaglandin G (PGG) or hydroperoxyeicosatetraenoic acid (HPETE)] or exogenous hydroperoxides to cause the co-oxidation of compounds present during the conversion of arachidonate to prostaglandins and related derivatives (22,27). In particular, xenobiotics such as the 7,8-diol of benzo(α)pyrene (BP) are converted to the 7,8-diol-9,10-epoxide, the most potent ultimate carcinogen from BP (35,51).

The P-450 system may involve radical reactions. The cytochrome P-450 enzyme system is an extremely important pathway to oxidize, solubilize, and thus detoxify xenobiotics. This system involves production of an oxenoid complex (23) that acts as an oxygen atom donor and is able, for example, to convert arenes to arene oxides (20). This oxygen-atom transfer in the P-450 system often may not involve isolatable radical intermediates; however, for some substrates, free radicals clearly are involved (11,28,30,39).

Tumor cells have anomalous rates of lipid peroxidation. Tissue from tumors autoxidizes at an anomalously slow rate relative to matched normal controls (14,48). The reasons for this appear to be a lower content of PUFAs in the tumor, higher concentrations of antioxidants such as tocopherol (perhaps because of their being spared by the lower rates of autoxidation), and an anomalous P-450 activity (48,53). Thus, tumors autoxidize slowly and have lower levels of LOOH, and also undergo cell proliferation at anomalously high rates; these characteristics are related. For example, Cornwell (53) has shown that certain types of lipid hydroperoxides actually retard cell proliferation.

Autoxidizing lipids are able to cause the co-oxidation of many types of xenobiotics. For example, autoxidizing PUFAs co-oxidize and convert BP to derivatives that are mutagenic as tested by sister chromatid exchange (49). Interestingly, as the BP is converted to oxygenated products, the yield of TBARMs from the PUFAs decreases (49). This is consistent with our suggestion that peroxyl radicals are the precursors of TBARMs (66). We have proposed that peroxyl radicals cyclize to form endoperoxides that are the "non-volatile precursors" of the malonaldehyde that is detected by the thiobarbituric acid (TBA) test (66). It appears likely that peroxyl radicals can react with polycyclic aromatic hydrocarbons (PAHs) to yield arene oxides and alkoxyl radicals as follows (59):

$$ROO\cdot + PAH \longrightarrow RO\cdot + \text{arene oxide}$$

Since peroxyl radicals that react in this equation would not be able to cyclize to yield endoperoxides (66), the reaction of peroxyl radicals with PAHs would be predicted to cause the yield of endoperoxides to drop and thus lower the yield of TBARMs, as is observed (49).

Superoxide itself causes DNA damage. Superoxide (released by superoxide-generating systems such as neutrophils) itself causes DNA strand scission. For example, KO_2 produces single-strand breaks in T7 DNA; the damage is completely inhibited by ethylene diaminetetraacetic acid, diethylenetriaminepentaacetic acid, catalase, and alcoholic hydroxyl radical scavengers (45). The promoter TPA produces DNA strand breaks in human leukocytes, and catalase, superoxide dismutase (SOD), and benzoate (a hydroxyl radical scavenger) all protect DNA against nicks (7,45). Superoxide causes chromosomal damage in human lymphocytes, and again catalase, SOD, and hydroxyl radical scavengers protect DNA against nicks (17,24). Phagocytic cells, such as leukocytes, that can release superoxide are mutagenic in the Ames Salmonella assay, and catalase, SOD, benzoate, and tocopherol all protect against these radical-induced mutations (79,80).

PAH metabolism and P-450 activity differ in various tissues. The processes by which PAHs such as BP are oxidized by nonradical-dependent processes to diol-epoxides cannot rationalize the carcinogenicity of all PAHs in all tissues of all animals. Some chemical carcinogenicity appears to be the result of radical-mediated oxidation of PAHs, and these oxidations can involve radicals (16). Tissues that are low in P-450 activity but high in prostaglandin synthetase activity oxidize PAHs and other procarcinogens by radical-dependent processes (16,35,51).

Cigarette smoke contains radicals and is tumorigenic. Cigarette smoke contains very high concentrations of two different populations of free radicals. Gas-phase smoke contains approximately 10^{17} free radicals per puff; these are small carbon- and oxygen-centered radicals that are strong oxidants (12,19,57,58,67). Cigarette tar contains about 10^{17} free radicals per gram (19); these radicals are semiquinones in nature and are reducing agents (69). Thus, the radicals in gas-phase smoke and in tar are very different, and they probably lead to differing physiological effects. For example, gas-phase smoke oxidizes a1PI, but tar does not (61a,64). [Smoke also increases the release of superoxide from pulmonary macrophages, and oxy-radicals derived from this also oxidize a1PI (34).]

The carcinogenic effects of smoke appear to be predominantly associated with tar, and there is reason to suspect a connection between the tumorigenicity of tar and the radicals in tar (9,19). We have shown that the principal organic free radical in cigarette tar is a quinone-hydroquinone-semiquinone ($Q/QH_2/QH\cdot$) redox system bound in a low molecular weight polymer (69). Aqueous extracts of cigarette tar reduce oxygen to produce superoxide, which dismutates to form hydrogen peroxide, and metals in the tar catalyze the production of hydroxyl radicals from this hydrogen peroxide (21). We suggest that it is the quinone system in tar that is the agent that is able to reduce dioxygen to superoxide (21). These aqueous solutions of tar also nick DNA (9). The scavengers (such as SOD and catalase) that block production of hydroxyl radicals in these solutions also protect the DNA against damage (9,21). We also have shown that the paramagnetic species in tar becomes associated with DNA when tar and calf thymus DNA are incubated for 1 day at 37°C (68), suggesting that a complex forms between the Q/QH_2 species in tar and DNA, possibly with both the quinone/hydroquinone radical system and the DNA chelated and "sandwiched" about a metal ion (9). Thus, aqueous extracts of tar can produce the hydroxyl radical, and may do so in the vicinity of the DNA. If this is so, site-directed hydroxyl damage to the DNA might result, as is proposed to be the case with certain of the anticancer drugs that complex with DNA, and then oxidizing free radicals might be produced (4a,9,44,47).

RADICAL INVOLVEMENT IN TUMORIGENESIS: A HYPOTHESIS

It is reasonable to ask: How can free radicals accomplish the modification of DNA bases that is involved in transforming pro-oncogenes to oncogenes? This question may be asked in such a way as to imply that free radicals are too reactive to perform specific chemical transformations in a selective and controlled way, and therefore that radicals would not be able to perform base substitutions and consequently could not cause point mutations. Reasoning in this manner might suggest that radicals could not be implicated in carcinogenesis. This clearly is an erroneous conclusion since, for example, radiation is tumorigenic and acts via radical intermediates.

One problem with this line of reasoning, I suggest, is that it omits possible mechanisms for cell transformation in which chemicals do not add to DNA to form discrete adducts. Table 2 shows five possible mechanisms for the initiation of a tumor. Of these five, radicals are involved in three. Note that the fifth mechanism is suggested not to involve the production of adducts between DNA and chemical carcinogens, and it is this mechanism I wish to discuss in some detail.

It is known that many cellular toxins cause enzyme systems to produce superoxide and hydrogen peroxide; these systems invariably lead to the formation of the hydroxyl radical (4). It is also known that hydroxyl radicals can attack DNA and that this is quite common in vivo (2,15). Much of the damage caused by this process undoubtedly is repaired without transforming the cell, but the level of damage appears to be so high that some

Tab. 2. Mechanisms for the chemical initiation of tumors, using BP as a typical carcinogen.

Mechanism number	Typical event	Are radicals involved in the mechanism of chemical transformation?	Are radicals involved in the mechanism of DNA binding?
1	BP $\xrightarrow{\text{P-450}}$ diol-epoxide	Generally not[a]	No[b]
2	BP $\xrightarrow{\text{PES}}$ diol-epoxide	Yes	No[b]
3	BP $\longrightarrow$ quinones	Probably	Possibly
4	Alkylation	No	No
5	Production of $O_2^{\overline{\cdot}}/H_2O_2$[c]	Yes	Binding does not occur and is not involved

[a] A radicaloid mechanism may be involved in which odd electron species are produced but never become free (39).
[b] Nucleophilic opening by DNA of the BP epoxide is involved.
[c] From redox cycling of a quinone, from electron leakage, from electron transport, or from other sources.

of the damage, under particular circumstances, may lead to the transformation of the cell. That is, the occurrence of nonerror-free DNA repair in response to the high and continuous levels of radical damage may lead to cellular transformations. If this were true, radicals would be the ultimate initiators of cellular transformations, but radical-derived adducts of carcinogens would not necessarily be present.

In summary, I am suggesting three mechanisms for the initiation of cellular transformations by radicals. The first is mechanism 2 in Table 2. In this mechanism BP is activated [e.g., by prostaglandin endoperoxide synthetase (PES)] to the 7,8-diol-9,10-epoxide by a radical-dependent process, and this electrophilic epoxide reacts with DNA in an ionic process to lead to mutations. The second is mechanism 3 (Tab. 2), in which a radical or radicaloid process (involving PGS or P-450) may convert BP to diols and quinones (16,74) that can bind to DNA, perhaps via radical intermediates (36). The third process is mechanism 5 in Tab. 2. This mechanism appears to apply to cigarette tar and may equally well apply to BP and other quinones (18). In this process, BP is converted to diols and quinones, and these Q/QH_2 systems do not bind to DNA, but instead undergo redox cycling to produce oxy-radicals that damage DNA (74).

When the hydroxyl radical attacks DNA or RNA, free radical sites are produced on the DNA/RNA backbone by this attack (25). This may lead to what I term "reverse labeling"; in this process, DNA/RNA radical sites attack an unmodified and unactivated substrate molecule to lead to binding (72a). For example, hydroxyl radicals might attack DNA and produce DNA radicals (DNA•) that then attack PAHs or their quinones, as shown in Scheme 2 (Fig. 2). Since PAHs themselves are too insoluble in water to diffuse through the cell, this process probably is not important for PAHs themselves. However, it could be important for water soluble species such as quinones that have high reactivities toward radicals and that might be near DNA or RNA at the time the nucleic acid radicals are produced.

The impact of these ideas was brought home to my group recently with regard to the chemistry of cigarette tar. As discussed above, we had shown that cigarette tar contains semiquinone radicals (69), and that the ESR signal from these radicals becomes incorporated into DNA when the tar radical is incubated with DNA in vitro (68). Our initial aim then was to attempt to identify the specific molecular fragment from tar and the DNA base to which binding occurred. We were unsuccessful in this effort, and we now no longer believe this is necessarily the mechanism for cigarette tar tumorigenicity, since the indirect mechanism discussed above seems more

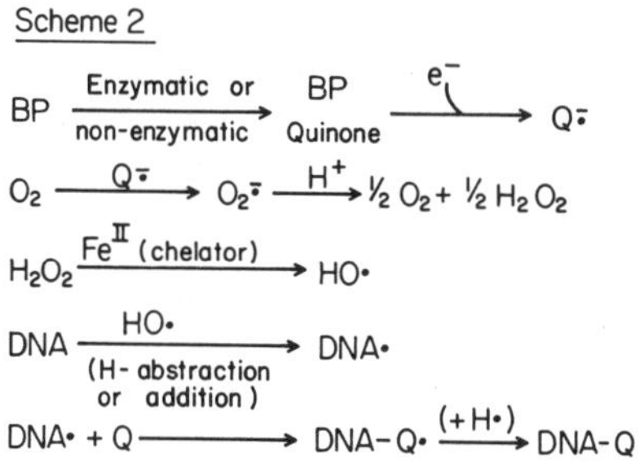

Scheme 2

$$\text{BP} \xrightarrow[\text{non-enzymatic}]{\text{Enzymatic or}} \text{BP Quinone} \xrightarrow{e^-} Q^{\bar{\cdot}}$$

$$O_2 \xrightarrow{Q^{\bar{\cdot}}} O_2^{\bar{\cdot}} \xrightarrow{H^+} \tfrac{1}{2}O_2 + \tfrac{1}{2}H_2O_2$$

$$H_2O_2 \xrightarrow{Fe^{II}\ (\text{chelator})} HO\cdot$$

$$\text{DNA} \xrightarrow[\text{(H- abstraction or addition)}]{HO\cdot} \text{DNA}\cdot$$

$$\text{DNA}\cdot + Q \longrightarrow \text{DNA-Q}\cdot \xrightarrow{(+H\cdot)} \text{DNA-Q}$$

Fig. 2. The "reverse labeling" mechanism for attachment of an unactivated substrate to activated DNA.

probable. That is, aqueous extracts of tar reduce oxygen to superoxide, which then dismutates to form hydrogen peroxide. Metal ions (either in the tar or in the biological system) then catalyze the production of hydroxyl radicals from the hydrogen peroxide (21), and these hydroxyl radicals damage DNA (9). Since the tar radical does become associated with DNA, we now believe that the tar radical-DNA complex (which may be held together by chelation to metal ions) leads to the redox cycling of oxygen and the production of hydroxyl radicals in the vicinity of the DNA. Thus, the mutagenic event (81) is not the result of covalent binding of an activated tar component such as a semiquinone to DNA, but rather of generalized radical-mediated damage to DNA caused by redox cycling of smoke-derived radicals.

ACKNOWLEDGEMENTS

The research reported here was performed by the co-workers acknowledged in the literature citations, to whom I am grateful. Specific mention, however, should be made of the contributions of Drs. Daniel F. Church, Edward T. Borish, and John P. Cosgrove. This work was supported in part by grants from the National Institutes of Health, and by a contract from the National Foundation for Cancer Research.

REFERENCES

1. Adelman, R.C., and E.E. Dekker, eds. (1985) Modification of Proteins During Aging, Alan R. Liss, Inc., New York, 120 pp.
2. Ames, B.N., R.L. Saul, E. Schwiers, R. Adelman, and R. Cathcart (1984) Oxidative DNA damage as related to cancer and aging: The assay of thymine glycol, thymidine glycol, and hydroxymethyluracil in human and rat urine. In Proceedings of the Symposium on Molecular Biology of Aging: Gene Stability and Gene Expression, Raven Press, New York.
3. Armstrong, D., ed. (1984) Free Radicals in Molecular Biology, Aging, and Disease, Raven Press, New York, 412 pp.
4. Aust, S.D., and B.A. Svingen (1982) The role of iron in enzymatic lipid peroxidation. In Free Radicals in Biology, Vol. V, W.A. Pryor, ed. Academic Press, New York, pp. 1-28.

4a. Bachur, N.R., S.L. Gordon, and M.V. Gee (1978) A general mechanism for microsomal activation of quinone anticancer agents to free radicals. Cancer Res. 38:1745-1750.

5. Bernstein, L., L.S. Gold, B.N. Ames, M.C. Pike, and D.G. Hoel (1985) Totologenic aspects of the comparison of carcinogenic potency in rats and mice. Fund. Appl. Tox. 5:79-86.
6. Birnboim, H.C. (1982) DNA strand breakage in human leukocytes exposed to a tumor promoter, phorbol myristate acetate. Science 215:1247-1249.
7. Birnboim, H.C. (1983) Importance of DNA strand-break damage in tumor promotion. In Radioprotectors and Anticarcinogens, O.F. Nygaard and M.G. Simic, eds. Academic Press, New York, pp. 539-556.
8. Borg, D.C., and K.M. Schaich (1984) Cytotoxicity from coupled redox cycling of autoxidizing xenobiotics and metals. Israel J. Chem. 24: 38-53.
9. Borish, E.T., J.P. Cosgrove, D.F. Church, W.A. Deutsch, and W.A. Pryor (1985) Cigarette tar causes single strand breaks in DNA. Biochem. Biophys. Res. Commun. (in press).
10. Brot, N., and H. Weissbach (1983) Biochemistry and physiological role of methionine sulfoxide residues in proteins. Arch. Biochem. Biophys. 223:271-281.

11. Burka, L.T., F.P. Guengerich, R.J. Willard, and T.L. Macdonald (1985) Mechanism of cytochrome P-450 catalysis. Mechanism of n-dealkylation and amine oxide deoxygenation. J. Am. Chem. Soc. 107:2549-2551.
12. Burkey, T., D.F. Church, and W.A. Pryor (1986) Initiation of lipid peroxidation by gas phase cigarette smoke models. (To be submitted.)
13. Burton, G.W., L. Hughes, and K.U. Ingold (1983) Antioxidant activity of phenols related to vitamin E. Are there chain-breaking antioxidants better than alpha-tocopherol. J. Am. Chem. Soc. 105:5950-5951.
14. Burton, G.W., K.H. Cheesman, K.U. Ingold, and T.F. Slater (1983) Lipid antioxidants and products of lipid peroxidation as potential tumour protective agents. Biochem. Soc. Trans. 11:261-262.
15. Cathcart, R., E. Schwiers, R.L. Saul, and B.N. Ames (1984) Thymine glycol and thymidine glycol in human and rat urine: A possible assay for oxidative DNA damage. Proc. Natl. Acad. Sci., USA 81:5633-5637.
16. Cavalieri, E.L., and E.G. Rogan (1984) One-electron and two-electron oxidation in aromatic hydrocarbon carcinogenesis. In Free Radicals in Biology, Vol. VI, W.A. Pryor, ed. Academic Press, New York, pp. 323-369.
17. Cerutti, P.A. (1985) Prooxidant states and tumor promotion. Science 227:375-381.
18. Chesis, P.L., D.E. Levin, M.T. Smith, L. Ernster, and B.N. Ames (1984) Mutagenicity of quinones: Pathways of metabolic activation and detoxification. Proc. Natl. Acad. Sci., USA 81:1696-1700.
19. Church, D.F., and W.A. Pryor (1985) The free radical chemistry of cigarette smoke and its toxicological implications. Env. Health Perspect. (in press).
20. Coon, M.J., et al., eds. (1980) Microsomes, Drug Oxidations, and Chemical Carcinogens, Vol. I, Academic Press, New York.
21. Cosgrove, J.P., E.T. Borish, D.F. Church, and W.A. Pryor (1985) The metal-mediated formation of hydroxyl radical by aqueous extracts of cigarette tar. Biochem. Biophys. Res. Commun. 132:390-396.
21a. Dipple, A., R.C. Moschell, and C.A.H. Bigger (1984) Polynuclear aromatic carcinogens. In Chemical Carcinogens (second edition, revised and expanded), Vol. 2, C.E. Searle, ed. American Chemical Society, Washington, D.C., 163 pp.
22. Dix, T.A., R. Fontana, A. Panthani, and L.J. Marnett (1985) Hematin-catalyzed epoxidation of 7,8-dihydroxyl-7,8-dihydrobenzo[a]pyrene by polyunsaturated fatty acid hydroperoxides. J. Biol. Chem. 260:5358-5365.
23. Dunford, H.B., and W.A. Pryor, in collaboration with S. Aust, W. Bors, I. Fridovich, W.H. Koppenol, T. Traylor, and V. Ullrich (1981) A list of recommended names and possible structures for various iron-oxygen complexes. In Oxygen and Oxy-radicals in Chemistry and Biology, M.A.J. Rogers and E.L. Powers, eds. Academic Press, New York, p. 196.
24. Emerit, I., and P. Cerutti (1984) Icosanoids and chromosome damage. In Icosanoids and Cancer, H. Thaler-Dao, A. Crastes de Paulet, and R. Paoletti, eds. Raven Press, New York, pp. 127-138.
25. Fitchett, M., and B.C. Gilbert (1985) ESR studies of the reactions of •OH with carbohydrates, sugar-phosphates and other nucleotide components. Life Chem. Rep. 3:57-61.
26. Fucci, L., C.N. Oliver, M.J. Coon, and E.R. Stadtman (1983) Inactivation of key metabolic enzymes by mixed-function oxidation reactions: Possible implication in protein turnover and aging. Proc. Natl. Acad. Sci., USA 80:1521-1525.
27. Gale, P.H., and R.W. Egan (1984) Prostaglandin endoperoxide synthetase-catalyzed oxidation reactions. In Free Radicals in Biology, Vol. VI, W.A. Pryor, ed. Academic Press, New York, pp. 1-38.

28. Ghezzi, P., M. Bianchi, L. Gianera, S. Landolfo, and M. Salmona (1985) Role of reactive oxygen intermediates in the interferon-mediated depression of hepatic drug metabolism and protective effect of n-acetylcysteine in mice. Cancer Res. 45:3444-3447.
29. Grollman, A.P., M. Takeshita, D.M.R. Pillai, and Francis Johnson (1985) Origin and cytotoxic properties of base propenals derived from DNA. Cancer Res. 45:1127-1131.
30. Groves, J.T., and D.V. Subramanian (1984) Hydroxylation by cytochrome P-450 and metalloporphyrin models. Evidence for allylic rearrangement. J. Am. Chem. Soc. 106:2177-2181.
31. Gutteridge, J.M.C. (1982) Iron-dependent free radical damage to DNA and deoxyribose. Separation of TBA-reactive intermediates. Int. J. Biochem. 14:891-893.
32. Gutteridge, J.M.C. (1984) Reactivity of hydroxyl and hydroxyl-like radicals discriminated by release of thiobarbituric acid-reactive material from deoxy sugars, nucleosides and benzoate. Biochem. J. 224: 761-767.
33. Huttermann, J., W. Kohnlein, and R. Teoule, eds.; A.J. Bertinchamps, coordinating ed. (1978) Effects of Ionizing Radiation on DNA, Springer-Verlag, Berlin, Heidelberg, New York, 650 pp.
34. Janoff, A., H. Capp, P. Laurent, and L. Raju (1983) The role of oxidative processes in emphysema. Am. Rev. Respir. Dis. 127:531-538.
35. Kalyanaraman, B., and K. Sivarajah (1984) The electron spin resonance study of free radicals formed during the archidonic acid cascade and cooxidation of xenobiotics by prostaglandin synthase. In Free Radicals in Biology, Vol. VI, W.A. Pryor, ed. Academic Press, New York, pp. 149-198.
36. Kennedy, K.A., E.G. Mimnaugh, M.A. Trush, and B.K. Sinha (1985) Effects of glutathione and ethylxanthate on mitomycin C activation by isolated rat hepatic or EMT6 mouse mammary tumor nuclei. Cancer Res. 45:4071-4076.
37. Kensler, T.W., and M.A. Trush (1984) Role of oxygen radicals in tumor promotion. Environ. Mutag. 6:593-616.
38. Kensler, T.W., D.M. Bush, and W.J. Kozumbo (1983) Inhibition of tumor promotion by a biomimetic superoxide dismutase. Science 221:75-77.
39. Korzekwa, K., W. Trager, M. Gouterman, D. Spangler, and G.H. Loew (1985) Cytochrome P450 mediated aromatic oxidation: A theoretical study. J. Am. Chem. Soc. 107:4273-4279.
40. Kozumbo, W.J., M.A. Trush, and T.W. Kensler (1985) Poly(adenosine diphosphate-ribosylation) of nuclear matrix proteins in alkylating agent resistant and sensitive cell lines. Chem. Biol. Interact. 54:199-208.
41. Lands, W.E.M. (1985) Interactions of lipid hydroperoxides with eicosanoid biosynthesis. J. Free Radicals Biol. Med. 1:97-101.
42. Lands, W.E.M., R.J. Kulmacz, and P.J. Marshall (1984) Lipid peroxide actions in the regulation of prostaglandin biosynthesis. In Free Radicals in Biology, Vol. VI, W.A. Pryor, ed. Academic Press, New York, pp. 39-61.
43. Laughrea, M. (1982) On the error theories of aging. Exp. Gerontology 17:305-317.
44. Lazo, J.S., and M.P. Hacker (1985) Organ-directed toxicology of antitumor agents. Fed. Proc. 44:2335-2338.
45. Lesko, S.A., R.J. Lorentzen, and P.O.P. Ts'o (1980) Role of superoxide in deoxyribonucleic acid strand scission. Biochemistry 19:3023-3028.
46. Lin, S.W., and K.J.A. Davies (1985) General aspects of protein damage by oxygen radicals. Fed. Proc. Vol. 44 (abstract of paper no. 3990).
47. Lown, J.W. (1985) Molecular mechanisms of action of anticancer agents involving free radicals. Adv. Free Radical Biol. Med. (in press).

48. McBrien, D.C.H., and T.F. Slater, eds. (1982) Free Radicals, Lipid Peroxidation and Cancer, Academic Press, New York.
49. McNeill, J.M., and E.D. Wills (1985) The formation of mutagenic derivatives of benzo[α]pyrene by peroxidising fatty acids. Chem. Biol. Interact. 53:197-207.
50. Man, E.H. (1983) Accumulation of d-aspartic acid with age in the human brain. Science 220:1407-1408.
51. Marnett, L.J. (1984) Hydroperoxide-dependent oxidations during prostaglandin biosyntheses. In Free Radicals in Biology, Vol. VI, W.A. Pryor, ed. Academic Press, New York, pp. 63-94.
52. Mead, J.F. (1976) Free radical mechanisms of lipid damage and consequences for cellular membranes. In Free Radicals in Biology, Vol. I, W.A. Pryor, ed. Academic Press, New York, pp. 51-67.
53. Morisaki, N., J.A. Lindsey, J.M. Stitts, H. Zhang, and D.G. Cornwell (1984) Fatty acid metabolism and cell proliferation. V. Evaluation of pathways for the generation of lipid peroxides. Lipids 19:381-394.
54. Myers, C.C., W.P. McGuire, R.H. Liss, L. Ifrim, K. Grotzinger, and R.C. Young (1977) Adriamycin. The role of lipid peroxidation in cardiac toxicity and tumor response. Science 197:165-167.
55. Oliver, C.N., R. Fulks, R.L. Levine, L. Fucci, A.J. Rivett, J.E. Roseman, and E.R. Stadtman (1984) Oxidative inactivation of key metabolic enzymes during aging. In Molecular Basis of Aging, A.K. Roy and B. Chatterjee, eds. Academic Press, New York, pp. 235-260.
56. Phillips, G.O., ed. (1968) Energetics and Mechanisms in Radiation Biology, Academic Press, New York, 527 pp.
57. Pryor, W.A. (1980) Mechanisms and detection of pathology caused by free radicals: Tobacco smoke, nitrogen dioxide, and ozone. In Environmental Health Chemistry, J.D. McKinney, ed. Ann Arbor Science Publishers, Ann Arbor, Michigan, pp. 445-467.
58. Pryor, W.A. (1980) Methods of detecting free radicals and free radical-mediated pathology in environmental toxicology. In Molecular Basis of Environmental Toxicity, R.S. Bhatnagar, ed. Ann Arbor Science Publishers, Ann Arbor, Michigan, pp. 3-36.
59. Pryor, W.A. (1982) Free radical biology: Xenobiotics, cancer and aging. In Vitamin E: Biochemical, Hematological and Clinical Aspects, B. Lubin and L.J. Machlin, eds. New York Academy of Sciences, New York, pp. 1-22.
60. Pryor, W.A. (1984) Free radicals in autoxidation and in aging. Part I. Kinetics of the autoxidation of linoleic acid in SDS micelles: Calculations of radical concentrations, kinetic chain lengths, and the effects of vitamin E. Part II. The role of radicals in chronic human diseases and in aging. In Free Radicals in Molecular Biology, Aging, and Disease, D. Armstrong, R.S. Sohal, R.C. Cutler, and T.F. Slater, eds. Raven Press, New York, pp. 13-41.
61. Pryor, W.A. (1986) Oxy-radicals and related species: Their formation, their lifetimes, and their reactions. Ann. Rev. Phys. 48:657-667.
61a. Pryor, W.A. (1986) Free radicals in cigarette smoke. In Oxidative and Proteolytic Damage to the Lung: Second International Symposium, C. Mittman and J.C. Taylor, eds. Academic Press, New York (in press).
62. Pryor, W.A., and M.M. Dooley (1985) Inactivation of human alpha-1-proteinase inhibitor by cigarette smoke: Effect of smoke phase and buffer. Am. Rev. Respir. Dis. 131:941-943.
63. Pryor, W.A., M.M. Dooley, and D.F. Church (1984) Inactivation of human alpha-1-proteinase inhibitor by gas-phase cigarette smoke. Biochem. Biophys. Res. Commun. 122:676-681.
64. Pryor, W.A., M.M. Dooley, and D.F. Church (1986) The mechanisms of the inactivation of human alpha-1-proteinase inhibitor by gas-phase cigarette smoke. Adv. Free Radical Biol. Med. (in press).

65. Pryor, W.A., M.J. Kaufman, and D.F. Church (1985) Autoxidation of micelle-solubilized linoleic acid. Relative inhibitory efficiencies of ascorbate and ascorbyl palmitate. J. Org. Chem. 50:281-283.
66. Pryor, W.A., J.P. Stanley, and E. Blair (1976) Autoxidation of polyunsaturated fatty acids: II. A suggested mechanism for the formation of TBA-reactive materials from prostaglandin-like endoperoxides. Lipids 11:370-379.
67. Pryor, W.A., M. Tamura, and D.F. Church (1984) ESR spon-trapping study of the radicals produced in NOx/olefin reactions: A mechanism for the production of the apparently long-lived radicals in gas-phase cigarette smoke. J. Am. Chem. Soc. 106:5073-5079.
68. Pryor, W.A., K. Uehara, and D.F. Church (1984) The chemistry and biochemistry of the radicals in cigarette smoke: ESR evidence for the binding of the tar radical to DNA and polynucleotides. In Oxygen Radicals in Chemistry and Biology, W. Bors, M. Saran, and D. Tait, eds. Walter de Gruyter & Co., Berlin, pp. 193-201.
69. Pryor, W.A., B.J. Hales, P.I. Premovic, and D.F. Church (1983) The radicals in cigarette tar: Their nature and suggested physiological implications. Science 220:425-427.
70. Slaga, T.J., ed. (1980) Modifiers of Chemical Carcinogenesis: An Approach to the Biochemical Mechanism and Cancer Prevention, Raven Press, New York, 275 pp.
71. Slaga, T.J., A.J.P. Klein-Szanto, L.L. Triplett, L.P. Yotti, and J.E. Trosko (1981) Skin tumor-promoting activity of benzoyl peroxide, a widely-used free radical-generating compound. Science 213:1023-1025.
72. Smith, C.V., H. Hughes, B.G. Lauterburg, and J.R. Mitchell (1983) Chemical nature of reactive metabolites determines their biological interactions with glutathione. In Functions of Glutathione: Biochemical, Physiological, Toxicological, and Clinical Aspects, A. Larsson, ed. Raven Press, New York, pp. 125-137.
72a. Steenken, S., and V. Jagannadham (1985) Reaction of 6-yl radicals of uracil, thymine, and cytosine and their nucleosides and nucleotides with nitrobenzenes via addition to give nitroxide radicals. OH^--catalyzed nitroxide hetrolysis. J. Am. Chem. Soc. 107:6818-6826.
73. Troll, W., G. Witz, B. Goldstein, D. Stone, and T. Sugimura (1982) The role of free oxygen radicals in tumor promotion and carcinogenesis. In Carcinogenesis: A Comprehensive Survey, Vol. 7, E. Hecker, N.E. Fusenig, W. Kunz, F. Marks, and H.W. Thielmann, eds. Raven Press, New York, pp. 593-597.
74. Ts'o, P.O.P., W.J. Caspary, and R.J. Lorentzen (1977) The involvement of free radicals in chemical carcinogenesis. In Free Radicals in Biology, Vol. III, W.A. Pryor, ed. Academic Press, New York, pp. 251-303.
75. Tsuruta, Y., P.D. Josephy, A.D. Rahimtula, and P.J. O'Brien (1985) Peroxidase-catalyzed benzidine binding to DNA and other macromolecules. Chem. Biol. Interact. 54:143-158.
76. Wattenberg, L.W. (1980) Inhibitors of chemical carcinogens. In Cancer and the Environment, H.B. Demopoulos and M.A. Mehlman, eds. Pathotox Publishers, Inc., Park Forest South, Illinois, pp. 35-52.
77. Wattenberg, L.W., and L.D.T. Lam (1983) Phenolic antioxidants as protective agents in chemical carcinogenesis. In Radioprotectors and Anticarcinogens, O.F. Nygaard and M.G. Simic, eds. Academic Press, New York, pp. 461-469.
78. Weinstein, I.B. (1980) Evaluating substances for promotion, cofactor effects and synergy in the carcinogenic process. In Cancer and the Environment, H.B. Demopoulos and M.A. Mehlman, eds. Pathotox Publishers, Inc., Park Forest South, Illinois, pp. 89-101.

79. Weitzman, S.A., and T.P. Stossel (1981) Mutation caused by human phagocytes. Science 212:546-547.
80. Weitzman, S.A., and T.P. Stossel (1982) Effects of oxygen radical scavengers and antioxidants on phagocyte-induced mutagenesis. J. Immun. 128:2770-2778.
81. Yamasaki, E., and B.N. Ames (1977) Concentration of mutagens from urine by adsorption with the nonpolar resin XAD-2: Cigarette smokers have mutagenic urine. Proc. Natl. Acad. Sci., USA 74:3555-3559.

CYTOCHROMES P-450 AND THE ACTIVATION AND INACTIVATION OF MUTAGENS AND CARCINOGENS*

Henry W. Strobel, Wan-Fen Fang, Richard S. Takazawa,**
Daniel J. Stralka, S.N. Newaz, Gary P. Kurzban,
David R. Nelson, and R. Scott Beyer

Department of Biochemistry and Molecular Biology
The University of Texas Medical School
Houston, Texas 77225

INTRODUCTION

The activation and inactivation of a wide variety of endogenous and exogenous compounds, including alkanes, fatty acids, steroids, drugs, polycyclic hydrocarbons, carcinogens, and mutagens, are catalyzed by the cytochrome P-450-dependent drug metabolism system (5). This system is found in the endoplasmic reticulum of various mammalian tissues and organs, e.g., lung, small intestines, etc. (20,24), but the greatest concentration and activity are in the liver (5). For this reason, many pioneering studies on the activation or inactivation of carcinogens, mutagens, and other drugs have been carried out using hepatic microsomes from various animal models. In turn, mammalian liver also was used in attempts to isolate, purify, and characterize components of the mixed-function oxidase system.

The first report on the solubilization of the components of the drug metabolism system and reconstitution of metabolic activity from solubilized components was published in 1968 by Lu and Coon (11). They showed that solubilized cytochrome P-450 and cytochrome P-450 reductase fractions could support the hydroxylation of lauric acid when reconstituted with a lipid factor later identified as phosphatidylcholine (26). In rapid order, the solubilized system was shown to catalyze the metabolism of many substrates (15,16) and carcinogens (14). The demonstration of a second form of cytochrome P-450 led to the recognition that substrate specificity was conferred by the cytochrome P-450 form present (14). Together with the earlier demonstrations of differential inducibility of cytochrome P-450 forms and consequently substrate preference (22,23), this work led the way to the demonstration of the importance of the cytochrome P-450 system in the activation and inactivation of carcinogens and other compounds. This is well

* Supported by Grant CA37148 from the National Cancer Institute and Grant Ag02081 from the National Institute of Aging, DHHS.

** Recipient of a Rosalie B. Hite Fellowship for Cancer Research.

illustrated by the Salmonella-microsome mutagenesis assay developed by Ames and colleagues (1).

The cytochrome P-450 system catalyzes the formal hydroxylation of its substrates. This process has been designated as "Phase I drug metabolism" to denote the oxidative phase of drug metabolism as separate from "Phase II drug metabolism," the conjugation reactions. The conjugation reactions include the sulfotransferases, glutathione-S-transferases, UDP-glucuronyl transferases, and other reactions, in general, between the products of Phase I reactions and conjugation acceptor molecules. The general reaction for the oxidative phase is described by the following equation:

$$RH + NADPH + H^{+} + O_2 \rightarrow ROH + HOH + NADP^{+}$$

where RH is any substrate and ROH its hydroxylated product. This reaction may be a direct hydroxylation or an indirect hydroxylation through the formation of an epoxide with subsequent enzymatic or nonenzymatic hydration.

Characteristics of Components of the Drug Metabolism System

The purification and reconstitution of activity from components of the drug metabolism system led to the recognition that cytochrome P-450 and cytochrome P-450 reductase are the two required protein components that carry out the overall oxidative reaction of drug metabolism. The interaction of these two enzymes in the presence of the phospholipid component (PL) is shown in Fig. 1. As shown in this scheme, electrons are conveyed from NADPH through the two flavin centers of cytochrome P-450 reductase to the cytochrome P-450 component. The cytochrome P-450 component binds substrate, as indicated by spectral shifts in the cytochrome P-450 absorption spectrum when substrate is added (15). The cytochrome P-450 component also binds O_2, as indicated by the inhibition of activity in the presence of carbon monoxide, which competes with oxygen for cytochrome P-450 (13).

Cytochrome P-450. Cytochrome P-450 has been purified from many sources and shown to exist as multiple forms with distinct but overlapping substrate specificity (6,12). The general properties of the isoenzymes of cytochrome P-450 are listed in Tab. 1. More than a dozen cytochrome P-450 mRNAs have been cloned and their amino acid sequences deduced (e.g., Ref. 17). Others have been sequenced by protein chemical methods (8,19). From the sequences, several structural features of cytochrome P-450 may be predicted. Of the total mass (M_r = 56,000) of rabbit liver cytochrome P-450

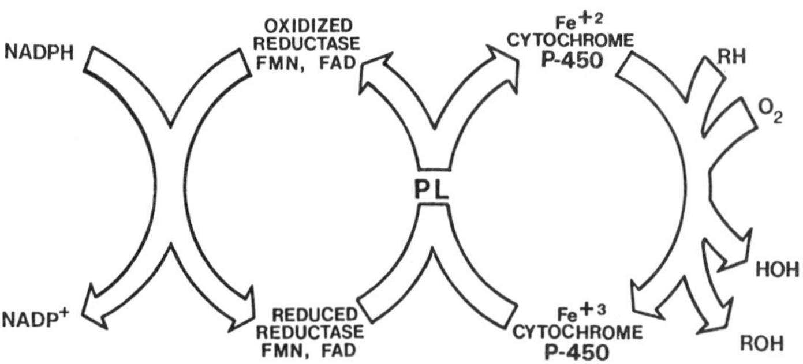

Fig. 1. Pathway of electron transfer in the microsomal mixed-function oxidase system.

Tab. 1. General properties of cytochromes P-450.

Prosthetic group	Protoporphyrin IX.
Absorption maxima	Soret 414-418. α ∿ 569 β ∿ 540 CO reduced 446-452.
Substrate interaction	Spectral shift in soret; shift in spin state.
Catalytic activity	Predominately hydrophobic compounds, e.g., alkanes, fatty acids, steroids, drugs, polycyclic hydrocarbons, and xenobiotics.
Molecular weight	48,000-60,000 daltons (by SDS gel).
Multiple forms	Yes, approximately 20 known.

Form 2 (LM2), about 20,000 daltons may be present as 8 α helical membrane-spanning segments, 6,000 daltons may be present in the endoplasmic reticulum lumen, and the remainder (30,000 daltons) probably forms a cytoplasmic domain (8). This may constitute a reasonable model for the structure of cytochrome P-450. One cysteine residue near the C-terminus is conserved in all of the cytochromes P-450 whose structures are known (17). This cysteine residue must contribute the thiolate (S^-) ligand recognized as the fifth ligand of the heme iron (29).

Cytochrome P-450 reductase. NADPH cytochrome P-450 reductase has been purified to electrophoretic homogeneity and shown to contain one mole of FMN and one mole of FAD per mole of enzyme (3). The general properties of the purified enzyme are summarized in Tab. 2. Structurally, the reductase is a stalked integral membrane protein anchored to the cytosolic surface of the endoplasmic reticulum by a short hydrophobic peptide of molecular

Tab. 2. Properties of cytochrome P-450 reductase.

Prosthetic group	1 mole FAD 1 mole FMN	
Absorption maxima	383, 456, and shoulder at 480 nm.	
Substrate	Electron donor:	NADPH
	Electron acceptor:	Cytochrome P-450
	Artificial acceptors:	Cytochrome *c* ferricyanide dichlorophenol-indophenol, etc.
Molecular weight	76,962 by sequence analysis (21). 678 amino acids.	
Structural features	Has a membrane binding domain at N terminus (6,400).	
Multiple forms	No.	

weight 6,400 (7). The membrane binding domain is required to support cytochrome P-450 dependent drug metabolism. The reductase has been sequenced (21), and the N-terminal amino acid is blocked. The flow of electrons through the flavin centers has been proposed to be from NADPH to FAD to FMN to acceptor (27).

Interaction Between Cytochrome P-450 and Cytochrome P-450 Reductase

During the course of substrate hydroxylation, NADPH cytochrome P-450 reductase serves as an electron donor, transferring electrons from its flavin centers one at a time to the heme iron of cytochrome P-450. The iron can pass the electron to an oxygen molecule which binds at the sixth axial ligand position (29). The 2-electron reduced iron oxygen complex can react with bound substrate and protons to form water and hydroxylated substrate. The interaction between cytochrome P-450 and cytochrome P-450 reductase which occurs during catalysis is represented in the model in Fig. 2. NADPH is shown near the FAD site and FMN, near the site of interaction with cytochrome P-450 in accord with current models (27). Charge complementarity between cytochrome P-450 and cytochrome P-450 reductase may impart some direction to the interaction of these molecules. Cytochrome P-450 is indicated as having regions of α helix in the membrane and the heme, with its thiolate ligand above the membrane in a cytoplasmic domain. In this model, both hydrophobic and ionic forces are suggested as playing a role in the interaction between the two proteins. Substrate-enhanced interaction between the two proteins remains an attractive idea based on enhanced electron flow in the presence of substrate. However, it remains to be shown whether this interaction is physically staged.

EXPERIMENTAL TECHNIQUES

The experimental techniques used in the studies presented have been fully described elsewhere and will only be listed for reference in this section. Purification of cytochrome P-450 or cytochrome P-450 reductase and preparation of antibodies to these proteins were as described (2,9). Assays of drug metabolism activities and carcinogen activation have been

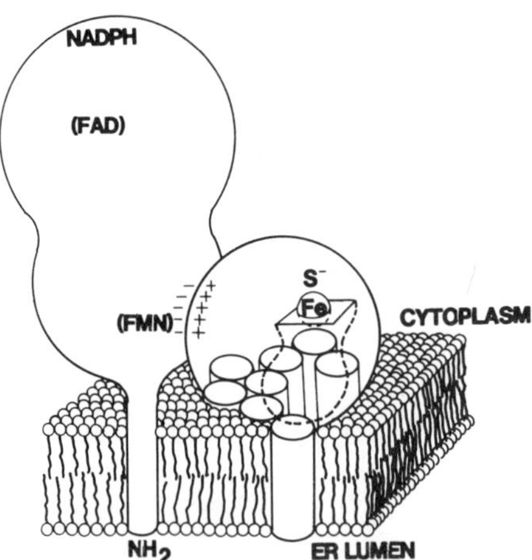

Fig. 2. Model for interaction of cytochrome P-450 and cytochrome P-450 reductase in the endoplasmic reticulum.

reported previously (15,16,18). Other assay techniques for determination of protein, cytochrome P-450 specific content, and cytochrome P-450 reductase activity, unless otherwise described, were identical to those used previously (2,4,9). The radial immunodiffusion assay was conducted according to Strobel and Lau (25).

RESULTS

A variety of factors affect the overall activity of the cytochrome P-450 system. Among these effectors are the tissue or organ in which the cytochrome P-450 system is studied, the interaction of other enzymes with cytochrome P-450, the age of the organism, and the amount of oxygen presented to the system. Our laboratory is studying these and other effectors of overall drug and carcinogen activation as they influence the distribution and catalytic function of the cytochromes P-450.

Effects of Tissue or Organ and Substrate on Cytochrome P-450-Dependent Drug Metabolism Activity

The interest of our laboratory in the origins of colon cancer led us to investigate the ability of human colon mucosal microsomes to support the activation of the organ-specific colon carcinogen 1,2-dimethylhydrazine (28). As shown in Tab. 3, microsomes prepared from the descending segments of human colons obtained at autopsy catalyze the hydroxylation of 1,2-dimethylhydrazine at a substantial rate. As is evident from the data, there is a certain individual variation in activity within the small number of samples available to us. The ascending and transverse segments of these same colons showed substantially lower activities when compared to the respective descending segments (data not shown). These findings are consistent with a gradient of activity through the colon with the highest levels of activity occurring in the same region where most tumors are located (10). Evidence is also presented in Tab. 3 that dimethylhydrazine

Tab. 3. Catalysis of 1,2-dimethylhydrazine metabolism by microsomes from human colon and human colon tumor.

Source of microsomes		Pretreatment	Activity (nmol HCHO/min/mg protein)
Human colon	Case 1[a]	None	95.3
	2	None	21.5
	3	None	64.6
	4	None	24.3
	5	None	55.4
Human colon tumor		None	15.2
		Phenobarbital and hydrocortisone	42.8

[a] Microsomes were prepared from the descending segment of the human colon.

metabolism is inducible with phenobarbital and hydrocortisone (4). These experiments were conducted using a cultured human colon tumor cell line (LS174T) and show a 3-fold induction of activity after treatment. Interestingly, these activities are higher than those catalyzed by rat liver microsomes. Thus, it appears that the effectiveness of activation of particular carcinogens varies with the tissue or organ of study.

In order to assess whether the variation in activity might be a reflection of a differential distribution of cytochromes P-450, we examined rat colonic microsomes for the presence of various forms of cytochrome P-450 by radial immunodiffusion. As shown in Tab. 4, different multiple forms of cytochrome P-450 could be detected in colonic microsomes from β-naphthoflavone-pretreated rats as compared with phenobarbital pretreated rats. These data indicate that, like liver, the colon responds differently to induction with phenobarbital than it does to induction with β-naphthoflavone. On the other hand, while liver microsomes from β-naphthoflavone-pretreated rats contain Forms 1 to 5, colon microsomes appear to contain detectable amounts of Forms 1, 4, and 5. While these data must be interpreted cautiously due to limits of immunochemical detection, the data seem consistent with a differential distribution of forms of cytochrome P-450 in liver and colon. A different profile of cytochromes P-450 for the colon provides a possible explanation for differing catalytic potentials.

Interaction of Other Enzyme Components with the Cytochrome P-450-Dependent Drug Metabolism System

While characterizing the components of the cytochrome P-450 system in rat and human colon and in human colon tumor cells we observed that the ratio of cytochrome b_5 to cytochrome P-450 is higher than in liver. In human colon tumor cells (LS174T), for instance, the ratio of cytochrome b_5 to cytochrome P-450 for control cells is 2.5 and for phenobarbital-treated cells, 2.7. In rat liver microsomes this ratio for control is 0.35, for phenobarbital-induced, 0.17. We examined human colon microsomes to determine whether cytochrome b_5 played a role in cytochrome P-450-mediated hydroxylation reactions. As shown in Fig. 3, antibody to cytochrome b_5 was as effective as antibody to cytochrome P-450 reductase in inhibiting dimethylhydrazine hydroxylation by colonic microsomes in the presence of NADPH. These data suggest that cytochrome b_5 may be involved in cytochrome P-450 reactions in the colon.

Tab. 4. Presence of cytochromes P-450 determined by radial immunodiffusion in rat colon microsomes.

Forms of cytochrome P-450	Pretreatment	
	β-Naphthoflavone	Phenobarbital
1	+	
2		
3		
4	+	+
5	+	
b		+

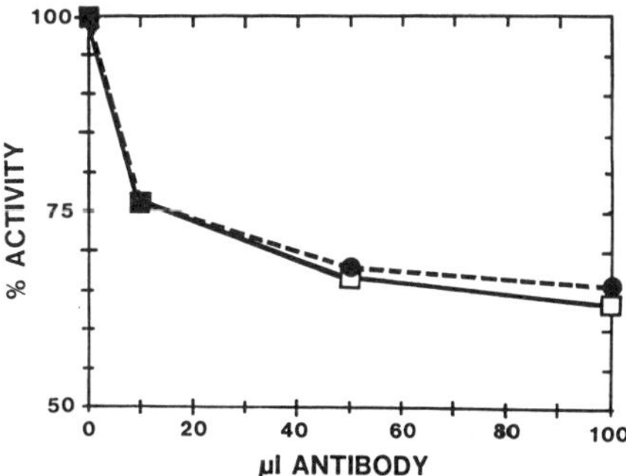

Fig. 3. Inhibition by antibodies of dimethylhydrazine metabolism catalyzed by human colon microsomes. Antibodies were preincubated with microsomes for 5 min at 30°C. Reaction time was 10 min. □-□-□ represents antibody to cytochrome P-450 reductase. ●-●-● represents antibody to cytochrome b_5.

Aging Affects Induction and Catalytic Activity of Cytochrome P-450

The response of the microsomal cytochrome P-450 systems of rat liver and colon to induction as a function of age is shown in Fig. 4. Benzo(α)-pyrene hydroxylation activity is detectable without induction in both tissues in each age group. In all except the earliest ages, the specific activity in liver is almost 10 times that of colon tissue. In control animals, peak activity is attained at 4 weeks in colon, but at 10 weeks in liver. With aging, activity in control animals drops to about 50% of peak activity. In both colon and liver, β-naphthoflavone induces benzo(α)pyrene hydroxylation activity. In liver, the highest induced levels are attained

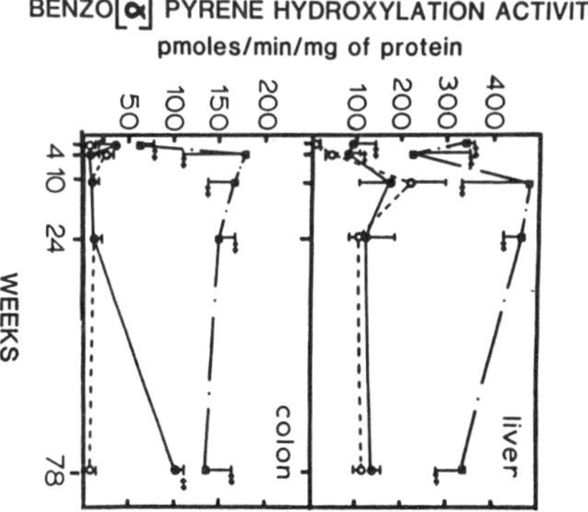

Fig. 4. Benzo(α)pyrene hydroxylation specific activity in liver and colon microsomes as a function of age. Values for control rats are designated by o-o-o; phenobarbital/hydrocortisone-pretreated rats by ●-●-●; and β-naphthoflavone-pretreated rats by ■-■-■. I's represent the standard deviation of each point. ◆ indicates that the point is significantly different from control (P = 0.05, Student t test); ◆◆ significantly different (P = 0.01).

at 10 weeks, but when 78-week-old animals are treated, only 67% of peak activity is produced. Inducibility of activity therefore declines with age in the liver. In the colon, by contrast, β-naphthoflavone induces hydroxylation activity at all ages. Inducibility in the colon does not significantly decline with age. Phenobarbital does not significantly induce benzo(α)pyrene hydroxylation in the liver at any age group. In the colon, however, phenobarbital dramatically induces hydroxylation in 78-week-old rats. The phenobarbital-induced levels are almost as high as the β-naphthoflavone-induced levels. Thus, aging has a markedly different effect on hydroxylation activity depending on the tissue examined.

We further examined the effects of aging by assessing the amount of cytochrome P-450 Form 5 as a function of age. As shown in Tab. 5, the amount of Form 5 in the colon at 78 weeks of age is almost double the amount present at 2 weeks. A similar relationship is seen for lung, but for liver the amount present at 78 weeks is only half that present in β-naphthoflavone-treated 2-week-old rats. This observation correlates well with the activity data in light of the fact that Form 5 has the highest known turnover number for benzo(α)pyrene (9). These changes in the amount of Form 5 would markedly affect benzo(α)pyrene hydroxylation activity.

Effects of Oxygen Concentration on Cytochrome P-450-Mediated Drug Metabolism Activity

When oxygen is excluded from cytochrome P-450 reaction mixtures, product formation is markedly reduced with most substrates (13). As shown in Tab. 6, however, the opposite appears true for the production of pentachlorobenzene from its substrate, hexachlorobenzene. In fact, product formation under aerobic conditions, the usual conditions associated with cytochrome P-450-mediated reactions, is only one-sixth the rate under anaerobic conditions (N_2 atmosphere). That pentachlorobenzene formation is cytochrome P-450-mediated is clearly demonstrated by the loss of activity when N_2 is replaced with CO or when metyrapone is added to the anaerobic reaction mixture. These data demonstrate that reductive dechlorination of hexachlorobenzene is catalyzed by cytochrome P-450. Under aerobic conditions, pentachlorophenol is observed as the major product. This is consistent with the production of pentachlorobenzene as a first product followed by

Tab. 5. Effects of age on amount of cytochrome P-450 Form 5 present after treatment of rats with β-naphthoflavone.

	Relative amount of Form 5[a]	
Tissue	2 weeks	78 weeks
Colon	1.0	1.80
Lung	1.0	1.90
Liver	1.0	0.45

[a] For each tissue the content of Form 5 at 2 weeks of age determined by radial immunodiffusion was set at unity (1.0).

the rapid formation of pentachlorophenol. Thus, the amount of oxygen present profoundly affects product formation by the cytochrome P-450-mediated system.

DISCUSSION

The existence of multiple forms or isoenzymes of cytochrome P-450 brings a certain measure of complexity to drug metabolism system activities. Further complicating the regulation of the system is the fact that the ratio of cytochrome P-450 reductase to cytochrome P-450 in the endoplasmic reticulum is 1 to 20. Thus, interaction (and, more important, productive interaction) between the two protein components of this system is by nature complex. We have shown a series of examples of effectors that impinge on and modify the productive interaction of cytochrome P-450 and cytochrome P-450 reductase.

Tissue and organ specificity is reflected in the distribution of cytochromes P-450 and consequently in the substrate that is preferentially activated. As we have indicated, dimethylhydrazine is metabolized efficiently in the tissue susceptible to dimethylhydrazine-induced carcinogenesis (colon). Tissues are affected in a differential manner by aging. In our work, we have shown that both human colon and rat colon show an elevated and sustained metabolic rate of benzo(α)pyrene hydroxylation which is stable with age. Of particular importance is the change in fidelity to induction observed in the colon of 78-week-old rats. The induction of benzo(α)pyrene hydroxylation activity by phenobarbital is unique. In all our other studies, phenobarbital has elicited either no change or a decrease in benzo(α)pyrene hydroxylation activity.

The importance of the effects of oxygen concentration on the formation of products from various carcinogens (e.g., hexachlorobenzene) is emphasized when considered in light of the reduced blood flow to various organs observed in aging. Reduced oxygenation to peripheral tissues and organs could have a marked effect on the metabolites produced by cytochrome P-450-mediated systems.

Tab. 6. Metabolism of hexachlorobenzene by microsomal cytochrome P-450.

System[a]	Percent activity[b]
+ Air + NADPH + HCB	16
+ N_2 + NADPH + HCB	100
+ N_2 - NADPH + HCB	--
+ N_2 + NADPH + HCB + 1.0 mM metyrapone	30
+ CO + NADPH + HCB	20

[a] System contained phenobarbital-induced microsomes (4 μM P-450). Source of microsomes: liver, male rats, 10 weeks. NADPH was present at 2.5 mM.

[b] Percent activity was calculated using the formation of pentachlorobenzene after 1 hr at 37°C under N_2 atmosphere (449 pmol) minus the product formed in the absence of NADPH (121 pmol) as 100%.

These effectors of the rate and mechanism of the cytochrome P-450-dependent drug metabolism system play a significant role in the activation and inactivation of carcinogens and mutagens. Continued deconvolution of these interactive effects will shed light on the regulation and control of carcinogenesis and mutagenesis.

REFERENCES

1. Ames, B.N., W.E. Durston, E. Yamasaki, and F.D. Lee (1973) Carcinogens are mutagens: A simple test system combining liver homogenates for activation and bacteria for detection. Proc. Natl. Acad. Sci., USA 70:2281-2285.
2. Dignam, J.D., and H.W. Strobel (1977) NADPH-cytochrome P-450 reductase from rat liver: Purification by affinity chromatography and characterization. Biochemistry 16:1116-1123.
3. Dignam, J.D., and H.W. Strobel (1975) Preparation of homogeneous NADPH-cytochrome P-450 reductase from rat liver. Biochem. Biophys. Res. Commun. 63:845-852.
4. Fang, W.F., and H.W. Strobel (1982) Effects of cyclophosphamide and polycyclic aromatic hydrocarbons on cell growth and mixed-function oxidase activity in a human colon tumor cell line. Cancer Res. 42:3676-3681.
5. Gillette, J.R. (1966) Biochemistry of drug oxidation and reduction by enzymes in hepatic endoplasmic reticulum. Adv. Pharmacol. 4:219-261.
6. Guengerich, F.P., G.A. Dannan, S.T. Wright, M.V. Martin, and L.S. Kaminsky (1982) Purification and characterization of liver microsomal cytochromes P-450: Electrophoretic, spectral, catalytic, and immunochemical properties and inducibility of eight isozymes isolated from rats treated with phenobarbital or β-naphthoflavone. Biochemistry 21:6019-6030.
7. Gum, J.R., and H.W. Strobel (1981) Isolation of the membrane-binding peptide of NADPH-cytochrome P-450 reductase. Characterization of the peptide and its role in the interaction of reductase with cytochrome P-450. J. Biol. Chem. 256:7478-7486.
8. Heinemann, F.S., and J. Ozols (1983) The complete amino acid sequence of rabbit phenobarbital-induced liver microsomal cytochrome P-450. J. Biol. Chem. 258:4195-4201.
9. Lau, P.P., and H.W. Strobel (1982) Multiple forms of cytochrome P-450 in liver microsomes from β-naphthoflavone pretreated rats: Separation, purification and characterization of five forms. J. Biol. Chem. 257:5257-5262.
10. Lipkin, M., E. Deschner, and F. Troncale (1970) Cell differentiation and the development of colonic neoplasm. Gastroenterology 59:303-309.
11. Lu, A.Y.H., and M.J. Coon (1968) Role of hemoprotein P-450 in fatty acid ω-hydroxylation in a soluble enzyme system from liver microsomes. J. Biol. Chem. 243:1331-1332.
12. Lu, A.Y.H., and S.B. West (1978) Reconstituted mammalian mixed function oxidases: Requirements, specificities and other properties. Pharmacol. Ther. A. 2:337-358.
13. Lu, A.Y.H., K.W. Junk, and M.J. Coon (1969) Resolution of the cytochrome P-450 containing ω-hydroxylation system of liver microsomes into three components. J. Biol. Chem. 244:3714-3721.
14. Lu, A.Y.H., R. Kuntzman, S.B. West, M. Jacobson, and A.H. Conney (1972) Reconstituted liver microsomal enzyme system that hydroxylates drugs, other foreign compounds, and endogenous substrates. II. Role of the cytochrome P-450 and P-448 fractions in drug and steroid hydroxylations. J. Biol. Chem. 247:1727-1734.

15. Lu, A.Y.H., H.W. Strobel, and M.J. Coon (1969) Hydroxylation of benzphetamine and other drugs by a solubilized form of cytochrome P-450 from liver microsomes: Lipid requirement for drug demethylation. Biochem. Biophys. Res. 36:545-551.
16. Lu, A.Y.H., H.W. Strobel, and M.J. Coon (1970) Properties of a solubilized form of the cytochrome P-450 containing mixed function oxidase of liver microsomes. Mol. Pharmacol. 6:213-220.
17. Morohashi, K., Y. Fujii-Kuriyama, Y. Okada, K. Sogawa, T. Hirose, S. Inayama, and T. Omura (1984) Molecular cloning and nucleotide sequence of cDNA for mRNA of mitochondrial cytochrome P-450 (SCC) of bovine adrenal cortex. Proc. Natl. Acad. Sci., USA 81:4647-4651.
18. Newaz, S.N., W.F. Fang, and H.W. Strobel (1983) Metabolism of the carcinogen 1,2-dimethylhydrazine by isolated human colon microsomes and human colon tumor cells in culture. Cancer 52:794-798.
19. Ozols, J., F.S. Heinemann, and E.F. Johnson (1985) The complete amino acid sequence of a constitutive form of liver microsomal cytochrome P-450. J. Biol. Chem. 260:5427-5434.
20. Philpot, R.M., E. Arinc, and J.R. Fouts (1975) Reconstitution of the rabbit pulmonary microsomal mixed function oxidase system from solubilized components. Drug Metab. Dispos. 3:118-126.
21. Porter, T.D., and C.B. Kasper (1985) Coding nucleotide sequence of rat NADPH cytochrome P-450 oxidoreductase cDNA and identification of flavin-binding domains. Proc. Natl. Acad. Sci., USA 82:973-977.
22. Sladek, N.E., and G.J. Mannering (1969) Induction of drug metabolism. I. Differences in the mechanisms by which polycyclic hydrocarbons and phenobarbital produce their inductive effects of microsomal N-demethylating systems. Mol. Pharmacol. 5:174-185.
23. Sladek, N.E., and G.J. Mannering (1969) Induction of drug metabolism. II. Qualitative differences in the microsomal N-demethylating systems stimulated by polycyclic hydrocarbons and by phenobarbital. Mol. Pharmacol. 5:186-199.
24. Stohs, S.J., R.C. Grafstrom, M.D. Burke, P.W. Moldeus, and S. Orrenuis (1976) The isolation of rat intestinal microsomes with stable cytochrome P-450 and their metabolism of benzo[α]pyrene. Arch. Biochem. Biophys. 177:105-116.
25. Strobel, H.W., and P.P. Lau (1982) Cytochromes P-450 in liver microsomes of rats pretreated with β-naphthoflavone. In Cytochrome P-450: Biochemistry, Biophysics and Environmental Implications, E. Hietanen, M. Laitinen, and O. Hanninen, eds. Elsevier Biomedical Press, Amsterdam, pp. 321-328.
26. Strobel, H.W., A.Y.H. Lu, J. Heidema, and M.J. Coon (1970) Phosphatidylcholine requirement in the enzymatic reduction of hemoprotein P-450 and in fatty acid, hydrocarbon, and drug hydroxylation. J. Biol. Chem. 245:4851-4854.
27. Vermilion, J.L., D.P. Ballou, V. Massey, and M.J. Coon (1981) Separate roles for FMN and FAD in catalysis by liver microsomal NADPH-cytochrome P-450 reductase. J. Biol. Chem. 256:266-277.
28. Weisburger, J.H. (1971) Colon carcinogens: Their metabolism and mode of action. Cancer 28:60-70.
29. White, R.E., and M.J. Coon (1982) Heme ligand replacement reactions of cytochrome P-450: Characterization of the bonding atom of the axial ligand trans to thiolate as oxygen. J. Biol. Chem. 257:3073-3083.

ALTERATION OF MUTAGENIC POTENTIALS BY PEROXIDASE, CATALASE, AND SUPEROXIDE DISMUTASE

M. Nagao,[1] K. Wakabayashi,[1] Y. Suwa,[2] and T. Kobayashi[2]

[1]National Cancer Center Research Institute
1-1, Tsukiji 5-chome
Chuo-ku, Tokyo 104, Japan

[2]Laboratory of Applied Biology and Biotechnology
Central Research Institute
Suntory Ltd.
1-1, Wakayamadai 1-chome
Shimamoto-cho, Mishima-gun, Osaka 618, Japan

INTRODUCTION

In studies of the mutagens/carcinogens in foods using Ames' *Salmonella* test, we found that peroxidase, catalase, and superoxide dismutase (SOD) affected the mutagenic potentials of the compounds present in various kinds of ordinary foods.

Horseradish peroxidase, in the presence of hydrogen peroxide, degraded heterocyclic amines: Trp-P-1, Trp-P-2, Glu-P-1, and 2-amino-α-carboline, which are produced by heating amino acids, proteins, or proteinaceous foods (17). The mutagenicity of methanol extracts of broiled fish to *Salmonella typhimurium* TA98 detected in the presence of S-9 mix was suppressed by horseradish peroxidase without addition of hydrogen peroxide.

Coffee induced base pair change mutations in *S. typhimurium* TA100. This mutagenic activity was suppressed by peroxidases and catalase, but was enhanced by SOD. Coffee has a hydrogen peroxide-generating system, and freshly prepared, normal strength instant coffee contains 3.5 ppm of hydrogen peroxide (2). Hydrogen peroxide itself showed only weak mutagenicity to TA100, but its mutagenicity was markedly enhanced by a small amount of methylglyoxal (2), which is also known to be a mutagen in coffee (4). Superoxide dismutase increased the mutagenicity of coffee, possibly due to the catalytic formation of hydrogen peroxide by SOD from the superoxide anion radical and hydrogen cation.

The mutagenicity of quercetin was enhanced by the cytosol fraction (S105) of rat liver. The presence of at least four enhancing factors in S105 was demonstrated by high performance liquid chromatography (HPLC) on a G2000SWG gel-filtration column. One component was identified as SOD (11).

Superoxide dismutase prevented the degradation of quercetin under aerobic conditions.

Aniline, diphenylamine, and 3-aminopyridine showed mutagenicity only in the presence of norharman and microsomes (12). This co-mutagenicity of norharman was enhanced by SOD.

The roles of peroxidase, catalase, and SOD in the mutagenic effects of some compounds are discussed in this chapter.

MUTAGENICITIES OF HETEROCYCLIC AMINES AND COOKED FISH AND MEAT: SUPPRESSION BY PEROXIDASE

In the presence of hydrogen peroxide, peroxidases suppress the mutagenicities of the heterocyclic amines: Trp-P-1, Trp-P-2, Glu-P-1, and 2-amino-α-carboline, which have been isolated from pyrolysates of tryptophan, glutamic acid, and soybean globulin (17). Horseradish peroxidase, myeloperoxidase, and lactoperoxidase have similar suppressive activities. The mutagenicities of 37 nmol of Trp-P-1, 19.5 nmol of Trp-P-2, 35.8 nmol of Glu-P-1, and 370 nmol of 2-amino-α-carboline were suppressed 80%, 99%, 76%, and 75%, respectively, by incubation for 1 hr at 37°C with 60 pmol of myeloperoxidase and 200 nmol of hydrogen peroxide. This suppressive effect was due to degradation of heterocyclic amines by peroxidase and hydrogen peroxide. On treatment with peroxidase and hydrogen peroxide, Glu-P-1 and 2-amino-3-methylimidazo(4,5-_f_)quinoline (IQ) were converted to their azo-dimer (13) and hydrazine-type N,N'-dimer (5), respectively, as shown in Fig. 1. The same azo-Glu-P-1 was produced from Glu-P-1 by treatment with hypochlorite (14,15). These two dimers showed almost no mutagenic activity, although the hydrazine-type N,N'-dimer of IQ showed a slight mutagenicity without S-9 mix, inducing 400 revertants of TA98 per μg (5).

The mutagenicities of cooked fish and meat were also suppressed by horseradish peroxidase. The mutagens in homogenates of fish and meat were extracted with 5 volumes of 0.05 M HCl, and the extracts were neutralized and loaded on XAD-2 columns. Eluates with methanol were subjected to the mutagenicity test after treatment with various amounts of horseradish

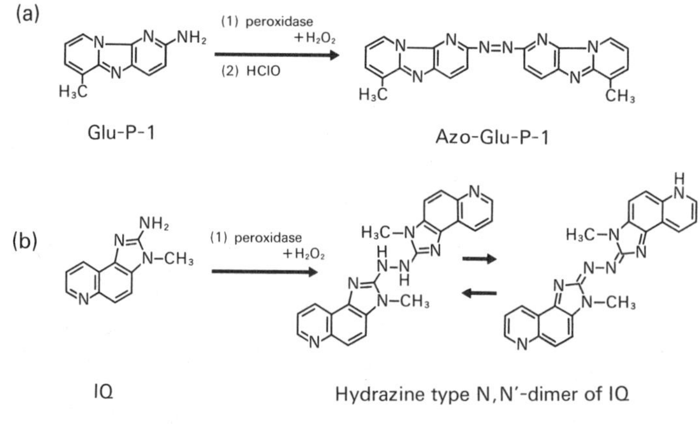

(a) Glu-P-1 → (1) peroxidase $+H_2O_2$ (2) HClO → Azo-Glu-P-1

(b) IQ → (1) peroxidase $+H_2O_2$ → Hydrazine type N,N'-dimer of IQ

Fig. 1. Structures of derivatives of Glu-P-1 (13) and 2-amino-3-methylimidazo(4,5-_f_)quinoline (IQ) (5) produced by horseradish peroxidase and hydrogen peroxide.

peroxidase. As shown in Fig. 2, the mutagenicities of broiled sardine and meat were suppressed by this treatment. In this case, external addition of hydrogen peroxide was not required. Since the main mutagens in cooked fish and meat were reported to be IQ, 2-amino-3,4-dimethylimidazo(4,5-f)quinoline (MeIQ), and 2-amino-3,8-dimethylimidazo(4,5-f)quinoxaline (MeIQx), these compounds must have been converted to nonmutagenic dimers by this treatment.

When an extract of broiled fish was applied to an XAD-2 column and the eluate with methanol was evaporated, hydrogen peroxide was detected in the residue by measurement with an oxygen electrode, indicating that the charred part should contain a hydrogen peroxide-generating system as described later for an extract of coffee. Peroxidases are widely distributed in the animal body, including the saliva, milk, and peroxisomes in almost all organs, and are also present in various vegetables. Therefore, peroxidases may be important in suppressing the effects of environmental mutagens.

MUTAGENICITY OF COFFEE: SUPPRESSION BY CATALASE AND PEROXIDASES, AND ENHANCEMENT BY SUPEROXIDE DISMUTASE

We have reported that instant coffee and freshly brewed coffee are mutagenic to S. typhimurium TA100, TA102, TA104, and TA98, and to Escherichia coli WP2 uvrA and E. coli WP2 uvrA/pKM101, without metabolic activation (6,8); under standard conditions, 15 mg of instant coffee induced 400-600 revertants of TA100. Instant coffee also induced lysogenic λ prophage in E. coli (6). Moreover, instant coffee and brewed coffee induced diphtheria toxin-resistant mutations in Chinese hamster lung cells (10). However, the mutagenicity of instant coffee to S. typhimurium TA100 was remarkably suppressed by S105 of rat liver homogenate. S105 obtained from livers of both

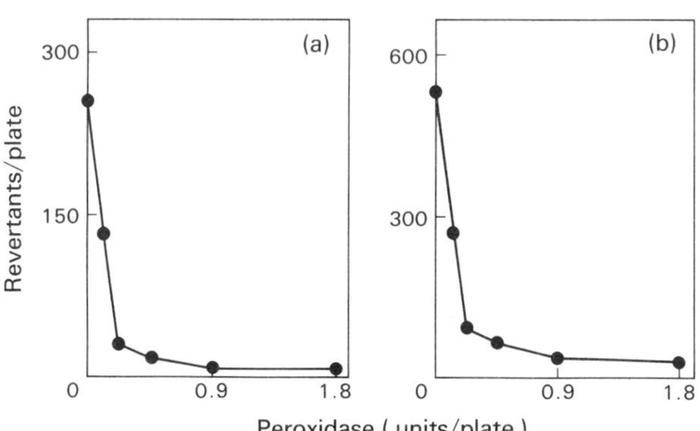

Fig. 2. Suppression of mutagenicity of broiled foods by horseradish peroxidase. Broiled foods were homogenized in 0.05 M HCl and supernatants obtained by centrifugation at 3,000 rpm were loaded on XAD-2 columns, washed with water, and eluted with methanol. Evaporated residues of methanol eluates were dissolved in dimethyl sulfoxide, and the mutagenicities of the solutions toward S. typhimurium TA98 were tested in the presence of S-9 mix and various amounts of horseradish peroxidase. (a) Broiled sardine, 0.4 g equivalent, was used for each point. (b) Broiled beef, 1.9 g equivalent, was used for each point.

polychlorinated biphenyl (PCB)-treated and untreated rats was equally effective. A factor suppressing the mutagenicity was partially purified from S105 of normal rat liver on a molecular sieve column of Sephadex G-100 and an ion-exchange column of CM-cellulose. The partially purified factor required no co-factor and showed absorption at 420 nm, which indicated that it had the nature of a hemoprotein (8).

The partially purified fraction showed catalase activity (8). The activity for suppression of mutagenicity was inactivated by rabbit anti-rat catalase immunoglobulin (kindly provided by Dr. K. Miyazaki, Osaka University). Bovine catalase (Sigma Chemical Co., St. Louis, Missouri) also suppressed the mutagenicity. Twenty units of catalase were enough to inactivate the mutagenicity of 15 mg of instant coffee (1). Horseradish peroxidase, lactoperoxidase, and glutathione peroxidase with glutathione also suppressed the mutagenicity.

These results prompted us to measure the hydrogen peroxide content of instant coffee. By measurements with an oxygen electrode in the presence of catalase, a linear dose-dependent increase of hydrogen peroxide was observed at concentrations of up to 10 mg/ml of instant coffee (2). A freshly prepared solution of instant coffee of ordinary concentration (15 mg/ml) contained about 3.5 ppm of hydrogen peroxide, and the concentration of hydrogen peroxide increased with time after preparation of the solution; instant coffee kept at 37°C for 24 hr contained 50 ppm of hydrogen peroxide. This means that coffee contains a hydrogen peroxide-generating system. When instant coffee was dissolved in water containing SOD, the formation of hydrogen peroxide in the solution increased. This suggests that the superoxide anion is a precursor of hydrogen peroxide.

Freshly brewed coffee prepared from roasted beans also contained hydrogen peroxide. However, coffee prepared from green coffee beans contained scarcely any hydrogen peroxide, and showed almost no mutagenicity. We previously reported that methylglyoxal is one of the mutagens in coffee

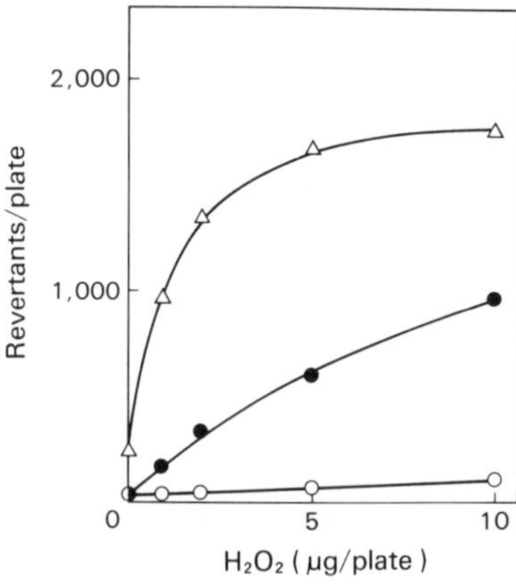

Fig. 3. Effect of hydrogen peroxide on the mutagenicity of methylglyoxal. The mutagenicities of methylglyoxal at 0 μg (o——o), 1.50 μg (●——●), and 3.75 μg (Δ——Δ) with various amounts of hydrogen peroxide were tested. Reprinted with permission from Elsevier Biomedical Press (2).

(4). However, methylglyoxal itself can account for only about 5% of the mutagenicity of instant coffee and 20% of that of freshly brewed coffee.

A sample of 1.5 μg of methylglyoxal, the amount present in 15 mg of instant coffee, showed scarcely any mutagenicity. However, this amount of methylglyoxal induced about 500 revertants, in the presence of 5 μg of hydrogen peroxide, which is about the amount of hydrogen peroxide present in 15 mg of instant coffee (Fig. 3). Furthermore, instant coffee dissolved in deoxygenated water was not mutagenic to S. typhimurium TA100 under anaerobic conditions.

These results suggest that the presence of both methylglyoxal and hydrogen peroxide accounts for most of the mutagenicity of coffee. Methylglyoxal is known to be converted to pyruvate by hydrogen peroxide, but pyruvate is not mutagenic with or without hydrogen peroxide. Thus, the mechanism of the enhancing effect of hydrogen peroxide on the mutagenicity of methylglyoxal is unknown.

ENHANCEMENT OF THE MUTAGENICITY OF QUERCETIN BY SUPEROXIDE DISMUTASE

Quercetin is a direct-acting mutagen to S. typhimurium TA98 and TA100. The mutagenicity of quercetin to TA98 was enhanced a maximum of 2-fold by rat liver cytosol. On gel-filtration on a column of TSKgel G2000SWG [21.5 mm inside diameter (ID) x 600 mm], the mutagenicity-enhancing activity of S105 was separated into 4 fractions with molecular weights of approximately 60,000; 35,000; 6,000; and less than 1,000 daltons, respectively (Fig. 4). Fraction II, which coincided in position with SOD, was further purified by DE-52 column chromatography. The mutagenesis-enhancing factor gave a single band with a molecular weight of 18,000 daltons on SDS polyacrylamide gel electrophoresis (11). Superoxide dismutase is known to have a molecular weight of 32,000-38,000 daltons and to consist of a homodimer.

In the presence of oxygen, quercetin is unstable in weakly alkaline solution and is rapidly degraded even at pH 7.8. Superoxide dismutase

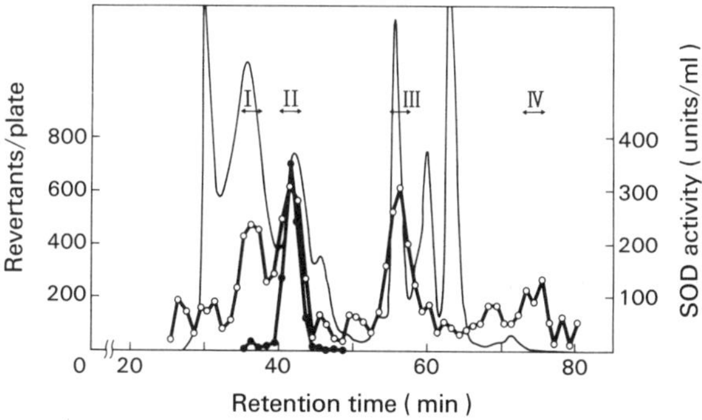

Fig. 4. Gel filtration pattern on HPLC on a G2000SWG column (21.5 mm ID x 600 mm, Toyo Soda) of S105 of normal rat liver. ——, absorbance at 280 nm; o—o, enhancement of quercetin mutagenesis on TA98; •—•, SOD activity. Mutagenesis-enhancing activity is shown as the revertant number with 50 μg of quercetin and 0.3 ml of each fraction minus the revertant number with 50 μg of quercetin alone (11). Reprinted with permission from Elsevier Biomedical Press.

partly prevented its degradation (11). Quercetin was also degraded under the conditions for mutagenesis assay in which 50 mM sodium phosphate buffer (pH 7.4) was used. These results strongly suggest that quercetin is autocatalytically degraded by a self-produced superoxide anion radical. Quercetin was recently reported to be degraded by the superoxide anion radical produced from xanthine and xanthine oxidase (16). Thus, SOD may enhance the mutagenicity of quercetin by preventing autodegradation of quercetin. Hydrogen peroxide, which is a reaction product of SOD, did not have any effect on the mutagenicity of quercetin. Quercetin was not mutagenic to *S. typhimurium* TA98 under anaerobic conditions, indicating that oxygen is necessary for exhibition of its mutagenicity.

IMPLICATION OF SUPEROXIDE DISMUTASE IN THE CO-MUTAGENESIS OF NORHARMAN

Norharman [9*H*-pyrido(3,4-*b*)indole, β-carboline] is a co-mutagen (9, 12); it is not itself mutagenic to *S. typhimurium* TA100 or TA98 with or without S-9 mix, but it shows mutagenicity to *S. typhimurium* TA98, a frameshift mutant, in the presence of some kinds of nonmutagenic amines, such as aniline, 3-aminopyridine, *o*-toluidine, yellow OB, or diphenylamine in the presence of microsomes. A microsomal enzyme is necessary for expression of the mutagenicity of norharman with amines. Addition of the cytosol fraction of rat liver caused a 7- to 8-fold increase in the mutagenicity of a mixture of aniline, norharman, and microsomes.

Phenylhydroxylamine, which is considered to be a proximate mutagen formed from aniline, also required a microsomal enzyme and reduced nicotinamide adenine dinucleotide phosphate (NADPH) to show mutagenicity with norharman. Of 2 types of PCB-P-450 (Ic and Id) and 2 types of PCB-P-448 (IIa and IId) of cytochrome P-450 (3), only Ic induced mutagenicity of a mixture of phenylhydroxylamine and norharman. In this case also, the cytosol fraction was effective for enhancing the mutagenicity. The mutagenicity-enhancing factor present in the cytosol fraction was purified by HPLC on a gel-filtration column of G2000SWG. As shown in Fig. 5, the active factor was eluted in one fraction, and its retention time coincided with that of SOD.

We tested the effect of SOD of bovine blood (Sigma Chemical Co.) on the co-mutagenicity of norharman with aniline, diphenylamine, and 3-aminopyridine in the presence of microsomes. Addition of SOD caused about a 5-fold increase in the mutagenicities of mixtures of amines and norharman. The mechanism of the co-mutagenic effect of norharman and the implication of SOD in mutagenic activation of these compounds require investigation.

DISCUSSION

Peroxidases were found to be important in inactivation of environmental mutagens. Hydrogen peroxide and peroxidase inactivate heterocyclic amines of both the IQ type and non-IQ type (14). The charred parts of fish and meat contain hydrogen peroxide and do not require exogenous addition of hydrogen peroxide for their inactivation by peroxidase. Since peroxidases are widely distributed in mammals and plants, and some peroxidases are heat stable, they may play important roles in modulating the activities of mutagens/carcinogens present in ordinary foods. The mutagenicity of coffee was also shown to be greatly decreased by peroxidases or catalase.

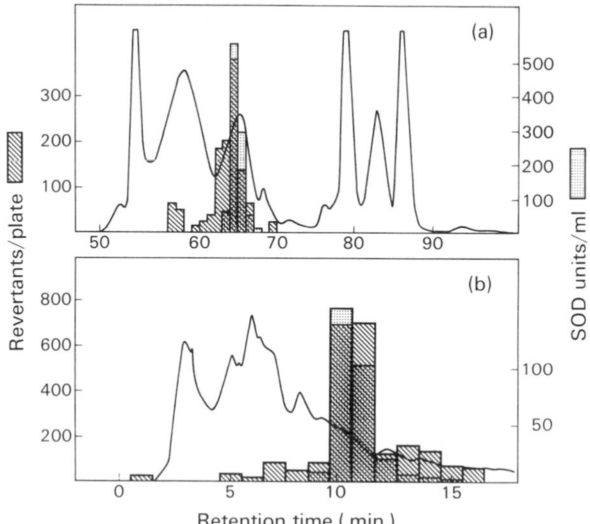

Fig. 5. Purification of SOD and the enhancing factor for norharman co-mutagenesis from the cytosol of normal rat liver. (a) S105 of rat liver was subjected to gel-filtration by HPLC on a G2000SWG column (21.5 mm ID x 600 mm, Toyo Soda) in 0.15 M KCl solution. (b) Fractions with retention times of 64-65 min were collected and purified further by HPLC on an ion exchange column of IEX545-DEAE (7.5 mm ID x 75 mm, Toyo Soda) in 2.5 mM sodium phosphate buffer (pH 7.4).

Superoxide dismutase enhanced the mutagenicities of some compounds. This finding was unexpected, since the superoxide anion radical induces DNA damage. We found that SOD enhanced the mutagenicity of quercetin and the co-mutagenicity of norharman. Recently, SOD was also shown to enhance the mutagenicity of hydroxyxanthone (7). The role of SOD in the genotoxicity of chemicals requires further study, although the superoxide anion radical is suspected to be involved in tumor promotion.

REFERENCES

1. Fujita, Y., K. Wakabayashi, M. Nagao, and T. Sugimura (1985) Characteristics of major mutagenicity of instant coffee. Mutat. Res. 142: 145-148.
2. Fujita, Y., K. Wakabayashi, M. Nagao, and T. Sugimura (1985) Implication of hydrogen peroxide in the mutagenicity of coffee. Mutat. Res. 144:227-230.
3. Kamataki, T., H. Yoshizawa, K. Ishii, and R. Kato (1982) Separation and purification of multiple forms of cytochrome P-450 from PCB-treated rat and human livers. In Microsomes, Drug Oxidation and Drug Toxicity, R. Sato and R. Kato, eds. Japan Scientific Societies Press, Tokyo, and Wiley Inter-Science, New York, pp. 99-100.
4. Kasai, H., K. Kumeno, Z. Yamaizumi, S. Nishimura, M. Nagao, Y. Fujita, T. Sugimura, H. Nukaya, and T. Kosuge (1982) Mutagenicity of methylglyoxal in coffee. Gann 73:681-683.
5. Kasai, H., M. Nakayama, Z. Yamaizumi, H. Saitô, and S. Nishimura (1985) Formation of mutagenic IQ dimer by peroxidase treatment of IQ.

In Proceedings of the Environmental Mutagen Society, Japan, 14th Annual Meeting, p. 97 (in Japanese).

6. Kosugi, A., M. Nagao, Y. Suwa, K. Wakabayashi, and T. Sugimura (1983) Roasting coffee beans produces compounds that induce prophage λ in E. coli and are mutagenic in E. coli and S. typhimurium. Mutat. Res. 116:179-184.
7. Matsushima, T., M. Muramatsu, O. Yagame, A. Araki, L. Tikkanen, and S. Natori (1985) Mutagenicity and chemical structure relations of naturally occurring mutagens from plants. In Proceedings of the 4th International Conference on Environmental Mutagens, Stockholm, I. Knudsen et al., eds. (in press).
8. Nagao, M., Y. Suwa, H. Yoshizumi, and T. Sugimura (1984) Mutagens in coffee. In Coffee and Health, Banbury Report No. 17, B. MacMahon and T. Sugimura, eds. Cold Spring Harbor Laboratory Press, Cold Spring Harbor, New York, pp. 69-77.
9. Nagao, M., T. Yahagi, M. Honda, Y. Seino, T. Matsushima, and T. Sugimura (1977) Demonstration of mutagenicity of aniline and o-toluidine by norharman. Proc. Japan. Acad. 53B:34-37.
10. Nakasato, F., M. Nakayasu, Y. Fujita, M. Nagao, M. Terada, and T. Sugimura (1984) Mutagenicity of instant coffee on cultured Chinese hamster lung cells. Mutat. Res. 141:109-112.
11. Ochiai, M., M. Nagao, K. Wakabayashi, and T. Sugimura (1984) Superoxide dismutase acts as an enhancing factor for quercetin mutagenesis in rat-liver cytosol by preventing its decomposition. Mutat. Res. 129: 19-24.
12. Sugimura, T., M. Nagao, and K. Wakabayashi (1982) Metabolic aspects of the comutagenic action of norharman. In Biological Reactive Intermediates-II: Chemical Mechanisms and Biological Effects, Part B, R. Snyder et al., eds. Plenum Press, New York and London, pp. 1011-1035.
13. Tsuda, M., K. Wakabayashi, M. Nagao, T. Hirayama, and T. Sugimura (1981) Structure determination of a nonmutagenic product obtained by the treatment of Glu-P-1 with horseradish peroxidase. In Proceedings of the Environmental Mutagen Society, Japan, 14th Annual Meeting, p. 75 (in Japanese).
14. Tsuda, M., C. Negishi, R. Makino, S. Sato, Z. Yamaizumi, T. Hirayama, and T. Sugimura (1985) Use of nitrite and hypochlorite treatments in determination of the contribution of IQ-type and non-IQ-type heterocyclic amines to the mutagenicities in crude pyrolyzed materials. Mutat. Res. 147:335-341.
15. Tsuda, M., K. Wakabayashi, T. Hirayama, and T. Sugimura (1983) Inactivation of potent pyrolysate mutagens by chlorinated tap water. Mutat. Res. 119:27-34.
16. Ueno, I., M. Kohono, K. Yoshihira, and I. Hirono (1983) Inactivation of quercetin, a mutagenic flavonoid, by superoxide radicals. In Proceedings of the Japanese Cancer Association, The 42nd Annual Meeting, p. 301.
17. Yamada, M., M. Tsuda, M. Nagao, M. Mori, and T. Sugimura (1979) Degradation of mutagens from pyrolysates of tryptophan, glutamic acid and globulin by myeloperoxidase. Biochem. Biophys. Res. Commun. 90:769-776.

ACTIVATION AND INACTIVATION OF MUTAGENS AND CARCINOGENS--PART II

EFFECTS OF VITAMINS C AND E ON CARCINOGEN FORMATION AND ACTION, AND RELATIONSHIP TO HUMAN CANCER

Sidney S. Mirvish

Eppley Institute for Research in Cancer
University of Nebraska Medical Center
Omaha, Nebraska 68105

To introduce this subject, I will briefly review the inhibition by vitamins C and E of the formation of carcinogenic nitrosamines and nitrosamides (N-nitroso compounds), as well as effects of these vitamins on carcinogenesis and human cancer. I have previously reviewed the inhibition by vitamins C and E of N-nitroso compound formation (4,5) and hope shortly to publish a more complete review on the subject (7).

In 1972 we found that vitamin C inhibited the acid-catalyzed formation of N-nitroso compounds from nitrite and amines or amides. The inhibition occurs when vitamin C competes successfully with amines or amides for nitrite, and vitamin C is more effective at pH 3-5 than most other reducing agents. The inhibition has been studied in rodents treated with nitrite and amines or amides, and has been demonstrated by chemical analysis of stomach contents and by observing an inhibition of acute toxicity and tumor induction (4,5). Inhibition has also been demonstrated in humans (see Bartsch et al., this Volume). Vitamin C is added to nitrite-preserved meat products to prevent nitrosamine formation and should, I believe, be administered to people who ingest readily nitrosatable drugs, such as piperazine.

Vitamin E inhibits nitrosamide formation in lipid media from amides and nitrogen oxides (5). Vitamin E emulsions in water can be used together with vitamin C to reduce nitrosamine formation in bacon (10) and inhibit in vivo nitrosation by nitrite (7,10). Vitamin E does not inhibit nitrosamine formation from a nitrogen dioxide-derived nitrosating agent in mouse skin lipids (8).

In most published studies, vitamins C and E also inhibited carcinogenesis by preformed carcinogens, including N-nitroso compounds and polycyclic aromatic hydrocarbons, when these compounds were administered to laboratory rodents and the vitamin was fed in the diet (7). There was a fairly consistent 30-70% inhibition by each vitamin, provided that it was given in the diet at an adequate level and the carcinogen dose was not excessive. Vitamin E may exert its effect synergistically with that of selenium (3). In a cell culture system, vitamin C prevented transformation (conversion to cancer-like cells) by 3-methylcholanthrene and also reversed transformed cells back to normal cells, provided that the transformed cells had not progressed too far towards malignancy (1).

In vivo formation of N-nitroso compounds may be involved in the induction of human cancer of the stomach, esophagus, mouth, larynx, and cervix. The basis of this statement is, first, that these cancers are negatively associated with the intake of fresh fruits and vegetables, which are rich in vitamin C (2). Polyphenols in these foods could also be involved, since most polyphenols are good nitrite scavengers. Other lines of evidence also support the N-nitroso compound hypothesis for the etiology of stomach cancer (6) and esophagus cancer. Dr. Bartsch will present some important new supporting data on these cancers (see Bartsch et al., this Volume). People in areas with high incidences of esophageal cancer usually consume a poor diet, which is deficient not only in vitamin C, but also in protein, vitamin A, riboflavin, thiamine, molybdenum, zinc, and magnesium (7). These other dietary constituents may also be involved in the etiology.

In conclusion, vitamins C and E are good examples of anticarcinogens that can act at several stages of the carcinogenic process and that may also inhibit cancer induction in people. Partly on the basis of these findings, people should be encouraged to increase their consumption of fresh fruits and vegetables (rather than that of vitamin supplements) as a method of reducing their risk of cancer (9).

ACKNOWLEDGEMENTS

My current research in this area is supported by National Cancer Institute grants R01-CA-30593, R01-CA-32192, R01-CA-35628, and P01-CA-36727.

REFERENCES

1. Benedict, W.F., W.L. Wheatley, and P.A. Jones (1980) Inhibition of chemically induced morphological transformation and reversion of the transformed phenotype by ascorbic acid in C3H/10T1/2 cells. Cancer Res. 40:2796-2801.
2. Graham, S. (1983) Toward a dietary prevention of cancer. Epidemiol. Rev. 5:38-40.
3. Ip, C. (1985) Attenuation of the anticarcinogenic action of selenium by vitamin E deficiency. Cancer Lett. 25:325-331.
4. Mirvish, S.S. (1980) Ascorbic acid inhibition of N-nitroso compound formation in chemical, food and biological systems. In Inhibition of Tumor Induction and Development, M.S. Zedeck and M. Lipkin, eds. Plenum Press, New York, pp. 101-126.
5. Mirvish, S.S. (1981) Inhibition of the formation of carcinogenic N-nitroso compounds by ascorbic acid and other compounds. In Cancer 1980: Achievements, Challenges, Projects, Vol. 1, J.H. Burchenal and H.P. Oettgen, eds. Grune and Stratton, Inc., New York, pp. 557-587.
6. Mirvish, S.S. (1983) The etiology of gastric cancer: Intragastric nitrosamide formation and other theories. J. Natl. Cancer Inst. 71:629-647.
7. Mirvish, S.S. (1985) Inhibition by vitamins C and E of N-nitroso compound formation and carcinogenesis, and relationship to human cancer. Cancer (ms. submitted for publ.).
8. Mirvish, S.S., and K. Laughlin (1985) Lack of vitamin E inhibition of nitrosamine formation from NO_2-derived nitrosating agent in mouse

skin. Abstract of presented poster at the International Conference on Mechanisms of Antimutagenesis and Anticarcinogenesis, Lawrence, Kansas (this Volume).

9. National Academy of Sciences (1982) Diet, Nutrition and Cancer, National Academy Press, Washington, D.C.

10. Newmark, H.L., and W.J. Mergens (1981) α-Tocopherol (vitamin E) and its relationship to tumor induction and development. In Inhibition of Tumor Induction and Development, M.S. Zedeck and M. Lipkin, eds. Plenum Press, New York, pp. 127-160.

MODIFIERS OF ENDOGENOUS NITROSAMINE SYNTHESIS AND METABOLISM

H. Bartsch, H. Ohshima, J. Nair,
B. Pignatelli, and S. Calmels

International Agency for Research on Cancer
Division of Environmental Carcinogenesis
69372 Lyon, France

INTRODUCTION

N-Nitroso compounds (NOCs), a class of versatile carcinogens (1,13), are formed in nature, most likely since mankind first existed on earth, whenever nitrosating agents such as nitrite or nitrosating gases encounter nitrosatable amines. To date, more than 300 NOCs have been tested in animals, and about 90% of them produced tumors in 40 animal species, including primates. Humans are exposed to NOCs from exogenous and endogenous sources through nitrosation of ingested/inhaled amino precursors. Nitrite is produced by bacterial reduction of nitrate, normally in the mouth, and the nitrosation reaction generally proceeds through an acid-catalyzed reaction in the stomach. Any nitrosation reaction occurring in vivo is, however, affected by many factors, such as the pH, precursor concentration, and the presence of catalysts and inhibitors. These various factors, some of which are difficult to measure in vivo, have complicated the estimation of nitrosation reactions occurring in humans.

Endogenous formation of NOCs has been suspected to be associated with an increased risk of cancer of the stomach, esophagus and bladder, but convincing epidemiological evidence is lacking (1). Because of the lack of noninvasive methods, our laboratory has recently developed a simple and sensitive method [N-nitrosoproline (NPRO) test] for quantitative estimation of endogenous nitrosation in man (2,20,21). It is based on the fact that certain noncarcinogenic N-nitrosamino acids (NAAs), such as NPRO, are excreted quantitatively in the urine (6,7,22) while concomitantly formed carcinogenic nitrosamines are not readily detectable, due to fast metabolism and reaction with cellular material. Thus, the amount of urinary NPRO and of other NAAs that is excreted during 24 hr per person, following ingestion of nitrate and/or proline, is an index of endogenous nitrosation.

The general objective of our study is to apply this test in clinical and field studies, with the aim of measuring nitrosamine exposure and of identifying dietary, life style, and host factors or disease states that modify endogenous nitrosation in man. Results from such studies are used to identify populations/individuals at high risk for cancers of the stomach, esophagus, and oral cavity possibly caused by endogenous NOCs, and to indicate preventive measures by which the body burden of endogenous NOCs can be efficiently lowered.

Ongoing studies and results obtained to date are summarized below. The role of ethanol as a modulator of nitrosamine carcinogenesis is briefly discussed.

RESULTS AND DISCUSSION

The NPRO Test

Human subjects were given the following (protocol A): (i) 200 ml beetroot juice (260 mg nitrate); (ii) 30 min later, L-proline (500 mg); (iii) the subjects fast for a further 2 hr; (iv) 24-hr urine samples were collected in plastic bottles containing 10 g NaOH (during urine collection, foodstuffs rich in nitrate and cured meat, smoked fish, and beer were avoided); (v) 100-ml urine aliquots were stored at -20°C prior to analysis; no artifact formation or degradation of NPRO and nitrate/nitrite was shown to occur (21,24). Alternatively, three 24-hr urine samples had been collected from each subject according to the following protocol (protocol B): (i) undosed; (ii) sample taken after intake of 100 mg L-proline 3 times a day after each meal; and (iii) sample taken after intake of 100 mg L-proline and 100 mg ascorbic acid 3 times a day. Whenever possible, information was collected from each study subject on demographic data, smoking, drinking and dietary habits, and clinical findings.

The rationale for applying this procedure in field or clinical studies is based on the following observations: (i) NPRO has been reported not to be carcinogenic and mutagenic (13,18); (ii) after gavage of rats with ^{14}C-NPRO, the $^{14}CO_2$ production and DNA alkylation were negligible (6), but urinary excretion of NPRO (as the unchanged compound) was rapid and almost complete (6,7,22); (iii) urinary levels of NPRO in rats gavaged with proline reflected well endogenous nitrosation of proline (7); and (iv) in humans, preformed NPRO ingested in the aqueous extract of broiled dried squids was also rapidly and almost quantitatively eliminated in the urine 24 hr after ingestion (20). Thus, the amount of NPRO excreted in the 24-hr urine was an indicator of daily endogenous nitrosation (21).

Human urine contains several NAAs among which the major ones have been identified as NPRO, N-nitrososarcosine (NSAR), N-nitrosothiazolidine 4-carboxylic acid (NTCA), and trans- and cis-isomers of N-nitroso-2-methylthiazolidine 4-carboxylic acid (NMTCA) (21,23,24,30), all of which are currently analyzed as indicators of human exposure to exogenously and endogenously formed N-nitroso compounds (see Fig. 1).

Recently, 3 more hitherto unknown NAAs in human urine were identified as 3-(N-nitroso-N-methylamino)propionic acid, N-nitrosoazetidine 2-carboxylic acid, and N-nitrosotetrahydro-4H-1,3-thiazine 4-carboxylic acid (Ohshima et al., ms. in prep.).

Analysis of N-Nitroso Compounds in the Urine

Urine samples were spiked with N-nitrosopipecolic acid as an internal standard and analyzed for NPRO, NSAR, and other NAAs according to a published procedure (21,24) after conversion to their methyl esters by diazomethane. Samples were analyzed on a Tracor 550 gas chromatograph (GC),

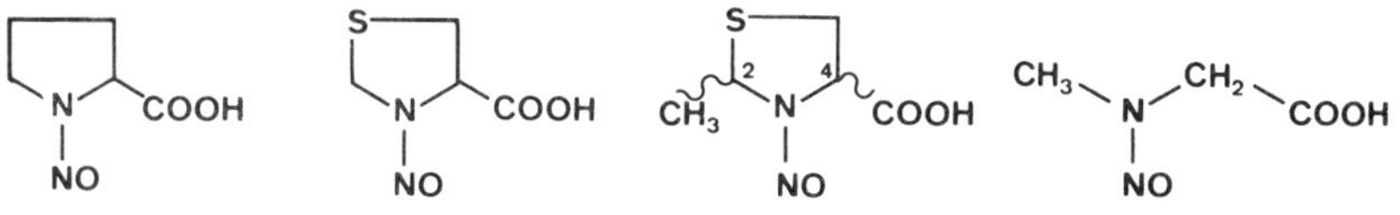

NPRO NTCA NMTCA NSAR

Fig. 1. Structures of 4 urinary N-nitrosamino acids (NAAs) used as indicators of human exposure to exogenously and endogenously formed N-nitroso compounds: N-nitrosoproline (NPRO), N-nitrosothiazolidine 4-carboxylic acid (NTCA), N-nitroso-2-methylthiazolidine 4-carboxylic acid (NMTCA), and N-nitrososarcosine (NSAR).

which was interfaced to a Thermal Energy Analyzer (TEA) (Thermo Electron Corporation, Waltham, Massachusetts), a nitrosamine-specific detector.

Kinetic Studies on the Endogenous Nitrosation in Human Subjects

In the absence of adverse biological effects of NPRO (13,18), a number of kinetic studies were previously carried out on the formation of NPRO in vivo in human volunteers who had ingested vegetable juice (as a source of nitrate) and an aqueous solution of proline with or without dietary nitrosation modifiers. Urinary excretion of NPRO was monitored as an index for endogenous nitrosation. Results allowed 3 major conclusions to be drawn: (i) an unequivocal demonstration that NOCs are formed in the human body, even after ingestion of amounts of precursors (amine, nitrate) that are considered as normal daily intake. The amounts of NAA formed in vivo were around 25 µg per day per person, when subjects were living on an uncontrolled western diet; (ii) it was found that the intake of nitrate above a dose of 260 mg per day per person led to a sharp increase in the amount of NPRO formed in vivo (21); and (iii) nitrosation inhibitors such as vitamins C and E markedly reduced the yield of NAA formed in healthy human subjects (see Fig. 2).

Another class of nitrosation modifiers, polyphenols, are ubiquitously present in the plant kingdom, and substantial exposure of humans to them occurs through intake of vegetables and fruits, coffee, tea, etc. It was noted previously in rats that, after co-administration of proline, nitrite, and the modifier, some polyphenols, like catechin, catalyzed in vivo nitrosation while others, like chlorogenic acid, strongly inhibited this reaction (26). To explain this different behavior, 3 determinants appear to be important: (i) the nature of the nitrosatable amine and the relative position of the hydroxy groups in the phenol ring; consequently the resorcinol-type compounds are often nitrosation catalysts; (ii) the molar ratio (R): nitrite to polyphenol; when $R > 1$, usually catalysis is observed for certain polyphenols; inhibition is seen when $R < 1$, a situation which may normally prevail in the human stomach; and (iii) the pH; with certain phenols catalysis is seen at pH 4 to pH 6, while inhibition normally occurs below pH 4.

Using the NPRO test, one can now examine the effects of some polyphenols after ingestion of proline and nitrate. As an example, we were able to examine in human volunteers some extracts of areca (betel) nuts, which are rich in polyphenols and tannins. Urinary NPRO excretion was strongly inhibited after ingestion of the extract (27). The studies have been extended to other polyphenols that are more commonly consumed, such as naturally occurring mixtures or pure compounds occurring in coffee and tea

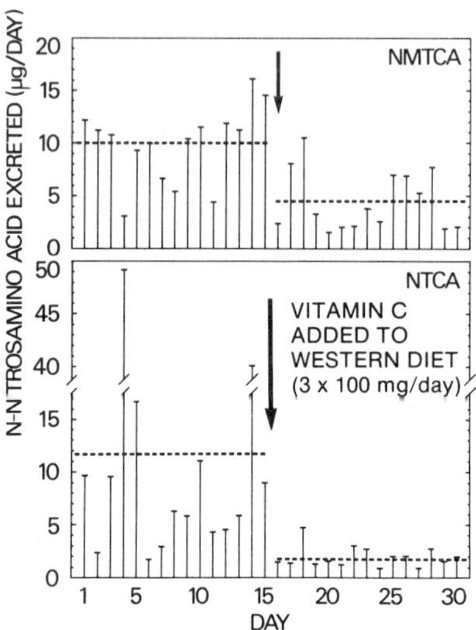

Fig. 2. Endogenous formation of N-nitrosothiazolidine 4-carboxylic acid (NTCA) and its 2-methyl derivative (NMTCA) in one human subject living on an uncontrolled western diet, and its inhibition by vitamin C. Twenty-four-hr urine samples were collected each day during a 30-day period from one healthy subject. From days 16 to 30 the diet was supplemented with vitamin C (as indicated by the arrow). (From Ref. 24.)

preparations (28). With ingestion of doses that are considered as normal daily intake, inhibition of endogenous nitrosation up to 80% was observed (Tab. 1). Although polyphenols exert a number of pharmacological actions, based on our results, their potential for inhibiting nitrosamine formation in vivo in humans cannot be neglected and should be considered when interpreting epidemiological data.

Uric acid and allantoin, which are present in human body fluids, were also studied as potential nitrosation modifiers (Tab. 2). Both compounds turned out to be extremely poor inhibitors of nitrosation of diethylamine in vitro, and their activity was only about 1 to 4% as compared to an equimolar ascorbic acid solution. Therefore, uric acid and allantoin probably cannot be very effective nitrite-trapping agents in vivo. However, upon prolonged incubation with nitrite, both compounds yielded direct-acting mutagenic nitroso derivatives in Salmonella typhimurium, an interesting finding as these compounds could be formed in the (infected) urinary bladder (Pignatelli et al., ms. in prep.).

N-Nitrosation Catalysis by Bacteria

Since endogenous nitrosation catalyzed by bacteria has been suggested to play a role in the stomach and the urinary bladder (9), studies on the microbial formation of NOCs have been initiated (5). A total of 38 strains of various species of bacteria and one strain of yeast (Candida) from human sources were examined for their ability to form N-nitrosomorpholine (NMOR) from morpholine and nitrite at neutral pH. Twenty-five bacterial strains exhibited nitrosation activity (Tab. 3). Kinetic studies were conducted with resting cells of E. coli A10. The formation of NMOR was optimal at pH

Tab. 1. Effect of phenolic compounds on urinary levels of N-nitrosoproline (NPRO) in human volunteers.

Nitrate + proline + modifying agent	NPRO excreted in urine	
	μg	Control (%)
None	6.4	100
Caffeic acid (3 x 300 mg)	0.9	14
Ferulic acid (2 x 300 mg)	1.8	28
Instant coffee decaffeinated (2 cups)	1.3	20
Instant coffee regular (1 cup)	0.9	14
Indian tea (1 cup)	2.6	41
Chinese tea (1 cup)	1.6	25

Note: Collated from Ref. 28. Nitrate (300 mg); proline (300 mg).

7.25, and the reaction followed Michaelis-Menten kinetics. Substrate specificity for several amines was found with proline being a poor substrate. These results indicate that nitrosation is catalyzed by a bacterial enzyme(s), and attempts to characterize this process in E. coli are in progress. Bacterial strains isolated and identified in gastric juice samples from patients with chronic gastritis, and urine samples from subjects having urinary bladder infections are now being tested (in collaboration with P. Vincent, Université Claude Bernard, Lyon, and H. Leclerc, INSERM, Villeneuve d'Ascq, France) for their ability to catalyze nitrosation.

A study reported an increased risk of bladder cancer among individuals with a history of urinary infections, and especially among cigarette smokers (15). Nitrate and thiocyanate are present in the urine of smokers together with the tobacco-derived alkaloids, nicotine and cotinine. Nitrosation in the infected urinary bladder could take place if nitrate-reducing bacteria are present together with bacteria possessing the ability to catalyze nitrosation.

The NPRO test is now applied in field studies in order to collect data on exposure to endogenous NOC in subjects and areas where nitroso carcinogens may be involved in certain human cancers. Studies are being done on tobacco smokers and chewers, betel quid chewers, and on inhabitants of areas at high risk of stomach and esophageal cancer in Japan and in some northern parts of the People's Republic of China. Some interim results of these studies, which have not yet been fully completed or evaluated, are presented below.

Endogenous Nitrosation in Cigarette Smokers and Tobacco Users

Tobacco smokers, chewers of tobacco and of betel quid (often containing tobacco), and snuff dippers are all exposed to increased levels of nitrosamine precursors (such as nicotine, nitrate, nitrite, and NOx) and nitrosation modifiers (thiocyanate). Results to date prove that in these

Tab. 2. Effect of vitamin C, uric acid, and allantoin on N-nitrosodiethylamine (NDEA) formation.

Inhibitor (I)	Incubation time (min)	% Inhibition of NDEA formation ratio: I/nitrite 10	2
Vitamin C	30, 60	100.0	100.0
Uric acid	30	1.2	0.6
	60	3.7	2.3
Allantoin	30	3.5	0.5
	60	3.7	3.0

Note: Diethylamine (20 mmol/l); I (4 mmol/l in citric acid/citrate buffer; pH: 2.85 at 37°C); from Pignatelli et al., manuscript in prep.

subjects, endogenous NOC synthesis occurs at a higher rate than in persons without such habits (3,10,16), thus adding to the body burden of carcinogens ingested or inhaled from exogenous intake. Endogenous NPRO formation in smokers could be inhibited by daily addition of 1 g ascorbic acid to their diet (10); its inhibition of endogenous nitrosation of proline in nonsmokers has been previously demonstrated (see Fig. 2). In a study in male subjects from Bombay, India, who were smokers of western-type cigarettes and bidis (native cigarettes) or chewers of betel quid containing Indian tobacco, the excretion of NPRO and NTCA in the 24-hr urine was measured (Fig. 3) (J. Nair et al., ms. in prep.). The means tended to be increased in the habit groups, most pronounced in cigarette and bidi smokers as compared to subjects without such habits. Ascorbic acid at a dose of 3 x 100 mg/day effectively reduced the levels of urinary NPRO in smokers and chewers (data not shown). Large interindividual variations were observed.

There is now good evidence available to assume that tobacco-specific nitrosamines (TSNAs) are major carcinogenic agents alone and in combination with other tobacco constituents, in tobacco-associated cancers, in both smokers and, to an even greater degree, in users of combusted tobacco products (12). A substantial fraction (as yet undetermined) of tobacco-related nitrosamines and other NOCs appears to be synthesized in vivo. Ascorbic acid, shown to be an effective inhibitor of nitrosation in humans, also inhibits endogenous NOC synthesis in smokers. Regular consumption of vegetables and fresh fruits, a potential source for ascorbic acid, has some protective effect against tobacco-associated malignancies (31), and it can be assumed that this protective effect is attributable mainly to the inhibitory action of ascorbate on nitrosamine formation, although other various modifiers of carcinogenesis are present in these food items.

Nitrosamine Exposure of Betel Quid Chewers

The correlation between oral cancer and chewing of betel quid (often containing tobacco in India and other Southeast Asian countries) is well established (4). Nitrosation in vitro of areca nut-specific alkaloids, in

Tab. 3. Activity of some microorganisms/strains to catalyze the nitrosation of morpholine at pH 7.2 (NMOR, N-nitrosomorpholine).

Species*		Nitrosation rate (range) (nmol NMOR formed per mg protein per hr)
Escherichia coli	(strain) A10	50
	(14 strains)	10-44
	(4 strains)	0-7
Pseudomonas aeroginosa (2 strains)		0-36
Proteus morganii		10
Proteus mirabilis		4
Salmonella typhimurium		0
Staphylococcus aureus		0
Acinetobacter		0
Lactobacillus		0
Candida		0

* Microorganisms or strains were isolated from human sources (urinary infections, wounds, hemocultures, nasotracheal secretions, vagina, feces) (collated from Ref. 5).

particular arecoline, leads to the formation of NOCs of which 3-(methylnitrosamino)propionitrile was found to be carcinogenic in experimental animals (33). The extent to which this nitrosation of betel nut constituents is taking place in vivo, i.e., in the oral cavity of chewing subjects, has not been investigated until recently.

Saliva and urine samples were collected from chewers of betel quid with or without tobacco, from tobacco chewers, from cigarette smokers, and from people with no such habits (in collaboration with S.V. Bhide, Cancer Research Institute, Tata Memorial Center, Bombay, India). In order to evaluate exposure of betel quid chewers to NOCs, these samples were analyzed for TSNAs such as N'-nitrosonornicotine (NNN), N'-nitrosoanatabine (NAT), and 4-(methylnitrosamino)-1-(3-pyridyl)-1-butanone (NNK), and for areca nut-specific nitrosamines (ASNAs) such as N-nitrosoguvacoline (NGCO) and N-nitrosoguvacine (NGCI) (see Fig. 4) (19).

Tobacco-specific nitrosamines and ASNAs were detected (ng/ml) in saliva of chewers of betel quid with tobacco (BQT); ASNAs were detected in chewers of betel quid without tobacco (BQ). Precursor tobacco alkaloids (i.e., nicotine) and betel nut alkaloids (i.e., arecoline), as well as nitrite and thiocyanate, were detected (µg/ml) in the saliva. A good positive correlation was found between TSNAs and nicotine in saliva. Urinary NPRO levels tended to be higher in chewers of BQT, BQ, and tobacco and in smokers as compared to the no-habit group, suggesting an increased endogenous nitrosation. Urinary levels of tobacco-related compounds were similar in chewers of BQT and tobacco when compared with smokers, implying similar uptake. Certain TSNAs and ASNAs can be formed while chewing betel quid with or without tobacco, as demonstrated by an increase in their levels after in vitro nitrosation of betel quid extracts at pH 7.4. Up to 250 times higher concentrations of TSNAs and ASNAs are formed at pH 2.1 (simulated

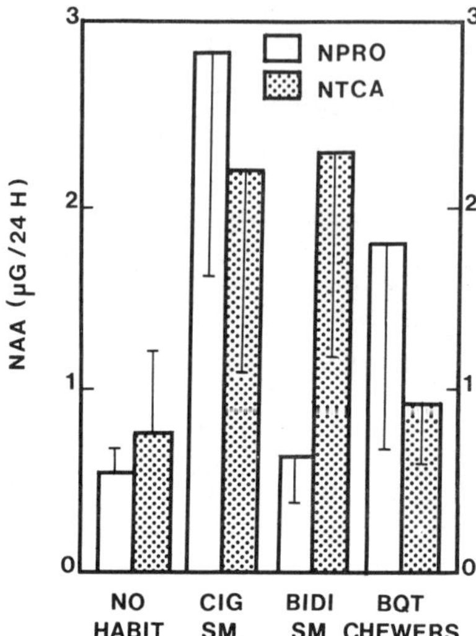

Fig. 3. Urinary excretion of N-nitrosamino acids (NAAs) (ng/24 hr) in Indian tobacco users and no-habit controls. Urinary excretion of NAAs in groups of subjects from Bombay, India who had ingested 100 mg proline one hour after each meal 3 times a day, and who had smoked one or more cigarettes (CIG SM) or one or more bidis (BIDI SM), or who had chewed one or more betel quids containing tobacco (BQT) after each meal. Controls (no-habit group) received proline only. Urine was collected for 24 hr and analyzed for NPRO, NTCA, NMTCA, and NSAR as previously described (24).

gastric condition) when compared with those formed at pH 7.4 after nitrosation of BQT and BQ. These levels of NOC could be formed intragastrically after ingestion of betel quid.

More recently, nitrosation was shown to occur in the oral cavity of betel quid chewers as evidenced by an increased NPRO formation during a 20-min period of chewing betel quid containing Indian tobacco to which proline had been added (J. Nair et al., ms. in prep.).

There is a firmly established association between buccal cancer and snuff dipping in humans (34); these tobacco products contain TSNAs at high concentrations (11), some of them known to be potent animal carcinogens. N-Nitroso compounds, therefore, may also be causative agents in cancers associated with betel quid chewing, either alone or in combination with other betel quid ingredients, some of which exert genetic and toxic effects.

Study in a High and Low Incidence Area for Gastric Cancer in northern Japan

In order to study the role of endogenous nitrosation and nutritional deficiency in the etiology of stomach cancer, a pilot study was initiated in 1983. Samples of 24-hr urine and blood were collected from 100 subjects living in high-incidence (Akita) and low-incidence (Iwate) areas for stomach cancer in the northern part of Japan. Three different urine samples were collected from each subject (see the NPRO test, protocol B, above).

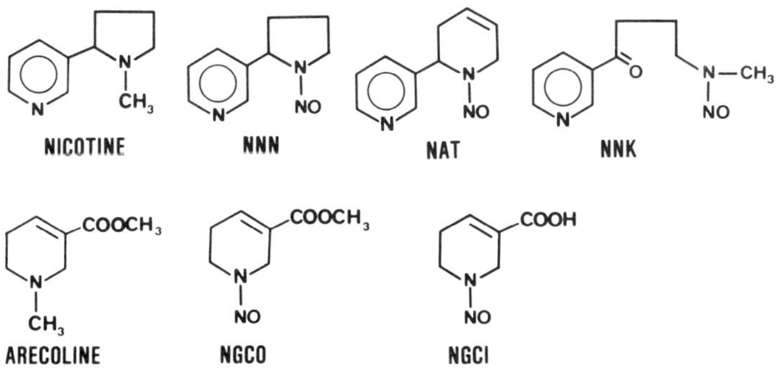

NICOTINE NNN NAT NNK

ARECOLINE NGCO NGCI

Fig. 4. Structure of some tobacco-specific nitrosamines (TSNAs), i.e., NNN, NAT, NNK, and areca nut-specific nitrosamines (ASNAs), i.e., NGCO, NGCI, and precursor alkaloids, detected in the saliva of tobacco and betel quid chewers.

These samples were analyzed for nitrate and chloride ions excreted and for some NAAs such as NPRO, NTCA, and NMTCA. Blood samples are being analyzed for the levels of vitamins (A, B_2, and E), trace elements (Zn, Fe, and Se), uric acid, and some enzyme activities as an index of nutritional status. In addition, dietary questionnaires were obtained from each subject. Initial results revealed that Akita subjects had a 2-fold higher exposure to NAA (median value: 47 μg NAA per day per person) compared to those in Iwate (25 μg per day per person). Vitamin C led to a reduction of endogenous NAA (16 μg per day). The potential of endogenous nitrosation in each individual is now being correlated with the results obtained from the analysis for nutritional status and food habits.

Evaluation of Human Exposure to N-Nitroso Compounds and Nitrate in Areas at High Risk of Esophageal Cancer

Esophageal cancer is a prevalent disease in northern China. The Linxian county in Henan province has the highest age-adjusted mortality rate with 151/100,000 in men and 115/100,000 in women. An extensive search for the causative factors of this cancer was begun in 1972 (35). Those studies suggested that environmental factors such as NOCs and their precursors (secondary amine, nitrate, nitrite) may be important.

In order to evaluate human exposure to NOCs through either ingestion of preformed compounds in foods or endogenous formation, 24-hr urine samples were collected (in collaboration with S.H. Lu, Cancer Institute, Chinese Academy of Sciences, Beijing, People's Republic of China) from about 250 subjects living in Linxian (high-risk area) and Fanxian (low-risk area) with a mortality rate for esophageal cancer of 35/100,000 in men and 16/100,000 in women, according to protocol B (see NPRO test above). The urine samples were analyzed for the presence of NPRO, NTCA, and NSAR (μg/24 hr per person) and nitrate (mg/24 hr per person).

The conclusions that can be drawn from this study are: (i) in the high-incidence area (Linxian), exposure to NPRO, NTCA, NSAR, and nitrate is higher than in the Fanxian country; (ii) intake of L-proline 3 times a day further increased the urinary excretion of NPRO in these subjects; the potential for endogenous nitrosation appears to be at least twice as high in the inhabitants of the high-risk area as in those of the low-risk area; and (iii) ingestion of vitamin C reduced the amount of NPRO and NTCA formed in vivo in the high-risk subjects to levels seen in the low-risk subjects. Although it remains to be demonstrated whether endogenous nitrosation is one of the risk factors related to the development of esophageal cancer in certain provinces in northern China, intervention studies can be considered, as our results (17) have demonstrated the efficiency of vitamin C in blocking N-nitrosation in vivo.

The significance of these findings is being verified in studies that involve different areas in the People's Republic of China. The excretion of urinary NAA is being correlated with mortality figures for cancer of the esophagus, stomach, and liver in different provinces; a number of risk factors for malignant diseases and nutritional indices that had also been analyzed will be included in the comparisons (in collaboration with Dr. J. Chen, National Center for Preventive Medicine, Beijing, People's Republic of China; and Dr. T.C. Campbell, Cornell University, Ithaca, New York).

Metabolic Modulators of Nitrosamine Carcinogenesis

Nitrosamines produce their adverse biological effects, including induction of tumors, following metabolic conversion into reactive intermediates (or nonenzymic decomposition, in the case of nitrosamides), that react with DNA, resulting in the formation of adducts at at least 12 alkylated sites. The interplay between enzymic activation of the (pro)carcinogen and the removal of DNA adducts appears to be a critical determinant of which organ- or cell-given nitrosamine results in its carcinogenic effect.

However, the tissue-specific carcinogenicity of a given nitrosamine is also conditioned by additional factors, e.g., the role of the liver in determining the extent to which extrahepatic tissues are exposed to nitrosamines, since the liver plays the major role in the process of activation or detoxification of most of these carcinogens. In fact, after oral administration of very low doses of N-nitrosodimethylamine (NDMA), very little or none of this nitrosamine reaches extrahepatic tissues because of the efficient metabolism of the liver (8,25). Thus, interactions of the nitrosamine are strongly determined by the dose of nitrosamine, the rate of absorption from the intestine, and by various factors that could modify the metabolic competence of the liver. Experimental studies have shown that a number of modifiers, including ethanol, can drastically change the organotropism of the carcinogenic effects of nitrosamines (see Fig. 5).

Ethanol, for example, has been shown to decrease the metabolism and toxicity of NDMA and some other nitrosamines in the liver and, at the same time, to increase tumor incidence in the esophagus, kidney, and nasal cavities. It should be stressed that this effect is observed with very low doses of ethanol and that some observations in man are consistent with the experimental data (see Ref. 29 for some details). Figure 5 shows some other agents that modify the various biological effects of nitrosamines in the liver and in extrahepatic tissues. Epidemiological studies have shown that tobacco smoking and alcohol drinking are risk factors for cancer of the upper digestive tract (14,32); it can be hypothesized on the basis of the available evidence that ethanol facilitates metabolic activation of TSNAs (29), notably in the target organs of nitrosamine carcinogenicity, while inhibiting their detoxifying metabolism in the liver.

Modulator	Liver				Extrahepatic tissues	
	Metabolism-nucleic acid alkylation *in vivo*	Toxicity	Mutagenicity	Carcinogenicity	Metabolism-nucleic acid alkylation *in vivo*	Carcinogenicity
Protein-restricted diet	↓	↓	↓		↑ Kidney	↑ Kidney
Ethanol	↓ =	↓	↓↑	=	↑ Kidney	↑ Nasal cavity Mice NDEA Oesophagus
	↓ NMBzA			↓ NDEA	↑ NMBzA Oesophagus, Lung	↑ NPYR Nasal cavity Trachea Hamster
Disulfiram	↓ NDMA Mouse NMBzA Rat NDMA, NDEA Rat	↓ NDMA Mouse Rat NDEA Rat	↓	↓ NDMA,	↑ NMBzA Lung, Oesophagus	↑ NDMA Nasal cavity NDEA Oesophagus NMBzA Oesophagus NDBA Lung
Carbon tetrachloride	↓	↓		↑ =	= Kidney	↑ Kidney
Aminoacetonitrile	↓	↓	↓	↓	↓	
Zinc deficiency						↑ NMBzA Oesophagus

Fig. 5. Modulators of biological effects of nitrosamines in liver and extrahepatic tissues. The results refer to N-nitrosodimethylamine (NDMA) and rat tissues, unless otherwise specified; for clarity, in some cases, the animal species, the tissue, and the nitrosamine are specified in full. "=" indicates that the modulators have no effect or contradictory results were observed. Mutagenicity experiments were carried out with postmitochondrial fractions that were prepared from livers of rats treated in vivo with the various modulators; the mutagenicity in vitro of NDMA was then determined in bacteria, with the exception of one study carried out in vivo as a host-mediated assay. NDEA, N-nitrosodiethylamine; NPYR, N-nitrosopyrrolidine; NDBA, N-nitrosodibutylamine; NMBzA, nitrosomethylbenzylamine. (From Ref. 1, reprinted with permission from IRL Press.)

CONCLUSIONS

By applying a sensitive method for monitoring in vivo nitrosation to more than 1,000 subjects, our results unequivocally demonstrated that NOCs are formed in the human body. Their yield increased markedly when the daily intake of nitrate was above 260 mg. Vitamins C and E and polyphenolic compounds inhibited the yield of NAAs excreted in the urine, while catalysts (salivary SCN^-) in smokers led to an increase. Sites where nitrosation occurs in vivo include the acidic stomach, the oral cavity, and possibly the infected urinary bladder.

Because of the effectiveness of the nitrosation modifiers, the individual monitoring of human subjects is required, rather than measuring the intake of precursors (amine and nitrate/nitrite), to associate endogenous formation of NOCs with cancer at specific sites.

When the NPRO test was applied in clinical or field studies, higher exposures to endogenous NOCs were found in the following subjects at higher risk for cancers at specific sites: cigarette smokers (respiratory tract); chewers of betel quid containing tobacco (oral cavity); inhabitants of a high-risk area for stomach cancer in northern Japan; and inhabitants of a high-risk area for esophageal cancer in the People's Republic of China.

Individual exposure was greatly affected by dietary components and modifying chemicals. Most important, the efficiency of moderate doses of vitamin C to lower the body burden of endogenous NOC was clearly demonstrated in tobacco users and in study subjects from Japan and China. Ethanol profoundly affects the metabolism and pharmaco-kinetics of nitrosamines, thus offering a possible explanation for its synergistic action with tobacco use in relation to cancer of the upper digestive tract.

ACKNOWLEDGEMENTS

We wish to thank J.C. Bereziat, M.C. Bourgade, and J. Michelon for technical assistance, and Y. Granjard for secretarial work. Part of these studies was supported by Contract No. 9304498 from the Délégation Générale à la Recherche Scientifique et Technique (DGRST), France. One of the TEA detectors was provided on loan by the National Cancer Institute of the United States under Contract NO1 CP-55715. Part of the work reported on betel quid was undertaken during the tenure of a Research Training Fellowship awarded to J. Nair by the International Agency for Research on Cancer.

The authors gratefully acknowledge the scientific contributions and collaborative efforts of S.H. Lu (Cancer Institute, Academy of Medical Sciences, Beijing, People's Republic of China), H.F. Stich (British Columbia Cancer Research Center, Vancouver, Canada), S. Kamiyama (Akita University, Akita, Japan), S.V. Bhide (Cancer Research Institute, Tata Memorial Center, Bombay, India), P. Vincent (Universite Claude Bernard, Lyon, France), H. Leclerc (INSERM, Villeneuve d'Ascq, France), J. Chen (China National Center for Preventive Medicine, Beijing, People's Republic of China), and T.C. Campbell (Cornell University, Ithaca, New York).

REFERENCES

1. Bartsch, H., and R. Montesano (1984) Relevance of nitrosamines to human cancer. Carcinogenesis 5:1381-1393.

2. Bartsch, H., H. Ohshima, N. Muñoz, and S.H. Lu (1983) Measurement of endogenous nitrosation in humans: Potential applications of a new method and initial results. In Human Carcinogenesis, C.C. Harris and H.N. Autrup, eds. Academic Press, New York, p. 833.
3. Bartsch, H., H. Ohshima, N. Muñoz, M. Crespi, V. Cassale, V. Ramazotti, R. Lambert, Y. Minaire, J. Forichon, and C.L. Walters (1984) In-vivo nitrosation, precancerous lesions and cancers of the gastrointestinal tract. On-going studies and preliminary results. In N-Nitroso Compounds: Occurrence, Biological Effects and Relevance to Human Cancer (IARC Scientific Publications No. 57), I.K. O'Neill, R.C. von Borstel, C.T. Miller, J. Long, and H. Bartsch, eds. International Agency for Research on Cancer, Lyon, France, p. 957.
4. Bhide, S.V., A.S. Shah, J. Nair, and D. Nagarajarao (1984) Epidemiological and experimental studies on tobacco related oral cancer in India. In N-Nitroso Compounds: Occurrence, Biological Effects and Relevance to Human Cancer (IARC Scientific Publications No. 57), I.K. O'Neill, R.C. von Borstel, C.T. Miller, J. Long, and H. Bartsch, eds. International Agency for Research on Cancer, Lyon, France, p. 851.
5. Calmels, S., H. Ohshima, P. Vincent, A.M. Gounot, and H. Bartsch (1985) Screening of microorganisms for nitrosation catalysis at pH 7 and kinetic studies on nitrosamine formation from secondary amine by E. coli strains. Carcinogenesis 6:911.
6. Chu, C., and P.N. Magee (1981) The metabolic fate of nitrosoproline in the rat. Cancer Res. 41:3653.
7. Dailey, R.E., R.C. Braunberg, and A.M. Blaschka (1975) The absorption, distribution and excretion of (^{14}C)-nitrosoproline by rats. Toxicology 3:23.
8. Diaz-Gomez, M.I., P.F. Swann, and P.N. Magee (1977) The absorption and metabolism in rats of small oral doses of dimethylnitrosamine. Implication for the possible hazard of dimethylnitrosamine in human food. Biochem. J. 164:497-500.
9. Hawksworth, G., and M. Hill (1971) Bacteria and the N-nitrosation of secondary amines. Br. J. Cancer 25:520.
10. Hoffmann, D., and K.D. Brunnemann (1983) Endogenous formation of N-nitrosoproline in cigarette smokers. Cancer Res. 43:5570.
11. Hoffmann, D., K.D. Brunnemann, J.D. Adams, and S.S. Hecht (1984) Formation and analysis of N-nitrosamines in tobacco products and their endogenous formation in consumers. In N-Nitroso Compounds: Occurrence, Biological Effects and Relevance to Human Cancer (IARC Scientific Publications No. 57), I.K. O'Neill, R.C. von Borstel, C.T. Miller, J. Long, and H. Bartsch, eds. International Agency for Research on Cancer, Lyon, France, p. 743.
12. Hoffmann, D., and S.S. Hecht (1985) Nicotine-derived N-nitrosamines and tobacco-related cancer: Current status and future directions. Cancer Res. 45:935.
13. International Agency for Research on Cancer (1978) Monographs on the Evaluation of the Carcinogenic Risk of Chemicals to Humans. Vol. 17. Some N-Nitroso Compounds, 365 pp.
14. International Agency for Research on Cancer (1985) Monographs on the Evaluation of the Carcinogenic Risk of Chemicals to Humans. Vol. 38. Tobacco Smoking (in press).
15. Kantor, A.F., P. Hartige, R.N. Hoover, A. Narayana, J.W. Sullivan, and J.F. Fraumeni, Jr. (1984) Urinary tract infection and risk of bladder cancer. Am. J. Epidemiol. 119:510.
16. Ladd, K.F., H.L. Newmark, and M.C. Archer (1984) N-nitrosation of proline in smokers and non-smokers. J. Natl. Cancer Inst. 73:83.
17. Lu, S.H., H. Ohshima, and H. Bartsch (1984) Recent studies on N-nitroso compounds as possible etiological factors in oesophageal cancer.

N-Nitroso Compounds: Occurrence, Biological Effects and Relevance to Human Cancer (IARC Scientific Publications No. 57), I.K. O'Neill, R.C. von Borstel, C.T. Miller, J. Long, and H. Bartsch, eds. International Agency for Research on Cancer, Lyon, France, p. 947.

18. Mirvish, S.S., O. Bulay, R.G. Runge, and K. Patil (1980) Study of the carcinogenicity of large doses of dimethylnitramine, N-nitroso-L-proline and sodium nitrite administered in drinking water to rats. J. Natl. Cancer Inst. 64:1435.
19. Nair, J., H. Ohshima, M. Friesen, M. Croisy, S.V. Bhide, and H. Bartsch (1985) Tobacco-specific and betel nut specific N-nitrosamino compounds: Occurrence in saliva and urine of betel quid chewers and formation in vitro by nitrosation of betel quid. Carcinogenesis 6: 295.
20. Ohshima, H., and H. Bartsch (1982) Quantitative estimation of endogenous nitrosation in humans by measuring excretion of N-nitrosoproline in the urine. In Environmental Mutagens and Carcinogens, T. Sugimura, S. Kondo, and H. Takabe, eds. University of Tokyo Press, Tokyo, p. 577.
21. Ohshima, H., and H. Bartsch (1981) Quantitative estimation of endogenous nitrosation in humans by monitoring N-nitrosoproline excreted in the urine. Cancer Res. 41:3658.
22. Ohshima, H. J.C. Bereziat, and H. Bartsch (1982) Monitoring N-nitrosamino acids excreted in the urine and feces of rats as an index for endogenous nitrosation. Carcinogenesis 3:115.
23. Ohshima, H., M. Friesen, I.K. O'Neill, and H. Bartsch (1983) Presence in human urine of a new N-nitroso compound, N-nitrosothiazolidine 4-carboxylic acid. Cancer Lett. 20:183.
24. Ohshima, H. I.K. O'Neill, M. Friesen, J.C. Bereziat, and H. Bartsch (1984) Occurrence in human urine of new sulphur-containing N-nitrosamino acids, N-nitrosothiazolidine 4-carboxylic acid and its 2-methyl derivative, and their formation. J. Cancer Res. Clin. Oncol. 108:121.
25. Pegg, A.E., and W. Perry (1981) Alkylation of nucleic acids and metabolism of small doses of dimethylnitrosamine in the rat. Cancer Res. 41:3128-3132.
26. Pignatelli, B., J.C. Bereziat, G. Descotes, and H. Bartsch (1982) Catalysis of nitrosation in vitro and in vivo in rats by catechin and resorcinol and inhibition by chlorogenic acid. Carcinogenesis 3:1045.
27. Stich, H.F., H. Ohshima, B. Pignatelli, J. Michelon, and H. Bartsch (1983) Inhibitory effect of betel nut extracts on endogenous nitrosation in humans. J. Natl. Cancer Inst. 70:1047.
28. Stich, H.F., B.P. Dunn, B. Pignatelli, H. Ohshima, and H. Bartsch (1984) Dietary phenolics and betel nut extracts as modifiers on N-nitrosation in rat and man. In N-Nitroso Compounds: Occurrence, Biological Effects and Relevance to Human Cancer (IARC Scientific Publications No. 57), I.K. O'Neill, R.C. von Borstel, C.T. Miller, J. Long, and H. Bartsch, eds. International Agency for Research on Cancer, Lyon, France, p. 213.
29. Swann, P.F. (1984) Effect of ethanol on nitrosamine metabolism and distribution. Implications for the role of nitrosamines in human cancer and for the influence of alcohol consumption on cancer incidence. In N-Nitroso Compounds: Occurrence, Biological Effects and Relevance to Human Cancer (IARC Scientific Publications No. 57), I.K. O'Neill, R.C. von Borstel, C.T. Miller, J. Long, and H. Bartsch, eds. International Agency for Research on Cancer, Lyon, France, p. 501.
30. Tsuda, H., T. Hirayama, and T. Sugimura (1983) Presence of N-nitroso-L-thioproline and N-nitroso-L-methylthioproline in human urine as major N-nitroso compounds. Gann 74:331.

31. Tuyns, A.J. (1983) Protective effect of citrus fruit on esophageal cancer. Nutr. Cancer 5:195.
32. Tuyns, A.J., G. Pequignot, and O.M. Jensen (1977) Le cancer de l'oesphage en Ille et Vilaine en fonction des niveaux de consommation d'alcool et de tabac. Bull. Cancer 64:45.
33. Wenke, G., and D. Hoffmann (1983) A study of betel quid carcinogenesis. 1. On the in vitro N-nitrosation of arecoline. Carcinogenesis 4:169.
34. Winn, D.M. (1984) Tobacco chewing and snuff dipping: An association with human cancer. In N-Nitroso Compounds: Occurrence, Biological Effects and Relevance to Human Cancer (IARC Scientific Publications No. 57), I.K. O'Neill, R.C. von Borstel, C.T. Miller, J. Long, and H. Bartsch, eds. International Agency for Research on Cancer, Lyon, France, p. 837.
35. Yang, C.S. (1980) Research on oesophageal cancer in China: A review. Cancer Res. 40:2633.

GLUTATHIONE TRANSFERASES AND CARCINOGENESIS

Brian Ketterer, David J. Meyer, Brian Coles,
John B. Taylor, and Sally Pemble

Courtauld Institute of Biochemistry
Middlesex Hospital Medical School
London W1P 7PN, United Kingdom

INTRODUCTION

Carcinogenesis has been regarded as being divisible into at least 3 major phases: (a) initiation, a largely irreversible process which has been compared to mutation; (b) promotion, a process during which cells undergo phenotypic changes, some of which are referred to as preneoplastic, and a proportion of which attain autonomy when they can be regarded as neoplastic; and (c) progression, a period of continuing phenotypic diversification during which a selection is made of those cells that possess autonomy and advantages over the host tissue. The result is increasing malignancy and, without intervention, the death of the host (22). The selective forces involved in carcinogenesis are reminiscent of those proposed for evolution, but in the case of cancer it is evolution within a lifetime and the result is biological chaos.

Glutathione (GSH) transferases have the potential to play a role in all 3 phases, although not necessarily an anticarcinogenic role. Thus it will be seen that they may have an anticarcinogenic role in protecting cells from initiating agents during initiation, and a possible anticarcinogenic role in promotion. But, at least in some cases, those cells that arise in promotion and progress to autonomy may be rich in GSH transferases when their protective role is procarcinogenic. In order to appreciate how this might be so, it is necessary to understand the reactions catalyzed by GSH and GSH transferases.

DEFINITION OF GLUTATHIONE TRANSFERASES

The term glutathione transferase defines a number of enzymes, all of which catalyze the reaction between the nucleophile GSH and a range of electrophiles, i.e., GSH conjugation. Although defined by their ability to catalyze GSH conjugation, GSH transferases also catalyze a number of other reactions involving GSH, including organic hydroperoxide reduction, GSH-dependent isomerization of certain double bonds, organic nitrate and thiocyanate reduction, and thioester formation (12). Glutathione transferases also bind lipophilic nonsubstrate ligands, perhaps at or near the substrate binding site, since all electrophilic substrates have lipophilic moieties.

These enzymes appear to occur in all tissues and, within the cell, in a number of compartments. Membrane-bound forms are found in the mitochondria and endoplasmic reticulum (53), soluble forms in the cytosol, mitochondria, and in the nucleoplasm (41,42). The most studied GSH transferases are the soluble forms of the cytosol, in particular, the rat cytosol. The rat soluble GSH transferases represent a family of isoenzymes consisting of homodimers and heterodimers derived from a series of different subunits. In the current nomenclature these subunits in the rat are classified numerically in the chronological order in which they have been characterized. So far, 7 have been named and shown to give rise to 10 dimers as demonstrated in Tab. 1. It is seen that they can be distinguished according to apparent subunit molecular weight and isoelectric point, and to their ability to form heterodimers. Thus, subunits 1 and 2 form both 2 homodimers and a heterodimer 1-2; subunits 3, 4, and 6 form 3 homodimers and also the three heterodimers 3-4, 3-6, and 4-6. So far, subunits 5 and 7 have been observed to form only homodimers (39,49,50).

Subunits are also distinguished by substrate specificity (47). Heterodimers have enzymic activity that is approximately the mean of the activities seen in the two homodimers to which they are related, suggesting that the subunits in a dimer express their activity independently of each other (see Tab. 2). Thus subunit 1 is specifically associated with Δ^5-3,17-androstenedione isomerase activity and high affinity binding for the lipophiles heme and bilirubin. Subunits 3 and 4 are associated with GSH conjugating activity towards the electrophiles 1,2-dichloro-4-nitrobenzene and trans-4-phenyl-3-buten-2-one, respectively. Subunits 1, 2, 5, and 7 have GSH peroxidase activity, while subunits 3, 4, and 6 do not. Among those with GSH peroxidase activity, subunit 7 is unique in catalyzing the reduction of endogenous fatty acid hydroperoxides, but not the model substrate cumene hydroperoxide (52). However, certain electrophiles are a good deal

Tab. 1. Classification of GSH transferases purified from rat tissues.

GSH transferases[a]	Subunit $\underline{M}r \times 10^{-3}$ [b]	pI
1-1	25.0/25.0 (25.587[c])	9.2
1-2	25.0/28.0	9.1
2-2	28.0/28.0	9.1
3-3	26.5/26.5	8.4
3-4	26.5/26.5	7.7
3-6	26.5/26.0	7.4
4-4	26.5/26.5	6.7
4-6	26.5/26.0	6.1
5-5	26.5/26.5	7.0
6-6	26.5/26.0	5.3
7-7	23.5/23.5 (23.307[d])	7.0

[a] Nomenclature according to Jakoby et al. (35).
[b] Approximate subunit $\underline{M}r$ determined by SDS PAGE.
[c] Molecular weight of a subunit 1 deduced from sequence (43).
[d] Molecular weight of a subunit 7 deduced from sequence (65).

Tab. 2. Substrates that specify GSH transferase subunits.

Substrate	1-1	2-2	3-3	4-4	5-5	6-6	7-7
Δ^5-Androstene-3,17-dione	0.23*	0.07	n.d.**	n.d.	n.d.	n.d.	0.001
Cumene hydroperoxide	1.4	3.0*	0.1	0.4	12.5	0.04	0.01
1,2-Dichloro-4-nitrobenzene	0.15	0.15	8.4*	0.7	n.d.	0.26	0
Trans-4-phenyl-3-buten-2-one	0.1	0.10	0.1	1.2*	n.d.	0.1	0.02
1,2-Epoxy-3(p-nitrophenoxy)propane	0.7	0.9	0.2	0.9	25.5*	0.1	1.0
1-Chloro-2,4-dinitrobenzene	34	32	44	16	trace	187*	20
Ethacrynic acid	0.3	2.1	0.4	1.0	n.d.	0.1	4.0*
Linoleate hydroperoxide	3.00*	1.6	0.2	0.2	5.3	0.06	1.5*

* The subunit specified by each substrate.
** No data.
Note: Activities are expressed as μmol/min/mg protein.

less specific. For example, the carcinogenic electrophile benzo(α)pyrene-7,8-diol-9,10-oxide is a substrate for all the subunits in the series 1 to 7 (36,49). A little more specific, but of considerable practical importance, is 1-chloro-2,4-dinitrobenzene (CDNB), which gives the highest rates so far obtained and is used as a general substrate for detecting GSH transferase activity.

Tissue Distribution of Soluble Glutathione Transferases from the Rat

Considerable variation of isoenzyme content is seen from tissue to tissue (39); indeed, the distribution of GSH transferase isoenzymes is often unique to a tissue (see Fig. 1). Interesting exceptions are the adrenals and the lactating mammary glands which seem to be almost identical. They both have an isoenzyme distribution in which GSH transferase 2-2 predominates. Why these tissues should be so similar is not known; they have little in common in either embryonic origin or function.

Not all subunits so far identified have been characterized, and there may yet be others which have still to be identified. A separation by chromatofocusing of the numerous acidic GSH transferases of skeletal muscle is shown in Fig. 2, which gives some idea of the problem. Of the isoenzymes separated in this study, only 2 previously characterized GSH transferases, namely, 4-4 (peak 1) and 4-6 (peak 4), have been identified (51).

It is interesting to note in Fig. 1 that the much studied subunit 1, which is so abundant in the liver and kidney and may also be important in the intestine, is otherwise absent or of low occurrence. Most of the other subunits have a wide occurrence, but taking into account the contribution to overall body mass, plus enzyme content of the liver and skeletal muscle, subunits 2 and 4 are the most abundant. Subunit 2 has GSH peroxidase activity, and the only tissue we have seen in which it appears to be absent is spermatogenic tissue; however, other subunits, such as 5 and 7, may be alternative providers of this activity in spermatogenic tissue.

Changes in GSH transferase isoenzyme distribution during differentiation have been studied very little so far. A little is known about the liver. It is interesting that the immature liver has more isoenzymes than the mature liver, including more acidic isoenzymes and a basic isoenzyme with very high GSH peroxidase activity which has yet to be observed elsewhere (52). Striking changes in the pattern of isoenzymes occur during chemical hepatocarcinogenesis, which can be regarded as abnormal differentiation. Thus, subunit 1 becomes a minor rather than a major component; subunit 2 remains at high levels; levels of subunit 3 increase several times; and subunit 7, not present in normal liver, is expressed as a major component. Subunit 7 makes its appearance early during hepatocarcinogenesis, appearing in hyperplastic nodules (61) and persisting into the primary hepatoma (49). Its abundance and its clear-cut separation from other liver proteins make it a useful marker (22).

GLUTATHIONE CONJUGATION

Glutathione conjugation is the reaction of electrophiles with the nucleophilic thiol of GSH. It is one of many reactions of electrophiles with nucleophiles that may occur in vivo.

The electrophiles are usually produced metabolically and may be endogenous or exogenous. The number of known endogenous electrophiles is small and includes 2-hydroxy-estradiol, leukotriene A_4, prostaglandin A_1, cholesterol-5,6-oxide, and 4-hydroxyalk-2-en-1-als. Exogenous electrophiles, on the other hand, are potentially very numerous. They may be: (a) specialized plant microbial and fungal metabolites; (b) pyrolysis products of normal nutrients such as protein, plant matter such as tobacco or fossil fuel; and (c) synthetic industrial chemicals.

The nucleophiles of importance are almost all endogenous. Glutathione is the most abundant single organic nucleophile, occurring at concentrations between 1 and 10 mM according to the tissue. Other nucleophilic centers of importance are in nucleic acids and proteins. Electrophilic attack on these macromolecules is frequently toxic, and it is the role of GSH and GSH transferases to compete with macromolecular nucleophiles for electrophiles and provide a means of detoxification.

In order to understand this route of detoxification it is important to appreciate that there are large differences in the spontaneous rates of reaction between electrophiles and nucleophiles which depend on the nature of the nucleophilic center involved (13). Classification of electrophiles and nucleophiles according to their chemical hardness or softness is the concept most useful in predicting the ease of reaction of an electrophilic/nucleophilic pair. Soft electrophiles and nucleophiles have readily polarizable reactive centers, whereas hard electrophiles and nucleophiles have polarized reactive centers. In practice, "like" tends to react with "like." Glutathione is a soft nucleophile because of the large atomic volume of sulphur which causes the electron density to be readily polarizable. As a consequence, it reacts most rapidly with soft electrophiles (e.g., Michael addition to a polarized double bond) and relatively slowly with hard electrophiles (e.g., carbonium ions). Nucleophilic targets within the cell upon which electrophilic attack is very likely to lead to toxicity are the "soft" cysteinyl thiols at the active centers of enzymes, etc., and the "hard" nitrogen and oxygen atoms in DNA (13). Other targets almost certainly exist and remain to be studied. One interesting hard nucleophilic center subject to an electrophilic attack which results in toxicity is pyrrole nitrogen in the heme prosthetic group of cytochrome P-450 (17).

When GSH transferases are present, catalysis by GSH transferases can: (a) increase the rate of conjugation of both soft and hard electrophiles, and (b) give the reaction independence of GSH concentration until it falls to levels in the region of the K_m for GSH (e.g., 0.1 mM in the case of GSH transferases 1-1 and 3-3). Catalysis is particularly beneficial for the detoxification of hard electrophiles, while independence of GSH concentration enables detoxification to occur as rapidly in cells with low GSH concentrations as it does in tissues with high GSH concentrations.

Glutathione Transferase Subunits and Specificity of Function

The association of a multiplicity of enzyme subunits with a range of enzymic functions is shown in Tab. 1. However, many of these substrates shown are not encountered in biological situations, but have been chosen as model substrates because of the ease with which GSH conjugation can be followed spectrophotometrically.

The number of exogenous electrophiles of biological importance that have been tested thus far is small and has yet to provide a sufficient basis to indicate what sort of exogenous substrates may have provided the assumed evolutionary pressure. Emphasis has been placed on drugs and carcinogens of which some examples are given below.

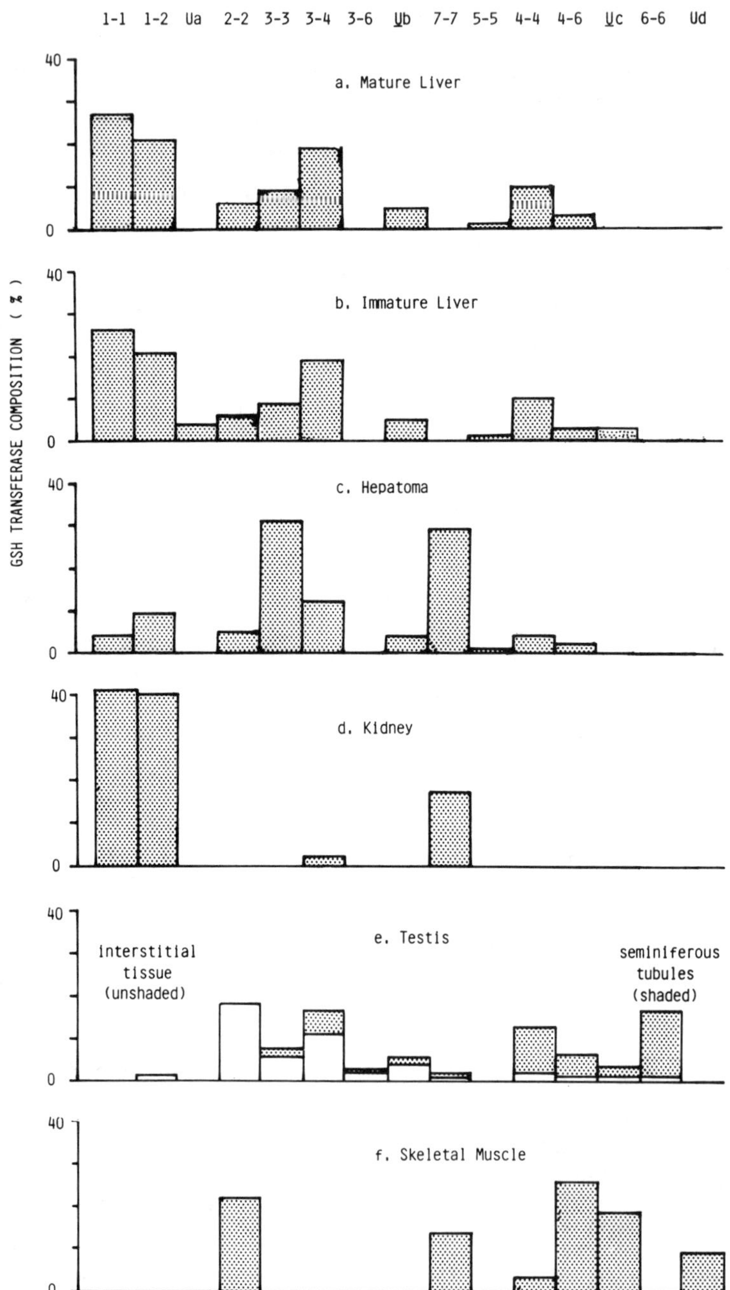
1-1 1-2 Ua 2-2 3-3 3-4 3-6 Ub 7-7 5-5 4-4 4-6 Uc 6-6 Ud
GSH TRANSFERASE COMPOSITION (%)
40
0
a. Mature Liver
b. Immature Liver
c. Hepatoma
d. Kidney
e. Testis
interstitial tissue (unshaded)
seminiferous tubules (shaded)
f. Skeletal Muscle

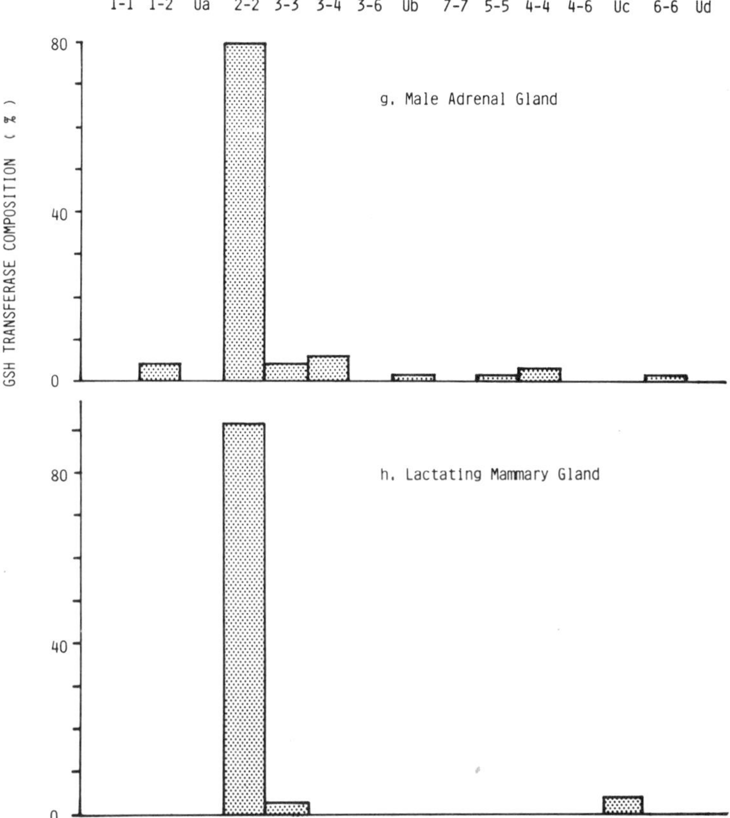

Fig. 1. Tissue distribution of GSH transferases in the rat. GSH transferases were prepared from tissue soluble fractions by affinity chromatography on S-hexyl glutathionyl Sepharose and finally separated either by isoelectric focusing or by fast protein liquid chromatography (FPLC) chromatofocusing. GSH transferase 5-5 was determined separately by gel filtration and isoelectric focusing (50). a, Mature liver; b, Immature (30 day) liver; c, Primary hepatoma induced by N,N-dimethyl-4-aminoazobenzene (DAB); d, Kidney; e, Testis (spermatogenic and steroidogenic tissues were separated by dissection); f, Skeletal muscle; g, Male adrenal gland; h, Lactating mammary gland. GSH transferase content is presented according to pI, the most basic enzymes being on the left. "U" signifies a GSH transferase fraction that is incompletely characterized and may contain more than one form.

Examples of Enzymatic and Spontaneous Glutathione Conjugation of Relevance to Carcinogenesis

Benzo(α)pyrene-7,8-diol-9,10-oxide. Benzo(α)pyrene-7,8-diol-9,10-oxide (BPDE) is a hard electrophile associated with DNA adduct formation in vitro and in vivo (involving principally exocyclic N^2 of guanine), mutagenesis in prokaryote and eukaryote cells in culture, and carcinogenesis in the mouse skin (19). As mentioned above, it is a substrate for all 7 GSH transferase subunits from the rat. Kinetic data for the homodimers 1-1,

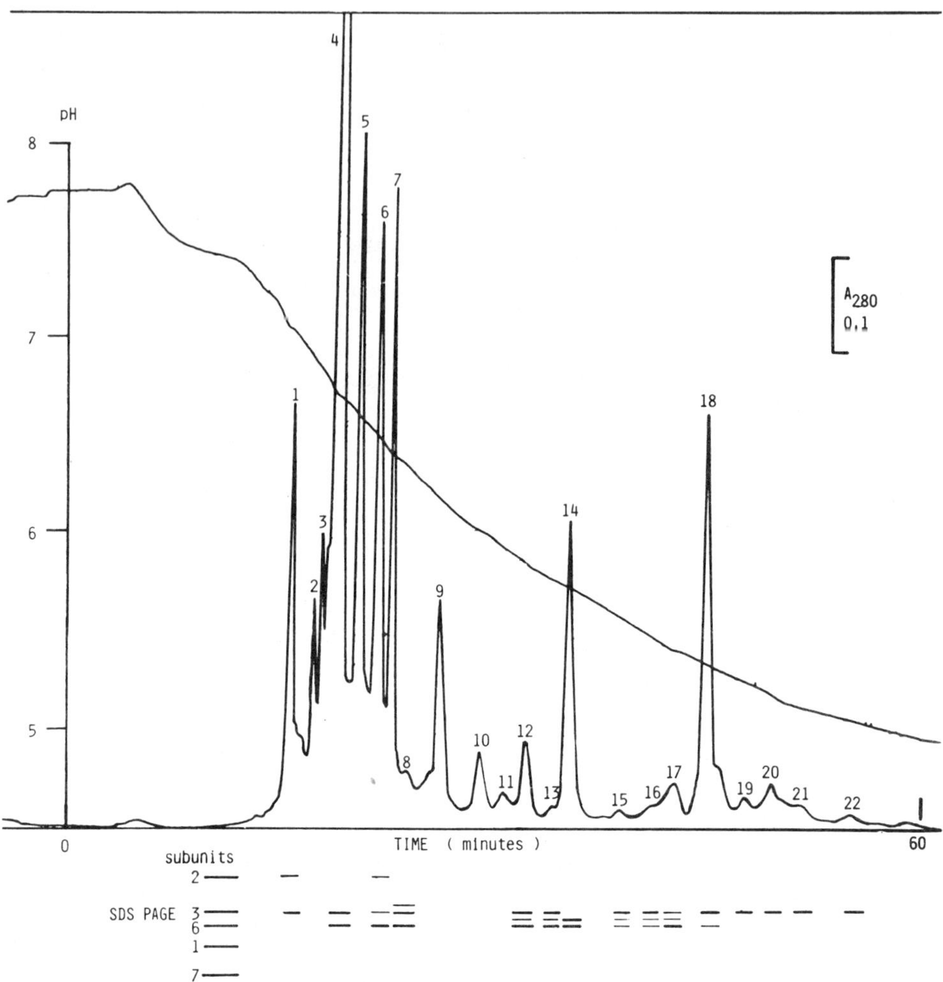

Fig. 2. Acidic GSH transferases of skeletal muscle separated by FPLC chromatofocusing. A GSH transferase fraction was obtained from skeletal muscle by affinity chromatography and separated by FPLC chromatofocusing on a Pharmacia mono P column at 0.5 ml/min in the pH range 7.8 to 4.8. GSH transferases 2-2 and 7-7 were not retained by the column. Peaks 1 and 4 contained GSH transferases 4-4 and 4-6, respectively; the other peaks remain to be fully characterized. Peaks not represented in SDS PAGE analysis do not appear to be GSH transferases.

2-2, 3-3, and 4-4 are shown in Tab. 3. Subunit 4 is the most active and subunit 2, the least (36). It can be deduced that in the liver the detoxification of BPDE is 300-fold the spontaneous rate, and this may be an important determinant of the very low levels of benzo(α)pyrene (BP) bound to DNA and the lack of its hepatocarcinogenicity.

In a similar vein, it is tempting to relate the low level of GSH transferases in the rat mammary gland (where the relatively inactive subunit 2 predominates) to the susceptibility of this organ to carcinogenesis

Tab. 3. Classification of GSH transferases purified from human tissues.

GSH transferase	Subunit $\underline{M}r \times 10^{-3}$	pI
α	26.0	7.80
β	26.0	8.25
γ	26.0	8.55
δ	26.0	8.75
ε	26.0	8.80
μ	26.3 27.0	6.60 6.10
ρ	24.0 24.5	4.50 4.80
π	22.5 23.5	4.65 4.80

Note: Data are taken from Ref. 2, 4, 29, 37, 48, and 72.

by BP (19). However, the physiology of the mammary gland is so different from that of the liver that other considerations, particularly those associated with promotion, may also be important.

Note on the metabolism of lipophilcs. When one is considering the intracellular behavior of compounds with lipophilic moieties such as metabolites of BP and other potential substrates for GSH transferases, it is important to take account of the fact that these metabolites are distributed throughout a system containing lipid and aqueous phases and hydrophobic binding sites on macromolecules (e.g., membrane lipid; aqueous phases of the cytosol, nucleus, mitochondria, etc.; the hydrophobic interior of the DNA double helix; and lipophile binding sites on protein and RNA). Expressed per liter, hepatocytes contain 0.1 moles lipid, 6.5×10^{-3} moles DNA nucleotide, 33×10^{-3} moles RNA nucleotide, 164 g protein, and 690 g water (10). The major soluble lipophile binding proteins are the GSH transferases, which have 2 binding sites per mole, and the fatty acid binding protein (FABP), which has 1. Their contribution to lipophile binding in cytosol is the provision of 0.4 and 0.1 mM binding sites, respectively.

In a model in which these components are homogenously arranged, it is possible to determine the theoretical distribution of BPDE provided appropriate binding constants are available. Unfortunately, only one such constant exists, namely, that of BPDE for DNA which is 8×10^3 M^{-1} (26,27) expressed per mole nucleotide. However, some binding constants are available for the proximate metabolite, BP-7,8-diol, which may be sufficiently similar to those for BPDE to be useful. These are 10^5 M^{-1} for lipid bilayers expressed per mole lipid and 3×10^5 M^{-1} for the binding sites of the cytosolic proteins (71). No data are available for RNA, but since binding to DNA depends considerably on intercalation in the double helix, binding to RNA may be expected to be several-fold lower.

On the basis of these various figures and assumptions, it can be calculated that 99% of BPDE is lipid bound, 0.001% is free in the aqueous phase (71), and the rest is divided more or less equally between DNA, RNA, and GSH transferases.

DNA itself partially detoxifies BPDE. Ninety-five percent of the BPDE it binds is hydrolyzed catalytically. This appears to be a fraction bound by intercalation, while the remainder appears to bind noncovalently at other DNA sites from which it forms adducts, mainly with the exocyclic amino group of guanine (27). It is this second fraction of DNA-bound BPDE with which it is assumed that competition with GSH transferases is most important. Our data indicated that BPDE will be detoxified by GSH transferases at a rate of 2.5 x [BPDE] per second. The low K_m's which GSH transferases possess for BPDE ensure that it will be efficiently detoxified at the low concentrations of substrate expected in vivo. Unfortunately, a rate constant for BPDE-DNA adduct formation is not available and therefore in the end it is not possible to predict how effectively GSH transferases might compete with DNA for BPDE in this simple model.

However, even if the binding constant for BPDE-DNA adduct formation were available, the above model would be misleading since it does not take into account the topography of the cell; notably, the important consideration that nuclear DNA resides in the center of the cell remote from the periphery where BP, the precursor for BPDE, enters. The consequence is the development of a gradient of BP and its metabolites from high values near the surface of the cell to low values near the nucleus. In the course of diffusion down this gradient BP is progressively detoxified, resulting in a much lower concentration of BPDE near the nucleus than near the point of entry (70). The fact that mitochondrial DNA, which is distributed throughout the cytoplasm, binds 40 times the BP bound to nuclear DNA seems to support this concept (5).

Thus, GSH transferases may have an important effect on the outcome of the administration of a carcinogen. GSH transferase is high in rat liver and low in mouse skin (28) and rat mammary gland. The amount of BP bound to hepatic DNA after a single intraperitoneal injection is 35 times lower than the amount bound to skin DNA by an equimolar dose of carcinogen (45). This, presumably, is one of the reasons why BP is a skin carcinogen in the mouse and not usually a hepatocarcinogen in the rat. It is possible to initiate hepatocarcinogenesis with BP, but only under special circumstances when BP activation is strongly induced by polychlorinated biphenyls (18) or when BP is administered after partial hepatectomy (40).

Aflatoxin B_1. Aflatoxin B_1 (AFB_1) is a widely occurring fungal metabolite possibly responsible for the high incidence of human liver cancer in parts of Africa and China. It is a potent experimental hepatocarcinogen and mutagen (9,25).

Like BP, AFB_1 exerts its carcinogenic effect through microsomal metabolism to a nonpolar oxide. This reacts with DNA principally at the N-7 of guanine bases in DNA. However, unlike BP, AFB_1 is a powerful hepatocarcinogen and hepatotoxin in the rat (9,25). Detoxification of the oxide is through spontaneous hydrolysis to the dihydrodiol and by GSH conjugation which appears to depend on the presence of GSH transferases (9,16,25). The GSH transferases would appear to play an important role in detoxification, since species that are resistant to the carcinogenic effects of AFB_1 have hepatic GSH transferases that are particularly effective at conjugation of the oxide (55).

For example, in the rat, the GSH conjugate forms only 6% of the AFB_1-oxide produced in vitro, whereas in the mouse, a relatively resistant species, it is 72%. The poor conjugation by rat liver cytosol is reflected in the narrow range of subunits that catalyze conjugation, namely 1 and 2. Rat GSH transferases are also poor competitors with DNA for the oxide [only 22-28% inhibition of covalent DNA binding in vitro (44)], nor do they affect the mutagenicity of microsomally activated AFB_1 towards Salmonella typhimurium (14). Interestingly, the increased resistance of the rat to AFB_1 toxicity which is induced by low-level feeding of the toxin causes an increase in GSH transferases (54) and a decrease in DNA binding in vitro in comparison with control animals.

Since, like BPDE, AFB_1-oxide is a nonpolar electrophile with a high affinity for DNA, the principles with respect to intracellular distribution discussed above will also apply to AFB_1. However, less data are available in this case since the oxide is too labile to be isolated. We do not know, for example, if DNA catalyzes hydrolysis of AFB_1-oxide. There are also differences in the chemistry of the two oxides (9,25), for the sites of reaction at guanine are different which may be related to different alignment of the two oxides within DNA, prior to reaction. Some idea of the complexity of the system can be gained by comparing the effect of mouse (57) and hamster liver cytosols (44) on the inhibition of covalent binding of AFB_1-oxide to DNA. Both species are relatively resistant to the carcinogenic effects of AFB_1, but whereas in the hamster covalent binding is inhibited in vitro by 50-80%, in the mouse inhibition is only 5-20% at similar cytosol concentrations.

Nitropyrene oxides. 1-Nitropyrene (NP) is one of a number of nitropolyaromatics that are of current interest. It is an environmental pollutant, particularly abundant in diesel exhaust, and is both mutagenic and carcinogenic (59). Mammalian metabolic systems give rise to an electrophilic metabolite which reacts with DNA to give an aminopyrene adduct at C-8 of guanine (34,64). Activation of this electrophile requires reduction. The liver also produces electrophiles by oxidative metabolism, namely, NP-4,5-oxide and NP-9,10-oxide, which are conjugated with GSH and can be recovered as major biliary excretion products (20). Studies with synthetic epoxides and purified GSH transferases show that they are both good substrates for GSH transferases 3-3 and 4-4, but much less good for GSH transferase 1-2. In preliminary studies values obtained for V_{max} and K_m for the catalysis of GSH conjugation of NP-4,5-oxide by GSH transferase 3-3 are 111 nmol/min/mg and 3.3 µM, respectively, and, for GSH conjugation of NP-9,10-oxide, 10.9 nmol/min/mg and 2.5 µM, respectively (20). In the case of GSH transferase 4-4, the V_{max} and K_m for the 4,5-oxide are 11.5 nmol/min/mg and 0.3 µM, respectively, and, for the 9,10-oxide, 10.0 nmol/min/mg and 0.4 µM, respectively. In the case of GSH transferase 3-3, the nitro group might have a steric effect on the reaction such that the 4,5-oxide is a better substrate than the rate of conjugation of the 9,10-oxide.

Like the values for BPDE, the K_m values are noteworthy for being very low, particularly in the case of GSH transferase 4-4.

Electrophiles Arising from N-Methyl-4-aminozobenzene and N-Acetyl-2-aminofluorene

N-Methyl-4-aminozobenzene (MAB) and N-acetyl-2-aminofluorene (AAF) are powerful liver carcinogens that give rise to a number of both hard and soft electrophiles. The hard electrophiles which react with DNA and which are believed to be essential for their carcinogenic activity are N-sulphonyloxy-MAB in one case and N-sulphonyloxy-AAF and N-acetoxyaminofluorene (N-acetoxy-AF) in the other. In addition, each carcinogen gives rise to reactive soft electrophiles. In the case of MAB, these are 4-aminoazobenzene methimine and its 4'-hydroxyderivative, which are excreted as GSH conjugates that account for at least 60% of the biliary detoxification products of the azodye carcinogen. In the case of AAF the soft electrophile produced is nitrosofluorene. This is believed to react with GSH to give a sulphinamide adduct which, with further GSH, is reduced to aminofluorene (AF). The significance of this pathway which could occur in both aromatic amino and aromatic nitro compounds is yet to be assessed (6,38).

None of these reactions with GSH have thus far been shown to be catalyzed by GSH transferases; however, these tests were made when GSH transferases 1-1 and 1-2 were the ones most commonly available (38), and it is necessary to reexamine the question using all GSH transferases. The need to reexamine this area is emphasized by the more recent observation that the N-hydroxy derivative of the heterocyclic polyaromatic amine Trp-P-2, a product of amino acid pyrolysis, gives rise to unstable GSH conjugates, the formation of which is catalyzed by GSH transferases (60). However, the lack of catalysis of the GSH conjugation of N-sulphonyloxy-MAB, N-sulphonyloxy-AAF, and N-acetoxy-AF would explain the low levels of the relevant GSH conjugates in the bile and the hepatocarcinogenicity of MAB and AAF.

N-Acetylbenzoquinone Imine

N-Acetylbenzoquinone imine (NAPQI) is a reactive, soft electrophile produced by the cytochrome P-450 oxidation of paracetamol and is also a metabolite of phenacetin. The carcinogenicity of paracetamol is still under question, but that of phenacetin is established (32). NAPQI reacts with soft nucleophilic sites in macromolecules with possible toxic consequences and also undergoes comproportionation to a toxic phenoxy radical (23). GSH can compete with both these reactions of NAPQI either by conjugating it or by reducing it back to paracetamol (32). The spontaneous rates of both these reactions with GSH are rapid but, in addition, GSH conjugation is catalyzed by GSH transferases. We have been interested in discovering to what extent this enzymic catalysis affects the toxicology of paracetamol and phenacetin.

N-Acetylbenzoquinone imine reacts so rapidly with GSH that stopped-flow techniques are required to measure rates. However, some indication of the efficacy of the GSH transferases has been obtained by incubating NAPQI with a mixture of GSH and cysteine in the presence and absence of GSH transferases. It is shown that the effect of GSH transferase isoenzymes is to increase the ratio of the GSH adduct to the cysteine adduct to a degree dependent on their catalytic effectiveness. GSH transferases 1-2, 3-3, 3-4, and 4-6 have been tested, with GSH transferase 3-3 having the greatest effect (33).

Preliminary kinetic data have been obtained using stopped-flow techniques. In order to obtain these data the rate of reaction was reduced by working at 25°C and pH 6.5. Under these conditions, the second-order rate constant for the spontaneous reaction was 6 x 10^5 M^{-1}/min and the V_{max} and K_m for GSH transferase 3-3 were 1.5 μmol/min/mg and 3 μM, respectively (75). If the rate constant for the spontaneous reaction is determined using a physiological value for GSH concentration (10 mM) and a notional NAPQI concentration (1 μM), the rate of conjugation can be calculated to be

6 mM/min. If the reaction is carried out in the presence of a physiological concentration of GSH transferase 3-3, e.g., 50 mM, the enzymic rate is calculated to be 4 mM/min.

At first sight it might be thought that enzymic and nonenzymic reactions make equal contributions to GSH conjugation, but such a conclusion does not take into account the effect of enzymes in lowering the free substrate concentration available to the enzyme. Under the above conditions, the concentration of substrate below the K_m of the enzyme and the concentration of the enzyme is at least an order of magnitude above the K_m. In these circumstances the rate of binding of substrate to enzyme compared with the combined nonenzymic reactions is such that at least 90% is enzyme-catalyzed conjugation.

When concentrations of NAPQI are well in excess of K_m the contribution of nonenzymic reactions should become more important. While GSH conjugation is purely detoxifying, NAPQI reduction and GSH oxidation may affect redox state. A combination of a high dose of paracetamol and inhibition of GSH reductase causes severe toxicity due to a gross alteration in redox state (1). However, at moderate doses there may be no toxicity in vivo since the enzymic reaction should predominate and the GSH reductase present should reduce the small amount of glutathione disulfide produced without an important effect on the redox state of the cell. GSH transferases are therefore relatively effective in detoxifying paracetamol and also play an important subsidiary role in detoxifying phenacetin. Whether or not there is significant electrophilic or free radical attack on macromolecules at ordinary dosage has yet to be determined.

GLUTATHIONE TRANSFERASES AND LIPID PEROXIDATION

Detoxification of Lipid Hydroperoxides

The generation of reactive oxygen species is a general feature of normal aerobic metabolism and may be enhanced by xenobiotics either during oxidative metabolism at the endoplasmic reticulum (e.g., carcinogens, paracetamol), through one electron redox-cycling (e.g., menadione, adriamycin), or by photoactivation (e.g., porphyrins, psoralens) (3). These reactive O_2 species may initiate lipid peroxidation by the abstraction of a hydrogen atom from an unsaturated phospholipid. The resulting acyl radical reacts rapidly with oxygen to yield a lipid peroxy radical, and the latter then propagates the chain reaction by hydrogen abstraction from a neighboring phospholipid to produce a lipid hydroperoxide and a new phospholipid radical.

Lipid hydroperoxides can undergo very damaging metal-catalyzed decomposition to a number of compounds, including yet more free radicals, electrophilic and mutagenic hydroxyalkenals, and inert hydrocarbons such as ethane which are exhaled and can be estimated (63). Several mechanisms exist to inhibit lipid peroxidation. One is the trapping of free radicals within the membrane by tocopherols (76). Another involves the removal and detoxification of fatty acyl hydroperoxide. The latter depends on the consecutive action of phospholipase A_2 and the reduction of the released free fatty acid hydroperoxide to fatty acid alcohol by Se-dependent or Se-independent GSH peroxidase activity (66). As has already been mentioned, GSH transferase subunits 1, 2, 5, and 7 are responsible for GSH peroxidase activity (52). Removal of lipid and lipid peroxy-radicals by α-tocopherol prevents the primary chain reaction but still yields lipid hydroperoxides which, through oxidative decomposition, yield alkoxy and alkyl-peroxy radicals capable of initiating a secondary phase of chain reactions. Reduction of hydroperoxides by GSH peroxidases prevents this, and in doing so has a dramatic inhibitory effect on overall lipid peroxidation. The substrates for GSH peroxidases are free fatty acid hydroperoxides, and inhibition of lipid peroxidation is dependent on phospholipase A_2 activity (66). It is likely that the turnover of phospholipids is accelerated by distortions in the bilayer brought about by the introduction of the hydroperoxy group in the previously nonpolar hydrocarbon chain of the phospholipid; distortions, as brought about by epoxidation of polyunsaturated fatty acyl residues in phospholipid bilayers, have already been shown to have the effect (62).

Both Se-dependent and Se-independent GSH peroxidases exist in all tissues thus far examined. Whether or not they have exactly the same action on lipid peroxidation is not yet known. There may be, for instance, differences in substrate specificity. In the liver they have different locations in the lobule. Thus, Se-dependent GSH peroxidase is more abundant in the more oxygenated periportal region (77), and the GSH transferases are more abundant in the centrilobular regions (58).

The importance of GSH in the inhibition of lipid peroxidation is demonstrated by the effect of GSH depletion in enhancing the exhalation of ethane in animals utilizing increased levels of oxygen in the course of drug metabolism (74). The importance of the Se-dependent GSH peroxidase is demonstrated by enhanced ethane exhalation in Se-deficient animals (74). Similar evidence has yet to be collected to demonstrate the importance of GSH transferases in the inhibition of lipid peroxidation in vivo, but in homogenates, depending on the species and the tissue, the percentage of GSH peroxidase activity due to GSH transferases is always significant and may approach 100%.

Hydroxyalkenals

The chief hydroxyalkenals produced during the decomposition of fatty acyl hydroperoxides are 4-hydroxy-non-2-en-1-al and 4-hydroxy-dec-2-en-1-al (21). The former is known to be a genotoxic electrophile that is effectively detoxified by GSH transferase isoenzymes, the subunits of which have activity in the order $4 > 3 = 1 > 2$ (46).

THE INDUCTION OF GLUTATHIONE TRANSFERASES

GSH transferases are induced by a number of agents of diverse structure (73) such as plant products (e.g., extracts of certain brassicas such as brussel sprouts); the narcotic phenobarbitone; the carcinogens 3-methylcholanthrene and MAB; the antioxidants BHA and BHT; the insecticide DDT; the polychlorinated biphenyls that are electrical insulators; aflatoxin B_1; the lupinosis toxin phomopsin; and otherwise unclassified substances (e.g., trans-stilbene oxide) (46). Se-deficiency is also an inducer as is the normal dietary constituent, cholesterol (31). Induction is subunit selective; for example, phenobarbitone induces GSH transferase subunits 1 and 3, whereas 3-methylcholanthrene induces only subunit 1. Phomopsin induces only subunit 3; Se-deficiency induces subunits 1 and 2. In the rat, phenobarbitone increases GSH transferase activity about 3-fold; in the mouse, BHA can increase it 10-fold, to the point that GSH transferases account for at least 20% of the soluble protein of the liver (56).

All these agents also induce isoenzymes in the P-450 system with the consequence that many xenobiotics are metabolized more rapidly, and those electrophilic metabolites produced which are substrates for GSH transferases are more rapidly detoxified by GSH.

THE STRUCTURE OF GLUTATHIONE TRANSFERASES

Dimers of subunits with similar molecular weights have been found in all soluble GSH transferases thus far isolated, be they from the plants, members of the animal kingdom as lowly as the platyhelminths or as diverse as insects, molluscs, and man. Thus far, soluble GSH transferases from rat, mouse, and man have been most studied.

The Rat Soluble GSH Transferases

As previously stated, subunit 1 forms dimers only with itself or subunit 2, and subunits 3, 4, and 6 form a similar associating group. Subunits 5 and 7 have been observed only to form homodimers. It is now known that subunits 1 and 2 have homology in their amino acid sequences determined either by analysis of tryptic peptides (7) or by analysis of cDNA sequences (69). Analysis of tryptic peptides of subunits 3, 4 (Ref. 8), and 6 also shows homology of amino acid sequence, and this is confirmed in detail for N-terminal sequences of subunits 3 and 4 by sequenation (24). Recently, cDNA sequences corresponding to mRNAs coding for subunits 1, 2, and 7 have become available, and the amino acid sequences for which they code are shown in Fig. 3 (43,65,67,69). The close homology between subunits 1 and 2 is immediately apparent. Very much less homology is seen with subunit 7. Even so, the same amino acid residue occupies a remarkable 26% of the equivalent positions in all 3 subunits and there are small areas of even higher homology, for example, residues 8-26 (58%), residues 80-101 (79%), and residues 180-199 (50%). The conservation of these domains may be related to a common property of all 3 subunits, for example, GSH binding. The divergence in amino acid sequence may be expected, not only because the enzymic activities of the subunits vary, but also because the sequences involved in heterodimer formation also differ since subunit 7 does not form a heterodimer with subunits 1 or 2.

This information alone indicates that the GSH transferases constitute an interesting multigene family. However, closer analysis of individual subunits gives evidence of sequence heterogeneity within the subunits described above. Thus in different laboratories, analysis of subunit 1 has shown the N-terminus to be either a blocked N-terminal serine or a free proline (24,68); also in our laboratory, two C-termini have consistently been detected in the same preparation, namely, Phe/Lys/Phe-OH and Lys/Arg/Lys-OH (7). In addition, analyses of tryptic peptides suggest that in the one preparation there may be heterogeneity of sequence in the region of cysteinyl residues where 4 cysteinyl peptides occur in subunit 1, whereas amino acid analysis indicates only 2 cysteinyl residues per subunit (7). A comparison of full length cDNA sequences shows that 2 mRNA species coding for subunit 1 must exist, between which there are 15 base changes in the coding region resulting in 8 conservative amino acid changes. Only 3 tryptic cysteinyl peptides and one C-terminal tripeptide would be generated from these sequences, which suggests that other full-length cDNA sequences encoding subunit 1 mRNAs have yet to be found.

Thus far, there have been no reports of these amino acid changes being linked with a variation in the properties of the two subunits. The

```
     1           10          20          30          40          50          60          70
1    SGKPVLHYFNARGRMECIRWLLAAAGVEFDEKFIQSPEDLEKLKKDGNLMFDQVPMVEIDGMKLAQTRAILNYIA

2    P.......DG.....P..............Q.LKTRD..AR.RN..S...Q............V..........

7    *PPYTIV..PV...C.AT.M...NQ.QSWK.EVVTIDVW.QGSL.STC.*YG.L.KF.DGDLT.Y.SN...RHLG

π    *PPYTVV..PV...C.AL.M...D


        80          90          100         110         120         130         140         150
1    TKYDLYGKDMKERALIDMYTEGILDLTEMIMQLVICPPDQKEAKTALAKDRTKNRYLPAFEKVL**KSHGQDYLV

2    ...N...............A..VA..D.IVLHYPYI..GE...SL.KI..KAR...F.......**.........

7    RSLG.....Q..A..V..VND.VE..RCKYGT.IYTNYENGKDDYVK.****LPGH.KP..TL.SQNQG.KAFI.


                 160         170         180         190         200         210         220
1    GNRLTRVDIHLLELLLYVEEFDASLLTSFPLLKAFKSRISSLPNVKKFLQPGSQRKLPMDAKQIEEARKIFKF

2    ....S.A.VY.VQV.YH...L.P.A.AN......LRT.V.N..T............PLE.E.CV.S.V.I.S*

7    ..QISFA.YN..D...VHQVLAPGC.DN....S.YVA.L.AR.KI.A..SSPDHLNR.********INGNG.Q
```

Fig. 3. Comparison of primary structure of GSH transferase subunits 1, 2, 7, and π. Amino acid sequences of subunits 1 (43,67), 2 (69), and 7 (65) deduced from cDNA sequences are aligned to produce maximum homology. The N-terminal portion of human GSH transferase subunit π determined by protein sequencing (2) is shown to emphasize its homology to subunit 7. . = Same amino acid residue as in subunit 1; * = space inserted to allow maximum homology.

nucleotide variations, scattered as they are along the coding sequences, suggest that there are at least 2 genes encoding subunit 1 polypeptides rather than alternative RNA splicing regimens of a single primary transcript. Such an interpretation is supported by substantial differences occurring in the 3' untranslated region of these two clones (α-actins) and suggests that subunit 1 may itself represent a multigene family.

This view gets more direct support from Southern blots of rat genomic DNA cleaved with Pstl, EcoRl, or Pvull restriction enzymes (68) and hybridized with a radiolabeled fragment of subunit 1 cDNA (Fig. 4). Even though cleavage sites for the chosen restriction enzymes do not exist in the cDNA probe, several hybridizing DNA fragments of varying sizes are detectable in each digest. Since the probe is fairly small, there is little likelihood of multiple restriction fragments arising from a single complementary genomic sequence containing several introns since each intron would need to include the low frequency cleavage sites recognized by Pstl, EcoRl, and Pvull. The varying of intensities of the bands may reflect either varying homology to the probe (for example, subunit 2 and other subunit 1 genomic sequences), or nonequimolar ratios of restriction fragments due to varying multiplication of highly homologous genes. In future experiments it will be possible to differentiate between genes for subunit 2 and both subunit 1 mRNAs by using probes complementary to the 3' untranslated sequences of the three cDNA clones.

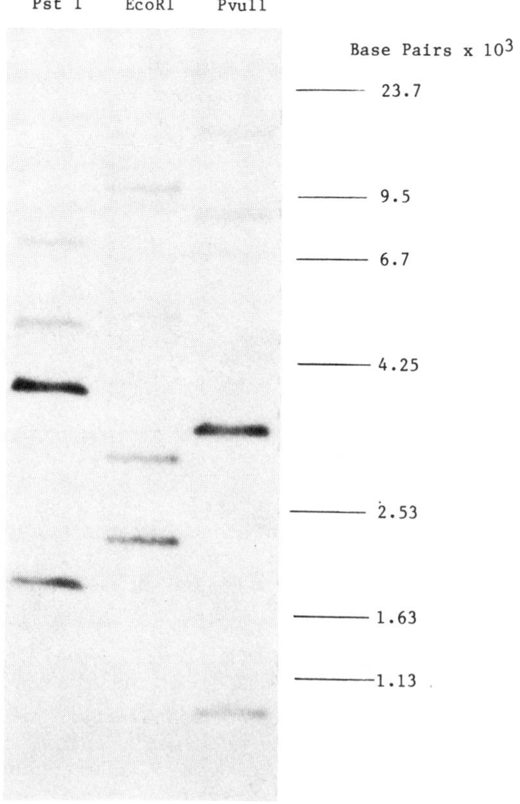

Fig. 4. Genetic complexity of GSH transferase subunit 1 determined by Southern blot. Rat liver DNA was digested to completion with the restriction enzymes as indicated. The DNA fragments were separated by electrophoresis on 0.55% agarose, transferred to Gene Screen Plus membrane, and probed with [^{32}P] cDNA. The probe was an isolated 360-bp fragment contained within plasmid pGSTr155 (67) which corresponds to the sequence coding for amino acid residues 26-129 of subunit 1. There are no cleavage sites for Pst1, EcoR1, or Pvu11 restriction enzymes within the probe.

It is probable that the genetic complexity seen in subunit 1 also occurs in other subunits. For instance, chemical analysis of subunit 2 shows heterogeneity around cysteinyl residues and at the C-terminus similar to that seen in subunit 1 (7).

Human GSH Transferases

Multiplicity of GSH transferases also occurs in the human. The nomenclature is based on the greek alphabet, and physical, chemical, and enzymic properties associated with the various human enzymes are shown in Tab. 3 and 4. In earlier times a trans-species relationship was not immediately apparent, but an examination of substrate specificity (72) and observations of immunological cross-reactivity (61) between antirat GSH transferases and human GSH transferases have enabled relationships to be seen between: (a) human GSH transferases α - ε and rat GSH transferase subunits 1 and 2; (b)

Tab. 4. Specific activities of human GSH transferases towards selected model substrates.

Substrate	α - ε	μ	π
Δ^5-Androsten-3,17-dione	8.0*	0.12	0.01
Cumene hydroperoxide	10.6*	0.63	0.03
Trans-4-phenyl-3-buten-2-one	0.001-2	0.36*	0.01
Ethacrynic acid	0.017-44	0.08	0.86*

* Substrates that relate human enzymes α - ε, μ and π to rat enzymes containing subunits 1 and 2, 4 and 7, respectively.
Note: Activity is expressed in μmol/min/mg protein. Data are taken from Ref. 72.

human GSH transferase μ and rat GSH transferase subunit 4; and (c) human GSH transferase π and rat GSH transferase subunit 7. The closeness of these relationships is apparent in a study of N-terminal amino acid sequences of rat GSH transferase 7-7 (65) and human GSH transferase π (2) which show considerable homology.

Soluble GSH Transferases in the Mouse

Multiplicity of genes encoding GSH transferases also occurs in the mouse. Of particular relevance to the multigenic nature of the rat subunit 1/2 family has been the selection of clones of mouse genomic DNA that hybridize to a radiolabeled probe of rat subunit 1 cDNA. It has been shown that these subunit 1/2-like genes are numerous, can occur in tandem, and therefore also constitute a multigene family (15). Mouse GSH transferases with close structural relationship with both rat subunits 3 and 4 have been demonstrated (56).

CONCLUSIONS

We have shown that GSH transferases are a large multigenic family of isoenzymes which collectively have the power to detoxify many electrophiles and, in addition, fatty acid hydroperoxides. Their tissue distribution varies enormously and according to the tissue in which they are found, some are inducible by a wide range of compounds, including carcinogens and tumor promoters.

Initiation has at least 2 important requirements. One is DNA damage resulting from reaction with electrophiles or attack by oxidizing free radicals. GSH transferases can prevent or limit the reaction of electrophiles with DNA and in this respect are anticarcinogenic. Their success requires that the electrophiles be substrates for the GSH transferases present and that the relevant GSH transferases be present in sufficient quantity. Induction can make a large difference. GSH transferases that are GSH peroxides can also prevent or limit oxidative free radical attack which might arise from the decomposition of lipid peroxides (11,22).

Another important requirement for initiation is cell turnover. Many tissues turn over by nature and those which do not require the induction of hyperplasia or regenerative cell division. The latter may be a consequence of the reaction of cytotoxins which, if they are electrophiles or derived from lipid hydroperoxides, are also susceptible to GSH transferases which are therefore once more anticarcinogenic.

Promotion is a complex process, still poorly understood. The phenotypic diversity with which it is associated may be induced by compounds that do not give electrophiles (e.g., phorbol esters, phenobarbitone) or possibly in the case of complete carcinogens, nonelectrophilic forms. It is, however, noteworthy that the action of phorbol esters, for example, has been associated with the release of oxidizing free radicals. GSH transferases may therefore be antipromoting if effects arising from lipid peroxides are important. Cell turnover is required to bring about the clonal expansion of potential tumor cells, and in as much as this involves regenerative hyperplasia such as in initiation, GSH transferases are anticarcinogenic (11,22).

Because they are protective, GSH transferases may also be procarcinogenic. In hepatic hyperplastic nodules resulting from carcinogen treatment, a well-orchestrated change in differentiation occurs which is highly advantageous since it enables these cells to avoid most of the toxic hazards dealt with by the liver. This change in differentiation involves the maintenance of high levels of GSH transferases, but with a new isoenzyme distribution: GSH itself is elevated. This pattern exists in the primary hepatoma (22,49).

GSH transferases may also be advantageous in malignant progression. Cells that develop resistance to chemotherapeutic drugs retain or increase GSH transferase levels and have elevated GSH (30).

Thus, GSH transferases are certainly an anticarcinogenic defense in normal tissue, but they may also be useful armaments in preneoplastic and neoplastic cells.

They are a double-edged sword in yet another respect. In liver carcinogenesis, GSH transferase inducers are often tumor promoters (22,73).

ACKNOWLEDGEMENTS

The authors thank the Cancer Research Campaign for which B.K. is a Fellow for generous support, and Miss C.A. Brough for help with the manuscript.

REFERENCES

1. Albano, E., M. Rundgren, P.J. Harvison, S.D. Nelson, and P. Moldeus (1985) Mechanisms of N-acetyl-p-benzoquinone imine cytotoxicity. Molec. Pharmacol. 28:306-311.
2. Ålin, P., B. Mannervik, and H. Jörnvall (1985) Structural evidence for three different types of glutathione transferases in human tissues. FEBS Lett. 182:319-322.
3. Ames, B.N. (1983) Dietary carcinogens and anticarcinogens. Oxygen radicals and degenerative diseases. Science 221:1256-1264.
4. Awasthi, Y.C., D.D. Dao, and R.P. Saneto (1980) Interrelationships between anionic and cationic forms of glutathione S-transferases of human liver. Biochem. J. 191:1-10.

5. Backer, J.M., and I.B. Weinstein (1980) Mitochondrial DNA is a major cellular target for dihydrodiol-epoxide derivative of benzo(a)pyrene. Science 209:297-299.
6. Bartsch, H., D. Dworkin, E.C. Miller, and J.A. Miller (1972) Electrophilic N-acetylaminoarenes by enzymatic deacetylation and trans acetylation in liver. Biochim. Biophys. Acta 286:272-298.
7. Beale, D., B. Ketterer, T. Carne, D. Meyer, and J.B. Taylor (1982) Evidence that Ya and Yc subunits of glutathione transferase B (Ligandin) are the products of separate genes. Eur. J. Biochem. 126:459-463.
8. Beale, D., D.J. Meyer, J.B. Taylor, and B. Ketterer (1983) Evidence that the Yb subunits of hepatic glutathione transferases represent two different but related families of polypeptides. Eur. J. Biochem. 137: 125-129.
9. Busby, Jr., W.F., and G.N. Wogan (1984) Aflatoxins. In Chemical Carcinogenesis, C.E. Searle, ed. ACS Monograph 182, American Chemical Society, Washington, D.C., pp. 945-1093.
10. Campbell, R.M., and H.W. Kosterlitz (1947) Ribonucleic acid as a constituent of labile liver cytoplasm. J. Physiol. 106:12P.
11. Cerutti, P.A. (1985) Prooxidant states in tumour promotion. Science 227:375-381.
12. Chasseud, L.F. (1979) The role of glutathione and glutathione S-transferases in the metabolism of chemical carcinogens and other electrophilic agents. Adv. Cancer Res. 29:176-274.
13. Coles, B. (1984/85) Effects of modifying structure on electrophilic reactions with biological nucleophiles. Drug Metab. Rev. 15:1307-1334.
14. Coles, B., D.J. Meyer, B. Ketterer, C.A. Stanton, and R.C. Garner (1985) Studies on the detoxication of microsomally-activated aflatoxin B_1 by glutathione and glutathione transferases in vitro. Carcinogenesis 6:693-697.
15. Czosnek, H., S. Sarid, P.E. Barker, F.H. Ruddle, and V. Daniel (1984) Glutathione S-transferase Ya subunit is coded by a multigene family located on a single mouse chromosome. Nucl. Acids Res. 12:4825-4833.
16. Degen, G.H., and H.G. Neumann (1978) The major metabolite of aflatoxin B_1 in the rat is a glutathione conjugate. Chem. Biol. Interact. 22: 239-255.
17. De Matteis, F., H.A. Gibbs, and C. Hollands (1985) N-alkylated porphyrins: Products of inactivation of cytochrome P-450 isoenzymes by suicide substrates. In Microsomes and Drug Oxidations, A.R. Boobis, F. De Matteis, and C.R. Elcombe, eds. Taylor and Francis, London, pp. 265-273.
18. Deml, E., D. Oesterle, and F.J. Wiebel (1983) Benzo(a)pyrene initiates enzyme-altered islands in the liver of adult rats following single pretreatment and promotion with polychlorinated biphenyls. Cancer Lett. 19:301-304.
19. Dipple, A., R.C. Moschel, and C.A. Bigger (1984) Polynuclear aromatic hydrocarbons. In Chemical Carcinogenesis, C.E. Searle, ed. ACS Monograph 182, American Chemical Society, Washington, D.C., pp. 41-126.
20. Djuric, Z., B.F. Coles, F.A. Beland, and B. Ketterer (unpublished information).
21. Esterbauer, H., K.H. Cheeseman, M.V. Dianzani, and T.F. Slater (1982) Separation and characterization of the aldehydic products of lipid peroxidation stimulated by ADP-Fe^{2+} in rat liver microsomes. Biochem. J. 208:129-140.
22. Farber, E. (1984) Cellular biochemistry of the stepwise development of cancer with chemicals. Cancer Res. 44:5463-5474.

23. Fischer, V., and R.P. Mason (1984) Stable free radical and benzoquinone imine metabolites of an acetaminophen analogue. J. Biol. Chem. 259:10284-10288.

24. Frey, A.B., T. Friedberg, F. Oesch, and G. Kreibich (1983) Studies on the subunit composition of rat liver glutathione S-transferases. J. Biol. Chem. 258:11321-11325.

25. Garner, R.C., and C.N. Martin (1979) Fungal toxins, aflatoxins and nucleic acids. In Chemical Carcinogens and DNA, Vol. I, CRC Press, Boca Raton, Florida, pp. 187-225.

26. Geacintov, N.E., H. Yoshida, V. Ibanez, and R.G. Harvey (1981) Noncovalent intercalative binding of 7,8-dihydroxy-9,10-epoxybenzo(α)pyrene to DNA. Biochem. Biophys. Res. Commun. 100:1569-1577.

27. Geacintov, N.E., H. Yoshida, V. Ibanez, and R.G. Harvey (1982) Noncovalent binding of 7β,8α-dihydroxy-9α,10α-epoxytetrahydrobenzo(α)pyrene to deoxyribonucleic acid and its catalytic effect on the hydrolysis of the diol epoxide to tetrol. Biochemistry 21:1864-1869.

28. Grover, P.L., and B. Jernström (unpublished information).

29. Guthenberg, C., and B. Mannervik (1981) Glutathione S-transferase (transferase π) from human placenta is identical or closely related to glutathione S-transferase (transferase P) from erythrocytes. Biochim. Biophys. Acta 661:255-260.

30. Hamilton, T.C., M.A. Winker, K.G. Louie, G. Batist, B.C. Behrens, T. Tsuruo, K.R. Grotzunger, W.M. McKoy, R.C. Young, and R.F. Ozols (1985) Augmentation of adriamycin, melphalan, and cisplatin cytotoxicity in drug-resistant and sensitive human ovarian carcinoma cell lines by buthionine sulphoximine mediated glutathione depletion. Biochem. Pharmacol. 34:2583-2586.

31. Hietanen, E., M. Ahotupa, A. Heikela, and M. Laitinen (1982) Dietary cholesterol-induced changes of xenobiotic metabolism in liver. II. Effects of phenobarbitone and carbon tetrachloride on activities of drug-metabolising enzymes. Drug-Nutrient Interact. 1:313-327.

32. Hinson, J.A. (1983) Reactive metabolites of phenacetin and acetaminophen: A review. Environ. Health Perspect. 49:71-79.

33. Hinson, J.A., B. Coles, S.D. Nelson, and B. Ketterer (unpublished information).

34. Howard, P.C., R.H. Heflich, F.E. Evans, and F.A. Beland (1983) Formation of DNA adducts in vitro and in Salmonella typhimurium upon metabolic reduction of the environmental mutagen 1-nitropyrene. Cancer Res. 43:2052-2058.

35. Jakoby, W.B., B. Ketterer, and B. Mannervik (1984) Glutathione transferases: Nomenclature. Biochem. Pharmacol. 33:2539-2540.

36. Jernström, B., M. Martinez, D.J. Meyer, and B. Ketterer (1980) Glutathione conjugation of the carcinogenic and mutagenic electrophile (±)-7β,8α-dihydroxy-9α,10α-oxy-7,8,9,10-tetrahydro-benzo(a)pyrene catalyzed by purified rat liver glutathione transferases. Carcinogenesis 6:85-89.

37. Kamisaka, K., W.H. Habig, J.N. Ketley, I.M. Arias, and W.B. Jakoby (1975) Multiple forms of human glutathione S-transferase and their affinity for bilirubin. Eur. J. Biochem. 60:153-161.

38. Ketterer, B., B. Coles, and D.J. Meyer (1983) The role of glutathione in detoxication. Environ. Health Perspect. 49:56-69.

39. Ketterer, B., D.J. Meyer, B. Coles, and J.B. Taylor (1985) Glutathione transferase isoenzymes in the rat: Their multiplicity and tissue distribution. In Microsomes and Drug Oxidations, A.R. Boobis, F. De Matteis, and C.R. Elcombe, eds. Taylor and Francis, London, pp. 167-177.

40. Kitagawa, T., T. Hirakawa, T. Ishikawa, M. Nemoto, and S. Takayama (1980) Induction of hepatocellular carcinoma in rat liver by initial treatment with benzo(a)pyrene after partial hepatectomy and promotion by phenobarbital. Toxicol. Lett. 6:161-171.
41. Kraus, P., and B. Gross (1979) Particle-bound glutathione S-transferases. Enzyme 24:205-208.
42. Kraus, P. (1980) Resolution, purification and some properties of three glutathione transferases from rat liver mitochondria. Hoppe-Seyler's Z. Physiol. Chem. 361:9-15.
43. Lai, H.-C.J., N.-Q. Li, M.J. Weiss, C.C. Reddy, and C.-P.D. Tu (1984) The nucleotide sequence of a rat liver glutathione S-transferase subunit cDNA clone. J. Biol. Chem. 259:5536-5542.
44. Lotlikar, P.D., E.C. Jhee, S.M. Insetta, and M.S. Clearfield (1984) Modulation of microsomal-mediated aflatoxin binding to exogenous and endogenous DNA by cytosolic glutathione S-transferases in rat and hamster livers. Carcinogenesis 5:269-276.
45. Lutz, W.K., A. Viviani, and C. Schlatter (1978) Non-linear dose-response relationship for the binding of the carcinogen benzo(a)pyrene to rat liver DNA in vivo. Cancer Res. 38:575-578.
46. Mannervik, B., P. Ålin, C. Guthenberg, H. Jensson, and M. Warholm (1985) Glutathione transferases and the detoxification of products of oxidative metabolism. In Microsomes and Drug Oxidations, A.R. Boobis, F. De Matteis, and C.R. Elcombe, eds. Taylor and Francis, London, pp. 221-228.
47. Mannervik, B., and H. Jensson (1982) Binary combinations of four protein subunits with different catalytic specificities explain the relationship between six basic glutathione S-transferases in rat liver cytosol. J. Biol. Chem. 257:9909-9912.
48. Marcus, C.J., W.H. Habig, and W.B. Jakoby (1978) Glutathione transferase from human erythrocytes. Nonidentity with the enzymes from liver. Arch. Biochem. Biophys. 188:287-293.
49. Meyer, D.J., D. Beale, K.H. Tan, B. Coles, and B. Ketterer (1985) Glutathione transferases in primary rat hepatomas: The isolation of a form with GSH peroxidase activity. FEBS Lett. 184:139-143.
50. Meyer, D.J., L.G. Christodoulides, K.H. Tan, and B. Ketterer (1984) Isolation, properties and tissue distribution of rat glutathione transferase E. FEBS Lett. 173:327-330.
51. Meyer, D.J., M. Hall, and B. Ketterer (unpublished information).
52. Meyer, D.J., K.H. Tan, L.G. Christodoulides, and B. Ketterer (1985) Recent studies of selenium-independent GSH peroxidases (GSH transferases) from the rat. In Free Radicals in Liver Injury, M.V. Dianzani, G. Poli, T.F. Slater, and K.H. Cheeseman, eds. IRL Press, Oxford, pp. 221-222.
53. Morgenstern, R., C. Guthenberg, and J.W. DePierre (1982) Microsomal glutathione S-transferase. Purification, initial characterization and demonstration that it is not identical to the cytosolic glutathione S-transferases A, B and C. Eur. J. Biochem. 128:243-248.
54. Neal, G.E., S.A. Metcalfe, R.F. Legg, D.J. Judah, and J.A. Green (1981) Mechanism of the resistance to cytotoxicity which precedes aflatoxin B_1 hepatocarcinogenesis. Carcinogenesis 2:457-461.
55. O'Brien, K., E. Moss, D. Judah, and G. Neal (1983) Metabolic basis of the species difference to aflatoxin B_1 induced hepatotoxicity. Biochem. Biophys. Res. Commun. 114:813-821.
56. Pearson, W.R., J.J. Windle, J.F. Morrow, A.M. Benson, and P. Talalay (1983) Increased synthesis of glutathione S-transferase in response to anticarcinogenic antioxidants. Cloning and measurement of messenger RNA. J. Biol. Chem. 258:2052-2062.

57. Rahimtula, A.D., and M. Martin (1984) Dietary administration of 3,3-di-t-butyl-4-hydroxyanisole elevates mouse liver microsome-mediated DNA binding and mutagenicity of AFB_1. Chem. Biol. Interact. 48:207-220.

58. Redick, J.A., W.B. Jakoby, and J. Baron (1982) Immunohistochemical localization of glutathione S-transferases in livers of untreated rats. J. Biol. Chem. 257:15200-15203.

59. Rosenkranz, H.S., and R. Mermelstein (1983) Mutagenicity and genotoxicity of nitroarenes: All nitro-containing compounds were not created equal. Mutat. Res. 114:217-267.

60. Saito, K., Y. Yamazoe, T. Kamataki, and R. Kato (1983) Activation and detoxication of N-hydroxy-Trp-P-2 by glutathione and glutathione transferases. Carcinogenesis 4:1551-1557.

61. Satoh, K., A. Kitahara, Y. Soma, Y. Inaba, I. Hatayama, and K. Sato (1985) Purification, induction and distribution of placental glutathione transferase: A new marker enzyme for preneoplastic cells in the rat chemical hepatocarcinogenesis. Proc. Natl. Acad. Sci., USA 82: 3964-3968.

62. Sevanian, A., R.A. Stern, and J.F. Mead (1982) Metabolism of epoxidized phosphatidylcholine by phospholipase A_2 and epoxide hydrolase. Lipids 16:781-789.

63. Slater, T.F. (1984) Free-radical mechanisms in tissue injury. Biochem. J. 222:1-15.

64. Stanton, C.A., F.L. Chow, D.H. Phillips, P.L. Grover, R.C. Garner, and C.N. Martin (1985) Evidence for N-(deoxyguanosin-8-yl)-1-aminopyrene as a major DNA adduct in female rats treated with 1-nitropyrene. Carcinogenesis 6:535-538.

65. Suguoka, Y., T. Kano, A. Okuda, M. Sakai, T. Kitagawa, and M. Muramatsu (1985) Cloning and the nucleotide sequence of rat glutathione S-transferase P. Nucl. Acids Res. 13:6049-6057.

66. Tan, K.H., D.J. Meyer, J. Belin, and B. Ketterer (1984) Inhibition of microsomal lipid peroxidation by glutathione and glutathione transferases B and AA--Role of phospholipase A_2. Biochem. J. 220:243-252.

67. Taylor, J.B., R.K. Craig, D. Beale, and B. Ketterer (1984) Construction and characterization of a plasmid containing complementary DNA to mRNA encoding the N-terminal amino acid sequence of the rat glutathione transferase Ya subunit. Biochem J. 219:223-231.

68. Taylor, J.B., S. Pemble, and B. Kettererr (unpublished information).

69. Telakowski-Hopkins, C.A., J.A. Rodkey, C.D. Bennett, A.Y.H. Lu, and C.B. Pickett (1985) Rat liver glutathione S-transferases. Construction of a cDNA clone complementary to a Yc mRNA and prediction of the complete amino acid sequence of a Yc subunit. J. Biol. Chem. 260: 5820-5825.

70. Tipping, E., and B. Ketterer (1981) The influence of soluble binding proteins on lipophile transport and metabolism in hepatocytes. Biochem. J. 195:441-452.

71. Tipping, E., B.P. Moore, C.A. Jones, G.M. Cohen, B. Ketterer, and J.W. Bridges (1980) The non-covalent binding of benzo(a)pyrene and its hydroxylated metabolites to intracellular proteins and lipid bilayers. Chem. Biol. Interact. 32:291-304.

72. Warholm, M., C. Guthenberg, and B. Mannervik (1983) Molecular and catalytic properties of glutathione transferase from human liver: An enzyme efficiently conjugating epoxides. Biochemistry 22:3610-3616.

73. Wattenberg, L.W. (1983) Inhibition of neoplasia by minor dietary constituents. Cancer Res. 43(Suppl.):2248s-2453s.

74. Wendel, A., and S. Feuerstein (1982) Drug induced lipid peroxidation in mice. I. Modulation by monooxygenase activity, glutathione and selenium status. Biochem. Pharmacol. 30:2513-2520.

75. Wilson, I., J. Hinson, B. Coles, S.D. Nelson, P. Wardman, and B. Ketterer (unpublished information).
76. Witting, L.A. (1980) Vitamin E and lipid antioxidants in free-radical-initiated reactions. In Free Radicals in Biology, Vol. IV, W.A. Pryor, ed. Academic Press, New York, pp. 295-319.
77. Yoshimura, S., N. Komatsu, and K. Watanabe (1980) Purification and immunohistochemical localization of rat liver glutathione peroxidase. Biochim. Biophys. Acta 621:130-137.

NATURAL ENVIRONMENTAL ANTIMUTAGENS

INTRODUCTION: NATURAL ENVIRONMENTAL ANTIMUTAGENS

Thomas S. Matney

University of Texas
Health Science Center at Houston
Houston, Texas 77225

The topic, Natural Environmental Antimutagens, is thoroughly and provocatively discussed by Dr. Ames in this Volume. I should like to make only one small addendum.

In 1979, Dr. Chiu-Nan Lai performed some experiments in my laboratory which related the antimutagenic activities of common vegetables to their chlorophyll content (4). The results are summarized in Fig. 1. Percent of mutagenic activity of 3-methylcholanthrene (3-MC) is plotted against the chlorophyll content of vegetables (closed circles) and purified chlorophyllin (open circles). The standard Ames *Salmonella* mutagenesis test with strain TA100 and S-9 metabolic activation was employed.

A previous pair of studies (2,3) had shown that aqueous fractions of wheat sprout leaves and roots selectively inhibited the mutagenic effect of carcinogens requiring metabolic activation. Formation of dihydrodiol metabolites of benzo(α)pyrene was significantly reduced, as seen in high-pressure liquid chromatographic profiles of metabolites. Chlorophyll was shown to be the major active antimutagenic factor, with vitamin E and carotene contributing marginally to the overall inhibitory activity; chlorophyllin, a water-soluble derivative of chlorophyll, similarly exhibited antimutagenic activities.

It would seem important to recognize the antimutagenic activity of chlorophyll, not only because it is a major constituent of plants and therefore should play a leading anticarcinogenic role in effects of diet on the epidemiology of certain types of cancer (1), but also as a possible contaminant of plant extracts purported to contain specific plant components with antimutagenic properties (Mitscher et al., this Volume).

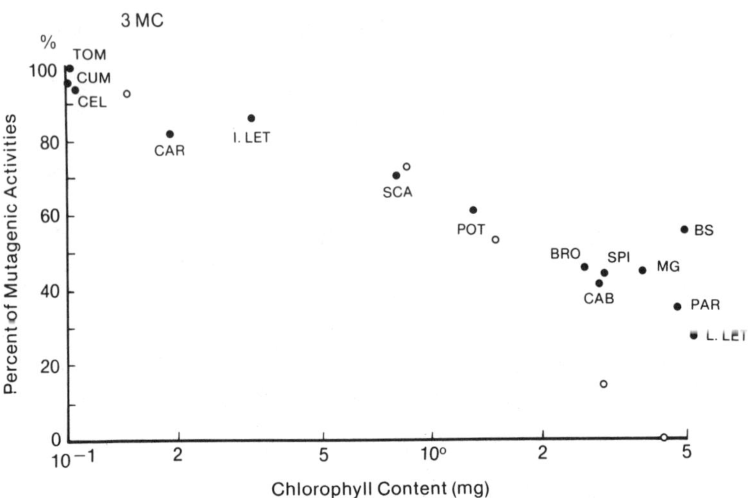

Fig. 1. Chlorophyll content of common vegetables. TOM = tomato, CUM = cucumber, CEL = celery, CAR = carrot, I. LET = iceberg lettuce, SCA = scallion, POT = potato, BRO = broccoli, SPI = spinach, CAB = cabbage, MG = mustard green, BS = brussels sprout, PAR = parsley, and L. LET = leafy lettuce. Reprinted with permission from Elsevier Biomedical Press (4).

REFERENCES

1. Graham, S.H., M. Dayal, A. Swanson, A. Mittleman, and G. Wilkinson (1975) Diet in the epidemiology of cancer of the colon and rectum. J. Natl. Cancer Inst. 61:709-714.
2. Lai, C.N. (1979) Chlorophyll: The active factor in wheat sprout extract inhibiting the metabolic activation of carcinogens in vitro. Nutr. Cancer 1:19-21.
3. Lai, C.N., B.J. Dabney, and C.R. Shaw (1978) Inhibition of the in vitro metabolic activation of carcinogens by wheat sprout extracts. Nutr. Cancer 1:27-30.
4. Lai, C.N., M.A. Butler, and T.S. Matney (1980) Antimutagenic activities of common vegetables and their chlorophyll content. Mutat. Res. 77:245-250.

ANTIOXIDANTS/ANTIMUTAGENS IN FOODS

Mitsuo Namiki and Toshihiko Osawa

Department of Food Science and Technology
Nagoya University
Chikusa, Nagoya 464, Japan

INTRODUCTION

Oxygen is indispensable for aerobic organisms including, of course, human beings, but it is believed that oxygen also may be responsible for undesired phenomena. In particular, oxygen species such as hydrogen peroxide, superoxide radical anion, and singlet oxygen are proposed as agents attacking polyunsaturated fatty acids in cell membranes, giving rise to lipid peroxidation. Several reports have suggested that lipid peroxidation may result in destabilization and disintegration of cell membranes, leading to liver injury and other diseases, and finally, to aging and susceptibility to cancer (4).

Recently, much attention has been focused on studies of cellular defenses against damage caused by oxygen radicals: such systems include enzymatic inactivation, such as superoxide dismutase, glutathione (GSH)-peroxidase and catalase, as well as nonenzymatic protection of polyunsaturated fatty acids by endogenous antioxidants like α-tocopherol, ascorbic acid, β-carotene, and uric acid (1).

In addition, several antioxidants have been reported to reduce tumor incidence in animals. For example, α-tocopherol, ascorbic acid, and selenium reduced skin tumors induced by 7,12-dimethylbenzo(α)anthracene (DMBA)-croton oil (13). The synthetic antioxidants, such as butylated hydroxyanisole (BHA) and butylated hydroxytoluene (BHT), were also reported to inhibit the carcinogenicity of DMBA and benzo(α)pyrene (BP) (14,16). These positive results suggested that some antioxidants may possibly play an important role in prevention of carcinogenesis related to active oxygen radicals. There are also indications that dietary antioxidants in foods may play an important role in protecting the cell against damage caused by oxygen radicals (15).

From these bases, we have undertaken studies on natural antioxidants. Our studies involve isolation and identification of antioxidants from plant materials, especially those used as foods. We have investigated their activities in inhibiting lipid peroxidation in cell membrane systems. In

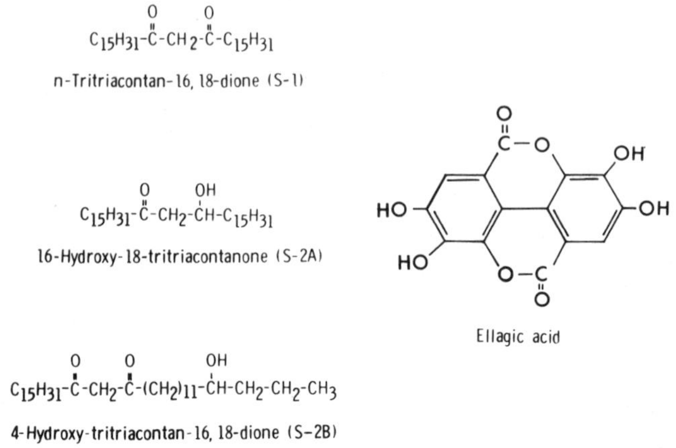

$C_{15}H_{31}-\overset{O}{\overset{\|}{C}}-CH_2-\overset{O}{\overset{\|}{C}}-C_{15}H_{31}$

n-Tritriacontan-16, 18-dione (S-1)

$C_{15}H_{31}-\overset{O}{\overset{\|}{C}}-CH_2-\overset{OH}{\overset{|}{C}H}-C_{15}H_{31}$

16-Hydroxy-18-tritriacontanone (S-2A)

$C_{15}H_{31}-\overset{O}{\overset{\|}{C}}-CH_2-\overset{O}{\overset{\|}{C}}-(CH_2)_{11}-\overset{OH}{\overset{|}{C}H}-CH_2-CH_2-CH_3$

4-Hydroxy-tritriacontan-16, 18-dione (S-2B)

Ellagic acid

Fig. 1. Structures of natural antioxidants isolated from Eucalyptus leaves.

the present study, natural antioxidants isolated from Eucalyptus leaf wax (7,9), sesame seeds, and tea leaves have been examined for inhibitory effects on in vitro lipid peroxidation. The systems used were an erythrocyte membrane ghost and a rat liver microsome with peroxidation induced by ADP-Fe^{++}/NADPH, by ADP-Fe^{+++}/EDTA-Fe^{+++}/NADPH (11), or by adriamycin (5). As several microbiological assay systems for detecting mutagenicity induced by oxygen radicals, hydroperoxides (1), and malonaldehydes (MDAs) (6) formed during lipid peroxidation have been developed in the recent past, preliminary investigations using such microbiological systems to determine antimutagenic effects of the isolated antioxidants are also presented.

NATURAL ANTIOXIDANTS ISOLATED FROM EUCALYPTUS LEAVES

In the course of our study on screening of natural antioxidants from plant materials, we have isolated and identified a novel type of the β-diketone which occurs as antioxidative constituents of Eucalyptus leaf waxes. The β-diketone type of antioxidant has a very strong antioxidative activity in food systems but showed no antioxidant activity in in vitro systems due to its poor solubility. Further investigation of other types of natural antioxidants in polyphenol fractions of Eucalyptus leaf extracts indicated that ellagic acid is one of the main antioxidants (8) (Fig. 1). Ellagic acid is also widely distributed in plant foods, and we decided to evaluate the antioxidative activity of ellagic acid together with other well-known plant polyphenols (shown in Fig. 2) using the in vitro lipid peroxidation systems.

Initial evaluation of the antioxidative activity of plant phenols, including ellagic acid, was carried out by the rabbit erythrocyte membrane ghost system. Lipid peroxidation was induced by t-butylhydroperoxide. Formation of thiobarbituric acid (TBA)-reactive substances was determined spectrophotometrically at 535 nm. As shown in Fig. 3, ellagic acid showed the strongest antioxidative activity among the plant polyphenols, but was still less active than α-tocopherol.

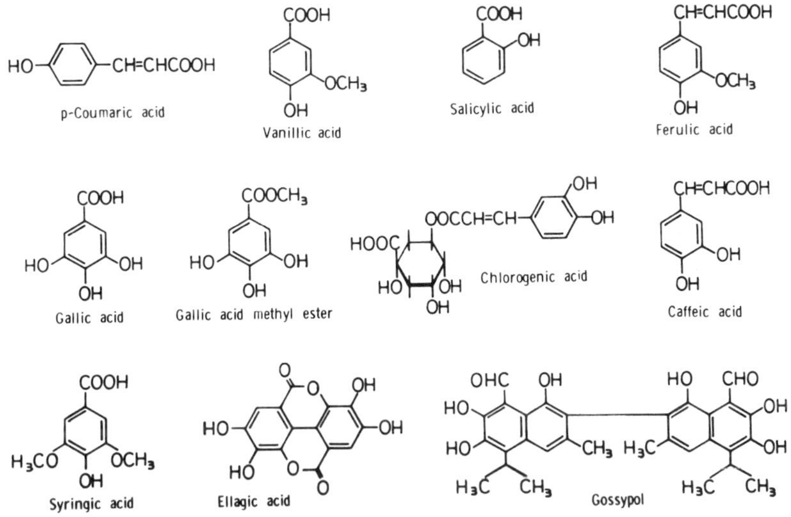

Fig. 2. Structures of plant polyphenols.

In vitro antioxidant activity was also examined using rat liver microsomal lipid peroxidation systems induced by ADP-Fe^{++}/NADPH, by ADP-Fe^{+++}/EDTA-Fe^{+++}/NADPH, or by adriamycin. The determination of NADPH-dependent, in vitro lipid peroxidation has been well studied and is an established method for determining inhibitory activity of antioxidants against radical-induced oxidative destruction of polyunsaturated fatty acids in cell membranes (12). Lipid peroxidation induced by ADP-Fe^{++} complex and NADPH has been considered to have a correlation with the perferryl ion-dependent initiation step, and addition of EDTA-Fe^{+++} may enhance the hydroperoxide-dependent initiation step of microsomal lipid peroxidation.

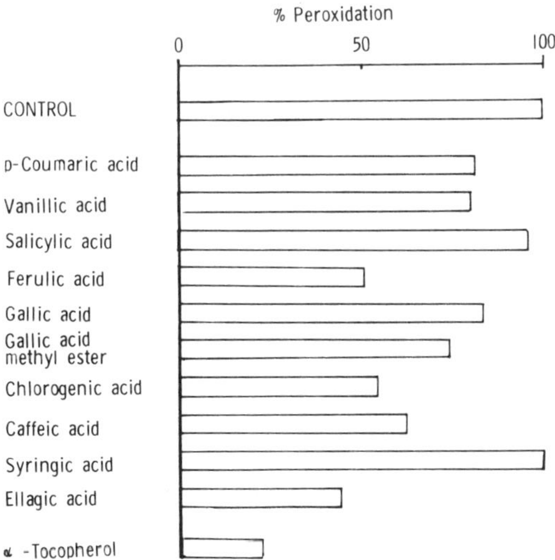

Fig. 3. Antioxidative assay of plant polyphenols by erythrocyte membrane ghost system. (Each sample was prepared as 100 μM final concentration.)

As shown in Tab. 1, ellagic acid inhibited lipid peroxidation induced by both systems, in particular, the perferryl ion-dependent initiation step of lipid peroxidation, and α-tocopherol inhibited the hydroperoxide-dependent initiation step of lipid peroxidation.

In order to confirm the structure-activity relationship, the in vitro antioxidant activity of ellagic acid and two of its derivatives (Fig. 4) were compared. In this connection it is essential to note that adriamycin, a member of the anthracycline group of antineoplastic agents widely used in the medical treatment of carcinoma, has been reported to cause several side effects. Cardiotoxicity, one of the severe side effects of adriamycin, has been reported to be caused by lipid peroxidation induced by oxygen radicals derived from the quinone structure of the antibiotic. If antioxidants inhibit lipid peroxidation induced by adriamycin, they could be of value in combination with chemotherapeutic agents such as adriamycin.

As seen in Tab. 2, ellagic acid strongly inhibited lipid peroxidation induced by adriamycin, but the two ellagic acid derivatives were much less effective. This difference was true of all three systems (Tab. 2).

We also examined, using model systems, the possibility that ellagic acid may scavenge superoxide anion radicals, hydroxy radicals, and singlet oxygen, but found that ellagic acid did not scavenge these oxygen radicals to an appreciable extent. At this stage, the evidence strongly suggests

Tab. 1. Inhibition of lipid peroxidation in the rat liver microsome system by polyphenols.

Compounds	IC_{50}*	
	ADP-Fe^{++}/NADPH	ADP-Fe^{+++}/EDTA-Fe^{+++}/NADPH
Ellagic acid	20	23.0
Hexahydroxydiphenic acid	>100	>100.0
Ellagic acid tetra acetate	>100	>100.0
p-Coumaric acid	330	330.0
Vanillic acid	400	>100.0
Salycilic acid	460	>100.0
Ferulic acid	>500	8.0
Gallic acid	>500	>100.0
Gallic acid methyl ester	>500	67.0
Chlorogenic acid	>500	80.0
Caffeic acid	340	75.0
Syringic acid	>500	>100.0
α-Tocopherol	320	0.8
Butylated hydroxyanisole	220	0.3
Propyl gallate	2,500	20.0

* IC_{50}: Concentration (μM) for 50% inhibition of lipid peroxidation in the model systems.

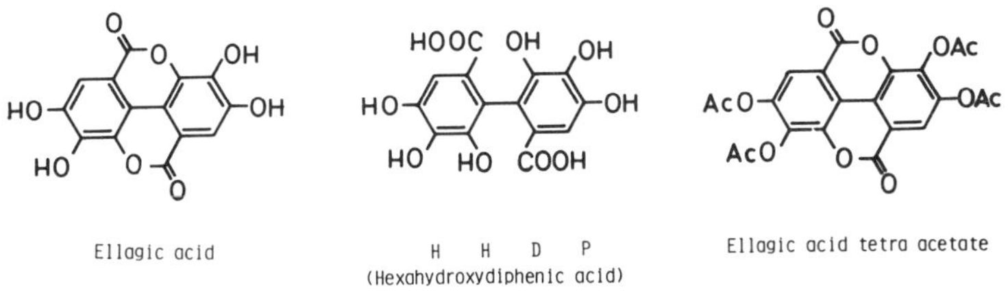

Fig. 4. Structures of ellagic acid and its derivatives.

that ellagic acid may inhibit the perferryl ion-dependent initiation of microsomal lipid peroxidation, but its role in the case of adriamycin-induced lipid peroxidation is not yet clear. However, these results suggest that ellagic acid may be useful not only as a food additive functioning as an antioxidant to inhibit auto-oxidation of fat in foods, but also for medicinal purposes.

ANTIOXIDATIVE GREEN TEA FACTORS

The custom of drinking green tea was established in Japan a long time ago and is still well established. Recently, one of our collaborators separated antioxidative green tea factors for examination as antioxidants in food systems. We have been involved in the determination of in vitro antioxidant activity of these green tea factors for evaluation of the wholesomeness of green tea.

Six different catechins, as shown in Fig. 5, have been isolated and identified. In vitro antioxidant activities of these catechins were also evaluated, using the erythrocyte ghost and rat liver microsome systems. The inhibitory effects on microsomal lipid peroxidation are summarized in Tab. 3. Epicatechin gallate and epigallocatechin gallate strongly inhibited microsomal lipid peroxidation in the ADP-Fe^{+++}/EDTA-Fe^{+++}/NADPH system, and also adriamycin-induced lipid peroxidation. Epigallocatechin exhibited

Tab. 2. Inhibition of lipid peroxidation in the rat liver microsome system by ellagic acid, 2 derivatives, and some other antioxidants.

Compounds	IC_{50}*		
	ADP-Fe^{++}/NADPH	ADP-Fe^{+++}/EDTA-Fe^{+++}/NADPH	Adriamycin
α-Tocopherol	320	0.8	5.5
Butylated hydroxyanisole	220	0.3	0.1
Propyl gallate	2,500	20.0	0.3
Ellagic acid	20	23.0	0.1
Hexahydroxydiphenic acid	>100	>100.0	7.0
Ellagic acid tetra acetate	>100	>100.0	31.0

*IC_{50}: Concentration (μM) for 50% inhibition of lipid peroxidation in the model systems.

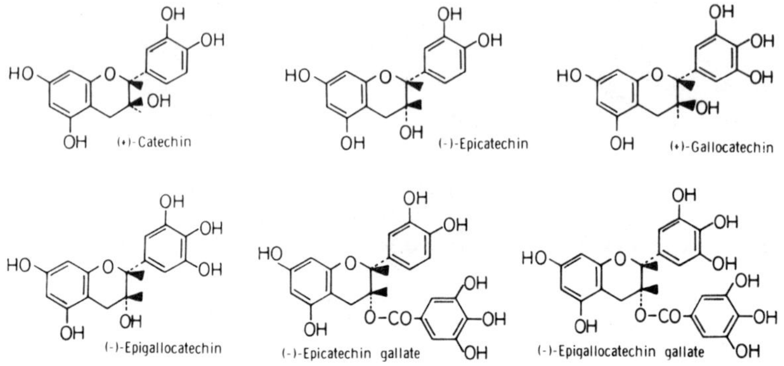

Fig. 5. Structures of catechins isolated from tea leaves.

a strong antioxidative activity against microsomal lipid peroxidation induced by the ADP-Fe^{++} complex and NADPH. Epicatechin gallate and epigallocatechin gallate also showed a strong inhibitory effect on lipid peroxidation in the erythrocyte ghost system (Fig. 6).

It is also very important to evaluate the role of these dietary antioxidants in the mechanisms of the antioxidation process. Consequently, we decided to confirm the role of dietary antioxidants in relation to superoxide anion radicals, singlet oxygen, and hydroxy radicals.

Effects of singlet oxygen formed by photochemical reactions using rose-bengal as the sensitizer and of hydroxy radicals produced by γ-irradiation were determined by measuring decomposition of dietary antioxidants

Tab. 3. Inhibition of lipid peroxidation in the rat liver microsome system by flavonoids: Comparison with some other antioxidants.

Compounds	IC_{50}*		
	ADP-Fe^{++}/NADPH	ADP-Fe^{+++}/EDTA-Fe^{+++}/NADPH	Adriamycin
α-Tocopherol	320	0.8	5.5
Butylated hydroxyanisole	220	0.3	0.1
Propyl gallate	2,500	20.0	0.3
Catechin	>500	20.0	1.6
Epicatechin	460	13.0	0.7
Gallocatechin	110	13.0	0.5
Epigallocatechin	70	15.0	1.7
Epicatechin gallate	>500	3.0	0.4
Epigallocatechin gallate	400	2.0	0.3

*IC_{50}: Concentration (μM) for 50% inhibition of lipid peroxidation in the model systems.

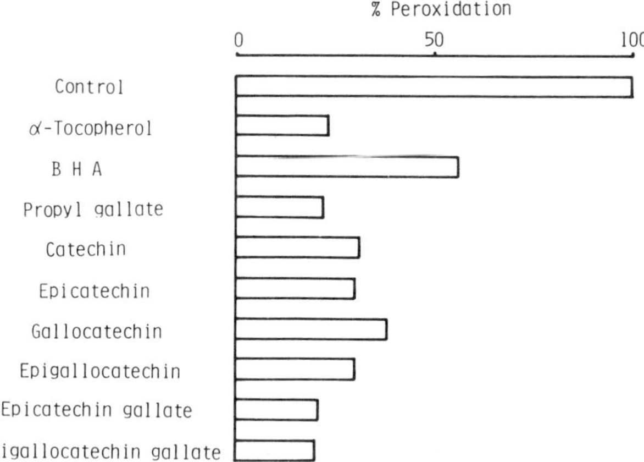

Fig. 6. Antioxidative assay of catechins by erythrocyte membrane ghost system. (Each sample was prepared as 100 μM final concentration.)

such as epicatechin gallate, epigallocatechin, and ellagic acid. However, none of them showed an appreciable effect as scavengers.

Effects on superoxide anion radicals in a KO_2 tetrahydrofuran system have also been studied. As shown in Fig. 7, ellagic acid had no effect as a radical scavenger against superoxide anion radical, but epicatechin gallate and epigallocatechin gallate had a very strong effect as scavengers against superoxide anion radicals.

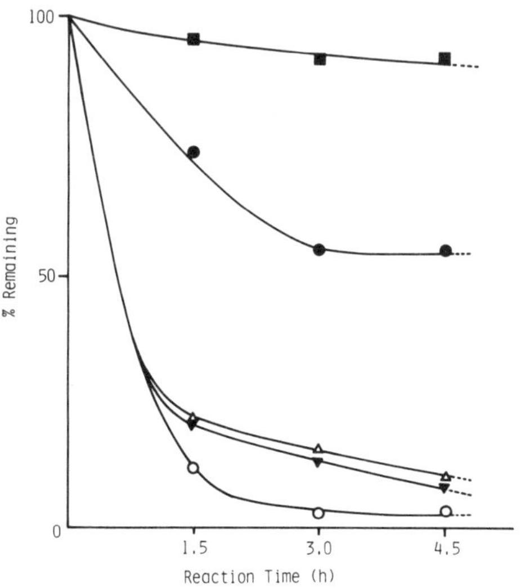

Fig. 7. Decomposition of antioxidants by superoxide anion radicals generated in the KO_2/THF system. (The amounts of residual polyphenols were calculated by reading maximal UV absorption of each compound.) ■, ellagic acid; ●, propyl gallate; Δ, epicatechin gallate; ▼, epigallocatechin gallate; o, α-tocopherol.

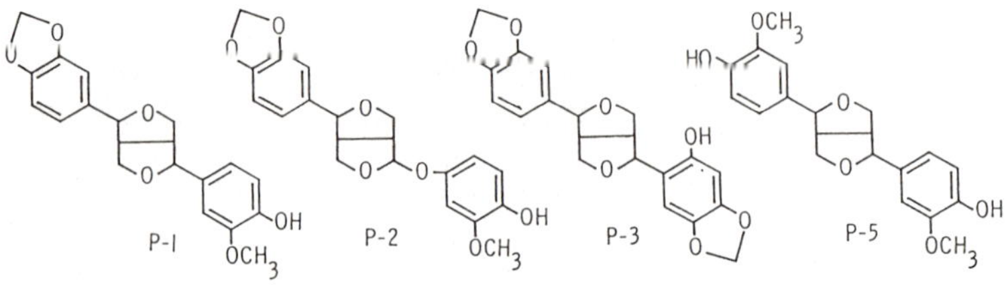

Fig. 8. Structures of antioxidative lignans isolated from sesame seed.

LIGNAN-TYPE ANTIOXIDANTS IN SESAME SEEDS

Sesame seeds have been used traditionally in Japan, China, and other Eastern Asia countries from the viewpoint of wholesomeness. Sesame oils, especially roasted sesame oils, are widely used in Chinese and Japanese dishes and have been evaluated as being highly antioxidative. However, correlations between the chemical constituents, their wholesomeness, and their antioxidative nature have not been clearly interpreted yet, and this prompted us to investigate the antioxidative components in sesame seeds in an attempt to evaluate this relationship.

Sesamol and γ-tocopherol have been reported as being present in sesame seeds, especially in sesame oils. Our data indicated that sesamol is only partly responsible for the antioxidative activity of the oil and that other types of antioxidative substances are probably present in sesame seeds. Therefore, we have made a large-scale isolation and identification of antioxidative substances from defatted sesame seeds. Several antioxidative lignans have been purified as shown in Fig. 8. P-1 and P-3 (sesaminol) were identified as compounds already isolated from Indian plant drugs (3), but P-2 was a new phenolic lignan and was named sesamolinol.

Recently, we completed the structural determination of sesamolinol and P-3 (sesaminol) by X-ray crystallographic analyses (10). The in vitro antioxidative activity of sesame lignans has also been evaluated using the rabbit erythrocyte ghost and the rat liver microsomal systems. As shown in Fig. 9, sesamolinol had a strong inhibitory effect on microsomal lipid

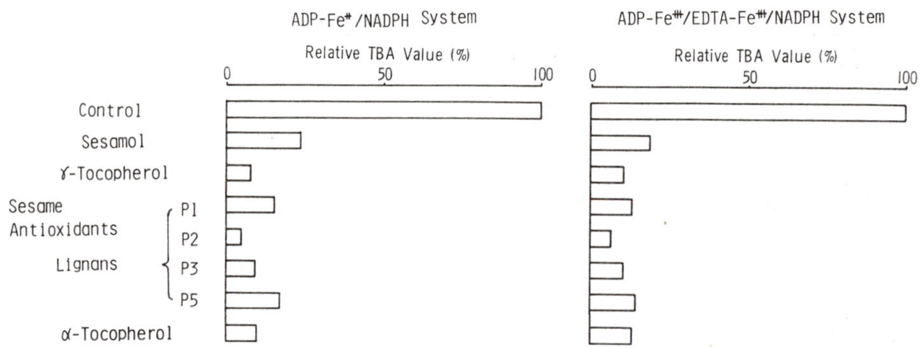

Fig. 9. Antioxidative assay of lignans isolated from sesame seed by rat liver microsome by ADP-Fe^{++}/NADPH and ADP-Fe^{+++}/EDTA-Fe^{+++}/NADPH systems.

peroxidation induced by the ADP-Fe^{++} complex and NADPH, although P-5 (pinoresinol) showed more activity in the erythrocyte ghost system. These lignans are also present as aglycones of glucosides in the defatted sesame seeds and may be responsible for the wholesomeness of sesame seeds. Further chemical investigation of these lignan glucosides is now in progress.

DESMUTAGENIC AND ANTIMUTAGENIC ACTIVITIES OF NATURAL ANTIOXIDANTS

Endogenous and dietary antioxidants have been proposed to have an important role in scavenging oxygen radicals and inhibiting lipid peroxidation in cell membranes and other organs. Moreover, the possible inhibitory effects of these antioxidants on chemical carcinogenesis appear to be a promising method for cancer inhibition.

Recently, several reports have indicated that some endogenous and dietary antioxidants may play an important role in suppressing carcinogen-induced mutations in several bacterial mutagenesis assay strains, including the *Salmonella*.

Also, Ames et al. (2) have recently introduced new *Salmonella* tester strains, such as *Salmonella* TA102 and TA104, which were reverted efficiently by oxygen radicals, in particular, by hydroperoxides. *Salmonella* TA1978 strain has already been reported to have been reverted by MDA, a substance assumed to be the most important degradation product of peroxidized lipid in mutagenicity.

Consequently, we began this project by evaluating the influence of the dietary antioxidants that we have isolated and identified from plant materials on *Salmonella* mutagenesis induced by hydroperoxides and MDA.

Initially, the influence of dietary antioxidants, mainly sesame lignans and ellagic acid, on t-butylhydroperoxide-induced mutagenesis in *Salmonella typhimurium* TA104 strain was determined. As shown in Fig. 10, most of the sesame lignans had no antioxidative activity, but ellagic acid showed antimutagenic activity.

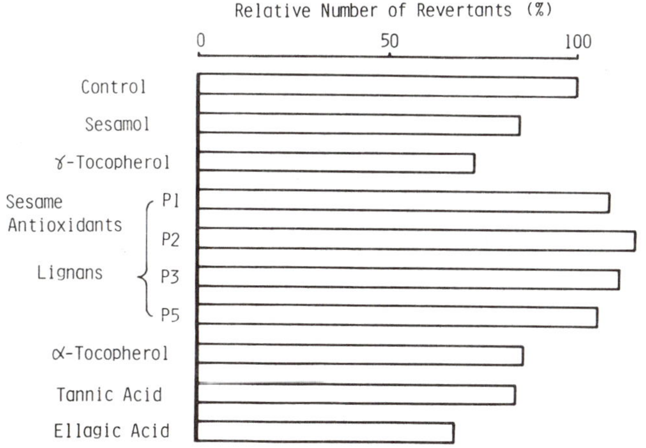

Fig. 10. Antimutagenic assay using *Salmonella typhimurium* TA104.

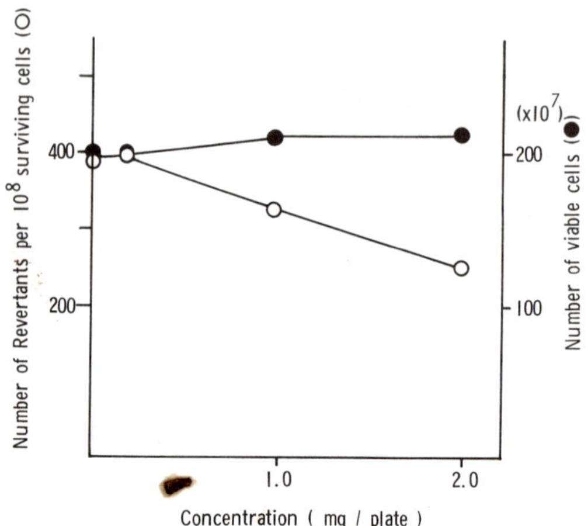

Fig. 11. Effects of ellagic acid on t-butylhydroperoxide-induced mutagenesis in Salmonella typhimurium TA104.

Figure 11 presents the dose-response curve of the effects of ellagic acid on t-butylhydroperoxide-induced mutagenesis and on cellular viability. A significant decrease in cell number was not observed up to the concentration of 2.0 mg/plate, and ellagic acid is not toxic for viable cells below this concentration.

Preliminary experiments using the S. typhimurium TA1978 strain induced by MDA have also been carried out to determine the antimutagenic activity of sesame lignans, especially P-2 (sesamolinol). P-2 was selected for study because it showed no antimutagenic effects on t-butylhydroperoxide-induced mutagenesis, but showed the strongest inhibitory effects against microsomal lipid peroxidation.

In Tab. 4, the data on effects of P-2, ellagic acid, and some known antioxidants on MDA-induced mutagenesis are shown. Ellagic acid also showed the most inhibitory effects on MDA-induced mutagenesis, and P-2

Tab. 4. Effects of dietary antioxidants on malonaldehyde-induced mutagenesis in Salmonella typhimurium TA1978 strain.

Antioxidants	% Inhibition	
	2 mg/plate	1 mg/plate
α-Tocopherol	35	10
γ-Tocopherol	39	36
δ-Tocopherol	45	27
Ascorbic acid	45	38
Glutathione	32	24
Sesamol	53	47
P-2	35	23
Ellagic acid	52	50

showed moderate activity. We have not been able to obtain any strong correlation between antioxidative activity and antimutagenic activity at this time; however, our data, especially on the effects of dietary antioxidants on bacterial mutagenesis, are preliminary. We believe that it is important to develop new in vitro biological systems to evaluate antimutagenicity of dietary antioxidants.

ACKNOWLEDGEMENTS

The authors wish to thank Professor Y. Fukuda, Mr. M. Nagata, and Mr. A. Ide for their cooperation. This work was supported by a Cancer Research Grant and a Scientific Research Grant from the Ministry of Education, and by the Research Fund from the Nissan Science Foundation.

REFERENCES

1. Ames, B.N., R. Cathcart, E. Schwiers, and P. Hochstein (1981) Uric acid provides an antioxidant defense in humans against oxidant- and radical-caused aging and cancer: A hypothesis. Proc. Natl. Acad. Sci., USA 78(11):6858-6862.
2. Levin, D.E., M. Hollstein, M.F. Christman, E.A. Schwiers, and B.N. Ames (1982) A new *Salmonella* tester strain (TA 102) with A•T base pairs at the site of mutation detects oxidative mutagens. Proc. Natl. Acad. Sci., USA 79:7445-7449.
3. Fukuda, Y., T. Osawa, M. Namiki, and T. Ozaki (1985) Studies on antioxidative substances in sesame seed. Agric. Biol. Chem. 49(2):301-306.
4. McBrien, D.C.H., and T.F. Slater (1982) *Free Radicals, Peroxidation and Cancer*, Academic Press, New York.
5. Mimnaugh, E.G., T.E. Gram, and M.A. Trush (1983) Stimulation of mouse heart and liver microsomal lipid peroxidation by anthracycline anticancer drugs: Characterization and effects of reactive oxygen scavengers. J. Pharm. Exper. Therapeu. 226(3):806-816.
6. Mukai, F.H., and Goldstein (1976) Mutagenicity of malonaldehyde, a decomposition product of peroxidized polyunsaturated fatty acids. Science 191:868-869.
7. Osawa, T., and M. Namiki (1981) A novel type of antioxidant isolated from leaf wax of eucalyptus leaves. Agric. Biol. Chem. 45(3):735-739.
8. Osawa, T., and M. Namiki (1983) Natural antioxidants isolated from plant leaf waxes. In *Research in Food Science and Nutrition*, Vol. 3, J.V. McLoughlin and B.M. McKenna, eds. Boole Press, Dublin, pp. 49-50.
9. Osawa, T., and M. Namiki (1985) Natural antioxidants isolated from eucalyptus leaf waxes. J. Agric. Food Chem. 33(5):77-80.
10. Osawa, T., M. Nagata, M. Namiki, and Y. Fukuda (1985) Sesamolinol, a novel antioxidant isolated from sesame seeds. Agric. Biol. Chem. 49(11):3351-3352.
11. Pederson, T.C., J.A. Buege, and S.D. Aust (1973) Microsomal electron transport. J. Biol. Chem. 248(20):7134-7141.
12. Pryor, W.A., ed. (1980) *Free Radicals in Biology*, Vol. IV, Academic Press, New York.
13. Shamberger, R.J., and G. Rudolph (1966) Protection against co-carcinogenesis by antioxidants. Experientia 22:116.
14. Shamberger, R.J. (1970) Relationship of selenium to cancer. I. Inhibitory effect of selenium on carcinogenesis. J. Natl. Cancer Inst. 44:931-936.

15. Simic, G.M., and M. Karel, eds. (1980) Autoxidation in Food and Biological Systems, Plenum Press, New York and London.
16. Wattenberg, L.W. (1972) Inhibition of carcinogenic and toxic effects of polycyclic hydrocarbons by phenolic antioxidants and ethoxyquin. J. Natl. Cancer Inst. 48:1425-1430.

ROLE OF ENZYMES IN ANTIMUTAGENESIS OF HUMAN SALIVA AND SERUM

H. Nishioka and T. Nunoshiba

Department of Biochemistry
Doshisha University
Kyoto, Japan 602

INTRODUCTION

In order to evaluate the risk of dietary environmental mutagens and/or carcinogens to man and to search for protection from them, it is important to study possible modifications in every part of the digestive tract.

In 1981, the author and colleagues, using bacterial assay systems, reported (7) that human saliva excreted from the salivary gland inactivates the mutagenicity of various types of carcinogens and suspected carcinogens which may possibly be contained in our daily foods. These include aflatoxins, benzo(α)pyrene, 2-(2-furyl)-3-(5-nitro-2-furyl)acrylamide, hydrogen peroxide, amino acid pyrolysates, the pyrolysates of beef and salmon, etc. This finding suggests that the habit of chewing food well may be important to prevent cancer.

In 1983, we found (5) that human serum also has a similar capacity. Our successive studies with saliva and serum obtained from a number of donors has made clear that these inactivating capacities are different among individuals and that the individual variation seems to relate to the activity of enzymes, such as peroxidase, catalase, superoxide dismutase, etc., in saliva and serum. It is hypothesized, therefore, that these enzymes play an important role in the mechanism of inactivation. In order to confirm this hypothesis, the activities of these enzymes in saliva and serum were determined by means of microanalytical methods using a sensitive oxygen electrode, and the relation between the enzymatic activity in individual samples of saliva and serum and their inactivating capacities has been studied. In our present report, the experimental results obtained up to now are reviewed.

MATERIALS AND METHODS

Bacterial Tester Strains

The bacterial strains used in these experiments were *Salmonella typhimurium* TA98 and TA100 derived by Ames (1) for detecting possible environmental carcinogens as mutagens. (These strains were kindly donated by Dr. B.N. Ames, University of California, Berkeley.) TA98 is an indicator strain for frameshift-type mutagens and TA100, an indicator strain for base pair substitution. Overnight cultures of bacteria in nutrient broth were used for mutagenicity assays.

Collection of Saliva and Serum

Whole human saliva was obtained individually from healthy donors (24 males and 16 females) of various ages who expectorated after stimulation with the odor of pickled Japanese apricot just before lunch. Approximately 2.0 ml of saliva from each donor was collected during a 5-min period. Immediately after collection, saliva was used to treat the mutagen samples tested.

Human serum was collected individually from healthy donors (6 males and 2 females) and cancer patients (5 males and 4 females) of various ages. All healthy donors and cancer patients were Japanese. (A cancerous portion of each patient is described in Fig. 5.) To avoid hemolysis, 2 ml of blood taken from a vein in the arm of each patient by syringe was allowed to stand for 30 min and then centrifuged at 3,000 r/min for 10 min at 25°C. The supernatant, which was almost transparent, was used in the experiment within 1 hr after collection. The other experimental procedures have been described elsewhere (7).

Mutagen Samples

The mutagen samples used in the experiment were 2-(2-furyl)-3-(5-nitro-2-furyl)acrylamide (AF-2), a nitrofuran compound used as a food additive from 1966 to 1975 in Japan (8), N-methyl-N'-nitro-N-nitrosoguanidine (MNNG), methyl methanesulfonate (MMS), 4-nitroquinoline 1-oxide (4NQO), quercetin, aflatoxin B_1, benzo(α)pyrene (BP), tryptophan pyrolysate-1 (Trp-P-1) (3) and glutamic acid pyrolysate-1 (Glu-P-1) (4), the pyrolysates of beef (2), salmon and sodium glutamate, and the condensate of cigarette smoke (9) with or without metabolic activation.

These samples were dissolved in dimethylsulfoxide (DMSO) at the concentrations indicated in the figures. For preparing the samples of pyrolysates, 10 g of fresh beef or salmon and sodium glutamate were broiled in an oven at 300°C for 10 min. Then they were suspended in DMSO (2 ml/g) with vigorous stirring for 30 min at 25°C. The suspensions were centrifuged at 10,000 r/min for 5 min at 25°C and the supernatants were subjected to the mutagenicity assay. Cigarette smoke was trapped in a commercial filter (Aquafilter®) containing approximately 1 ml of 1% ethanol during the smoking of Japanese cigarettes (Seven Stars®) with this filter. A cigarette was smoked, for about 4 min, to two-thirds butt length. After smoking 20 cigarettes, the sample of cigarette smoke condensate was obtained by squeezing out the filter. (All chemicals were purchased from Nakarai Chemicals Company, Kyoto, Japan unless stated otherwise.)

Treatment with Saliva or Serum

First, 1.8 ml of saliva or serum was mixed with 0.2 ml of each test material at the concentration indicated in Fig. 1 and incubated for 5 min at 37°C with gentle shaking. Then, 0.1-ml samples, with or without treatment with saliva or serum, were assayed for mutagenicity.

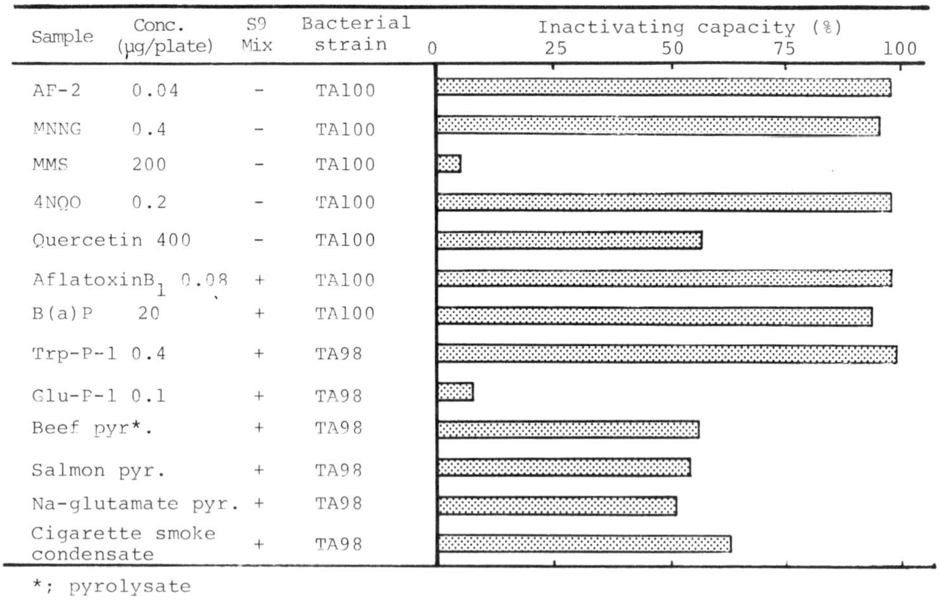

Sample	Conc. (µg/plate)	S9 Mix	Bacterial strain
AF-2	0.04	-	TA100
MNNG	0.4	-	TA100
MMS	200	-	TA100
4NQO	0.2	-	TA100
Quercetin	400	-	TA100
AflatoxinB_1	0.08	+	TA100
B(a)P	20	+	TA100
Trp-P-1	0.4	+	TA98
Glu-P-1	0.1	+	TA98
Beef pyr*.		+	TA98
Salmon pyr.		+	TA98
Na-glutamate pyr.		+	TA98
Cigarette smoke condensate		+	TA98

*; pyrolysate

Fig. 1. Inactivating capacity (%) of human saliva (from donor No. 20 indicated in Fig. 3) for mutagenicity of the mutagen samples.

Mutagenicity Assay

The mutagenicity assay was carried out by the method of Ames et al. (1) with some modifications. For metabolic activation, we used 0.5 ml of an S-9 mix derived from the livers of Sprague-Dawley rats weighing 180 to 200 g pretreated with phenobarbital and 5,6-benzoflavone for enzyme induction. The S-9 mix contained reduced nicotinamide adenine dinucleotide (NADH), nicotinamide adenine dinucleotide phosphate (NADPH), ATP, and cofactors. Samples of 0.1 ml were preincubated with 0.1 ml of the suspension of bacterial tester strain and 0.5 ml of Na-P buffer (pH 7.4) or 0.5 ml of S-9 mix (in the case of metabolic activation) for 20 min at 37°C with gentle shaking. They were then poured, with 2 ml of soft agar, into a plate of minimal-glucose agar containing ampicillin (20 µg/ml) in the case of the saliva experiment in order to inhibit the growth of the bacteria contaminating the saliva. There was no significant effect of ampicillin (at the concentration used) on survival and mutation of Salmonella strains used. Colonies of histidine prototrophs in triplicate plates were counted after incubation for 2 days.

Calculation of Inactivating Capacity

Inactivating capacity (%) was obtained by calculation as follows:

$$\text{Inactivating capacity (\%)} = \frac{A \quad B}{A}$$

Here, A is the number of revertant colonies of the cells treated with the sample without saliva or serum, and B is the number of those with saliva or serum. Numbers of spontaneous mutations were previously subtracted from both.

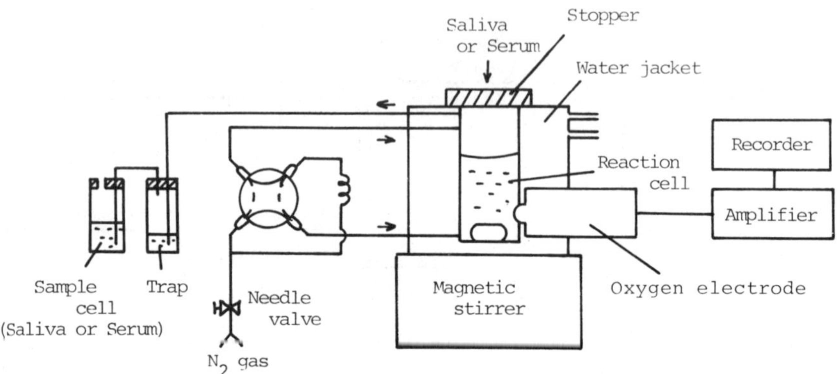

Fig. 2. Scheme of the Oritector Model III for the determination of peroxidase and catalase activities in individual samples of human saliva and serum.

Measurement of Peroxidase and Catalase Activity

The activity of peroxidase and catalase in saliva and serum was determined by the microanalytical method using Oritector Model III (11) (Oriental Electric Company, Ltd., Niiza, Japan) with a sensitive oxygen electrode. The scheme of the apparatus is shown in Fig. 2. This apparatus was originally developed for the determination of a small quantity of hydrogen peroxide in daily foods, such as heat-sterilized milk. Applying this method, the activity of the enzymes in saliva and serum was measured. Then 2 ml of 10-ppm H_2O_2 and the saliva or serum sample were put into the reaction cell and the sample cell, respectively (Fig. 2). The bubbling of N_2 gas was continued until most of the dissolved oxygen was removed from the analytical system. Then, 10 μl of saliva or serum was injected into the reaction cell, and the concentration of the generated oxygen due to H_2O_2-decomposition catalyzed by peroxidase or catalase in saliva or serum was determined with the sensitive oxygen electrode.

RESULTS AND DISCUSSION

Inactivating capacity (%) of saliva for mutagenicity of the samples is shown in Fig. 1. The saliva used in the experiment was obtained from donor No. 20 (30 year-old, male) indicated in Fig. 3. It is clear that human saliva inactivates mutagenicity of AF-2, MNNG, 4NQO, aflatoxin B_1, BP, and Trp-P-1 by nearly 100%.

Mutagenicities of quercetin, the pyrolysates of beef, salmon and sodium glutamate, and the condensate of cigarette smoke were reduced to approximately 50%. However, no significant inactivation was seen for MMS and Glu-P-1.

It has been observed (7) that (a) the inactivation was dependent on the temperature during treatment, (b) the diluted saliva showed lower capacity, (c) no inactivating capacity was seen in the boiled saliva, and (d) the filtrate of saliva filtered by millipore (0.45 μm) still had inactivating capacity. These observations suggest that certain components in saliva may be involved in the inactivation.

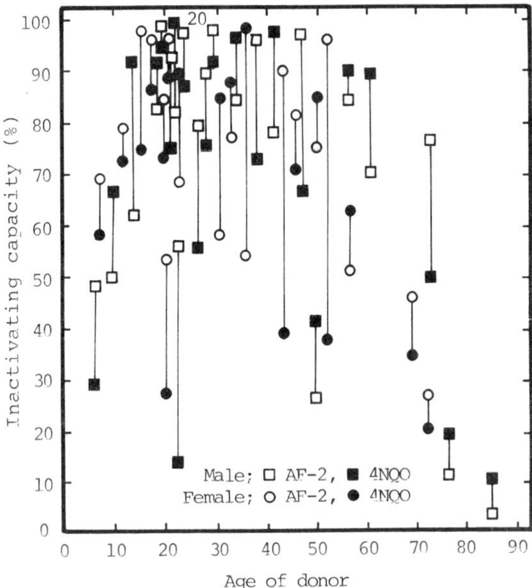

Fig. 3. Relation between the ages of the donors and the inactivating capacity (%) of individual saliva for mutagenicity of AF-2 (0.04 μg/plate) and 4NQO (0.2 μg/plate) in S. typhimurium TA100.

The inactivating capacities (%) in the individual saliva samples collected from 40 donors were compared. The relationship between the age of the donors and the inactivating capacity of the individual saliva from those donors for AF-2 and 4NQO mutagenicity is shown in Fig. 3. It is obvious that there is a considerable difference in the inactivating capacity of saliva among individuals. Generally, it seems that the majority of donors with low capacity saliva is observed among the very young (younger than 10 years old) or among those of advanced age (older than 55 years old), though there are some exceptions.

It is also clear that human serum inactivates the mutagenicity of various types of mutagens and/or carcinogens, such as AF-2, 4NQO, and BP, to a great extent as shown in Fig. 4. The inactivating capacity for quercetin

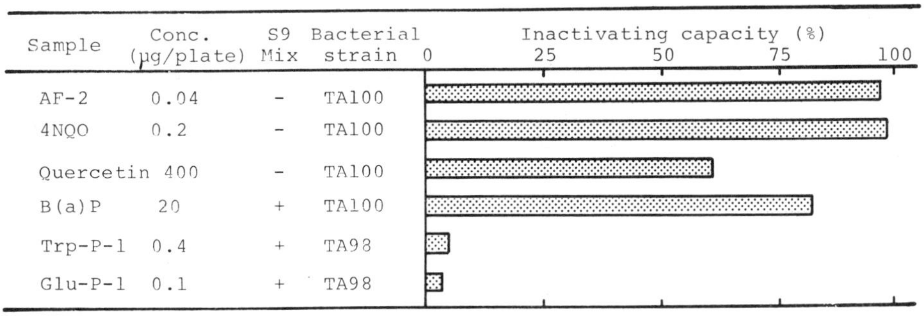

Sample	Conc. (μg/plate)	S9 Mix	Bacterial strain	Inactivating capacity (%) 0 25 50 75 100
AF-2	0.04	-	TA100	
4NQO	0.2	-	TA100	
Quercetin	400	-	TA100	
B(a)P	20	+	TA100	
Trp-P-1	0.4	+	TA98	
Glu-P-1	0.1	+	TA98	

Fig. 4. Inactivating capacity (%) of human serum (from donor N8 indicated in Fig. 5) for mutagenicity of the mutagen samples.

was approximately 60%. However, no significant inactivating capacity of serum for two pyrolysates, Trp-P-1 and Glu-P-1, was seen. It is interesting that a difference was seen between saliva and serum in regard to their inactivating capacities for Trp-P-1. It is possible to assume that the difference is due to the different components, such as enzymes, vitamins, etc., in saliva and serum.

In the experiment on individual serum, the serum samples were obtained from 8 healthy donors and 9 cancer patients as shown in Fig. 5. Serum from the healthy donors showed very high inactivating capacity for AF-2 and 4NQO without exception. It may be important to collect data from more donors in order to know whether there is any healthy person with a lower capacity in his serum.

On the other hand, there were distinct variations in the inactivating capacity of the serum from the cancer patients. Many had a lower inactivating capacity by nature. Since those patients were comparably older than healthy donors, had reached the last stage, and were under strong medication, many of their vital functions may have been in a weakened state. Therefore, it may still be impossible to determine whether a certain person is prone to cancer as a result of the lack of inactivating capacity of his saliva or serum for mutagenicity of carcinogens and suspected carcinogens.

Generally, for the mechanism of inactivation of mutagens, the following reactions can be considered:

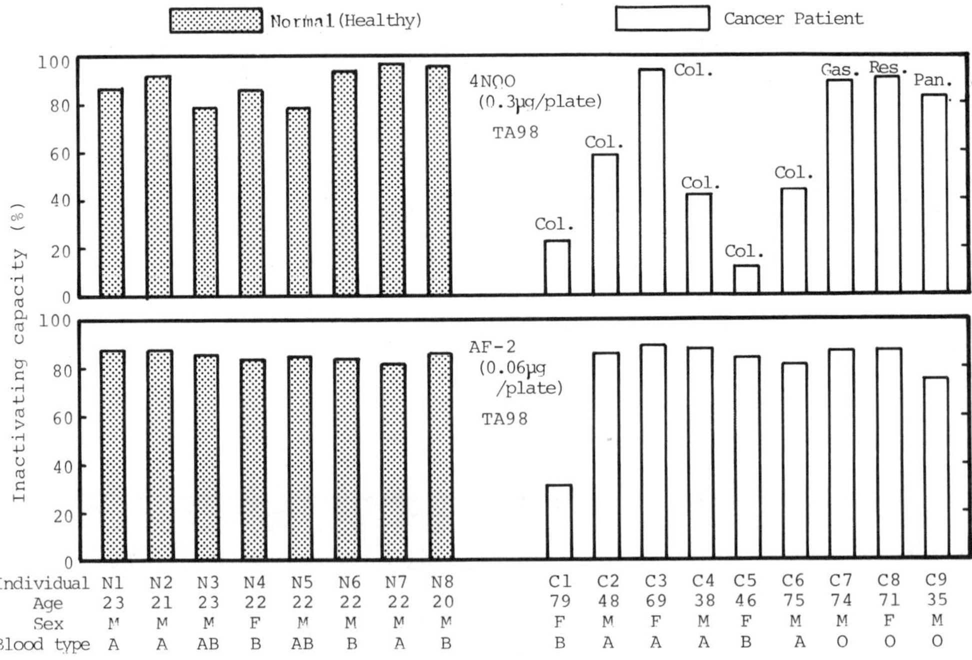

Fig. 5. Comparison of inactivating capacity (%) of individual samples of serum from healthy donors and cancer patients for mutagenicity of 4NQO (0.3 μg/plate) and AF-2 (0.06 μg/plate) in Salmonella typhimurium TA98.

I. Direct Inactivation

a. Chemical reaction.
b. Enzymatic reaction.
c. Physical adsorption.

II. Indirect Inactivation

a. Inhibition of mutagen formation from precursor.
b. Inhibition of metabolic activation.

Furthermore, according to Shannon et al. (10), it has been confirmed that both human saliva and serum contain a large number of enzymes, vitamins, hormones, other organic and inorganic compounds, etc. Among these components, especially, the enzymes, such as peroxidase, catalase and superoxide dismutase, seem to play a role in the mechanism of inactivation, as evidenced by our previous studies (5,6).

To confirm this possibility, the inactivating capacities of peroxidase and catalase for the various mutagens were determined. As shown in Fig. 6, it is quite clear that peroxidase almost completely inactivates the mutagenicity of the pyrolysate mutagens, such as Trp-P-1, Trp-P-2, Glu-P-1, and Glu-P-2.

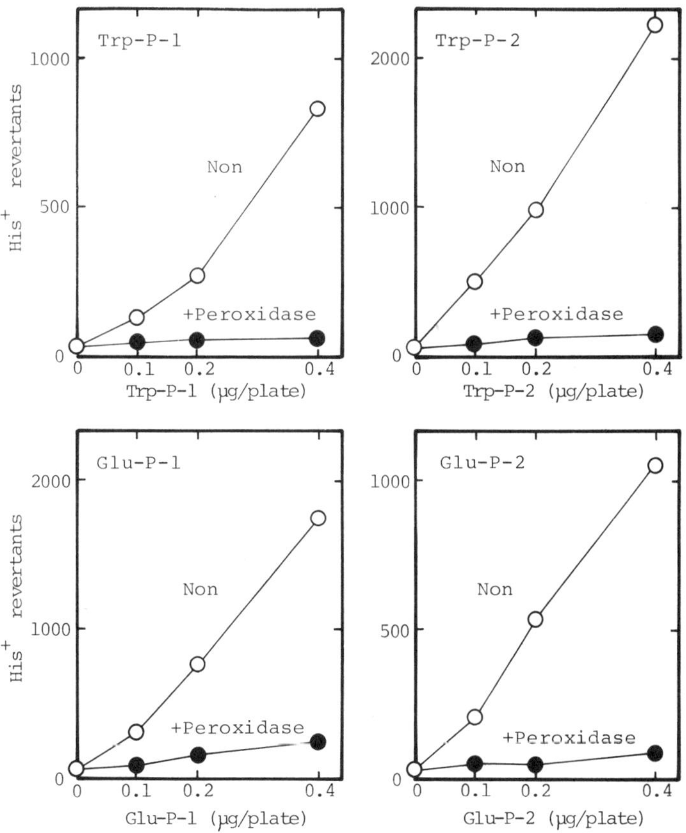

Fig. 6. Effect of peroxidase on mutagenicity of amino acid pyrolysates.

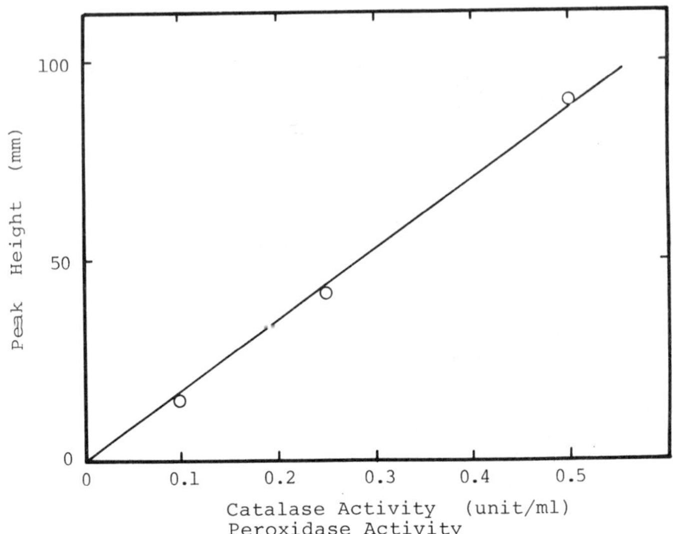

Fig. 7. Standard curve for the determination of peroxidase and catalase activities in individual samples of human saliva and serum.

In general, the catalyzation of peroxidase requires hydrogen peroxide as an oxidizing agent. In our experiment, however, hydrogen peroxide was not added into the inactivation test system. This fact may suggest that hydrogen peroxide is produced by metabolic activation in bacterial cells or that some different functions of peroxidase, besides oxidative catalyzation, are involved in the mechanism of the inactivation.

On the contrary, no effect of catalase on these pyrolysates was seen. It has been observed (6), however, that the mutagenicities of AF-2 and 4NQO are inactivated by the enzyme to some extent. These results support the hypothesis that the difference in inactivating capacities between saliva and serum depends on the different kinds of enzymes obtained.

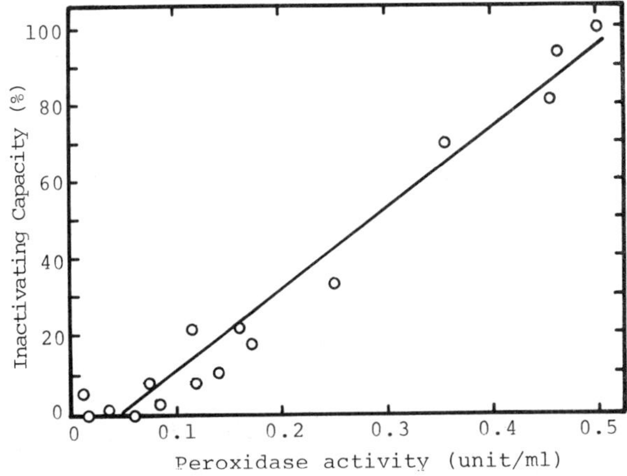

Fig. 8. Relation between peroxidase activity of individual samples of saliva and inactivating capacity for Trp-P-1 mutagenicity.

In order to clarify the role of peroxidase and catalase in the inactivation of mutagens, the activities of these enzymes in individual saliva and serum samples were determined. For this purpose, the microanalytical method with a sensitive oxygen electrode (as described above; see Fig. 2) was adopted. Using this technique, it was possible to sensitively determine the activity of peroxidase and catalase in individual samples of human saliva and serum. The standard curves are shown in Fig. 7.

The correlation between the activity of these enzymes and the inactivating capacity of saliva for Trp-P-1 and of serum for AF-2 was investigated. As shown in Fig. 8, the clear correlation (correlation factor 95%) between peroxidase activity and the inactivating capacity for Trp-P-1 was observed in saliva. In serum, however, such a relationship was not seen (data not shown). From these results, it seems that most of the inactivation can be explained by the activity of peroxidase in the case of saliva. On the other hand, the inactivation by serum may be due to some other mechanisms excluding a role of catalase. Further study is required to elucidate the whole mechanism of inactivation.

REFERENCES

1. Ames, B.N., J. McCann, and E. Yamasaki (1975) Methods for detecting carcinogens and mutagens with the Salmonella/mammalian-microsome mutagenicity test. Mutat. Res. 31:347.
2. Nagao, M., M. Honda, Y. Seino, T. Yahagi, and T. Sugimura (1977) Mutagenicities of smoke condensates and the charred surface of fish and meat. Cancer Lett. 2:221.
3. Nagao, M., M. Honda, Y. Seino, T. Yahagi, T. Kawachi, and T. Sugimura (1977) Mutagenicities of protein pyrolysates. Cancer Lett. 2:334.
4. Nebert, D.W., S.W. Bigelow, A.B. Okey, T. Yahagi, Y. Mori, M. Nagao, and T. Sugimura (1979) Pyrolysis products from amino acids and protein: Highest mutagenicity requires cytochrome P_1-450. Proc. Natl. Acad. Sci., USA 76:5929.
5. Nishioka, H. (1983) Antimutagenic activity of human saliva and serum. In Carcinogens and Mutagens in the Environment. Vol. 2. Naturally Occurring Compounds, CRC Press, Inc., Boca Raton, Florida, pp. 77-83.
6. Nishioka, H., and K. Nishi (1981) Effect of human saliva and its components on mutagenicity of carcinogens. Sci. Eng. Rev. Doshisha Univ. 22:62.
7. Nishioka, H., K. Nishi, and K. Kyokane (1981) Human saliva inactivates mutagenicity of carcinogens. Mutat. Res. 85:323.
8. Odashima, S., and Y. Hashimoto (1970) Carcinogenicity and target organs of methoxyl derivatives of 4 aminoazo-1 benzene in rats. Gann 61:153.
9. Sato, S., Y. Seino, T. Ohka, T. Yahagi, M. Nagao, T. Matsushima, and T. Sugimura (1977) Mutagenicity of smoke condensates from cigarettes, cigars and pipe tobacco. Cancer Lett. 3:1.
10. Shannon, I.L., R.P. Suddick, and F.J. Dowd, Jr. (1974) Saliva: Composition and Secretion (Monograph in Oral Science), S. Karger, Basel, p. 2.
11. Toyoda, M., Y. Ito, M. Iwaida, and M. Fujii (1982) Rapid procedure for the determination of minute quantities of residual hydrogen peroxide in food by using a sensitive oxygen electrode. J. Agric. Food Chem. 30: 346-349.

ISOLATION AND IDENTIFICATION OF HIGHER PLANT AGENTS ACTIVE IN ANTIMUTAGENIC ASSAY SYSTEMS: GLYCYRRHIZA GLABRA

Lester A. Mitscher,[1] Steven Drake,[1] Sitaraghav Rao Gollapudi,[1] Jane A. Harris,[2] and Delbert M. Shankel[2]

[1]Department of Medicinal Chemistry
[2]Department of Microbiology
University of Kansas
Lawrence, Kansas 66045

The reproduction and maintenance of identity of species are biological imperatives, and it is not surprising that there exist mechanisms to minimize or repair the deleterious influence of noxious chemicals in the environment on DNA. Animals tend to defend themselves through the use of enzymes which intercept aggressive chemicals and convert them to less dangerous substances. Animal cells also contain a variety of preformed smaller molecular weight chemicals which can react with oxidized species and free radicals and convert them to less virulent electrophiles (1). It is now becoming clear that higher plants also contain a variety of preformed secondary metabolites which represent a structurally diverse array of antimutagenic and desmutagenic compounds (6). Many, but not all, would appear to be enzyme inhibitors or antioxidants. Since a number of plant constituents are mutagenic (18), it seems reasonable that higher plants should also contain molecules capable of antimutagenicity so as to survive the effects of their own metabolism. Study of such substances has the potential of revealing much interesting molecular detail about the processes of mutagenesis and antimutagenesis. A rather more distant hope is that such substances might be safe enough to provide protection for individuals perceived to be at risk. This would appear to be the case with a number of minor anticarcinogenic constituents consumed as part of our diet (4,20).

Using short-term mutagenicity assays modified to include a mutagenic insult (UV, free radicals, alkylating agents, oxidants, etc.), a number of studies, several of which are described elsewhere in this Volume, have implicated extracts from various higher plants as antimutagenic or desmutagenic agents (8,10,14,15,17). Since not all plant extracts possess activity in such screens, and since those that are active are not active against all mutagens, it seems unlikely that there is a single universal mechanism at work (6).

Those studies that have led to the isolation and structural characterization of responsible higher plant agents show them to be structurally diverse. A few examples are given in Fig. 1. These include cinnamaldehyde, protoanemonin, hypoglycin, coumarins, furanocoumarins, quinacrine, a number

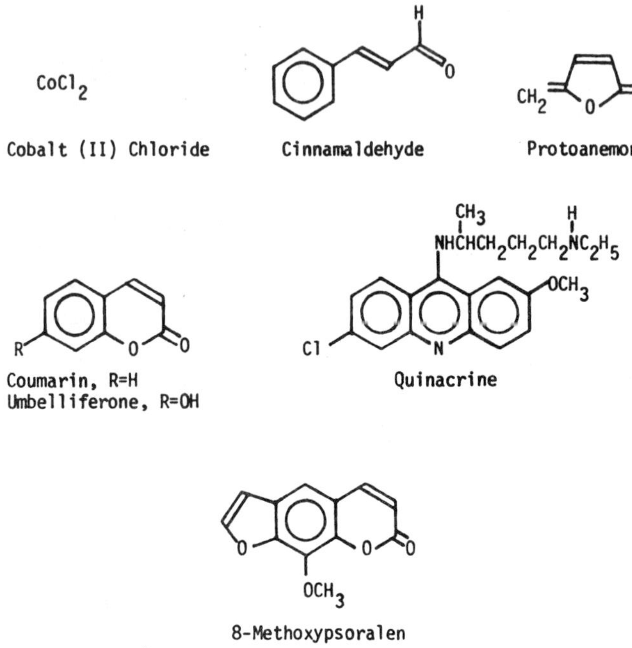

CoCl$_2$

Cobalt (II) Chloride

Cinnamaldehyde

Protoanemonin

Coumarin, R=H
Umbelliferone, R=OH

Quinacrine

8-Methoxypsoralen

Fig. 1. Some known antimutagens.

of plant phenolics and peptides, etc. (3). These studies suggest that such agents are relatively readily obtainable provided that suitable screening procedures are developed for their detection and evaluation.

In our laboratory we have utilized the simple modified Ames test (12) for this purpose. A screen has, by definition, holes in it, and the necessary limiting assumptions which have to be made in order to get started dictate whether one will be successful or not and to a major extent dictate the kinds of compounds that will be found. Generally speaking, there are about eight characteristics that an optimal screen will possess. These are listed in Tab. 1. Whether or not the elements are meaningfully related to the phenomenon in question can only be guessed at in the beginning. The screen must run a few cycles and discrete compounds be produced before

Tab. 1. Some important characteristics of an ideal screening assay.

- Meaningfully related to phenomenon in question.
- Tolerant of impurities.
- Can be performed locally.
- Sample-sparing.
- Rapid turnaround.
- Robust.
- Quantitative and precise.
- Simple to interpret.

questions relating to meaningful relevance can be asked. Since the extracts will be crude mixtures, a useful screen must reliably pick up activity even when a large number of inactive impurities contaminate the sample. It is frustrating in the extreme to have to wait for significant periods of time for test results before knowing what experiments to perform next. Consequently, an assay simple enough to be performed quickly and reliably in or near the chemist's laboratory is best.

The samples will also be precious depending upon the source of the plant material and the amount of work expended upon the sample prior to analysis, so that a sample-sparing assay is of greatest value.

The assays should also be robust. That is, the tests should produce relatively constant results regardless of changes in analyst or subtle variations in analytical techniques. If the results are easily interpreted by laboratory workers of reasonable intelligence and are quantitative, then rapid progress will usually follow. Few screens possess all of these characteristics. In our experience with antimutagenic screening of the type that we will describe next, lack of sufficient robustness has been the most troublesome feature. Our colleagues doing similar work in other laboratories suggest that they have encountered some of the same difficulties. Mutagenesis is a complex phenomenon, so optimization of the relevant screening parameters cannot be done until we collectively have more experience with these kinds of studies.

The essential features of the screen that we presently employ are set forth in Fig. 2. This consists of a relatively minor modification of the now classical Ames *Salmonella typhimurium* assay. Each of the Ames tester

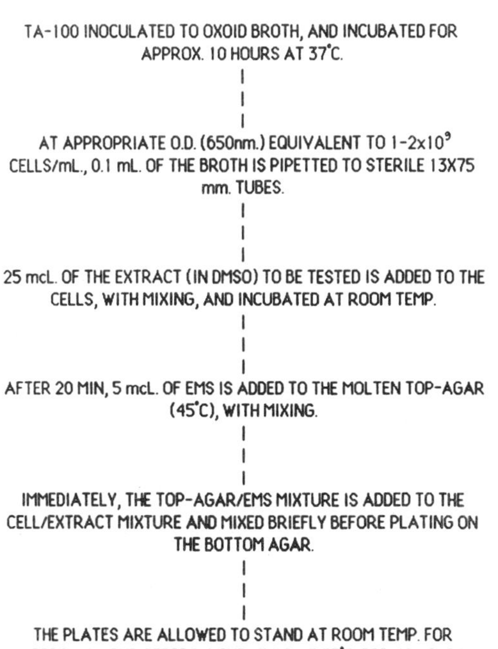

Fig. 2. Brief outline of the antimutagenesis assay.

strains contains a mutation in the histidine operon and other mutations that enhance its ability to detect mutagens (12). This procedure is widely employed and has been optimized in many laboratories.

The TA100 strain was found to be most responsive to the effect of *Glycyrrhiza glabra* extracts. The extract is diluted with the cell suspension of fixed population density and preincubated for 20 min to allow for sample uniformity and cellular uptake. Perhaps some cellular preconditioning takes place also during this time. In this assay a significant simplification of the overall procedure is obtained through the use of a complete mutagen, i.e., one not requiring metabolic activation. Thus the complex mixture of biopolymers present in rat liver post-mitochondrial supernatant preparations need not be added because oxidative activation is not required.

Of the various mutagens screened (ultraviolet irradiation, acriflavine, and 9-aminoacridine, for example), the most sensitive results were obtained with ethyl methanesulfonate (EMS). This substance boils at over 200°C, and the concentration in the molten top agar (5 μL) is adjusted so as to give at least three times the spontaneous mutation rate seen in the control plates. A notable feature of this protocol is that the plant extract and the mutagen are both diluted separately and homogenized before mixing. This minimizes the possibility of a direct reaction between them which would produce a desmutagenic, rather than an antimutagenic, effect. The rest of the procedure follows the standard Ames protocol.

Ethyl methanesulfonate is an aggressive nucleophile inducing mutations by damaging DNA at numerous electron-rich sites by introducing N- and O-ethyl groups. The effect of EMS on various nucleic acid bases is illustrated in Fig. 3 (16).

In using this assay for screening our library of crude plant extracts, *G. glabra* L. was among the more potent early findings. The Spanish vari-

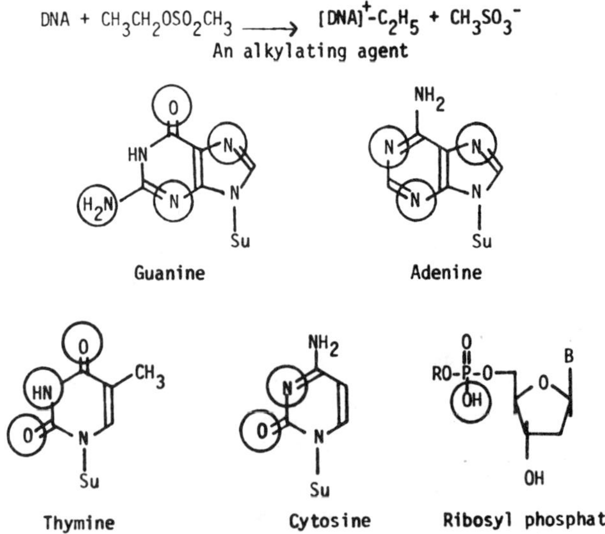

Fig. 3. Known reactions of DNA with ethyl methanesulfonate [G.A. Saga (1984) *Mutat. Res.* 134:113–142].

ety, known as licorice, is an article of commerce widely used as a flavoring agent. It has been used medicinally for flavoring medications and for soothing sore throats. It is also used for a variety of lay purposes such as in candies, chewing tobacco, cigars, soft drinks, and the like. In fact, it has a history of human consumption going back for at least 6,000 years, and it also has the advantage of being inexpensive and readily available.

On the other hand, extracts of *G. glabra* are tricky to evaluate for antimutagenic activity. As we had demonstrated in an earlier study, this plant produces a number of fairly potent antimicrobial agents (13). *Dead cells do not mutate* (19)! Thus we had always to determine that the concentrations under test did not measurably decrease the viability of the tester strain. Tests were thus performed at a maximum of 0.5 times the maximum tolerated dose and then in decade series less than that so as to construct a dosage-response curve. In practice, antimicrobial activity is relatively easy to detect on these plates. It is not only detected in the normal way by determining viable colony-forming units, but one can also see suppression of the background lawn haze and formation of microcolonies among the survivors when toxic doses are approached.

A battery of different test systems were evaluated in a preliminary way before settling on a protocol and beginning fractionation. For example, the extracts were screened using a modification of the *Bacillus subtilis* repair test (Rec-assay). The *B. subtilis* strain M45 rec$^-$ isolated by Kada et al. (7) contains the rec-45 mutation which renders it deficient in genetic recombination. Its heightened sensitivity to various mutagens clearly distinguishes it from the *B. subtilis* H17 rec$^+$ strain to which it is compared in the Rec-assay. The radial streak method (9) was used and a modification was made from the procedure described in that the cells were treated with a mutagen. This was accomplished simply by the addition of the antimutagen directly to the same filter paper disc to which the mutagen was also added. In different experiments the antimutagen was added either at the same time as the mutagen, or added up to 24 hr prior to the mutagen. The *G. glabra* extract was used at a concentration of 100 μg/plate and EMS at a concentration of 4 μL/plate.

In the Rec-assay, zones of inhibition are measured as indicators of the degree of sensitivity of the Rec$^+$ or Rec$^-$ strain to the chemical being tested. In this investigation, addition of the *G. glabra* extract greatly decreased the zone of inhibition of Rec$^-$ by EMS in two different situations. The cells were either co-treated with the mutagen and putative antimutagen (both EMS and extract added to the disc simultaneously), or the Rec$^-$ cells were preincubated with the extract, washed twice with phosphate buffer, and then streaked radially onto plates containing only EMS.

When Rec$^+$ and Rec$^-$ strains were grown in the presence of EMS, the inhibition of Rec$^-$ growth was quite extensive; the zone of inhibition was greater than 1 cm. However, when the *G. glabra* extract was added to the disc along with the EMS, the survival of the Rec$^-$ strain, as measured by the size of the zone of inhibition, increased greatly. In this situation, the zone of inhibition was only 3-5 mm. Thus, *G. glabra* possesses a definite antimutagenic effect in this assay system. The results of these experiments are shown in Tab. 2.

In other preliminary experiments, the *G. glabra* extract has not appeared to be antimutagenic when plated together with the frameshift mutagens 9-aminoacridine and acriflavine HCl. However, the fact that it

Tab. 2. Results of the Rec-assay.

Mutagen	Amount of Glycyrrhiza glabra extract (μg/disc)	Diameter of inhibition zone (mm): H17 Rec$^+$ Bacillus subtilis	Diameter of inhibition zone (mm): M45 Rec$^-$ Bacillus subtilis
Ethyl methanesulfonate	0	1	15
		2	17
		1	15
Ethyl methanesulfonate	100	0	5
	(co-treatment	0	4
	with EMS)	0	3
Ethyl methanesulfonate	100	0	3
	(pretreatment)	0	4
		0	3
Dimethyl sulfoxide	0	0	0
control		0	0
		0	0

decreased mutation rates induced by EMS in the microbial mutagenicity assays used in this study led us to think that the root extract of G. glabra might be acting as an antimutagen either by enhancing a DNA repair response or by blocking the effects of the mutagen in some fashion. Support for this inference was obtained when the S. typhimurium strains TA100 and TA1535, which are both sensitive to base-pair substitutions, showed a greater than 2-fold decrease in reversion frequency when cells were exposed to both the mutagen and the extract.

Bacillus subtilis cells, as well as Escherichia coli cells, become more resistant to both killing and mutagenesis by a challenge dose of N-methyl-N'-nitro-N-nitrosoguanidine (MNNG) after adaptation with lower doses of the alkylating agent (5). Since nonadapted B. subtilis cells contain higher levels of a methyltransferase with an activity similar to the enzyme in E. coli and therefore have a greater enzyme concentration available at the time of invocation of the adaptive response, the possibility that the crude G. glabra extract may act as a low-level alkylating agent and thereby induce the adaptive response was considered.

In order to test this hypothesis, preliminary experiments were conducted on the effects of G. glabra against E. coli K-12 AB1157 and GW5352. Strain GW5352 is distinguished from AB1157 only by the insertion of a transposon (Tn 10) within the ada locus (11). Indeed, in contrast to GW5352, the survival of AB1157 was dramatically enhanced when the cells were preincubated with the extract prior to exposure to the mutagen EMS. These results are illustrated in Tab. 3. Possibly G. glabra crude extracts exert an antimutagenic effect in certain assays by inducing an adaptive response.

Thus, G. glabra extracts demonstrated antimutagenic effects in several test systems. The most reproducible and easiest to quantitate was the modified Ames Salmonella test, and the remainder of the data to be presented were obtained with this system. The results with the total extract are illustrated in Fig. 4. The viability data indicate that loss in viability

Tab. 3. Effects of Glycyrrhiza glabra extract and ethyl methanesulfonate on survival in E. coli AB1157 and GW5352 strains.

Treatment	Percent survival in	
	E. coli AB1157	E. coli GW5352
Glycyrrhiza glabra extract only (100 μg/plate)	100	95 ± 2
Glycyrrhiza glabra extract + ethyl methanesulfonate	75 ± 2	13 ± 3
Ethyl methanesulfonate only	28 ± 3	20 ± 2

was caused by 500 μg of extract per plate (Fig. 5), and signs of toxicity were apparent down to 250 μg per plate in the complete system. The extract was active against spontaneous mutations in a linear fashion from 3.9 to 125 μg per plate. This is definitely below the toxic range, and these results indicate that the extract is not exerting a desmutagenic effect as no EMS is present in these plates. Under the conditions used, EMS induces more than a 3-fold mutation increase. Glycyrrhiza glabra extract profoundly decreases the number of induced mutations over a wide range of concentrations well below the toxic level. The activity, however, is a complex function of concentration. Perhaps this reflects the crudity of the mixture being examined.

With these data in hand, fractionation studies were initiated in order to identify the constituent(s) responsible for the effects noted. The crude extract contains dozens of constituents. The first step in simplifying the mixture is to fractionate by bulk transfer methods into subclasses of grossly related substances. The fractionation scheme is adapted from

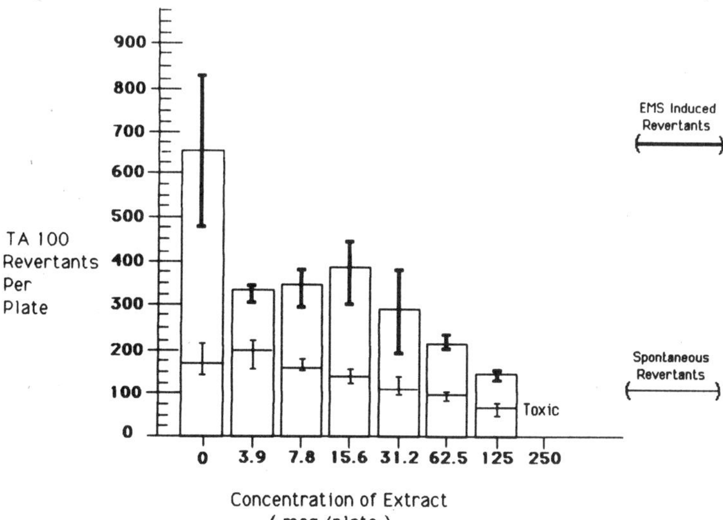

Fig. 4. Antimutagenic activity of Glycyrrhiza glabra extracts in the modified Ames tests.

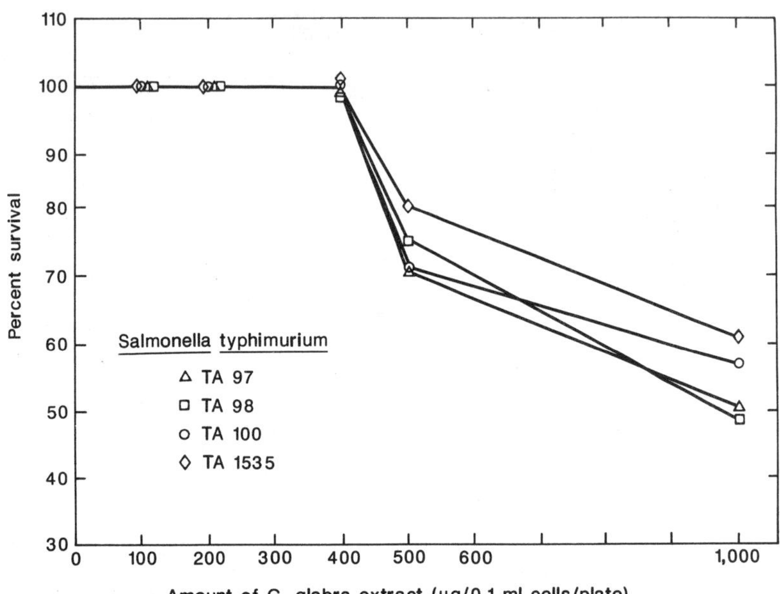

Fig. 5. Percent survival of Ames Salmonella typhimurium tester strains as a function of Glycyrrhiza glabra extract concentration.

that used by antitumor and antibiotic chemists to deal with similar mixtures. The scheme is illustrated in Fig. 6. Each fraction was tested for antimutagenic activity, and this chapter will discuss the bioactive polar-lipid-containing fraction.

The dose-response curve obtained with the semipurified polar-lipid fraction is illustrated in Fig. 7. This mixture shows activity against spontaneous mutations only at higher concentrations relatively near the toxic doses, so no significance can be placed upon this small effect. On the other hand, a definite and linear decrease in effect against the EMS-induced mutations was obtained. The difference between the numbers of spontaneous mutations and the numbers of EMS-induced mutations was almost abolished before toxic doses were reached, and a definite decrease in mutagenic effect was observable at concentrations 10,000 times less than the dose where toxic effects could be discerned.

Whereas this fraction is rather significantly purified as compared with the total extract of the plant, many compounds are still present. These were separated effectively by silica gel chromatography into individual pure components. The separated substances were then examined for activity. Figure 8 gives the structural formulae of the 11 compounds isolated in pure form from this fraction to date.

Glabrene is a known isoflavene shown by us in an earlier study to be active in vitro against Staphylococcus aureus and Mycobacterium smegmatis (13). It is now found to be a highly active antimutagenic agent against both spontaneous and EMS-induced mutations at doses well below those showing signs of toxicity and at levels that do not decrease the colony-forming units of S. typhimurium TA100 strain (Fig. 9). It is not an aggressively functionalized compound, and its action against spontaneous mutations would

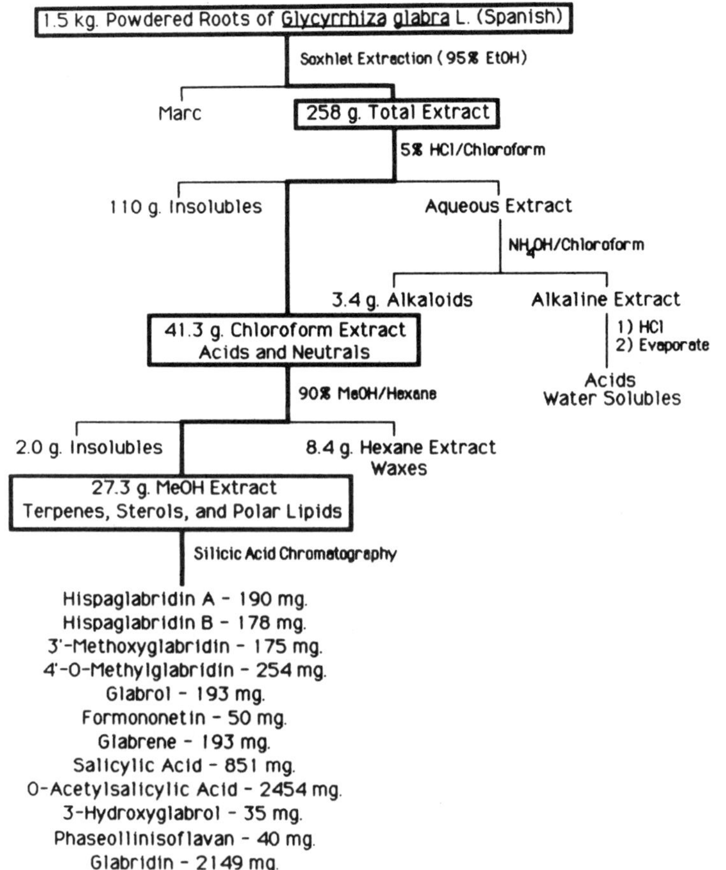

Fig. 6. Bioassay-directed fractionation of Glycyrrhiza glabra extract.

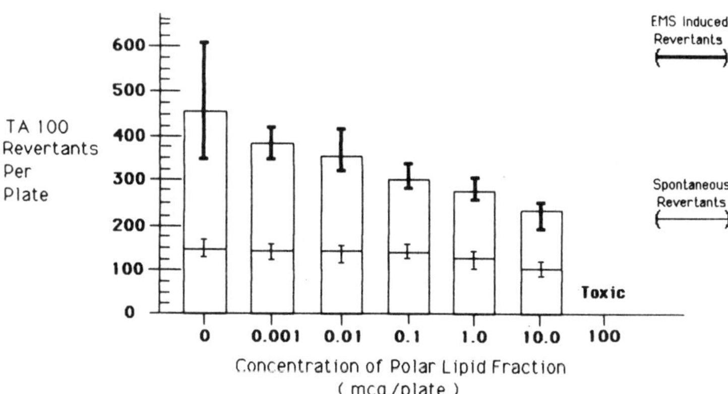

Fig. 7. Antimutagenic activity of the polar-lipid-containing portion of Glycyrrhiza glabra extracts.

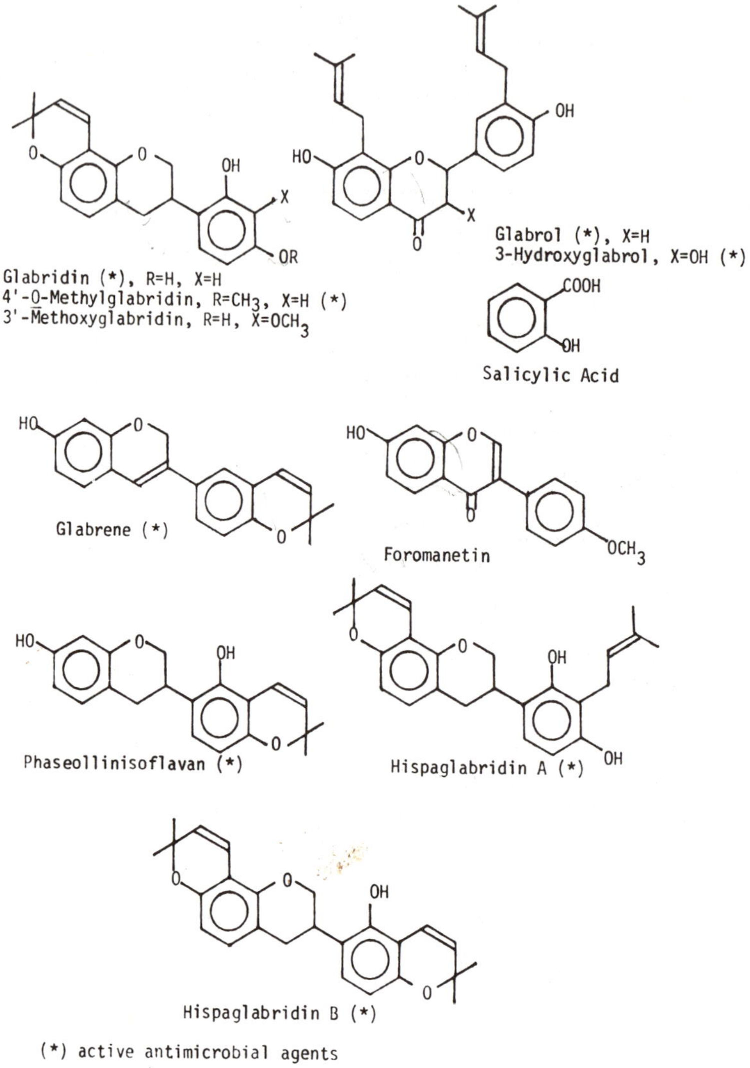

Glabridin (*), R=H, X=H
4'-O-Methylglabridin, R=CH_3, X=H (*)
3'-Methoxyglabridin, R=H, X=OCH_3

Glabrol (*), X=H
3-Hydroxyglabrol, X=OH (*)

Salicylic Acid

Glabrene (*)

Foromanetin

Phaseollinisoflavan (*)

Hispaglabridin A (*)

Hispaglabridin B (*)

(*) active antimicrobial agents

Fig. 8. Structural formulae for pure components isolated from the polar-lipid-containing fraction of Glycyrrhiza glabra extracts [L.A. Mitscher et al. (1980) J. Natural Prod. 43:259 and (1983) Phytochem. 22:573].

appear to rule out a desmutagenic effect. No toxic effects towards the tester strain were apparent in these studies. The survival rate did not decrease until at least 10 μg per plate of glabrene was employed.

Glabridin is also an antibacterial phenolic constituent. It belongs to the isoflavan class of natural products. It shows relatively little activity against spontaneous mutations and, while it shows modest activity against EMS-induced mutations, it is much less active than glabrene. It is interesting to note that it is by far the most abundant phenolic component that we have isolated from this fraction.

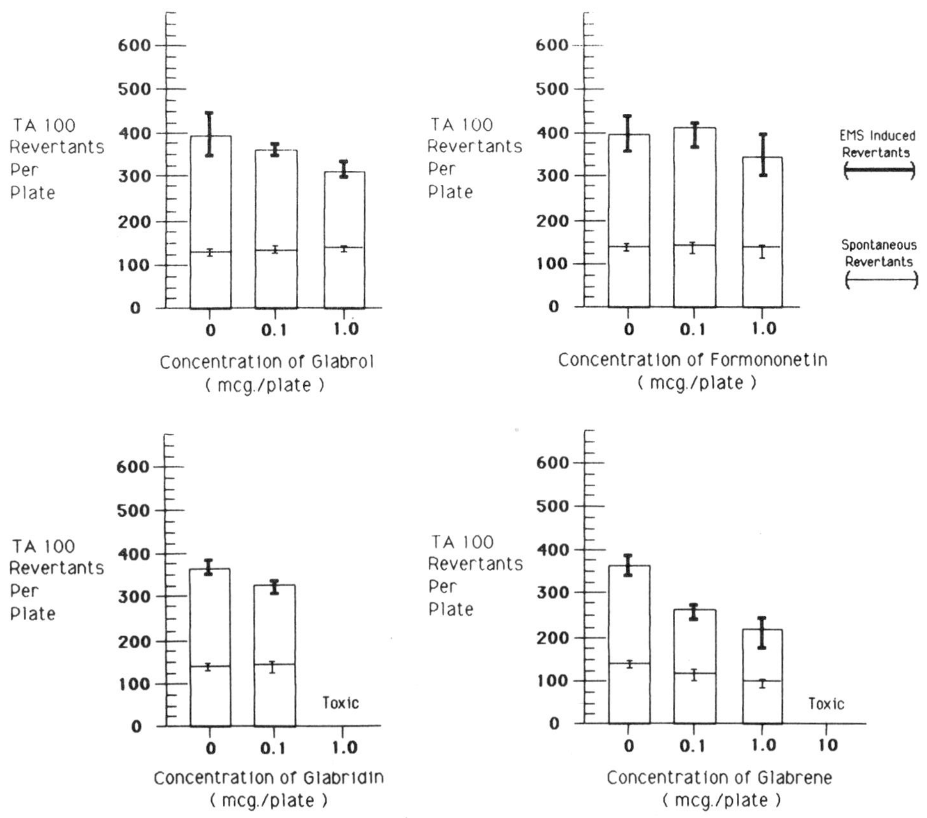

Fig. 9. Antimutagenic activity of glabrol, formononetin, glabridin, and glabrene.

Glabrol, an antimicrobially active flavanone, shows no detectable antimutagenic activity against spontaneous mutations and only modest activity against the EMS-induced mutations.

Formononetin, a phytoestrogen, has no antimicrobial, antimutagenic, or mutagenic properties. The lack of mutagenic activity confirms the earlier findings of Bartholomew and Ryan (2).

Hispaglabridin A is an antimicrobial flavan whose structure is somewhat similar to that of glabridin. It is active as an antimicrobial agent but shows no significant antimutagenic activity (Fig. 10). As a phenolic compound it is chemically quite capable of reacting with EMS, but these results suggest strongly that the precautions taken in separately diluting the mutagen and the putative antimutagens have prevented this, for no desmutagenic activity is seen with hispaglabridin A.

It is interesting to note from these results that there is no direct correlation between antimicrobial and antimutagenic effects with these compounds. It is also clear that structural specificity is present in this series because structurally related compounds are either active or inactive depending upon their intrinsic properties.

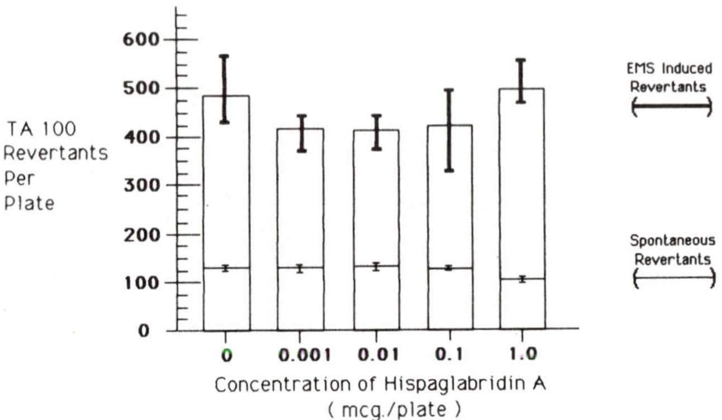

Fig. 10. Antimutagenic activity of hispaglabridin A.

Thus, the screen devised for this purpose has proven to be useful in revealing constituents with a reproducible antimutagenic effect in these assay systems. Use of the assay has allowed us to fractionate the active extracts systematically by a bioassay-directed scheme and has led to the isolation of individual pure compounds with the intrinsic activity of the total extracts.

With these substances in hand, many further experiments are now suggested with which we can challenge the parameters of the assay and determine whether the compounds isolated are in fact intrinsically active and potentially useful antimutagenic compounds. It will be particularly interesting to explore whether the individual pure components act through an adaptive mechanism. In this regard, it is important to bear in mind that *S. typhimurium* tester strains have not been shown to possess the adaptive response to alkylating agents which is known to be present in *E. coli* and *B. subtilis* strains.

In light of the fact that *G. glabra* extract is not apparently mutagenic itself in any of the mutagenicity assays described herein, and that it is sometimes a component of the human diet, we hope that the results of this study can lead to a further understanding of the mechanisms by which cells protect themselves from environmental insult, the role that certain dietary factors, especially from plants, may have in antimutagenesis, and the development of methodology by which new agents can be detected and isolated.

ACKNOWLEDGEMENTS

We thank Bruce Ames for providing the *S. typhimurium* tester strains, Tsuneo Kada for the *B. subtilis* Rec-assay strains, and Graham Walker for the *E. coli* ada$^+$ and ada$^-$ strains. This research was supported in part by a grant from the Univ. of Kansas Biomedical Support Grant Fund (to D.M.S.) and Grant No. AI 13155 from the National Institutes of Health (to L.A.M.).

REFERENCES

1. Ames, B.N. (1983) Dietary carcinogens and anticarcinogens: Oxygen radicals and degenerative diseases. *Science* 221:1256-1264.

2. Bartholomew, R.M., and D.S. Ryan (1980) Lack of mutagenicity of some phytoestrogens in the Salmonella, mammalian microsome assay. Mutat. Res. 78:317-321.
3. Clarke, C.H., and D.M. Shankel (1975) Antimutagenesis in microbial systems. Bacteriol. Rev. 39:33-56.
4. Fiala, E.S., B.S. Reddy, and J.H. Weisburger (1985) Naturally occurring anticarcinogenic substances in foodstuffs. Am. Rev. Nutr. 5:295-321.
5. Hadden, C.T., R.S. Foote, and S. Mitra (1983) Adaptive response of Bacillus subtilis to N-methyl-N'-nitro-N-nitrosoguanidine. J. Bacteriol. 153:756-762.
6. Ishii, R., K. Yoshikawa, H. Minakata, H. Komura, and T. Kada (1984) Specificities of bio-antimutagens in plant kingdom. Agric. Biol. Chem. 48:2587-2591.
7. Kada, T., K. Tuitkawa, and Y. Sadaie (1972) In vitro and host mediated "Rec-assay" procedures for screening chemical mutagens; and phloxine a mutagenic red dye detected. Mutat. Res. 16:165-170.
8. Kada, T., T. Inoue, and M. Namiki (1982) Environmental desmutagens and antimutagenic agents. In Environmental Mutagenesis, Carcinogenesis and Plant Biology, E.S. Klekowski, Jr., ed. Praeger Publishers, Inc., New York, pp. 135-151.
9. Kada, T., Y. Sadaie, and Y. Sakamoto (1984) Bacillus subtilis repair test. In Handbook of Mutagenicity Test Procedures, B.J. Kilbey, M. Legator, W. Nichols, and C. Ramel, eds. Elsevier Science Publishers, Amsterdam, pp. 13-31.
10. Kakinuma, K., J. Koike, K. Kotani, N. Ikekawa, T. Kada, and M. Nomoto (1984) Cinnamaldehyde: Identification of an antimutagen from a crude drug, Cinnamoni Cortex. Agric. Biol. Chem. 48:1905-1906.
11. LeMotte, P.K., and G.C. Walker (1985) Induction and autoregulation of ada, a positively acting element regulating the response of Escherichia coli K-12 to methylating agents. J. Bacteriol. 161:888-895.
12. Maron, D.M., and B.N. Ames (1983) Revised methods for the Salmonella mutagenicity test. Mutat. Res. 113:173-215.
13. Mitscher, L.A., Y.H. Park, D. Clark, and J.L. Beal (1980) Antimicrobial agents from higher plants. Antimicrobial isoflavanoids and related substances from Glycyrrhiza glabra L. var. Typica. J. Natural Prod. 43:259-269.
14. Morita, K., T. Kada, and M. Namiki (1984) A desmutagenic factor isolated from burdock (Arctium lappa Linne). Mutat. Res. 129:25-31.
15. Ohta, T., K. Watanabe, M. Moriya, Y. Shirasu, and T. Kada (1983) Antimutagenic effects of coumarin and umbelliferone on mutagenesis induced by 4-nitroquinoline-1-oxide or UV irradiation in E. coli. Mutat. Res. 117:135-138.
16. Saga, G.A. (1984) A review of the genetic effects of ethyl methanesulfonate. Mutat. Res. 134:113-142.
17. Sato, T., Y. Suzuki, Y. Ose, T. Youki, and T. Ishikawa (1984) Desmutagenic substance in water extract of grass-wrack pondweed (Potamogeton oxyphylus Miquel). Mutat. Res. 129:33-38.
18. Sugimura, T., and S. Sato (1983) Mutagens-carcinogens in foods. Cancer Res. 43(Suppl.):2415s-2421s.
19. Thilly, W.G. (1985) Dead cells do not form mutant colonies: A serious source of bias in mutation assays. Environ. Mutag. 7:255-258.
20. Wattenberg, L.W. (1983) Inhibition of neoplasia by minor dietary constituents. Cancer Res. 43(Suppl.):2448s-2459s.

MECHANISM OF ACTION OF ANTIMUTAGENS AND ANTICARCINOGENS

INTERCEPTION OF TOXIC AGENTS/MUTAGENS/CARCINOGENS: SOME OF NATURE'S NOVEL STRATEGIES

Philip E. Hartman

Department of Biology
The Johns Hopkins University
Baltimore, Maryland 21218

INTRODUCTION

In this section focused on mechanisms of action of antimutagens and anticarcinogens, it might be well to note that there are a wide variety of levels at which defense mechanisms operate. Nature provides numerous diverse examples at each level. The following descriptions center on the flow of interests of my own laboratory over a period of years. The descriptions will highlight a few of the many mechanisms that serve to "intercept" potential mutagens and carcinogens before these deleterious molecules can attack critical cells ("stem cells," see Ref. 35) or, once in a critical cell, before the DNA can be reached, to elicit a potentially mutagenic lesion. It will be tacitly assumed that lesions in DNA are a most important component in formation of focal lesions during the lifespan. As pointed out elsewhere, such focal lesions are certain to be of critical importance not only in carcinogenesis but also in a wide array of additional situations that compromise the human condition and collectively constitute an important aspect of physiological "aging" (27,29).

There certainly are multiple mechanisms that operate at the organismal level to protect against toxic substances; examples are avoidance responses and regurgitation. Protective devices also operate at the level of organ systems. One example is the role in the mammal of the portal vein in delivering assimilated molecules to the liver for processing before dissemination to the rest of the organism. In this chapter we look at lower but equally important levels of interception that operate at the tissue level, that concern detoxification by enzymes, and that involve direct interaction of important but often neglected cellular molecules with potential mutagens/carcinogens.

TISSUE ORGANIZATION

Carcinomas involving the external and internal epithelia account for the vast majority of human cancers worldwide (reviewed in Ref. 14 and 23). Study of the origins of carcinomas, then, ought to provide material for analyses of the relation between tissue organization and induction of the more common forms of cancer. Carcinoma of the glandular stomach will be surveyed here because there exists a wealth of information combined from laboratory animal experiments as well as data from man, mainly through the efforts of Japanese workers with substantial contributions by researchers in other nations (reviewed in Ref. 24 and 29).

Progression to Gastric Cancer

Figure 1 diagrams some key steps in the development of gastric carcinomas as they occur in man. One key point is that carcinomas can appear, but with extremely low frequency, in regions of what seem to be virgin,

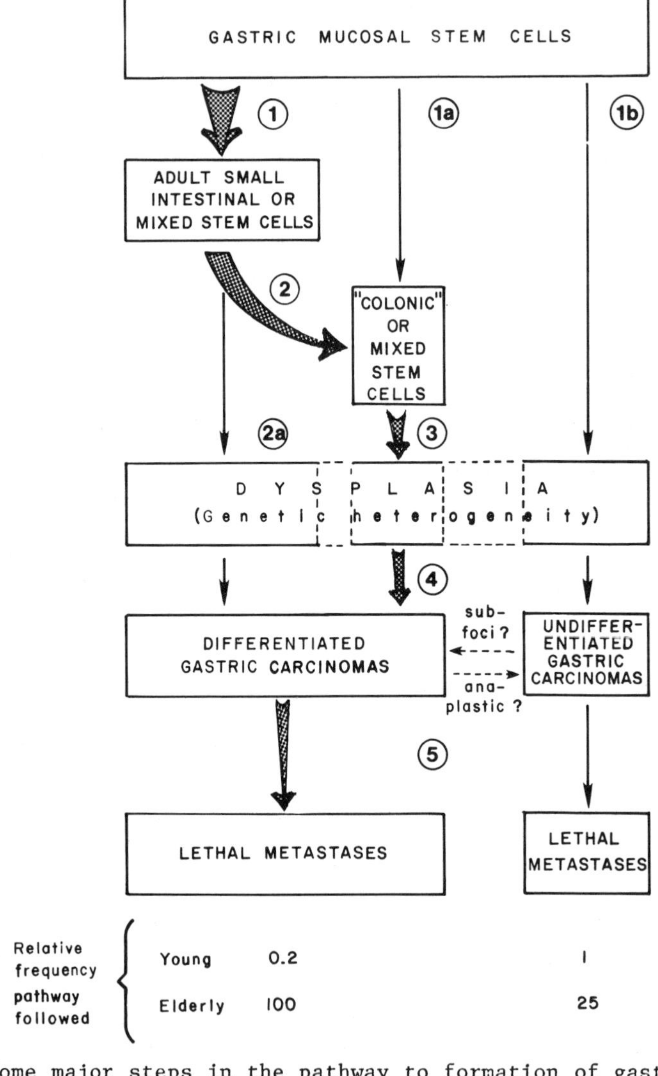

Fig. 1. Some major steps in the pathway to formation of gastric carcinomas in the stomach of man and the glandular stomach of rodents. Further explanation is in the text and in Ref. 29, from which the figure is taken with permission from Raven Press.

undisturbed mucosa. These are mainly undifferentiated carcinomas, largely exhibiting features of adult gastric tissue with regard to enzyme and mucin contents. Such tumors predominate in the young, but gastric cancer is relatively rare in young persons (bottom, Fig. 1). A second key point is that the frequency of such undifferentiated carcinomas increases considerably with age. In addition, in older persons undifferentiated carcinomas are often, although not exclusively, found at sites in conjunction with intestinal metaplasia (IM) of the background mucosal region. Finally, the large class of differentiated gastric carcinomas typical of the elderly possess many features (enzymes, mucins, structure) that characterize adult small or colonic intestinal tissue. This type of carcinoma also predominantly originates at patches of gastric mucosa that seem to have been previously intestinalized. Even in mutagen-treated rats, carcinomas preferentially occur at foci of IM (43,44).

How do we account for: (a) the possession of intestinal properties by many gastric carcinomas, a characteristic that typifies a gradually increasing proportion of carcinomas as people grow older; (b) the propensity for both types of carcinomas to increase in frequency with age; and (c) the preferential locations of both types of fully developed carcinomas at sites in the stomach where IM prevails in the surrounding mucosa? One very tenable proposition is that IM precedes and predisposes to gastric cancer.

Tissue Organization and Barrier Breakdown

Figure 2 presents a model that depicts multiple roles for IM in disruption of the integrity of the gastric mucosa. This disruption (barrier breakdown) sensitizes nearby critical stem cells, competent in DNA synthesis and cell division, to continual bombardment by mutagens and carcinogens throughout the remainder of the lifespan. Many spots of IM are gradually formed, accumulate, and expand through life, particularly in gastric cancer high-risk populations; however, only an occasional site progresses to gastric cancer (reviewed in Ref. 24, 27, and 29).

A gastric gland is diagrammed on the left side of Fig. 2. The native structure is shown with critical stem cells hidden in the neck of the gland

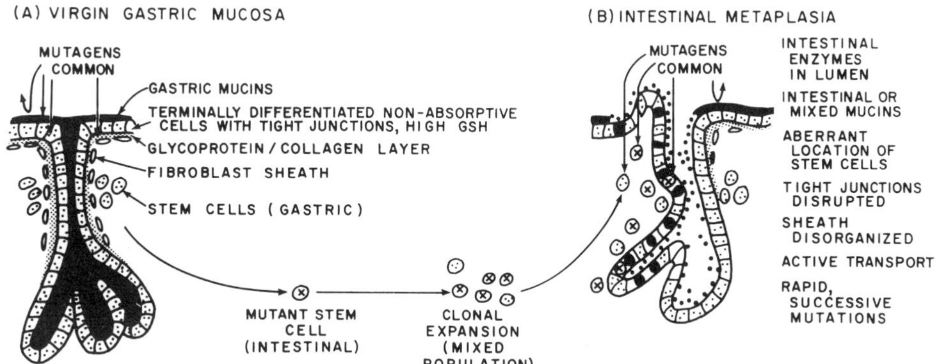

Fig. 2. "Barrier breakdown" of the coordinated structure and synthetic activities of a normal gastric gland (left) during development of intestinal metaplasia (right). Further explanation is in the text and in Ref. 24, from which the figure is taken with permission from Plenum Press.

and protected from mutagens in the gastric lumen. Protective devices include the following.

Gastric mucins. These prevent bacterial adhesion and, like the cell walls of *Salmonella*, block permeation of some species of toxic molecules. Each gastric mucin molecule contains about 100 cysteinyl residues in the exposed 28-kDa portions of the polypeptide backbones of the tetrameric mucoprotein (1). These cysteines could play a defensive role against electrophiles besides holding the mucoprotein subunits together through disulfide bridges.

Surface epithelial layer. Terminally differentiated cells incompetent in DNA synthesis or further division (21,30,37) are joined by very tight junctions (16). Cells of the glandular gastric tissue contain millimolar levels of reduced glutathione (GSH) (9). Furthermore, the most rapidly renewed surface cells slough off and are completely replaced over a 4-day period (21,37); they would carry reactive chemical species and the products of chemical reactions with them into the gastric lumen and then through the digestive system. Presumably, the vast majority of transformed surface epithelial cells, should there be any, would undergo the same fate before tumor progression could ensue.

Basement glycoprotein/collagen/fibroblast layer. A specially organized layer (21,31) further separates the underlying stem cell population from mutagens in the gastric lumen.

The right side of Fig. 2 illustrates how the above-mentioned defenses, due to stringent organization in native gastric mucosa, may be disrupted at sites of IM. Intercellular communication, tight junctions, coordinated cellular locations, and cell products befitting a gastric environment are all disrupted (resulting in barrier breakdown). Active intestinal-type transport ensues. Some of the many possible consequences of barrier breakdown are outlined elsewhere with references to the original literature (24).

Clonal Expansion and "Self"

Two properties of IM foci may be critical to their role in barrier breakdown. First, the mitotic rate of the stem cells is increased, indicating a replacement time of 2 days versus the 4 days typical of gastric surface epithelium (21,37,52). This gives the IM foci a proliferative advantage (12), allowing continuous and progressive clonal expansion once a site has been initiated. Second, the tissue is essentially "self" and thus less susceptible to immune surveillance and subsequent elimination than would be a more foreign type of cellular alteration.

Metaplasias and Barrier Breakdown

Metaplasias are exceedingly common, although curiously overlooked by many hematoxylin-eosin oriented pathologists. A number of reports of metaplasias in various epithelial tissues have been catalogued by Hartman and Morgan (29), and there is good reason to believe that numerous additional cases have been repeatedly overlooked in the literature. Metaplasias may play important roles in the generation of cancers in organs other than the stomach, and they also may participate importantly in numerous other diseases that, like cancer, increase with age (27,29). A hypothesis has been presented that attributes the formation of metaplasias to somatic mutations (27,29). That is, mutant stem cells are considered to channel the very terminal stages of differentiation into a pathway mimicking that found in a very closely related tissue. This process seems analogous to the phenomenon of transdetermination in Drosophila, a process that also is unresolved as to cause, genetic or epigenetic (48).

DETOXIFICATION BY ENZYMES

Other chapters in this Volume describe some important, widely distributed interceptor detoxification reactions mediated by enzymes such as superoxide dismutases and glutathionyl S-transferases. An additional enzymologically mediated detoxification mechanism recently has come to our attention. It is worth considering because it appears to be a rather specialized reaction in the sense that its activity varies widely among mammalian species. This indicates that some animals may have strategies for dealing with entire classes of mutagens that seem completely lacking in other species. One can guess that numerous other species-dependent enzymological mechanisms await discovery. Their elucidation may be important in understanding the differential responses of various animals to particular environmental mutagens.

Rodent (rat, mouse) and rabbit blood plasma samples contain an enzyme, apparently a carboxylesterase, that hydrolyzes and thus inactivates direct-acting mutagens of the nitrosocarbamate and nitrosamide classes without affecting homologous nitrosoureas (6,7,10). The distribution in mammals of blood plasma carboxylesterases varies greatly among species (5), and the purpose of their presence in plasma has remained conjectural. Hydrolyzing activity is almost entirely absent in human plasma (7). Perhaps we are seeing here a novel strategy for dealing with potent mutagens, a strategy that has not been granted to man during evolution. Yet we find that testing of nitrosamides, nitrosocarbamates, and nitrosoureas for carcinogenicity, for germline mutations, and for antitumor activity is all done in rodents (references in Ref. 6, 7, 10, and 11) with extrapolation to the plasma carboxylesterase-deficient human!

ANTIMUTAGENIC ACTIVITIES OF INDIVIDUAL COMPOUNDS

Kada et al. (Ref. 32 and this Volume) have pioneered the detection of antimutagens in natural products, particularly in plant extracts, through the use of bacterial mutagenesis screening. The work of Wattenberg (54,55) has focused attention on anticarcinogens and most cogently on the variety of mechanisms by which anticarcinogens may exert their influence in the mammal. One purpose of these proceedings is to hasten convergence of these two types of approaches in antimutagenesis/anticarcinogenesis.

There are really three quite different ways to find antimutagens. One way is to screen complex mixtures and isolate protective components. In this regard, we need to examine places where antimutagens might be expected to be present in abundance, for example, in seminal fluid, ova, seeds, spores, conidia, etc. A second way to find antimutagens is to look at novel niches that lack a prevalent defense mechanism. Archibald and Fridovich (4) did this, and they discovered that millimolar Mn^{+2} in Lactobacillus could replace the micromolar levels of superoxide dismutase present in most other aerobes. The third general way to find antimutagens is to look at prevalent molecules, known compounds that possess chemical attributes making them attractive prospects as antimutagens/anticarcinogens. Examples are the discovery by the Ames group (2) that uric acid, abundant in the human, possesses antioxidant properties, and our own observation (25) that the 4 mM (saliva) to 600 mM (urine) levels of urea in man might serve to lessen formation of N-nitroso compounds by scavenging nitrite. Spread through nature are a variety of strategies for intercepting mutagens, several of which we have encountered in our own laboratory as noted below.

Polyamines

A number of years ago we were alerted to antimutagenic effects of specific compounds by the observation that polyamines inhibited the toxicity and mutagenicity of hycanthone in *Salmonella* (28). We found that this observation was not entirely unexpected for a class of intercalating, ionically binding mutagens, as reviewed very perceptively a full decade ago by two colleagues among those primarily responsible for the existence of these proceedings (15). Polyamines presumably compete with hycanthone for DNA binding sites rather than, for example, competing for transport into the bacteria, although precise mechanisms of protection have not been elucidated. Polyamines also have been found to lower alkylation of DNA by N-methyl-N-nitrosourea, again by mechanisms that are not clearly delineated (47).

In any event, polyamines are ubiquitous and present at high levels in many, but not all, organisms. In *E. coli*, polyamines are normally present in the millimolar range, but they are still luxury molecules in the sense that mutants have been obtained that grow in the complete absence of polyamines (51). Like many molecules, polyamines may be two-edged swords. That is, they may be beneficial under many sets of conditions but may be possible substrates for nitrosation reactions, with formation of mutagens, under other circumstances (53). Polyamines are deserving of further study, particularly in light of their increased synthesis during tumor promotion and relative abundance in a number of tumors.

Do cells overproducing polyamines have a selective advantage during tumor promotion and progression by virtue of their ability to intercept toxic molecules? Since polyamine synthesis is coupled to DNA synthesis in eukaryotes, selection for polyamine production may constitute one step in lessening cellular restrictions on DNA synthesis.

Glutathione

Modification of reduced GSH levels and activity of GSH in detoxification reactions are the subjects of numerous investigations summarized in this Volume. *Escherichia coli* mutants blocked in GSH biosynthesis are more sensitive to inhibition of growth by a variety of chemicals than are wild-type bacteria (3). Cellular GSH also protects bacteria against inactivation by γ-irradiation (45) by blocking fixation of lethal damage (40).

Recently, Roland Owens has found that some laboratory strains of both *Salmonella typhimurium* and *E. coli* export into the medium about 50% of their synthesized GSH in an energy-dependent process (Ref. 46 and ms. in prep.). Exported GSH accumulates in the medium to micromolar levels in contrast to the millimolar levels found intracellularly. However, micromolar quantities of extracellular GSH are adequate to serve as a prominent defense when bacteria are challenged with micromolar levels of compounds that otherwise are active in this range of concentrations against *Salmonella* and *E. coli*, namely N-methyl-N'-nitro-N-nitrosoguanidine (MNNG), mercuric chloride, methyl mercuric chloride, silver nitrate, cisplatin, cadmium chloride, cadmium sulfate, and iodoacetamide (R.A. Owens, ms. in prep.). Glutathione-deficient E. coli cells are able to survive and grow on standard media in the absence of GSH; thus, GSH appears to be an example of a luxury molecule with, perhaps, detoxification as one primary benefit of its presence.

While internal GSH and export of GSH may comprise important mechanisms for protection of enteric bacterial DNA and other macromolecules from exogenous insults, nature has apparently devised alternate strategies for other organisms. Glutathione is not a predominant low molecular weight thiol in numerous biological systems, being replaced by other molecules in various species (17,18).

Ergothioneine

L-Ergothioneine (thioneine or 2-thiol-L-histidine betaine) is predominantly synthesized by fungi, where it is accumulated in levels of up to 2 mM (22,39,50). It is concentrated in conidia (18). Its function in fungi has been proposed to be as a cellular protector against hydrogen peroxide (39). Dietary ergothioneine is assimilated by animals and stored predominantly in the red blood cells, the central nervous system, and the liver (39,50). It can reach a concentration of 1 mM and inhibit lipid peroxidation in the liver of rats (33,34).

In ongoing explorations of ergothioneine as an antioxidant and as an antimutagen in this laboratory, Thomas Dahl (unpubl. results) has preliminary evidence in chemical experiments in vitro that ergothioneine at an equimolar concentration (1 mM) is an even better blocker of radical damage mediated by illuminated rose bengal than is the classical singlet oxygen quencher, sodium azide. Since ergothioneine does not intercept pure singlet oxygen in an isolated system (41), we speculate that ergothioneine reacts with excited dye to block production of singlet oxygen. This would be an entirely new type of defense mechanism (T. Dahl, unpubl. data). Martin Citardi (unpubl. results) has found strong protection by 2 mM ergothioneine against γ-irradiation inactivation of Salmonella P22 bacteriophage. Zlata Hartman (unpubl. data) has found ergothioneine to protect Salmonella hisG46 uvrB (strain TA1950) against direct-acting mutagens formed from nitrosated spermidine. The imidazole sulfhydryl of ergothioneine possesses quite different properties from the cysteinyl sulfhydryl of GSH; for example, ergothioneine does not activate MNNG (R.A. Owens, unpubl. results).

Intensive studies of ergothioneine as an antioxidant and an antimutagen are in progress. We are excited about ergothioneine because the literature indicates that it is relatively nontoxic to animals on mineral-sufficient diets. It also appears to supplement GSH in the animal liver without perturbing drug-metabolizing enzyme systems or GSH concentrations (33). It seems possible that ergothioneine will be found to be safe enough to reach the vitamin and mineral section of stores on the shelf next to ascorbic acid.

Carnosine

Singlet oxygen preferentially degrades deoxyguanosine residues in DNA (13) with genetic impacts that remain to be thoroughly assessed. L-Histidine is one of the more effective scavengers of singlet oxygen (8,36,49), but free L-histidine is not commonly found in nature at levels that would be protective. On the other hand, the dipeptide carnosine (β-alanyl-L-histidine) is present in human and rodent skeletal muscle at levels up to 20 mM, and is perhaps present at even higher concentrations in the sarcoplasm where the mitochondria are located.

In ongoing experiments, Thomas Dahl (unpubl. results) has found that millimolar concentrations of carnosine can, in fact, effectively scavenge singlet oxygen and could be slightly better than L-histidine in this regard. This is true both in a pure singlet oxygen-generating system (41) and in irradiated aqueous solutions of the photosensitizer, rose bengal. Carnosine reactivity may not be specific for singlet oxygen, because carnosine also appears to be a very effective protector of phage P22 inactivation by γ-irradiation (Martin Citardi, unpubl. observ.). Freifelder (19, 20) had previously shown L-histidine to be a very effective protecting agent with regard to bacteriophage T7 inactivation by X-rays.

One idea among muscle physiologists is that carnosine affords a convenient buffer because of the favorable pKa of the imidazole ring. Our experiments indicate that the high level of carnosine typical of striated muscle also may be just one more of nature's novel defense mechanisms against otherwise potential mutagens/carcinogens.

AGRICULTURAL PRACTICES

We are finding out enough about mutagens and antimutagens so that perhaps it is time to ask agriculture for a helping hand. Clearly, toxic molecules <u>can</u> be lessened in just a few years through standard plant breeding practices. Just one example is nitrate or, better, the ratio of nitrate to ascorbate concentrations in various plants. In spite of extensive searches over a period of years, no beneficial effects of nitrite or nitrate ingestion were detected (24-26). High ascorbate-containing breeding lines of cabbage, muskmelon, tomato, and white potato exist (cited in Ref. 25), and many others could probably be found if searched for. Economically feasible agricultural practices can decrease the nitrate load of market vegetables considerably (38; summarized in Ref. 25). Such steps would seem to be important because nitrate ingestion is related to gastric cancer incidence (24-26,42), possibly due to the formation of nitroso compounds in the stomach (reviewed in Ref. 24 and 42). Ascorbate, on the other hand, reduces nitrosation in the human, as reviewed by H. Bartsch and colleagues in this Volume. Because we eat traditional foods is no reason we have to continue eating traditional mutagens.

ACKNOWLEDGEMENTS

This minireview is fondly dedicated to my wife, Zlata Hartman, who has helped my various colleagues in the laboratory through the years with patience and understanding, while at the same time heading a happy household and home. Dr. W. Robert Midden has supplied both cogent advice and facilities during the initiation of our experiments in blocking photodynamic activities with ergothioneine, carnosine, and related compounds. Thanks also to Chris Wertz for typing the manuscript and to the National Institute of Environmental Health Sciences for Research Grant ES03217 which has supported, in part, our more recent studies described herein. This is contribution number 1302 of the Department of Biology, The Johns Hopkins University.

REFERENCES

1. Allen, A., A. Bell, M. Mantle, and J.P. Pearson (1982) The structure and physiology of gastrointestinal mucus. In Mucus in Health and Disease-II, E.N. Chantler, J.B. Elder, and M. Elstein, eds. Plenum Press, New York, pp. 115-133.
2. Ames, B.N., R. Cathcart, E. Schwiers, and P. Hochstein (1981) Uric acid provides an antioxidant defense in humans against oxidants and radical-caused aging and cancer: A hypothesis. Proc. Natl. Acad. Sci., USA 78:6858-6862.
3. Apontoweil, P., and W. Berends (1975) Isolation and initial characterization of glutathione-deficient mutants of Escherichia coli K12. Biochim. Biophys. Acta 399:10-22.
4. Archibald, F.S., and I. Fridovich (1982) Investigations of the state of the manganese in Lactobacillus plantarum. Arch. Biochem. Biophys. 215:589-596.
5. Augustinsson, D. (1968) The evolution of esterases in vertebrates. In Homologous Enzymes and Biochemical Evolution, N.V. Thoai and J. Roche, eds. Gordon and Breach, New York, pp. 299-311.
6. Aukerman, S.L., R.B. Brundrett, J. Hilton, and P.E. Hartman (1983) Effect of plasma and carboxylesterase on the stability, mutagenicity, and DNA cross-linking activity of some direct-acting N-nitroso compounds. Cancer Res. 43:175-181.
7. Aukerman, S.L., R.B. Brundrett, and P.E. Hartman (1984) Detoxification of nitrosamides and nitrosocarbamates in blood plasma and tissue homogenates. Environ. Mutag. 6:835-849.
8. Bellus, D. (1979) Physical quenchers of singlet molecular oxygen. Adv. Photochem. 11:105-205.
9. Boyd, S.C., H.A. Sasame, and M.R. Boyd (1979) High concentrations of glutathione in glandular stomach: Possible implications for carcinogenesis. Science 205:1010-1012.
10. Brundrett, R.B., and S.L. Aukerman (1985) Enzymatic inactivation of N-nitroso compounds in murine blood plasma. Cancer Res. 45:1000-1004.
11. Brundrett, R.B., M. Colvin, E.H. White, J. McKee, P.E. Hartman, and D.L. Brown (1979) Comparison of mutagenicity, antitumor activity, and chemical properties of selected nitrosoureas and nitrosamides. Cancer Res. 39:1328-1333.
12. Burnet, M. (1965) Somatic mutation and chronic disease. Brit. Med. J. 1:338-342.
13. Cadet, J., C. Decarroz, S.Y. Wang, and W.R. Midden (1983) Mechanisms and products of photosensitized degradation of nucleic acids and related model compounds. Israel J. Chem. 23:420-429.
14. Cairns, J. (1978) Cancer: Science and Society, W.H. Freeman & Co., New York.
15. Clarke, C., and D.M. Shankel (1975) Antimutagenesis in microbial systems. Bacteriol. Rev. 39:33-53.
16. Claude, P., and D.A. Goodenough (1973) Fracture faces of zonulae occludentes from "tight" and "leaky" epithelia. J. Cell Biol. 58:390-400.
17. Fahey, R.C., W.C. Brown, W.B. Adams, and M.B. Worsham (1978) Occurrence of glutathione in bacteria. J. Bacteriol. 133:1126-1129.
18. Fahey, R.C., and G.L. Newton (1983) Occurrence of low molecular weight thiols in biological systems. In Functions of Glutathione: Biochemical, Physiological, Toxicological and Clinical Aspects, A. Larsson, S. Orrenius, A. Holmgren, and B. Mannervik, eds. Raven Press, New York, pp. 251-260.
19. Freifelder, D. (1965) Mechanisms of inactivation of coliphage T7 by X rays. Proc. Natl. Acad. Sci., USA 54:128-134.

20. Freifelder, D. (1966) Lethal changes in bacteriophage DNA produced by X-rays. Radiat. Res. 6(Suppl.):80-96.
21. Fujita, S., and T. Hattori (1977) Cell proliferation, differentiation and migration in the gastric mucosa: A study on the background of carcinogenesis. In Pathophysiology of Carcinogenesis in Digestive Organs, E. Farber, T. Kawachi, T. Nagayo, H. Sugano, T. Sugimura, and J.H. Weisburger, eds. University Park Press, Baltimore, pp. 21-34.
22. Genghof, D.S. (1970) Biosynthesis of ergothioneine and hercynine by fungi and actinomycetales. J. Bacteriol. 103:475-478.
23. Hartman, P.E. (1982) Announcement. Environ. Mutag. 4:529-530.
24. Hartman, P.E. (1982) Nitrates and nitrites: Ingestion, pharmacodynamics and toxicology. In Chemical Mutagens: Principles and Methods for Their Detection, Vol. 7, F.J. de Serres and A. Hollaender, eds. Plenum Press, New York, pp. 211-294.
25. Hartman, P.E. (1982) Overview: Nitrite load in the upper gastrointestinal tract--Past, present and future. In Banbury Conference: Nitrosamines and Human Cancer, Banbury Report No.12, P.N. Magee, ed. Cold Spring Harbor Laboratory, Cold Spring Harbor, New York, pp. 415-431.
26. Hartman, P.E. (1983) Review: Putative mutagens and carcinogens in foods. I. Nitrate/nitrite ingestion and gastric cancer mortality. Environ. Mutag. 5:111-121.
27. Hartman, P.E. (1983) Mutagens: Some possible health impacts beyond carcinogenesis. Environ. Mutag. 5:139-152.
28. Hartman, P.E., and P.B. Hulbert (1975) Genetic activity spectra of some antischistosomal compounds, with particular emphasis on thioxanthenones and benzothiopyranoidazoles. J. Toxicol. Environ. Health 1:243-270.
29. Hartman, P.E., and R.W. Morgan (1985) Mutagen-induced focal lesions as key factors in aging: A review. In Molecular Biology of Aging: Gene Stability and Gene Expression, R.S. Sohal, L. Birnbaum, and R.G. Cutler, eds. Raven Press, New York, pp. 93-136.
30. Hollander, F. (1962) Recent advances in the physiology of gastric secretions. Ann. N.Y. Acad. Sci. 99:4-8.
31. Ishikawa, M. (1977) Study of stromal aspect of stomach cancer in rat induced by MNNG (N-methyl-N'-nitro-N-nitrosoguanidine) in special reference of morphological change of basement membrane. Okayama Igakkai Zasshi 89:589-607.
32. Kada, T., T. Inoue, and M. Namiki (1982) Environmental desmutagens and antimutagens. In Environmental Mutagenesis, Carcinogenesis and Plant Biology, Vol. I, E. Klekowski, Jr., ed. Praeger, New York, pp. 133-151.
33. Kawano, H., K. Cho, Y. Haruna, Y. Kawai, T. Mayumi, and T. Hama (1983) Studies on ergothioneine. X. Effects of ergothioneine on the hepatic drug metabolizing enzyme system and on experimental hepatic injury in rats. Chem. Pharm. Bull. 31:1676-1681.
34. Kawano, H., H. Murata, S. Iriguchi, T. Mayumi, and T. Hama (1983) Studies on ergothioneine. XI. Inhibitory effect on lipid peroxide formation in mouse liver. Chem. Pharm. Bull. 31:1682-1687.
35. Lajtha, L.G. (1979) Stem cell concepts. Differentiation 14:23-34.
36. Lindig, B.A., and M.A.J. Rodgers (1981) Rate parameters for the quenching of singlet oxygen by water-soluble and lipid-soluble substrates in aqueous and micellar systems. Photochem. Photobiol. 33:627-634.
37. Lipkin, M. (1977) Neoplastic transformation of cells in the intestinal tract of man and rodent. In Pathophysiology of Carcinogenesis in Digestive Organs, E. Farber, T. Kawachi, T. Nagayo, H. Sugano, T.

Sugimura, and J.H. Weisburger, eds. University Park Press, Baltimore, pp. 413-428.

38. Maynard, D.N., and A.V. Barker (1979) Regulation of nitrate accumulation in vegetables. Acta Horticulturae 93:153-162.
39. Melville, D.G. (1958) Ergothioneine. Vitamins and Hormones 17:155-204.
40. Michael, D., H.A. Harrop, and K.D. Held (1981) Timescale and mechanisms of the oxygen effect in irradiated bacteria. In Oxygen and Oxy-Radicals in Chemistry and Biology, M.A.J. Rodgers and E.L. Powers, eds. Academic Press, New York, pp. 285-291.
41. Midden, W.R., and S.Y. Wang (1983) Singlet oxygen generation for solution kinetics: Clean and simple. J. Am. Chem. Soc. 105:4129-4135.
42. Mirvish, S.S. (1983) The etiology of gastric cancer. Intragastric nitrosamide formation and other theories. J. Natl. Cancer Inst. 71:630-647.
43. Morgan, R.W., J.M. Ward, and P.E. Hartman (1981) Aroclor 1254-induced intestinal metaplasia and adenocarcinoma in the glandular stomach of F344 rats. Cancer Res. 41:5052-5059.
44. Morgan, R.W., J.M. Ward, and P.E. Hartman (1981) Detection of mutagens-carcinogens: Carcinogen-induced lesions pinpointed by alkaline phosphatase activity in fixed gastric specimens from rats. J. Natl. Cancer Inst. 66:941-945.
45. Morse, M.L., and R.H. Dahl (1978) Cellular glutathione is a key to the oxygen effect in radiation damage. Nature 271:660-662.
46. Owens, R.A., and P.E. Hartman (1985) Export of glutathione (GSH) and other sulfur-containing compounds by Salmonella typhimurium and Escherichia coli. Environ. Mutag. 7(Suppl. 3):47.
47. Rajalakshmi, S., P.M. Rao, and D.S.R. Sarma (1980) Carcinogen-DNA interaction: Differential effects of distamycin-A and spermine on the formation of 7-methylguanine in DNA by N-methyl-N-nitrosourea, methylmethanesulfonate, and dimethylsulfate. Terat. Carcin. Mutagen. 1:97-104.
48. Shearn, A. (1985) Analysis of transdetermination. In Comprehensive Insect Physiology, Biochemistry, and Pharmacology, G.A. Kerkut and L.I. Gilbert, eds. Pergamon Press, New York, pp. 183-199.
49. Singh, A. (1982) Chemical and biochemical aspects of superoxide radicals and related species of activated oxygen. Can. J. Physiol. Pharmacol. 60:1330-1345.
50. Stowell, E.C. (1961) Ergothioneine. In Organic Sulfur Compounds, N. Kharasch, ed. Pergamon Press, New York, pp. 462-490.
51. Tabor, C.W., and H. Tabor (1985) Polyamines in microorganisms. Microbiol. Rev. 49:81-99.
52. Teir, H., and T. Räsänen (1961) A study of mitotic rate in renewal zones of nondiseased portions of gastric mucosa in cases of peptic ulcer and gastric cancer, with observations on differentiation and so called "intestinalization" of gastric mucosa. J. Natl. Cancer Inst. 27:949-971.
53. Thomas, H.F., P.E. Hartman, M. Mudryj, and D.L. Brown (1979) Nitrous acid mutagenesis of duplex DNA as a three-component system. Mutat. Res. 61:129-151.
54. Wattenberg, L.W. (1978) Inhibition of chemical carcinogenesis. J. Natl. Cancer Inst. 60:11-18.
55. Wattenberg, L.W. (1982) Inhibition of chemical carcinogens by minor dietary components. In Molecular Interrelations of Nutrition and Cancer, M.S. Arnott, J. van Eys, and Y.-M. Wang, eds. Raven Press, New York, pp. 43-56.

ANTIMUTAGENS AND THEIR MODES OF ACTION

Tsuneo Kada,[1] Tadashi Inoue,[1], Toshihiro Ohta,[2]
and Yasuhiko Shirasu[2]

[1]Department of Molecular Genetics
National Institute of Genetics
Yata 1111, Mishima, Shizuoka-ken 411, Japan

[2]Institute of Environmental Toxicology
Suzuki-cho 2-772, Kodaira, Tokyo 187, Japan

INTRODUCTION

Agents suppressing cellular mutagenesis have been known for some time, and their modes of action have been analyzed in the field of bacterial genetics (8,49). More recently, a number of mutagens have been detected in our environment and their genotoxicities recognized (45). Therefore, it is necessary to have knowledge about antimutagens and their modes of action in order to assess the genotoxic nature of our environment. Crow (9) considers that keeping a low mutation rate in germ cells is important to mankind. For the past several years, we have focused our attention on factors that suppress cellular mutagenesis (19,22). The word antimutagen has an old origin and has been adopted for factors that reduce the rates of spontaneous and induced mutagenesis by different modes of action. We recently proposed a distinction among categories of antimutagens (20,22) (Fig. 1) as follows:

(a) Desmutagens: The frequency of induced mutations will be reduced if mutagens are inactivated by desmutagens in vitro before reaching the cells. There also exist factors that inhibit metabolic activation of chemicals or formation of active forms of mutagens from precursors. "Desmutagens" refers to agents that cause chemical or biochemical modifications of mutagens outside cells (Fig. 2).

(b) Bioantimutagens: Factors that interfere with cellular functions which produce genetically stable informative genes from primary damage to DNA should be distinguished from desmutagens and named "bioantimutagens." Since the word "antimutagens" can be used for factors that reduce the apparent frequencies of mutations, including "desmutagens," the term bioantimutagens is more specific to those factors that are biologically active (29).

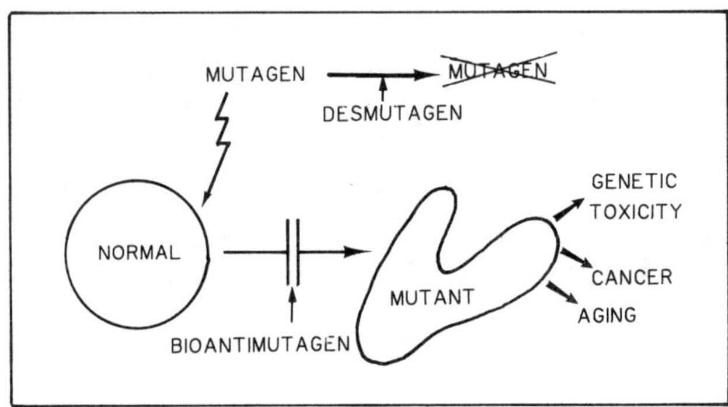

Fig. 1. Scheme of two distinct modes of action of antimutagens.

In this chapter we provide examples of desmutagens and bioantimutagens and discuss their modes of action.

DESMUTAGENS

Agents that Chemically React with Mutagens

Artificially synthesized mutagens, such as 4-nitroquinoline 1-oxide (4NQO), 2-(2-furyl)-3-(5-nitro-2-furyl)acrylamide (AF-2), etc., are inactivated simply by warming the mixture with cysteine or with cysteamine (55, 56). Moriya et al. (43) showed that Captan, a pesticide, was inactivated by treating it with cysteine. Namiki and Osawa (47,56) showed that a reaction product named Y from 2 food additives, sorbic acid and sodium nitrite, was inactivated by treatment with vitamin C. There are a number of examples where mutagens are inactivated through direct actions of desmutagens. Hydrogen oxide and active oxygen that are formed in vivo through peroxidation of fat may be captured and inactivated in vivo by reducing agents such as vitamin C, vitamin E, or uric acid as shown by Ames (1,2) and by Matsushita et al. (37). Antioxidants are generally excellent desmutagens (7,46).

Agents that Enzymatically Reduce Mutagen Activities

A series of food mutagens were identified and studied extensively in Japan and include the pyrolysis products of amino acids and protein such as Trp-P-1, Trp-P-2, Glu-P-1, and Glu-P-2 (64,65). In our attempt to identify vegetable desmutagenic factors active against tryptophan pyrolysis products (27,39,77), we first found a hemoprotein in cabbage (16). This protein had peroxidase and NADPH-oxidase activities. A similar protein was isolated from broccoli (41). Peroxidases, which are desmutagenic on heterocyclic amines, were also found in human sources such as saliva and myeloma cells (48,74). Chemical inactivation phenomena were also shown for Trp-P-1 and related pyrolysate mutagens (63,66,76).

In a broad sense, liver enzymes active in detoxification of xenobiotic chemicals in vivo can be considered desmutagenic enzymes. However, we would like to limit the designation of desmutagenic enzymes to those working in vitro.

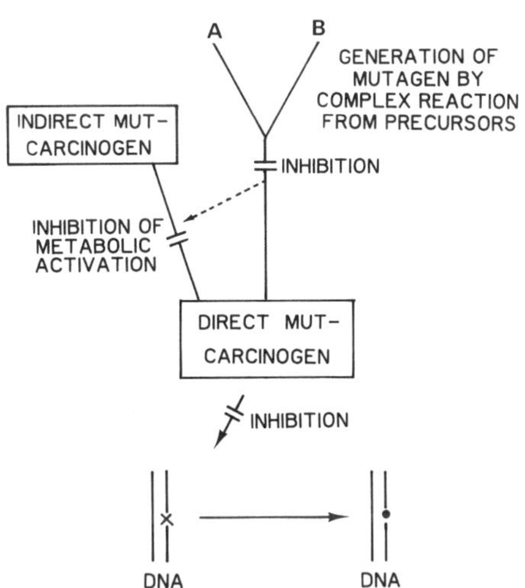

Fig. 2. Scheme of desmutagens, inhibitors of mutagens.

High Molecular Weight Compounds that Adsorb Mutagens

We recently found that the heat-resistant desmutagenic factors of vegetables consist of certain macromolecular substances and fibers (26,40). When purified fibers prepared from different kinds of vegetables, such as burdock, green pepper, cabbage, etc., were immersed in aqueous solutions containing either Trp-P-1, Trp-P-2, Glu-P-2, or 2-amino-methylimidazo(4,5)-quinoline (IQ), the solutions quickly lost their mutagenicity. We concluded that each of the pyrolysate mutagens was adsorbed, irreversibly, onto the fiber. Similar adsorption phenomena with mutagens on cereal fibers and on cotton containing an artificial blue-colored compound have been reported (4,14).

Inhibitors of Metabolic Activation of Mutagens

For indirect-acting mutagens requiring metabolic activation, inhibition of metabolic activation obviously can result in a reduction of apparent mutagenicities (13,34,35). We also observed that various vegetables contain inhibitors of metabolic activation which abolished apparent mutagenicities of Trp-P-1, Trp-P-2, etc. (16,22,27). However, when the above mixture of the mutagen, S-9, and the inhibitor was extracted with an appropriate organic solvent, intact mutagens were often recovered. We have thus hesitated to classify these inhibitors as antimutagens. However, some other possibilities exist. Suppose an indirect mutagen is in close proximity to the target cell. If the mutagen is not activated because of inhibition by a certain agent and is lost without affecting the cell, such an inhibitor of metabolic activation is obviously antagonistic to the cellular mutability in question. It is important to consider that certain inhibitors of metabolic activation of carcinogens reduce chemical incidences of cancer in animals (69).

On the other hand, a very important implication for metabolic activation exists. Obviously, metabolically activated forms of chemicals are often very reactive with certain other chemicals. For example, a metabolically active form of benzo(α)pyrene (BP) is very reactive to ellagic acid

(73). It has also been shown by Arimoto et al. (3) that metabolically activated forms of Trp-P-1 and Trp-P-2 are reactive with certain hemins and are thus inactivated.

Inhibitors that Prevent Chemical Generation of Mutagens

It is obvious that inhibition of mutagen generation from inactive precursors causes the suppression of induced mutations. We would like to classify these agents as one class of antimutagens, since we believe that these factors often operate in the bodies of man and other mammals and cause suppression of induced mutabilities. We are all aware of the famous example of vitamin C, which suppresses the generation of dimethyl nitrosamine from Na-nitrite and secondary amines, as shown by Mirvish et al. (38).

BIOANTIMUTAGENS

It has long been known that induction of mutations is influenced by physiological conditions of the test organisms (8,10,71,72). There are also a number of "artificial" factors which modulate mutagenicity in microbial tests (36). In this section, we discuss chemical antimutagens that work at the cellular level.

Agents that Increase the Fidelity of DNA Replication

The bioantimutagenicity of cobaltous chloride was observed early in our studies on N-methyl-N'-nitro-N-nitrosoguanidine (MNNG)-induced mutations in Escherichia coli B/r WP2, trp^- (25) (Fig. 3). Cobaltous chloride also reduced the high rate of spontaneous mutations in Bacillus subtilis strain NIG1125, a high rate due to a defective DNA polymerase III (17). We supposed that the high frequency of errors made during DNA replication was reduced in some way by the presence of cobaltous chloride, and that a similar mechanism is involved in reducing the frequency of MNNG-induced mutations in E. coli. Later studies showed a possible effect of cobaltous chloride on the functioning of the E. coli recA protein (32).

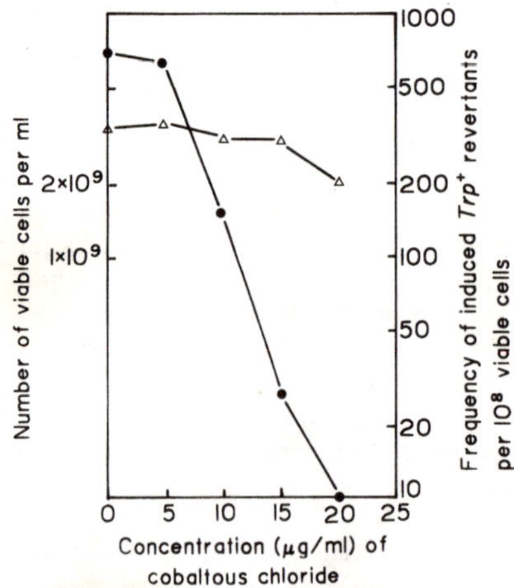

Fig. 3. Effects of cobaltous chloride on lethality and mutation induction in Escherichia coli B/r WP2 trp^- cells previously treated with MNNG (25). Survival, Δ; revertants, ●.

The relationship between this and the proofreading (3' to 5' exonuclease) of erroneous DNA replication is not clear. However, the problem might be clarified if inducible error-prone repair after UV and gamma-irradiation involves DNA polymerase III (5,6).

Recently a bioantimutagen was isolated from Japanese green tea (leaves of *Camellia sinensis*) which also reduced spontaneous mutation frequencies in *B. subtilis* strain NIG1125 (24). Chemical studies showed that the factor was epigallocatechin gallate (EGCg) (Fig. 4). It is supposed that such a tannin derivative might be able to interact directly with a protein such as DNA polymerase III.

Repair Promotion of DNA Damage

We believe that certain bioantimutagens reduce frequencies of induced mutabilities by promoting repair of DNA damage. There exist 2 categories of these bioantimutagens: (a) those promoting repair which work nonspecifically with regard to the nature of the DNA damage involved, and (b) those others that are effective with regard to specific DNA damage.

Cinnamaldehyde was identified from cinnamon as an active agent using UV-exposed *E. coli* B/r WP2 *trp* (30). Its bioantimutagenicity was very specific to UV, or to UV-mimetic mutagens such as 4NQO or AF-2; it is not active against gamma-ray or MNNG-induced mutations (Fig. 5) (51). Chemicals that are structurally related to cinnamaldehyde, such as coumarin, umbelliferone, and vanillin (Fig. 4), possess bioantimutagenic activities with similar specificity (Fig. 6) (52,54). On the contrary, methyl cinnamate derivatives enhanced UV-induced mutagenesis, probably by inhibiting a cellular excision repair process (62). Ohta et al. (50,54) found that the survival of 4NQO-treated cells increased in the presence of cinnamaldehyde

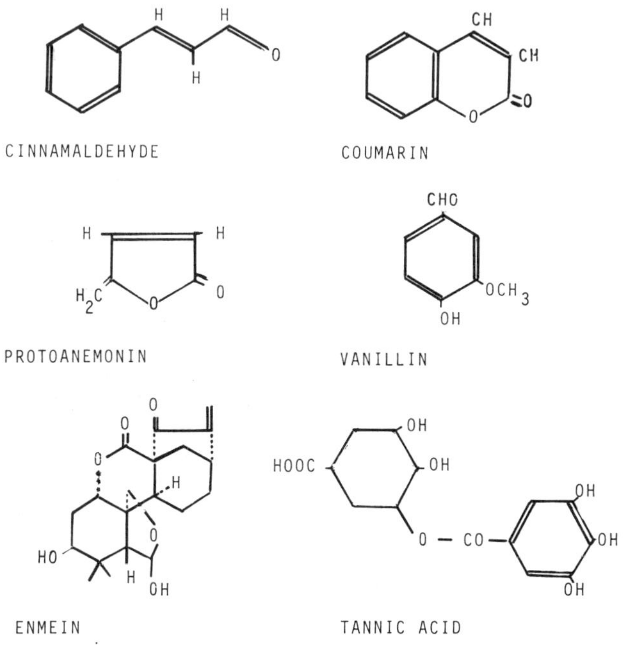

CINNAMALDEHYDE

COUMARIN

PROTOANEMONIN

VANILLIN

ENMEIN

TANNIC ACID

Fig. 4. Chemical structures of bioantimutagens.

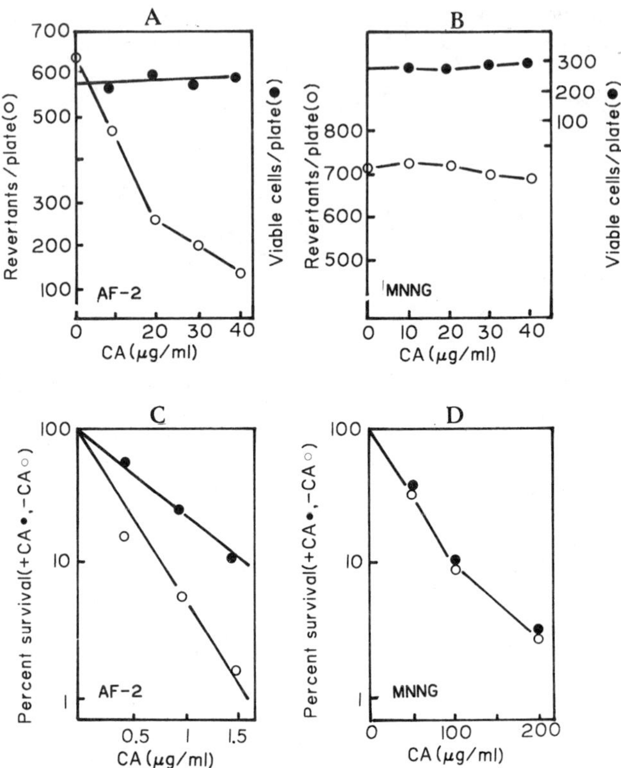

Fig. 5. Effect of cinnamaldehyde (CA) on mutation induction (A, B) and cellular viability (C, D) of Escherichia coli WP2 uvrA trpE (50).

or vanillin at appropriate concentrations (Fig. 5 and Fig. 7). They also showed that this response is strictly dependent on the recA function in E. coli K12 or B strains possessing different genetic repair deficiencies (Tab. 1). A similar increase in survival rate was observed with cobaltous chloride in UV- exposed cells of E. coli grown in the presence of the bio-antimutagen.

Cobaltous chloride is an extremely potent antimutagen in bacteria for mutations induced by irradiation (gamma-ray, UV) (Fig. 8) and chemicals (MNNG, etc.) (Fig. 3) (20,23,25). In earlier studies we thought that it was particularly involved in DNA replication (as noted above in the first section of "Bioantimutagens"). Later, because of its wide activity spectrum, we concluded that cobaltous chloride also enhances recombination repair of a variety of DNA damages. Detailed analyses were recently carried out on mutation induction by double exposures to UV irradiations. We reported some 20 years ago that cells previously exposed to UV showed high mutabilities (after postirradiation incubation) resulting from a second UV exposure; this increase was dependent on post-irradiation synthesis of protein and RNA (21). We more recently found that the presence of cobaltous chloride for 5 min after the first UV exposure totally abolishes the induced potentiality in the UV-exposed and incubated cells (unpubl. data). Since induced SOS functions are mediated by an activation of recA protein, we concluded that the cobaltous chloride effect might be related to the function of recA protein.

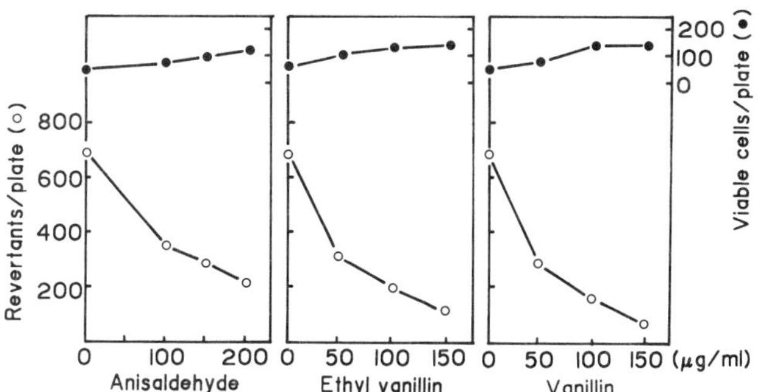

Fig. 6. Antimutagenic effects of anisaldehyde, ethyl vanillin, and vanillin on 4NQO-induced mutagenesis in Escherichia coli WP2s. Cells were treated with 4NQO at 2 μg/ml for 15 min (54).

We carried out in vitro studies in a model system of genetic recombination (Fig. 9) (60), using the recA protein of E. coli in collaboration with Shibata and colleagues. It was clearly shown that cobaltous chloride enhances D-loop formation in vitro (32) (Tab. 2).

Shimoi et al. (61) recently screened about 150 kinds of medicinal plants to detect the factors that show bioantimutagenic effects on UV-mutagenesis in bacteria. Special attention was paid to tannins as active principles. They found that tannic acid enhanced the excision-repair system in E. coli B/r WP2 trp and suppressed mutagenesis induced by UV or 4NQO, but not mutagenesis induced by gamma-rays or MNNG. We propose that tannic acid interacts directly with enzymes involved in UV-excision repair. This contrasts to the inhibition of dark repair that can be elicited at the level of DNA (59).

Inhibition of Error-Prone Repair

Ohta et al. (53) recently reported that 5-fluorouracil (5-FU) markedly inhibited induction of SOS repair in a system where colorimetrically

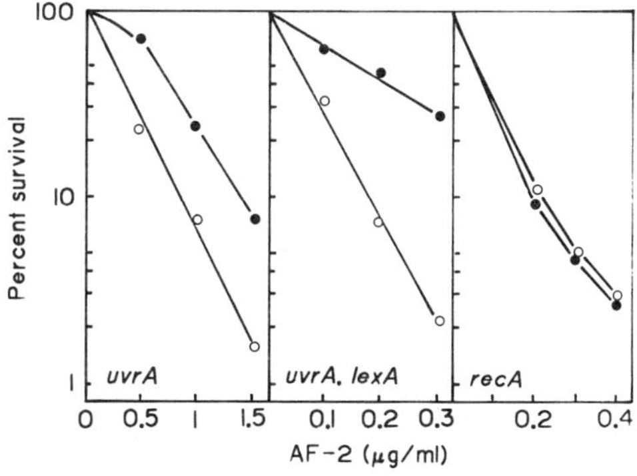

Fig. 7. Effect of vanillin on cellular viability of WP2s (uvrA), CM611 (uvrA, lexA) and CM571 (recA) after treatment with AF-2. Cells were treated with AF-2 at indicated doses for 15 min. Treated cells were washed, diluted, and then spread on SEM plates (o) and on those containing vanillin at 150 μg/ml (●) (54).

Tab. 1. Effect of cinnamaldehyde on cellular viability of DNA repair mutants of Escherichia coli after treatment with 4NQO. [For details of experiments, see Ohta et al. (50).]

Strain	4NQO treatment	Cinnamaldehyde in plate	% Survival	Increased ratio at 1% survival
WP2s (uvrA155)	10 μg/ml (15 min)	-	3.0	8.50
		+	15.0	
CM571 (recA56)	10 μg/ml (15 min)	-	5.8	1.00
		+	5.9	
TK603 (uvrA6)	10 μg/ml (15 min)	-	2.5	2.70
		+	5.7	
AB2463 (recA13)	30 μg/ml (15 min)	-	3.1	0.95
		+	2.9	

measurable expression of a structural gene is connected with a signal coupled to SOS induction (Fig. 10). They showed that this inhibitory activity of 5-FU is associated with a potent bioantimutagenic action against UV-induced mutations. Inhibitors of SOS DNA repair, such as protease inhibitors, have also been studied in connection with chemical carcinogenesis (15,67).

By studying the chemical inhibition of repair of potential lethal damage (PLD) in Chinese hamster cells (V79) cultured in vitro, Yokoiyama et al. (75) observed that gamma-ray-induced mutability decreased in the presence of cordycepin (3'-dA). The survival level also was lowered due to the presence of this drug. They proposed that PLD repair has a considerable fraction of error-prone repair which is inhibited by cordycepin.

OTHER BIOANTIMUTAGENS

We have discussed above some possible mechanisms of bioantimutagens whose modes of action are partially elucidated. On the other hand, we have detected several additional bioantimutagens whose modes of action are not yet so well studied.

Experiments have been conducted in several laboratories to attempt to find bioantimutagens in plant ingredients. Minakata et al. (42) identified protoanemonin from Ranunculus and Anemone plants as showing bioantimutagenic activities against UV- and MNNG-induced mutations. Kakinuma and his collaborators (31) isolated antimutagenic diterpenoides from a crude drug, Isodonis Herba (Enmei-so), and chemically identified them as Enmein, Nodosin and Oridonin, which were all active against UV-induced mutations in E. coli B/r trp. An extensive survey of bioantimutagens in the plant kingdom was recently published (18).

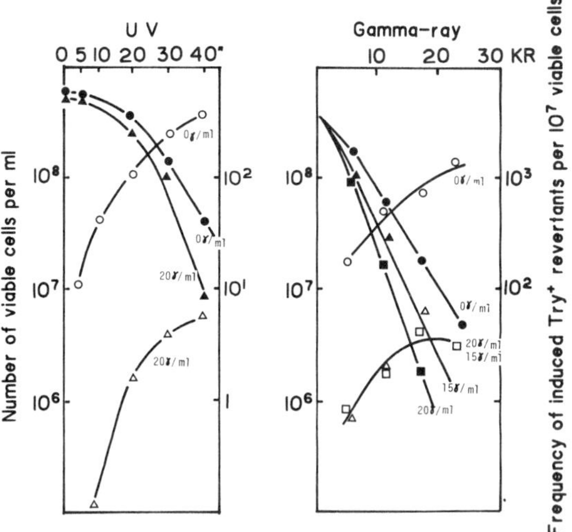

Fig. 8. Bioantimutagenic effects of cobaltous chloride on UV- and gamma-ray-induced mutations in Escherichia coli B/r WP2 trp. Irradiated cells were appropriately diluted and plated on semi-enriched minimal agar containing cobaltous chloride at different levels to determine their viability and mutability (20,23).

We have found that human placental extract works as a strong bioantimutagen against UV-, gamma-ray-, and MNNG-induced mutations in E. coli WP2 trp (28). Komura et al. (33) confirmed that the active principle in the placental extract is the cobaltous ion, which could explain an important part of its antimutagenic potency (see first two sections under "Bioantimutagens," above). This is reasonable because of similarities in the action spectra between cobaltous chloride and placental extracts. We are currently trying to find additional factors that might also be potent antimutagens.

There are a number of other possibilities for bioantimutagenicity. In E. coli, exposure to low levels of alkylating chemicals leads to induction

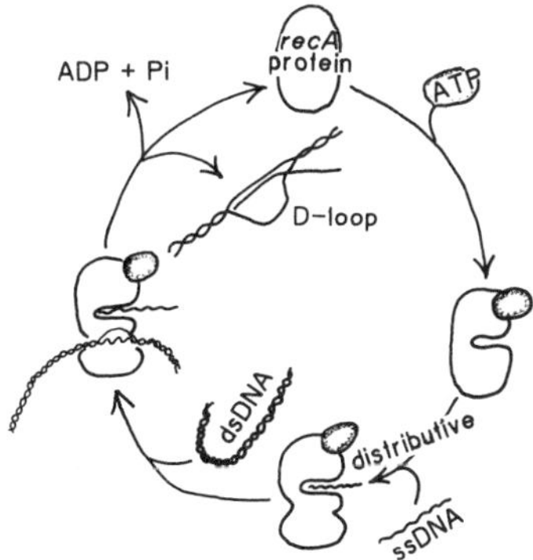

Fig. 9. Scheme of a step of genetic recombination indicating the D-loop formation involving recA protein (60).

Tab. 2. Effect of cobaltous chloride on the in vitro formation of D-loop molecules [Kiuchi et al. (32)]. The assay system was based on that described by Shibata et al. (60) and consisted of ^{3}H-labeled form-1 DNA of phage fd, its single-stranded fragments, ATP, $MgCl_2$, $CaCl_2$ and <u>recA</u> protein of <u>E. coli</u> DR1453.

Cobaltous chloride (mM)	Amount of D-loop formed (% of total)		
	Minutes of incubation		
	0	1.5	4.0
0	0.4	1.2	10.9
1	0.4	8.2	28.1
2	0.4	13.3	32.5

of error-free cellular repair functions. Mechanisms of this adaptive response are well analyzed (58). Promotion of such a response should result in reduction of mutability. In addition, certain antimutator genes have been well analyzed (11,57), and it may be interesting to study them in the future from the viewpoint of antimutagenicity (12,68).

CONCLUSIONS

We have described several different possible mechanisms for antimutagenic activities of chemicals. We proposed the term "desmutagens" to indicate factors directly reacting with mutagens before affecting the cells, and to distinguish them from "bioantimutagens," which modulate mutagenesis at the cellular level in vivo.

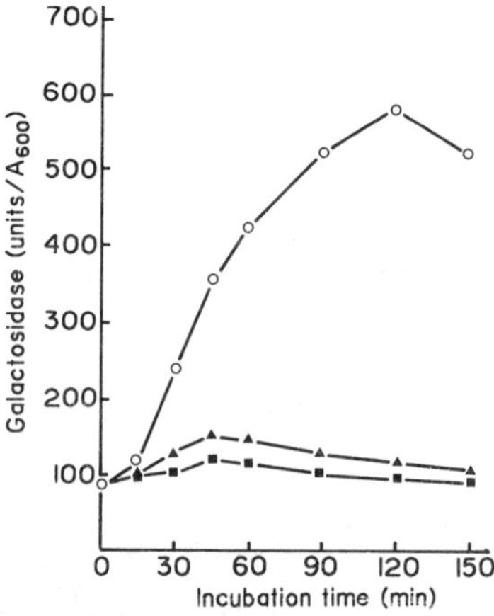

Fig. 10. Inhibition of SOS-induction by 5-fluorouracil (5-FU) and 5-fluorodeoxyuridine (FUDR). Cells were irradiated with UV at a dose of 1.5 J/m^2 and then incubated in LB medium (o), in LB medium containing 5-FU at 5 μg/ml (▲), or in LB medium containing FUDR at 10 μg/ml (■) (53).

Implications of the desmutagenic events taking place in vitro are of similar importance for other biological systems and for humans. For example, foods would become more safe if their carcinogenic ingredients were irreversibly inactivated before consumption. It is interesting to note that desmutagenic activities detected in bacterial systems often coincide with the capability to suppress cancer formation in experimental animals (69). For example, it has been shown that ellagic acid is highly reactive with metabolites of BP (73).

On the other hand, it is obvious that bioantimutagens which are effective in bacteria are not necessarily active in mammals. We are now examining the activities of bacterial bioantimutagens in cultured mammalian cells as well as in animals. While it is too early to state any conclusion, we expect that certain bioantimutagenic substances might possess universal characteristics if they are involved in fundamental and common pathways of DNA repair and DNA replication in different biological species. Also of interest is the fact that some of the bioantimutagens that we have detected are reported to show anticarcinogenic activities in experimental animal systems. For example, Wattenberg et al. (70) reported that coumarin and structurally related chemicals inhibited 7,12-dimethylbenz(α)anthracene (DMBA)- and BP-induced neoplasia in rats. They postulate that a reduction in microsomal enzyme activity might be responsible for the protective effect of coumarin and related compounds, or that the ring-opened products of these anticarcinogenic chemicals may react with certain metabolites of the carcinogens. A possibility remains that these chemicals may be antagonistic to "initiation" of malignant cells. It was recently shown that coumarin reduced DMBA-induced formation of micronuclei in bone marrow cells of mice (44). This effect could be closely allied to the repair of cellular DNA.

When we work in animal systems, it is very difficult to know if the antimutagens work as desmutagens or as bioantimutagens. Suppose an animal is treated with a mutagenic or a carcinogenic chemical. This chemical usually is metabolized and remains in the body for some time before it is totally excreted. Effects of a modulating chemical are estimated only by using "contaminated" animals. When a secondarily administered chemical reduces the incidence of tumors in experimental animals, one must first suppose that this chemical might suppress, directly or indirectly, activating enzymes of the primarily administered carcinogen. It also is often suspected that the modulator may directly capture the carcinogen. These possibilities must be eliminated before concluding, for any modulating chemicals, that they are really affecting cellular mutagenesis events involving DNA repair and DNA replication. These problems must be solved in the future.

ACKNOWLEDGEMENTS

This study was aided in part by grants for basic sciences from the Ministry of Education and the Ministry of Health and Welfare, Japan, as well as from the Nissan Science Foundation, Japan. Contribution No. 1670 from the National Institute of Genetics, Mishima, Shizuoka-ken 411, Japan.

REFERENCES

1. Ames, B.N., R. Cathcart, E. Schwiers, and P. Hochstein (1981) Uric acid provides an antioxidant defense in humans against oxidant- and

radical-caused aging and cancer: A hypothesis. Proc. Natl. Acad. Sci., USA 78:6858-6862.

2. Ames, B.N. (1983) Dietary carcinogens and anticarcinogens. Science 21:1256-1263.
3. Arimoto, S., Y. Ohara, T. Namba, T. Negishi, and H. Hayatsu (1980) Inhibition of the mutagenicity of amino acid pyrolysis products by hemin and other biological pyrrole pigments. Biochem. Biophys. Res. Commun. 9:662-669.
4. Barnes, W.S., J. Maiello, and J.H. Weisburger (1983) In vitro binding of the food mutagen 2-amino-3-methylimidazo(e,5-f)quinoline to dietary fibers. J. Natl. Cancer Inst. 70:757-760.
5. Bridges, B.A. (1980) The involvement of E. coli DNA polymerase III in repair and mutation induction by ionizing radiation. Int. J. Radiat. Biol. 37:93-96.
6. Bridges, B.A., and R.P. Mottershead (1978) Mutagenic DNA repair in Escherichia coli. VIII. Involvement of DNA polymerase III in constitutive and inducible mutagenic repair after ultraviolet and gamma irradiation. Mol. Gen. Genet. 162:35-41.
7. Calle, L.M., and P.D. Sullivan (1982) Screening of antioxidants and other compounds for antimutagenic properties towards benzo(α)pyrene-induced mutagenicity in strain TA98 of Salmonella typhimurium. Mutat. Res. 101:99-114.
8. Clarke, C.H., and D.M. Shankel (1975) Antimutagenesis in microbial systems. Bacteriol. Rev. 39:33-53.
9. Crow, J. (1986) Population consequences of mutagenesis and antimutagenesis (this Volume).
10. Doudney, C.O., and F.L. Haas (1960) Some biological aspects of the postirradiation modification of ultraviolet induced mutation frequency in bacteria. Genetics 45:1481-1502.
11. Drake, J.W., and E.F. Allen (1968) Antimutagenic DNA polymerases of bacteriophage T_4. Cold Spring Harbor Symp. Quant. Biol. 33:339-344.
12. Glickman, B. (1986) Mechanisms of spontaneous mutagenesis: Clues from altered mutational specificity in DNA repair defective species (this Volume).
13. Hayatsu, H., S. Arimoto, K. Togawa, and M. Makita (1981) Inhibitory effect of the ether extract of human feces on activation of mutagens, inhibition by oleic and linoleic acids. Mutat. Res. 81:287-293.
14. Hayatsu, H., T. Oka, A. Wakata, Y. Ohara, T. Hayatsu, H. Kobayashi, and S. Arimoto (1983) Adsorption of mutagens to cotton bearing covalently bound trisulfo-copper-phthalocyanine. Mutat. Res. 119:234-238.
15. Ichikawa-Ryo, H., and S. Kondo (1980) Differential antimutagenic effects of caffeine and the protease inhibitor antipain on mutagenesis by various mutagens in Escherichia coli. Mutat. Res. 72:311-322.
16. Inoue, T., K. Morita, and T. Kada (1980) Purification and properties of a plant desmutagenic factor for the mutagenic principle of tryptophan pyrolysate. Agric. Biol. Chem. 45:345-353.
17. Inoue, T., Y. Ohta, Y. Sadaie, and T. Kada (1981) Effect of cobaltous chloride on spontaneous mutation induced in a Bacillus subtilis mutator strain. Mutat. Res. 91:41-45.
18. Ishii, R., K. Yoshikawa, H. Minakata, H. Komura, and T. Kada (1984) Specificities of bio-antimutagens in plant kingdom. Agric. Biol. Chem. 48:2587-2591.
19. Kada, T. (1982) Mechanisms and genetic implications of environmental antimutagens. In Environmental Mutagens and Carcinogens, T. Sugimura, S. Kondo, and H. Takebe, eds. University of Tokyo Press, Tokyo, and Alan R. Liss, Inc., New York, pp. 355-359.
20. Kada, T. (1984) Desmutagens and bio-antimutagens: Their action mechanisms and possible roles in the modulation of dose-mutation relation-

ships. In Problems of Threshold in Chemical Mutagenesis, Y. Tazima et al., eds. The Environmental Mutagen Society of Japan, Mishima, Shizuoka, pp. 73-82.

21. Kada, T., C.O. Doudney, and F.L. Haas (1956) Mutation frequency response of irradiated bacteria to a second radiation exposure. Mutat. Res. 3:118-128.

22. Kada, T., T. Inoue, and M. Namiki (1981) Environmental mutagenesis and plant biology. In Environmental Desmutagens and Antimutagens, E.J. Klekowski, Jr., ed. Praeger, New York, pp. 134-151.

23. Kada, T., T. Inoue, A. Yokoiyama, and L.B. Russell (1980) Combined genetic effects of chemicals and radiation. In Radiation Research, S. Okada et al., eds. The 6th International Congress on Radiation Research, Tokyo, pp. 711-720.

24. Kada, T., K. Kaneko, S. Matsuzaki, T. Matsuzaki, and Y. Hara (1985) Detection and chemical identification of natural bio-antimutagens. A case of the green tea factor. Mutat. Res. 150:127-132.

25. Kada, T., and N. Kanematsu (1978) Reduction of N-methyl-N'-nitro-N-nitrosoguanidine-induced mutations by cobalt chloride in Escherichia coli. Proc. Japan Acad. 54:234-237.

26. Kada, T., M. Kato, K. Aikawa, and S. Kiriyama (1984) Adsorption of pyrolysate mutagens by vegetable fibers. Mutat. Res. 141:149-152.

27. Kada, T., K. Morita, and T. Inoue (1978) Anti-mutagenic action of vegetable factor(s) on the mutagenic principle of tryptophan pyrolysate. Mutat. Res. 53:351-353.

28. Kada, T., and H. Mochizuki (1981) Antimutagenic activities of human placental extract on ultraviolet light- and gamma-ray-induced mutation in Escherichia coli WP2 B/r trp. J. Radiat. Res. 22:297-302.

29. Kada, T., Y. Shirasu, N. Ikekawa, and M. Nomoto (1985) Detection of natural bio-antimutagens and in vivo and in vitro analysis of their actions. In Proceedings of the 4th International Conference on Environmental Mutagens, Stockholm (in press).

30. Kakinuma, K., J. Koike, K. Kotani, N. Ikekawa, T. Kada, and M. Nomoto (1984) Cinnamaldehyde: Identification of an antimutagen from a crude drug, Cinnamoni cortes. Agric. Biol. Chem. 48:1905-1906.

31. Kakinuma, K., Y. Okada, N. Ikekawa, T. Kada, and M. Nomoto (1984) Antimutagenic diterpenoids from a crude drug isodonis herba (Enmeiso). Agric. Biol. Chem. 48:1647-1648.

32. Kiuchi, H., T. Inoue, T. Kada, O. Makino, T. Shibata, and T. Ando (1984) Effect of an antimutagenic metal compound, cobaltous chloride, on E. coli recA protein in vitro. Mutat. Res. 130:367-368.

33. Komura, H., H. Minakata, H. Mochizuki, and T. Kada (1983) Antimutagenicity of human placenta extract: Chemical studies. In Carcinogens and Mutagens in the Environment, Vol. II, H.F. Stich, ed. CRC Press, Inc., Boca Raton, Florida, pp. 85-99.

34. Lai, C.N., M.A. Butler, and T.S. Matney (1980) Antimutagenic activities of common vegetables and their chlorophyll content. Mutat. Res. 77:245-250.

35. Martin, S.E., G.H. Adams, M. Schillaci, and J.A. Milner (1981) Antimutagenic effect of selenium on acridine orange and 7,12-dimethylbenz(α)anthracene in the Ames Salmonella/microsomal system. Mutat. Res. 82:41-46.

36. Matsushima, T., T. Sugimura, M. Nagao, T. Yahagi, A. Shirai, and M. Sawamura (1980) Factors modulating mutagenicity in microbial tests. In Short-term Test Systems for Detecting Carcinogens, K.H. Norpoth and R.C. Garner, eds. Springer-Verlag, Berlin, pp. 273-285.

37. Matsushita, S., F. Ibuki, and A. Aoki (1963) Chemical reactivity of the nucleic acid bases. I. Antioxidative ability of the nucleic acids

and their related substances on the oxidation of unsaturated fatty acids. Arch. Biochem. Biophys. 102:446-451.

38. Mirvish, S.S., I. Wallcave, M. Eagen, and P. Shubik (1972) Ascorbate-nitrite reaction: Possible means of blocking the formation of carcinogenic N-nitroso compounds. Science 177:65-68.
39. Morita, K., M. Hara, and T. Kada (1978) Studies on natural desmutagens: Screening for vegetable and fruit factors active in inactivation of mutagenic pyrolysis products from amino acids. Agric. Biol. Chem. 42:1235-1238.
40. Morita, K., T. Kada, and M. Namiki (1984) A desmutagenic factor isolated from burdock (Arctium lappa Linne). Mutat. Res. 129:25-31.
41. Morita, K., H. Yamada, S. Iwamoto, M. Sotomura, and A. Suzuki (1982) Purification and properties of desmutagenic factor from broccoli (Brassica olerancea var. Italica Plenck) for mutagenic principle of tryptophan pyrolysate. J. Food Safety 4:139-150.
42. Minakata, H., H. Komura, K. Nakanishi, and T. Kada (1981) Protoanemonin, an antimutagen isolated from plants. Mutat. Res. 116:317-322.
43. Moriya, M., K. Kato, and Y. Shirasu (1978) Effects of cysteine and a liver metabolic activation system on the activities of mutagenic pesticides. Mutat. Res. 57:259-263.
44. Morris, D.L., B.H. Harper, and M.S. Legator (1985) Coumarin inhibits micronucleus formation induced by benzo(α)pyrene in male ICR mice. 4th International Conference on Environmental Mutagens, Book of Abstracts, Stockholm, p. 55.
45. Nagao, M., T. Sugimura, and T. Matsushima (1978) Environmental mutagens and carcinogens. Ann. Rev. Genet. 12:117-159.
46. Namiki, M., and T. Osawa (1985) Antioxidants/antimutagens in foods (this Volume).
47. Namiki, M., S. Udaka, T. Osawa, K. Tsuji, and T. Kada (1980) Formation of mutagens by sorbic acid-nitrite reaction: Effects of reaction conditions on biological activities. Mutat. Res. 73:21-28.
48. Nishioka, H., K. Nishi, and K. Kyokane (1981) Human saliva inactivates mutagenicity of carcinogens. Mutat. Res. 85:323-333.
49. Novick, A. (1956) Mutagens and antimutagens. Brookhaven Symposium Biol. 8:201-215.
50. Ohta, T., K. Watanabe, M. Moriya, Y. Shirasu, and T. Kada (1983) Analysis of the antimutagenic effect of cinnamaldehyde on chemically induced mutagenesis in Escherichia coli. Mol. Gen. Genet. 192:309-315.
51. Ohta, T., K. Watanabe, M. Moriya, Y. Shirasu, and T. Kada (1983) Antimutagenic effects of cinnamaldehyde on chemical mutagenesis in Escherichia coli. Mutat. Res. 107:219-227.
52. Ohta, T., K. Watanabe, M. Moriya, Y. Shirasu, and T. Kada (1983) Antimutagenic effects of coumarin and umbelliferone on mutagenesis induced by 4-nitroquinoline 1-oxide or UV irradiation in E. coli. Mutat. Res. 117:135-138.
53. Ohta, T., M. Watanabe, R. Tsukamoto, Y. Shirasu, and T. Kada (1985) Antimutagenic effects of 5-fluorouracil and 5-fluorodeoxyuridine on UV-induced mutagenesis in Escherichia coli. Mutat. Res. (in press).
54. Ohta, T., M. Watanabe, K. Watanabe, Y. Shirasu, and T. Kada (1986) Inhibitory effects of flavourings on mutagenesis induced by chemicals in bacteria. Food Chem. Toxicol. (in press).
55. Onitsuka, S., N.Y. Chanh, S. Murakawa, and T. Takahashi (1978) Report of Special Grants from Ministry of Education of Japan, Environmental Sciences, A-4, Tokyo.
56. Osawa, T., H. Ishibashi, M. Namiki, and T. Kada (1980) Desmutagenic actions of ascorbic acid and cysteine on a new pyrrole mutagen formed by the reaction between food additives: Sorbic acid and sodium nitrite. Biochem. Biophys. Res. Commun. 95:835-841.

57. Quah, S.K., R.C. von Borstel, and P.J. Hastings (1980) The origin of spontaneous mutation in Saccharomyces. Genetics 96:819-839.
58. Sekiguchi, M., H. Kondo, K. Sakumi, and Y. Nakabeppu (1986) Molecular mechanism of adaptive response to alkylating agents (this Volume).
59. Shankel, D.M. (1972) Inhibition of dark repair of ultraviolet damage by ethidium bromide. Radiat. Res. 51:Gd-5.
60. Shibata, T., T. Ohtani, M. Iwabuchi, and T. Ando (1982) D-loop cycle. A circular reaction sequence which comprises for formation and dissociation of D-loop and inactivation and reactivation of superhelical closed circular DNA promoted recA protein of Escherichia coli. J. Biol. Chem. 257:13981-13986.
61. Shimoi, K., Y. Nakamura, I. Tomita, and T. Kada (1985) Bio-antimutagenic effects of tannic acid on UV and chemically induced mutagenesis in Escherichia coli B/r. Mutat. Res. 149:17-23.
62. Shimoi, K., Y. Nakamura, T. Noro, I. Tomita, S. Fukushima, T. Inoue, and T. Kada (1985) Methyl cinnamate derivatives enhance UV-induced mutagenesis due to the inhibition of DNA excision repair in Escherichia coli B/r. Mutat. Res. 146:15-22.
63. Stich, H.F., C. Wu, and W. Powrie (1982) Enhancement and suppression of genotoxicity of food by naturally occurring components in these products. In Environmental Mutagens and Carcinogens, T. Sugimura, S. Kondo, and H. Takebe, eds. University of Tokyo Press, Tokyo, and Alan R. Liss, Inc., New York, pp. 347-353.
64. Sugimura, T. (1982) Mutagens, carcinogens, and tumor promoters in our daily food. Cancer 49:1970-1984.
65. Sugimura, T. (1985) Carcinogenicity of mutagenic heterocyclic amines formed during the cooking process. Mutat. Res. 150:33-41.
66. Tsuda, M., Y. Takahashi, M. Nagao, T. Hirayama, and T. Sugimura (1980) Inactivation of mutagens from pyrolysates of tryptophan and glutamic acid by nitrite in acidic solution. Mutat. Res. 78:331-339.
67. Umezawa, K., T. Matsushima, and T. Sugimura (1977) Antimutagenic effect of elastatinal, a protease inhibitor from Actinomycetes. Proc. Japan Acad. 53(ser.B):30-33.
68. Walker, G.C. (1979) Mutagenesis- and repair-enhancing activities associated with the plasmid pKM101. Cold Spring Harbor Symp. Quant. Biol. 43:893-896.
69. Wattenberg, L.W. (1978) Inhibitors of chemical carcinogenesis. Adv. Cancer Res. 26:197-223.
70. Wattenberg, L.W., L.K.T. Lam, and A.V. Fladmoe (1979) Inhibition of chemical carcinogen-induced neoplasis by coumarins and α-angelicalactone. Cancer Res. 39:1651-1654.
71. Witkin, E.M. (1956) Time, temperature, and protein synthesis: A study of ultraviolet-induced mutation in bacteria. Cold Spring Harbor Symp. Quant. Biol. 21:123-140.
72. Witkin, E.M., and E.L. Farquharson (1969) Enhancement and diminution of ultraviolet light-initiated mutagenesis by post-treatment with caffeine in Escherichia coli. In Mutation as a Cellular Process, CIBA Foundation Symposium, G.E.W. Wolstenholme and M. O'Connor, eds. J. and A. Churchill Ltd., London, pp. 36-49.
73. Wood, A.W., M.T. Huang, R.L. Chang, H.L. Newmark, R.E. Lehr, H. Yagi, J.M. Sayer, D.M. Jerina, and A.H. Conney (1982) Inhibition of the mutagenicity of bay-region diol epoxides of polycyclic aromatic hydrocarbons by naturally occurring plant phenols: Exceptional activity of ellagic acid. Proc. Natl. Acad. Sci., USA 79:5513-5517.
74. Yamada, M., M. Tsuda, M. Nagao, M. Mori, and T. Sugimura (1979) Degradation of mutagens from pyrolysates of tryptophan, glutamic acid and globulin by myeloperoxidase. Biochem. Biophys. Res. Commun. 90:769-776.

75. Yokoiyama, A., T. Kada, and T. Sugahara (1985) 3-dA (3'-deoxyadenosine), inhibitor of potentially lethal damage (PLD) repair, reduced the frequency of mutations in cultured Chinese hamster cells *in vitro*. *Ann. Rep. Natl. Inst. Genet.* No. 35 (in press).

76. Yoshida, D., and T. Matsumoto (1978) Changes in mutagenicity of protein pyrolysates by reaction with nitrate. *Mutat. Res.* 58:35-40.

77. Yoshikawa, K., R. Ishii, and T. Kada (1981) Desmutagenic action of vegetable extracts on the mutagenic pyrolysis products of foods. In *Proceedings of the 3rd International Conference on Environmental Mutagens*, Tokyo, 3P19.

AVOIDANCE OF ERRORS IN DNA

INTRODUCTION: AVOIDANCE OF SPONTANEOUSLY OCCURRING ERRORS IN DNA

Lawrence A. Loeb

The Joseph Gottstein Memorial
Cancer Research Laboratory
University of Washington
Seattle, Washington 98195

In order to perpetuate species homogeneity, DNA replication must be an exceptionally accurate process. Yet, a small number of mistakes in DNA replication must also be permitted to promote species evolution. The mechanism for achieving high fidelity with bacterial systems has been investigated extensively. The major contributors to accuracy are the DNA polymerases themselves. Alterations in DNA polymerases and associated proteins can modulate fidelity and effect mutagenesis. In animal cells, the accuracy of DNA replication is likely to be even higher than in bacteria. Measurements of spontaneous mutation frequencies in animal cells suggest that the average frequency of errors in DNA replication is in the range of 10^{-9} to 10^{-11} errors per base pair replicated. This value is the net result of errors made during DNA replication and repair minus the correction of these mistakes by processes such as mismatch correction.

So far, two methods have been used to analyze the fidelity of DNA polymerases and replicating complexes. The first, simplest, and most extensively used method involves the copying of a synthetic polynucleotide template composed of only 1 or 2 nucleotide species. An error is defined as the incorporation of any nucleotide that is noncomplementary to the base(s) in the template. This assay is limited by the chemical purity of the DNA template and substrates. At best, one can detect 1 error in 10^5 nucleotides polymerized.

The second approach is to use a natural biologically active DNA as a template, copy with DNA polymerase, and measure the reversion of a single base mutation to wild type. The most extensively used assay is with single-stranded ØX174 DNA containing an amber 3 mutation. After copying by a polymerase in the test tube, the partially double-stranded product is transfected into <u>Escherichia coli</u> spheroplasts and the error rate of the polymerase is determined from the reversion rate of progeny phage after plating on appropriate bacteria. The sensitivity of this assay approaches 10^{-6} in reactions with all four nucleoside triphosphates at equal concentration. By extrapolating from the results of experiments in which the concentration of the noncomplementary nucleotide is progressively varied, error rates as low as 10^{-8} to 10^{-9} can be ascertained.

Using these assays, several laboratories have been able to copy DNA in the test tube with purified bacterial components and achieve a fidelity similar to that exhibited by bacteria in vivo. With prokaryotic DNA polymerases, an accuracy of 10^{-7} can be achieved by at least two sequential steps: the first involves selection of the complementary nucleotide by the polymerase, and the second involves proofreading by the associated 3' → 5' exonuclease. The fact that one can achieve an accuracy in the test tube approaching that in vivo is not an argument against the involvement of other replicating components in enhancing accuracy in vivo. Studies carried out in vitro are under optimal conditions, and might not be those occurring in the cell.

In contrast, most purified animal DNA polymerases are highly error-prone and after purification lack any associated exonucleolytic activity. The error rates of purified homopolymeric animal DNA polymerases vary from 1 in 2,000 to 1 in 40,000. When one compares these numbers to the spontaneous mutation rates in animal cells, one must conclude that our gap in knowledge may be as great as 6 orders of magnitude. Even though it seems likely that 3' → 5' exonucleolytic activities are lost during purification of eukaryotic DNA polymerases, there is no strong evidence for proofreading by DNA polymerase-associated exonucleases in animal cells. Based on energy considerations, it also seems likely that proofreading, if operative in animal cells, cannot account for enhancing fidelity by more than 2 or 3 orders of magnitude. Thus, a mechanism for postsynthetic correction, like mismatch repair, is likely to occur.

However, it should be noted that enzyme activities that preferentially excise noncomplementary base pairs remain to be demonstrated in eukaryotic cells. Also, cells have to establish a mechanism to distinguish the newly synthesized daughter DNA strand from the parental DNA strand. Thus, a major goal of future research in this area is to establish an in vitro system, using purified components from animal cells, that approaches the fidelity achieved during eukaryotic DNA replication in vivo, and then to determine how environmental factors might modulate this fidelity.

INTRODUCTION: AVOIDANCE OF ERRORS AFTER DNA DAMAGE

D.G. MacPhee

Department of Microbiology
La Trobe University
Bundoora, Victoria, Australia

Two of the following chapters are concerned with adaptive responses in *Escherichia coli*. Dr. Demple discusses possible mechanisms involved in the adaptive response to alkylation of DNA, and then describes several ways in which the response of *E. coli* cells to oxidative damage can be measured. In responding to an oxidative challenge, a novel type of adaptive response, different from either the SOS response or the heat shock response, appears to be involved.

Dr. Sekiguchi outlines the molecular mechanisms that appear to be involved in the adaptive response to alkylation damage and, in particular, refers to studies of the control and expression of the *ada* and *alkA* genes. It is perhaps noteworthy that in their abstract, Dr. Demple and colleagues refer to how little we actually know about the biochemical mechanisms of repair of ionizing radiation damage in the yeast *Saccharomyces cerevisiae*. However, very little is actually known about the biochemical mechanisms of repair of any kind of DNA damage in yeast, even including UV damage.

This is made clear in Dr. Friedberg's chapter, which is essentially a progress report on a systematic (and very probably long-term) attempt to remedy the situation for UV, although much useful information about mechanisms of repair of ionizing radiation damage will almost certainly emerge along the way.

Before proceeding, I would like to present a few brief remarks and observations specifically about bacterial mutagenesis. The topic area implies that cells possess mechanisms which permit avoidance of errors after sustaining DNA damage, as indeed they do with, for example, the adaptive response to alkylation damage and the inducible response to oxidative damage discussed by Dr. Demple. Thus, agents or systems that assist in error avoidance would tend to be considered antimutagenic, while agents whose effects are manifested in the opposite direction, i.e., inhibition of error avoidance mechanisms, or promotion of fixation of premutational lesions, would tend to be considered co-mutagenic. Indeed, in bacterial systems in which mutation yields following DNA damage can be measured at reasonable levels, the actual yields of mutants observed could be influenced in a number of different ways:

	Error Avoidance	Error Fixation	Mutagenesis Level
(a)	unchanged	unchanged	intermediate
(b)	promoted	promoted	how strong is each effect?
(c)	promoted	unchanged	decreased
(d)	unchanged	promoted	increased
(e)	inhibited	unchanged	increased
(f)	unchanged	inhibited	decreased

The well-known capacity of caffeine to enhance UV-induced mutation yields in _E. coli_ and other species (4) would thus be an example of situation (e); in this case the error-avoidance system implicated is almost certainly the _uvrA,B,C_-dependent excision repair pathway. However, the decreased mutation yields observed when UV-irradiated cells of an excision-deficient strain of _E. coli_ are plated on caffeine-containing media (1) would be an example of situation (f). Similarly, the decrease (to zero) of 9-aminoacridine-induced frameshift revertants of the _hisC3076_ marker in a _uvr_ derivative of _Salmonella typhimurium_ (3) could be an example of situation (f), whereas the slight enhancement in 2-aminopurine-induced mutation yields observed with the missense _trpE8_ marker [situation (d)] would therefore tend to implicate different types of avoidance/fixation systems in the two types of mutagenesis (3).

Another possible way of influencing mutation yields might involve changes in the relative amounts of avoidance/fixation enzymes in response to normal cellular regulatory mechanisms. Results recently obtained in a study of spontaneous revertant yields with a missense mutant strain of _S. typhimurium_ (2) can be explained on this basis if it is assumed that: (i) there is an inducible (error-prone) gene product involved in error fixation (in this case, the errors would be naturally arising), and (ii) induction of this gene product is subject to normal catabolite (glucose) repression. Given these assumptions, the reduced yields of spontaneous mutants observed when cells are plated on glucose rather than glycerol can be explained, as can the reduced yields observed when both carbon sources are present. [Thus, both would be examples of situation (f).] A simple extension of these suggestions would imply that bacterial cells are intrinsically more mutable under adverse conditions (see Ref. 2 for more detailed discussion).

Recent observations in my laboratory by Susan Thomas and George Kopsidas indicate that 9-aminoacridine mutagenesis is greatly reduced during liquid treatment in the presence of glucose (as opposed to treatment in the presence of glycerol, or even no carbon source at all), and suggest that glucose may act either by strongly promoting error avoidance [situation (c)], or else by strongly inhibiting error fixation [situation (f)]. Other data we have obtained suggest that glucose may, in fact, be acting by inhibiting error fixation, although the precise mechanism is not yet clear.

REFERENCES

1. Clarke, C.H. (1967) Caffeine- and amino acid-effects upon try$^+$ revertant yield in UV-irradiated di-auxotrophs of _E. coli_ B/r. _Molec. Gen. Genet._ 99:97-108.
2. MacPhee, D.G. (1985) Indications that mutagenesis in Salmonella may be subject to catabolite repression. _Mutat. Res._ 151:35-41.

3. Podger, D.M., G.W. Grigg, and D.G. MacPhee (1983) Ethionine abolishes mutagenesis by 9-aminoacridine (but not by 2-aminopurine) in Salmonella plate tests. Mutat. Res. 119:113-120.
4. Shankel, D.M. (1962) Mutational synergism of ultraviolet light and caffeine in Escherichia coli. J. Bacteriol. 84:410-415.

ALKYLATION AND OXIDATIVE DAMAGES TO DNA: CONSTITUTIVE AND INDUCIBLE REPAIR SYSTEMS

Bruce Demple, Yasmin Daikh, Jean Greenberg, and Arlen Johnson

Department of Biochemistry and Molecular Biology
Harvard University
Cambridge, Massachusetts 02138

INTRODUCTION

Simple chemical agents that damage DNA are of both environmental and intracellular origin. Among these are carcinogen-related compounds, including simple alkylating agents (e.g., S-adenosyl methionine) and simple oxidizing agents (e.g., oxygen radicals). Living cells have therefore to counter the potentially mutagenic and carcinogenic effects of physiological compounds on a daily basis, and not just when they encounter an occasional environmental challenge.

The alkylating agents form an important class of chemicals, some of which are very potent mutagens and carcinogens, such as N-methyl-N'-nitro-N-nitrosoguanidine (MNNG) and N-methyl-N-nitrosourea (MNU). The latter chemical can cause mammary carcinomas in rats following a single injection, evidently by means of a single G:C → A:T transition in the twelfth codon of the Ha-*ras*-1 gene (48). The discovery of the adaptive response to alkylating agents in *Escherichia coli* (39), combined with the existing extensive knowledge of the chemistry of DNA alkylation (43), allowed a spectacularly rapid advance in the understanding of the molecular details of alkylation mutagenesis and the biochemistry of alkylated DNA repair. Recent results regarding the biochemistry and control of the adaptive response are presented in the next section.

Ionizing radiation was the first environmental agent that was shown to be capable of causing mutations (36). A large number of common compounds and some cellular metabolites appear to cause DNA damages by mechanisms similar or identical to ionizing radiations, namely, via the production of active oxygen species (1). Despite the considerable effort that has been expended analyzing the chemistry of radiation damages to DNA (see Ref. 19 for a review), our understanding of the mechanisms of radiation mutagenesis and carcinogenesis remains poor. It is therefore of primary importance to understand the natures of radiation and oxidative DNA damages and the biochemical mechanisms of their repair.

Escherichia coli bacteria can mount an adaptive resistance to the lethal effects of oxidative stress, as produced by the simple oxidizing agent hydrogen peroxide (11), and a similar phenomenon has recently been described in *Salmonella* (8). The response is induced specifically by oxidizing agents, and peroxide-adapted bacteria show enhanced resistance specifically towards oxidative stresses (8,11) and heat (8), but not towards other damaging agents, such as MNNG or UV light (11). This inducible response is accompanied by enhanced reactivation of oxidatively damaged phages (10,11,17) and an induced resistance to the lethality caused by γ-rays, indicating that a component of the response is an increased ability of the bacteria to handle the types of damaged DNA that arise from oxidation or ionizing radiation.

The adaptive response to oxidative damages is global. The synthesis of some 30 proteins is induced in response to H_2O_2 (8), including the protective activities catalase (11) and an alkyl hydroperoxide reductase (8). The destruction of active oxygen species by these enzymes would seem an important component of adapted cell survival, but such a conclusion is not always warranted (see section on "Adaptive Response to Oxidizing Agents").

The biochemical pathways of repair of oxidative and radiation DNA damages in eukaryotes have not been thoroughly explored. We have addressed this situation by beginning a systematic identification and characterization of enzymes in the yeast *Saccharomyces cerevisiae* that act on oxidatively damaged DNA, the subject of the section on "Repair of Oxidation and Radiation Damages in Yeast."

ADAPTIVE RESPONSE TO ALKYLATING AGENTS

Some years ago, it was discovered that *E. coli* bacteria have the capacity to activate specific defenses against alkylation mutagenesis (21,39) by inducing, in response to agents such as MNNG or MNU, two cellular repair activities specific for alkylated bases in DNA. A DNA glycosylase of relatively broad specificity, encoded by the *alkA* gene (16), is elevated about 20-fold during adaptation and removes N3-alkylated purines (22) and O^2-alkylated pyrimidines (31). The AlkA glycosylase appears to remove potentially lethal damages that contain methyl groups in the minor groove of DNA (16,31).

The mutagenic damages O^6-methylguanine (O^6MeG) and O^4-methylthymine (O^4MeT) are repaired by an inducible activity that operates by an unusual suicide mechanism: the methyl (alkyl) groups of O^6MeG and O^4MeT are transferred by a protein to one of its own cysteine residues in a self-inactivating reaction (14,38). This inducible suicide repair enzyme is the product of the *ada* gene (15,32,37,44), previously identified as the key regulatory locus controlling the adaptive response to alkylating agents (20,40,41). The Ada protein also contains a separate suicide methyltransferase site that scavenges the methyl groups from methyl phosphotriesters, i.e., alkylations of the DNA backbone (30,32). In addition to its repair activities, the Ada protein is a positive regulator of *ada*, *alkA*, and *aidB* transcription (25,41,45).

Escherichia coli mutants defective in the adaptive response to alkylating agents were originally isolated by virtue of their high sensitivity to alkylation mutagenesis (20). Such mutants are expected to fall into at least three categories based on the known functions of the *ada* gene: (a) *ada* alleles encoding a defective methyltransferase for O^6MeG;

(b) alleles encoding an Ada protein that is impaired in positive control of transcription; and (c) promoter mutants in ada. Two of the mutants isolated by Jeggo (20) synthesize significant amounts of Ada protein in response to MNNG [the ada3 and ada5 mutants (44)], but are nevertheless sensitive to MNNG mutagenesis even after such an induction (20). Thus, ada3 and ada5 are candidates for alleles encoding defective O^6MeG methyltransferase.

A Kinetic Defect in ada Mutants

Cell-free extracts of adapted PJ3 (ada3) exhibited a small amount of O^6MeG-DNA methyltransferase activity in the standard assay (14), suggesting that the Ada3 methyltransferase is active but impaired. To test this hypothesis, kinetic measurements were made of the induced methyltransferase activities in crude extracts of adapted wild-type and ada mutant cells.

Purified wild-type Ada proteins effect a complete reaction in <15 sec (26,32). Similar rapid kinetics are observed using crude extracts of an adapted ada^+ strain (14,34) (Fig. 1). In contrast, the extract from induced PJ3 cells exhibits much slower repair kinetics, with the reaction being complete only after about 35 min (Fig. 1). Repeated kinetic analysis of the Ada3 activity confirms a 5 to 10 times slower initial rate for the mutant compared to the wild-type enzyme. The possibility that ada5 also encodes an active but retarded O^6MeG methyltransferase was tested by

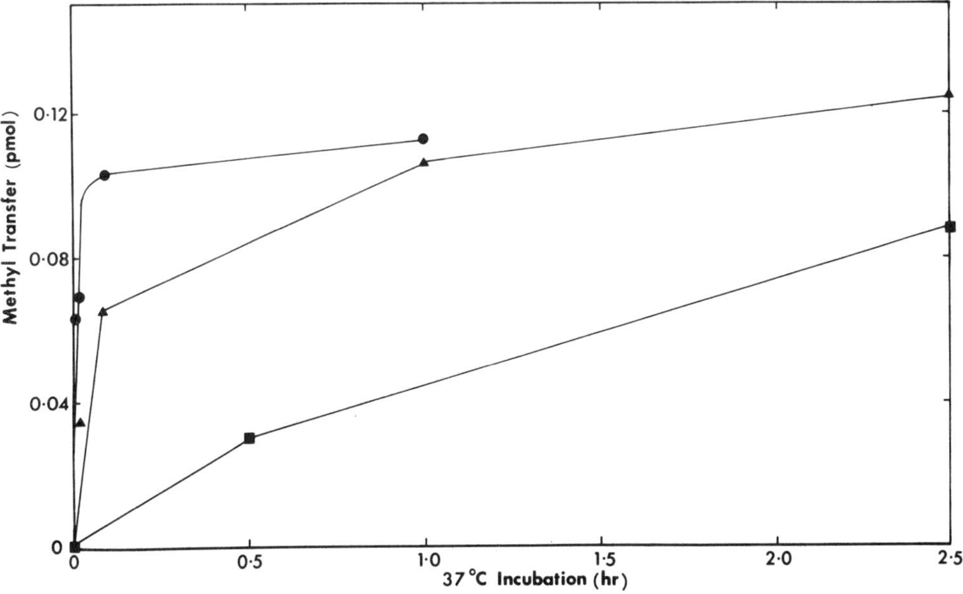

Fig. 1. O^6MeG-methyltransferase kinetics in ada^+ and ada mutants. Extracts of AB1157, PJ3, and PJ5 cultures treated with MNNG (0.1 µg/ml for 30 min, followed by 0.5 µg/ml for an additional 30 min) were made as previously described (11), except that methyltransferase extraction buffer (14) was used. Standard O^6MeG-methyltransferase reactions (26) containing samples of the individual extracts were incubated at 37°C, and aliquots removed at the indicated times for analysis of methyl groups remaining in the DNA (14). Extracts: ●, ada^+; ▲, ada3; ■, ada5.

kinetic analysis (Fig. 1). The induced PJ5 methyltransferase is even more impaired than the Ada3 activity, requiring at least 150 min to complete the methyl transfer reaction, and with an initial rate about 100-fold slower than the Ada^+ activity (Fig. 1). If the length of time required for the reactions involving the mutant methyltransferases is taken into account, the levels of activity (pmol methyl transfer per mg of protein) induced by MNNG are comparable in <u>ada</u>$^+$, <u>ada3</u>, and <u>ada5</u> strains (data not shown).

The location within the Ada3 protein of the defect that results in a decreased methyl transfer efficiency was deduced by taking advantage of the fortuitous limited proteolysis of the 39-kDa Ada protein to a 19-kDa domain that retains O^6MeG methyltransferase activity (14,44). A concentrated extract of adapted PJ3 cells was subjected to gel filtration chromatography on Ultrogel AcA54 (14). Two peaks of methyltransferase activity for O^6MeG were located, corresponding to the expected proteins of 39 and 19 kDa, respectively. The active fractions were pooled and concentrated by precipitation with ammonium sulfate.

Both the intact Ada3 protein and its 19-kDa subfragment have O^6MeG methyltransferase activity that is significantly slower than the wild-type enzyme by a factor of at least 10 (B. Demple, ms. in prep.). Thus, the sluggishness of the Ada3 methyltransferase is a result of a mutational change in the 19-kDa domain that corresponds to the carboxy-terminal 50% of Ada (15) and cannot be ascribed to a defect in the N-terminal portion of the 39-kDa protein.

The impaired ability of the Ada3 methyltransferase to correct O^6MeG rapidly is not a result of a replacement of the active cysteine residue by another amino acid. S-Methylcysteine is the only methylated amino acid observed in proteolytic hydrolysates following reaction of the 19-kDa domain with alkylated DNA (data not shown). It is possible that other residues in the active site (15) are altered to reduce the nucleophilicity of the cysteine sulfhydryl. The 19-kDa Ada3 methyltransferase could also be defective in its ability to bind to DNA at all, or to recognize the damaged bases correctly.

<u>Regulation in ada3 and ada5</u>

The sensitivity to mutagenesis of the adapted <u>ada3</u> and <u>ada5</u> mutants is evidently a result of the inability of the induced proteins to repair O^6MeG and O^4MeT in parental DNA before replication has fixed mispairs at these sites (17a,26a). It has been calculated that this repair must occur within 1 sec in vivo to account for the induced resistance to MNNG mutagenesis in adapted wild-type cells (5), and the wild-type Ada protein is certainly sufficient to this task in vitro (26). To a first approximation, one would also expect the induction of the <u>ada</u> and <u>alkA</u> genes in these mutants to be normal if the O^6MeG methyltransferase and positive regulation are separate functions of the Ada molecule. However, it should be noted that these mutants also exhibit abnormal sensitivity to the toxicity of MNNG even when adapted, as though they have a defect in regulation of the response (20), i.e., diminished induction of the AlkA protein.

The question of regulation in PJ3 and PJ5 was examined in 3 independent ways. Measurements of the rate of induction of methyltransferase activity in PJ3 were made by making a series of extracts of PJ3 cells removed from a culture at various times following the addition of MNNG, and using a prolonged incubation time (45 min) in the assay. Elevated methyltransferase levels could be detected in AB1157 (<u>ada</u>$^+$) after only 5 to 10 min, and increased at an initial rate of 0.17 Units/mg protein/min. In the case of PJ3 (ada3), increased transferase activity could only be detected after 45 min, increasing then at an initial rate of only 0.019 Units/mg protein/min (B. Demple, ms. in prep.).

Further confirmation of this result was sought by analyzing the series of extracts using immunoblotting. Antibodies raised against the purified 19-kDa methyltransferase can be used to quantitate sensitively the intact Ada protein (44). Such an analysis shows rapid induction of Ada protein in the ada^+ strain (<10 min) (44), while the increase is significantly retarded in the ada3 mutant, in which up to 45 min are required to increase Ada to detectable levels (B. Demple, ms. in prep.).

Induction of ada transcription can be measured by using gene fusions. LeMotte and Walker (25) have constructed a hybrid gene in which the ada^+ promoter is fused to the lacZ structural gene. They have shown that all of the mutants isolated by Jeggo (20) are leaky with respect to regulation, but they did not examine the early kinetics of induction. We have used these same fusions (located on a plasmid) to focus on the activation of ada transcription at early times during exposure to 0.25 μg/ml MNNG. The data reveal that, as seen when monitoring Ada activity or protein levels, induction of transcription under the control of either ada3 or ada5 shows a longer lag than under ada^+ regulation--25 min vs 10 min, respectively. Following this lag, β-galactosidase activity is induced at the same slow rate in both mutants, reduced by a factor of 7- to 8-fold (B. Demple, ms. in prep.). A similar kinetic defect in the induction of the ada::lacZ fusion was also observed at the higher level of MNNG used by LeMotte and Walker (25) (data not shown).

Repair and Control by Ada

The ada3 and ada5 alleles fall into an expected category of ada mutants in that both genes encode defective O^6MeG methyltransferase activities. What is surprising is that these same alleles give rise to defective regulation of the adaptive response. The degree of impairment of the methyl transferring capacity of Ada is not directly reflected in the degree by which positive regulation is defective: both ada3 and ada5 strains induce the ada::lacZ fusion at the same slow rate, but differ by a factor of 10 in their rates of O^6MeG methyl transfer. Thus, self-methylation of Ada at the O^6MeG-DNA methyltransferase active site is evidently not a limiting event for activation of the adaptive response genes.

Suicide methylation of the methyl phosphotriester repair site might activate positive control by Ada. However, in the case of ada3, the rate of methylphosphotriester repair in crude extracts is indistinguishable from the wild-type strain (B. Demple, ms. in prep.). It therefore appears that the 19-kDa O^6MeG methyltransferase domain of the Ada protein is required for efficient activation of the adaptive response in *E. coli*. The nature of this requirement (e.g., in Ada dimerization or interaction with RNA polymerase), apparently distinct from repair activity per se, should help to reveal the mechanism of Ada-mediated positive control.

The ada3 and ada5 alleles turn out to have doubly deleterious consequences for the cell. Not only are mutants containing these genes slow to induce the O^6MeG-DNA methyltransferase that is needed to counteract most alkylation mutagenesis, but the proteins that they do eventually synthesize are incapable of acting rapidly enough to neutralize the miscoding capacity of O^6MeG and O^4MeT during DNA synthesis. On another level, the defective (but not inactive) Ada3 and Ada5 methyltransferase proteins should be informative for understanding the mechanism of this unusual DNA repair enzyme.

As a final note, it is of interest that at least two mutant Ada proteins retain the ability to carry out the specific methyl transfer from O^6MeG, albeit slowly. Indeed, no ada mutant has yet been reported that has been convincingly demonstrated to be devoid of any O^6MeG methyltransferase activity. One possibility that we are currently testing is that an active methyltransferase is essential for viability in *E. coli* (12), perhaps by performing some as yet unsuspected function different from O^6MeG repair.

ADAPTIVE RESPONSE TO OXIDIZING AGENTS

Processing of oxidatively damaged DNA is enhanced in *E. coli* pretreated with low doses of H_2O_2 (11). We have addressed the possibility that this enhancement is due to specific DNA repair enzymes by assaying cell-free extracts of peroxide-adapted bacteria for several known activities. In this manner, we have shown that exonuclease III and endonuclease III (11), 5-hydroxymethyluracil-DNA glycosylase, formamidopyrimidine-DNA glycosylase, and endonuclease IV (unpubl. data) levels are all unchanged in response to H_2O_2. It seems possible that novel DNA repair activities could be induced by peroxide. However, no DNA glycosylase or oxidoreductase activity specific for adenine-N1-oxide residues in DNA has been detected (B. Demple and T. Lindahl, unpubl. data).

The state of the bacterial genome following oxidative damage was assessed by measuring the strand breaks formed during a challenge with hydrogen peroxide. Somewhat surprisingly, the number of single-strand breaks formed by a given level of H_2O_2 is similar in naive and adapted cells (Tab. 1). Thus, the DNA in *E. coli* pretreated with hydrogen peroxide appears nearly equally susceptible to oxidative damages as that in unadapted bacteria, despite the extra protective mechanisms that are operating in the adapted cells.

Tab. 1. H_2O_2-induced nicks in chromosomal DNA.

Strain	H_2O_2 challenge (min)	Nicks per genome Nonadapted	Adapted
AB1157	0	<0.1	<0.1
	15	2.2	1.8
	30	3.4	2.4
821	0	<0.1	<0.1
	15	4.8	3.0
	30	5.4	4.8

Bacterial DNA was labeled by growth of the cells for 90 min at 37°C in M9 medium containing 10 μCi/ml [^{3}H]methyl thymidine. After treatment with 10 mM H_2O_2 for 30 min, the number of nicks per bacterial genome was measured by alkaline sucrose gradient sedimentation as described by Krasin and Hutchinson (23).

An analysis of DNA synthesis rates following an oxidative challenge reveals two major features. Both naive and H_2O_2-adapted bacteria shut down most DNA synthesis within 5 min upon challenge with millimolar levels of hydrogen peroxide (Fig. 2). The rate at which DNA synthesis ceases during an oxidative challenge is a function of the peroxide concentration. For a given level of H_2O_2, synthesis stops more rapidly in unadapted cells (Fig. 2). As little as 0.5 mM H_2O_2 abolishes DNA synthesis in naive bacteria for periods up to 2.5 hr, while the same level interrupts replication in adapted cells only briefly, about 10 min.

These observations are in line with the previously demonstrated increases in protective activities that are induced by H_2O_2 (Ref. 8 and 11; Tab. 2). However, the very large survival differences between naive and adapted cells, and the similar levels of DNA nicking that both types of cell suffer, suggest that replication of oxidatively damaged DNA might occur more efficiently in adapted bacteria. This interpretation is borne out by the observation that the resumption of DNA synthesis occurs much more rapidly in adapted bacteria than in naive cells, even after peroxide challenges that are about equally toxic to both types of cell (Fig. 2).

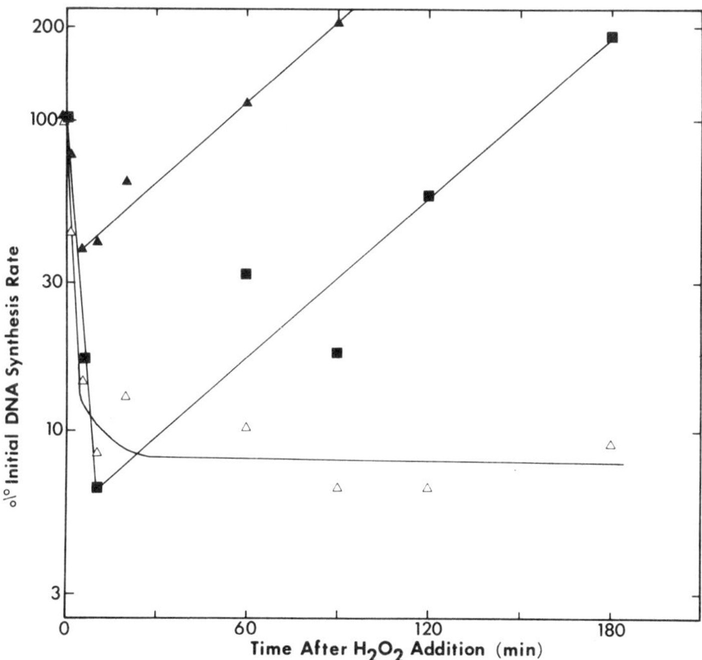

Fig. 2. DNA synthesis in naive and adapted E. coli challenged with H_2O_2. AB1157 was grown at 37°C in K medium (13) to ∿5 x 10^7 cells/ml, and half of the culture adapted with 50 μM H_2O_2 for 20 min at 37°C. Then both cultures were shifted to 30°C for an additional 20 min. The cultures were then challenged with 0.5 mM H_2O_2 (Δ, not pretreated; ▲, adapted) or 2.5 mM H_2O_2 (■, adapted). At the indicated times, 100-μl aliquots were labeled with [^{3}H]methyl-thymidine for 2 min at 30°C, and the acid-insoluble radioactivity measured by filtration on glass-fiber filters.

Tab. 2. Catalase induction by hydrogen peroxide.

Strain	Genotype	Fold induction by H_2O_2
AB1157	gshA$^+$ F$^-$	4.0
821	gshA$^-$ F$^-$	1.3
821/KL711	gshA$^-$/F$^+$ gshA$^+$	3.1
JG10	gshA::Tn10	2.8
TJ5	gshA$^+$ gshB$^-$	5.4

Strains were grown in M9 medium to 2 x 10^8 to 3 x 10^8 cells/ml and adapted with 50 μM H_2O_2 for 35 min at 37°C. The cells were collected by centrifugation, lysates prepared, and catalase activity determined as described (11).

The role of protective activities in promoting cell survival was addressed in another way by taking advantage of the availability of mutants deficient in catalase-peroxidase activities. A strain, UM2 (katE2 katG15) (29), that lacks any detectable catalase activity in vitro (data not shown), was about twice as sensitive to hydrogen peroxide as its wild-type parent. Following exposure to adapting conditions, both the Kat$^-$ and Kat$^+$ strains attained a similar high-level resistance to H_2O_2 (data not shown), although extracts of the adapted kat$^-$ mutant still lacked any detectable catalase activity. Thus, catalase plays a role in destroying hydrogen peroxide before it can damage critical cellular components, but the enzyme is also quite dispensable for the resistance that is induced by H_2O_2. The assumption that such protective activities will automatically lead to increased resistance to oxidative damages is thus open to question.

Glutathione is another cellular component that is thought to be important for protection from oxidative damages (2,33,35). We have examined the role of this intracellular reducing agent in the constitutive and inducible resistances of E. coli against hydrogen peroxide. An MNNG-induced mutant was already available (42), and we obtained another by Tn10 insertion (47) and selection for MNNG-resistance (42; J. Greenberg and B. Demple, ms. in prep.). Both mutants have considerably reduced intracellular thiol levels; the transposon-containing mutant has no detectable glutathione. Neither strain is abnormally sensitive to killing by H_2O_2, and both strains exhibit the same inducible resistance to peroxide toxicity that is observed in wild-type bacteria (data not shown). Rather surprisingly, therefore, glutathione does not seem to play an important role in either the constitutive or the inducible defenses against oxidative damages in E. coli.

In the course of examining glutathione-deficient mutants for their ability to adapt to hydrogen peroxide, we found that one of the original mutants (strain 821) isolated by Apontoweil and Berends (2) is incapable of inducing H_2O_2 resistance, and does not induce catalase activity (Tab. 2). On the other hand, this same strain without adaptation exhibits an increased resistance (as much as 4-fold) to peroxide at early challenge times (10 mM H_2O_2, <20 min), but at later times the mutant experiences as much or more killing than its wild-type parent. This entire phenotype can be reversed by the introduction of an episome covering the region around the gshA gene (56-61 min on the E. coli chromosome) (Fig. 3; Tab. 2). Since such a phenotype does not occur in any of the other gshA mutants that we have tested, and the episome is dominant (indicating a loss of function by the chromosomal mutation), it appears that the unadaptable phenotype we have observed is due to a mutation in a gene that is linked to gshA. Such multiple, linked mutations appear frequently following MNNG mutagenesis (7). More precise mapping of this mutation is underway.

The mutation responsible for the unadaptable phenotype appears to be distinct from previously discovered genes (Fig. 3). The oxyR gene maps near 90 min, and noninducible oxyR mutants are hypersensitive to H_2O_2 in agar plates (8). The inducible catalase genes katE and katG map at 37.8 and 89.2 min, respectively, far away from the gshA gene (27,29). Interestingly, the katF gene maps at 59 min (28) and should be included on the episome introduced into strain 821. However, this enzyme has a very high K_m for H_2O_2 and is not inducible by H_2O_2 (28), which is in contrast to the other catalase activities.

A number of other known genes have been tested for a role in inducible oxidation resistance. The adaptation to hydrogen peroxide occurs normally in recA and xthA mutants (11), and in bacteria harboring the lexA3, ΔpolA, nth, gshB, or gor mutations. Two of the proteins induced in Salmonella by low levels of H_2O_2 are heat-shock proteins, and peroxide-adapted bacteria are heat-resistant (8). Adaptation against peroxide toxicity occurs normally in 2 rpoH (formerly called htpR) (24) amber mutants (data not shown), indicating that oxidation resistance can be regulated distinctly from heat shock. It is noteworthy that strain 821 is still able to induce heat resistance in response to H_2O_2 (J. Greenberg and B. Demple, ms. in prep.).

We are currently working to isolate additional mutants that are specifically unable to adapt to hydrogen peroxide. We are examining the potential roles of other known genes and enzymes in adaptive resistance to hydrogen peroxide, and will also test for the induction of novel activities for oxidized DNA.

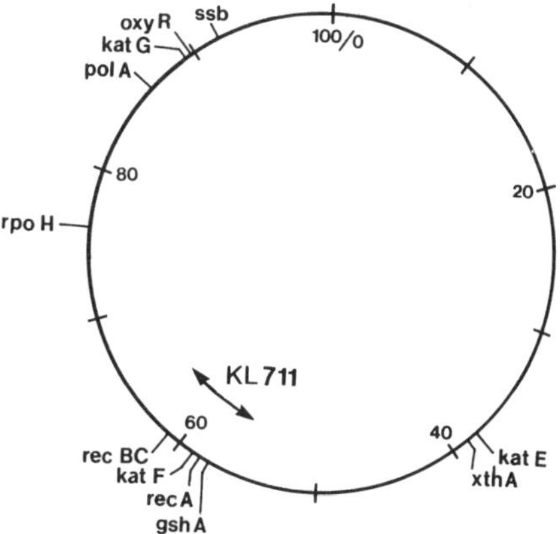

Fig. 3. Map of E. coli genes involved in constitutive and inducible resistances to hydrogen peroxide and other oxidizing agents.

REPAIR OF OXIDATION AND RADIATION DAMAGES IN YEAST

We have focused on the repair of two damages that are formed in significant amounts by oxidation and by ionizing radiation. Thymine glycol (5,6-dihydroxydihydrothymine) is the major thymine damage formed by γ-irradiation of DNA in vivo (4). In E. coli this lesion is repaired by an unusual DNA glycosylase activity that has an associated AP endonuclease (12a). A similar glycosylase has been extensively purified from calf thymus (3). Disruptions of the gene encoding the E. coli enzyme cause an elevated spontaneous reversion rate of an ochre mutation (9), indicating a role for this interesting enzyme in the repair of mutagenic lesions formed spontaneously in DNA.

We have found a DNA glycosylase activity for thymine glycol in cell-free extracts of the yeast Saccharomyces cerevisiae. This enzyme is similar to its prokaryotic and bovine counterparts in having no apparent requirement for a divalent cation (data not shown). Using the synthetic substrate developed by Breimer (3), the only detectable radioactive species released under our standard conditions is free thymine glycol. The yeast thymine glycol glycosylase has been partly purified by concentration with ammonium sulfate and gel filtration on an Ultrogel Aca54 column. The activity is recovered as a single, symmetrical peak corresponding to a globular protein of molecular weight about 35,000. The protein is thus slightly larger or more asymmetric than previously described thymine glycol glycosylases. The significance of this difference has yet to be established.

The glycosylase is found at normal levels in yeast mutants of the rad52 epistasis group, genes that affect the sensitivity of yeast to ionizing radiation. The purification of this glycosylase is underway in order both to understand its precise mechanism of action, and also with an aim to eventually identify the yeast gene that encodes it.

Considerable numbers of strand breaks are formed in DNA by ionizing radiation. The 3' termini left at such breaks bear either 3'-phosphate or 3'-phosphoglycolate ester groups, and can be excised in vitro by E. coli exonuclease III (18). This observation probably explains the exceptional sensitivity of E. coli xth mutants (which lack exonuclease III) to killing by hydrogen peroxide (13). We are currently working to construct synthetic substrates that have as the only damage 3'-phosphoglycolate esters. Synthetic polymers have been made containing modified termini that are susceptible to the action of exonuclease III, and that also are substrates for an activity or activities in crude extracts from RAD^+ yeast. Thus, yeast appears to contain easily detectable enzymes that can act at fragmented deoxyribose residues, as found after radiation damage. Such an activity would be in keeping with the efficient repair of radiation-induced strand breaks in wild-type yeast. We are refining the assay we have developed to detect 3' end repair enzymes, and will proceed with the purification of the yeast repair proteins that act on such lesions.

CONCLUSION

The strong correlation between the mutagenicity and the carcinogenicity of a great many agents (1) underscores the importance of understanding the formation, fate, and consequences of various DNA damages. The ability of living cells to repair a variety of lesions in the genome is an important factor in the ability of organisms to escape mutagenesis and of vertebrates to counteract carcinogenesis. The physiological changes brought about by the persistence of damages in the DNA no doubt also play an important epigenetic role in the causation of mutations (46) and of cancer (6). It is our overall aim to understand the effects of chemical and physical attacks on DNA at all these levels as an eventual approach to determining the events that lead to mutations and disease.

ACKNOWLEDGEMENTS

We thank Leona Samson for helpful discussions, and Eric Alani for performing the experiment shown in Fig. 2. This work was supported by start-up funds from the Department of Biochemistry and Molecular Biology, Harvard University, and by a grant from the National Cancer Institute.

REFERENCES

1. Ames, B.N. (1983) Dietary carcinogens and anticarcinogens. Oxygen radicals and degenerative diseases. Science 221:1256-1264.
2. Apontoweil, P., and W. Berends (1975) Isolation and initial characterization of glutathione-deficient mutants of Escherichia coli K12. Biochim. Biophys. Acta 399:10-22.
3. Breimer, L.H. (1983) Urea-DNA glycosylase in mammalian cells. Biochemistry 22:4192-4197.
4. Breimer, L.H., and T. Lindahl (1985) Thymine lesions produced by ionizing radiation in double-stranded DNA. Biochemistry 24:4018-4022.
5. Cairns, J. (1980) Efficiency of the adaptive response of Escherichia coli to alkylating agents. Nature 286:176-178.
6. Cairns, J. (1981) The origin of human cancers. Nature 289:353-357.
7. Cerda-Olmedo, E., P.C. Hanawalt, and N. Guerola (1968) Mutagenesis of the replication point by nitrosoguanidine: Map and pattern of replication of the E. coli chromosome. J. Mol. Biol. 33:705-719.
8. Christman, M.F., R.W. Morgan, F. Jacobson, and B.N. Ames (1985) Positive control of a regulon for defenses against oxidative stress and some heat-shock proteins in Salmonella typhimurium. Cell 41:753-762.
9. Cunningham, R.P., and B. Weiss (1985) Endonuclease III (nth) mutants of Escherichia coli. Proc. Natl. Acad. Sci., USA 82:474-478.
10. Demple, B. (1984) Repair of oxidative DNA damage: Constitutive and inducible systems in E. coli. In Oxidative Damage and Related Enzymes (Life Chemistry Reports, Suppl. 2), G. Rotilio and J.V. Bannister, eds. Harwood Academic Publishers, London, pp. 422-429.
11. Demple, B., and J. Halbrook (1983) Inducible repair of oxidative DNA damage in Escherichia coli. Nature 304:466-468.
12. Demple, B., and P. Karran (1983) Death of an enzyme: Suicide repair of DNA. Trends in Biochem. Sci. 8:137-139.
12a. Demple, B., and S. Linn (1980) DNA N-glycosylases and UV repair. Nature 287:203-208.
13. Demple, B., J. Halbrook, and S. Linn (1983) Escherichia coli xth mutants are hypersensitive to hydrogen peroxide. J. Bacteriol. 153:1079-1082.
14. Demple, B., A. Jacobsson, M. Olsson, P. Robins, and T. Lindahl (1982) Repair of alkylated DNA in Escherichia coli. Physical properties of O^6-methylguanine-DNA methyltransferase. J. Biol. Chem. 257:13776-13780.
15. Demple, B., B. Sedgwick, P. Robins, N. Totty, M.D. Waterfield, and T. Lindahl (1985) Active site and complete sequence of the suicidal methyltransferase that counters alkylation mutagenesis. Proc. Natl. Acad. Sci., USA 82:2688-2692.

16. Evensen, G., and E. Seeberg (1982) Adaptation to alkylation resistance involves induction of a DNA glycosylase. Nature 296:773-775.
17. Farr, S.B., D.O. Natvig, and T. Kogoma (1985) Toxicity and mutagenicity of plumbagin and the induction of a possible new DNA repair pathway in Escherichia coli. J. Bacteriol. 164:1309-1316.
17a. Hall, J., and R. Saffhill (1983) The incorporation of O^6-methyldeoxyguanosine and O^4-methyldeoxythymidine monophosphates into DNA by DNA polymerases I and α. Nucl. Acids Res. 11:4185-4193.
18. Henner, W.D., S.M. Grunberg, and W.A. Haseltine (1983) Enzyme action at 3' termini of ionizing radiation-induced DNA strand breaks. J. Biol. Chem. 258:15198-15205.
19. Hutchinson, F. (1985) Chemical changes induced in DNA by ionizing radiation. Prog. Nucl. Acids Mol. Biol. 32:115-154.
20. Jeggo, P. (1979) Isolation and characterization of Escherichia coli K12 mutants unable to induce the adaptive response to simple alkylating agents. J. Bacteriol. 139:783-791.
21. Jeggo, P., M. Defais, L. Samson, and P. Schendel (1977) An adaptive response of E. coli to low levels of alkylating agent: Comparison with previously characterized DNA repair pathways. Mol. Gen. Genet. 157:1-9.
22. Karran, P., T. Hjelmgren, and T. Lindahl (1982) Induction of a DNA glycosylase for N-methylated purines is part of the adaptive response to alkylating agents. Nature 296:770-773.
23. Krasin, F., and F. Hutchinson (1977) Repair of DNA double-strand breaks in Escherichia coli, which requires recA function and the presence of a duplicate genome. J. Mol. Biol. 116:81-98.
24. Krueger, J.H., and G.C. Walker (1984) groEL and dnaK genes of Escherichia coli are induced by UV irradiation and nalidixic acid in an htpR -dependent fashion. Proc. Natl. Acad. Sci., USA 81:1499-1503.
25. LeMotte, P., and G.C. Walker (1985) Induction and autoregulation of ada, a positively acting element regulating the response of Escherichia coli K12 to methylating agents. J. Bacteriol. 161:888-895.
26. Lindahl, T., B. Demple, and P. Robins (1982) Suicide inactivation of the E. coli O^6-methylguanine-DNA methyltransferase. EMBO J. 1:1359-1363.
26a. Loechler, E.L., C.L. Green, and J.M. Essigmann (1984) In vivo mutagenesis by O^6-methylguanine built into a unique site in a viral genome. Proc. Natl. Acad. Sci., USA 81:6271-6275.
27. Loewen, P.C. (1984) Isolation of catalase-deficient Escherichia coli mutants and genetic mapping of katE, a locus that affects catalase activity. J. Bacteriol. 157:622-626.
28. Loewen, P.C., and B.L. Triggs (1984) Genetic mapping of katF, a locus that with katE affects the synthesis of a second catalase species in E. coli. J. Bacteriol. 160:668-675.
29. Loewen, P.C., B.L. Triggs, C.S. George, and B.E. Hrabarchuk (1985) Genetic mapping of katG, a locus that affects synthesis of the bifunctional catalase-peroxidase I in Escherichia coli. J. Bacteriol. 162:661-667.
30. McCarthy, J.G., B.V. Eddington, and P.F. Schendel (1983) Inducible repair of phosphotriesters in Escherichia coli. Proc. Natl. Acad. Sci., USA 80:7380-7384.
31. McCarthy, T.V., P. Karran, and T. Lindahl (1984) Inducible repair of O-alkylated DNA pyrimidines in Escherichia coli. EMBO J. 3:545-550.
32. McCarthy, T.V., and T. Lindahl (1985) Methyl phosphotriesters in DNA are repaired by the Ada regulatory protein of E. coli. Nucl. Acids Res. 13:2683-2698.
33. Meister, A., and M.E. Anderson (1983) Glutathione. Ann. Rev. Biochem. 52:711-760.

34. Mitra, S., B.C. Pal, and R.S. Foote (1982) O^6-Methylguanine-DNA methyltransferase in wild-type and ada mutants of E. coli. J. Bacteriol. 152:534-537.

35. Morse, M.L., and R.H. Dahl (1978) Cellular glutathione is a key to the oxygen effect in radiation damage. Nature 271:660-662.

36. Muller, H.J. (1927) Artificial transmutation of the gene. Science 66: 84-87.

37. Nakabeppu, Y., H. Kondo, S. Kawabata, S. Iwanaga, and M. Sekiguchi (1985) Purification and structure of the intact Ada regulatory protein of Escherichia coli K12, O^6-methylguanine-DNA methyltransferase. J. Biol. Chem. 260:7281-7288.

38. Olsson, M., and T. Lindahl (1980) Repair of alkylated DNA in Escherichia coli. Methyl group transfer from O^6-methylguanine to a protein cysteine residue. J. Biol. Chem. 255:10569-10571.

39. Samson, L., and J. Cairns (1977) A new pathway for DNA repair in Escherichia coli. Nature 267:281-283.

40. Sedgwick, B. (1982) Genetic mapping of ada and adc mutations affecting the adaptive response of Escherichia coli to alkylating agents. J. Bacteriol. 150:984-988.

41. Sedgwick, B. (1983) Molecular cloning of a gene which regulates the adaptive response to alkylating agents in E. coli. Mol. Gen. Genet. 191:466-472.

42. Sedgwick, B., and P. Robins (1980) Isolation of mutants of Escherichia coli with increased resistance to alkylating agents: Mutants deficient in thiols and mutants constitutive for the adaptive response. Mol. Gen. Genet. 180:85-90.

43. Singer, B., and J.T. Kusmierek (1982) Chemical mutagenesis. Ann. Rev. Biochem. 51:655-693.

44. Teo, I., B. Sedgwick, B. Demple, B. Li, and T. Lindahl (1984) Induction of resistance to alkylating agents in E. coli: The ada^+ gene product serves both as a regulatory protein and as an enzyme for repair of mutagenic damage. EMBO J. 3:2151-2157.

45. Volkert, M.R., and D.C. Nguyen (1984) Induction of specific Escherichia coli genes by sublethal treatments with alkylating agents. Proc. Natl. Acad. Sci., USA 81:4110-4114.

46. Walker, G.C. (1984) Mutagenesis and inducible responses to deoxyribonucleic acid damage in Escherichia coli. Microbiol. Rev. 48:60-93.

47. Way, J.C., M.A. Davis, D. Morisato, D.E. Roberts, and N. Kleckner (1984) New Tn10 derivatives for transposon mutagenesis and for construction of lacZ operon fusions by transposition. Gene 32:369-379.

48. Zarbl, H., S. Sukumar, A.V. Arthur, D. Martin-Zanca, and M. Barbacid (1985) Direct mutagenesis of Ha-ras-1 oncogenes by N-nitroso-N-methylurea during initiation of mammary carcinogenesis in rats. Nature 315:382-385.

MOLECULAR MECHANISM OF ADAPTIVE RESPONSE TO ALKYLATING AGENTS

Mutsuo Sekiguchi, Hidemasa Kondo, Kunihiko Sakumi, and Yusaku Nakabeppu

Department of Biology, Faculty of Science
and Department of Biochemistry, Faculty of Medicine
Kyushu University 33
Fukuoka 812, Japan

INTRODUCTION

Alkylating agents are potent mutagens and carcinogens, and organisms respond in a complex manner to these agents. Growth of *Escherichia coli* in the presence of low levels of simple alkylating agents, such as N-methyl-N'-nitro-N-nitrosoguanidine (MNNG), results in a marked increase in resistance of cells to both the mutagenic and the lethal effects of challenging doses of the same agents (24). This adaptive response is distinct from previously characterized pathways of DNA repair, particularly from the SOS response, another inducible effect resulting from DNA damage. The adaptation does not lead to expression of the SOS functions, and *recA* and *lexA* mutant cells that are unable to perform SOS repair can be adapted to MNNG (8,27).

The adaptive response seems to be related to the induced synthesis of certain enzymes that repair alkylation lesions in DNA. It has been shown that O^6-methylguanine-DNA methyltransferase and 3-methyladenine-DNA glycosylase II are formed when *E. coli* cells are exposed to low doses of MNNG (4,5,9,20). The former enzyme transfers the methyl group of the O^6-methylguanine residue in DNA to one of its cysteine residues (20), while the latter liberates 3-methyladenine and other alkylated bases from DNA (30). Thus, it would be of interest to know how the formation of these enzymes is controlled.

CLONING OF GENES INVOLVED IN REPAIR OF ALKYLATED DNA

There are *E. coli* mutations that cause an abnormal sensitivity to alkylating agents. These include *alkA*, *ada*, *alkB*, and *tag*, and the locations on the chromosome are shown in Fig. 1. These genes may be responsible for repair of DNA damaged by alkylating agents, and some may control the process of adaptive response. As a first step toward elucidating the control mechanisms of the adaptive response, we cloned these genes and identified the products (Tab. 1).

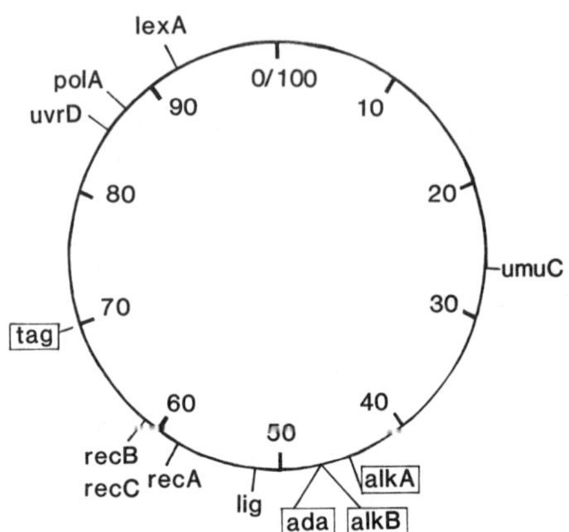

Fig. 1. The chromosomal location of Escherichia coli genes involved in repair of alkylated DNA.

The alkA Gene

The alkA mutants are specifically sensitive to simple alkylating agents, such as methyl methanesulfonate (MMS) and MNNG, and the mutation responsible was mapped at 44 to 45 min on the E. coli genetic map (32,33). By in vitro recombination, we constructed hybrid plasmids which can suppress the increased MMS sensitivity caused by the alkA mutation in E. coli. Since the cloned DNA fragment was mapped at 45 min of the genetic map, we concluded that the cloned DNA fragment contains the alkA gene itself but not other genes that suppress the alkA mutation (17). Cloning of the alkA gene was also done by Clarke et al. (1).

Specific labeling of plasmid-encoded proteins by the maxicell method revealed that the alkA gene codes for a polypeptide whose molecular weight is about 30,000. When cells harboring the $alkA^+$ plasmids were grown in the presence of low doses of a simple alkylating agent (adapted condition), the activity of 3-methyladenine-DNA glycosylase II was increased. Taking advantage of this overproduction, 3-methyladenine-DNA glycosylase II was purified to apparent physical homogeneity (17). The enzyme activity was co-purified with the molecular weight 30,000 polypeptide.

The nucleotide sequence of a DNA fragment containing the alkA gene and its control region has been determined (19). Only one open reading frame responsible for 3-methyladenine-DNA glycosylase II was found. The hypothetical polypeptide deduced from the DNA sequence, with a molecular weight of 31,400, has an amino-terminal sequence and total amino acid composition identical to that of purified glycosylase II. From these results we concluded that alkA is the structural gene for 3-methyladenine-DNA glycosylase II.

The ada Gene

The ada mutants are specifically sensitive to alkylating agents and are unable to induce both O^6-methylguanine-DNA methyltransferase and 3-methyladenine-DNA glycosylase II (4,7.15). Thus, the ada gene appears to be a regulatory one for the adaptive response. There is another class of mutants, adc, and these are more resistant to killing and mutagenic effects

Tab. 1. Cloned genes and their protein products.

Gene	Number of nucleotides in coding region	Molecular weight of protein deduced from nucleotide sequence	Enzyme activity of protein
alkA	864	31,400	3-Methyladenine-DNA glycosylase II
ada	1,062	39,385	O^6-Methylguanine-DNA methyltransferase
alkB	648	23,900	
tag	561	21,100	3-Methyladenine-DNA glycosylase I

of alkylating agents than is the wild-type strain, and constitutively synthesize a large amount of methyltransferase but only a small amount of glycosylase II (26). Both ada and adc map at almost the same position (47 min) on the chromosome and, thus, adc appears to be a constitutive mutation arising in the ada gene.

The ada gene has been cloned from E. coli strains B and K12 (13,14,18, 25). We placed the cloned ada gene under the control of the lac regulatory region on a multicopy runaway plasmid, pUC9, thereby yielding a hybrid plasmid pYN3059 (16). Ada protein with a molecular weight of about 38,000 was overproduced when cells harboring pYN3059 were incubated at 42°C in the presence of a lac inducer, isopropyl-β-D-thiogalactopyranoside (IPTG). Taking advantage of the overproduction of Ada protein, we purified the protein to apparent physical homogeneity. The purified 38-kDa Ada protein transferred the methyl group from the O^6-methylguanine residue of alkylated DNA to the Ada protein, per se. Amino acid composition of the purified Ada protein was in accord with the nucleotide sequence of the ada gene, as determined by the dideoxy method using M13 phage. It was deduced that Ada protein comprises 354 amino acids and that the molecular weight is 39,385.

Recently, Demple et al. (3) determined the nucleotide sequence of the ada gene of E. coli B. The data are roughly in accord with ours, which was made with the gene derived from E. coli K12. The total number of amino acid residues and the first 74 amino acid sequences are essentially the same for the two fraternal genes; however, there are 25 base substitutions and 6 amino acid alterations in the central and carboxy-terminal region. A partial nucleotide sequence of the ada gene from E. coli K12 was also determined by Lemotte and Walker (13).

Teo et al. (29) showed that the 38,000-dalton Ada protein is processed to smaller polypeptides, including an 18,000-dalton protein which may be identical to the previously characterized E. coli O^6-methylguanine-DNA methyltransferase (2). We found that this processing is dependent on both the pH and the ionic strength of the extraction buffer; the 38,000-dalton Ada protein was fairly stable in low ionic strength buffer at pH 8.5, while it was rapidly degraded when the extract was incubated in the presence of salts at pH 7.4 (16). The purified Ada protein, however, remained stable even under the latter condition, thereby indicating that the Ada protein may not undergo autodegradation but may be processed by a proteinase(s) present in the extract.

As will be shown later, Ada protein acts as a positive regulator for transcription of the ada and alkA genes. Thus, the Ada protein has dual functions and seems to play key roles in the adaptive response.

The alkB Gene

The alkB mutant exhibits the same phenotype as the alkA mutant but their genetic locations differ; the alkB gene is located at 47 min, while the alkA gene is mapped at 45 min on the genetic map. Although the ada gene is located in the region where the alkB is mapped, the alkB mutant exhibits a normal adaptive response to MNNG (12). Introduction of a small deletion into the alkB region of the bacterial chromosome resulted in inactivation of both the alkB and ada genes, thereby suggesting that the two genes are adjacent on the E. coli chromosome (11).

The alkB gene was cloned in a multicopy plasmid, pBR322. Specific labeling of plasmid-encoded proteins by the maxicell method revealed that the alkB gene codes for a polypeptide with a molecular weight of about 27,000 (11). Since the enzyme activity associated with ALkB protein has not been characterized, AlkB protein has been purified by monitoring the 27,000-dalton polypeptide band on the gel. The amino acid composition and amino-terminal sequence of the most purified preparation of AlkB protein were determined.

The nucleotide sequence of the alkB fragment has been determined by the dideoxy method using M13 phage. An open reading frame responsible for the molecular weight 27,000 polypeptide was found, and the amino acid sequence deduced from the DNA sequence was in accord with the data obtained by analyses of the purified AlkB protein. One of the interesting findings is that the first initiation codon for AlkB protein overlaps with the termination codon for Ada protein (Fig. 2). Thus, the alkB gene is indeed adjacent to the ada gene. Recently, we obtained evidence to show that expression of the alkB gene is controlled by the ada promoter, indicating that the ada and alkB genes constitute an operon (H. Kondo et al., ms. in prep.).

The tag Gene

The 3-methyladenine-DNA glycosylase I is a constitutive enzyme that

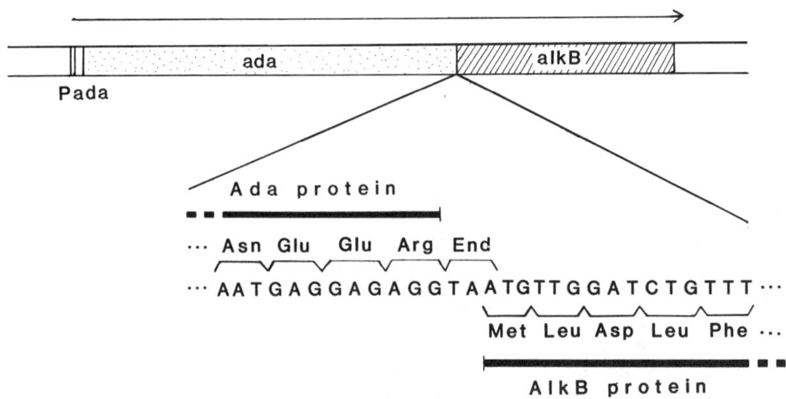

Fig. 2. Organization of the ada and alkB genes.

removes one of the major alkylation products, 3-methyladenine, from DNA (22). The tag mutants, which are deficient in this enzyme, are moderately sensitive to MMS (10). Recombinant plasmids carrying the tag gene have been constructed (1,31).

To overproduce 3-methyladenine-DNA glycosylase I, a 0.8-kb DNA fragment that carries the tag gene was placed under the control of the lac promoter in plasmid vector pUC8. The plasmid (pCY5) overproduced the glycosylase to a level of 450-fold that of the wild-type strain, on exposure to IPTG. From an extract of such cells, glycosylase I (approximate molecular weight, 21,000) was purified to apparent physical homogeneity. With the purified preparations, the amino acid composition and the N-terminal amino acid sequence were determined.

The nucleotide sequence of the tag gene was determined by using the dideoxy method. The open reading frame found would code for a polypeptide consisting of 187 amino acids. The molecular weight of the purified protein is 21,100, which is almost equal to that of glycosylase I. Furthermore, the amino acid sequence and composition, deduced from the nucleotide sequence, were in good agreement with the data obtained from direct protein analyses. Thus, it is evident that tag is the structural gene for 3-methyladenine-DNA glycosylase I.

POSITIVE CONTROL MECHANISM FOR ADAPTIVE RESPONSE

When plasmid pYN3028 carrying the ada^+ gene was introduced into ada^- strains (ada-3, ada-5, and ada-136), the occurrence of MNNG-induced mutations in these strains was suppressed to almost the level of the wild-type strain. To determine the effects of the cloned ada^+ gene on the synthesis of DNA repair enzymes that are known to be inducible as a part of the adaptive response, DNA methyltransferase and 3-methyladenine-DNA glycosylase II activities were assayed on cells with these plasmids (Fig. 3). Constitutive levels of the two enzymes were significantly higher in ada^+ or ada^- cells harboring pYN3028, compared with those in the cells harboring the vector plasmid. When such cells were adapted with a low dose of MMS, even higher levels of the enzyme activities were produced; the levels attained with ada^+ and ada^- cells were essentially the same. These results are consistent with the findings in the case of pCS33, which carries the ada^+ gene from E. coli strain B (25).

Inactivation of the ada gene by transposon insertion or by partial deletion resulted in the loss of synthesis of Ada protein. In such cells no adaptive response took place (11,13,25). These results suggest that the ada^+ gene codes for a positive regulator of the adaptive response.

To study the mechanism of control of expression of the ada and alkA genes, we constructed the ada'-lacZ' and alkA'-lacZ' fused genes and analyzed their expression under various conditions (18,19). First, we found that the expression of these genes is inducible by low concentrations of methylating agents such as MNNG and MMS, while ethylating agents, even at lethal doses, have little effect on the gene expression. Second, it appeared that the expression of the ada gene was regulated autogenously by its own product because ada^- mutant cells harboring the ada'-lacZ' fused gene are defective in the induction of β-galactosidase activity with MMS. Although Ada protein synthesized in the absence of alkylating agents is effective for induction of expression of both ada and alkA genes, the full

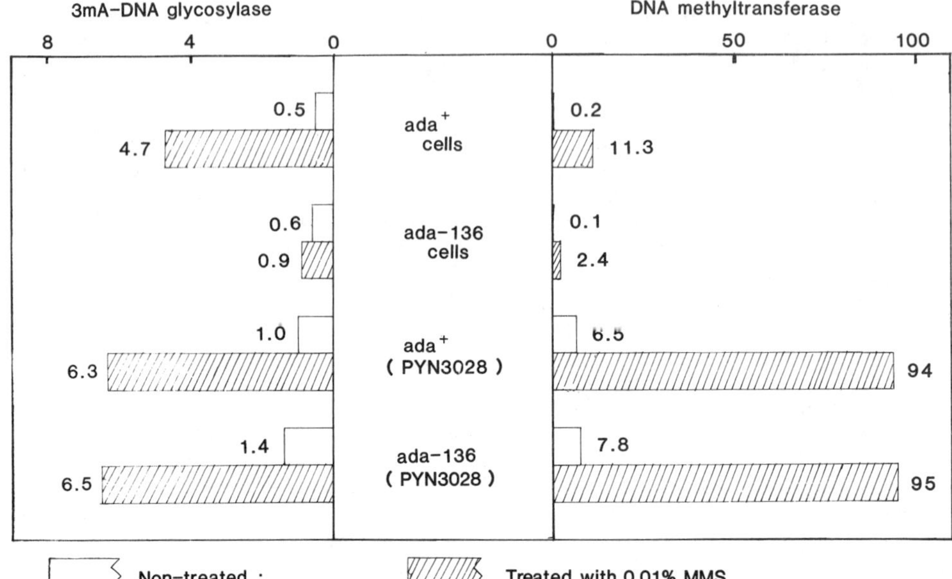

Fig. 3. DNA methyltransferase and 3-methyladenine-DNA glycosylase II activities in $\underline{ada}^+$ and $\underline{ada}^-$ cells, with or without $\underline{ada}^+$ plasmid, pYN3028.

expression of these genes requires not only Ada protein but also treatment with methylating agents.

Ada Protein as a Positively Acting Transcriptional Regulator

To specify the promoter region of the ada gene, we determined the transcription initiation site of ada by S1 nuclease mapping (6,21,23). The 175-bp HindIII/EcoRI fragment containing the ada promoter region was labeled at the 5'-ends, and hybridized with RNA extracted from normal and adapted cells of YN3 cells harboring pYN3028. After S1 nuclease digestion, the protected DNA was determined by comparison with the products of Maxam-Gilbert sequence reactions. A major band was present at the same position for both normal and adapted cells, although the intensity of the band for adapted cells exceeded that of normal cells. Thus, we concluded that the major transcription initiation site for ada is adenine at 22 bases upstream from the translation initiation site, in either normal or adapted conditions. A similar result was obtained with primer extension experiments.

Immediately upstream of the transcription initiation site, a DNA sequence was found that is similar to the consensus sequence for E. coli promoters (6,21,23), thereby suggesting that sequences TTGCGT (for -35 region) and TAAAGG (for -10 region), separated by 18 bases, comprise the functional promoter for the ada gene. There was a typical Shine-Dalgarno sequence GGAG at 7 bases upstream from the translational initiation site (28).

In a similar manner, the transcription initiation site of alkA was determined. The major initiation site was found to be guanine at 19 bases upstream from the translation initiation site, in either normal or adapted conditions.

Having accurate in vivo transcription initiation sites for the two genes in hand, we proceeded to in vitro transcription experiments using a reconstituted system. A 262-bp MluI/AccI fragment containing the alkA promoter and a 175-bp HindIII/EcoRI fragment containing the ada promoter region were used as DNA templates. As a control, a 205-bp EcoRI fragment containing the lac UV5 promoter was also used. To either one or a mixture of DNA fragments, E. coli RNA polymerase holoenzyme was added and the preparation incubated at 37°C for 1 hr in the presence or absence of a purified preparation of Ada protein, to produce a transcription initiation complex. Then a mixture of ATP, CTP, GTP, and α-^{32}P-UTP was added together with heparin, which prevents further initiation. The reaction was performed at 37°C for 5 min. If accurate transcription occurs in this reconstituted system, 23-, 98-, and 63-nucleotide-long RNAs would be produced from the alkA, ada, and lac promoters, respectively.

When the products were analyzed by polyacrylamide gel electrophoresis, the results shown in Fig. 4 were obtained. The 23-b alkA transcript and the 98-b ada transcript were produced from the corresponding promoters when Ada protein was present in the reaction mixture. No band or only a faint band was found when the reaction was performed without Ada protein. In the case of ada a 90-b RNA, which is not produced in vivo and may be an in vitro artifact product, was formed in addition to the 98-b transcript. Formation of this 90-b RNA as well as of the control 63-b lac transcript was not significantly affected by the presence of Ada protein. These results clearly indicate that Ada protein acts on the ada and alkA promoters as a positive regulator to enhance their transcription.

We next examined effects of a methylating agent on the reconstituted system (Fig. 4). When 100 pmol of methylnitrosourea were added to the reaction mixture, formation of the 98-b ada transcript was enhanced, while formation of the 23-b alkA and the 63-b lac transcripts was rather slightly suppressed. This result was confirmed in separate experiments in which various concentrations of methylnitrosourea were used (data not shown). This result is consistent with our in vivo data, and further implies that a methylated form of Ada protein acts primarily on the ada promoter, thereby producing Ada protein, which in turn acts on the ada and alkA promoters to overproduce both proteins. Recently, Lindahl and his associates found that a methylated form of Ada protein promotes transcription of the ada gene (T. Lindahl, pers. comm.).

Based on these findings, we propose a model for regulation of the adaptive response (Fig. 5). In normal cells, the ada gene and alkA gene are expressed weakly but at significant levels. When methylating agents attack the cells, the DNAs of the cells are rapidly methylated, and O^6-methylguanine, which is scavenged by the Ada protein, is formed. This repair reaction, which gives rise to the methylated form of the Ada protein, may be an initiating signal for the adaptive response. It seems that the methylated form of the Ada protein preferentially binds to the ada promoter to enhance transcription of the ada gene. Once the Ada protein is overproduced in this way, the protein, in either methylated or unmethylated form, could bind to both the alkA and ada promoters, thereby yielding a large amount of AlkA protein (3-methyladenine-DNA glycosylase II) as well as of Ada protein (O^6-methylguanine-DNA methyltransferase).

The Ada protein is processed to a smaller polypeptide that continues to possess methyltransferase activity (16,29). We are doing experiments to determine whether the processed form of the Ada protein is active as a transcriptional activator. If the processed protein is devoid of such

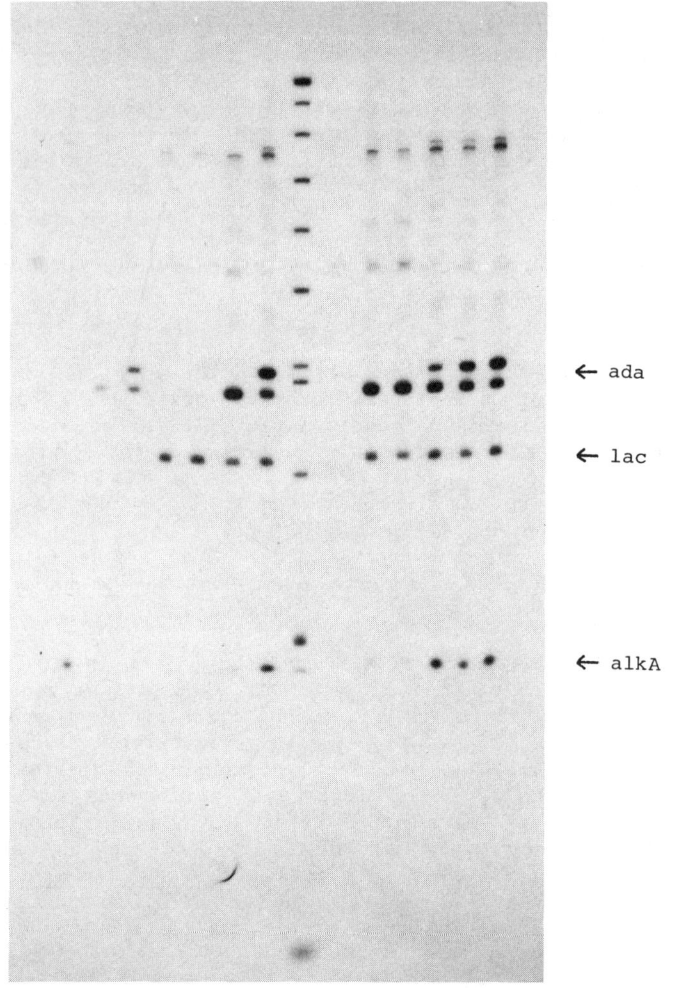

Fig. 4. Formation of alkA, ada, and lac transcripts in an in vitro reconstituted system. Either one or a mixture of DNA fragments (0.3 pmol each) containing alkA, ada, or lacUV5 promoter and 10-fold molar excess of E. coli RNA polymerase was incubated with or without Ada protein in 35 μl of the reaction mixture at 37°C for 60 min. Then 15 μl of a solution containing 0.05 mM α-^{32}P-UTP (2 μCi per reaction), 0.16 mM each of ATP, GTP, and CTP, and 200 μg/ml of heparin were added, and the mixture was incubated at 37°C for 5 min. The samples were precipitated with ethanol, applied to an 8% polyacrylamide gel containing 50% urea, and run at 65 W for 1.5 hr. Samples 1 and 2, alkA DNA; 3 and 4, ada DNA; 5 and 6, lac DNA; 7 to 13, alkA, ada, and lac DNA. Three pmol of Ada protein were added to samples 2, 4, 6, 11, and 12, and 9 pmol of Ada protein were added to samples 8 and 13, while no Ada protein was added to other samples. Samples 10 and 12 were treated with 100 pmol of methylnitrosourea. M is HpaII-digested pUC9 DNA for size markers.

I. Non-induced

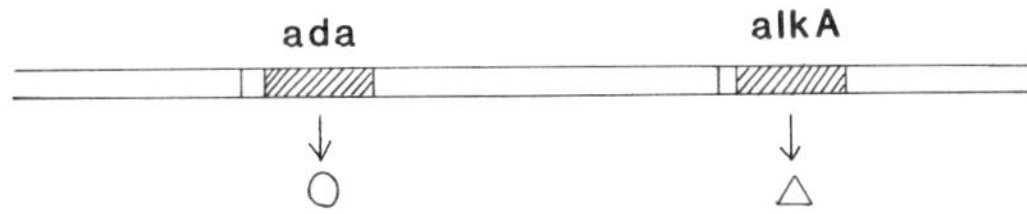

II. Induced by MMS

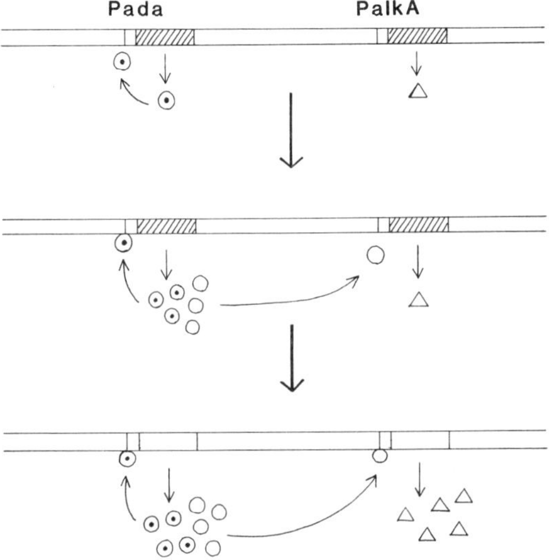

Fig. 5. A model for adaptive response. In the noninduced state, only small amounts of *ada* (o) and *alkA* (Δ) gene products were formed. After treatment with methylating agents such as MMS, Ada protein is methylated by accepting the methyl group from DNA. The methylated form of Ada protein (⊙) acts on the *ada* promoter, thereby producing Ada protein which, in turn, acts on the *ada* and *alkA* promoters to overproduce both proteins.

activity, the cell would revert to the normal state in which expression of the *ada* and *alkA* genes is suppressed.

Another area requiring attention is how the Ada protein is activated by methylation. It has been shown that the methyl acceptor site of the Ada protein is cysteine^{321} that is located near the C-terminus (3). We recently produced an altered form of the Ada protein (by site-directed mutagenesis of the cloned *ada* gene) in which this cysteine residue is replaced by alanine (Nakabeppu et al., ms. in prep.). In cells carrying such a mutant *ada* gene, the Ada protein was overproduced without adaptive treatment. Thus, activation of the Ada protein seems to be induced by replacement of the cysteine residue by alanine as well as by formation of S-methylcysteine, at the same site. This implies that such amino acid changes would cause a local alteration of conformational structure of the Ada protein so that the altered form of the protein could bind more readily to the *ada* promoter region than could the original form.

ACKNOWLEDGEMENTS

We extend special thanks to Drs. Akira Ishihama and Sadaaki Iwanaga for pertinent advice, and to Mariko Ohara for comments on the manuscript. This work was supported by grants for scientific and cancer research and a grant-in-aid for special project research (Cancer-Bioscience) from the Ministry of Education, Science and Culture and by grants from the Japan Society for the Promotion of Science.

REFERENCES

1. Clarke, N.D., M. Kvaal, and E. Seeberg (1984) Cloning of *Escherichia coli* genes encoding 3-methyladenine DNA glycosylase I and II. *Molec. Gen. Genet.* 197:368-372.
2. Demple, B., A. Jacobsson, M. Olsson, P. Robins, and T. Lindahl (1982) Repair of alkylated DNA in *Escherichia coli*: Physical properties of O^6-methylguanine-DNA methyltransferase. *J. Biol. Chem.* 257:13776-13780.
3. Demple, B., B. Sedgwick, P. Robins, N. Totty, M.D. Waterfield, and T. Lindahl (1985) Active site and complete sequence of the suicidal methyltransferase that counters alkylation mutagenesis. *Proc. Natl. Acad. Sci., USA* 82:2688-2692.
4. Evensen, G., and E. Seeberg (1982) Adaptation to alkylation resistance involves the induction of a DNA glycosylase. *Nature* (London) 296:773-775.
5. Foote, R.S., S. Mitra, and B.C. Pal (1980) Demethylation of O^6-methylguanine in synthetic DNA polymer by an inducible activity in *Escherichia coli*. *Biochem. Biophys. Res. Commun.* 97:654-659
6. Hawley, D.K., and W.R. McClure (1983) Compilation and analysis of *Escherichia coli* promoter DNA sequences. *Nucl. Acids Res.* 11:2237-2255.
7. Jeggo, P. (1979) Isolation and characterization of *Escherichia coli* K-12 mutants unable to induce the adaptive response to simple alkylating agents. *J. Bacteriol.* 139:783-791.
8. Jeggo, P., M. Defais, L. Samson, and P. Schendel (1977) An adaptive response of *E. coli* to low levels of alkylating agent: Comparison with previously characterized DNA repair pathways. *Molec. Gen. Genet.* 157:1-9.
9. Karran, P., T. Hjelmgren, and T. Lindahl (1982) Induction of a DNA glycosylase for N-methylated purines is part of the adaptive response to alkylating agents. *Nature* (London) 296:770-773.
10. Karran, P., T. Lindahl, I. Ofsteng, G.B. Evensen, and E. Seeberg (1980) *Escherichia coli* mutants deficient in 3-methyladenine DNA glycosylase. *J. Molec. Biol.* 140:101-127.
11. Kataoka, H., and M. Sekiguchi (1985) Molecular cloning and characterization of the *alkB* gene of *Escherichia coli*. *Molec. Gen. Genet.* 198:263-269.
12. Kataoka, H., Y. Yamamoto, and M. Sekiguchi (1983) A new gene (*alkB*) of *Escherichia coli* that controls sensitivity to methylmethane sulfonate. *J. Bacteriol.* 153:1301-1307.
13. Lemotte, P.K., and G.C. Walker (1985) Induction and autoregulation of *ada*, a positively acting element regulating the response of *Escherichia coli* K-12 to methylating agents. *J. Bacteriol.* 161:888-895.
14. Margison, G.P., D.P. Cooper, and J. Brennand (1985) Cloning of the *E. coli* O^6-methylguanine and methylphosphotriester methyltransferase gene using a functional DNA repair assay. *Nucl. Acids Res.* 13:1939-1952.

15. Mitra, S., B.C. Pal, and R.S. Foote (1982) O^6-methylguanine-DNA methyltransferase in wild-type and ada mutants of Escherichia coli. J. Bacteriol. 152:534-537.
16. Nakabeppu, Y., H. Kondo, S. Kawabata, S. Iwanaga, and M. Sekiguchi (1985) Purification and structure of the intact Ada regulatory protein of Escherichia coli K12, O^6-methylguanine-DNA methyltransferase. J. Biol. Chem. 260:7281-7288.
17. Nakabeppu, Y., H. Kondo, and M. Sekiguchi (1984) Cloning and characterization of the alkA gene of Escherichia coli that encodes 3-methyladenine DNA glycosylase II. J. Biol. Chem. 259:13723-13729.
18. Nakabeppu, Y., Y. Mine, and M. Sekiguchi (1985) Regulation of expression of the cloned ada gene in Escherichia coli. Mutat. Res. 146:155-167.
19. Nakabeppu, Y., T. Miyata, H. Kondo, S. Iwanaga, and M. Sekiguchi (1984) Structure and expression of the alkA gene of Escherichia coli involved in adaptive response to alkylating agents. J. Biol. Chem. 259:13730-13736.
20. Olsson, M., and T. Lindahl (1980) Repair of alkylated DNA in Escherichia coli: Methyl group transfer from O^6-methylguanine to a protein cysteine residue. J. Biol. Chem. 255:10569-10571.
21. Pribnow, D. (1975) Nucleotide sequence of an RNA polymerase binding site at an early T7 promoter. Proc. Natl. Acad. Sci., USA 72:784-788.
22. Riazuddin, S., and T. Lindahl (1978) Properties of 3-methyladenine-DNA glycosylase from Escherichia coli. Biochemistry 17:2110-2118.
23. Rosenberg, M., and D. Court (1979) Regulatory sequences involved in the promotion and termination of RNA transcription. Ann. Rev. Genet. 13:319-353.
24. Samson, L., and J. Cairns (1977) A new pathway for DNA repair in Escherichia coli. Nature (London) 267:281-283.
25. Sedgwick, B. (1983) Molecular cloning of a gene which regulates the adaptive response to alkylating agents in Escherichia coli. Molec. Gen. Genet. 191:466-472.
26. Sedgwick, B., and P. Robins (1980) Isolation of mutants of Escherichia coli with increased resistance to alkylating agents: Mutants deficient in thiols and mutants constitutive for the adaptive response. Molec. Gen. Genet. 180:85-90.
27. Schendel, P.F., M. Defais, P. Jeggo, L. Samson, and J. Cairns (1978) Pathway of mutagenesis and repair in Escherichia coli exposed to low levels of simple alkylating agents. J. Bacteriol. 135:466-475.
28. Shine, J., and L. Dalgarno (1975) Determinant of cistron specificity in bacterial ribosomes. Nature (London) 254:34-38.
29. Teo, I., B. Sedgwick, B. Demple, B. Li, and T. Lindahl (1984) Induction of resistance to alkylating agents in E. coli: The ada^+ gene product serves both as a regulatory protein and as an enzyme for repair of mutagenic damage. EMBO J. 3:2151-2157.
30. Thomas, L., C.-H. Yang, and D.A. Goldthwait (1982) Two DNA glycosylases in Escherichia coli which release primarily 3-methyladenine. Biochemistry 21:1162-1169.
31. Yamamoto, Y., H. Kataoka, Y. Nakabeppu, T. Tsuzuki, and M. Sekiguchi (1983) The genes involved in the repair of alkylated DNA in Escherichia coli K-12. In Celluar Responses to DNA Damage, E.C. Friedberg and B.A. Bridges, eds. Alan R. Liss, Inc., New York, pp. 271-278.
32. Yamamoto, Y., M. Katsuki, M. Sekiguchi, and N. Otsuji (1978) Escherichia coli gene that controls sensitivity to alkylating agents. J. Bacteriol. 135:144-152.
33. Yamamoto, Y., and M. Sekiguchi (1979) Pathways for repair of DNA damaged by alkylating agent in Escherichia coli. Molec. Gen. Genet. 171:251-256.

NUCLEOTIDE EXCISION REPAIR GENES FROM THE YEAST *SACCHAROMYCES CEREVISIAE*

Errol C. Friedberg, Reinhard Fleer, Louie Naumovski, Charles M. Nicolet, Gordon W. Robinson, William A. Weiss, and Elizabeth Yang

Department of Pathology
Stanford University School of Medicine
Stanford, California 94305

ABSTRACT

The genetics of nucleotide excision repair in the yeast *Saccharomyces cerevisiae* is complex, apparently requiring at least 10 genes. We have isolated 5 of these genes (designated *RAD1*, *RAD2*, *RAD3*, *RAD4*, and *RAD10*) by molecular cloning and plan to overexpress them in order to generate proteins for biochemical study. We have sequenced four of these five genes and have noted regions of homology with other proteins in the predicted amino acid sequence of some of them. In particular, there is striking homology between Rad3 protein and a number of prokaryotic and eukaryotic proteins that bind nucleotides and hydrolyze ATP or GTP. Mutations in this region of the *RAD3* gene render cells defective in the nucleotide excision repair function. In addition to its role in nucleotide excision repair, the *RAD3* gene is essential for the viability of haploid cells in the absence of DNA damage. The nature of the essential function is unknown. The *RAD1* and *RAD3* genes are not inducible by DNA damaging agents. However, exposure of cells to UV radiation, 4-nitroquinoline 1-oxide, or γ radiation results in 4- to 6-fold enhanced expression of the *RAD2* gene.

INTRODUCTION

The phenomenon of excision repair of DNA was discovered over 20 years ago (see Ref. 7). However, despite intensive study, a detailed understanding of this ubiquitous cellular response to DNA damage is far from complete. Early cellular biological experiments resulted in simple working models of excision repair, many of which continue to pervade the literature. In fact, excision repair of DNA has proven to be surprisingly complicated, and the details of this phenomenon in prokaryotes are just beginning to emerge.

Complexity is apparent at a number of levels. First, there are at least two distinct biochemical pathways by which damaged or inappropriate bases are removed from DNA (7). Base excision repair is so called because some damaged moieties in DNA (e.g., simple alkylated bases) are excised as free bases by the action of a repair-specific class of enzymes called DNA glycosylases. The repair of the resulting sites of base loss (apurinic or apyrimidinic sites) involves the action of another repair-specific class of enzymes called apurinic/apyrimidinic (AP) endonucleases. In contrast, base damage that results in the formation of bulky adducts, e.g., pyrimidine dimers, is repaired by an alternative mechanism in which the damaged bases are removed as part of a nucleotide structure, hence the designation nucleotide excision repair.

A second level of complexity concerns the biochemistry of each of these distinct excision repair modes. Recent studies on *Escherichia coli* suggest that the specific degradation of DNA during nucleotide excision repair normally involves the action of at least 5 genes, designated *uvrA*, *uvrB*, *uvrC*, *uvrD*, and *polA* (5,12,14). Studies with purified proteins encoded by these genes suggest that it may be incorrect to consider the biochemistry of nucleotide excision repair as a series of specific sequential steps, each of which results from the action of a distinct protein. It seems more likely that under normal conditions most, if not all, phases of nucleotide excision repair, i.e., specific incision, unwinding, excision, repair synthesis, and ligation of DNA, require the concerted action of a complex of proteins (5). There are indications from both in vivo and in vitro studies on *E. coli* that some elements of this putative multiprotein complex (repairosome) are indispensable for nucleotide excision repair. Thus, for example, a defect in the UvrA, UvrB, or UvrC proteins, which collectively constitute the so-called uvrABC endonuclease required for specific incision of DNA, completely abolishes nucleotide excision repair. However, defects in UvrD and PolA proteins reduce the rate and extent of DNA incision, but do not abolish the process completely (8).

Very little is known about the molecular mechanism of nucleotide excision repair in eukaryotes. The two eukaryotic systems that have been genetically most informative are the yeast *Saccharomyces cerevisiae* and human cells through the study of the human disease xeroderma pigmentosum (XP). Both systems reveal a greater genetic complexity for nucleotide excision repair than in *E. coli* (7). In yeast, 10 genetic loci have been implicated in the process, while to date 9 complementation groups have been identified in the classical form of XP. An analysis of the phenotypes of mutants in both systems suggests interesting parallels with *E. coli* (8). Defects in some yeast and human genes completely abolish the capacity for the excision of pyrimidine dimers in UV-irradiated cells, reminiscent of the phenotype of *E. coli* *urvA*, *uvrB*, or *uvrC* strains. Other mutants have a significant residual capacity for nucleotide excision repair, as is true of *uvrD* and *polA* strains.

Our long-term goal is to isolate and study the proteins encoded by the 10 genes apparently involved in nucleotide excision repair in yeast. Initially, we have focused our attention on five of these genes (*RAD1*, *RAD2*, *RAD3*, *RAD4*, and *RAD10*). Mutations in any of these genes completely abolish detectable nucleotide excision repair in vivo. In this chapter, we review the characterization of these five genes, with an emphasis on their roles in the molecular mechanism of nucleotide excision repair.

MOLECULAR CLONING AND PHYSICAL CHARACTERIZATION OF YEAST *RAD* GENES

We isolated the *RAD1*, *RAD2*, *RAD3*, and *RAD10* genes by screening a yeast genomic library for recombinant plasmids that complement the UV sensitivity of mutants in these genes (19,21,32,34). Such screening failed to yield plasmids containing RAD4. The RAD4 gene has been genetically mapped to a position very close to the SPT2 gene of *S. cerevisiae* (33). We have recently shown that some rad4 mutant strains, transformed with an integrating plasmid containing SPT2 as well as flanking DNA, show normal levels of UV resistance. This observation suggested that part (and possibly all) of the RAD4 gene is present on the plasmid. Using a gene rescue strategy, we isolated a DNA fragment from the genome of a UV-resistant integrant strain and showed by complementation and by genetic analysis that this fragment contains the RAD4 gene (R. Fleer, G. Pure, and E.C. Friedberg, ms. in prep.).

The failure to isolate RAD4 by screening a yeast library is apparently due to the inability to propagate the functional gene in *E. coli*. Current evidence suggests that during passage through *E. coli* one or more defects are generated, perhaps because expression of RAD4 is lethal to this bacterium (R. Fleer and E.C. Friedberg, unpubl. observ.). However, if the defective gene is integrated into the genome of rad4 mutants, then, depending on the position of the integration events relative to that of the defect in the cloned gene and of the mutation in the particular host rad4 allele, some fraction of the integrants contain a wild-type copy of the gene. Following propagation in *E. coli*, nonintegrating (i.e., centromeric) plasmids contain a defective RAD4 gene and fail to complement the UV sensitivity of rad4 mutant alleles. If the defective region is deleted and the resulting linear plasmid is transformed into yeast cells, the gap created by the deletion can be repaired by gene conversion from homologous wild-type chromosomal sequences. When these gap-repaired plasmids are propagated in yeast (rather than in *E. coli*), they fully complement the UV sensitivity of all rad4 mutants tested (R. Fleer, G. Pure, and E.C. Friedberg, ms. in prep.).

The RAD1, RAD2, RAD3, RAD4, and RAD10 genes are all present in the yeast genome in single copies. There is no evidence for introns in any of these genes, and this is consistent with the observation that the majority of yeast genes transcribed by RNA polymerase II are free of intervening sequences. The sizes of the coding regions in 4 of these genes have been established by DNA sequencing (Tab. 1) (22,23,32,34). Based on this information, the expected sizes of the polypeptides encoded by the RAD1, RAD2, and RAD3 genes are quite large (Tab. 1). Rad10 protein is considerably smaller. The molecular weights of the Rad3 and Rad10 proteins have been confirmed by gel electrophoresis following their expression in *E. coli* and in yeast (L. Naumovski, W. Weiss, and E.C. Friedberg, unpubl. observ.). Thus, a multiprotein complex possibly formed by these gene products is expected to be very large (>350 kDa).

Tab. 1. Sizes of some of the yeast RAD genes.

RAD1	RAD2	RAD3	RAD10
Coding region = 2,916 bp	Coding region = 2,925 bp	Coding region = 2,334 bp	Coding region = 588 bp
No introns	No introns	No introns	No introns
Calculated size of protein = 110 kDa	Calculated size of protein = 111 kDa	Calculated size of protein = 89.9 kDa	Calculated size of protein = 22.6 kDa

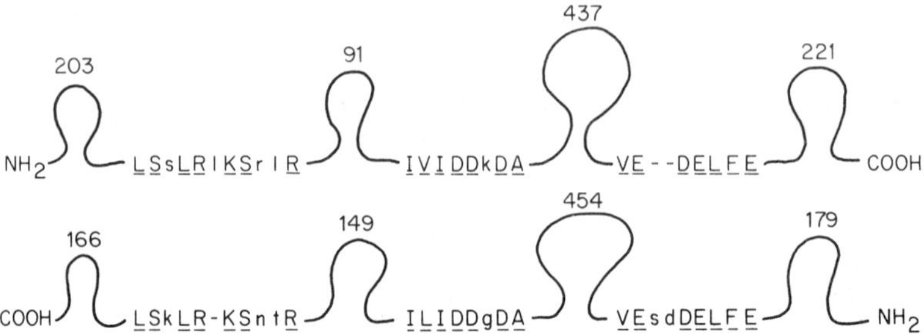

Fig. 1. Regions of amino acid sequence homology between the Rad2 (top) and Rad1 (bottom) polypeptide chains. The numbers refer to the distances (in amino acids) that separate the regions of homology from each other and from the ends of the polypeptide chains.

Three limited regions of homology have been observed between the predicted amino acid sequences of the Rad1 and Rad2 polypeptides (23). Curiously, the order of these sequences is reversed on the two polypeptide chains (Fig. 1). Nonetheless, in each polypeptide, the distances separating the regions of homology from one another and from the ends of the respective polypeptide chains are similar (Fig. 1). Their reversed order makes it unlikely that the two proteins evolved from a common ancestral polypeptide. However, this observation is not inconsistent with convergent evolution of similar catalytic, structural, or regulatory functions.

Comparisons with other protein sequences have revealed a number of interesting observations, the significance of which remains to be determined. The yeast Rad10, Rad1, and Rad3, and the E. coli UvrA, UvrC, UvrD, and PolA excision repair proteins share a limited region of sequence homology in which the tripeptide Gly-Lys-Thr-(Ser or Gly) is invariant (Fig. 2). Both UvrA and UvrD proteins are DNA-dependent ATPases (1,15,24,30), and the

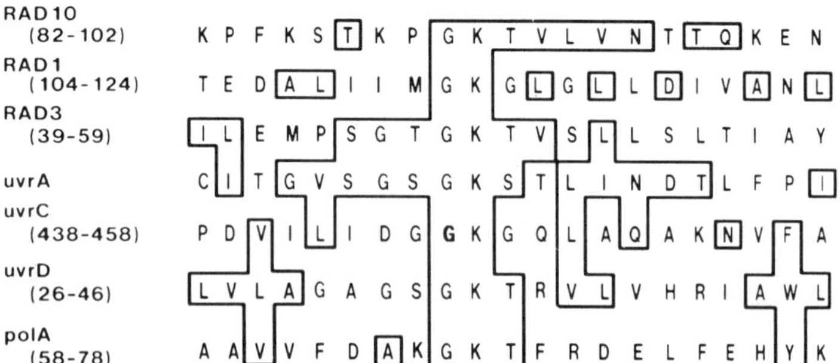

Fig. 2. Homology between some yeast and E. coli excision repair proteins. Regions of amino acid sequence homology in the Rad1 (34), Rad10 (32), and Rad3 (22) yeast proteins and the UvrA, UvrC (29), UvrD (6), and PolA (13) proteins of E. coli. The numbers refer to amino acid positions in the respective polypeptide chains. The nucleotide and predicted amino acid sequence of the uvrA gene and the UvrA protein were generously communicated to us by Dr. Aziz Sancar. (The exact position of the region shown is not known since the uvrA sequence is not yet complete.)

sequence Gly-Lys-Thr (Ser) is common to a number of other prokaryotic and eukaryotic ATPases and adenine nucleotide-binding proteins, as well as a number of GTPases and guanine nucleotide-binding proteins (L. Naumovski and E.C. Friedberg, ms. submitted for publ.). UvrD protein also has DNA helicase activity (1,15, 24). In addition to the homology suggested in Fig. 2, Rad1 protein shows homology with a second region identified in a number of ATPases, in which the amino acid Asp is preceded by a string of 4 hydrophobic amino acids and is often followed by Asp or Glu (10) (Fig. 3).

A particular dnaB mutant of E. coli shows a decreased rate of thymine excision and is hypermutable by UV radiation; however, this strain is not abnormally sensitive to killing by UV light and its role in excision repair of DNA is unknown (4). It is interesting to note that DnaB protein is also a DNA-dependent ATPase, and the region between amino acids 226-243 includes the Gly-Lys-Thr tripeptide discussed above. Homology between this region and the Gly-Lys-Thr region of Rad3 protein is shown in Fig. 4. Curiously, this region of Rad3 also shows homology with a second region in DnaB just 12 amino acids downstream (Fig. 4).

Sequences even further downstream in the Rad3 and DnaB polypetides are also homologous with each other and with a region of Sir3 protein (Fig. 5). Sir3 is one of a number of yeast proteins involved in the regulation of yeast mating type (9). Its precise function is unknown; however, its postulated role in mating type gene expression is not inconsistent with a DNA-binding activity (3). In addition to its ATPase activity, DnaB protein is also a DNA-binding protein that is required for DNA replication in E. coli (9). Thus, these comparisons suggest that Rad3 protein may be a purine ribonucleoside triphosphatase and a DNA-binding protein.

The latter possibility is supported by yet another amino acid sequence comparison. Structural analysis of λ repressor, λ Cro, and E. coli Cap

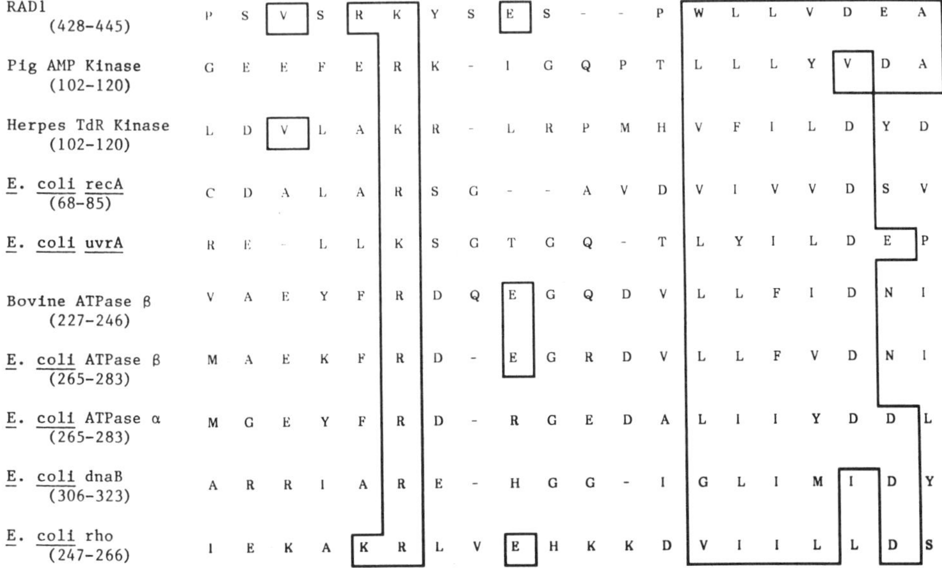

Fig. 3. Regions of amino acid sequence homology between Rad1 protein and various ATPases (see Ref. 6).

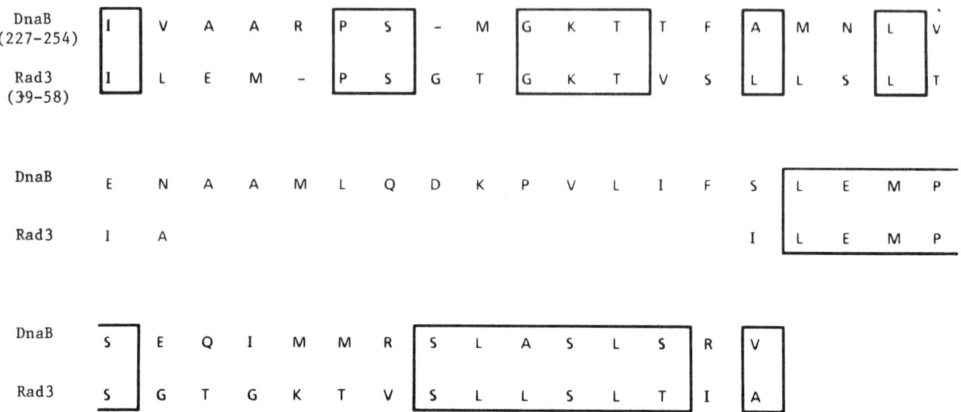

Fig. 4. Sequence homology between Rad3 and DnaB proteins. The region of the Rad3 polypeptide between amino acids 39-58 shows homology with two regions of DnaB protein (18) separated by 12 amino acids.

proteins has shown that the putative specific DNA-binding domains of all three proteins are characterized by αhelix-turn-αhelix motifs (26). A number of other proteins known to bind to specific regions of DNA contain amino acid sequences that are good candidates for αhelix-turn-αhelix domains (26). A comparison of Rad3 protein with a consensus sequence derived from these proteins (including λ repressor, λ Cro, and E. coli Cap proteins) suggests that Rad3 may contain an αhelix-turn-αhelix domain (L. Naumovski and E.C. Friedberg, ms. submitted for publ.).

Collectively, these homologies suggest the following possibilities: (a) some E. coli and yeast excision repair proteins may have evolved from a common ancestral protein; (b) one or more yeast Rad proteins may have nucleotide-binding, purine ribonucleoside triphosphatase, and/or DNA helicase activity; and (c) Rad3 protein may be a DNA-binding protein. These possibilities will be systematically explored when purified proteins are available for study.

RAD3 IS AN ESSENTIAL GENE IN SACCHAROMYCES CEREVISIAE

Disruptions of the chromosomal RAD1, RAD2, RAD4, and RAD10 genes do not affect the viability of haploid cells, indicating that these genes are not essential (8; R. Fleer and E.C. Friedberg, ms. in prep.). However, disruption of the RAD3 gene is lethal to haploid cells, and the recessive lethal mutation generated in diploid cells can be inferred from the demonstration that only two of the four spores that result from each cell during meiosis are viable (11,20).

The observation that a gene required for early events in nucleotide excision repair is also essential in the absence of DNA damage is unexpected. The additional finding that many rad3 mutant strains are perfectly viable but are completely defective in the incision of DNA (20; L. Naumovski and E.C. Friedberg, ms. submitted for publ.) led to the suggestion (20) that RAD3 has two distinct functions. One explanation for such a phenomenon is that the gene encodes two distinct products as a result of

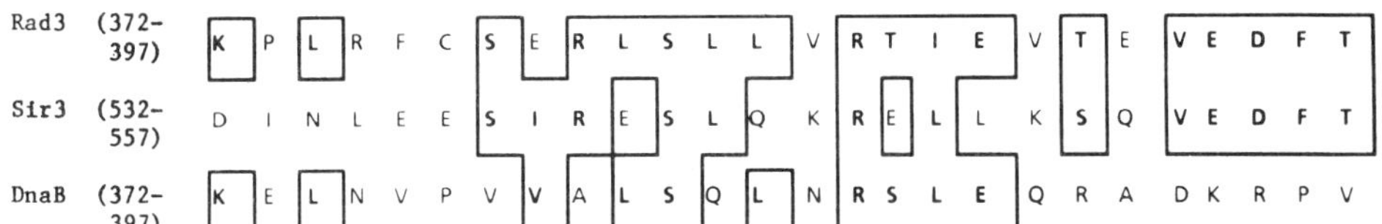

Fig. 5. Sequence homology between Rad3, Sir3, and DnaB proteins. Rad3 and DnaB proteins show a second region of homology (see Fig. 4) and share this homology with a region in the yeast Sir3 protein (R. Doolittle, pers. comm.).

transcriptional and/or translational regulation. However, as indicated previously, the RAD3 gene does not contain introns, and examination of RAD3 mRNA by blot hybridization reveals a single transcript. Furthermore, transcript mapping reveals single major transcriptional start and stop sites. Thus, there is no evidence for more than one form of RAD3 mRNA. There is also no indication of translational regulation. Rad3 protein has been overexpressed in yeast cells following the tailoring of the coding region next to one of several strong yeast promoters (L. Naumovski and E.C. Friedberg, unpubl. observ.). In all cases, only a single major species has been identified, with a molecular weight expected from the size of the coding region of the gene.

There are at least four mechanisms by which a single gene product could participate both in excision repair of DNA and in an essential function:

(a) A single species of Rad3 protein may contain two distinct functional (catalytic and/or regulatory) domains.

(b) Some Rad3 protein may undergo a form of post-translational modification to yield a species not observed by gel electrophoresis.

(c) Rad3 protein may contain a single activity that is required for both the excision repair and the essential functions, but the relative affinity of Rad3 protein for its substrate may differ for each of these functions. The excision repair function may require a high affinity (low K_m) interaction with its substrate, while the essential function may involve a much lower affinity interaction (higher K_m). Hence, a single amino acid substitution in Rad3 protein may alter its conformation such that the excision repair function is defective while the essential function remains intact. This model predicts that most, if not all, mutants defective in the essential function would also be defective in the excision repair function.

(d) The relative amounts of Rad3 protein required for the essential and excision repair functions may differ.

We are attempting to decipher the functions of the RAD3 gene by directing studies at two levels: the gene itself and the protein. With respect to the former, we have investigated the effect of mutations in the two regions of the gene that, based on amino acid sequence homology with other proteins, have been tentatively identified as ribonucleoside triphosphatase (and/or nucleotide-binding) and DNA-binding domains (see section "Molecular Cloning and Physical Characterizations of Yeast RAD Genes").

Two independent mutations generated by site-directed mutagenesis in the putative nucleotide binding box (codons 32-55) (Fig. 6) resulted in loss of only the excision repair function of the RAD3 gene. A third mutant gene contains two mutations in adjacent codons in the putative DNA binding box (codons 592-611) (Fig. 6) and is defective in both the excision repair and the essential function of RAD3 (Fig. 6). Other mutants in this region are defective only in the excision repair function.

Further mutants were generated by random mutagenesis of the cloned gene. After screening plasmids for the loss of the excision repair function, five mutations were mapped between codons 462-594 (Fig. 6). However, none of these mutations affects the essential function of RAD3. Following random mutagenesis, we isolated no plasmids in which only the essential function of RAD3 was inactivated (L. Naumovski and E.C. Friedberg, ms. submitted for publ.).

It is evident from these studies that mutations at multiple sites can affect the excision repair function of the RAD3 gene. Furthermore, these sites are not confined to a single domain on the linear polypeptide chain, but can be as far apart as codons 48-604. Thus it would appear that both the putative nucleotide- and the DNA-binding domains of the gene are involved in the excision repair function. However, in light of the observation that mutations affecting this function can be widely distributed in the gene, further site-directed mutagenesis is required to confirm the specificity of mutations in these domains.

The essential function of RAD3 is more resistant to mutagenesis, and thus far only a single mutant has been isolated in which this phenotype is affected. This is consistent with (but certainly does not prove) the idea that the essential function is determined by a limited domain of the gene. On the other hand, the apparent resistance of the essential function relative to the excision repair function to inactivation by point mutations is also consistent with the notion that the former activity requires a low-affinity protein-substrate interaction, whereas the affinity of this interaction during excision repair is higher. Clearly, it would be extremely useful to generate conditional lethal rad3 mutants to explore the nature of the essential function more fully. Such studies are in progress in our laboratory.

The identification of possible nucleotide- and DNA-binding domains in the Rad3 polypeptide has prompted specific biochemical studies on the isolated protein. We have identified Rad3 protein of Mr ∿90 kDa in extracts of yeast cells transformed with a multicopy plasmid in which expression of RAD3 is under control of the yeast ADH1 promoter (L. Naumovski and E.C. Friedberg, unpubl. data). The purification of this protein and its physical and biochemical characterization are in progress. A basically similar strategy is being used to isolate proteins encoded by the other cloned RAD genes.

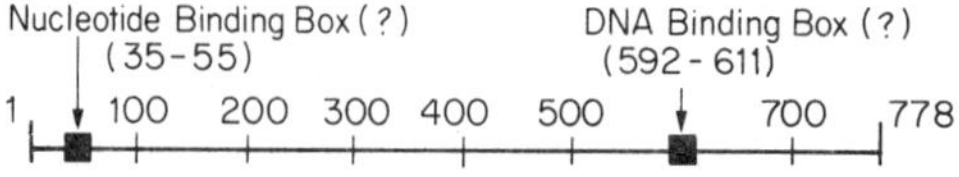

Fig. 6. Diagrammatic representation of the Rad3 polypeptide chain (778 codons) showing the location of the regions that show homology with known nucleotide-binding and DNA-binding proteins.

THE EXPRESSION OF YEAST RAD GENES

Measurements of the amount of RAD3 mRNA relative to URA3 mRNA in constitutive, untransformed Rad$^+$ cells indicate the presence of only 1 to 5 RAD3 transcripts per cell (19). The amounts of RAD1 and RAD2 mRNA are apparently even lower, since these messengers are extremely difficult to identify by northern hybridization (21; E. Yang and E.C. Friedberg, unpubl. observ.). Direct comparisons of the relative amounts of RAD1, RAD2, RAD3, and RAD10 mRNA have not been carried out. However, comparisons have been made with respect to a β-galactosidase activity expressed from RAD-lacZ gene fusions. Such comparisons must be interpreted with caution, since different fusion proteins may have varying stability and specific activity.

Nonetheless, it is interesting to note that, relative to the amount of β-galactosidase expressed by fusion of the E. coli lacZ gene to HIS4 (a strongly expressed yeast gene), RAD1-, RAD2-, RAD3-, and RAD10-lacZ fusions are all weakly expressed (Tab. 2). It is also interesting to note that all four RAD genes thus far sequenced show no detectable bias for codon usage, i.e., codons generally considered rare in yeast are well-represented in these genes (23). This is consistent with the view that strongly expressed genes have evolved a bias in favor of codons that correspond to the major tRNA isoacceptor species in yeast, while weakly expressed genes show little, if any, such bias (2).

In view of the low level of expression of these genes, we were prompted to investigate their possible induction following exposure of cells to DNA damage. Single copy RAD-lacZ gene fusions were integrated into the chromosome of wild-type cells. When cells containing RAD1-lacZ or RAD3-lacZ integrant genes were exposed to 4-nitroquinoline 1-oxide (4NQO), no enhanced expression of β-galactosidase was observed (22; G. Robinson, C. Nicolet, D. Kalainov, and E.C. Friedberg, ms. submitted for publ.). Similar results have been reported by others (17). However, cells containing a RAD2-lacZ fusion showed approximately a 4-fold increase in β-galactosidase activity after treatment with 4NQO. This response is inhibited in the presence of cycloheximide, indicating a requirement for protein synthesis. Furthermore, direct measurement of RNA levels in untransformed cells showed about a 3-fold increase in RAD2 mRNA following treatment with 4NQO (G. Robinson, C. Nicolet, D. Kalainov, and E.C. Friedberg, ms. submitted for publ.). Thus, the enhanced expression of β-galactosidase reflects induction of the RAD2 gene rather than the modification of extant RAD2-lacZ fusion protein.

Tab. 2. Expression of β-galactosidase from gene fusions.

Fusion	β-Galactosidase activity ($U/10^7$ cells)
HIS4::lacZ	28.20
HIS3::lacZ	2.70
RAD3::lacZ	0.88
RAD1::lacZ	0.20
RAD2::lacZ	0.01
RAD10::lacZ	<0.01

A response of similar magnitude to that observed with 4NQO is obtained following treatment of RAD2-lacZ integrant strains with UV or γ radiation, or with naladixic acid. Other agents that damage DNA, including the alkylating agent methyl methanesulfonate, result in lower levels of induction. In addition, disrupting normal DNA replication by treating cells with methotrexate (which results in loss of thymidylate synthetase activity) also induces weakly. The response to agents such as UV radiation or 4NQO, on the one hand, and agents such as methyl methanesulfonate, on the other, can also be kinetically distinguished, i.e., stronger inducing agents result in a more rapid peak response than do weaker ones (G. Robinson, C. Nicolet, D. Kalainov, and E.C. Friedberg, ms. submitted for publ.).

The broad spectrum of inducing signals for RAD2, the magnitude of the induction response, and the differential kinetics with different agents are reminiscent of the SOS response in *E. coli* (25). In addition, both in yeast and in *E. coli*, some (but not all) genes involved in nucleotide excision repair of DNA are inducible. These correlations notwithstanding, there is currently no direct experimental evidence relating induction of RAD2 in yeast to an SOS response reminiscent of that in *E. coli*. Furthermore, in *E. coli* not all the genes induced in response to DNA damage are under recA-lexA control (31). The ada gene regulates the response to agents that alkylate the O^6 position of guanine in DNA. In addition, the groEL and dnaK genes of *E. coli* are inducible by DNA damage and by heat shock and are thought to be regulated by the htpR gene. Finally, there is recent evidence suggestive of the induction of genes in *E. coli* in response to oxidative damage to DNA (31).

Aside from RAD2, the CDC9 gene of *S. cerevisiae* has been reported to be inducible by DNA damage (27). Furthermore, two groups have independently isolated a series of damage-inducible (DIN) yeast genes of unknown function (16,28). A comparison of the kinetics and level of induction of one of these genes (DIN1) with that of RAD2 shows significant differences, suggesting that at least these two genes may be differentially regulated (G. Robinson, C. Nicolet, D. Kalainov, and E.C. Friedberg, ms. submitted for publ.). The regulation of RAD2 induction, its possible relationship to the SOS response in *E. coli*, and the response of other RAD genes to DNA-damaging agents are currently under active study.

ACKNOWLEDGEMENTS

We thank Dr. Aziz Sancar for the partial sequence of uvrA and Dr. Russell Doolittle for pointing out the homology between Rad3 and Sir3 proteins. These studies are supported by research grant #CA 12428 from the U.S. Public Health Service. W.A.W. is supported by Training Grant #CA09302, L.N. by MSTP Training Grant #GM 07365, and G.W.R. by National Research Service Award #CA 07437 from the USPHS. R.F. is supported by a postdoctoral fellowship from Deutscher Akademischer Austauschdienst, West Germany.

REFERENCES

1. Arther, H.M., D. Bramhill, P.B. Eastlake, and P.T. Emmerson (1982) Cloning of the uvrD gene of *E. coli* and identification of the product. *Gene* 19:285-295.
2. Bennetzen, J.L., and B.D. Hall (1982) Codon selection in yeast. *J. Biol. Chem.* 257:3026-3031.

3. Brand, A.H., L. Breeden, J. Abraham, R. Sternglanz, and K. Nasmyth (1985) Characterization of a "silencer" in yeast: A DNA sequence with properties opposite to those of a transcriptional enhancer. Cell 41: 41-48.
4. Bridges, B.A., R.P. Mottershead, and A. Lehmann (1976) Error-prone DNA repair in Escherichia coli. IV. Excision repair and radiation-induced mutation in a dnaB strain. Biol. Zentralblat. 95:393-403.
5. Caron, P.R., S.R. Kushner, and L. Grossman (1985) Involvement of helicase II (uvrD gene product) and DNA polymerase I in excision mediated by the uvrABC protein complex. Proc. Natl. Acad. Sci., USA 82:4925-4929.
6. Finch, P.W., and P.T. Emmerson (1984) The nucleotide sequence of the uvrD gene of E. coli. Nucl. Acids Res. 12:5789-5799.
7. Friedberg, E.C. (1985) DNA Repair, W.H. Freeman and Co., New York.
8. Friedberg, E.C. (1985) Nucleotide excision repair of DNA in eukaryotes: Comparisons between human cells and yeast. Cancer Surveys (in press).
9. Herskowitz, I., and Y. Oshima (1981) Control of cell type in Saccharomyces cerevisiae. In The Molecular Biology of the Yeast Saccharomyces, J.N. Strathern, E.W. Jones, and J.R. Broach, eds. Cold Spring Harbor Laboratory, Cold Spring Harbor, New York, pp. 181-209.
10. Higgins, C.F., I.D. Hiles, K. Whalley, and D.J. Jamieson (1985) Nucleotide binding by membrane components of bacterial periplasmic binding protein-dependent transport systems. EMBO J. 4:1033-1040.
11. Higgins, D.R., S. Prakash, P. Reynolds, R. Polakowska, S. Weber, and L. Prakash (1983) Isolation and characterization of the RAD3 gene of Saccharomyces cerevisiae and inviability of rad3 deletion mutants. Proc. Natl. Acad. Sci., USA 80:5680-5684.
12. Husain, I., B. Van Houten, D.C. Thomas, M. Abdul-Monem, and A. Sancar (1985) Effect of DNA polymerase I and DNA helicase II on the turnover rate of UvrABC excision nuclease. Proc. Natl. Acad. Sci., USA (in press).
13. Joyce, C.M., W.S. Kelley, and N.D.F. Grindley (1982) Nucleotide sequence of the Escherichia coli polA gene and primary structure of DNA polymerase I. J. Biol. Chem. 257:1958-1964.
14. Kumara, K., M. Sekiguchi, A.-L. Steinum, and E. Seeberg (1985) Stimulation of the UvrABC enzyme-catalyzed reactions by the UvrD protein (DNA helicase II). Nucl. Acids Res. 13:1483-1492.
15. Maples, V.F., and S.R. Kushner (1982) DNA repair in Escherichia coli: Identification of the uvrD gene product. Proc. Natl. Acad. Sci., USA 79:5616-5620.
16. McClanahan, T., and K. McEntee (1984) Specific transcripts are elevated in Saccharomyces cerevisiae in response to DNA damage. Mol. Cell. Biol. 4:2356-2363.
17. Nagpal, M.L., D.R. Higgins, and S. Prakash (1985) Expression of the RAD1 and RAD3 genes of Saccharomyces cerevisiae is not affected by DNA damage or during the cell division cycle. Mol. Gen. Genet. 199:59-63.
18. Nakayama, N., N. Arai, M.W. Bond, Y. Kaziro, and K. Arai (1984) Nucleotide sequence of dnaB and primary structure of the dnaB protein from Escherichia coli. J. Biol. Chem. 259:97-101.
19. Naumovski, L., and E.C. Friedberg (1982) Molecular cloning of eukaryotic genes required for excision repair of UV-irradiated DNA: Isolation and partial characterization of the RAD3 gene of Saccharomyces cerevisiae. J. Bacteriol. 152:323-331.
20. Naumovski, L., and E.C. Friedberg (1983) A DNA repair gene required for the incision of damaged DNA is essential for viability in Saccharomyces cerevisiae. Proc. Natl. Acad. Sci., USA 80:5680-5684.

21. Naumovski, L., and E.C. Friedberg (1984) Saccharomyces cerevisiae RAD2 gene: Isolation, subcloning and partial characterization. Mol. Cell. Biol. 4:290-295.
22. Naumovski, L., G. Chu, P. Berg, and E.C. Friedberg (1985) RAD3 gene of Saccharomyces cerevisiae: Nucleotide sequence of wild-type and mutant alleles, transcript mapping and aspects of gene regulation. Mol. Cell. Biol. 5:17-26.
23. Nicolet, C.M., J.M. Chenevert, and E.C. Friedberg (1985) The RAD2 gene of Saccharomyces cerevisiae: Nucleotide sequence and transcript mapping. Gene 36:225-334.
24. Oeda, K., T. Horiuchi, and M. Sekiguchi (1982) The uvrD gene of E. coli encodes a DNA-dependent ATPase. Nature 298:98-100.
25. Oishi, M., R.M. Irbe, and L.M.E. Morin (1981) Molecular mechanism for the induction of "SOS" functions. Prog. Nucl. Acid Res. Mol. Biol. 26:281-301.
26. Pabo, C.O., and R.T. Sauer (1984) Protein-DNA recognition. Ann. Rev. Biochem. 53:293-321.
27. Peterson, T.A., L. Prakash, S. Prakash, M.A. Osley, and S.I. Reed (1985) Regulation of CDC9, the Saccharomyces cerevisiae gene that encodes DNA ligase. Mol. Cell. Biol. 5:226-235.
28. Ruby, S., and J.W. Szostak (1985) Specific Saccharomyces cerevisiae genes are expressed in response to DNA-damaging agents. Mol. Cell. Biol. 5:75-84.
29. Sancar, G.B., A. Sancar, and W.D. Rupp (1984) Sequences of the E. coli uvrC genes and protein. Nucl. Acids Res. 12:4593-4608.
30. Seeberg, E., and A.-L. Steinum (1982) Purification and properties of the uvrA protein from Escherichia coli. Proc. Natl. Acad. Sci., USA 79:988-992.
31. Walker, G. (1984) Mutagenesis and inducible responses to deoxyribonucleic acid damage in Escherichia coli. Bacteriol. Rev. 48:60-93.
32. Weiss, W.A., and E.C. Friedberg (1985) Molecular cloning and characterization of the yeast RAD10 gene and expression of RAD10 protein in E. coli. EMBO J. 4:1575-1582.
33. Winston, F., D.T. Chaleff, B. Valent, and G.R. Fink (1984) Mutations affecting Ty-mediated expression of the HIS4 gene of Saccharomyces cerevisiae. Genetics 107:179-197.
34. Yang, E., and E.C. Friedberg (1984) Molecular cloning and nucleotide sequence analysis of the Saccharomyces cerevisiae RAD1 gene. Mol. Cell. Biol. 4:2161-2169.

THE FIXATION OF ERRORS

INTRODUCTION: MOLECULAR BASIS OF GENOMIC STABILITY AND CHANGE

Robert H. Haynes

Department of Biology
York University
Toronto, Canada

In 1952, just before publication of the epochal work of Watson and Crick on the DNA double helix, H.J. Muller argued that mutation is the result of "some biochemical disorganization in which processes _normally tending to hold mutation frequencies in check_ are to some extent interfered with" (8). This prescient statement crisply summarizes our contemporary picture (Fig. 1) of the molecular basis of genetic stability and change (5). There are great and persistent pressures on genomes to erode as a result of the physicochemical "decay" of DNA structure under normal physiological conditions, and to mutate as a result of replication errors and attack on DNA by endogenous and exogenous mutagens. Were it not for the existence of a complex array of biochemical and other mechanisms that counteract the natural tendency for DNA to mutate and disintegrate, cells could not survive, even under "normal" conditions, and genomes could not have evolved beyond a few hundred nucleotides in length (3,9). As indicated in Fig. 1, the _potential_ mutation rate per nucleotide (PMR) that would be associated with nonenzymatic, purely template-directed polymerization (3) of nucleic acids is very high ($\sim 10^{-2}$), whereas observed _residual_ mutation rates (RMRs) are very low ($\sim 10^{-9}$). In this context, it is interesting to note that the most skillful human keypunch operators make about one mistake per 5,000 symbols copied.

The genetic information of cells is stabilized biochemically from one generation to the next despite the chronic pressure toward mutation. The great stability of genes does not arise from static physical properties of the genetic material, or lack of mutagens in the genomic environment. Rather, it is a consequence of the biochemical dynamics of DNA replication and repair, together with the existence of enzymes and other materials, of both inter- and extracellular origin, which are capable of neutralizing many common genotoxic agents, including those that arise as byproducts or intermediaries of normal oxidative metabolism (1). These antimutagens provide a "first line of defense" for genomes in the hostile environments in which they must exist, although processes that act to inhibit or reverse any biochemical step in the mutational pathway can be regarded, in a formal sense, as being antimutagenic.

In 1953, Dancoff and Quastler pointed out that, in the context of information theory, if very great fidelity of operation is to be achieved with equipment of poor precision, extensive checking procedures must be built into the system (2). For optimum economy, the energy cost of such procedures should be just sufficient to reduce the error rate to a tolerable level. It would appear that this "principle of maximum error" is exemplified in the genetic machinery of cells. The evolution of long genetic messages has been made possible, in large part, by the fact that they encode extensive instructions for their own correction. The energy cost obviously is not prohibitive for the general economy of the cell, and the residual error rate is consistent both with the genetic integrity of the individual and genomic expansion in evolution.

The fantastic, though not perfect, accuracy of DNA synthesis is achieved in bacteria through a combination of three devices that actively promote replicational fidelity (7). These are, first, the utilization of specific DNA polymerases that discriminate against the insertion of incorrect nucleotides at the point of polymerization in the nascent DNA strand. Such enzymes reduce the error rate to about 10^{-4}. Second, the "editing" or "proofreading" 3'-nucleolytic function of bacterial polymerase acts to catalyze the removal of mispaired nucleotides immediately after their insertion. This allows a further decrease in error rate to about 10^{-7}. Third, there exists a mismatch repair system capable of excising mispaired nucleotides from newly replicated, unmethylated DNA strands. Genetic or other interference with these "antimutagenic" enzyme activities can lead to increased error rates in DNA synthesis.

There are four basic biochemical requirements for the polymerization of nucleotides in double-stranded DNA synthesis. First, an appropriate template from which the newly synthesized strand can be copied; second, the enzymes necessary to catalyze the various reactions involved; third, a suitable concentration of magnesium ions; and fourth, suitable concentrations of the four deoxynucleotide precursors. A further battery of enzymes is necessary for biosynthesis of the nucleotides. These requirements are all rather stringent, and if any fail to be met, the cell may die, or become genetically unstable in the sense of exhibiting pathological levels of mutation and/or recombination (5). Thus, mutation rates in cells may be increased by damage to DNA templates caused by various mutagenic chemicals or radiation, by substituting other metal ions for magnesium, and by altering the relative concentrations of the nucleotide precursors required for DNA synthesis. In addition, genetically heritable alterations in spontaneous mutation and recombination frequencies are observed in mutant lines of organisms in which enzymes involved in DNA replication, repair, or nucleotide biosynthesis are structurally abnormal. Conversely, any agent or situation that serves to stabilize the biochemical conditions required for normal DNA synthesis can be regarded formally as being antimutagenic.

The chemical vulgarity of DNA makes it prey to all the horrors and misfortune that might befall any molecule in a warm aqueous medium. Bits and pieces fall away, and even its backbone may be broken, at alarming rates under normal physiological conditions (9). Furthermore, the immediate intracellular environment swarms with nasty items ever ready to chew upon genes and break their chemical bones. A partial list of these incubi would include certain highly reactive free radicals, peroxides, singlet oxygen, reducing agents, alkyl nitrosamines, nitrites and alkylating agents, plus low, but sometimes significant, exposures to ionizing radiation and ultraviolet light (1). It is hardly surprising that, in addition to mechanisms that promote replicative fidelity, a complex battery of DNA repair

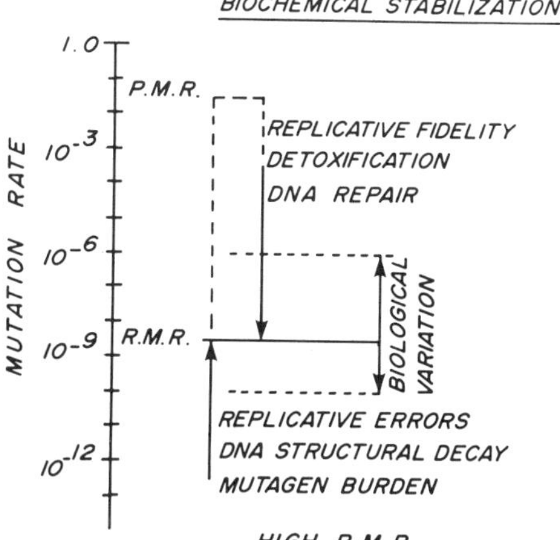

Fig. 1. Schematic view of the molecular mechanisms involved in genetic stability and change. Observed residual mutation rates in cells (RMRs) are extremely low. They are kept low through the efficient, combined actions of a variety of biochemical mechanisms that stabilize the genome against damage and normal physicochemical "decay." Were it not for the existence of various antimutagenic mechanisms that promote replicative fidelity, detoxification of mutagens, and repair of damaged DNA, the genetic information would soon erode (because the potential mutation rate, PMR, is very high) and organisms as such could not survive. However, the few random mutations that do escape these protective devices are the ultimate source of the genetic variation necessary for evolution. Residual mutation rate (per nucleotide) is not a finely controlled quantity; its magnitude can vary over three or four orders of magnitude among organisms, although the rate per genome is relatively constant.

systems has evolved to reverse or bypass potentially mutagenic or lethal damage to DNA (4,6). The three classical modes of DNA repair are photoreactivation of UV damage, base and nucleotide excision repair, and postreplication recombinational repair. The two latter mechanisms are rather versatile, and a variety of different types of DNA damage can be repaired by them. These rather "general" DNA damage detecting and correcting systems could not exist were it not for the redundancy of the genetic information that is inherent in the complementary double-stranded structure of DNA. Indeed, this structural feature may account, in large part, for the ubiquity of double-stranded DNA as the genetic material of organisms.

DNA repair processes serve to reverse genetic damage after it has occurred. However, as noted above, there also exist enzymes and antimutagens, in the strict sense of the word, that neutralize many mutagens, both within the cell and in extracellular environments. These substances are, of course, the primary object of concern in this conference. I cannot resist relating an amusing anecdote about my first encounter with the fact that many antimutagens are to be found in fresh vegetables and other foods. In 1979, I visited Dr. Kada in Mishima. It was then that I heard of his fascinating discoveries in this area. This made me breathe a deep sigh of relief, as earlier, in Tokyo, I had learned from Dr. Sugimura of the potent mutagenicity of tryptophan pyrolysate, and related materials, found in broiled fish. The identification of these genotoxic products in ordinary cooked foods seemed to deepen even further the dread sea of mutagens in which we are supposedly immersed. On the train from Tokyo to Mishima, I began to wonder how any species could evolve, let alone individuals survive, when exposed to so many mutagens. In Mishima, Dr. Kada's work seemed to provide the answer. In order to camouflage my genuine amazement at his results, I commented, in a cool, offhand way, "Your results beautifully exemplify Haynes' Law: in any well-balanced diet, for every mutagen, there is an equal and opposite antimutagen." Kada immediately, and with much excitement, asked, "Where published?" Well, I can now answer your question, Dr. Kada: in the proceedings of the International Conference on Mechanisms of Antimutagenesis and Anticarcinogenesis!

Upon reflection, there may be more truth than humor in "Haynes' Law," especially in light of Fig. 1. The many pressures toward mutation are in fact balanced, to a remarkable degree, by antimutagenic devices and agents. However, being themselves composed of ordinary molecules, such systems cannot, for thermodynamic and other reasons, operate with 100 percent efficiency. Still, the net result of this Manichean conflict between genetic order and decay is that mutation rates in normal organisms are extraordinarily low. However, even though most mutations are deleterious, natural selection has not, and indeed cannot, drive mutation rates to zero. A background of random genetic noise will be with us always, no matter how much we do to reduce exposure to environmental mutagens.

Unfortunately, we do not know what fraction of the total mutagen burden on the human population is effectively avoidable by appropriate public health legislation, or personal changes in diet and lifestyle. Until we learn much more about the magnitude of the net mutagen burden on various human populations, what proportion of this burden is avoidable, and what fraction of the genetic alterations that occur in germinal and somatic cells is induced by this avoidable burden, it will be difficult to give legislators and the public a quantitatively accurate assessment of the public health benefits that can be expected to flow from the control of environmental mutagens. Quantitative evaluation of the role played by antimutagenic agents and processes in the maintenance of genetic stability in humans will contribute significantly to the resolution of the important practical problems of genetic toxicology.

REFERENCES

1. Ames, B.N. (1983) Dietary carcinogens and anticarcinogens. Science 221:1256-1264.
2. Dancoff, S.M., and H. Quastler (1953) The information content and error rate in living things. In Information Theory in Biology, H. Quastler, ed. University of Illinois Press, Urbana, Illinois, pp. 263-273.
3. Eigen, M., and P. Schuster (1979) The Hypercycle, Springer-Verlag, Berlin.
4. Hanawalt, P.C., P.K. Cooper, A.K. Ganesan, and C.A. Smith (1979) DNA repair in bacteria and mammalian cells. Ann. Rev. Biochem. 48:783-836.
5. Haynes, R.H. (1985) Molecular mechanisms in genetic stability and change: Role of deoxyribonucleotide pool balance. In Genetic Consequences of Nucleotide Pool Imbalance, F. de Serres, ed. Plenum Press, New York, pp. 1-23.

6. Haynes, R.H., and B.A. Kunz (1981) DNA repair and mutagenesis in yeast. In Molecular Biology of the Yeast Saccharomyces: Life Cycle and Inheritance, J.N. Strathern, E.W. Jones, and J.R. Broach, eds. Cold Spring Harbor Laboratory, Cold Spring Harbor, New York, pp. 371-414.
7. Loeb, L.A., and T.A. Kunkel (1982) Fidelity of DNA synthesis. Ann. Rev. Biochem. 52:429-457.
8. Muller, H.J. (1954) The nature of the genetic effects produced by radiation. In Radiation Biology, Vol. I, A. Hollaender, ed. McGraw-Hill Publishing Co., New York, p. 417.
9. Tice, R.R., and R.B. Setlow (1985) Repair and replication in aging organisms and cells. In Handbook of the Biology of Aging, 2nd ed., C.E. Finch and E.L. Schneider, eds. Van Nostrand Reinhold Co., New York, pp. 173-224.

GENETIC ANALYSES OF THE ROLES OF UmuDC AND MucAB IN MUTAGENESIS

Lorraine Marsh, Lori A. Dodson, Christine Dykstra,
David Sobell, and Graham C. Walker

Biology Department
Massachusetts Institute of Technology
Cambridge, Massachusetts 02139

INTRODUCTION

Mutagenesis by some alkylating agents and base analogs in *Escherichia coli* appears to involve a simple mispairing during replication. In contrast, mutagenesis by ultraviolet (UV) light and many chemicals in *E. coli* requires the participation of an inducible process that is dependent on the products of the *umuD*, *umuC*, and *recA* genes. The requirement for induced cellular functions for UV mutagenesis was first noted by Weigle (37), who found that UV was not mutagenic for bacteriophage lambda unless the bacteria had been preirradiated with UV. The induction of this process was later shown by Defais et al. (7) to be blocked by *recA*(Def) or *lexA*(Ind$^-$) mutations. The fact that the same *recA*(Def) and *lexA*(Ind$^-$) mutations rendered *E. coli* nonmutable by UV and many chemicals (16,24,39) indicated that mutagenesis of the bacterial chromosome required the participation of the same *recA*$^+$ *lexA*$^+$-controlled process, and the inducibility of the system was demonstrated by subsequent physiological experiments (40).

The RecA and LexA proteins are now known to control the expression of a network of more than 17 genes that are responsible for the physiologically observable SOS responses. The existence of this regulatory network was first postulated by Radman (28), and the specific induced component involved in UV and chemical mutagenesis has been referred to by a number of terms, including "error-prone repair" and "SOS processing." Through the efforts of many groups, the mechanistic aspects of SOS regulation are understood in considerable detail and have been extensively reviewed (19,34,35,41).

In brief, LexA functions as the repressor of *recA* and at least 16 other genes. Activation of the RecA protein in response to an SOS-inducing treatment permits RecA to mediate a proteolytic cleavage of the LexA protein at an ala-gly bond. Induction of bacteriophages lambda, 434, and P22 is similarly induced by SOS-inducing treatments and involves a RecA-mediated proteolytic cleavage of an ala-gly bond of their respective repressors (29). In vitro experiments have suggested that activation of the RecA protein requires single-stranded DNA and a nucleoside triphosphate. As the pools of LexA become depleted, the *recA* gene and the other SOS genes are expressed at higher levels; the strength of the SOS-inducing treatment required for full induction varies between different SOS genes and is influenced by a number of factors including the Kd's of LexA binding to the operators of the individual genes.

A significant advance in our understanding of UV and chemical mutagenesis in *E. coli* was achieved by the independent isolation of *umuD* and *umuC* mutants by Kato and Shinoura (14) and Steinborn (31). These mutants, which were isolated by an arduous screen of mutagenized populations for nonmutable derivatives, are (i) nonmutable with a large variety of mutagens including UV, methyl methanesulfonate, and 4-nitroquinoline 1-oxide, (ii) deficient in Weigle-mutagenesis of UV-irradiated lambda, and (iii) slightly UV-sensitive. However, unlike the highly pleiotropic *recA*(Def) and *lexA*(Ind$^-$) mutations, *umuDC* mutations did not affect the induction of other SOS functions such as filamentous growth. The properties of these mutants suggest that the UmuD and UmuC proteins are the best candidates for proteins that play a unique role in UV and chemical mutagenesis. Furthermore, the fact that *umuC* mutations can be generated by the insertion of Mu *d1*(Ap *lac*) (1) or Tn5 (9) suggests that their phenotype arises from the loss of a cellular function. The isolation of the *umuC*::Mu *d1*(Ap *lac*) mutants allowed us to show that the *umuDC* locus is induced by DNA-damaging treatments and regulated by the SOS control system (1).

The RecA protein seems to have at least one other role in UV mutagenesis besides mediating the cleavage of LexA. This is implied by the fact that *lexA*(Def) mutants (which lack functional LexA protein and consequently constitutively express the SOS genes, including *umuDC*) are rendered nonmutable by the introduction of *recA*(Def) mutations (1,3,11,19). Moreover, this requirement for RecA seems to be related to its ability to be activated rather than to its ability to promote homologous recombination, since the introduction of a *recA430* mutation (which blocks the ability of RecA to mediate lambda repressor cleavage and impairs its ability to mediate LexA cleavage but does not impair its recombination proficiency) into a *lexA*(Def) background also yields a strain which is nonmutable with UV (3, 11).

IDENTIFICATION OF THE UmuD/UmuC and MucA/MucB GENE PRODUCTS

As part of an effort to understand the molecular mechanism of UV and chemical mutagenesis, our group (9) and Shinagawa et al. (32) cloned the *umuDC* locus. The cloning was made more difficult by the fact that the phenotype conferred on a cell by a plasmid carrying *umuD*$^+$*C*$^+$ is dependent on the plasmid copy number. We finally obtained the *umuD*+*C*+ locus by generating a probe from a *umuC*::Tn5 mutant and using this to screen a lambda library of *E. coli* DNA, and then subcloning it onto a relatively low copy number vector. Analysis of the recombinant plasmid revealed that the *umuDC* locus consists of two genes, *umuD* and *umuC*, coding for proteins of 16 and 45 kDa, respectively. The products of both genes are required for UV and chemical mutagenesis. The two genes are organized in an operon, *umuDC*, that is induced by DNA damage and regulated by SOS circuitry.

At this point our work on *umuDC* converged with our analyses of the mutagenesis-enhancing plasmid pKM101 (25,38). The pKM101 was derived from the clinically isolated plasmid R46 by an in vivo deletion catalyzed by IS46 that removed a 14-kb fragment encoding several drug resistances and metal resistance (5,17). Both pKM101 and R46 are members of a subset of naturally occurring plasmids that increase both the resistance of cells to killing by UV and their susceptibility to mutagenesis by UV and various chemicals. Because of this latter property, pKM101 was incorporated into Ames *Salmonella* strains, where it has played a major role in the success of the system (22,23). The effects of pKM101 on UV resistance and mutagenesis are *recA*$^+$ *lexA*$^+$-dependent (33), and pKM101 will suppress the nonmutability of *umuDC* mutants (36). Analysis of pKM101 indicated that a 2,000-bp region was both necessary (30) and sufficient (26) for pKM101's effects on UV and chemical mutagenesis. This region was shown to encode two genes, *mucA* and *mucB*, with products of 16 and 45 kDa, respectively. The genes are organized in an operon, *mucAB*, and the products of each gene are required for the effects on mutagenesis and UV resistance (26). Moreover, although the *mucAB* locus is located on a broad host range IncN plasmid, it is repressed by LexA and regulated by the SOS circuitry (10).

Both the *umuDC* locus (15,27) and the *mucAB* locus (27) have been sequenced. The deduced amino acid sequences of the UmuD and MucA proteins are 41% homologous and those of the UmuC and MucB proteins are 55% homologous. Thus, although the two loci have undergone considerable divergence, it is apparent that they share a common evolutionary origin. Furthermore, the fact that UV and chemical mutagenesis is not observed unless each gene is paired with its cognate partner provides a genetic suggestion that the 16 and 45 kDa proteins physically interact (27). The UmuC and MucB proteins are each quite basic and therefore might interact with DNA, although this has not yet been demonstrated. A LexA binding site is located upstream of the *umuDC* operon, and a putative LexA binding sequence has been located in an analogous position upstream of the *mucAB* operon. A single G:C to A:T mutation in this sequence, that moves the putative LexA binding site away from consensus and the putative promoter towards consensus, causes a phenotype similar to that caused by cloning the *mucAB* genes onto a multicopy plasmid (Perry and Walker, unpubl. data).

THE UmuD, MucA, AND LexA PROTEINS SHARE HOMOLOGY

Analysis of their deduced amino acid sequences revealed that the UmuD and MucA proteins share homology with the carboxy terminal domain of the LexA protein (27). This domain of LexA contains the ala-gly cleavage site, the presumed recognition site for activated RecA protein and possibly the dimerization site; it does not contain the DNA-binding sequences. The MucA protein has an ala-gly pair at the site corresponding to the LexA ala-gly cleavage site, while the UmuD protein has the chemically closely related cys-gly pair. A number of the amino acids conserved between UmuD, MucA, and LexA are also conserved between the repressors of lambda, 434, and P22. The homology of these mutagenesis proteins to proteins that are cleaved when they interact with activated RecA is particularly intriguing, as it suggests possible models for the additional role(s) of RecA in mutagenesis referred to above.

We have recently been able to obtain genetic evidence that those elements of UmuD and MucA homologous to LexA are functionally significant and that these proteins may interact with activated RecA. We have used site-directed oligonucleotide mutagenesis in cloned *mucAB* to change the gly at codon 16 of MucA to a glu. The change alters the ala-gly homologous to the LexA cleavage site to ala-glu; a corresponding change in lambda repressor has been shown to prevent cleavage (12). Three sets of experiments, as follows, have been carried out comparing the phenotypes of strains carrying the mucAB(ala-glu) genes to the phenotypes of strains carrying the wild-type mucA+B+(ala-gly) genes (Marsh and Walker, unpubl. data):

(i) Reversion of the argE3 or his-4 ochre mutations of *E. coli* provides a crude but useful system for looking at additional spectrum changes, since "true" revertants of these mutations are easily distinguished from ochre suppressor mutations. Strains carrying the mucAB(ala-glu) genes exhibit an altered ratio of true to suppressor mutants (21). It is interesting to note that potential hairpins can form in tRNA genes; Glickman has suggested that such potential hairpins might be hotspots for mutagenesis (13). In any case, the result suggests that the umu and muc gene products may contribute to the specificity of mutagenesis at at least some sites, and thus that they may interact with the lesion itself or the DNA around it. It is also interesting to note that a similar change in spectrum was noted when the wild-type mucA+B+ genes were introduced into a recA430 mutant which is defective in lambda repressor cleavage (Perry and Walker, unpubl. data). The similarity of the UV mutational spectrum observed in a recA[mucAB(ala-glu)] strain, compared to that in a recA430(mucA+B+) strain, suggests that MucA and activated RecA may interact.

(ii) Both our lab (33) and Doubleday et al. (8) have observed that a recA441(mucA+B+) strain exhibits an elevated spontaneous mutational rate under conditions that led to an activation of RecA (raising the temperature or adding adenine). We have observed that a recA441[mucAB(ala-glu)] strain fails to exhibit this induced increase, suggesting that MucA is influenced, directly or indirectly, by the degree of activation of RecA (Marsh and Walker, unpubl. data).

(iii) We have also recently discovered that the induction of an SOS gene by UV is impaired by the modest overproduction of the mucA+B+ genes caused by cloning them into the multicopy vector pBR322. Such an inhibition was not observed with mucAB(ala-glu). The simplest interpretation of the results is that MucA competes with LexA for activated RecA in the cell and thereby inhibits SOS induction. In such a model, the mucAB(ala-glu) mutants would not compete effectively.

Taken together, these three types of experiments all suggest that MucA, and presumably UmuD, undergo a direct physical interaction with activated RecA, an inference consistent with the observed homology of these proteins to LexA and the phage repressors. The results do not necessarily imply that MucA and UmuD subsequently undergo a proteolytic cleavage, but this remains a possibility.

COLD SENSITIVITY CAUSED BY UmuD AND UmuC OVERPRODUCTION

We had experienced difficulty in introducing a derivative of multicopy vector pBR322 carrying the umuDC genes into lexA(Def) cells, which lack functional LexA and therefore constitutively express the SOS genes. However, we recently discovered that such a strain could be constructed if it were maintained at 43°C. Shifting the strain to 30°C resulted in a rapid cessation of DNA replication and eventually in cell death (20). In other words, the strain had properties similar to those of a "quick-stop" DNA replication mutant. This phenomenon requires both umuD+ and umuC+ and so would not appear to result from a nonspecific interaction of one of the proteins with DNA, but rather from some specific interaction of UmuD-UmuC with the replication apparatus. The fact that the phenomenon is recA+-independent, unlike UV mutagenesis which exhibits the additional requirement(s) for RecA function discussed above, reinforces the view that the cessation of DNA replication is due to a direct effect of UmuD-UmuC on the replication process.

We have also observed that the cold sensitivity of UmuD-UmuC overproduction can be suppressed by lon mutations which share the common property of causing defects in proteolysis (2,6). Since the stabilities of UmuD and UmuC do not appear to be significantly altered by a lon mutation, these observations raise the possibility that accumulation of UmuD-UmuC titrates out some normally labile limiting component of the replication apparatus.

A considerable body of genetic evidence is consistent with models in which "error-prone repair" acts by allowing by-pass of lesions that would otherwise block DNA replication. However, the role of UmuDC and MucAB in this process is still unclear. One of the most common models that is often proposed for their action is that they somehow modify DNA polymerase III holoenzyme, perhaps by inhibiting editing functions, to make a less fastidious enzyme that exhibits decreased fidelity. Though our data would be consistent with such a model, there are other possible models that are also attractive. These include models in which bulky or noninformational lesions do not block DNA replication but rather slow it down; UmuDC might then act by inhibiting mismatch repair or some similar system that might normally interfere with continued replication past the lesion. Bridges and Woodgate (4) have recently proposed a specific model in which UmuDC is not required for misincorporation of bases opposite UV-induced pyrimidine dimers but rather for continued chain elongation past the misincorporated bases.

RELEVANCE TO HIGHER ORGANISMS

It is clear that in bacteria, bulky and noninformational lesions in DNA require the active participation of UmuDC-like or MucAB-like proteins in order to cause mutations. Similar systems may operate in higher cells. For example, the yeast REV genes bear at least a formal resemblance to the umu locus. Mutations in these genes make yeast strains unmutable by UV light and other DNA-damaging agents without greatly affecting UV-resistance (18). Though the *E. coli umu* genes are inducible by DNA damage, inducibility is not necessarily an essential part of their function. It may be that in some organisms or tissues similar functions are expressed constitutively, as in *E. coli lexA*(Def) strains.

ACKNOWLEDGEMENTS

This work was supported by Public Health Service Grants CA21615 from the National Cancer Institute and GM28988 from the National Institute of General Medical Sciences.

REFERENCES

1. Bagg, A., C.J. Kenyon, and G.C. Walker (1981) Inducibility of a gene product required for UV and chemical mutagenesis in *Escherichia coli*. *Proc. Natl. Acad. Sci., USA* 78:5749-5753.
2. Baker, T.A., A.D. Grossman, and C.A. Gross (1984) A gene regulating the heat shock response in *Escherichia coli* also affects proteolysis. *Proc. Natl. Acad. Sci., USA* 81:6779-6783.

3. Blanco, M., G. Herrera, P. Collado, J. Rebollo, and L.M. Botella (1982) Influence of RecA protein on induced mutagenesis. Biochimie 64:633-636.
4. Bridges, B.A., and R. Woodgate (1984) Mutagenic repair in Escherichia coli. X. The umuC gene product may be required for replication past pyrimidine dimers but not for the coding error in UV mutagenesis. Mol. Gen. Genet. 196:364-366.
5. Brown, A.M.C., and N.S. Willetts (1981) A physical and genetic map of the Inc N plasmid R46. Plasmid 5:188-201.
6. Charette, M.F., G.W. Henderson, and A. Markovitz (1981) ATP hydrolysis-dependent protease activity of the lon(capR) protein of Escherichia coli K-12. Proc. Natl. Acad. Sci., USA 78:4728-4732.
7. Defais, M., P. Fauquet, M. Radman, and M. Errera (1971) Ultraviolet reactivation and ultraviolet mutagenesis of lambda in different genetic systems. Virology 43:495-503.
8. Doubleday, O.P., M.H.L. Green, and B.A. Bridges (1977) Spontaneous and UV-induced mutation in Escherichia coli: Interaction between plasmid and tif-1 mutator effects. J. Gen. Microbiol. 101:163-166.
9. Elledge, S.J., and G.C. Walker (1983) Proteins required for ultraviolet light and chemical mutagenesis: Identification of the products of the umuC locus of Escherichia coli. J. Mol. Biol. 164:175-192.
10. Elledge, S.J., and G.C. Walker (1983) The muc genes of pKM101 are induced by DNA damage. J. Bacteriol. 155:1306-1315.
11. Ennis, D.G., B. Fisher, S. Edmiston, and D.W. Mount (1985) Dual role for Escherichia coli RecA protein in SOS mutagenesis. Proc. Natl. Acad. Sci., USA 82:3325-3329.
12. Gimble, F.S., and R.T. Sauer (1985) Mutations in bacteriophage lambda repressor that prevent RecA-mediated cleavage. J. Bacteriol. 162:147-154.
13. Glickman, B. (1983) Mutational specificity of UV light in E. coli: Influence of excision repair and the mutator plasmid pKM101. In Induced Mutagenesis, C.W. Lawrence, ed. Plenum Press, New York, pp. 135-170.
14. Kato, T., and Y. Shinoura (1977) Isolation and characterization of mutants of Escherichia coli deficient in induction of mutations by ultraviolet light. Mol. Gen. Genet. 156:121-131.
15. Kitagawa, Y., E. Akaboshi, H. Shinagawa, T. Horii, H. Ogawa, and T. Kato (1985) Structural analysis of the umu operon required for inducible mutagenesis in Escherichia coli. Proc. Natl. Acad. Sci., USA 82:4336-4340.
16. Kondo, S. (1969) Mutagenicity versus radiosensitivity in Escherichia coli. In Proceedings of the XIIth International Congress of Genetics, Vol. II, pp. 126-127.
17. Langer, P.J., and G.C. Walker (1981) Restriction endonuclease cleavage map of pKM101: Relationship to parental plasmid R46. Mol. Gen. Genet. 182:268-272.
18. Lawrence, C.W. (1982) Mutagenesis in Saccharomyces cerevisiae. Adv. Genet. 21:173-254.
19. Little, J.W., and D.W. Mount (1982) The SOS regulatory system of Escherichia coli. Cell 29:11-22.
20. Marsh, L., and G.C. Walker (1985) Cold sensitivity induced by overproduction of UmuDC in Escherichia coli. J. Bacteriol. 162:155-161.
21. Marsh, L., and G.C. Walker (1985) Mutagenic DNA repair in bacteria: The role of UmuDC and MucAB. In Mechanisms of DNA Damage and Repair, M.G. Simic, L. Grossman, and A.D. Upton, eds. Plenum Press, New York (in press).
22. McCann, J., and B.N. Ames (1976) Detection of carcinogens as mutagens in the Salmonella/microsome test:assay of 300 chemicals: Discussion.

Proc. Natl. Acad. Sci., USA 73:950-954.

23. McCann, J., N.E. Spingarn, J. Kobori, and B.N. Ames (1975) Detection of carcinogens as mutagens: Bacterial tester strains with R factor plasmids. Proc. Natl. Acad. Sci., USA 72:979-983.

24. Miura, A., and J. Tomizawa (1968) Studies on radiation-sensitive mutants of E. coli. III. Participation of the Rec system in induction of mutation by ultraviolet irradiation. Mol. Gen. Genet. 103:1-10.

25. Mortelmans, K.E., and B.A.D. Stocker (1979) Segregation of the mutator property of plasmid R46 from its ultraviolet-protecting property. Mol. Gen. Genet. 167:317-328.

26. Perry, K.L., and G.C. Walker (1982) Identification of plasmid (pKM-101)-coded proteins involved in mutagenesis and UV resistance. Nature (London) 300:278-281.

27. Perry, K.L., S.J. Elledge, B.B. Mitchell, L. Marsh, and G.C. Walker (1985) umuDC and mucAB operons whose products are required for UV light- and chemical-induced mutagenesis: UmuD, MucA, and LexA proteins share homology. Proc. Natl. Acad. Sci., USA 82:4331-4335.

28. Radman, M. (1975) SOS repair hypothesis: Phenomenology of an inducible DNA repair which is accompanied by mutagenesis. In Molecular Mechanisms for Repair of DNA, Part A, P. Hanawalt and R.B. Setlow, eds. Plenum Press, New York, pp. 355-367.

29. Sauer, R.T., R.R. Yocum, R.F. Doolittle, M. Lewis, and C.O. Pabo (1982) Homology among DNA-binding proteins suggests use of a conserved super-secondary structure. Nature (London) 298:447-451.

30. Shanabruch, W.G., and G.C. Walker (1980) Localization of the plasmid (pKM101) gene(s) involved in $recA^+ lexA^+$-dependent mutagenesis. Mol. Gen. Genet. 179:289-297.

31. Steinborn, G. (1978) Uvm mutants of Escherichia coli K12 deficient in UV mutagenesis. I. Isolation of uvm mutants and their phenotypical characterization in DNA repair and mutagenesis. Mol. Gen. Genet. 165: 87-93.

32. Shinagawa, H., T. Kato, T. Ise, K. Makino, and A. Nakata (1983) Cloning and characterization of the umu operon responsible for inducible mutagenesis in Escherichia coli. Gene 23:167-174.

33. Walker, G.C. (1977) Plasmid (pKM101)-mediated enhancement of repair and mutagenesis: Dependence on chromosomal genes in Escherichia coli K-12. Mol. Gen. Genet. 152:93-103.

34. Walker, G.C. (1984) Mutagenesis and inducible responses to deoxyribonucleic acid damage in Escherichia coli. Microbiol. Rev. 48:60-93.

35. Walker, G.C. (1985) Inducible DNA repair systems. Ann. Rev. Biochem. 54:425-457.

36. Walker, G.C., and P.P. Dobson (1979) Mutagenesis and repair deficiencies of Escherichia coli umuC mutants are suppressed by the plasmid pKM101. Mol. Gen. Genet. 172:17-24.

37. Weigle, J.J. (1953) Induction of mutation in a bacterial virus. Proc. Natl. Acad. Sci., USA 39:628-636.

38. Winans, S.C., and G.C. Walker (1985) Conjugal transfer system of the IncN plasmid pKM101. J. Bacteriol. 161:402-410.

39. Witkin, E.M. (1969) Ultraviolet-induced mutation and DNA repair. Ann. Rev. Microbiol. 23:487-514.

40. Witkin, E.M. (1974) Thermal enhancement of ultraviolet mutability in a tif-1 uvrA derivative of Escherichia coli B/r: Evidence that ultraviolet mutagenesis depends upon an inducible function. Proc. Natl. Acad. Sci., USA 71:1930-1934.

41. Witkin, E.M. (1976) Ultraviolet mutagenesis and inducible DNA repair in Escherichia coli. Bacteriol. Rev. 40:869-907.

MECHANISMS OF SPONTANEOUS MUTAGENESIS: CLUES FROM ALTERED MUTATIONAL SPECIFICITY IN DNA REPAIR-DEFECTIVE STRAINS

Barry W. Glickman, Philip A. Burns,
and Douglas F. Fix

Department of Biology
York University
Downsview, Ontario M3J 1P3, Canada

INTRODUCTION

Spontaneous mutation may be described as the net result of all that can go wrong with DNA during the life cycle of an organism. The student of mutation, however, views the world of mutation through a myriad of filters and sees only a fraction of the molecular events surrounding mutation. To begin with, the student sees only those changes that produce a selectable and hence observable alteration in phenotype. Moreover, the majority of errors made during DNA replication and the errors produced by the accumulation of DNA damage are corrected by the plethora of repair mechanisms that have evolved to maintain the accurate transmission of genetic material. Hence, the study of mutagenesis in strains defective in DNA repair can be expected to yield information about the sources of spontaneous mutation, both with respect to the errors avoided as well as to those errors made during attempts at repair. Our increasing knowledge about mechanisms of DNA repair and their influence on mutation, coupled with the newly developed ability to clone and sequence DNA containing mutations, provides an opportunity to explore the sources of spontaneous mutation.

The sources of spontaneous mutation remain largely unknown. Early speculation on spontaneous mutation suggested that its source might be natural background radiation. However, Muller and Mott-Smith (43) showed that only a negligible fraction of spontaneous mutation could have arisen in this manner. Current views suggest that spontaneous mutation is essentially the product of intracellular events. A role for DNA replication in spontaneous mutation was first indicated by Speyer et al. (58) and Drake et al. (13) in their descriptions of bacteriophage T4 mutator and antimutator polymerases. Similarly, a DNA polymerase III mutant has been found to have a mutator phenotype (33). Other errors that occur during DNA replication (including repair replication) include the incorporation of unusual nucleotides (base analogs) or mispairings generated by tautomeric shifts of the natural bases (61,65). Additional sources of mutation can be expected to include DNA lesions resulting from naturally occurring cellular components. These include free radical species resulting from normal cellular metabolism (22) or natural methyl group donors.

A number of observations indicate a significant role for DNA repair in producing spontaneous mutation. Kondo et al. (32) demonstrated that a recA mutation reduced spontaneous mutation by 50%. While this was originally interpreted to mean that recombination is an error-prone process (32), this observation is more consistent with the role of recA in error-prone repair as suggested by Sargentini and Smith (50), who demonstrated that other mutations inactivating the inducible error-prone repair pathway in *Escherichia coli* also drastically reduce the level of spontaneous mutation. In addition, Sargentini and Smith (50) also demonstrated a role for excision repair in avoiding mutation, since uvrA and uvrB mutations increased the frequency of spontaneous mutation several-fold. Similar increases in spontaneous mutation have been observed in some excision repair-deficient strains of, for example, *Saccharomyces cerevisiae* (2,28,30). In addition, we have described some inherent properties of DNA sequences that may contribute to spontaneous mutation (26,48). These ideas are reviewed by von Borstel (63), Loeb and Kunkel (39), Drake et al. (14), and Sargentini and Smith (51).

In this chapter, we begin by considering the sequence alterations of 270 spontaneous lacI mutants recovered in a wild-type, repair-proficient strain of *E. coli*. These include examples of frameshift, deletion, duplication, and base substitution events. Seven insertion mutations due to the transposition of IS1 were also recovered. By examining the specificity of these mutational events, some inferences can be made concerning the origin of several specific mutational events. The analysis of mutation in the wild-type strain is followed by an examination of spontaneous mutation in Ung^- (defective in uracil-N-glycosylase) and $PolA^-$ (defective in DNA polymerase I) strains. The analysis of spontaneous mutation in an Ung^- background provides information on the role of cytosine deamination (and uracil misincorporation) in spontaneous mutation. The analysis of spontaneous mutation in the $PolA^-$ strain provides insights into the role(s) of DNA replication and repair during spontaneous mutagenesis. The recovered mutational spectra are discussed with respect to the origins of specific classes of mutation.

THE RATIONALE

The lacI gene of *E. coli* codes for the repressor of the lac operon. It has been used extensively for the analysis of mutational specificity. Mutants are selected by their ability to grow on a lactose analog, phenyl-B-D-galactose (P-GAL), which, while not an inducer, is a substrate for the metabolic pathway coded for by the lac operon. Hence, lacI mutants are able to grow on minimal medium containing P-GAL as the sole carbohydrate source. Initially, nonsense mutations were analyzed by classical genetic techniques to provide information on base substitution mutation at more than 70 sites within the gene (9,15,20,37,42,60). This same gene has been the subject of direct analysis by DNA sequencing (1,6).

Recently, we developed a rapid approach to the cloning and sequencing of lacI mutants (53). The approach involves the recovery of lacI mutations by recombination in vivo onto a bacteriophage M13 derivative which is $lacI^+$-$lacZ^-$. The double selection for $lacI^-$-$lacZ^+$ recombinants insures that the desired lacI mutation is recovered. Sequencing is done by the dideoxy method using synthetic oligomers as primers. The methodology is described in detail by Schaaper et al. (53,54).

SPONTANEOUS MUTATION IN A REPAIR-PROFICIENT STRAIN

The results of sequencing 270 spontaneous mutants obtained in the wild-type strain. NR3835, are given in Tab. 1, 2, and 3. About 66% of all the mutations recovered occur at the hotspot site previously reported by Farabaugh et al. (19). This site involves position 620-631 where the sequence 5'-CTGG-3' is repeated 3 times (Fig. 1). The hotspot event involves either the addition of a CTGG sequence (53%) or the loss of a CTGG sequence (13%). The fraction of hotspot mutations as well as the ratio of addition versus deletion events determined here by DNA sequencing are identical to the values obtained from mapping and reversion studies (19).

DETAILED ANALYSIS OF MUTATION IN THE WILD-TYPE STRAIN

The sequencing of 270 spontaneous mutations in the *lacI* gene of a repair-proficient strain of *E. coli* yielded representatives of 6 different classes of mutational events, presumably representing a wide variety of mutational mechanisms. The spectrum is dominated by addition and deletion events at the site of a repeated sequence (Tab. 1), the spontaneous *lacI* hotspot first reported by Farabaugh et al. (19). Together, these account for about 66% of all spontaneous mutations in the *lacI* gene. The question arises of how representative the prevalence of such a hotspot is for mutation in an average gene. Roughly calculated, the probability of a direct triple repeat of a 4-base sequence is only 1 in about 65,000. Thus, for genes of the size of *lacI*, only 1 out of roughly 50 might be expected to have a similar hotspot sequence. However, considerable evidence indicates that DNA sequences generally are not random and, therefore, similar kinds of repeated sequences may exist more often than might be expected.

The Frameshift Hotspot

As described above, a large fraction of spontaneous *lacI* mutations arise as the consequence of the gain or loss of one copy of the 4-bp sequence, 5'-CTGG-3', tandemly repeated 3 times at nucleotides 620-631 (Fig. 1). The ratio of the addition of 4 bp compared to their loss was 4:1, similar to that previously determined by genetic analysis (19). Frameshift

Tab. 1. The distribution of spontaneous *lacI*$^-$ mutants by class.

Class of mutation	Number of occurrences	Frequency X 10^{-7}
Base substitutions	43 (16.0%)	3.2
Single base frameshifts	12 (4.5%)	0.9
Deletions	28 (10.5%)	2.1
Duplications	3 (1.1%)	0.2
Insertions (IS1)	7 (2.6%)	0.5
(+) 5'-CTGG-3'*	142 (53.0%)	10.6
(-) 5'-CTGG-3'*	35 (13.0%)	2.6
TOTAL:	270	20.0 X 10^{-7}

* Frameshift hotspot at position 620-631.

Tab. 2. Description of all nonhotspot mutations obtained in the wild-type strain of *Escherichia coli*.

A. Base substitutions

Position	Change	Occurrences	Amino acid change
23	G → C	1	Val → Met
56	G → A	1	Ala → Thr
80	C → T	1	Gln → Amb (am5)
83	A → G	1	Thr → Ala
93	G → A	3	Arg → His
104	C → T	1	Gln → Amb (am6)
108	C → T	1	Ala → Val
140	G → A	1	Val → Met
141	T → A	1	Val → Glu
150	C → A	1	Ala → Glu
158	G → T	1	Glu → Amb (am7)
183	T → A	1	Val → Glu
213	T → A	1	Leu → Amb (am11)
222	G → A	1	Gly → Asp
293	C → A	1	Gln → Och (oc13)
419	C → T	1	Gln → Amb (am15)
582	T → G	1	Leu → Opal
768	A → T	1	Asp → Ser
777	C → A	1	Ala → Glu
845	T → G	1	Try → Asp
885	T → G	1	Ser → Opal
926	A → C	2	Ser → Arg
1085	G → C	1	Val → Leu
+6 (lacO)	T → C	17	O^C

B. Nonhotspot frameshifts

Position	Alteration	Occurrences	Sequence alteration
52	(-)T	1	CGA T GTC → CGAGTC
125-128	(-)A	1	CG AAA GC → CGAAGC
286-287	(+)C	1	TCG C GCC → TCGCCGCC
290	(-)G	1	GCC G ATC → GCCATC
388	(-)T	1	CAG T GGG → CAGGGG
394	(-)G	1	GCT G ATC → GCTATC
397	(-)C	1	GAT C ATT → GATATT
562	(+)TC	1	GG TC ACC → GGTCTCACC
581-582	(-)TT	1	CTG TT AG → CTGAG
620-621	(+)C	1	CGT C TGG → CGTCCTGG
646	(-)T	1	CAC T CGC → CACCGC
855-856	(-)C	1	ATA CC GA → ATACGA

C. Insertion Elements

Insertion element	Position[1]	Orientation[2]
IS1	42-50	II
IS1	53-61	I
IS1	148-156	II
IS1	222-230	II
IS1	488-496	II
IS1	496-504	I
IS1	670-678	I

Tab. 2, continued.

D. Deletions

Class I

Position	Size (bp)	Repeated bases	Sequence
-33- 18	52	none	
-10-137	148	none	
91-177	87	8	CCGCGTGG
114-127	14	3	ACG
146-268	123	8	GCGGCGAT
218-262	45	4	ATTG
272-286	15	2	GC
267-282	16	5	TCGCG
281-369	89	8	TCTCGCGC
313-334	22	7/8	GG$^{T}_{C}$GTCGA
507-524	18	none	
610-615	6	6	GCGTCT
691-768	78	none	
536-617	81	7/8	CGCG$^{A}_{T}$CTG
790-806	117	none	
814-1046	233	9	GCGCGTTGG
919-1029	111	7	GCAAACC
928-1014	87	10/12	AAACCA$^{G}_{C}$C$^{G}_{C}$TGG
928-1013	86	10/12	AAACCA$^{G}_{C}$C$^{G}_{C}$TGG

Class II

Position	Size (bp)	Occurrences	Repeated bases	Sequence
824-(+)8	380	1	7	GCGGATA
900-(+)4	300	1	none	
916-(+16)	296	1	none	
929-(+20)	287	1	none	
954-(+9)	251	1	none	
1108-(+4)	92	2	7	GTGAGCG
(-64)-(+7)	71	1	none	
(-64)-(+12)	76	1	none	

E. Tandem Duplications

Mutant	Duplicated bases	Size (bp)
1-8	362-619	258
1-28	368-824	457
3-25	380-534	155

[1] Nine base-pair repeat sequence generated upon IS1 insertion (39); only the 3' end of the IS1 insertion was sequenced (37).
[2] Orientation I is that of S114 (39).

Tab. 3. The nature of spontaneous base substitutions.

Transitions	Occurrences	Transversions	Occurrences
A:T → G:C*	19	A:T → T:A	4
G:C → A:T	9	A:T → C:G	5
		G:C → T:A	4
		G:C → C:G	2
TOTAL:	28		15

* Includes 17 mutations at (+6) of lacZ messenger (O^C mutation).

mutations involving repeated sequences are thought to occur via slipped or misaligned intermediates mediated by the repeats as proposed by Streisinger et al. (59). The model, therefore, predicts the occurrence of these mutations.

However, as we have discussed elsewhere (54), additional factors may also play an important role. The region neighboring the frameshift hotspot is extremely rich in both palindromic and repeated sequences. For example, the 4-bp repeats are part of two, more extensive, 9-bp repeats which predict the same mutational outcome. Moreover, two 7-bp repeats are immediately 3' to, and partially overlapping with, the 9-bp repeats; these neighboring repeats jointly form nearly perfect repeats of 19 and 17 bp also capable of templating the gain or loss of the 4 bases at the hotspot site. Palindromic DNA sequences are also available that are capable of stabilizing certain misaligned intermediates. Hence, the predominance of this hotspot site may reflect the accumulative effect of several structural components acting cooperatively to enhance the formation and stability of the predicted structural misalignments. A discussion of how misalignments can generate mutation is found in Ripley and Glickman (48).

Base Substitutions

Base substitutions accounted for 16% of the LacI$^-$ mutants and were recovered at a frequency of 2×10^{-7} (Tab. 2). Taking into account the number of generations the bacterial population had undergone, and estimating the number of sites in the gene that can yield a lacI mutation as roughly 200, the mutation rate can be estimated as 10^{-10} mutations per bp per cell per round of replication. Since 17 of the mutations occurred at an apparent hotspot site in the operator region of lacZ, the actual mutation rate in lacI is slightly lower. The estimated error rate per nucleotide per replication agrees well with earlier estimates by Drake (12).

The nature of the base substitutions was also examined. The ratio of transversion to transition mutations is shown in Tab. 3. Subtracting the 17 lacO^c mutations (A:T → G:C), no single event seems grossly favored. There are 4 possible transversion and 2 transition pathways. The ratio of transition and transversion events may not necessarily reflect their occurrence during DNA replication. Instead, they are more likely to be a reflection of the specificity of DNA repair. For example, mismatch repair may preferentially correct transitional mismatch intermediates (24,25). In addition, the apparent recovery of transversion events may also be enhanced

```
600        610        620        630        640
 |          |          |          |          |
5'-CTGTCTCGGCGCGTCTGCGTCTGGCTGGCTGGCATAAATAT-3'
```

Fig. 1. The sequence surrounding the frameshift hotspot. The mutations are the result of the gain or loss of this 4-bp repeat which is repeated three times (positions 620-632). The repeated sequences are underscored.

by the action of uracil DNA-glycosylase. The removal from DNA of uracil, the deamination product of cytosine, specifically reduces the frequency of G:C → A:T events (6). Not only is the frequency of transitions reduced by DNA repair mechanisms, but attempts at repair can also alter a mutational spectrum. Indeed, mutation at depurinated sites favors transversions (36,55). This makes any simple correlation between the relative recovery of transitions and transversions and their formation during DNA replication impossible.

The A:T → G:C transition at position +6 of the lacZ messenger has been shown previously to result in an O^c (operator constitutive) phenotype (41). The frequency of this base substitution is 23 x 10^{-8}. This is at least 50-fold greater than the frequency of base substitution at an average site within the lacI gene, and greater even than the frequency of mutation at the deamination hotspots (9,25). We note that Cheung et al. (7), in an NMR study, observed an enhanced imino proton exchange for this particular base pair compared to other base pairs in this region. This would be expected to enhance the deamination of the adenine residue. In the absence of repair, the resulting hypoxanthine subsequently could mispair with cytosine and A:T → G:C transition would result. We suggest that this mechanism is the basis for this base substitution hotspot.

Single-Base Frameshifts

Twelve independent, spontaneous frameshift mutations have been recovered in the wild-type strain. Since frameshifts severely disrupt protein structure, they are more likely to be recovered than, for example, base substitution mutations. Thus, correcting for effective target size, single-base pair frameshift events appear to occur at least one order of magnitude less frequently than single-base pair substitutions. As can be seen in Tab. 2, 8 of the frameshifts involved a single base deletion and 2, a single base addition. There was one occurrence each of a frameshift involving the gain or loss of 2 base pairs.

The DNA sequence information provides details of two critical aspects of mutation: the identification of the sequence change at the site of the mutation, and the larger DNA context in which the mutational site resides. This information provides an opportunity to examine the mechanisms by which these frameshift mutations may have occurred. Such an analysis provided the basis for the proposal of Streisinger et al. (59), who suggested that frameshift mutations were the consequence of slippage events during strand extension at sites of repetitive sequences. This model was later supported by several instances where the DNA sequence of frameshift sites could be resolved (44,45), and by the sequencing of the lacI frameshift hotspot (19). The Streisinger model has been extended to slippage schemes involving nonadjacent repeats. For example, the reversion of the T4 rIIB frameshift mutation, FC47, is thought to be templated by a 16-bp repeat 256 bp downstream from the mutated site (45,47).

More recent models of frameshift mechanisms provide a role for not only directly repeated sequences, but also palindromic sequences in producing misaligned intermediates for frameshift mutagenesis (46,48). In the case of direct repeats, interstrand misalignments result from the hydrogen bonding of one copy of the repeat onto the complement of the second copy of the repeat, whereas, in the case of inverted repeats, intrastrand misalignments are potentiated by the palindromic nature of the sequence. An interesting aspect of mutation directed by such misalignments is that not only frameshift mutations, but also base substitution mutations can result (10,46, 48).

An analysis of the origin of the frameshift mutations is presented in Tab. 4. Structural intermediates could be proposed for most of the mutations (10/12). Of these, in only 3 cases are repeated sequences available that could have directed a slippage event in the manner proposed by Streisinger (59). Most (10/12) of the sequenced frameshifts could have been templated by a nearby sequence, either a direct or an inverted repeat. In two cases, templates capable of producing the mutations and located within 25 bp were lacking. It is possible that these may reflect events templated by more remote sequences. In any case, our results indicate that diverse sources of sequences are available to direct the templating of frameshift events. Figure 2 includes an example of each of the subclasses of frameshift events described above.

Frameshift mutations, however, do not appear to occur at random throughout the gene. Even though there are only 3 _dam_ sites (5'-GATC-3') in the _lacI_ gene, three of the 12 mutations involve the loss of a G:C base pair within the _dam_ recognition sequence. The prevalence of frameshift mutations at _dam_ methylation sites is unlikely to be due to chance. More likely, they reveal a "DNA context" effect, reflecting metabolic activities involving the 5'-GATC-3' sequence.

Indeed, these sequences may represent metabolic hotspots. Several years ago, Gomez-Eichelmann and Lark (27) concluded that _dam_ methylation sites were nonrandom with respect to Okazaki fragments. They suggested that the 5'-GATC-3' sequences were at the ends of the Okazaki pieces. More

Tab. 4. Properties of spontaneous frameshift mutations recovered in the wild-type strain of _Escherichia coli_.

Origin of frameshift	Occurrences
Mediated by direct repeats	3*
Mediated by palindromic repeats	4
In runs	3
5'-GATC-3' (_dam_ methylation sites)	3
Without obvious template	2

* The total is greater than the number of frameshifts recovered since some fall into more than one category. Repeats and palindromes that template these events are all within 25 bp of the site of mutation.

recently, evidence from the study of the mechanism of methylation-instructed mismatch repair (40) and from mutational spectra in mismatch repair-defective strains (Glickman and Moessen, ms. in prep.) indicates that during mismatch repair, the incision occurs not at the site of the mismatch but rather at the 5'-GATC-3' sequences. An example of a frameshift at a dam methylation site is given in Fig. 2D.

The identification of 5'-GATC-3' sites as sites predisposed to single-strand breakage likely accounts for the DNA contextual effect. Single-strand breaks can be expected to provide potential substrates for the formation of misaligned intermediates, or sites for the action of DNA exonucleases leading to frameshift errors. The prediction that DNA context could modulate frameshift frequencies has been suggested by Ripley and

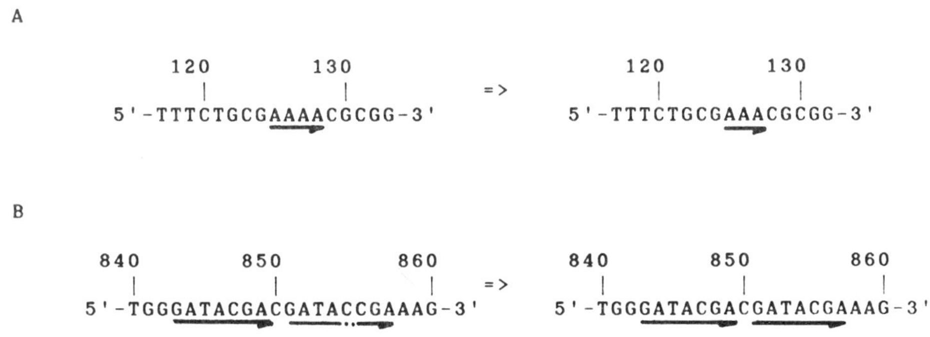

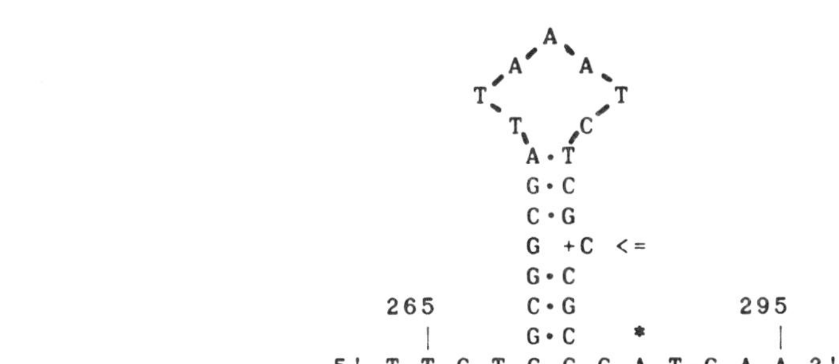

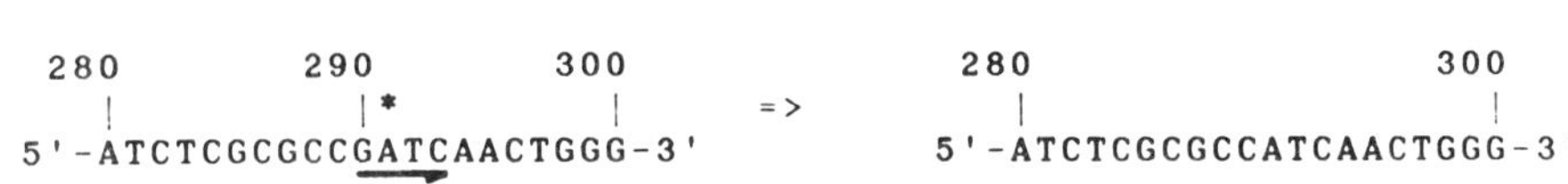

Fig. 2. DNA sequences surrounding frameshift mutations. (A) The loss of an A from a repeat of four A's. (B) A frameshift (loss of a C from position 855/856) templated by a nearby direct repeat. (C) The insertion of a C directed by a palindromic DNA sequence. (D) The loss of a G at a dam methylation site. Underscoring defines direct repeats. The asterisk indicates 6-methyladenine residues.

Shoemaker (49) and Ripley and Glickman (47) in the case of bacteriophage T4, where the reversion rate of different runs of A's appears to be related to their proximity to the dinucleotide AC, the sequence where Okazaki fragments are known to be initiated in the case of bacteriophage T4. The potential role for discontinuities in the DNA due to the initiation or termination of Okazaki fragments is discussed in detail in a later section.

Deletions

The mechanism of deletion formation has been the subject of much discussion. In many cases, short stretches of repeated bases are found at the deletion endpoints (1,19). This observation has lent support to the hypothesis that deletion mutation occurs via structural intermediates in which one copy of the repeat is misaligned on the complement of the other copy of the repeated sequence. The deletion then encompasses one copy of the repeat and the sequence between the repeats. This model is an extension of the Streisinger model for the formation of frameshift mutations by misalignment on repeated sequences (59). Secondary structure may also play a role in deletion formation, either by bringing together otherwise greatly separated sequences (1), or by juxtaposing deletion endpoints in deletion intermediates in which misaligned hydrogen-bonding is mediated by inverted or quasipalindromic DNA sequences (26,48).

At the time of writing, a total of 28 deletion mutations had been sequenced in this study. One deletion was isolated twice, and another (146-268 in Section D of Tab. 2) is identical to S23 isolated earlier (19). The recovery of several mutants more than once among a relatively small sample confirms that deletion formation is not a random event. For convenience, deletion events have been classified into two distinct groups. Class I deletion events (19/28) lie within the lacI gene, while Class II events (9/28) extend into the lacO region. Class II events are $lacO^c$.

Let us first consider Class I events. Class I events tend to be relatively short and are most often flanked by repeated sequences (14/19). An analysis of the sequences surrounding these endpoints indicates that in three of the five cases where repeated sequences do not flank the deletion, quasipalindromic sequences are available which could have directed the mutation. In the case of the two deletions for which neither direct nor inverted repeats were identified, it should be pointed out that another mechanism need not be invoked, as the templating sequence may be further away than the rather arbitrary 50 bp scanned.

The most interesting aspect of deletion specificity is just how often both palindromes and direct repeats cooperate to produce a deletion. In most of the Class I deletions analyzed, the misalignments potentiated by direct repeats were further stabilized by inverted repeats. In 12 of these, the quasipalindromic sequences alone could account for the specificity of the deletion endpoints. These included both deletions which have now been recovered more than once (19) and those for which structural intermediates have been described (36). The deletion presented in Fig. 3 is an example of a deletion event in which both repeated and palindromic sequences play a role. We have suggested that the presence of both repeated and palindromic sequences enhances the stability of the intermediate, hence making it more likely that such mutations will be recovered (26).

How significant is the recovery of deletions whose endpoints can be juxtaposed by inverted repeat sequences as well as by directly repeated sequences? In this study, 12 out of 14 mutants lying within the lacI gene

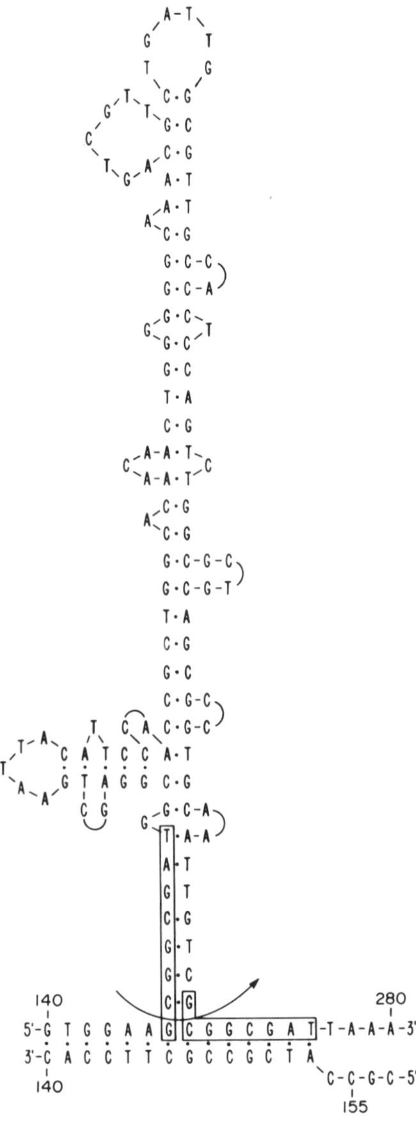

Fig. 3. Sequence-directed deletion mutation isolated more than once in the lacI gene. This 123-bp deletion is flanked by direct repeats (boxed) and inverted repeats (i.e., quasipalindromic sequences). The direct or inverted repeats are capable of accounting for the specificity of the deletion endpoints. However, both repeats can cooperate in the formation and stabilization of the proposed structural intermediate.

(i.e., Class I mutants) fell into this category, while 5 of the 7 from the study of Farabaugh et al. (19) demonstrated the concurrent occurrence of both palindromic and repeated sequences at their endpoints (26). In contrast, an analysis of 25 simulated deletions, ranging from 10 to 100 bp and contained entirely within the lacI gene, revealed only one case where their termini could be juxtaposed by palindromic sequences of 4 or more base pairs (26). Clearly, the endpoints of the deletions formed in vivo are not random. They must reflect the mechanism responsible for their creation.

An examination of the Class II deletions revealed a number of commonalities. Class II events are generally longer and often lack repeated sequences at their endpoints. Few of the Class II mutants can be explained on the basis of misalignments. This may be related to the fact that selection for the lacI mutants requires that the cell remain lacZ$^+$. In addition, many of the Class II deletions have their endpoints in the CAP-binding site. We do not yet know what influence the binding of the CAP protein to this region might have on mutation.

Duplications

The origin of duplication mutations is not well understood. They can potentially be created by the processing of structural intermediates similar to those envisioned for the production of deletion events (see Ref. 48). In this study, only 3 independent duplications were recovered. They are thus relatively rare events. The duplications ranged in size from 155 to 457 bp (Tab. 2, Section E). None of the duplications recovered in this study, however, were flanked by repeated sequences capable of directing their formation. Nor were palindromic sequences readily able to account for the duplication endpoints. A similar situation was noted by Calos et al. (3), who recovered a spontaneous duplication of 88 bp in the lacI gene.

While we are unable to account for the specific endpoints of these mutants, each of the duplicated sequences contains within it a region of very extensive secondary structure potential. As a consequence, structural intermediates formed in a single-strand region including the duplication would be expected to be stabilized. It is also possible that such structures may interfere with normal repair and replication processes and thereby enhance the opportunity for errors during subsequent repair or replication. DNA context effects may also play an important role. Duplications would be expected to occur where DNA breakage was frequent. Potential candidates might be sites of Okazaki primer synthesis or sites of frequent mismatch repair. The role of these contextual aspects of DNA sequence in the formation of duplications will be addressed later.

Transposable Elements

Transposable elements are normal constituents of most bacterial genomes and many extrachromosomal elements. The E. coli chromosome contains between 8 and 14 copies of IS1 (c.f. Ref. 31). They can alter the organization and expression of prokaryotic genes at frequencies similar to, and often greater than, spontaneous mutation. Thus, among the collection of spontaneous lacI$^-$ mutations, representatives of this class of events can be expected. Indeed, 7 independent occurrences of insertion elements were observed. In each case, the insertion was of IS1, a 768-bp insertion element characterized by short, inverted terminal repeats and producing a 9-bp direct repeat of the 9-bp segment at the site of insertion (5,29).

The seven IS1 insertions occurred at different sites in the lacI gene (Tab. 2). Two insertions of IS1 into lacI reported earlier by Calos et al. (4) occurred at sites identical to our mutants 3-19 and 3-24. However, in each case, the insertion had the reverse orientation to that found here. The insertion of IS1 is apparently not random (35) but demonstrates what has been termed regional, rather than site, specificity (31). The recovery of IS1 at 7 different sites in the lacI gene would confirm the relaxed aspect of integration specificity. G:C base pairs were found at one end, and often at both ends (10 out of 14), of the 9-bp duplication, a feature of IS1-mediated insertion noted previously by Galas et al. (23). It has been reported that insertions of IS1 and also transposon Tn9, which carries IS1 as flanking elements, occur preferentially in A/T-rich regions (23). The sites at which IS1 insertions were recovered in the lacI gene were not particularly rich in A/T. In fact, 64% of the bases in the duplicated 9-bp repeated sequence at the sites of insertion were G:C base pairs.

The seven IS1 insertions recovered in this study represent 12% of the nonhotspot mutations. These insertions are, therefore, not rare events. In addition, due to the lack of A/T-rich regions, the contribution of IS1-mediated mutagenesis in lacI might be an underestimate of the general role IS1-insertional mutagenesis plays in *E. coli*.

MUTATIONAL SPECIFICITY IN AN Ung$^-$ STRAIN

Independent spontaneous lacI mutants were collected from a strain defective in uracil-N-glycosylase, the enzyme responsible for the removal of uracil from DNA (64). The mutator phenotype expected of the ung mutation (16,17) was very modest and not readily seen in the lacI forward mutation system (Tab. 5). However, when the effect of the ung mutation on the frequency of base substitutions was determined, a 3.3- to 5.7-fold increase in transitions was noted (Tab. 5). In Tab. 5, the distribution of spontaneous amber mutants resulting from G:C → A:T transitions in the Ung$^-$ strain (NR8009) is compared to those isolated in the Ung$^+$ strain (NR3835). In total, 461 amber mutants from over 80,000 screened LacI$^-$ mutants were analyzed genetically and the base substitutions responsible for the lacI mutation determined.

The ung mutation increased the frequency of transition events but not transversions (Tab. 5). This is consistent with the role of uracil-N-glycosylase in the removal of uracil from DNA. Since uracil codes as thymine, uracil misincorporated in place of thymine (i.e., opposite adenine), and remaining in the DNA because of the ung defect, does not produce a mutation during DNA replication. (Note, however, that A:T → G:C transitions do not give rise to de novo nonsense mutants and, hence, cannot be detected in the lacI nonsense system.) In contrast, where uracil is present in the DNA as a consequence of either its misincorporation or the spontaneous deamination of cytosine, a G:C → A:T transition is expected. The increase in mutation frequency at a given G:C → A:T transition site can then be interpreted as being related to the frequency of deamination at that site.

The increase in mutation frequency varies considerably (Tab. 6), showing considerable site-to-site variation in deamination rates. Some sites show an order of magnitude or greater increase in mutation frequency in the Ung$^-$ strain as compared to the Ung$^+$ strain, whereas in some cases no increase was seen. At three such sites (A6, A15, and A34) no increase should be expected. This is because each of these three sites involves a site where the cytosine is naturally methylated to 5-methylcytosine. The deamination product of this modified base is thymine and, hence, not removed by uracil-N-glycosylase. These sites are normally recovered as mutational hotspots (9) and, in the absence of repair in the Ung$^-$ strain, the result is that other potential G:C transition sites now show mutational levels close to those seen at the 5-methylcytosine deamination hotspot sites (Tab. 5). A similar observation was made by Duncan and Miller (15).

Tab. 5. The effect of the ung mutation on spontaneous mutagenesis in the lacI gene of *Escherichia coli*.

	Ung+		Ung-	
Amber site	Number	Frequency[1]	Number	Frequency
5	6	0.040	14	0.90
6	99	0.670	7	0.50
9	12	0.080	45	3.00
15	71	0.480	7	0.50
16	21	0.140	7	0.50
19	22	0.150	5	0.35
21	13	0.090	2	0.14
23	15	0.100	3	0.20
24	21	0.140	1	0.07
26	12	0.080	16	1.00
31	1	0.007	3	0.20
33	9	0.060	2	0.14
34	29	0.200	3	0.20
35	2	0.014	2	0.14
Totals:				
Transitions[2]	333	2.200	117	7.20
Minus hotspots[3]	163	1.100	102	6.30
Transversions	113	0.800	5	0.40
Total:		300		330

[1] Mutation frequencies X 10^{-8}.
[2] Only G:C → A:T transitions are monitored.
[3] Hotspots are 5'-CMeCAGG-3' sites of A6, A15, and A34.
Note: Spontaneous mutation in NR 8009 (ung⁻) and NR3835 (ung⁺).

The ranking of the transition sites by the magnitude of the increase seen in the Ung⁻ strain compared to the Ung⁺ strain is given in Tab. 6. In order to understand the origin of the site-to-site variation, the bases neighboring the mutated sites were examined. A correlation between A/T richness and the relative increase in mutation frequency emerged. The biggest increases were observed at sites where 3 or more of the adjacent bases were A:T base pairs, and the least increases were observed where A:T base pairs were scarce. We suggest that the enhancement of mutation at A/T-rich sites may reflect the fact that deamination is more than 1,000-fold more rapid in single-stranded DNA (38). The weaker hydrogen bonding at the A/T-richer sites is more prone to disruption by DNA breathing and, as a consequence, deamination occurs preferentially at these sites.

Multiply-Recovered G:C → A:T Transitions in the Ung⁻ Strain

Mutant collections were made in strains carrying the ung mutation (NR3958 and NR8006). These were then cloned and their DNA sequences determined. Besides base substitution events, deletion and frameshift mutations were also recovered. An analysis of these mutations is currently underway

Tab. 6. A/T richness and the enhancement of mutation in an Ung^- strain.

Rank order		Fold-increase	A/T richness* 0	1	2	>3
1	A9	38.00				X
2	A31	29.00				X
3	A5	23.00				X
4	A26	12.00			X	
5	A35	10.00				X
6	A16	4.00		X		
7	A33	2.30			X	
8	A19	2.30		X		
9	A23	2.00				X
10	A21	1.50	X			
11	A15	1.00		X		
12	A34	1.00		X		
13	A6	0.75		X		
14	A24	0.50	X			

* Number of A:T base pairs adjacent to site of G:C → A:T transition event.

(Fix and Glickman, ms. in prep.). In addition, an analysis of mutants recovered when an excess of uracil is in the DNA has been carried out by Sedwick, Brown, and Glickman (ms. in prep.). The question we wish to answer, however, is how much of the spontaneous mutation in the wild-type strain grown under normal laboratory conditions is due to deamination? Our answer is based on the following analysis.

As in the case of the nonsense mutations recovered in the Ung^- strain, the G:C → A:T transitions recovered in the Ung^- strain by DNA sequencing are not random. In Tab. 7 we have listed all the transition sites that have been recovered 3 or more times. These sites can therefore be viewed as hotspots (or at least warm spots) for cytosine deamination. If these same sites are frequently seen among the spontaneous transition mutations, it would be reasonable to argue that their predominance reflects the contribution of deamination to the spontaneous spectrum. However, if mutations are rarely encountered at these sites, it would argue quite strongly that the spontaneous deamination of cytosine does not make a major contribution to the spontaneous spectrum.

Table 2 lists the nonhotspot mutants recovered in the Ung^+ strain. None of the sites multiply-recovered in the Ung^- background were recovered even once in the Ung^+ strain. In addition, the only site recovered more than once in the Ung^+ strain was not observed among those sequenced from an Ung^- strain. From these observations we are forced to conclude that the normal uracil-N-glycosylase activity in a cell is sufficient to avoid the fixation of mutation that would have occurred via the cytosine deamination pathway.

Tab. 7. Comparison of the multiply-recovered G:C → A:T transitions in an Ung^+ and an Ung^- strain.

Position	Ung^-	Ung^+	Amino acid change
42	4	0	Thr → Met
92	5	0	Arg → Cys
93	0	3	Arg → His
120	3	0	Ser → Phe
188	3	0	Gln → Ochre
191	5	0	Gln → Ochre

* Among NC^+ mutations in *lacI*.

SPONTANEOUS MUTATION IN A STRAIN DEFECTIVE IN DNA POLYMERASE I

A collection of 268 independent spontaneous *lacI* mutants was made in NR8035, a *polA1* derivative of NR3835. All classes of mutation, with the exception of the IS1 insertions, were recovered (see Tab. 8). However, since only 3 (or possibly 4) insertion events would be expected, we do not know if their absence in this collection is significant. In addition, since the IS1 insertions effectively disrupt homology, these mutations may be among those not successfully cloned. The spontaneous mutation frequency was approximately double that normally observed for the wild-type strain (compare Tab. 1 and 8).

A comparison of the frequencies of the various classes of mutation in the $PolA^-$ and $PolA^+$ strains indicated that only base substitution and the nonhotspot frameshift events were substantially enhanced. In the case of the nonhotspot frameshifts, the $PolA^-$ strain showed a greater than 4-fold enhancement over the $PolA^+$ strain. In the case of base substitutions, an overall enhancement of approximately 2.5-fold was observed. The increase in other classes of mutation was less than a factor of 2. The frameshift hotspot at position 620-631 (Fig. 1) is well represented, and the ratio of the +4 to the -4 event was not different than that seen in the $PolA^+$ strain. Among the base substitutions, the fraction of mutants at the +6 position in the *lac* operator region was substantially increased. The details of the spontaneous mutations are to be discussed in detail elsewhere (Fix, Burns, and Glickman, ms. in prep.).

$PolA^-$ strains are known to have high mutation rates. This was originally described by Coukell and Yanofsky (8), who reported an enhanced frequency of deletions in the *trp-tonB* region of the *E. coli* chromosome in strains carrying the *polA1* mutation. Later studies (57,62), in which the effect of *polA* mutations was assessed in *trp* frameshift reversion assays, indicated a propensity for $PolA^-$ mutants to revert "plus" frameshifts. Savic and Romac (52) provided additional evidence for the specificity of spontaneous mutation in a $PolA^-$ strain. They examined 122 spontaneous His^- mutants in a strain carrying the *polA1* mutation. Of these, 109 were thought to be "minus" frameshifts while the remaining 9 were thought to be deletions. As will be seen below, "plus" frameshifts were also recovered.

Tab. 8. The distribution of spontaneous lacI$^-$ mutants in a PolA$^-$ strain by class.

Class of mutation	Number of occurrences	Frequency X 10^{-7}
Base substitutions	84 (31.0%)	14.0
Single base frameshifts	13 (4.8%)	2.2
Deletions	13 (4.8%)	2.2
Duplications	2 (0.7%)	0.3
Insertions (IS1)	0	--
(+) 5'-CTGG-3'*	97 (36.0%)	17.0
(-) 5'-CTGG-3'*	17 (6.3%)	2.9
Unknown**	42 (16.0%)	7.2
	TOTAL: 268	46.0 X 10^{-7}

* Frameshift hotspot at position 620-631.
** Mutations not yet recovered onto M13 recovery vector.

Frameshift Events

The 13 frameshift mutants recovered in the PolA$^-$ strain are presented in Tab. 9. It is these events that demonstrate the most intriguing insights into the molecular events surrounding mutagenesis. Of the 13 frameshifts, three involve the gain, rather than the loss, of base pairs. The recovery of "plus" frameshifts in the polA1 strain was not predicted on the basis of work by Vaccaro and Siegel (57,62) and Savic and Romac (52), who concluded that frameshifts in the polA1 background would be the result of the loss of base pairs. As can be seen from Tab. 9, this is not the case. Several instances of the gain of one or even two base pairs are recorded. The recovery of the "plus" frameshift mutants effectively illustrates the limitation of the use of the genetic sign, i.e., "plus" and "minus," to predict the specificity of reversion events.

As illustrated earlier, the identification of the sequence change at the site of the mutation, and the larger DNA context in which the site resides, provide essential information from which the origin of the mutation can be inferred. This is particularly true with respect to frameshift mutations. An examination of the 13 frameshift mutations recovered in the PolA$^-$ strain revealed differences from the wild-type spectrum and suggested some interesting features of DNA context. About one-half of the frameshift mutations can be explained by mechanisms similar to those proposed for frameshift events in the wild-type strain.

For example, one mutation involves the loss of an A in a run of 5 A's (1006-1010), and four others can potentially be explained as being templated by very short (2-bp) repeats. These repeats are considerably shorter than those seen in the PolA$^+$ strain (see, e.g., Fig. 2). With respect to the 5'-GATC-3' sequence, only a single frameshift recovered was adjacent to this sequence. This frameshift is unique, however, as it involves the loss of 3 base pairs! Remarkably, 7 of the 13 frameshift mutations are adjacent to the sequence 5'-GTGG-3'.

Tab. 9. Nature of the nonhotspot frameshift mutations in the PolA⁻ strain.

Positions	Change	Occurrences	Sequence
46	+ A	1	ACGTTA(A)TACGAT
312	- T	1	GCGTGG(T)GGTGTC
357	- C	1	AAGCGG(C)GGTGCA
386	- A	3	CGCGTC(A)GTGGGC
390	+ G	1	CAGTGG(G)GCTGATC
440-441	- GC	1	GTGGAA(GC)TGCCT
543	- A	1	GCGTGG(A)GCATCT
610	+ GC	1	CGGCGC(GC)GTCTG
768	- ATC	2	CCAACG(ATC)AGAT
1006-1010	- A	1	AAAGAAAA(A)CCAC
		TOTAL: 13	

Note: The sequence 5'-GTGG-3' is underlined. We have proposed that these sequences denote sites of persistent breaks in the PolA⁻ strain. We suggest that these sequences are the sites at which Okazaki fragments are either initiated or terminated during DNA replication and propose their involvement in the formation of frameshift and deletion mutations.

We now comment on the 5'-GTGG-3' sequence and its role in mutation in the PolA⁻ strain. It is our belief that the high incidence of frameshift mutations adjacent to a defined sequence was not accidental. Rather, we prefer an argument analogous to that offered to explain the enhanced recovery of frameshift mutations at or nearby 5'-GATC-3' sequences in the PolA⁺ strain. This latter sequence is the site of active DNA metabolism and the model for frameshift mutation can be simplified as reflecting a requirement for DNA strand breakage. In other words, the increase in strand breakage at a specific site will be observed as a preferred site for frameshift events.

Frameshifts spawned by such events appear not necessarily to rely on mistemplating for their occurrence. The break may provide an opportunity for the action of any of a number of exonucleases present in the cell. Hence, frameshifts at such sites need not lie in runs of repeated bases and, thus, may appear untemplated. Such is the case for many of the frameshifts recovered in the PolA⁻ strain. A parallel situation is seen in an Ung⁻ strain during thymine deprivation (37). In that case, frameshift mutations were not observed, suggesting that the breaks resulting from the removal of uracil and the subsequent cleavage of the apyrimidinic sites result in the induction of frameshift mutations.

What metabolic operation does the 5'-GTGG-3' sequence presage in the PolA⁻ strain? The polA1 mutation demonstrates a major delay in the joining of Okazaki fragments during DNA synthesis [see Kornberg (34) for a review]. The delay reflects the role of the 5' → 3' exonucleolytic activity, along with RNase H, in the removal of the RNA primers that initiate the Okazaki fragments. In addition, polymerase I normally carries out the repair replication (short patch) that, together with ligase, converts the low molecular weight, nascent DNA to the high molecular weight DNA contained in the bacterial genome. Indeed, in the PolA$^-$ strain, the sealing of Okazaki fragments is approximately 10 times slower than in the wild-type. Hence, we propose that the frequent recovery of frameshift events in the neighborhood of the 5'-GTGG-3' sequence reflects the role that this sequence may play in the replication of DNA. We propose that the 5'-GTGG-3' sequences represent the sites of the initiation (or termination) of Okazaki fragments during normal DNA synthesis in *E. coli*. Our hypothesis is that the enhanced frequency with which frameshift events are recovered at these sequences reflects a delay in strand closure at these sites in a PolA$^-$ strain. A potential role for Okazaki fragments in the origin of frameshifts has been suggested earlier (47,49).

DNA Contextual Effects and the 5'-GTGG-3' Sequence

The identification of the 5'-GTGG-3' sequence as a favored site of DNA metabolism permits a retrospective examination of the role of this sequence in mutagenesis. Since we are proposing that the 5'-GTGG-3' sequence is the site of frequent metabolic activity, specifically involving single-strand breakage, it follows that sequences surrounding such preferred breakpoints would have the opportunity more often to form the misaligned structures such as have been proposed as intermediates of deletion.

An examination of the endpoints of the 28 deletions in the PolA$^+$ strain is intriguing. Nine of the 28 deletions (32%) have one or both of their endpoints adjacent to a 5'-GTGG-3' sequence. Considering that there are only 10 such sequences in the *lacI* gene, the choice of endpoints is unlikely to be coincidental. Indeed, if one assumes that short deletions are the consequence of local slippage events and may not require strand breakage (note that none of the seven deletions of less than 50 bp have the 5'-GTGG-3' sequence at their endpoints), then 7 of the 12 Class I deletions of greater than 50 bp have 5'-GTGG-3' at one or both of their termini. In the presence of the *polA1* mutation, 6 of the 12 Class I deletions have a 5'-GTGG-3' at one or both of their endpoints. During the course of this analysis we noticed that, in several other deletions, the sequence 5'-CTGG-3' was present at the terminus of a deletion. Since this is the sequence involved at the frameshift hotspot site, we can not help but speculate on whether the 5'-CTGG-3' sequence is not also a potential, though perhaps less-favored, site of breakage.

Finally, we should point out an additional observation which suggests the 5'-GTGG-3' sequence to be unusual, again in the context of DNA replication. The *mutD* mutation affects the epsilon or proofreading subunit of DNA polymerase III (11,18,56). This mutation is a powerful conditional mutator. Among *lacI* mutants sequenced (21), six A:T → T:A transversions were observed. All six occurred within the 5'-GTGG-3' sequence! While the relevance of this observation with respect to the action of the *mutD* gene product remains unknown, the involvement of these sites in some specific stage of DNA replication seems certain.

CONCLUSIONS

The analysis of the mutagenic specificity of spontaneous mutation in the *lacI* gene of *E. coli* has provided insights into the relative roles of the various pathways of mutagenesis. The analysis of the sequences surrounding mutation has provided some insights into the origin of deletion, insertion, duplication, frameshift, and base substitution events. Within the various classes of mutational events, often evidence for different mechanisms emerged. For example, frameshifts appear to be the product of one of several diverse mechanisms. Most intriguingly, insights into DNA "context" effects are emerging. Sites of DNA repair, e.g., sites of Okazaki initiation/termination, appear to be important components in frameshift, deletion, and duplication mutagenesis. Studies of spontaneous mutation in a PolA$^-$ strain reinforce these conclusions. Other studies, involving Ung$^-$ strains, indicate that the spontaneous deamination of cytosine does not normally contribute significantly to spontaneous mutation. The ability to clone and sequence large numbers of mutants opens the door to a radical new approach to our understanding of DNA repair and mutagenesis.

ACKNOWLEDGEMENTS

We sincerely appreciate the technical assistance of F.L. Allen and C.L. Bell in the laboratory and M.S. Bergart in the preparation of this manuscript. Special appreciation goes to Drs. J.A. Drake, R.M. Schaaper, and G.B. Golding for the continuing discussions on spontaneous mutation, and specifically to Dr. L.S. Ripley for discussions that helped give context to "DNA context" in relation to mutagenesis.

REFERENCES

1. Albertini, A.M., M. Hofer, M.P. Calos, and J.H. Miller (1982) On the formation of spontaneous deletions: The importance of short sequence homologies in the generation of large deletions. Cell 29:319-328.
2. Brychcy, T., and R.C. von Borstel (1977) Spontaneous mutability in U.V.-sensitive excision-defective strains of *Saccharomyces*. Mutat. Res. 45:185-194.
3. Calos, M.P., D. Galas, and J.H. Miller (1978) Genetic studies on the *lac* repressor. VIII. DNA sequence change resulting from an intragenic duplication. J. Mol. Biol. 126:865-869.
4. Calos, M.P., L. Johnsrud, and J.H. Miller (1978) DNA sequence at the integration sites of the insertion element IS1. Cell 13:411-418.
5. Calos, M.P., and J.H. Miller (1980) Transposable elements. Cell 20: 579-595.
6. Calos, M.P., and J.H. Miller (1981) Genetic and sequence analysis of frameshift mutations induced by ICR-191. J. Mol. Biol. 153:39-66.
7. Cheung, S., K. Arndt, and P. Lu (1984) Correlation of *lac* operator DNA imino protein exchange kinetics with its function. Proc. Natl. Acad. Sci., USA 81:3665-3669.
8. Coukell, M.B., and C. Yanofsky (1970) Increased frequency of deletions in DNA polymerase mutants of *Escherichia coli*. Nature (London) 228: 633-636.
9. Coulondre, C., J.H. Miller, P.J. Farabaugh, and W. Gilbert (1978) Molecular basis of base substitution hotspots in *Escherichia coli*. Nature 274:775-780.
10. DeBoer, J.G., and L.S. Ripley (1984) Demonstration of the production of frameshift and base substitution mutation by quasipalindromic DNA sequences. Proc. Natl. Acad. Sci., USA 81:5528-5531.
11. DiFrancesco, R., S.K. Bhatnagar, A. Brown, and M.J. Bessman (1984) The interaction of DNA polymerase III and the product of the *Escherichia coli* mutator gene, *mutD*. J. Biol. Chem. 259:5567-5573.
12. Drake, J.W. (1969) Comparative rates of spontaneous mutation. Nature 221:1182.

13. Drake, J.W., E.F. Allen, S.A. Forsberg, R.M. Preparata, and E.O. Greening (1969) Spontaneous mutation: Genetic control of mutation rates in bacteriophage T4. Nature 221:1128-1132.
14. Drake, J.W., B.W. Glickman, and L.S. Ripley (1983) Updating the theory of mutation. Am. Scientist 71:621-630.
15. Duncan, B.K., and J.H. Miller (1980) Mutagenic deamination of cytosine residues in DNA. Nature 287:560-561.
16. Duncan, B.K., and B. Weiss (1978) Uracil-DNA glycosylase mutants are mutators. In DNA Repair Mechanisms, P.C. Hanawalt, E.C. Friedberg, and C.F. Fox, eds. Academic Press, Inc., New York, p. 183.
17. Duncan, B.K., and B. Weiss (1982) Specific mutator effect of ung (uracil-DNA-glycosylase) mutations in Escherichia coli. J. Bacteriol. 151:750-755.
18. Echols, H., C. Lu, and P.M.J. Burgers (1983) Mutator strains of Escherichia coli, mutD and dnaQ, with defective exonucleolytic editing by DNA polymerase III holoenzyme. Proc. Natl. Acad. Sci., USA 80:2189-2192.
19. Farabaugh, P.J., U. Schmeissner, M. Hofer, and J.H. Miller (1978) Genetics studies of the lac repressor. VII. On the molecular nature of spontaneous hotspots in the lacI gene of Escherichia coli. J. Mol. Biol. 126:847-863.
20. Foster, P.L., E. Eisenstadt, and J.H. Miller (1983) Base substitution mutations induced by metabolically activated aflatoxin B1. Proc. Natl. Acad. Sci., USA 80:2695-2698.
21. Fowler, R.G., R.M. Schaaper, and B.W. Glickman (1986) The role of 3' → 5' exonuclease (proofreading activity) in DNA replication fidelity: A characterization of mutational specificity within the lacI gene for a mutD5 mutator strain of Escherichia coli. J. Bacteriol. (in press).
22. Fridovich, I. (1976) Oxygen radicals, hydrogen peroxide, and oxygen toxicity. In Free Radicals in Biology, Vol. 1, W.A. Pryor, ed. Academic Press, Inc., New York, pp. 239-277.
23. Galas, D.J., M.P. Calos, and J.H. Miller (1980) Sequence analysis of Tn9 insertions in the lacZ gene. J. Mol. Biol. 144:19-41.
24. Glickman, B.W. (1979) Spontaneous mutagenesis in Escherichia coli strains lacking N6-methyladenine residues in their DNA. Mutat. Res. 61:153-159.
25. Glickman, B.W., and M. Radman (1980) Escherichia coli mutator deficient in methylation-instructed DNA mismatch correction. Proc. Natl. Acad. Sci., USA 77:1063-1067.
26. Glickman, B.W., and L.S. Ripley (1984) Structural intermediates of deletion mutagenesis: A role for palindromic DNA. Proc. Natl. Acad. Sci., USA 81:512-516.
27. Gomez-Eichelmann, M.C., and K.G. Lark (1977) Endo R Dpn I restriction of Escherichia coli DNA synthesized in vitro. Evidence that the ends of Okazaki pieces are determined by template deoxyribonucleotide sequence. J. Mol. Biol. 117:621-635.
28. Hastings, P.J., S.K. Quah, and R.C. von Borstel (1976) Spontaneous mutation by mutagenic repair of spontaneous lesions in DNA. Nature 264:719-722.
29. Johnsrud, L. (1979) DNA sequence of the transposable element Is1. Mol. Gen. Genet. 169:213-218.
30. Kern, R., and F.K. Zimmermann (1978) The influence of defects in excision and error prone repair on spontaneous and induced mitotic recombination and mutation in Saccharomyces cerevisiae. Mol. Gen. Genet. 161:81-88.
31. Kleckner, N. (1981) Transposable elements in prokaryotes. Ann. Rev. Genet. 15:391-404.

32. Kondo, S., H. Ichikawa, K. Iwo, and T. Kato (1970) Base change mutagenesis and prophage induction in strains of Escherichia coli with different DNA repair capacities. Genetics 66:187-217.
33. Konrad, E.B. (1978) Isolation of an Escherichia coli K-12 dnaE mutation as a mutator. J. Bacteriol. 133:1197-1202.
34. Kornberg, A. (1980) DNA Replication, W.H. Freeman and Company, San Francisco.
35. Kuhn, S., H.J. Fritz, and P. Starlinger (1979) Close vicinity of IS1 integration sites in the leader sequence of the gal operon of E. coli. Mol. Gen. Genet. 167:235-242.
36. Kunkel, T.A. (1984) The mutational specificity of depurination. Proc. Natl. Acad. Sci., USA 81:1494-1498.
37. Kunz, B.A., and B.W. Glickman (1984) The role of pyrimidine dimers as premutagenic lesions: A study of targeted vs. untargeted mutagenesis in the lacI gene of Escherichia coli. Genetics 106:347-364.
38. Lindahl, T. (1979) DNA-glycosylases-endonucleases for apurinic/apyrimidic sites and base excision-repair. Prog. Nucl. Acid Res. Mol. Biol. 22:135-191.
39. Loeb, L.A., and T.A. Kunkel (1982) Fidelity of DNA synthesis. Ann. Rev. Biochem. 51:429-457.
40. Lu, A.L., K. Welsh, S. Clark, S.S. Su, and P. Modrich (1984) Repair of DNA basepair mismatches in extracts of Escherichia coli. Cold Spring Harbor Symp. Quant. Biol. 49:589-596.
41. Macquat, L., K. Thornton, and W. Reznikoff (1980) lac promoter mutations located downstream from the transcription start site. J. Mol. Biol. 139:537-549.
42. Miller, J.H. (1983) Mutational specificity in bacteria. Ann. Rev. Genet. 17:215-239.
43. Muller, H.J., and L.M. Mott-Smith (1930) Evidence that natural radioactivity is inadequate to explain the frequency of "natural" mutations. Proc. Natl. Acad. Sci., USA 16:277-285.
44. Okada, Y., G. Streisinger, J. Owen, J. Newton, A. Tsugita, and M. Inouye (1972) Molecular basis of a mutational hotspot in the lysozyme gene of bacteriophage T4. Nature 263:338-341.
45. Pribnow, D., D.C. Sigurdson, L. Gold, B.S. Singer, J. Brosius, T.J. Dull, and M.F. Noller (1981) rII cistrons of bacteriophage T4: DNA sequence around the intercistronic divide and positions of genetic landmarks. J. Mol. Biol. 149:337-376.
46. Ripley, L.S. (1982) Model for the participation of quasipalindromic DNA sequences in frameshift mutation. Proc. Natl. Acad. Sci., USA 79:4128-4132.
47. Ripley, L.S., and B.W. Glickman (1984) DNA secondary structure and mutation in cellular responses to DNA damage. In UCLA Symposia on Molecular and Cellular Biology. A New Series, E.C. Friedberg and B.A. Bridges, eds. Alan Liss, Inc., New York, pp. 521-540.
48. Ripley, L.S., and B.W. Glickman (1983) Unique self-complementarity of palindromic sequences provides DNA structural intermediates for mutation. Cold Spring Harbor Symp. Quant. Biol. 47:851-861.
49. Ripley, L.S., and N.B. Shoemaker (1983) A major role for bacteriophage T4 DNA polymerase in frameshift mutagenesis. Genetics 103:353-366.
50. Sargentini, N.J., and K.C. Smith (1981) Much spontaneous mutagenesis in Escherichia coli is due to error-prone DNA repair. Carcinogenesis 2:863-872.
51. Sargentini, N.J., and K.C. Smith (1985) Spontaneous mutagenesis: The roles of DNA repair, replication, and recombination. Mutat. Res. 154:1-27.

52. Savic, D.J., and S.P. Romac (1982) Powerful mutator activity of the polA1 mutation within the histidine region of Escherichia coli K-12. J. Bacteriol. 149:955-960.
53. Schaaper, R.M., B.N. Danforth, and B.W. Glickman (1985) Rapid repeated cloning of mutant lac repressor genes. Gene 39:181-189.
54. Schaaper, R.M., B.N. Danforth, and B.W. Glickman (1986) Mechanisms of spontaneous mutagenesis: An analysis of spectrum of spontaneous mutation in the E. coli lacI gene. J. Mol. Biol. (in press).
55. Schaaper, R.M., T.A. Kunkel, and L.A. Loeb (1983) Infidelity of DNA synthesis associated with bypass of apurinic sites. Proc. Natl. Acad. Sci., USA 80:487-491.
56. Scheuermann, R., S. Tam, P.M.J. Burgers, C. Lu, and H. Echols (1983) Identification of the E-subunit of Escherichia coli DNA polymerase III holoenzyme as the dnaQ gene product: A fidelity subunit for DNA replication. Proc. Natl. Acad. Sci., USA 80:7085-7089.
57. Siegel, E.C., and K.K. Vaccaro (1978) The reversion of trp frameshift mutations in mut, polA, lig, and dnaE mutant strains of Escherichia coli. Mutat. Res. 50:9-17.
58. Speyer, J.F., J.D. Karam, and A.B. Lenny (1966) On the role of DNA polymerase in base selection. Cold Spring Harbor Symp. Quant. Biol. 31:693-697.
59. Streisinger, G., Y. Okada, J. Emrich, J. Newton, A. Tsugita, E. Terzaghi, and M. Inouye (1966) Frameshift mutations and the genetic code. Cold Spring Harbor Symp. Quant. Biol. 33:77-84.
60. Todd, P.A., and B.W. Glickman (1982) Mutational specificity of UV-light in Escherichia coli: Indications for a role of DNA secondary structure. Proc. Natl. Acad. Sci., USA 79:4123-4127.
61. Topal, M.D., and J.R. Fresco (1976) Complementary base pairing and the origin of substitution mutations. Nature 263:285-289.
62. Vaccaro, K.K., and E.C. Seigel (1975) Increased spontaneous reversion of certain frameshift mutations in DNA polymerase I deficient strains of Escherichia coli. Mol. Gen. Genet. 141:251-262.
63. von Borstel, R.C. (1969) On the origin of spontaneous mutations. Japan. J. Genet. 44:102-105.
64. Warner, H.R., B.K. Duncan, C. Garett, and J. Neuhard (1981) Synthesis and metabolism of uracil-containing deoxyribonucleic acid in Escherichia coli. J. Bacteriol. 145:687-695.
65. Watson, J.D., and F.H.C. Crick (1953) The structure of DNA. Cold Spring Harbor Symp. Quant. Biol. 18:123-131.

GENETICS AND MOLECULAR BIOLOGY OF MAMMALIAN CELLS--PART I

INTRODUCTION: COMPARATIVE RESPONSES TO DNA DAMAGE IN BACTERIA AND MAMMALIAN CELLS

Philip C. Hanawalt

Department of Biological Sciences
Stanford University
Stanford, California 94305

Much of our basic understanding of cellular repair mechanisms began with a thorough analysis of the processing of cyclobutane pyrimidine dimers in the DNA of *Escherichia coli* (6,7). Furthermore, it is now well-documented in this Volume that there are still important discoveries to be made using bacteria as model systems for studying the mechanisms of antimutagenesis. The recent advances in our understanding of DNA damage processing in *E. coli* attest to the complexity of such mechanisms in even the simplest prokaryote systems (15). These advances include the discovery of new inducible responses to lesions in DNA, new insights into the mechanism of mismatch repair, and the development of defined sequence DNA probes for assessing the molecular spectrum of mutagenic actions. It is essential that basic research on important prokaryote model systems continues to receive adequate emphasis and support even though our long-term goal may be to understand the mechanisms of antimutagenesis and anticarcinogenesis in humans. The results from research on *E. coli* still serve to guide our exploration of DNA damage processing in mammalian cells (4,10).

There are, of course, important qualitative differences in the complexity of genomic organization and replication in mammalian cells when compared to bacteria, and these add new dimensions to the problems of dealing with genetic damage (2,7). These include multiple tandem replicons and many replication forks in actively replicating genomic domains in mammalian cells, in contrast to the situation in *E. coli* in which only 2 forks must traverse the entire chromosome bidirectionally from a unique origin to replicate the genome. The mechanisms for dealing with replication-blocking lesions could be very different in these respective systems, and it would, in fact, appear that mammalian cells are able to tolerate much higher levels of persisting lesions than are bacteria (7). Thus, for example, some rodent cell lines exhibit 50% survival of colony-forming ability while removing less than 20% of the ultraviolet (UV)-induced pyrimidine dimers from their genomes (5). Obviously an efficient and generally unfailing mechanism must exist in these cells for circumventing lesions, such as dimers, that pose blocks to the progression of DNA polymerases along the damaged template strands. We have obtained evidence for both replicative and recombinational modes for bypassing dimers, using simian virus 40 DNA as a probe in mammalian cells (3,13).

Another important difference in complexity between mammalian cells and bacteria is in their chromatin structure and gene organization (12). The protein-DNA structure in mammalian chromatin may necessitate special functional components to carry out DNA repair processes in addition to those required for the less encumbered bacterial chromosome (2,7). Furthermore, the highly specific protein-DNA associations in chromatin may render eukaryotic cells susceptible to DNA-protein cross-linking reactions in a fashion not found in bacteria. The same proteins may also protect the DNA and render it less reactive to some chemical agents, even at the level of vulnerable sites within a given base.

Since chromatin structure obviously plays a large role in the regulation of transcription and replication of eukaryotic DNA, it is likely that it is an important factor in DNA repair processes as well (12). The accessibility of sites of DNA damage to repair systems may be affected by their positions in the chromatin structure at the most basic level. For example, in UV-irradiated mammalian cells rendered permeable to enzymes under conditions of low ionic strength, the pyrimidine dimer-specific enzyme, T4 endonuclease V, incises DNA at only about half the dimer sites, while incisions can be made at nearly all of the sites if the permeabilized cells are first subjected to a brief exposure to high ionic strength that relaxes the chromatin structure in a nonreversible way (14).

At a higher level of organization, certain large domains of the genome may be more or less accessible to repair systems depending upon their state of expression or function in the cell. It is likely that the chromatin structure in a local region undergoing excision repair must be modified to allow the necessary enzymatic events to occur; perhaps the organization of higher-order domains, such as replicating units, must be altered as well (2). Finally, the sequence of repair reactions cannot be considered complete until the original chromatin structure at the site of damage is restored.

In attempts to correlate biological endpoints such as survival, mutagenesis, and transformation to DNA repair levels in mammalian cells, repair proficiency has usually been assumed to be uniform throughout the genome. However, there is now evidence from studies in our laboratory that DNA damage in some domains of the mammalian genome may be processed much more efficiently than that in others (1,9,11,16,17). The genome includes many different functional classes of DNA of which some are "silent" (e.g., repetitive sequences in heterochromatin, unexpressed genes) while others are active (e.g., expressed genes, elements required for the translation machinery, and various regulatory regions). The consequences of unrepaired or misrepaired damage in DNA will most certainly depend upon the precise location of the damage with respect to these functional classes, a fact that is well-documented at the nucleotide sequence level in bacteria by correlations of the spectrum of particular lesions with "hot spots" for mutagenesis. Differences in the repair response to damage in selected genomic regions may account for some of the profound differences seen in the carcinogenic response in different tissues or when different organisms are compared. Therefore, it is important to understand the "fine structure" of DNA repair in mammalian genomes in order to assess carcinogenic risks.

We have been developing technology for the analysis of DNA damage and repair in defined nucleotide sequences in mammalian genomes (1). Our basic strategy involves using ^{32}P-labeled hybridization probes to quantitate specific restriction fragments located in or around genes of interest. Typically, this involves alkaline gel electrophoresis to separate restriction fragments of isolated and purified genomic DNA, Southern transfer of the DNA to a support membrane, and hybridization to the probes. For quantitation of pyrimidine dimer frequencies in such fragments, identical DNA samples treated or not treated with T4 endonuclease V are analyzed on alkaline gels in parallel lanes. The amount of hybridization of a specific probe to the respective bands of unit-length restriction fragments is measured either by scintillation counting (for amplified genes) or by densitometry of the autoradiograms (for unamplified genes). Using the Poisson relation, the frequency of sites sensitive to the endonuclease (in this case, pyrimidine dimers) can be calculated. This technology is being applied to a number of different genes to learn more about the intragenomic heterogeneity of excision repair and the rules that govern accessibility of the DNA to repair.

We have quantified the levels of pyrimidine dimers and their repair in the active dihydrofolate reductase (DHFR) gene of Chinese hamster ovary (CHO) cells in which the gene is amplified 50-fold. Roughly 70% of the pyrimidine dimers were removed from a restriction fragment within the gene in 24 hr, while little repair was detectable in a fragment 30 kb upstream from the gene, and only 15% of the dimers were lost from the genome overall (1). We have now confirmed these findings in the parental, nonamplified CHO cells (Bohr, Okumoto, and Hanawalt, ms. submitted for publ.). On the basis of our results, we have suggested that the preferential repair of vital DNA sequences facilitates the UV resistance of the CHO cells in spite of their low overall repair levels. This is a plausible explanation, but it leaves unanswered the question of how the replication machinery in these cells is able to overcome the large number of persisting lesions in the DNA. Since viable progeny arise it is clear that bypass replication or some recombinational process must have occurred to generate functional daughter genomes.

We have also begun to look at the comparative DNA repair levels in several proto-oncogenes in rodent cells. The processing of damage in the relevant proto-oncogenes and in their flanking regions will undoubtedly affect the efficiency of transformation. In a preliminary study we have examined the formation and removal of pyrimidine dimers in 2 proto-oncogenes, c-*abl* and c-*mos*, in mouse Swiss 3T3 cells to assess the role of DNA damage processing in their activation. We have demonstrated 85% repair of dimers in 24 hr in a 19.5-kb BamH1 fragment in the expressed c-*abl* gene, but less than 20% repair in a 15-kb EcoR1 fragment containing the unexpressed c-*mos* gene (and in fact no detectable repair in a 6.2-kb fragment containing the gene) in the same cells. For comparison, the DHFR gene was 60 to 70% repaired in 24 hr in these cells which exhibit overall repair levels of less than 20% for pyrimidine dimers (Madhani et al., ms. in prep.).

The following chapters in this Volume deal with additional problems of importance to our understanding of the processing of damage in mammalian genomes. Dr. Michael Lambert's chapter addresses the question of inducible responses through the direct analysis of major new protein species by 2-dimensional protein gel electrophoresis of extracts from rat fibroblasts exposed to various chemical carcinogens. This approach should permit the isolation and eventual characterization of some of the unique proteins involved in inducible responses to damaged DNA in mammalian systems. It also facilitates comparative analysis of the responses in these cells to other types of stress, such as heat shock.

Dr. Angel Pellicer discusses approaches to understanding the specific nucleotide changes necessary to activate particular proto-oncogenes by different carcinogens. It will be important to correlate the results of these studies with those discussed above in which the repairability of defined lesions is analyzed in the respective proto-oncogenes.

Dr. Tadashi Inoue presents a possible animal model for the hereditary human disease, ataxia telangiectasia (AT), which confers extreme sensitivity to ionizing radiation. While the deficiency in the classical form of xeroderma pigmentosum is known to be at the level of damage recognition and incision, we still do not understand the nature of the defect, if any, in DNA repair in AT (8). The advancement of our understanding of human hereditary diseases with putative deficiencies in DNA repair could be immensely enhanced by the discovery and development of appropriate animal models to supplement the bacterial models that have heretofore served us so well.

REFERENCES

1. Bohr, V.A., C.A. Smith, D.S. Okumoto, and P.C. Hanawalt (1985) DNA repair in an active gene: Removal of pyrimidine dimers from the DHFR gene of CHO cells is much more efficient than in the genome overall. Cell 40:359-369.
2. Bohr, V.A., and P.C. Hanawalt (1984) Factors that affect the initiation of excision-repair in chromatin. In DNA Repair and Its Inhibition, A. Collins, R.T. Johnson, and C.S. Downes, eds. IRL Press, Oxford, pp. 109-125.
3. Clark, J.M., and P.C. Hanawalt (1984) Replicative intermediates in UV-irradiated simian virus 40. (DNA Repair Reports.) Mutat. Res. 132:1-14.
4. Ganesan, A.K., P.C. Hanawalt, P.K. Cooper, and C.A. Smith (1979) What can bacteria tell us about the responses of mammalian cells to DNA damage? In Radiation Research, Proceedings of the 6th International Congress on Radiation Research, O. Okada, M. Imamura, T. Terashima, and H. Yamaguchi, eds. Tokyo, Toppon, pp. 439-445.
5. Ganesan, A.K., G. Spivak, and P.C. Hanawalt (1983) Expression of DNA repair genes in mammalian cells. In Manipulation and Expression of Genes in Eukaryotes, P. Nagley, A.W. Linnane, W.J. Peacock, and J.A. Pateman, eds. Academic Press, Australia, pp. 45-54.
6. Hall, J.D., and D.W. Mount (1981) Mechanisms of DNA replication and mutagenesis in ultraviolet-irradiated bacteria and mammalian cells. Prog. Nucl. Acid Res. & Molec. Biol. 25:53-126.
7. Hanawalt, P.C., P.K. Cooper, A.K. Ganesan, and C.A. Smith (1979) DNA repair in bacteria and mammalian cells. Ann. Rev. Biochem. 48:783-836.
8. Hanawalt, P.C., and R.B. Painter (1985) On the nature of a DNA processing defect in ataxia telangiectasia. In Ataxia Telangiectasia: Genetics, Neuropathology, and Immunology of a Degenerative Disease of Childhood, R. Gatti, ed. A.R. Liss, Inc., New York, pp. 137-142.
9. Hanawalt, P.C. (1985) Intragenomic heterogeneity in DNA damage processing: Potential implications for risk assessment. In Mechanisms of DNA Damage and Repair, M. Simic, L. Grossman, and A. Upton, eds. Plenum Press, New York (in press).
10. Hanawalt, P.C., and S. Kondo (1979) Modes of DNA repair and replication. In Radiation Research, Proceedings of the 6th International Congress on Radiation Research, O. Okada, M. Imamura, T. Terashima, and H. Yamaguchi, eds. Tokyo, Toppon, pp. 434-438.
11. Mansbridge, J.N., and P.C. Hanawalt (1983) Domain-limited repair of DNA in ultraviolet irradiated fibroblasts from xeroderma pigmentosum complementation group C. In Cellular Responses to DNA Damage, E.C.

Friedberg and B.R. Bridges, eds. UCLA Symp. on Molec. and Cell. Biol., New Series Vol. II, Alan R. Liss, Inc., New York, pp. 195-207.

12. Reeves, R. (1984) Transcriptionally active chromatin. Biochim. Biophys. Acta 782:343-393.
13. Sarasin, A.R., and P.C. Hanawalt (1980) Replication of ultraviolet irradiated simian virus 40 in monkey kidney cells. J. Molec. Biol. 138:299-319.
14. van Zeeland, A.A., C.A. Smith, and P.C. Hanawalt (1981) Sensitive determination of pyrimidine dimers in DNA of UV irradiated mammalian cells: Introduction of T4 endonuclease V into frozen and thawed cells. Mutat. Res. 82:173-189.
15. Walker, G.C., L. Marsh, and L.A. Dodson (1985) Genetic analyses of DNA repair: Inference and extrapolation. Ann. Rev. Genet. 19:103-126.
16. Zolan, M.E., G.A. Cortopassi, C.A. Smith, and P.C. Hanawalt (1982) Deficient repair of chemical adducts in alpha DNA of monkey cells. Cell 28:613-619.
17. Zolan, M.E., C.A. Smith, and P.C. Hanawalt (1984) Formation and repair of furocoumarin adducts in alpha DNA and bulk DNA of monkey cells. Biochemistry 23:63-69.

INDUCIBLE CELLULAR RESPONSES TO DNA DAMAGE IN MAMMALIAN CELLS

Michael E. Lambert,[1] James I. Garrels,[1]
John McDonald,[2] and I. Bernard Weinstein[3]

[1]Cold Spring Harbor Laboratory
Cold Spring Harbor, New York 11724

[2]Department of Genetics
University of Georgia
Athens, Georgia 30602

[3]Comprehensive Cancer Center
Columbia University
New York, New York 10032

INTRODUCTION

The mechanisms by which chemical carcinogens induce cellular transformation are unknown. Polycyclic aromatic hydrocarbons, such as benzo(α)pyrene (BP), have been extensively studied, and their chemical properties, metabolic activation, and DNA-binding properties have been elucidated in great detail (34). There is also evidence that covalent binding of metabolites of these and of other carcinogens to cellular DNA is a critical event in their action as carcinogens (77). The subsequent series of biochemical and genetic events, however, that lead to the conversion of a normal cell to a cancer cell are not known.

There has been a tendency to think of the initiating event in chemical carcinogenesis in terms of simple random point mutations resulting from errors in the replication of damaged DNA. The simplest explanation is that the covalent binding of carcinogens to genes involved in growth control leads to point mutations at these sites, for example the types of base substitutions seen in mutated ras oncogenes (10). Several features of the carcinogenic process, however, including the high efficiency of carcinogen-induced cell transformation when compared to specific locus mutations, the lengthy latency period needed for the expression of the transformed state, and the multistep nature of the carcinogenic process, are not consistent with this mechanism (78,79).

An alternative mechanism would be that carcinogen-induced DNA damage might induce complex genomic changes, including, for example, gene rearrangements or gene amplification. Furthermore, it is possible that certain tumors are caused by synergistic interactions between viruses and environmental chemicals, although the mechanism of this synergy is not known.

Here, we provide an overview of known biological actions of chemical carcinogens in cellular transformation, including a review of previous studies of inducible cellular responses to DNA damage in mammalian cells. A series of three interrelated experimental strategies are presented to further define and characterize the roles of inducible responses to cellular stress, including DNA damage and heat shock, in molecular and cellular mechanisms of multistage chemical carcinogenesis.

BACKGROUND

Although the mechanism by which chemical carcinogens such as BP contribute to cellular transformation has not been specified, there has been an overriding tendency to think of these compounds in terms of their ability to induce somatic mutations. Such mutations are thought to arise from errors in the repair and replication of carcinogen-damaged cellular DNA, on the site, or in the immediate vicinity of chromosomal damage. This view has been supported by the finding that many carcinogens are active in the Ames bacterial mutation assay (1). Carcinogen-induced somatic mutation has been specifically invoked to explain the initiating action of carcinogens in multihit and multistep models of tumor initiation and promotion (6,61), and accumulation of somatic mutations with age has been hypothesized in models which describe cumulative age-specific patterns of certain human cancers, including lung cancer, that increase in incidence directly with time (17).

Recent studies of the mechanism of activation of the cellular _ras_ oncogenes, in both human and rodent tumors, have tended to strengthen the view that point mutation is important in chemical carcinogenesis. In the human bladder carcinoma-derived T24-_ras_ oncogene, a single point mutation distinguishes the normal from the transforming allele (10). Further studies of carcinogen-induced transformation in vivo in rodent systems have shown a highly reproducible pattern of induced point mutation within the _ras_ gene, including the same base substitution as is seen in T24-_ras_, and these mutations appear to be diagnostic for the transforming potential of _ras_ in these chemically derived tumors (31,55).

Despite these findings concerning the mutagenic potential of chemical carcinogens, and the limited examples in which cellular transformation can be reliably attributed, in part, to a specific point mutation, many features of the transformation process are not consistent with the somatic mutation theory of cancer.

In studies of carcinogen-induced transformation both in vitro and in vivo, the frequency of foci formation has been found to be as high as 100% of surviving cells (37), suggesting that the initiating event by chemical carcinogens in transformation may be a high-frequency, common event. Recently, it has also been found that X-ray-induced transformation in vitro can be effectively suppressed by addition of specific protease inhibitors (82). Since protease inhibitors have been found to suppress induction of the prokaryotic "SOS" inducible cellular response to DNA damage (81), it is possible that the suppression of transformation observed in this mammalian cell system may operate through an analogous biological mechanism. In fact, an attractive alternative hypothesis to the somatic mutation theory is that specific types of carcinogen-induced DNA damage are capable of inducing a high-frequency cellular response which is directly involved in the initiation process.

It is now also apparent that tumor cells can display numerous derangements in the organization of their genetic material, in addition to point mutations. These include: sequence amplifications; sequence deletions; chromosomal translocations; and the rearrangement of endogenous retroviral elements (26,78). Each of these rearrangements can, in turn, result in an alteration in the pattern of expression of specific cellular genes. Human neoplasms, for example, often display amplifications of oncogenes and other cellular genes (29,39). Though it is likely that gene amplification is a late step in tumor progression, DNA damage has previously been shown to increase the likelihood of amplification of specific genes during selection for drug resistance in defined cell culture systems (71), providing indirect evidence that exposure to chemical carcinogens may also contribute to late stages in multistep carcinogenesis. The exact mechanism by which chemical carcinogens enhance gene amplification in somatic cells is not known.

Retroviral elements have been found to be overexpressed in certain transformed cell types (38,78), and there is evidence for transposition of specific retroviral elements, including the murine intracisternal type A particle (IAP), in mouse tumors, including plasmocytomas (67). Both overexpression and rearrangement of endogenous retroviral-like elements can result in alterations in the patterns of expression of adjacent cellular genes. Likewise, an increase in retrotransposition (3) secondary to increased expression of retroviral elements could contribute to increased genomic instability during tumor progression (79). There is evidence, as well, that specific retroviral elements are transcriptionally responsive to specific types of cellular stress, including heat shock and DNA damage (58,73), and there is indirect evidence that damage to cellular DNA can both increase the probability of excision of retroviral elements from chromosomal loci (69) and act as a general stimulus to transposition (81). The molecular and cellular mechanisms, however, by which DNA damage or heat stress might lead to retroviral expression and rearrangement are not known.

Chemical carcinogens can act synergistically with both DNA and RNA tumor viruses in cellular transformation in rodent systems (19). Pretreatment of target cells with chemical carcinogens, both in vivo and in vitro, can enhance the frequency of cellular transformation during subsequent infection with DNA tumor viruses. This synergy may involve either increases in the uptake, replication or integration of viruses, or alterations in the pattern of expression of cellular genes, which enhance the probability of viral transformation. In contrast to DNA tumor viruses, synergy between carcinogens and RNA tumor viruses requires addition of the carcinogen to previously infected cells. In this case, enhancement of transformation more likely involves alterations in the pattern of expression of cellular genes which interact cooperatively with viral gene products in inducing stable transformation. There is evidence as well, from studies of human skin cancer, that ultraviolet light can act as a co-carcinogen with certain human papillomaviruses (62).

There is considerable evidence that pretreatment of cells with DNA-damaging agents can enhance the reactivation and mutagenesis of carcinogen-pretreated viruses, including herpes simplex, SV40, adenovirus, and parvovirus (13,15,70,83). Presumably these effects reflect the induction of DNA repair enzymes, although the putative genes and proteins involved in these responses have not been identified.

The above considerations and others suggest the possibility that DNA damage in mammalian cells might induce a cellular response analogous to the "SOS" response in bacteria. Indeed, in bacteria, carcinogens are potent inducers of the "SOS" response (32,81). Presumably the components of this response in mammalian cells would differ from those in bacteria, and may include the induction of factors that produce more complex changes in the genome than random point mutations.

The induction of several new proteins has previously been detected by two-dimensional gel electrophoresis in ultraviolet light or carcinogen-treated mammalian cell culture (66,72). It is not clear, however, that these proteins are induced specifically as a consequence of DNA damage, since they are also induced merely by arresting DNA synthesis. There are a limited number of other examples of carcinogen-induced increases in gene expression in eukaryotic systems, including ultraviolet light induction of plasminogen activator in mammalian cells (51), and induction of specific new transcripts after DNA damage in yeast (49). The biological significance of any of these induced proteins and transcripts is not known. In addition, there are recent reports of molecular cloning of specific eukaryotic repair genes (68), though these have not yet been studied in the wider context of mechanisms of cellular transformation.

Chemical carcinogens have also been found to induce changes in the patterns of methylation in cellular DNA and alterations in transcription at specific loci. Treatment of cells with chemical carcinogens, such as BP, has been found to reduce the overall content of 5-methylcytosine in cellular DNA (7,80), and demethylation of cellular genes after treatment with ultraviolet light is associated with an increase in the expression of specific genes, including metallothionein I (45). Alternatively, chemical carcinogens have also been reported to induce an increase in the activity of endogenous methyltransferases and to inactivate cellular gene expression through an increase in DNA methylation (33). Overall, however, there has been little progress in identifying inducible cellular responses to DNA damage that may be relevant, specifically, to molecular and cellular mechanisms in chemical carcinogenesis.

In order to further define and characterize the role of acute cellular stress responses, including both DNA damage and heat shock, in multistage carcinogenesis, we have developed three interrelated approaches. First, we have studied the effects of chemical carcinogens on the replication of latent papovavirus in transformed mammalian cells. Second, we have analyzed changes in patterns of cellular protein expression during the acute response to chemical carcinogens, in a well-defined cell culture system, by use of high resolution, two-dimensional isoelectric focusing protein gels. Third, we have studied the transcriptional responsiveness of a model *Drosophila* retroviral-like element, *copia*, to heat shock and cellular stress, after transfection into mammalian cells.

VIRAL-CHEMICAL SYNERGY AND GENE AMPLIFICATION

One of the most striking examples of a complex cellular response to DNA damage is the ability of chemical carcinogens and radiation to induce the replication of the papovaviruses SV40 or polyoma in transformed cells that contain integrated copies of the viral DNAs (20,41,42).

Infection of rat fibroblasts by a temperature-sensitive (ts-a) mutant of polyoma will lead to stable transformation of a small percentage of infected cells. The integrated virus DNA present in these cells can subsequently undergo spontaneous amplification in situ after a shift down of the culture temperature from 39°C to 33°C, the permissive temperature for the function of virally encoded, large T antigen (4). Continuous passage of these transformants at the permissive temperature also leads to a small amount of excision and asynchronous replication of free polyoma DNA species, as well as occasional curing of viral genomes from the rat chromosome (64). The intrinsic instability, therefore, of polyoma DNA in transformed rat embryo fibroblasts provides a semipermissive system for examining the consequences of exogenous stress, such as that resulting from treatment with chemical carcinogens, on the behavior of a strong viral replicon in mammalian cells.

As a specific approach to this question we have studied the effect of the ubiquitous chemical carcinogen BP on the replication of polyoma DNA in a well-characterized cell line designated ts-a H3 (25). Ts-a H3 cells contain 1.3 viral equivalents of polyoma DNA integrated in the cell genome and, because the ts-a mutant produces a temperature-sensitive, large T antigen, these cells produce free polyoma DNA at 33°C but not at 39°C. The integrated polyoma DNA also has three deletions, which provide useful markers for the two major species of free viral DNA that can be induced in this cell line. In these studies, we have also utilized the ultimate carcinogenic metabolite of BP, benzo(α)pyrene-trans-7,8-dihydrodiol-9,10-epoxy (anti) (BPDE), which has a very short half-life (41), in cell fusion experiments, to determine whether carcinogens might influence asynchronous DNA replication via a direct or indirect effect.

Experiments were carried out to determine whether exposure of cells to a low dose of BP (0.25 μg/ml) enhanced the production of extrachromosomal copies of polyoma DNA, and whether this process was dependent on large T antigen (see Fig. 1). As predicted from previous studies (25), simply shifting the temperature from 39°C to 33°C, in the absence of BP, resulted in the appearance of increasing amounts of free polyoma DNA molecules. The addition of BP, however, at the time of temperature shift led to a marked enhancement in production of free viral DNA. By contrast, when cells were maintained at 39°C, either in the absence or presence of BP, no free viral DNA was detected. These results provide evidence that both the spontaneous production of free polyoma DNA and the production induced by BP are dependent on the function of T antigen.

Previous studies on the mechanism of production of free polyoma DNA in transformed rat cells (4,60) are consistent with an onion skin model of viral replication (9) by which there is in situ replication of the integrated viral DNA, followed by excision of these copies and further extrachromosomal replication. Presumably, exposure of cells to BP enhances one or more stages of this process.

We have also addressed the possibility that the induction of polyoma DNA replication in cells exposed to BP or BPDE might be mediated by an inducible factor rather than by direct binding of the carcinogen to the integrated polyoma DNA. The induction of such a factor might reflect a general response of cells to DNA damage, and this factor might then act in trans to enhance polyoma DNA replication.

To examine this, normal rat fibroblasts (REF) lacking polyoma DNA were pretreated with BPDE and fused to cells containing integrated polyoma DNA (Py-REF) (see Fig. 2). Extrachromosomal DNA was then extracted from the heterokaryons 48 hr after cell fusion and analyzed for the level of free

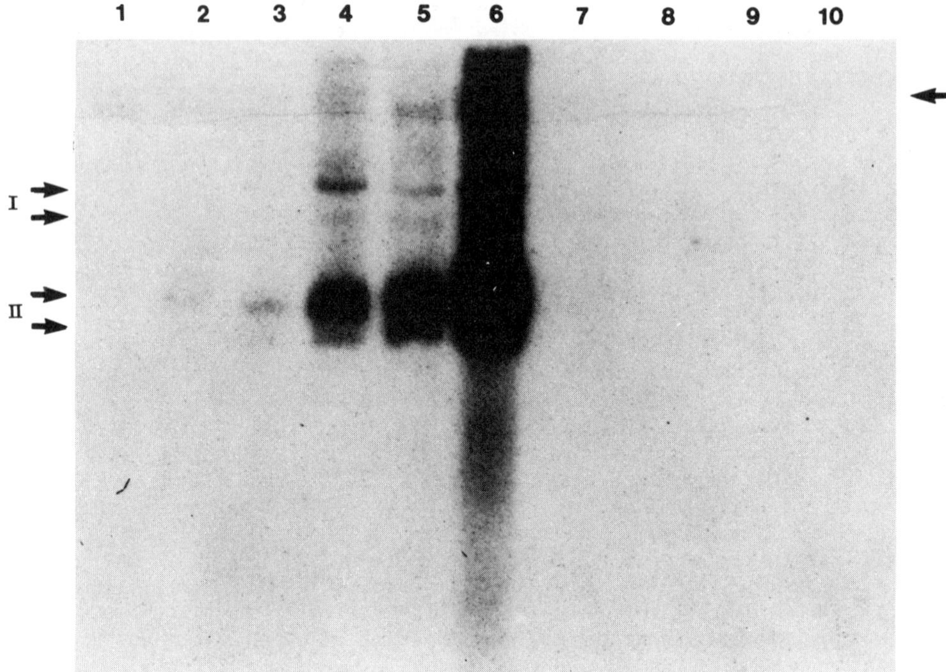

Fig. 1. Effect of benzo(α)pyrene (BP) on polyoma replication in ts-a H3 cells. Each lane contains 5 μg of extrachromosomal DNA, extracted at the indicated times, gel electrophoresed (without restriction enzyme cleavage), blotted, and hybridized to a ^{32}P-labeled polyoma probe (41). In lanes 1, 3, and 5 the cells were shifted from 39°C to 33°C and the DNA extracted at 24, 48, and 72 hr after the temperature shift, respectively. In lanes 2, 4, and 6 the procedure was the same except that BP (0.25 μg/ml) was added to the cultures at the time of the temperature shift. In lane 7 cells were grown continuously at 39°C. In lanes 8, 9, and 10 cells were grown at 39°C in the presence of BP (0.25 μg/ml) for 24, 48, and 72 hr, respectively. The arrows on the left side of the figure indicate the positions of supercoiled (Form I, lower arrows) and relaxed open circle (Form II, upper arrows) molecules of the two major species of free polyoma DNA produced by ts-a H3 cells. The arrow on the right indicates the position of the small amount of chromosomal DNA that contaminates the Hirt extract.

viral DNA. The results indicated that fusion of the BPDE-pretreated REF with untreated Py-REF resulted in a marked increase in extrachromosomal polyoma DNA, when compared with the results obtained when untreated REF cells were fused with untreated Py-REF cells. This suggests that the DNA damage produced by BPDE need not be on the same genome as the integrated copy of polyoma DNA in order to enhance asynchronous replication of polyoma DNA.

We have also studied the effects of known biochemical inhibitors of eukaryotic DNA repair on the ability of BPDE to induce polyoma viral DNA replication. The nuclear enzyme poly(ADP) ribosyl transferase is induced

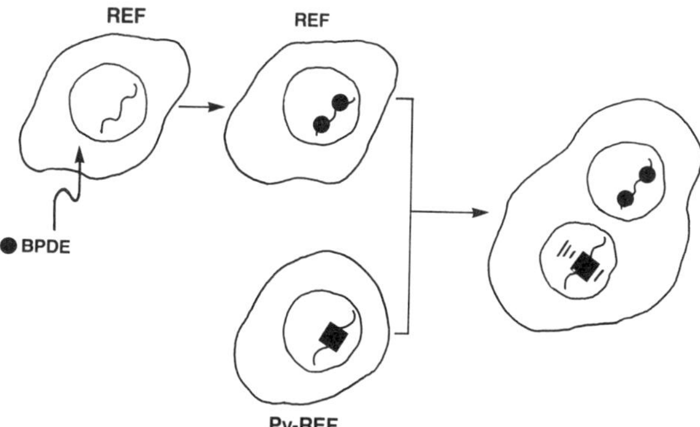

Fig. 2. Carcinogen-induced "trans" activation. Normal rat embryo fibroblasts (REF) were exposed to benzo(α)pyrene trans-7,8-dihydrodiol-9,10-epoxide (anti) (BPDE) for 3 hr and then washed extensively with PBS. Rat embryo fibroblasts containing integrated polyoma DNA (Py-REF) which had not been exposed to BPDE were then co-cultivated and fused with the BPDE-treated REF cells using polyethylene glycol. Extrachromosomal DNA extracted from the heterokaryons 48 hr after cell fusion and analyzed by Southern blot hybridization showed a marked increase in extrachromosomal polyoma DNA when compared with fused, untreated controls (41).

by several types of agents that produce DNA damage, and its function appears to be required for efficient repair (14). Increased poly(ADP) ribosylation is also associated with polyoma DNA replication (65) and control of gene expression and cell differentiation (18). We examined, therefore, whether addition of an inhibitor of poly(ADP) ribosyl transferase, 3-aminobenzamide, would inhibit the process of BPDE-induced polyoma DNA replication. We found that 5 mM 3-aminobenzamide inhibited BPDE-induced polyoma DNA replication in ts-a H3 cells but produced little or no inhibition of the constitutive level of polyoma DNA replication in these cells (see Fig. 3). We also found that the chemotherapeutic antibiotic bleomycin (20 μg/ml), which specifically damages DNA and not RNA or protein, is a potent inducer of polyoma DNA replication, supporting the view that DNA damage is the critical event in the ability of these agents to induce viral synthesis.

Finally, we examined the role of DNA polymerase alpha in the replication of polyoma virus by use of the specific inhibitor aphidicolin. Polymerase alpha has been shown to play a role in the replication of SV40 virus DNA (35). It is also induced by DNA damage and has been implicated in cellular DNA repair processes (11). We found that aphidicolin (0.1 to 10.0 μg/ml) caused a dose-dependent inhibition of constitutive polyoma DNA replication in the absence of BPDE, and that aphidicolin (0.1 μg/ml) completely inhibited BPDE induction of viral DNA synthesis. Thus it appears that carcinogen-enhanced polyoma DNA replication is strictly dependent on the function of DNA polymerase alpha.

What is the nature of the putative trans-acting agent that accumulates in carcinogen-treated cells? Time course studies suggest that there is a lag of several hours in its appearance and that it can then persist for several days after initial DNA damage (41). Recent studies by others (54) have extended our original observation and provide evidence for a trans-

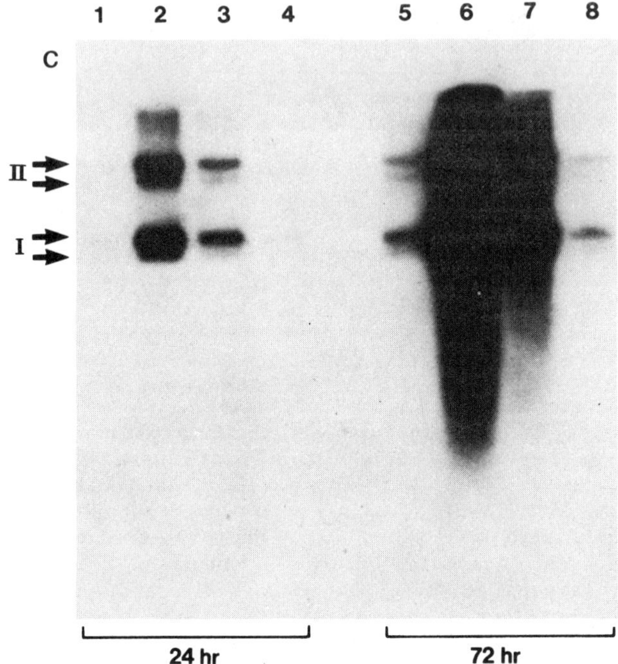

Fig. 3. Effect of 3-aminobenzamide on BPDE-induced polyoma DNA replication in H3 cells. Each lane contained 5 μg of low molecular weight DNA isolated from H3 cells 72 hr after the shiftdown in temperature. After electrophoresis all samples were hybridized to a polyoma probe. The positions of migration of Forms I and II of the two major species of free viral DNA are indicated on the left. Lanes 1 and 5, control cells; lanes 2 and 6, cells treated with BPDE (0.1 μg/ml); lanes 3 and 7, cells treated with BPDE (0.1 μg/ml) in the presence of 3-aminobenzamide (5 mM); and lanes 4 and 8, cells treated only with 3-aminobenzamide. DNA samples were collected at either 24 or 72 hr after the shiftdown in temperature and/or drug treatment.

acting factor in the process of ultraviolet light-induced SV40 virus DNA synthesis in Chinese hamster ovary (CHO) cells. Nomura and Oishi (54) also found that this response was eliminated by cycloheximide. Thus, it seems likely that the trans-acting factor is a protein, or that its accumulation requires de novo protein synthesis. This factor may be the same as, or similar to, putative permissivity factors constitutively made by certain cell types that support continuous replication of polyoma or SV40 virus (murine and simian cells, respectively), since the action of these factors can also be demonstrated by fusion of such cells with virus-infected non-permissive cells.

This putative trans-acting factor could be an enzyme which plays a general role in DNA replication, whose induction relieves otherwise rate-limiting steps for viral DNA synthesis. This might include certain enzymes known to play a general role in DNA synthesis, such as thymidine kinase or topoisomerases. Our observation that aphidicolin can completely inhibit carcinogen-induced polyoma DNA replication suggests, further, that DNA polymerase alpha plays an integral role in this process. It is of interest that herpes simplex virus (HSV-1) encoded polymerase has previously been shown to be required for HSV-1 mediated SV40 DNA synthesis in CHO cells (48), and that DNA polymerase alpha has been implicated in developmentally regulated ribosomal gene amplification in Xenopus laevis (84).

The nature of the inductive signal for the synthesis of this factor is not known, but it is clear that a variety of DNA-damaging agents are effective in inducing asynchronous polyoma DNA synthesis and SV40 DNA synthesis (12,41,42). The signal for induction could be the presence of blocked DNA replication forks, as postulated for the "SOS" response in bacteria (81), or the transient arrest of overall cellular DNA synthesis. We have found, in fact, that BPDE induction of polyoma DNA replication was enhanced by pretreating the cells with agents that cause transient arrest of the cell cycle. On the other hand, this pretreatment itself caused only a slight enhancement of polyoma DNA replication in the absence of BPDE.

Our results with 3-aminobenzamide and bleomycin provide direct evidence that DNA damage and repair play an important role in the induction process. Since a variety of DNA-damaging agents, which create diverse types of lesions in DNA, including single- and double-stranded breaks, thymidine dimers, bulky adducts, and DNA cross-links, are all capable of inducing polyoma and SV40 DNA synthesis, it is also likely that the signal is not lesion-specific. Since poly(ADP) ribosyl transferase is preferentially activated by single-strand DNA breaks, triggering of the response may occur after endonucleolytic cleavage during excision of DNA adducts or thymidine dimers; whereas agents that cause direct scission to the DNA strand (i.e., ionizing radiation, bleomycin, etc.) could induce the response directly.

An alternative possibility is that induction is the consequence of perturbations in nucleotide pools resulting from cellular response to DNA damage. Such perturbations are in fact known to induce a broad spectrum of deleterious genetic consequences (40).

DETECTION OF CARCINOGEN-INDUCIBLE PROTEINS BY TWO-DIMENSIONAL PROTEIN GEL ELECTROPHORESIS

A better appreciation of the significance of inducible responses to carcinogen treatment depends on systematic identification of specific proteins or sets of proteins which are expressed at altered levels as a consequence of DNA damage, and an understanding of their roles in regulation of DNA repair, DNA replication, and other cellular processes. In light of our findings from the study of inducibility of latent polyoma virus by chemical carcinogens, and in order to gain a deeper understanding of inducible cellular responses to carcinogen exposure, we have undertaken studies to identify changes in patterns of protein expression in rat embryo fibroblasts by use of the QUEST system for high resolution, two-dimensional protein gel electrophoresis.

The QUEST system (22-24) has been developed as a tool for the routine quantitative analysis of proteins detectable in radiolabeled cultured cells. It has previously been applied to a series of experiments involving normal and virally transformed secondary rat embryo fibroblast (REF52) cells (21). Shown in Fig. 4 is a gel obtained after [^{35}S]methionine labeling of an exponential culture of the rat embryo fibroblast cell line REF52. On average, such labeling studies can detect greater than 1,200 individual proteins, although the identity of only a very small fraction of these is definitely known. Since the gel system is standardized and since gels run over an extended period can be matched to a common standard pattern, it has also been possible to build a species-specific rat protein database. This database, in turn, provides a valuable reference against which to compare changes in patterns of cellular protein expression observed as a consequence specifically of carcinogen treatment.

Exposure of mammalian cells in vitro to low doses of physical or direct-acting chemical carcinogens rapidly induces pleiotropic effects which can broadly be divided into three distinct responses: (a) effects on cell growth and DNA replication, (b) induction of DNA repair systems, and (c) induction of other as yet unspecified cellular stress responses. Significant differences, however, among species and cell types also exist in the extent and character of such responses, with important consequences for the probability of such specific outcomes as cell viability, mutation, cytogenetic alterations, and cellular transformation. Expectations about the complexity of acute cellular responses to carcinogen treatment in vitro will depend, therefore, in large part on the choice of cell line, the carcinogen used, and the toxicity associated with the delivered dose.

In the current study we have limited ourselves to: (a) the use of the secondary rat embryo fibroblast cell line REF52; (b) the use of direct-acting chemical carcinogens; and (c) low dose ranges which produce minimal cytotoxicity. Selection of REF52 was based on the availability of an extensive protein database which permits comparisons to be made of the specificity of carcinogen treatment on changes in patterns of cellular protein expression. In addition, we have previously tested REF52 and found it to be capable of sustaining BPDE-enhanced polyoma viral replication. The putative trans-acting factor(s) responsible for asynchronous viral replication may, therefore, be detectable among the changes in protein expression seen in this line after BPDE treatment.

We have also limited the survey of carcinogens to well-characterized, direct-acting chemical carcinogens and avoided compounds that require metabolic activation to reactive intermediates. This has greatly simplified the changes in protein profiles by minimizing the changes expected due to activation of endogenous drug metabolizing systems. The carcinogens tested were selected to create different DNA lesions, thus permitting comparisons to be made of the possible specificity of various agents in inducing protein profile changes, as well as possible commonalities in the types of elicitable responses.

We have also stayed within low dose ranges, thus avoiding confounding changes in cellular protein expression which can be expected as a consequence of cell toxicity and cell death.

Shown in Fig. 5 are portions of gel patterns from cells treated with 0.5 μg/ml of BPDE for 2 hr, followed by incubation and labeling with [^{35}S]-methionine. As indicated by the arrows, a major protein species of approximately 30,000 dalton molecular weight and I.E.F. of pH 5.0 is readily observed during the early labeling period from 4 to 8 hr after treatment. By 20 to 24 hr the level of expression of this protein has returned to the basal level seen in parallel untreated cultures radiolabeled during the same time window. The protein, which we designate as Carcinogen Inducible Protein 1 (CIN 1), is actually only the most prominent spot in a small cluster within the same region of the gel which is induced coordinately. The induction of CIN 1 has not been detected in any previous experiment in which the REF52 cell line was studied under a variety of growth conditions. These studies include experiments designed specifically to examine changes

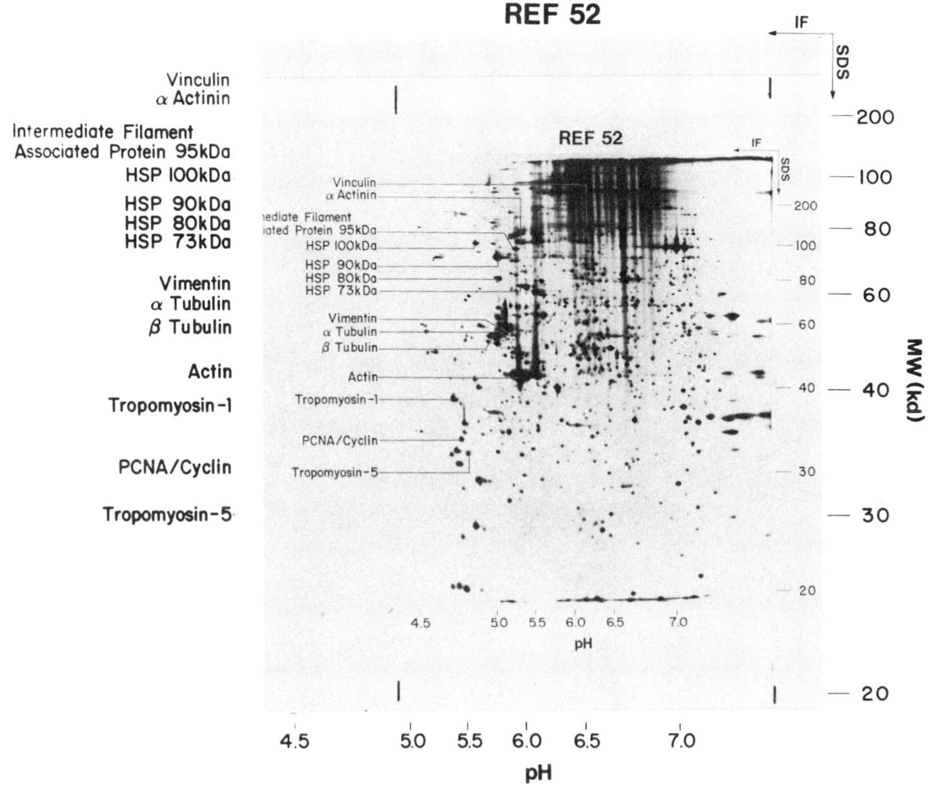

Fig. 4. Protein pattern of [^{35}S]methionine-labeled REF52. One x 10^5 REF52 cells, grown in DMEM supplemented with 10% FCS, were plated in a 60-cm^2 dish and 24 hr later (in the exponential phase of growth) were radiolabeled with 500 μCi of high specific activity [^{35}S]methionine for 4 hr. Cells were subsequently harvested and cell lysates prepared (28). The pattern shows an autoradiographic exposure of a broad range pH 3.5 to 10.0, 10% acrylamide protein gel loaded with 180,000 dpm. pH and molecular weight markers are shown on the bottom and right side, respectively. Known cellular proteins are indicated on the left.

in profiles of cellular proteins resulting from growth arrest and arrest of DNA synthesis. Thus, the CIN 1 protein would appear to be specifically induced as a consequence of exposure to BPDE.

We next examined the changes in patterns of cellular protein expression in REF52 which result from treatment with a different chemical carcinogen, N-acetoxy-2-acetyl aminofluorene (NAAF), the activated form of N-2 acetyl aminofluorene (AAF), which induces a characteristic DNA lesion which is distinct from that induced by BPDE (30). Cells were treated for 2 hr and then radiolabeled during the same subsequent time windows, in a manner identical to that used in labeling BPDE-treated REF52. As with BPDE, there is a rapid increase in the level of expression of this protein within 4 to 8 hr, and by 20 to 24 hr the level of expression has returned to the basal level observed in parallel cultures of control cells radiolabeled during the same time window. Thus at least two different chemical carcinogens can induce CIN 1 with similar time kinetics over the first 24 hr after treatment.

Studies are now in progress to further characterize the range of physical and chemical carcinogens that may induce CIN 1, and to obtain information on its regulation, including half-life, subcellular localization, post-translational modifications, and transcriptional regulation.

HEAT STRESS AND RETROVIRAL ACTIVATION

Heat shock rapidly induces a characteristic and highly regulated cellular stress response marked by a decrease in the transcription of many cellular genes and in overall protein synthesis, and the rapidly increased expression of a specific subset of proteins. This physiological response, as well as the major inducible heat-shock proteins themselves, are present in divergent species and are highly conserved (36,46). The induction of heat-shock proteins is not limited to exposure to elevated temperature, and a wide variety of different types of exogenous stress are capable of inducing a similar response. These include, in *Drosophila*, anoxic stress, amino acid starvation, and exposure to specific chemicals, including heavy metals, hydrogen peroxide, and ethanol (2).

The exact functions of the major heat-shock proteins are not known. In *Drosophila*, the hsp70 protein is rapidly translocated, under both heat shock and anoxic stress, from the cytoplasm to the nucleus, where it can bind to open chromatin structures and heteronuclear RNA, and possibly act to protect these from degradation (76). In addition, priming of an organism or cell with an initial pulse of heat increases thermotolerance to a subsequent challenge with higher or more protracted heat regimens (44), and can protect against X-irradiation-induced lethality (52). Heat-shock protein expression is also developmentally regulated. A highly specific pattern of hsp70 protein expression is observed early in mouse embryogenesis, and development of thermotolerance is acquired only at the blastocyst stage (5). Heat shock can also play an important role in life stage-specific differentiation in certain parasitic protozoa (75), and early genetic studies in *Drosophila* (27) indicated that mutant-like morphological alterations (phenocopies) could be induced by imposition of specific stresses, including heat shock and chemical carcinogens, at specific stages of embryological development. Together, these findings suggest that heat-shock proteins may play an important role, not only in cellular responses to a specific class of environmental stress, but in the regulation of gene expression during development and in differentiation.

Alterations in temperature and heat-shock mimetic regimens also affect the transcription and replication of several endogenous retroviral-like elements in a variety of species. The *Drosophila melanogaster* retroviral-like element *copia* is transcriptionally responsive in vivo to heat shock and heat-shock mimetic regimens, including exposure to sodium arsenite and hydrogen peroxide (73), and in *Dictyostelium*, the transposon-like element *DIR1* is transcriptionally responsive to heat shock, and heat-shock conditions mimic the effect of early development on the induction of *DIR1* expression in growing cells (85). Further, in yeast, the rate of *Ty* element transposition has been shown to be increased 100-fold by hyperthermia (8, 56) although it is not known whether this reflects an increase in the rate of retro-transposition secondary to stress-induced increases in the level of retroviral-specific mRNA.

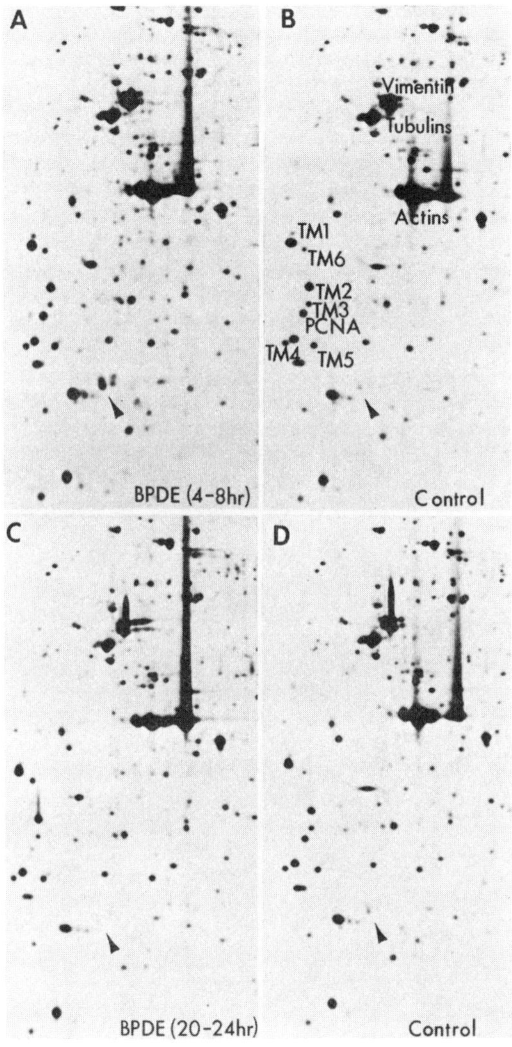

Fig. 5. Protein patterns for control and BPDE-treated REF52 cells. Proliferating REF52 cells were treated with 0.5 µg/ml BPDE for 2 hr and radiolabeled with [^{35}S]methionine from either 4 to 8 hr (Panel A), or 20 to 24 hr (Panel C). Untreated cells (Panels B and D) were labeled during the same intervals as BPDE-treated cells. The patterns shown are portions of pH 3.5 to 10.0, 10% acrylamide gels loaded with 180,000 dpm and exposed to film for 3 weeks. Indicated proteins are the tropomyosin forms (TM1-TM6) and proliferating cell nuclear antigen (PCNA). The upward pointing arrow in the lower left side of each panel points toward Carcinogen Inducible Protein 1 (CIN 1).

In light of the above findings, and in order to explore in greater detail the relationship between heat stress and the expression of retroviral- like elements, we have designed a model in vitro system to study the transcriptional responsiveness of the *D. melanogaster copia* retroviral-like element to heat shock, after introduction into a heterologous system.

In order to investigate whether *copia* retained its transcriptional responsiveness to heat shock after introduction into a mammalian cell system, we developed rat fibroblast cell lines stably transformed by a full-length 5.0-kb proviral copy of this element. The *copia*-containing plasmid cDM5002 (63) was co-transfected with pSV2-gpt (53) into the rat fibroblast cell line Rat 6, and cultures were selected for growth of Gpt^+ colonies. After subcloning, the chromosomal DNA from a number of these clones was isolated and checked for the presence of integrated *copia* DNA. A number of *copia*-positive colonies, designated Cl. 1, 2, and 3, were then grown in mass culture for RNA analysis.

The results of experiments to determine the transcriptional responsiveness of these clones to two cellular stress regimens are shown in Fig. 6. The first treatment involved addition of bromodeoxyuridine (BrdUrd) (50 µg/ml for 48 hr) to the culture medium. We included this drug since previous studies had indicated that this regimen of BrdUrd induces increased

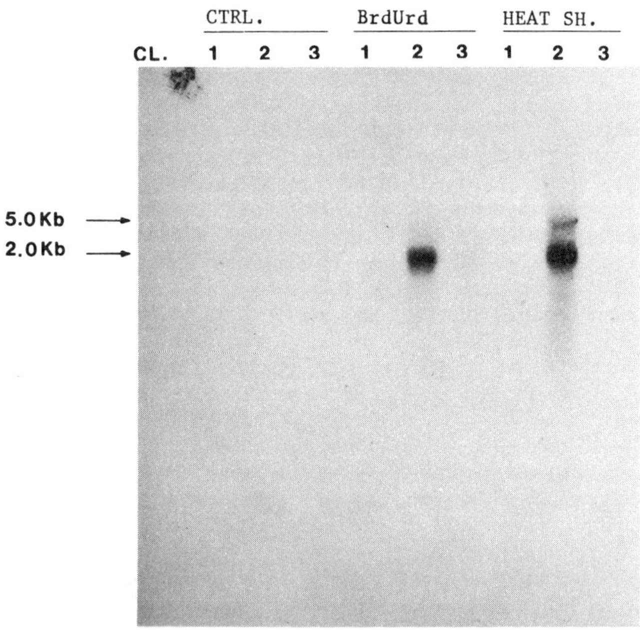

Fig. 6. Expression of *copia* after stable introduction into Rat 6 cells. Each lane contains 10 µg of Poly A^+ RNA isolated from *copia* transformed Gpt^+ Rat 6 Cl. 1, 2, and 3 cells, electrophoresed in 1% agarose gels, and blot-hybridized with a *copia* LTR-specific probe derived from the plasmid cDM5002 (63). CTRL, untreated cells collected at 48 hr after plating; BrdUrd, cells treated with bromodeoxyuridine (50 µg/ml) for 48 hr; HEAT SH, cells heat-shocked at 42.5°C for 30 min and processed immediately thereafter.

transcription of another retroviral element, Moloney murine leukemia virus, in vitro, in mouse cells (47). Timing of the heat-shock regimen (shift up from 37.5°C to 42.5°C for 30 min) was derived from in vivo experiments in Drosophila in which a 30-min heat shock (shift up from 28°C to 37.5°C) was found to induce the maximal increase in copia-specific mRNA.

Using a 5' long terminal repeat (LTR)-specific probe, we observed only an extremely low level of specific hybridization to PolyA$^+$ RNA in untreated Cl. 1, 2, or 3 cells. In contrast, in Cl. 2 cells, we observed the appearance of LTR-specific transcripts after both treatment with BrdUrd and heat shock. In the heat-shocked Cl. 2 cells, we observed the induction of two LTR-specific transcripts, 5.0 and 2.0 kb, which correspond in length to the two major copia-specific transcripts induced in vivo in Drosophila (73). The predominance of the smaller 2.0 kb LTR-specific transcript also parallels the profile observed under stress in vivo in Drosophila.

Since the increase in the steady state level of copia LTR-specific mRNAs after heat shock in Cl. 2 cells could be due to a decrease in mRNA stability, and not reflect a primary increase in the rate of initiation of transcription, we undertook to directly assess the transcriptional responsiveness of the copia LTR region, which contains promoter and transcriptional control sequences (73), utilizing a recombinant plasmid in which the LTR has been linked to the bacterial indicator gene, chloramphenicol acetyl transferase (CAT) (16). The results of experiments to assay the heat-shock responsiveness of the copia-CAT construct in transient expression assays in Rat 6 cells are shown in Fig. 7. Cells were heat-shocked for various times and in some instances permitted to recover at 37°C before assaying for the level of CAT enzyme. As in the in vivo induction profile in Drosophila, the major increase occurs at 30 min and has returned to the baseline line by 120 min. In both cases the induction is highly ephemeral and limited to a specific narrow time window. The molecular basis for this transient phenomenon is not known, but may reflect the induction of a short-lived heat-shock transcription factor (57), induced specifically in response to heat shock or heat-shock mimetic regimens, which is capable of interacting with the heat-shock consensus promoter sequence (59), which we have identified as being present within the copia LTR sequence (73).

DISCUSSION

We have presented evidence that specific types of cellular stress, including DNA damage and heat shock, are capable of inducing changes in the replicative and transcriptional state of certain papovaviral- and retroviral-like elements in mammalian cells. These effects are significant, not only because they are markers for inducible cellular responses to stress which can function in trans, but because they can contribute, through viral gene amplification and increased transposition, respectively, to the creation of genetic variation that is quantitatively and qualitatively unlike that produced by point mutation alone.

The ability of chemical carcinogens to induce a cellular response which enhances the constitutive process of polyoma viral DNA synthesis in transformed rat cells may also be of relevance in understanding the molecular and cellular basis of cellular gene amplification, including amplification and subsequent over-expression of genes involved in growth control, such as oncogenes and growth factor receptor genes. One possibility is that cellular genes located within responsive replicons might undergo asynchronous replication and amplification. Pretreatment of mouse cells with

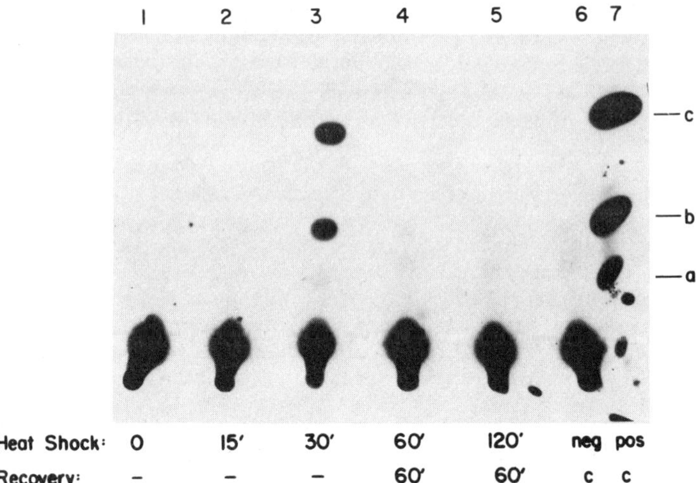

Fig. 7. Effect of heat shock on copia expression in Rat 6 cells in transient expression assays. In each lane 1 x 10^6 Rat 6 cells were transfected with 10 μg of Copia-CAT (16). Copia-CAT was derived by inserting a 1.66-kb EcoRl-Apal fragment from cDM5002 containing the 5' LTR and 1.25 kb of 5' cellular flanking sequences, after modification of the ends, into the Smal site of pSVO-CAT (16). Forty-eight hours later the cells were subjected to heat shock (42.5°C) for the indicated periods, followed in some instances by recovery at 37°C. Cultures were then processed for the level of CAT activity, as described by Gorman et al. (28). Neg C, incubation of ^{14}C-chloramphenicol alone; pos C, incubation of purified CAT enzyme alone in a standard reaction mixture; a, l-acetyl-chloramphenicol; b, 3-acetyl-chloramphenicol; c, 1,3-diacetyl-chloramphenicol.

ultraviolet light enhances the yield of methotrexate-resistant colonies that carry amplified dhfr genes (74), and Lavi (43) found that carcinogens induce transient amplification of the dhfr gene at both the in situ and extrachromosomal levels in CHO cells. Even if carcinogen-induced asynchronous DNA synthesis is largely restricted to certain viral sequences, however, it may be relevant to multistage and multifactor carcinogenesis. In the induction of certain tumors, for example, carcinogens may act synergistically with both DNA and RNA tumor viruses, by producing a change in the state of integration or the copy number of these viruses within the cell (19).

The transcriptional responsiveness of certain retroviral-like elements, such as the copia element of D. melanogaster, to specific types of cellular stress, including heat shock and oxidative stress, represents a distinct and novel mechanism for creation of genetic variation. An increase in the frequency of retroviral transposition secondary to an increase in retroviral expression (3) could result in dramatic phenotypic alterations. Retrovirus-like elements such as copia could function as a mobile class of heat-shock promoters and, through promoter insertion adjacent to specific cellular loci, could result in the acquisition of novel cis- and trans-acting regulatory controls at these loci (50).

A better understanding of the initiating action of chemical carcinogens and of genetic and karyotypic instability during tumor progression may require, therefore, a more detailed molecular understanding of inducible responses to cellular stress, including both DNA damage and heat shock.

ACKNOWLEDGEMENTS

This research was supported by National Cancer Institute Grant CA 02111 to I.B.W., CA 13106 and RR 02188 to J.I.G., and GM 07088 to M.L.

REFERENCES

1. Ames, B.N., W.E. Durston, E. Yamasaki, and F.D. Lee (1973) Carcinogens are mutagens: A simple test system combining liver homogenates for activation and bacteria for detection. Proc. Natl. Acad. Sci., USA 70:2281-2285.
2. Ashburner, M., and J.J. Bonner (1979) The induction of gene activity in Drosophila by heat shock. Cell 17:241-254.
3. Baltimore, D. (1985) Retroviruses and retrotransposons: The role of reverse transcription in shaping the eukaryotic genome. Cell 40:481-482.
4. Basilico, C., D. Zouzias, G. Della Valle, S. Gattoni, V. Colantuoni, R. Fenton, and L. Dailey (1980) Integration and excision of polyoma virus genomes. Cold Spring Harbor Symp. Quant. Biol. 44:611-620.
5. Bensaude, O., C. Babinet, M. Morange, and F. Jacob (1983) Heat shock proteins, first major products of zygotic gene activity in mouse embryo. Nature 305:331-332.
6. Berenblum, I. (1982) Sequential aspects of chemical carcinogenesis: Skin. In Cancer: A Comprehensive Treatise, Vol. 1, F.F. Becker, ed. Plenum Press, New York.
7. Boehm, L.J., and D. Drahovsky (1983) Alteration of enzymatic methylation of DNA cytosines by chemical carcinogens: A mechanism involved in the initiation of carcinogenesis. J. Natl. Cancer Inst. 71:429-434.
8. Boeke, J.D., D.J. Garfinkel, C.A. Styles, and G.R. Fink (1985) Ty elements transpose through an RNA intermediate. Cell 40:491-500.
9. Botchan, M., W. Topp, and J. Sambrook (1978) Studies on SV40 excision from cellular chromosomes. Cold Spring Harbor Symp. Quant. Biol. 43:709-719.
10. Capon, D.J., E.Y. Chen, A.D. Levinson, P.H. Seeburg, and D.V. Goeddel (1983) Complete nucleotide sequences of the T24 human bladder carcinoma oncogene and its normal homologue. Nature 302:33-37.
11. Collins, A. (1983) DNA repair in ultraviolet-irradiated HeLa cells is disrupted by aphidicolin. Biochim. Biophys. Acta 741:341-347.
12. Coohill, T.P., and S.P. Moore (1983) An SV40 mammalian inductest for putative carcinogens. Mutat. Res. Lett. 113:431-440.
13. Cornelis, J.J., Z.Z. Su, and J. Rommelaire (1982) Direct and indirect effects of ultraviolet light on the mutagenesis of parvovirus H-1 in human cells. EMBO J. 1:693-699.
14. Criessen, D., and S. Shall (1982) Regulation of DNA ligase activity by poly (ADP) ribose. Nature 296:271-272.
15. DasGupta, U.B., and W.C. Summers (1978) Ultraviolet re-activation of herpes simplex virus is mutagenic and inducible in mammalian cells. Proc. Natl. Acad. Sci., USA 75:2378-2381.

16. Di Nocera, P.P., and I.B. Dawid (1983) Transient expression of genes introduced into cultured cells of Drosophila. Proc. Natl. Acad. Sci., USA 80:7095-7098.
17. Doll, R., and R. Peto (1978) Cigarette smoking and bronchial carcinoma: Dose and time relationships among regular and lifelong non-smokers. J. Epidem. Comm. Health 32:303-313.
18. Farzaneh, F., R. Zalin, D. Brill, and S. Shall (1982) DNA strand breaks and ADP-ribosyl transferase activation during cell differentiation. Nature 300:362-366.
19. Fisher, P.B., and I.B. Weinstein (1980) Chemical-viral interactions and multistep aspects of cell transformation. In Molecular and Cellular Aspects of Carcinogen Screening Test, R. Montesanto, L. Bartsch, and L. Tomatis, eds. IARC Scient. Publ., 37, Lyon, France.
20. Fogel, M. (1972) Induction of viral synthesis in polyoma transformed cells by DNA anti-metabolites and by irradiation after pre-treatment by 5-bromodeoxyuridine. Virology 49:12-22.
21. Franza, B.R., and J.I. Garrels (1984) Transformation-sensitive proteins of REF52 cells detected by computer-analyzed two-dimensional gel electrophoresis. In Cancer Cells I/ The Transformed Phenotype, A. Levine, G.F. Vande Woude, W.C. Topp, and J.D. Watson, eds. Cold Spring Harbor Laboratory, Cold Spring Harbor, New York, pp. 137-145.
22. Garrels, J.I. (1979) Two dimensional gel electrophoresis and computer analysis of proteins synthesized by clonal cell lines. J. Biol. Chem. 254:7961.
23. Garrels, J.I. (1983) Quantitative two-dimensional gel electrophoresis of proteins. Methods Enzymol. 100:411-417.
24. Garrels, J.I., J.T. Farrar, and C.B. Burwell (1984) The QUEST system for computer analyzed two-dimensional gel electrophoresis of proteins. In Two-dimensional Gel Electrophoresis of Proteins: Methods and Applications, J.E. Celis and R. Bravo, eds. Academic Press, New York, pp. 37-91.
25. Gattoni, S., V. Colantuoni, and C. Basilico (1980) Relationship between integrated and non-integrated viral DNA in rat cells transformed by polyoma virus. J. Virol. 34:615-626.
26. Gattoni-Celli, S., W.L. Hsiao, and I.B. Weinstein (1983) Re-arranged c-mos locus in a MOPC-21 murine myeloma cell line and its persistence in hybridomas. Nature 306:795-796.
27. Goldschmidt, R., and L.K. Peternick (1947) The genetic background of chemically induced phenocopies in Drosophila. J. Exp. Zool. 135:127-202.
28. Gorman, C.M., L.F. Moffat, and B.H. Howard (1982) Recombinant genomes which express chloramphenicol acetyltransferase in mammalian cells. Mol. Cell. Biol. 2:1044-1051.
29. Groudine, M., and S. Collins (1982) Amplification of endogenous myc-related DNA sequences in a human myeloid leukemia cell line. Nature 298:679-681.
30. Grunberger, D., and R.M. Santella (1983) Conformational changes in DNA induced by chemical carcinogens. In Genes and Proteins in Oncogenesis, I.B. Weinstein and H. Vogel, eds. Academic Press, New York, pp. 13-40.
31. Guerrero, I., A. Villasante, V. Corces, and A. Pellicer (1984) Activation of a c-K-ras oncogene by somatic mutation in mouse lymphomas induced by gamma radiation. Nature 304:1159-1162.
32. Ivanovic, V., and I.B. Weinstein (1980) Genetic factors in Escherichia coli that affect cell killing and mutagenesis induced by benzo(a)pyrene 7,8-dihydrodiol,9,10-oxide. Cancer Res. 40:3508-3511.
33. Ivarie, R.D., and J.A. Morris (1982) Induction of prolactin-deficient variants of G3 rat pituitary tumor cells by ethyl methanesulfonate:

Reversion by 5-azacytidine, a DNA methylation inhibitor. Proc. Natl. Acad. Sci., USA 79:2967-2970.

34. Jeffrey, A.M., T. Kinoshita, R.M. Santella, D. Grunberger, L. Katz, and I.B. Weinstein (1980) The chemistry of polycyclic aromatic hydrocarbon DNA adducts. In Carcinogenesis: Fundamental Mechanisms and Environmental Effects, B. Pullman, P.O.P. Ts'O, and H. Gelboin, eds. Reidel Publishing Co., Amsterdam, pp. 565-579.

35. Jones, C., and R.T. Su (1982) DNA polymerase alpha from the nuclear matrix of cells infected with simian virus 40. Nucl. Acids Res. 10: 5517-5532.

36. Kelley, P.M., and M.J. Schlesinger (1982) Antibodies to two major chicken heat shock proteins cross-react with similar proteins in widely divergent species. Mol. Cell. Biol. 2:267-274.

37. Kennedy, A.R. (1985) Evidence that the first step leading to carcinogen induced malignant transformation is a high frequency, common event. In Cell Transformation Assays: Application to Studies of Mechanisms of Carcinogenesis and to Carcinogen Testing, J.C. Barrett, ed. Raven Press, New York (in press).

38. Kirschmeier, P., S. Gattoni-Celli, D. Dina, and I.B. Weinstein (1982) Carcinogen and radiation transformed C3H 10T1/2 cells contain RNA's homologous to the long terminal repeat sequences of a murine leukemia virus. Proc. Natl. Acad. Sci., USA 79:2773-2777.

39. Kohl, N.E., N. Kanda, R.R. Schreck, G. Bruns, S. Latt, F. Gilbert, and F.W. Alt (1983) Transposition and amplification of oncogene related sequences in human neuroblastomas. Cell 35:359-367.

40. Kunz, B.A. (1982) Genetic effects of deoxyribonucleotide pool imbalances. Environ. Mutag. 4:695-725.

41. Lambert, M.E., S. Gattoni-Celli, P. Kirschmeier, and I.B. Weinstein (1983) Benzo(a)pyrene induction of extrachromosomal viral DNA synthesis in rat cells transformed by polyoma virus. Carcinogenesis 4:587-593.

42. Lavi, S. (1981) Carcinogen-mediated amplification of viral DNA sequences in simian virus 40 transformed Chinese hamster embryo cells. Proc. Natl. Acad. Sci., USA 78:6144-6148.

43. Lavi, S., N. Kohn, T. Kleinberger, Y. Berko, and S. Etkin (1983) Amplification of SV40 and cellular genes in SV40 transformed Chinese hamster embryo cells treated with chemical carcinogens. In Cellular Responses to DNA Damage, E. Friedberg and B. Bridges, eds. Alan R. Liss, Inc., New York, pp. 659-670.

44. Li, G.C., N.S. Petersen, and H.K. Mitchell (1982) Induced thermal tolerance and heat shock protein synthesis in CHO cells. Int. J. Rad. Oncol. Biol. Phys. 8:63-67.

45. Lieberman, M.W., L.R. Beach, and R.D. Palmiter (1983) Ultraviolet radiation-induced metallothionein-I gene activation is associated with extensive DNA demethylation. Cell 35:207-214.

46. Lowe, D.G., W.D. Fulford, and L.A. Moran (1983) Mouse and Drosophila genes encoding the major heat shock protein (hsp70) are highly conserved. Mol. Cell. Biol. 3:1540-1543.

47. Lowy, D.R., W.P. Rowe, N. Teich, and J. Hartley (1971) Murine leukemia virus: High frequency activation in vitro by 5-iododeoxyuridine and 5-bromodeoxyuridine. Science 174:155-156.

48. Matz, B., J.R. Schlehofer, and H. zurHausen (1984) Identification of a gene function of Herpes Simplex Virus Type 1 essential for amplification of simian virus 40 DNA sequences in transformed hamster cells. Virology 134:328-337.

49. McClanahan, T., and K. McEntee (1984) Specific transcripts are elevated in Saccharomyces cerevisiae in response to DNA damage. Mol. Cell. Biol. 4:2356-2363.

50. McDonald, J. (1983) The molecular basis of adaptation: A critical review of relevant ideas and observations. Ann. Rev. Ecol. Systm. 14: 77-102.
51. Miskin, R., and E. Reich (1980) Plasminogen activator: Induction of synthesis by DNA damage. Cell 19:217-224.
52. Mitchell, R.E.J., and D.P. Morrison (1982) Heat shock induction of ionizing radiation resistance in Saccharomyces cerevisiae and correlation with stationary growth phase. Radiat. Res. 90:284-291.
53. Mulligan, R.C., and P. Berg (1981) Selection for animal cells that express the Escherichia coli gene coding for xanthine-guanine phosphoribosyltransferase. Proc. Natl. Acad. Sci., USA 7:2072-2076.
54. Nomura, S., and M. Oisha (1984) UV irradiation induces an activity which stimulates simian virus 40 rescue upon cell fusion. Mol. Cell. Biol. 4:1159-1162.
55. Notario, V., S. Sukumar, E. Santos, and M. Barbacid (1984) A common mechanism for the malignant activation of ras oncogenes in human neoplasia and in chemically induced animal tumors. In Cancer Cells, Vol. II, G. Vande Woude and A.J. Levine, eds. Cold Spring Harbor Laboratory, Cold Spring Harbor, New York, pp. 425-432.
56. Paquin, C.E., and V.M. Williamson (1984) Temperature effects on the rate of Ty transposition. Science 226:53-55.
57. Parker, C.S., and J. Topol (1984) A Drosophila RNA polymerase II transcription factor binds to the regulatory site of an hsp 70 gene. Cell 37:273-283.
58. Pearson, M.N., J.J. Karchesy, A. Deeney, M.L. Dwinzer, and G.S. Beaudreau (1984) Induction of endogenous avian tumor virus gene expression by pyrrolizidine alkaloids. Chem. Biol. Interact. 49:341-350.
59. Pelham, H.R.B. (1982) A regulatory upstream promoter element in the Drosophila hsp70 heat shock gene. Cell 3:517-528.
60. Pellegrini, S., L. Dailey, and C. Basilico (1984) Amplification and excision of integrated polyoma DNA sequences require a functional origin of replication. Cell 36:943-949.
61. Peterson, A.R., and V. Klement (1985) Oncogenic transformation of C3H/10T1/2 cells: Methods and mechanisms. In Cell Transformation Assays, J.C. Barrett and R. Tennant, eds. Raven Press, New York (in press).
62. Pfister, H., A. Gassenmaier, F. Nurnbergen, and G. Stuggen (1983) HPV-5 DNA in a carcinoma of an epidermodysplasia verruciformis patient infected with various human papillomavirus types. Cancer Res. 43:1436-1441.
63. Potter, S.S., W.J. Brorein, Jr., P. Dunsmuir, and G.M. Rubin (1979) Transposition of elements of the 412, copia and 297 dispersed repeated gene families in Drosophila. Cell 17:415-427.
64. Prasad, I., D. Zouzias, and C. Basilico (1976) State of the viral DNA in rat cells transformed by polyoma virus I. Virus rescue and the presence of non-integrated viral DNA molecules. J. Virol. 18:436-444.
65. Prieto-Soto, A., B. Gourlie, M. Miwa, V. Pigiet, T. Sugimura, N. Malik, and M. Smulson (1983) Polyoma virus minichromosomes: Poly ADP-ribosylation of associated chromatin proteins. J. Virol. 45:600-606.
66. Rahmsdorf, H.J., U. Mallick, H. Ponta, and P. Herrlich (1982) A B-lymphocyte specific high turnover protein: Constitutive expression in resting B cells and induction of synthesis in proliferating cells. Cell 29:459-468.
67. Rechavi, G., A. Givol, and L. Canaani (1982) Activation of a cellular oncogene by DNA re-arrangement: Possible involvement of an IS-like element. Nature 300:607-611.
68. Rubin, J.S., V.R. Prideaux, H.F. Willard, A.M. Dulhanty, G.F. Whitmore, and A. Bernstein (1985) Molecular cloning and chromosomal local-

ization of DNA sequences associated with a human DNA repair gene. Mol. Cell. Biol. 5:398-405.

69. Ryo, H., M.A. Yoo, K. Fujikawa, and S. Kondo (1985) Comparison of somatic reversions between the ivory allele and transposon-caused mutant alleles at the white locus of Drosophila melanogaster after larval treatment with X-rays and ethyl methanesulfonate. Genetics 110:441-451.
70. Sarasin, A., F. Bourre, and A. Benoit (1982) Error prone replication of ultraviolet-irradiated simian virus 40 in carcinogen treated monkey kidney cells. Biochemie 64:815-821.
71. Schimke, R.T. (1984) Gene amplification in cultured animal cells. Cell 37:705-713.
72. Schorpp, M., U. Mallick, H.J. Rahmsdorf, and J. Herrlich (1984) UV-induced extracellular factor from human-fibroblasts communicates the UV response to non-irradiated cells. Cell 37:861-868.
73. Strand, D.J., and J.F. McDonald (1985) Copia is transcriptionally responsive to environmental stress. Nucl. Acids Res. 13:4401-4410.
74. Tlsty, T.D., P.C. Brown, and R.T. Schimke (1984) UV radiation facilitates methotrexate resistance and amplification of the dihydrofolate reductase gene in cultured 3T6 mouse cells. Mol. Cell. Biol. 4:1050-1056.
75. Van der Ploeg, L.H.T., S.H. Giannini, and C.R. Cantor (1985) Heat shock genes: Regulatory role for differentiation in parasitic protozoa. Science 228:1443-1446.
76. Velazquez, J.M., and S. Lindquist (1984) Hsp70: Nuclear concentration during environmental stress and cytoplasmic storage during recovery. Cell 36:655-662.
77. Weinstein, I.B. (1981) Current concepts and controversies in chemical carcinogenesis. J. Supramol. Struct. Cell. Biochem. 17:99-120.
78. Weinstein, I.B., J. Arcoleo, M. Lambert, W. Hsiao, S. Gattoni-Celli, A. Jeffrey, and P. Kirschmeier (1985) Molecular mechanisms of multistage chemical carcinogenesis. In Molecular Biology of Tumor Cells, B. Wahren et al., eds. Raven Press, New York, pp. 55-70.
79. Weinstein, I.B., S. Gattoni-Celli, P. Kirschmeier, M. Lambert, W. L.-W. Hsiao, J. Backer, and A. Jeffrey (1984) Multistage carcinogenesis involves multiple genes and multiple mechanisms. In Cancer Cells I/The Transformed Phenotype, A. Levine, G.F. Vande Woude, W.C. Topp, and J.D. Watson, eds. Cold Spring Harbor Laboratory, Cold Spring Harbor, New York, pp. 229-237.
80. Wilson, V.L., and P.A. Jones (1983) Inhibition of DNA methylation by chemical carcinogens in vitro. Cell 32:329-346.
81. Witkin, E. (1985) The SOS response: Implications for cancer. In Genes and Cancer, J.M. Bishop, J. Rowley, and M. Greaves, eds. Alan R. Liss, Inc., New York, pp. 99-115.
82. Yavelow, J., M. Collins, Y. Birk, W. Troll, and A.R. Kennedy (1985) Nanomolar concentrations of Bowman-Birk soybean protease inhibitor suppress X-ray induced transformation in vitro. Proc. Natl. Acad. Sci., USA (in press).
83. Yu, L.Z., J.J. Cornelis, J.M. Vos, and J. Rommelaire (1982) U.V.-enhanced reactivation of capsid protein synthesis and infectious centre formation in mouse cells infected with U.V.-irradiated minute-virus-of-mice. Int. J. Radiat. Biol. 41:119-126.
84. Zimmerman, W., and A. Weissbach (1981) Participation of deoxyribonucleic acid polymerase-alpha in amplifications of ribosomal deoxyribonucleic acid in Xenopus laevis. Mol. Cell. Biol. 1:680-686.
85. Zuker, C., J. Capello, R.L. Chisholm, and H. Lodish (1983) A repetitive Dictyostelium gene family that is induced during differentiation and by heat shock. Cell 34:997-1005.

SINGLE-BASE MUTATIONS ASSOCIATED WITH MOUSE LYMPHOMAS

I. Guerrero, A. Villasante, L.E. Diamond, E.W. Newcomb, J.W. Berman, R. Lake, L. McMorrow, and A. Pellicer

Department of Pathology and Kaplan Cancer Center
New York University Medical Center
New York, New York 10016

INTRODUCTION

A few years ago we began to study a model system for carcinogenesis for two main reasons: first, because the molecular tools to approach the system were starting to become available and, second, because we thought that oncogene studies in humans would eventually need a model system since the development of human tumors is not easy to manipulate experimentally and cannot be studied in depth. With that in mind, we began to study the induction of mouse lymphomas after treatment with a chemical carcinogen, nitrosomethyl urea (NMU), or with gamma irradiation.

We chose two well-known protocols for induction of mouse lymphomas. One is the Kaplan fractionated gamma radiation scheme (8), while the other is also an established protocol for NMU tumor formation (7) (Fig. 1). In both cases, the mice developed thymic lymphomas with up to 90% incidence. In NMU induction, the latency period is a little shorter than in radiation, but when the tumors develop, the histological characteristics are identical.

DETECTION OF ONCOGENES

We decided to do what, at the time, was being initiated in the human system, which was to look for activation of oncogenes. The assay more readily available was foci formation in cultured rodent fibroblasts. Figure 2A illustrates how a culture of normal NIH 3T3 cells looks under microscopic observation. The cells have a flat morphology, are attached to the substrate, and are also contact-inhibited. When DNA was taken from the tumors, co-precipitated with calcium phosphate, and introduced into the culture (2,17), some of the cells took up DNA; when the DNA contained an oncogene, the cell developed into a transformed focus (3,12). The oncogene changed the morphology of these cells. This is illustrated in Fig. 2B, which shows that the cells are rounder, more refractile, and somewhat detached from the plate, in comparison with the ones in the periphery, which are the normal counterparts not affected by the uptake of the oncogene.

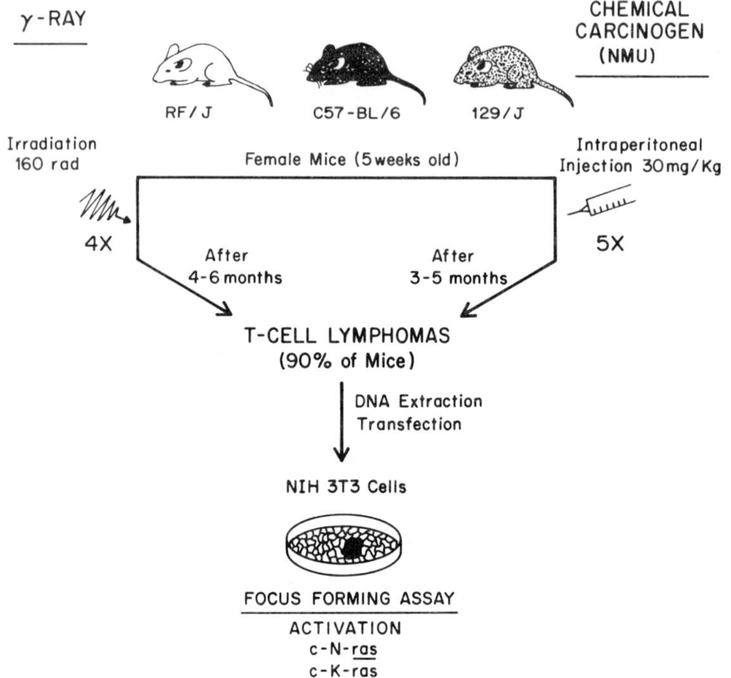

Fig. 1. Transforming potential of thymic lymphomas induced in several mouse strains by two different agents. (Reprinted from Leukemia Research with permission from Pergamon Press, Oxford.)

An even more definitive test for oncogenicity is the introduction of these cells into nude mice. Although parental cells of the same number do not cause any alteration, the transformed cells produce large tumors which appear very quickly (within a week after the inoculation) and eventually kill the animal. Nevertheless, it must be emphasized that the tumors do not frequently metastasize.

IDENTIFICATION OF THE ONCOGENES

Up to this point, we knew that these tumors contained some activated oncogene which produces tumors in nude mice, and that these tumors were certainly malignant. However, we did not know the identity of this oncogene. At the time, in the human system, it was starting to become apparent that when the focus formation assay was positive, the genes most likely involved were members of the ras family. The ras family is mainly composed of 3 oncogenes called H-ras, K-ras, and N-ras. The H-ras and K-ras genes were initially identified in viruses isolated from rat tumors (10,13), and the N-ras gene was first identified in a human neuroblastoma cell line by Shimizu et al. (14). The product of each of these genes is called p21. Although the genes are structurally different, the protein product of all three genes is quite similar.

Using DNA probes for these ras genes, we screened the transformants. In contrast to the investigators working with human tumors, we were faced with an additional problem since they were using human DNA to transform

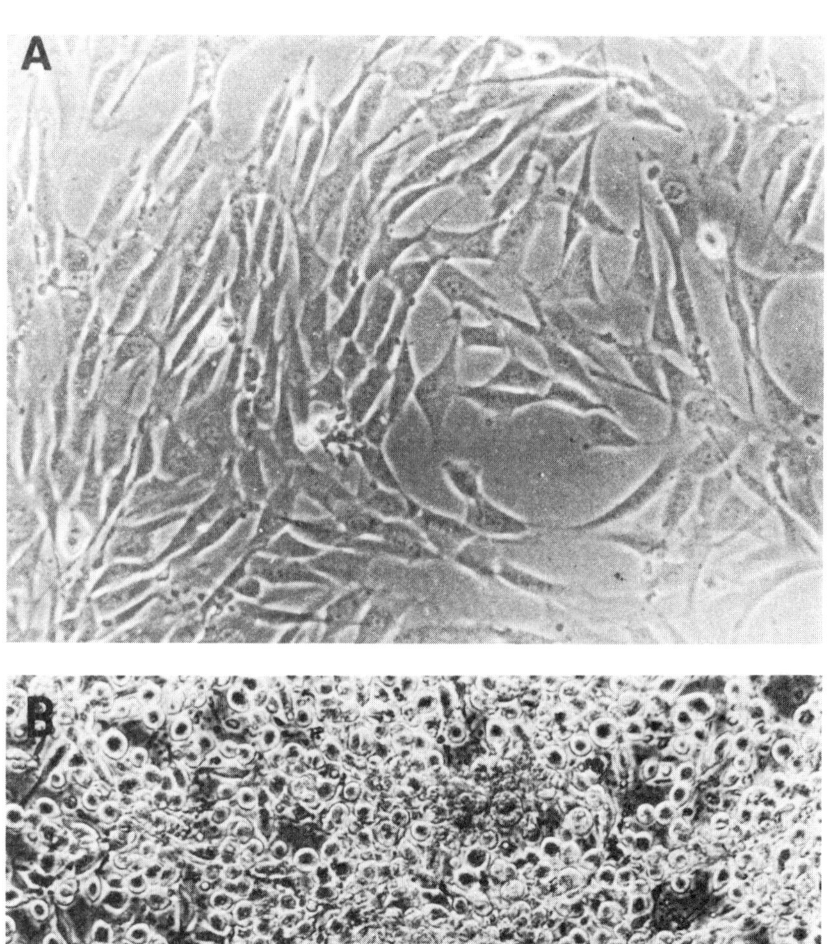

Fig. 2. Normal and transformed 3T3 cells. Panel A: 100X magnification of normal NIH 3T3 cells. Panel B: 100X magnification of transformed foci obtained by treating normal 3T3 cells with DNA from a chemically induced thymic lymphoma.

mouse cells while we were using an intraspecies DNA transfer. Since we had to look for the introduction of additional oncogene copies of mouse origin into a mouse cell, it was difficult to identify the transfer. We were looking for amplified bands or for something more definitive, such as an extra band. We interpreted the presence of new bands as coming from the additional oncogene that, upon introduction into the mouse genome and subsequent rearrangement which usually occurs during gene transfer, had acquired a different restriction site which was apparent in the Southern analysis.

To reaffirm this point, we took these DNAs and transformed rat cells. In the rat transformants, we obtained the rat pattern plus the superimposed mouse bands; this indicated that the oncogenic phenotype was segregating together with the presence of these mouse ras bands (3). To confirm that these were the oncogenes specifically activated in the system, we also did a northern analysis of these transformants.

We showed that in normal cells N-ras gives mainly 2 transcripts, whereas we observed an enormous increase in the transcription of N-ras in the transformed cells which were already assigned as being N-ras transformants. A parallel observation was made in the case of K-ras-activated clones (3). It was important to show that there was increased expression of these oncogenes at the protein level; moreover, there also seemed to be a change in p21 mobility compared with the homologous protein from the parental cell line (3). This latter observation suggested that some structural difference was present in the polypeptide with oncogenic potential.

A recapitulation of the data summarized shows that we have used 3 different strains of mice (RF/J, C57BL/6, and 129/J) and have treated them with 2 inducing agents (NMU and γ-rays). We obtained thymic lymphomas in 90% of the mice. DNA extracted and assayed in the 3T3 focus formation experiments scored positive 50% of the time. We found N-ras and K-ras to be the oncogenes activated in these tumors (Fig. 1). Thus far, we have analyzed over 50 tumors and we have not found any of them with an activated H-ras, which is the third member of the ras family. This suggests some kind of tissue specificity, because in other systems, such as mouse skin or rat breast carcinomas, H-ras is the only gene found to be activated.

MECHANISM OF ACTIVATION

We knew which oncogenes were responsible for focus formation and for oncogenic potential, but we also wanted to know the mechanism of activation. In order to analyze mechanisms of activation we had to clone the oncogenes. For the K-ras gene we postulated, due to the results in the human system (15), that activation would occur in the first exon. Therefore, we identified which fragment contained the first exon and isolated it both from normal and from transformed cells. Here normal DNA means DNA from the brain of the mouse that developed the tumor, and transformant DNA means DNA from a focus obtained in rat cells. When the sequence of the first exon from the normal gene, isolated from the brain of the animal that developed the tumor, and the gene isolated from the transformant were compared, they only differed by a single base, a G → A change in the 12th codon of the Kirsten ras gene. This substitution changed the normal amino acid, glycine, to aspartic acid (4).

Since we wanted to analyze all the tumors, the easiest way to do that, without having to sequence the genes individually from each of the tumors, would be to find a polymorphism for a restriction enzyme that would distinguish this particular mutation, a G → A substitution in the second base of the 12th codon. Unfortunately, no enzyme was available that recognized the normal or the mutated sequence that would be appearing in the tumor. For this reason we had to resort to another technique, namely the oligonucleotide mismatch hybridization technique (9).

In this particular case, we used a "nineteenmer" oligonucleotide which was labeled at the 5' end with ^{32}P by a kinase reaction. This oligonucleotide was exactly complementary to the sequence of the mutated gene identified as such in the previous experiments. In this technique, it is most important to work out the proper hybridization and washing conditions to distinguish among the DNAs present in the gel in order to differentiate the sequence that is absolutely identical to the one in the oligonucleotide from the one that has a single base pair difference. The base that we wanted to probe was located in the middle of the nineteenmer. The system is very sensitive because even when up to 100 copies of the cloned gene are analyzed it is still possible to clearly distinguish the one that has the appropriate sequence from the one that has only one base difference.

As shown in Fig. 3, two different classes of mutations were identified using this powerful approach. That was important to us for several reasons: (a) it meant that we could now screen numerous other tumors, and (b) since experiments were done using DNA from transformants, we could rule out the possibility that the mutation had occurred during gene transfer or in the gene manipulations for cloning. These experiments demonstrated the presence of the indicated alterations in tumor DNAs (4).

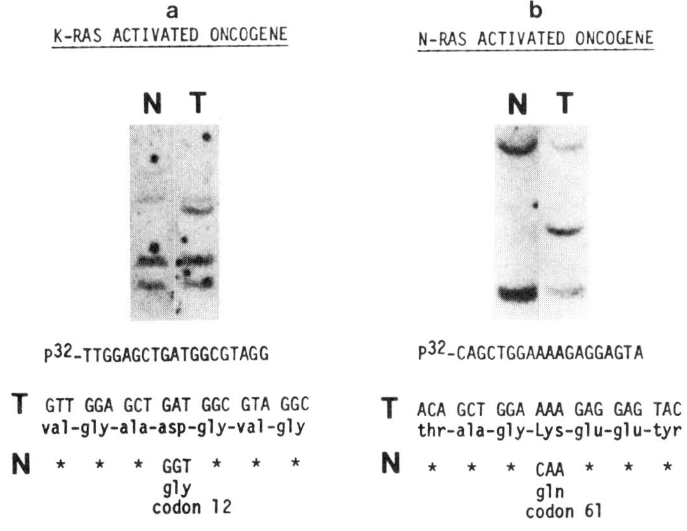

Fig. 3. Detection of single-base mutations by oligonucleotide mismatch hybridization. DNAs from tumors and brains were digested with restriction endonucleases, electrophoresed in an agarose gel, and hybridized with an oligonucleotide, the sequence of which is shown below the panels. Panel A: Mutation in codon 12 of the K-ras gene. Panel B: Mutation in codon 61 of the N-ras gene. N: Brain DNA. T: Tumor DNA.

ALLELIC COMPOSITION

We will now discuss the other gene that is activated in this system, namely, the N-ras gene. In humans the N-ras gene is smaller than K-ras (it is about 10 kb), so it was possible to isolate N-ras in an intact form using a λ-phage library. When we isolated the clone, we used the gene transfer assay to prove that the gene was functional. When the efficiency of transformation was compared between the clone DNA and the DNA from the original tumor, we found that there was a difference of 5 orders of magnitude in favor of the cloned gene (5). This means that the gene in question has actually been isolated, since a mammalian genome is 5 orders of magnitude more complex than the DNA of a 20-kb clone.

We next sequenced the four coding exons. We found only a single base change: a C → A transversion in the first base of codon 61 of the N-ras gene. This base change would change glutamine to lysine in the protein. No other base changes were found when the DNA from the transformant was compared to the DNA of the brain of the mouse that developed the tumor (6). Again, we looked for a restriction enzyme polymorphism that would give us the possibility of screening all the other tumors with just Southern analyses. Unfortunately, that was not the case, and so we again had to rely on the oligonucleotide mismatch hybridization technique in our analyses.

When we probed the different brains and tumors, we found that the only tumor that actually contained the mutation complementary to the oligonucleotide we made was the tumor that we had already cloned and sequenced (6) (Fig. 3). All the other tumors tested were negative, in spite of the fact that we knew from all the other criteria that they contained activated N-ras genes.

An interesting finding from these experiments was that when we probed the same DNAs with the oligonucleotide that is exactly complementary to the normal sequence in this portion of the gene, we observed that the tumor with the C → A mutation lacked the normal allele. Thus, it looked as though the normal allele was deleted in this particular tumor. From densitometric experiments, we concluded that the mutated gene was present in a diploid form; so there were probably 2 copies of the mutated gene and none of the normal gene in this tumor (6). It is possible that the deletion or absence of the normal allele might give these tumors an additional growth advantage. This is interesting because, so far, the ras genes have been considered to be primarily dominant-acting genes.

A summary analysis of several of these tumors shows that, in the tumors in which the K-ras gene has been identified as activated, 17 out of 18 contained the same mutation as identified in the original activated K-ras gene, which is the G → A mutation in the second base of the 12th codon, changing glycine to aspartic acid.

In the N-ras gene we do not have enough examples to compile statistics but the opposite picture seems to be emerging. We have identified one tumor that has a C → A mutation in codon 61 and two tumors that have a G → A mutation in the 12th codon. There are still another three with as yet unidentified mutations. The three tumors certainly do not contain the mutations mentioned above since we have probed them with the same oligonucleotides. Thus, in the activation of the N-ras oncogene, we are finding that different tumors contain different mutations.

CHROMOSOMAL ALTERATIONS

We also wanted to analyze the chromosomal composition of these tumors in comparison to the chromosomal composition of preleukemic lymphoma tissues. In a very preliminary study, we found that instead of 40 chromosomes, the normal complement for mice, there was some aneuploidy in the tumor tissues. This aneuploidy frequently was directed towards the presence of 41 chromosomes. When karyotypes were obtained, we saw trisomy for chromosome 15. Trisomy for chromosome 15 has been reported in similar systems by George Klein and by other investigators (1,16). Additionally, we saw the presence of a new marker chromosome, the prevalence of which is currently being investigated.

INTRODUCTION OF THE ACTIVATED ONCOGENE INTO MICE

The main problem in assessing the role of the *ras* oncogene in tumorigenesis is the difficulty in distinguishing between a causal event and a secondary event. One of the best ways to address this issue is to reintroduce the mutated oncogene into target cells of a live animal to see if tumors appear in the predicted fashion.

Since the cells that are believed to be the targets for thymic lymphoma development originate in the bone marrow, these are the best recipients for gene transfer experiments. From a technical point of view this represents a formidable task, because only 0.01% of the potential bone marrow recipients are stem cells. In addition, the efficiency of transfer using classical methods, such as calcium phosphate, is, at maximum, only around 1%. Therefore, the chances of success using this approach seem minimal.

On the other hand, efficient retroviral vectors provide an excellent approach to the problem. For this reason, we made the appropriate constructs with the mutated and normal mouse *N-ras* genes using a vector derived from the Moloney leukemia virus, namely, tll-pVHC, kindly provided by S. Goff.

We introduced the genes into 3T3 cells and showed that they were expressed, and that the transforming construct also induced foci with very high efficiency. To avoid the presence of helper virus, we obtained psi-2 cells (11) from R. Mann and transfected them with the constructs. We are now analyzing different transformed clones. Many of the clones are producing virus as judged by the ability of their supernatants to produce transformed foci in 3T3 cells in an infection protocol (Fig. 4). We are optimizing the conditions to undertake the bone marrow infection experiments to try to answer the central question: is the mutated *N-ras* gene sufficient to cause thymic lymphomas?

CONCLUSIONS

We have been working with a model system to induce thymic lymphomas in mice by NMU and gamma radiation; DNA from these tumors oncogenically transforms cultured rodent fibroblasts. The genes responsible for this transformation are the *N-ras* and *K-ras* genes. We have cloned them and found the alterations to be point mutations. The mutations in the *K-ras* gene are very consistent with G → A in the second base of the 12th codon. Meanwhile, in *N-ras* we found a variety of mutations. Chromosomal studies show trisomy 15 and some other abnormalities which we are exploring. We

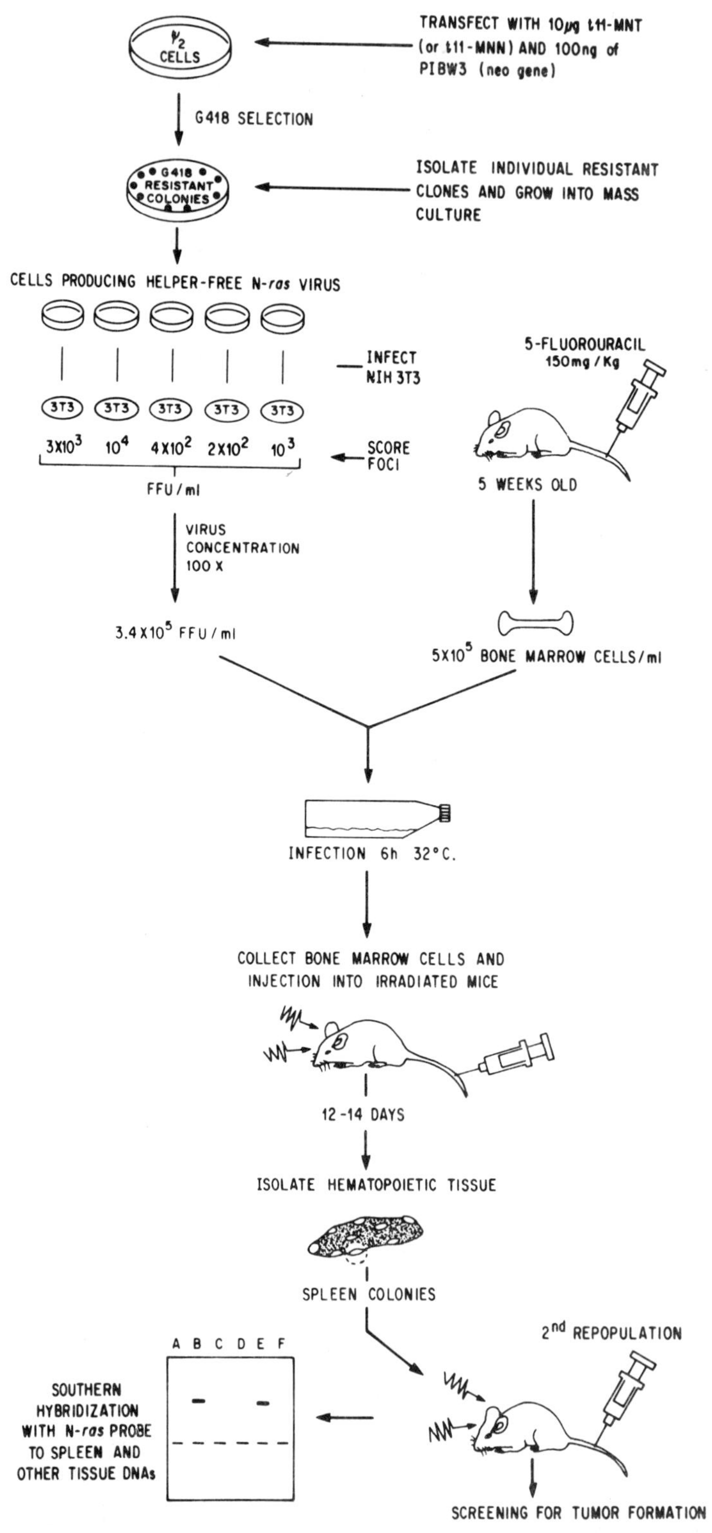

ψ2 CELLS
TRANSFECT WITH 10μg t11-MNT (or t11-MNN) AND 100ng of PIBW3 (neo gene)
G418 SELECTION
G418 RESISTANT COLONIES
ISOLATE INDIVIDUAL RESISTANT CLONES AND GROW INTO MASS CULTURE
CELLS PRODUCING HELPER-FREE N-ras VIRUS
INFECT NIH 3T3
3T3
3X10^3 10^4 4X10^2 2X10^2 10^3
SCORE FOCI
FFU/ml
5-FLUOROURACIL 150mg/Kg
5 WEEKS OLD
VIRUS CONCENTRATION 100 X
3.4X10^5 FFU/ml
5X10^5 BONE MARROW CELLS/ml
INFECTION 6h 32°C.
COLLECT BONE MARROW CELLS AND INJECTION INTO IRRADIATED MICE
12-14 DAYS
ISOLATE HEMATOPOIETIC TISSUE
SPLEEN COLONIES
2nd REPOPULATION
A B C D E F
SOUTHERN HYBRIDIZATION WITH N-ras PROBE TO SPLEEN AND OTHER TISSUE DNAs
SCREENING FOR TUMOR FORMATION

←

Fig. 4. Cloning of the N-ras oncogene in a retroviral vector and its introduction into mouse bone marrow cells.

have cloned the N-ras oncogene and its normal counterpart in a retroviral vector. We hope that the use of these constructs will facilitate critical investigation of the precise role of ras oncogenes in the origin of these tumors.

ACKNOWLEDGEMENTS

This research has been supported by NIH grants CA36327 and CA16239. I.G. is a Fulbright Fellow, L.D. was supported by NIH grant CA09161, and A.P. is a recipient of an Irma T. Hirschl-Monique Weill/Caulier Award.

REFERENCES

1. Dofuku, T., J.L. Biedler, B.A. Spengler, and L.J. Old (1975) Trisomy of chromosome 15 in spontaneous leukemia of AKR mice. Proc. Natl. Acad. Sci., USA 72:1515.
2. Graham, F.L., and A.J. van der Eb (1973) A new technique for the assay of infectivity of human adenovirus 5 DNA. Virology 52:456.
3. Guerrero, I., P. Calzada, A. Mayer, and A. Pellicer (1984) A molecular approach to leukemogenesis: Mouse lymphomas contain an activated c-ras oncogene. Proc. Natl. Acad. Sci., USA 81:202.
4. Guerrero, I., A. Villasante, V. Corces, and A. Pellicer (1984) Activation of a c-k-ras oncogene by somatic mutation in mouse lymphomas by γ-radiation. Science 225:1159.
5. Guerrero, I., A. Villasante, P. D'Eustachio, and A. Pellicer (1984) Isolation, characterization, and chromosome assignment of mouse N-ras gene from carcinogen-induced thymic lymphoma. Science 225:1041.
6. Guerrero, I., A. Villasante, V. Corces, and A. Pellicer (1985) Loss of the normal N-ras allele in a mouse thymic lymphoma induced by a chemical carcinogen. Proc. Natl. Acad. Sci., USA 82:7810.
7. Joshi, V.V., and J.V. Frei (1970) Effects of dose and schedule of methylnitrosourea on incidence of malignant lymphoma in adult female mice. J. Natl. Cancer Inst. 45:335.
8. Kaplan, H.S. (1967) On the natural history of the murine leukemias: Presidential address. Cancer Res. 27:1325.
9. Kidd, V.J., R.B. Wallace, K. Itakura, and S.L.C. Woo (1983) α_1-Antitrypsin deficiency detection by direct analysis of the mutation in the gene. Nature 304:230.
10. Langebeheim, H., T.Y. Shih, and E.M. Scolnick (1980) Identification of a normal vertebrate cell protein related to the p21 Src of Harvey murine sarcoma virus. Virology 106:292.
11. Mann, R., R.C. Mulligan, and D. Baltimore (1983) Construction of a retrovirus packaging mutant and its use to produce helper free defective retrovirus. Cell 33:153.
12. Shih, C., B.-Z. Shilo, M.P. Goldfarb, A. Dannenberg, and R.A. Weinberg (1979) Passage of phenotypes of chemically transformed cells via transfection of DNA and chromatin. Proc. Natl. Acad. Sci., USA 76:5714.

13. Shih, T.Y., D.R. Williams, M.O. Weeks, J.M. Maryak, W.C. Vass, and E.M. Scolnick (1978) Comparison of the genomic organization of Kirsten and Harvey sarcoma viruses. J. Virol. 27:45.
14. Shimizu, K., M. Goldfarb, M. Perucho, and M. Wigler (1983) Isolation and preliminary characterization of the transforming gene of a human neuroblastoma cell line. Proc. Natl. Acad. Sci., USA 80:383.
15. Varmus, H.E. (1984) The molecular genetics of cellular oncogenes. Ann. Rev. Genet. 18:553.
16. Wiener, F., J. Spira, S. Ohuno, N. Haran-Ghera, and G. Klein (1978) Chromosome changes (trisomy 15) in murine T-cell leukemia induced by 7,12-dimethylbenz(a)anthracene. Int. J. Cancer 22:447.
17. Wigler, M., A. Pellicer, S. Silverstein, R. Axel, G. Urlaub, and L. Chasin (1979) DNA mediated transfer of the adenine phosphoribosyltransferase locus into mammalian cells. Proc. Natl. Acad. Sci., USA 76:1373.

THE MOUSE MUTANT "WASTED": AN ANIMAL MODEL FOR ATAXIA-TELANGIECTASIA

Tadashi Inoue,[1] Hideo Tezuka,[1] Tsuneo Kada,[1] Katsuhiro Aikawa,[2] and Leonard D. Shultz[3]

[1]National Institute of Genetics
Mishima, Shizuoka-ken 411, Japan

[2]Institute for Animal Industry
Kukizaki, Ibaraki-ken 305, Japan

[3]The Jackson Laboratory
Bar Harbor, Maine 04609

INTRODUCTION

Ataxia-telangiectasia (AT) is a human autosomal recessive disease characterized by an increased predisposition to cancer, a progressive neurological disorder, a marked immune deficiency, and a spontaneous chromosomal instability. The frequency of AT heterozygotes in the human population has been estimated to be around 0.01. Considering that the heterozygotes are predisposed to cancer (23), it seems very important, from both a therapeutic and a preventive point of view, to elucidate the molecular mechanisms controlled by the AT gene that determine the susceptibility to cancer.

Hypersensitivity of cells from AT patients to ionizing radiation, expressed as reduced viability and increased cytogenetic damage in cultured fibroblasts, has suggested that defective DNA repair occurs in this disease (for review, see Ref. 2 and 17). Ionizing radiations produce many kinds of lesions in DNA, including strand breaks and modifications of base and sugar moieties. However, the living cell can usually repair most of these lesions efficiently.

We previously carried out a series of systematic biochemical studies on the repair mechanisms operating on DNA damage induced by ionizing radiation. Analysis of the biochemical requirements for DNA repair in bacterial cells indicated that some repair processes are necessary prior to those involving DNA polymerase and DNA ligase (13). Further studies revealed the existence of enzymes (primer activating enzymes) that convert some of the gamma-ray-induced lesions to effective priming sites for repair replication by DNA polymerase (14). These processes are schematically shown in Fig. 1.

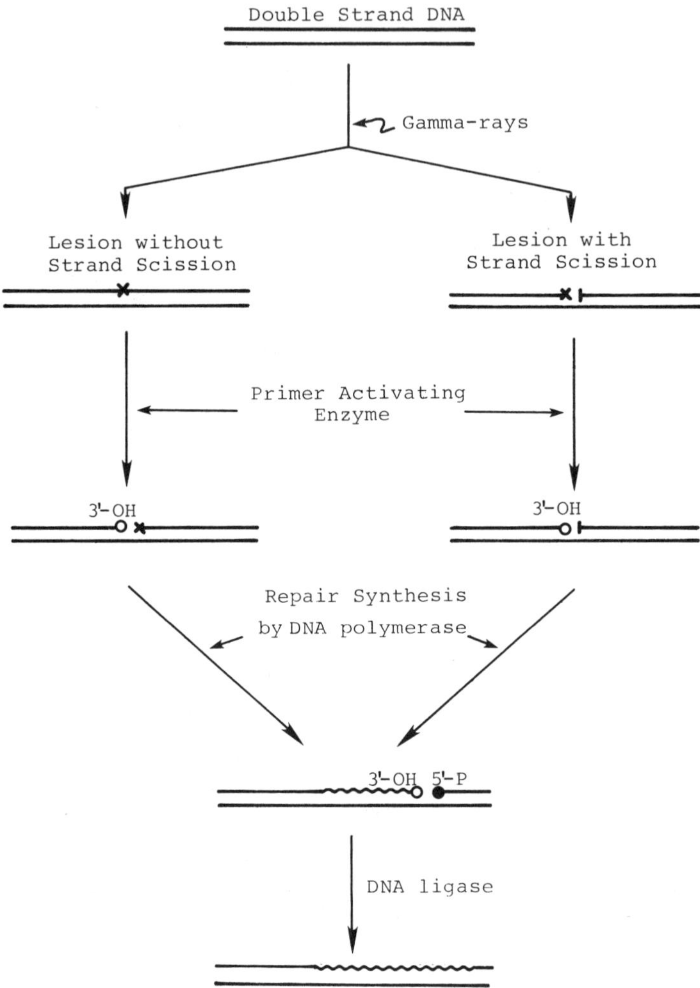

Fig. 1. Schematic representation of the action of the primer activating enzymes. The cross represents the lesion without a 3'-OH end, and the open circle represents a 3'-OH end. For details, see text.

Using the assay system for the primer activating enzyme, we have successfully identified in bacterial extracts 2 kinds of enzymes that may function in the repair of gamma-irradiated DNA. One of them was shown to be an apurinic/apyrimidinic site-specific endonuclease (AP endonuclease) (8), and the other was a "cleaning" exonuclease (6). We recently applied this assay system to studies on the repair properties of AT cells and found them to have a decreased activity of the primer activating enzyme (7,9,10). Similar results were independently reported by Edwards et al. (4).

Although the clinical and laboratory manifestations of AT have been rather well documented, the underlying mechanisms of this disease have not been well understood. Many aspects of this disease, including impairment of tissue differentiation, are not amenable to direct study of skin fibroblasts derived from patients; thus, an animal model for AT would be very

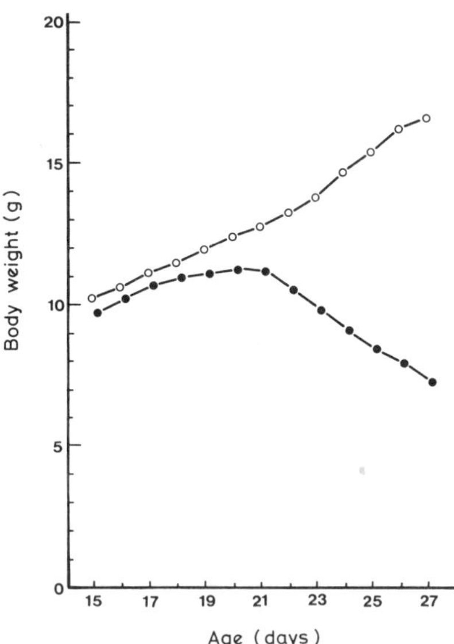

Fig. 2. Body weight gain of wasted and control mice. Mice were maintained in a heterozygous state in a closed colony system. Homozygotes (wst/wst) were first recognized at 20 to 22 days of age by low body weights and neurological abnormalities, including tremor and uncoordinated body movements. This recognition was later confirmed by absence of body weight gain and progressive paralysis. Affected homozygotes died at 28 to 30 days of age. Mice with a normal appearance in the same litter with affected homozygotes served as controls. Open circles, control mice (7 males and 6 females); closed circles, wasted mice (4 males and 4 females).

valuable in elucidating the underlying mechanisms and in developing effective preventive measures.

Recently, Shultz et al. (21) described a mouse mutant, "wasted" (wst), which shows pathological changes in both the central nervous and lymphoid systems, and exhibits increases in the frequency of spontaneous, as well as gamma-ray-induced, chromosomal aberrations. Since these symptoms are very similar to those of AT, we conducted cytogenetic and biochemical experiments to evaluate the wasted mouse as a model for AT. The results of the present study indicate that the wasted mouse may provide a useful model for investigating the relationships among neuropathology, immune deficiency, and defective DNA metabolism, although it may not be an exact animal homolog for AT.

EXPERIMENTAL RESULTS

Tissue-Specific and Age-Dependent Expression of the wst Gene

Body weights and organ weight ratios. A significant difference in mean body weight between wasted mice and controls became apparent at 23 days of age ($P < 0.05$) and continued from 24 to 27 days of age ($P < 0.001$) (Fig. 2). The difference in organ weight changes between wasted and con-

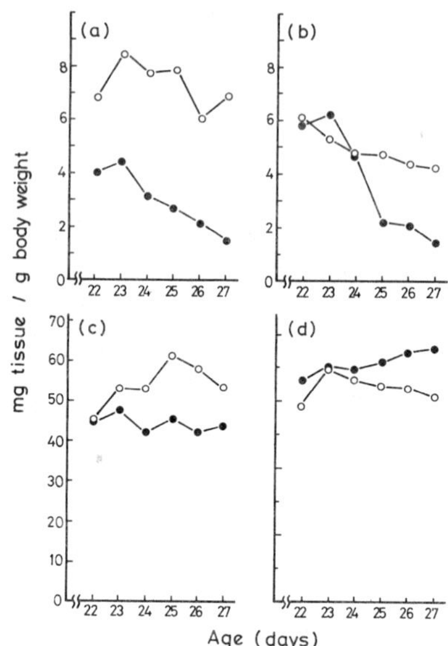

Fig. 3. Ratios of organ weights to body weights in wasted and control mice. Each point represents the mean value obtained from 5 to 23 animals. a, spleen; b, thymus; c, liver; d, kidney. Open circles, control mice; closed circles, wasted mice.

trol mice was evident when the ratios of organ weights to body weights were analyzed. As shown in Fig. 3a, the ratio of spleen to body weight in wasted mice was significantly lower than that in control mice ($P < 0.001$), even at 22 days of age. This ratio further decreased with age in wasted mice, while in controls it remained constant. The thymus-to-body weight ratio gradually decreased in controls (Fig. 3b), while the weights themselves were constant, suggesting that hyperplastic development may stop and natural involution may begin in this organ in control animals. The ratio of thymus to body weight in wasted mice was not lower than that in controls before 24 days of age, but showed a significant decrease after 25 days of age ($P < 0.01$).

In contrast to the above-mentioned lymphoid organs, organ-to-body weight ratios of livers and kidneys showed no marked changes in *wst/wst* mice (Fig. 3c and 3d). The value for the livers in wasted mice was constant, but significantly lower than that in controls, after 23 days of age ($P < 0.01$). The value for kidneys was rather high when compared to controls.

From these data, we suggest that, in wasted homozygotes, the effects of the *wst* gene may appear in specific organs, such as the spleen or the thymus, in an age-dependent manner. Alterations observed in the liver and in the kidney may be ascribed to secondary effects such as reduced food and water intake.

Chromosomal aberrations. The frequency of spontaneous chromosomal aberrations in wasted mice was not significantly different from that in controls at 22 days of age (Tab. 1). The difference in aberration frequency increased significantly and age-dependently at and after 24 days of age

($P < 0.05$). The chromosomal aberrations observed in the wst/wst homozygotes were chromatid gaps, breaks, fragments, and acentric fragments. There were 2 abnormal cells with chromosome-type aberrations in wasted mice, one each at 22 and 24 days of age, suggesting the possible existence of abnormal clones in homozygotes.

The frequency of chromosomal aberrations in bone marrow cells of irradiated wasted mice at 22 days of age was 0.6 per cell, which was comparable to that in age-matched controls (Tab. 1). There were few chromatid-type aberrations at this age in cells from either wasted mice or controls. The frequency of this type of aberration in homozygotes, however, increased markedly in an age-dependent manner after 24 days of age, and at 26 days of age it was 13-fold greater than at 22 days of age. The types of aberrations were chromatid and isochromatid gaps, breaks, fragments, and exchanges (mainly triradials). In contrast to the homozygotes, the aberration frequency in control groups remained rather constant.

Repair enzymes. Primer activating enzyme activity in cellular extracts of spleens and livers from both wasted and control mice at various ages is shown in Fig. 4. As clearly seen in this figure, activities in liver cells of wasted mice were similar to those in controls from 22 days to 27 days of age. In spleen cells from wasted mice, little, if any, activity was detected at 27 days of age, although levels of activity comparable to those in controls were detected at younger stages. In contrast,

Tab. 1. Number of spontaneous and gamma-ray-induced chromosomal aberrations in bone marrow cells of wasted mice as a function of age.

Age (days)	Animal group	Number of metaphases observed	Number of aberrations	
			Chromatid-type	Chromosome-type
	Nonirradiated groups			
22	Control	400	7	0
	Wasted	400	9	9
24	Control	400	7	0
	Wasted	400	33	7
26	Control	400	10	0
	Wasted	400	42	0
	2.0 Gy-irradiated groups			
22	Control	200	4	78
	Wasted	200	26	99
24	Control	200	78	127
	Wasted	200	160	107
26	Control	200	29	92
	Wasted	200	330	69

Note: Groups of mice were irradiated with 2.0 Gy of gamma-rays, and sacrificed 24 hr later. The mice were injected with 4 mg colchicine/kg for 2.0 to 2.5 hr prior to sacrifice. Bone marrow cells were harvested and prepared with a routine air-drying method including hypotonic treatment with 0.075 M KCl, followed by fixation with a mixture of acetic acid and methanol (1:3). Slides were stained with 2.5% Giemsa solution.

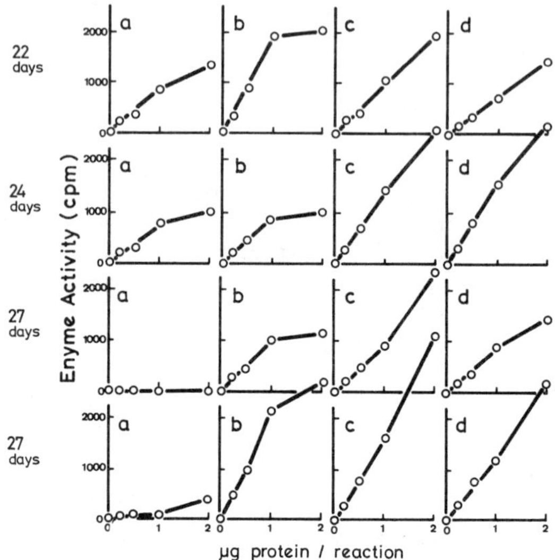

Fig. 4. Primer activating enzyme activity in the cellular extract of spleens and livers from wasted and control mice. The extracts were prepared from the organs at the indicated age. The assays were carried out using colicin E1 DNA as a template as described previously (6,7). a, wasted mice spleens; b, control mice spleens; c, wasted mice livers; d, control mice livers.

differences in apurinic DNA endonuclease activity between wasted and control animals could not be detected in either organ throughout the developmental stages examined (Fig. 5). This shows that the decrease in the activity of the primer activating enzyme in the spleen cells of wasted mice at 27 days of age is not due to nonspecific inactivation of the cellular activity.

Repair Properties of Isolated Spleen Cells

As described above, the spleen cells of wasted mice seemed to have defects in DNA metabolism. We therefore examined other biochemical properties of the spleen using isolated cells.

DNA synthesis inhibition. The extent of DNA synthesis in spleen cells from wasted and littermate control mice after exposure to increasing concentrations of bleomycin for 2 hr is shown in Fig. 6a. Under these conditions, DNA synthesis in normal spleen cells was inhibited to 20-23% of the control value after exposure to the drug at 200 µg/ml, whereas in spleen cells from wasted mice, synthesis was inhibited only to 65-70% of the control value. Similar results were obtained with another DNA-damaging chemical, 4-nitroquinoline 1-oxide (4NQO) (Fig. 6b). In Fig. 6c, the effect of gamma-irradiation on DNA synthesis is shown. Although the difference in sensitivity between the two types of cells was not as large as in the case of bleomycin or 4NQO, a significantly reduced DNA synthesis inhibition in wasted mouse cells was reproducibly observed.

In order to test whether the diminished DNA synthesis inhibition in spleen cells of wasted mice occurs when the cells are exposed to other types of DNA-damaging agents, we examined the effects of UV-light on DNA

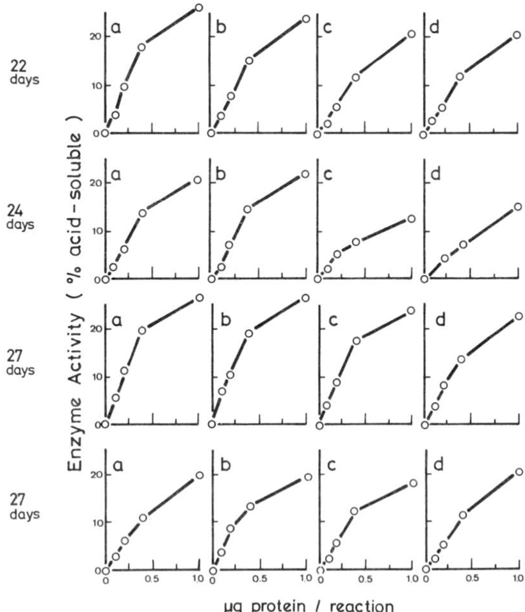

Fig. 5. Apurinic DNA endonuclease activity in the cellular extract of spleens and livers from wasted and control mice at various ages. Extracts to be assayed were portions of those prepared for the assay of the primer activating enzyme. Endonuclease activity was measured as the capacity to introduce nicks specific for apurinic sites in heavily depurinated DNA resulting in the release of acid-soluble materials. The assay conditions were the same as those described elsewhere (8). a, wasted mice spleens; b, control mice spleens; c, wasted mice livers; d, control mice livers.

synthesis. In contrast to the results obtained with gamma-irradiation or with the above-noted chemicals, an equal extent of DNA synthesis inhibition was observed in cells from wasted mice and from controls after exposure to UV-light (Fig. 6d), suggesting that any anomaly in DNA metabolism of the spleen cells of wasted mice was specific for certain types of DNA damage.

DNA repair synthesis after bleomycin-treatment. DNA repair synthesis in isolated spleen cells was measured by the bromodeoxyuridine density label method which enables the separation of DNA synthesis due to repair from that due to extensive chain elongation by semiconservative replication. Measurements of repair synthesis, which is determined as a specific activity of DNA of normal density, are shown in Fig. 7. This figure clearly demonstrates that the cells from wasted mice are quite similar to those from normal mice with respect to their ability to repair damages produced by bleomycin. The trace amount of incorporation in the cells without bleomycin treatment may be ascribed to DNA damage produced during the isolation and/or incubation of the spleen cells in vitro.

Examination of Gamma-Ray-Sensitivity of Wasted Mouse Cells

Hypersensitivity of fibroblasts to the lethal effect of gamma-rays is one of the most characteristic phenotypes of AT. Representative experiments performed to compare the sensitivity of lung fibroblasts of wasted

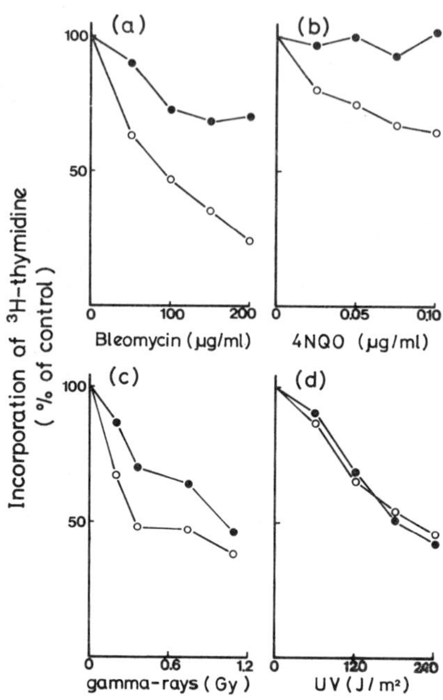

Fig. 6. Effect of various kinds of DNA-damaging agents on DNA synthesis in isolated spleen cells of wasted and control mice at 24 days of age. For the preparation of isolated cells, the spleens were disrupted by passing them through a fine stainless steel mesh with a rubber policeman, and the cells were gently suspended, with a Pasteur pipette, in Eagle's minimal essential medium (EMEM). The cells were washed twice with EMEM by centrifugation and finally suspended in EMEM. The isolated spleen cells were treated with DNA-damaging agents in one of 3 ways: (i) exposure for 2 hr at 37°C to various concentrations of bleomycin or 4NQO; (ii) exposure to gamma-rays of various doses; or (iii) exposure to UV light of various doses from a germicidal lamp. Following treatment, the cells were incubated with [^{3}H]thymidine for 20 min and chilled in an ice-water bath. Duplicate 40-µl samples from the reaction mixture were withdrawn and spotted on glass filters which had been soaked with 1% SDS. The filters were washed with 5% trichloroacetic acid, dried, and counted with a liquid scintillation spectrometer. The ratio of radioisotope incorporation in treated and corresponding control samples was taken as a measure of the inhibition of DNA synthesis. a, bleomycin; b, 4NQO; c, gamma-rays; d, UV light. Open circles, control mice; closed circles, wasted mice.

mice with that of normal mice are shown in Fig. 8. As shown in this figure, a difference in sensitivity to the killing effect of gamma-rays was not detectable between the two types of cells that had been derived from mice at 24 or 26 days of age, at which time degeneration of spleen cells as well as elevated frequency of chromosomal aberrations in bone marrow cells became apparent in wasted mice (see Fig. 2 and 3). Lung fibroblasts from younger (22-day) animals also showed no difference compared with control fibroblasts in their sensitivity to gamma-rays (data not shown).

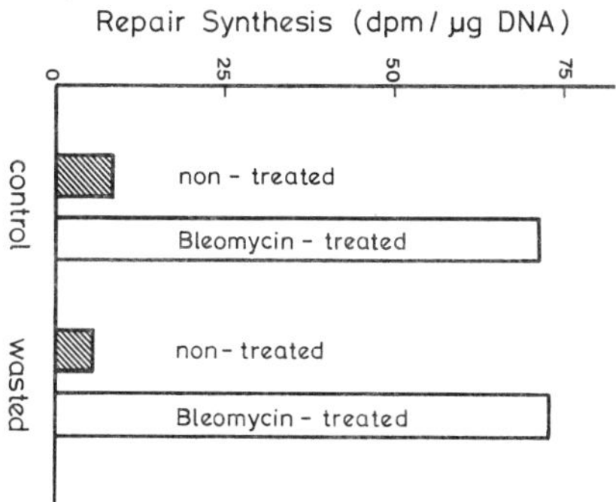

Fig. 7. DNA repair synthesis in isolated spleen cells from wasted and normal mice at 24 days of age after treatment with bleomycin. Spleen cells were preincubated with 10 µM of bromodeoxyuridine for 10 min at 37°C. After the preincubation, bleomycin and [^{3}H]thymidine were added, and the cells were incubated at 37°C for another 2 hr. Cellular DNA was purified by extraction with chloroform/phenol followed by column chromatography on hydroxyapatite and then subjected to a centrifugation in alkaline CsCl gradients. The amount of repair synthesis was determined from patterns of alkaline CsCl density equilibrium centrifugation by quantitating the amounts of [^{3}H]thymidine radioactivity and DNA under the normal density DNA peak (peak fraction only).

DISCUSSION

The results of the present study clearly show the pleiotropic effects of the wasted mutation on various tissues in affected homozygotes. Analysis of changes in organ-to-body weight ratios showed abnormal development of certain tissues in an age-dependent manner. The order of the degree of change was from thymus and spleen (severe degeneration), to liver (mild), and to kidney (seemingly not affected). The marked changes in lymphoid organs observed in the present study confirmed previous results by Shultz et al. (21).

In human AT, hypoplasia or aplasia of thymus and/or lymph nodes, and mild dysfunction of liver in some cases, have been observed (11,18). Such degeneration in human organs is similar to that observed in the present study on wasted mice. These changes in lymphoid organs of AT patients may be closely related to deficiencies found in both the cellular and humoral immune systems (Ref. 11; for review, see Ref. 25).

Concerning the molecular mechanism of the wasted mutation in mice, evidence from a preliminary cytogenetic study by Shultz et al. (21) suggests an underlying DNA repair deficiency. The present study revealed age-dependent increases in the frequency of spontaneous chromosomal aberrations in the bone marrow cells of affected homozygote mice. After exposure of cultured AT cells to ionizing irradiation of radiomimetic chemicals, a high frequency of chromatid-type aberrations has been observed. In wasted mice, the frequency of this type of aberration after irradiation showed age-dependent increases becoming 13-fold of that in normal mice. These data

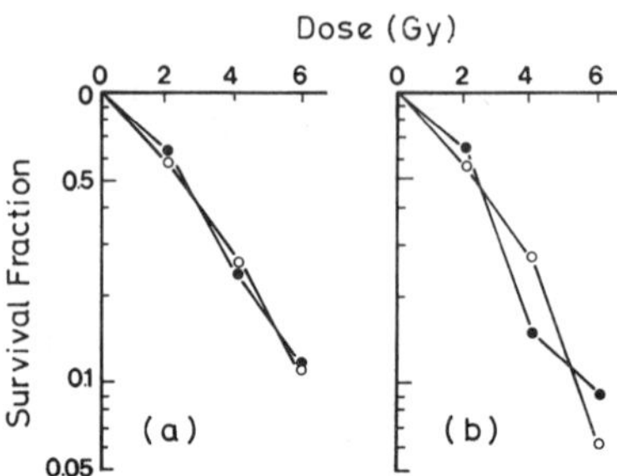

Fig. 8. Survival of gamma-irradiated primary lung fibroblasts from wasted and control mice. Cultures of primary fibroblasts were initiated from lungs of mice at 24 days (a) and 26 days (b) of age. Minced lung tissue was digested with collagenase. Cells were suspended by pipetting, washed by centrifugation, and incubated for several days in the presence of 20% fetal bovine serum. Propagated cells were collected by trypsinization and exposed to gamma-irradiation. Colonies were visually scored by staining with Giemsa after 4 weeks incubation. Open circles, control mice; closed circles, wasted mice.

suggest that a DNA repair deficiency similar to that in AT cells, causing specific chromosomal breakage after irradiation, may appear in an age-dependent manner in wasted mice.

The absence of the normal inhibition of replicative DNA synthesis following gamma-irradiation (5,22,26) or treatment with radiomimetic drugs, including bleomycin and neocarzinostatin (3,12,15,19), is an inherent characteristic of AT cells. This may have some relationship with the cytological observation by Zampetti-Bosseler and Scott (27) that AT cells exhibited less mitotic delay after X-irradiation than did normal cells. A possibility has been suggested that the primary defect in A-T is a disturbance in the regulation of DNA replication on damaged templates (26).

As described in the section "Experimental Results," DNA synthesis is more resistant to gamma-irradiation in wasted mouse spleen cells than in control cells. Bleomycin was also less effective in inhibiting DNA synthesis in the wasted mouse cells than in control cells. On the other hand, UV light, to which AT cells exhibit normal responses, did not distinguish wasted mouse cells from those of normal animals as to their sensitivity to DNA synthesis. These results indicate that any defect in DNA metabolism of spleen cells of wasted mice may be specific for damage induced by gamma-rays or by an agent that mimics gamma-rays. Our results with 4NQO, which is primarily a UV-mimetic DNA-damaging agent, are different from those of Smith and Paterson (22) who found that this agent inhibited DNA synthesis in AT and normal cells to a similar extent. This discrepancy may occur because 4NQO does not necessarily produce a unique lesion, but rather multiple types of damage to DNA, some of which may resemble those produced by gamma-irradiation.

The assays of the repair enzymes also revealed an age-dependent appearance of the effect of the wasted mutation. The primer activating enzyme, which enhances the priming activity of gamma-irradiated DNA for purified DNA polymerase, represents one of the cellular functions of DNA repair (13,14). Edwards et al. (4) and Inoue et al. (9) reported that activity of this enzyme in extracts from AT cells was lower than that in normal fibroblasts, indicating directly that one of the causal factors of AT is a DNA repair defect. As described in the section "Experimental Results," the enzyme activity in spleen cells of wasted mice was also significantly lower at 27 days of age. However, we could not detect a decrease in the enzyme activity at 24 days of age when the elevated frequencies of both spontaneous and gamma-ray-induced chromosomal aberrations were observed. This may be because we prepared the extracts from whole organs that contained cells at various degenerative stages.

Another line of evidence that supports the idea of abnormal repair in AT cells is that the cells exhibit reduced repair replication following treatment with DNA-damaging agents including gamma-rays (1,16,24). Scudiero (20) reported that N-methyl-N'-nitro-N-nitrosoguanidine-stimulated repair synthesis was also reduced in some, but not all, AT fibroblast lines compared with controls. In the present report, the data from experiments designed to determine the level of repair replication after treatment with bleomycin showed no difference between spleen cells from wasted and control mice (Fig. 7). As described, the primer activating enzyme activity decreased age-dependently in wasted mouse spleen; however, the decline of the activity at 24 days of age, when the spleen cell preparation examined in the present study was prepared, was negligible. A decrease in repair replication, therefore, may be detected in cells isolated from older animals.

One of the other important approaches to evaluating the wasted mouse as a model of AT is to compare the relative colony-forming ability of fibroblasts from wasted and control mice after gamma-irradiation. Conclusive results have not been obtained because of extremely poor plating efficiency of primary mouse fibroblasts (15). By using improved conditions for initiation and maintenance of primary lung fibroblasts, we concluded that sensitivity of lung fibroblasts from wasted mice to the killing effect of gamma-rays does not differ from that of controls throughout the ages examined.

The results described in this chapter show that although the "wasted" mutation shares many characteristics with AT, certain abnormalities seen in cells from AT patients are not observed in wasted mice. Among the dissimilarities, the most important may be that the lung fibroblasts of wasted mice were not hypersensitive to the killing effect of gamma-rays, although bone marrow cells in wasted mice exhibited many more radiation-induced chromosomal aberrations. It is noteworthy here that anomalies in wasted mice developed in a tissue-specific and age-dependent manner. It is thus reasonable to speculate that certain characteristics that are observed only in AT, but not in wasted mouse cells, may become evident in wasted mice if other types of cells from mice of different ages are examined.

In conclusion, both AT and the wasted mutation cause developmental abnormalities in the lymphoid system and a similar DNA repair deficiency, suggesting a close relationship between DNA repair and tissue differentiation in the lymphoid system. While wasted mice may not be an exact homolog for AT, further investigation of wasted mice may help us to understand some of the underlying mechanisms of AT and the genetic backgrounds that determine the susceptibility to cancer.

ACKNOWLEDGEMENTS

We thank Miss Junko Takahashi for excellent experimental assistance. T.I. and H.T. were supported by a grant-in-aid for scientific research

(#59480448) from the Ministry of Education. L.D.S. was supported by a grant from the National Institutes of Health (CA20408).

REFERENCES

1. Agarwal, S.S., D.Q. Brown, E.J. Katz, and L.A. Loeb (1977) Screening for deficits in DNA repair by the response of irradiated human lymphocytes to phytohemagglutinin. Cancer Res. 37:3594-3598.
2. Bridges, B.A., and D.G. Harnden, eds. (1982) Ataxia-Telangiectasia, a Cellualar and Molecular Link between Cancer, Neuropathology, and Immune Deficiency, John Wiley and Sons, New York.
3. Cohen, M.M.., and S.J. Simpson (1982) The effect of bleomycin on DNA synthesis in ataxia telangiectasia lymphoid cells. Environ. Mutag. 4:27-36.
4. Edwards, M.J., A.M.R. Taylor, and G. Duckworth (1980) An enzyme activity in normal and ataxia-telangiectasia cell lines which is involved in the repair of gamma-irradiation induced DNA damage. Biochem. J. 188:677-682.
5. Houldsworth, J., and M.F. Lavin (1980) Effect of ionizing radiation on DNA synthesis in ataxia telangiectasia cells. Nucl. Acids Res. 8: 3709-3720.
6. Inoue, T., and T. Kada (1977) Studies on DNA repair in Bacillus subtilis. III. Identification of an exonuclease which enhances the priming activity of γ-irradiated DNA by "cleaning" damaged ends. Biochim. Biophys. Acta 478:234-243.
7. Inoue, T., K. Hirano, A. Kokoiyama, T. Kada, and H. Kato (1977) DNA repair enzymes in ataxia-telangiectasia and Bloom's syndrome fibroblasts. Biochim. Biophys. Acta 479:497-500.
8. Inoue, T., and T. Kada (1978) Purification and properties of a Bacillus subtilis endonuclease specific for apurinic sites in DNA. J. Biol. Chem. 253:8559-8563.
9. Inoue, T., A. Yokoiyama, and T. Kada (1981) DNA repair enzyme deficiency and in vitro complementation of the enzyme activity in cell-free extracts from ataxia telangiectasia fibroblasts. Biochim. Biophys. Acta 655:49-53.
10. Inoue, T., M.S. Sasaki, A. Yokoiyama, and T. Kada (1982) Primer activating enzyme deficiency and in vitro complementation of the enzyme activity in cell-free extracts from ataxia-telangiectasia fibroblasts. In Ataxia-Telangiectasia, a Cellular and Molecular Link between Cancer, Neuropathology, and Immune Deficiency, B.A. Bridges and D.G. Harnden, eds. John Wiley and Sons, New York, pp. 305-317.
11. McFarlin, D.E., W. Strober, and T.A. Waldmann (1972) Ataxia-telangiectasia. Medicine 51:281-314.
12. Morris, C., R. Mohamed, and M.F. Lavin (1983) DNA replication and repair in ataxia telangiectasia cells exposed to bleomycin. Mutat. Res. 112:67-74.
13. Noguti, T., and T. Kada (1975) Studies on DNA repair in Bacillus subtilis. I. A cellular factor acting on γ-irradiated DNA and promoting its priming activity for DNA polymerase I. Biochim. Biophys. Acta 395:284-293.
14. Noguti, T., and T. Kada (1975) Studies on DNA repair in Bacillus subtilis. II. Partial purification and mode of action of an enzyme enhancing the priming activity of γ-irradiated DNA. Biochim. Biophys. Acta 395:294-305.
15. Nordeen, S.K., V.G. Schaefer, M.H. Edgell, C.A. Hutchson III, L.D. Shultz, and M. Swift (1984) Evaluation of wasted mouse fibroblasts and

SV-40 transformed human fibroblasts as models of ataxia telangiectasia in vitro. Mutat. Res. 140:219-222.

16. Paterson, M.C., B.P. Smith, P.H.M. Lohman, A.K. Anderson, and L. Fishman (1976) Defective excision repair of γ-ray-damaged DNA in human (ataxia telangiectasia) fibroblasts. Nature 260:444-447.
17. Paterson, M.C., and P.J. Smith (1979) Ataxia-telangiectasia: An inherited human disorder involving hypersensitivity to ionizing radiation and related DNA-damaging chemicals. Ann. Rev. Genet. 13:291-318.
18. Peterson, R.D.A., W.D. Kelly, and R.A. Good (1964) Ataxia-telangiectasia, its association with a defective thymus, immunological-deficiency disease, and malignancy. Lancet i:1189-1193.
19. Povirk, L.F., and I.H. Goldberg (1982) Inhibition of mammalian deoxyribonucleic acid synthesis by neocarzinostatin: Selective effect on replicon initiation in CHO cells and resistant synthesis in ataxia telangiectasia fibroblasts. Biochemistry 21:5857-5862.
20. Scudiero, D.A. (1980) Decreased DNA repair synthesis and defective colony-forming ability of ataxia telangiectasia fibroblast cell strains treated with N-methyl-N'-nitro-N-nitrosoguanidine. Cancer Res. 40:984-990.
21. Shultz, L.D., H.O. Sweet, M.T. Davisson, and D.R. Coman (1982) "Wasted," a new mutant of the mouse with abnormalities characteristic of ataxia telangiectasia. Nature 297:402-404.
22. Smith, P.J., and M.C. Paterson (1980) Gamma-ray induced inhibition of DNA synthesis in ataxia telangiectasia fibroblasts is a function of excision repair capacity. Biochem. Biophys. Res. Commun. 97:897-905.
23. Swift. M., L. Sholman, M. Perry, and C. Chase (1976) Malignant neoplasms in the families of patients with ataxia-telangiectasia. Cancer Res. 36:209-215.
24. Vincent, Jr., R.A., A.J. Fink, and P.C. Huang (1980) Unscheduled DNA synthesis in cultured ataxia telangiectasia fibroblast-like cells. Mutat. Res. 72:245-249.
25. Waldmann, T.A. (1982) Immunological abnormalities in ataxia-telangiectasia. In Ataxia-Telangiectasia, a Cellular and Molecular Link between Cancer, Neuropathology, and Immune Deficiency, B.A. Bridges and D.G. Harnden, eds. John Wiley and Sons, New York, pp. 37-51.
26. de Wit, J., N.G.J. Jaspers, and D. Bootsma (1981) The rate of DNA synthesis in normal human and ataxia telangiectasia cells after exposure to X-irradiation. Mutat. Res. 80:221-226.
27. Zampetti-Bosseler, F., and D. Scott (1981) Cell death, chromosome damage, and mitotic delay in human, ataxia telangiectasia and retinoblastoma fibroblasts after X-irradiation. Int. J. Radiat. Biol. 39:547-558.

GENETICS AND MOLECULAR BIOLOGY OF MAMMALIAN CELLS--PART II

INTRODUCTION: OXYGEN METABOLISM, DNA REPAIR, AND THE ORIGIN OF SPONTANEOUS GENETIC INSTABILITY

Hans Joenje

Institute of Human Genetics
Free University
Amsterdam, The Netherlands

An important issue in cancer research concerns the following questions: (a) How do living cells maintain the integrity of their genome? (b) What causes the genome to become destabilized? and (c) How can this destabilization be prevented? Under laboratory conditions it has been relatively easy to identify agents that efficiently destabilize the genome. Many of these agents occur in the natural environment, but it is uncertain whether they are major contributors to the background of genetic instability which occurs spontaneously in living cells. It is increasingly recognized that a substantial part of spontaneous genetic instability may be due to endogenous mechanisms.

In analyzing the origin of spontaneous genetic instability, several components may be distinguished. First, there is a base level of genetic instability due to the intrinsic chemical instability of the DNA (e.g., as a result of spontaneous depurination) and to the intrinsic infidelity of the DNA replication/repair process. Cellular physiology may increase this base level by generating potentially mutagenic changes in the DNA and/or by increasing the infidelity of the DNA replication/repair process. What physiological mechanisms could modulate genetic instability? One of the possible mechanisms that is currently receiving increased attention is based on the metabolism of molecular oxygen as a potential driving force in genetic destabilization, as summarized in Fig. 1.

Oxygen molecules that diffuse into cells are mainly used to meet the energy demand of the cell through tetravalent reduction to water by the cytochrome oxidase reaction in the mitochondria, but a significant fraction are activated to reactive species such as $O_2\cdot^-$, H_2O_2, $OH\cdot$, and singlet oxygen, generally called "oxygen radicals." These radicals pose a potential threat to the genome, either by acting directly on the nuclear chromatin or by interfering with the DNA replication/repair machinery of the cell.

Another indirect pathway is based on the peroxidation of polyunsaturated fatty acids. This is an autocatalytic process that is readily initiated by oxygen radicals. This process may lead to reactive products like

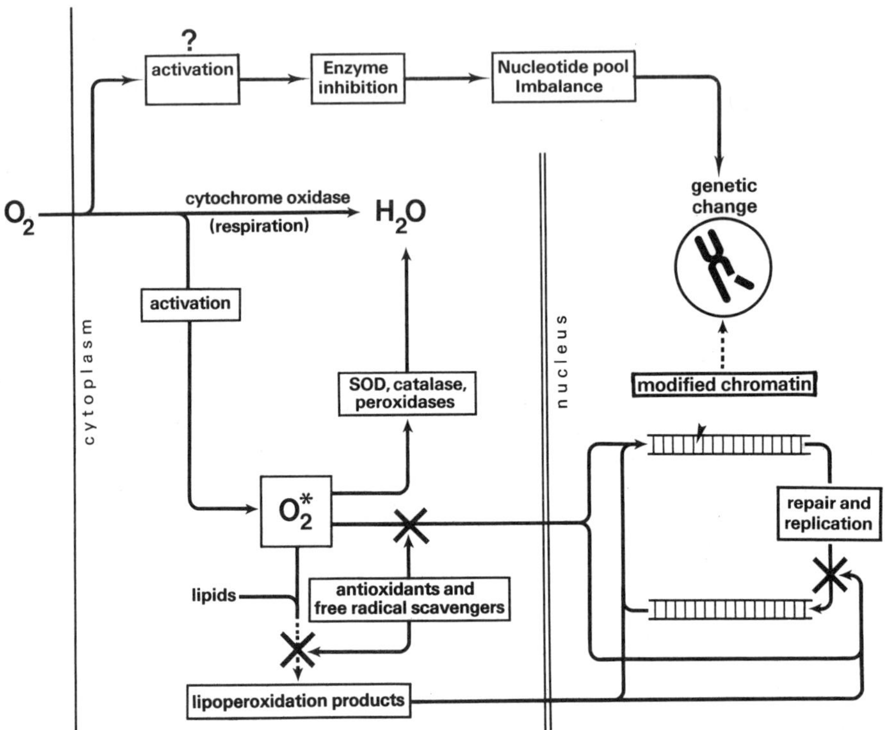

Fig. 1. Scheme summarizing some possible pathways in oxygen-dependent genetic instability.

lipid peroxides and aldehydes, which are capable of inducing genetic damage, either by interacting directly with the DNA or by interfering with the fidelity of DNA replication/repair.

Though it is currently largely hypothetical, there may be a third pathway. This pathway is based on the assumption that lack of oxygen may cause nucleotide pool imbalances, which are known to be potentially mutagenic; this mechanism does not necessarily depend on the production of reactive oxygen metabolites.

The steady-state level of oxygen radicals is thought to be kept under strict control by the presence of protective enzymes like superoxide dismutase, catalase, and peroxidases, as well as low molecular weight antioxidants and free radical scavengers. These factors all help to reduce the level of oxygen toxicity.

An interesting extrapolation can be made from the scheme in Fig. 1. It may be supposed that a primary defect in any of the protective mechanisms would lead to an increased level of physiological toxicants. This would result in an overproduction of DNA damage and/or in a malfunction of the DNA replication/repair machinery. In other words, a primary defect in a protective mechanism may manifest itself phenotypically as a DNA repair deficiency. DNA metabolizing enzymes are synthesized in the cytoplasm. Thus, prior to their arrival and function in the nucleus, such enzymes might have been exposed to potentially harmful metabolites. If some repair enzyme (e.g., an ultraviolet-endonuclease or a DNA polymerase) happens to be particularly vulnerable to some endogenous toxicant, it may become partially inactivated or may lose part of its fidelity.

This consideration, along with the fact that a surprisingly large number of complementation groups is found in the human DNA repair-deficiency syndromes and in experimental DNA repair mutants of mammalian cell lines, suggests that it may be rewarding to search for primary defects in the oxygen-protective mechanisms in addition to searching for defects in the DNA repair mechanism itself. Cloning of the responsible genes and identification of their products should help to reveal the origin of spontaneous and induced genetic instability in mammalian cells.

MULTIPLE INDUCIBLE CYTOCHROMES P-450 IN YEAST

G. Bronzetti and R. Del Carratore

Istituto di Mutagenesi e Differenziamento CNR
Vie Svezie 10, 56100
Pisa, Italy

ABSTRACT

A form of cytochrome P-450 is produced in *Saccharomyces cerevisiae* during the logarithmic growth phase (14). This form is inhibited by metyrapone, tetrahydrofurane, and carbon oxide which are inhibitors of ethanol-inducible cytochromes P-450 of mammals. Addition of Na-phenobarbital to the 0.5% glucose liquid medium induced a form of cytochrome P-450 inhibited only by metyrapone. Thus, yeast possess multiple forms of cytochrome P-450 that are individually inducible.

INTRODUCTION

Strains of *S. cerevisiae* produce cytochromes P-450 during the logarithmic growth phase in media containing 1-20% glucose (3,15). These P-450 systems of *S. cerevisiae* exhibit oxidation, hydroxylation, and demethylation activities on many compounds (2,9). There are multiple forms of cytochromes P-450 in animals with differential inducibility (5). Yeast also may possess more than two forms of cytochromes P-450. The level of cytochrome P-450 activity in the D7 strain of yeast is affected by ethanol produced by glycolysis during cell growth (4).

In the present investigation we have tested the hypothesis that the form produced during growth in 20% glucose might be different from the form induced by Na-phenobarbital. Specific inhibitors of drug metabolism, such as methyrapone, tetrahydrofurane, and carbon oxide were used to differentiate cytochrome P-450 species.

MATERIALS AND METHODS

Saccharomyces cerevisiae strain D7, obtained from Prof. F.K. Zimmermann, detects mitotic gene conversion at the *trp5* locus, point mutations of the mutant allele *ilv1-92*, and mitotic recombination between the centromere and the *ade2* locus (1,16).

Media

To 1 liter of deionized water were added 10 g of yeast extract (Difco), 20 g of Bacto Peptone (Difco), 200 g of glucose for the logarithmic growth phase cultures and 20 g for the cultures to be used in stationary phase.

Growth Conditions and Determination of Cytochrome P-450 Levels

To obtain a high level of cytochrome P-450, 5 x 10^6 cells/ml were inoculated into 20% glucose liquid growth medium and incubated at 30°C for 6 hr until a concentration of 5 x 10^7 cells/ml was attained. Cytochrome P-450 was determined directly on whole cells according to the method of Omura and Sata (8) as previously described (2-6,9).

Suspension Test Procedure

Samples of 5 x 10^8 cells were harvested during the logarithmic growth phase (when in the culture, density was about 5 x 10^7 cells/ml) and incubated in 4 ml of 0.1 M phosphate buffer, pH 7.4, in the presence of various concentrations of mutagens for 2 hr at 37°C. The incubation mixture was then plated on selective media to ascertain ade^+ recombinants, trp^+ convertants, and ilv^+ revertants (1). In assays for metabolic activation, we used an incubation mixture of 1 ml of post mitochondrial supernatant (9,000 x g) (S-9) of mouse liver induced with 100 mg/kg Na-phenobarbital and 80 mg/kg β-naphthoflavone.

Assay of Deethylation

The deethylation of 7-ethoxycoumarin was measured on whole cells by a modified method of Ullrich and Weber (12). Samples corresponding to about 4 x 10^9 cells were washed 2 times and resuspended in 15 ml of 0.05 M Tris HCl buffer pH 7.4 containing 1.15% KCl. Incubation of 2.5 ml of cell suspension with 250 λ of 1 mM ethoxycoumarin in Tris HCl was performed at 37°C. After incubation for 5 min the reaction was terminated by the addition of 0.6 ml 15% trichloroacetic acid and 2.2 ml of 1.6 M glycine-NaOH buffer, pH 10.3. The fluorescent umbelliferon was measured at 440 nm with excitation at 390 nm in a Perkin-Elmer 650-10S fluorescence spectrophotometer. Different concentrations of inhibitors were added directly to the incubation mixtures 10 min before assay.

Protein Determination

Protein concentrations were determined according to Lowry et al. (7) with bovine serum albumin as the standard.

RESULTS

4-Nitroquinoline 1-oxide (4NQO) (0.01 mM) gave a strong increase of ade recombinants, trp convertants, and ilv revertants on cells from stationary phase cultures (Tab. 1). The addition of S-9 fractions to the incubation mixtures decreased the 3 genetic effects 8, 10, and 13 times, respectively. These values were twice as high with logarithmic growth phase cells containing a level of cytochrome P-450 corresponding to 14 pM/mg total proteins. Specific inhibitors of monooxygenase systems, such as tetrahydrofurane and metyrapone, produced a dose-dependent increase in

Tab. 1. Effect of inhibitors on induction of ade recombinants, trp convertants, and ilv revertants in strain D7 of *Saccharomyces cerevisiae* by 4-nitroquinoline 1-oxide (4NQO) in cells from stationary growth phase with metabolic activation (S-9) and in cells in logarithmic growth phase. Values, means of 5 experiments, are expressed as colonies on selective medium/total colonies surviving ± standard deviation.

	Survival (%)		ade recombinants/ 10^4 survivors		trp convertants/ 10^5 survivors		ilv revertants/ 10^6 survivors	
	Stat. phase + S-9	Log. phase	Stat. phase + S-9	Log. phase	Stat. phase + S-9	Log. phase	Stat. phase + S-9	Log. phase
Control	100.0	100.0	5.1 ± 1.1	8.6 ± 1.4	0.6 ± 0.1	0.9 ± 0.1	0.2 ± 0.1	0.7 ± 0.3
Control + THF 80.0 mM	100.0	100.0	9.0 ± 1.2	4.5 ± 0.9	0.6 ± 0.1	0.8 ± 0.2	0.3 ± 0.1	0.6 ± 0.3
Control + MET 10.0 mM	100.0	100.0	4.1 ± 0.6	6.3 ± 0.5	0.7 ± 0.1	0.8 ± 0.1	0.3 ± 0.1	0.5 ± 0.1
4NQO	55.0 ± 3.3*	73.3 ± 4.5	357.8 ± 15.0*	19.4 ± 0.9	526.1 ± 14.7*	24.3 ± 1.7	55.0 ± 2.1*	2.3 ± 0.2
	56.2 ± 4.5	-	46.0 ± 3.1	-	49.8 ± 7.1	-	4.8 ± 1.3	-
4NQO + THF 0.1 mM	55.2 ± 5.7	62.1 ± 5.4	86.4 ± 4.2	25.2 ± 1.9	106.8 ± 9.1	38.3 ± 2.9	10.9 ± 1.9	3.4 ± 0.7
4NQO + THF 1.0 mM	61.1 ± 6.3	61.5 ± 3.4	103.9 ± 6.6	31.7 ± 1.8	115.0 ± 2.6	49.2 ± 3.6	12.3 ± 0.9	4.2 ± 0.8
4NQO + THF 10.0 mM	57.6 ± 5.7	61.7 ± 9.1	118.2 ± 6.9	65.4 ± 4.6	126.2 ± 3.5	109.0 ± 10.1	16.8 ± 2.5	5.5 ± 0.7
4NQO + MET 0.1 mM	58.6 ± 4.8	67.5 ± 9.3	55.3 ± 5.8	21.5 ± 3.9	98.4 ± 2.1	38.1 ± 6.5	4.8 ± 0.2	2.7 ± 0.3
4NQO + MET 1.0 mM	56.0 ± 7.7	65.6 ± 7.5	72.0 ± 3.4	17.5 ± 2.2	109.0 ± 0.9	45.8 ± 4.2	5.4 ± 0.5	3.0 ± 0.6
4NQO + MET 10.0 mM	50.5 ± 3.9	71.1 ± 8.7	110.5 ± 12.3	49.5 ± 4.9	145.6 ± 1.1	67.6 ± 5.5	7.4 ± 0.4	4.8 ± 0.6

* No metabolic activation (S-9) was used.
Abbreviations: 4NQO, 4-nitroquinoline 1-oxide; THF, tetrahydrofurane; MET, metyrapone.

Tab. 2. Ethoxycoumarin O-deethylase by whole cells of *Saccharomyces cerevisiae* D7 strain in the presence of different concentrations of inhibitors. The cells were harvested from 20% glucose and 0.5% glucose plus 0.2% Na-phenobarbital liquid medium during logarithmic growth phase when the cell concentration was about 5×10^7 cell/ml. The activity is expressed as pmol umbelliferone/mg total protein x minutes. Data represent mean ± standard deviation for 5 experiments.

	pmol Cytochrome P-450	Specific activity	Inhibition by metyrapone			Inhibition by tetrahydrofurane		
			0.1 mM	1.0 mM	10 mM	0.1 mM	1.0 mM	10.0 mM
20% Glucose	11.0 ± 1.6	7.6 ± 1.37	7.2 ± 1.32	3.6 ± 0.95	1.6 ± 0.12	7.1 ± 1.81	4.8 ± 0.75	2.1 ± 0.61
0.5% Glucose plus 0.2% Na-phenobarbital	3.5 ± 0.7	2.8 ± 0.81	2.0 ± 0.71	1.1 ± 0.26	0.6 ± 0.01	2.3 ± 0.21	2.7 ± 0.15	2.0 ± 0.33

genetic effects both in stationary cells with S-9 and in logarithmic growth phase cells (Tab. 1). These results were confirmed with the enzymatic assay.

Ethoxycoumarin O-deethylase activity measured on whole cells in the presence of inhibitors decreased in activity with respect to controls without inhibitor. A decrease of 79% of the O-deethylation activity was found in the presence of the maximal concentration of metyrapone and a 73% decrease in the presence of tetrahydrofurane (Tab. 2). Corresponding levels of cytochrome P-450 are also reported. When the induction was performed with 0.2% Na-phenobarbital added to 0.5% glucose liquid medium, a different form of cytochrome P-450, inhibited only by methyrapone, was obtained (Tab. 2).

DISCUSSION

4-Nitroquinoline 1-oxide appears to be transformed in cells by an enzymatic reduction and an acylation to the ultimate mutagenic-carcinogenic form, 4-aminoacyloxyaminoquinoline 1-oxide (10,11). The action of yeast cytochrome P-450 produced a detoxifying effect by metabolizing 4NQO to phenol. The detoxification by cytochrome P-450 was more efficient in cells from the logarithmic growth phase than in cells from stationary phase cultures with external metabolic activation (Tab. 1). This is in agreement with results obtained for the activation of styrene (3). Inhibitors of cytochrome P-450 caused a block of metabolic detoxification of the 4NQO and a dose-dependent increase of <u>ade</u> recombinants, <u>trp</u> convertants, and <u>ilv</u> revertants.

Metyrapone and tetrahydrofurane exhibited inhibitory activity both in cells grown in 20% glucose and in activation by liver S-9. In the case of cells grown in 20% glucose, the action of inhibitors mimics inhibition of mammalian isoenzymes that are induced by ethanol (13). Hepatic fraction enzymes (external metabolic activation used with cells from the stationary growth phase) were affected by both inhibitors. Inhibition was confirmed in measurements demonstrating decreases in 7-ethoxycoumarin O-deethylase activity (Tab. 2). Decreases of 79% and 73% of the deethylase activity were observed, corresponding to the maximal concentrations of metyrapone and tetrahydrofurane, respectively. Deethylation activity observed in yeast microsomal fractions was about 10% lower under the same experimental conditions.

After addition of 0.2% Na-phenobarbital in 0.5% glucose liquid medium (where no cytochrome P-450 is detectable) a form inhibited by metyrapone but not by tetrahydrofurane was obtained. 4NQO mutagenesis is yeast cells grown in the presence of Na-phenobarbital also showed an influence only of metyrapone (data not shown). The presence of two differently inducible forms of cytochromes P-450 is evident.

REFERENCES

1. Bronzetti, G., C. Bauer, C. Corsi, C. Leporini, R. Nieri, and R. Del Carratore (1981) Genetic activity of vinylidene chloride in yeast. <u>Mutat. Res.</u> 89:179-185.
2. Callen, D.E., C. Wolf, and R.M. Phylpot (1980) Cytochrome P-450 mediated genetic activity of seven halogenated aliphatic hydrocarbons in <u>Saccharomyces cerevisiae</u>. <u>Mutat. Res.</u> 77:55-63.

3. Del Carratore, R., G. Bronzetti, C. Bauer, C. Corsi, R. Nieri, M. Paolini, and P. Giagoni (1983) Cytochrome P-450 factors determining synthesis in strain D7 Saccharomyces cerevisiae. Mutat. Res. 121:117-123.
4. Del Carratore, R., C. Morganti, A. Galli, and G. Bronzetti (1984) Cytochrome P-450 inducibility by ethanol and 7-ethoxycoumarin O-deethylation in Saccharomyces cerevisiae. Biochem. Biophys. Res. Commun. 123:186-193.
5. Guengerich, F.P. (1978) Separation and purification of multiple forms of microsomal cytochrome P-450. J. Biol. Chem. 253:7931-7939.
6. King, D.J., M.R. Azari, A. Wiseman (1982) The induction of cytochrome P-450 dependent benzo(a)pyrene hydroxylase in Saccharomyces cerevisiae. Biochem. Biophys. Res. Comm. 105:1115-1121.
7. Lowry, O.H., N.J. Rosenbrough, A.L. Far, and R.J. Randall (1951) Protein measurement with the Folin phenol reagent. J. Biol. Chem. 193:265-275.
8. Omura, T., and R. Sato (1964) The carbon monoxide-binding pigment of liver microsomes. J. Biol. Chem. 239:2370-2378.
9. Sato, R., and T. Omura (1978) Cytochrome P-450, Kodonisha, Tokyo and Academic Press, New York.
10. Tada, M., and M. Tada (1975) Seryl-tRNA synthetase and activation of the carcinogen 4-nitroquinoline 1-oxide. Nature 255:510-512.
11. Takahashi, K., K. Huang-Khoda, and Y. Kawazoe (1983) Studies on chemical carcinogens XXIV. Mutagenic potency of alkylated 4-nitroquinoline 1-oxide: Its dependence on the rate of metabolic activation. Chem. Pharm. Bull. 31(3):959-965.
12. Ullrich, V., and P. Weber (1972) The O-deethylation of 7-ethoxycoumarin by liver microsomes. Z. Physiol. Chem. 353:1171-1177.
13. Ullrich, V., P. Weber, and P. Wollemberg (1975) Tetrahydrofuran--An inhibitor for ethanol-induced liver microsomal cytochrome P-450. Biochem. Biophys. Res. Commun. 64:808-813.
14. von Borstel, R.C., D.F. O'Connell, R.D. Mehta, and U.G.G. Hennig (1985) Modulation in cytochrome P-420 and P-450 content in Saccharomyces cerevisiae according to physiological conditions and genetic background. Mutat. Res. 150:217-224.
15. Wiseman, A., T.K. Lim, and C. McLoud (1975) Relationship of cytochrome P-450 to growth phase of brewer's yeast in 1% or 20% glucose medium. Biochem. Soc. Trans. 3:276-278.
16. Zimmerman, F.K., R. Kern, and H. Rosemberger (1975) A yeast strain for simultaneous detection of induced mitotic crossing-over, mitotic gene conversion and reverse mutation. Mutat. Res. 28: 381-388.

DNA REPAIR GENES OF MAMMALIAN CELLS

L.H. Thompson, K.W. Brookman, E.P. Salazar,
J.C. Fuscoe, and C.A. Weber

Biomedical Sciences Division
Lawrence Livermore National Laboratory
Livermore, California 94550

ABSTRACT

In the Chinese hamster ovary (CHO) cell line, various mutations affecting DNA repair have been obtained. Mutants that belong to 5 genetic complementation groups for ultraviolet (UV) sensitivity and resemble the cells from individuals having the cancer-prone genetic disorder xeroderma pigmentosum (XP) were previously identified. Each mutant is defective in the incision step of nucleotide excision repair and hypersensitive to bulky DNA lesions. These UV mutants can be divided into two subgroups; only Groups 2 and 4 are extremely sensitive to mitomycin C and other DNA cross-linking agents.

The clear-cut phenotypes of the CHO mutants have allowed us to construct hybrid cells by fusion with human lymphocytes and thereby identify which human chromosomes carry genes that correct the CHO mutations. The first two mutations analyzed, UV20 (excision-repair deficient; UV Group 2) and EM9, which has a very high frequency of sister chromatid exchange (SCE), are both corrected by chromosome 19.

Efforts are underway to isolate complementing repair genes by DNA-mediated gene transfer. The human gene that corrects mutant EM9 and the hamster gene that corrects UV135 (UV Group 5) have been introduced by co-transfer of genomic DNA and the dominant selectable marker <u>gpt</u> (guanine phosphoribosyltransferase) gene. In each case, the DNA repair function was co-selected based on resistance to 5-chlorodeoxyuridine (CldUrd) or repeated UV irradiation, respectively. The presence of a functional human repair gene in the EM9 transformants is shown by the presence of common human DNA sequences on some fragments produced by restriction enzyme cleavage. In UV135, transfer of a repair gene is indicated by a colony distribution containing "jackpots" and by instability of the resistant phenotype.

INTRODUCTION

In mammalian cells, as well as other organisms, multiple pathways exist for the repair of DNA lesions caused by insult from endogenous or environmental agents. An understanding of how these processes operate at the molecular level should help explain how the initial lesions are converted into both point mutations and chromosomal rearrangements, which may eventually produce a malignant phenotype or germ cell alteration. The study of repair has been difficult because each agent produces an array of DNA lesions, which evoke biochemically complex events governed by multiple genes. In mammalian cells, the repair proteins of several major pathways, such as bulky adduct and cross-link removal, have not been identified. In general, one can expect repair-pathway proteins to overlap with those involved in DNA replication and recombination.

In humans, the genetic disease XP has provided an important model for studying the major nucleotide excision-repair pathway, which acts on bulky DNA lesions produced by chemicals or far UV irradiation (4). In XP cells, there is either a partial or total defect in the repair of this class of damage. The fact that XP individuals have an elevated risk for cancer (4, 29), along with the finding that XP cells in culture are both hypermutable and hypersensitive to malignant transformation (21,22), has provided a strong argument for believing that somatic mutations arising from unrepaired lesions have oncogenic potential.

The genotypes underlying the XP phenotype are complex, as evidenced by the fact that the number of genetic complementation groups identified on the basis of unscheduled DNA synthesis has steadily increased. There are now 9 groups for excision-deficient XP (7). One simple interpretation of these results is that each of the complementation groups corresponds to a different genetic locus that has homozygous recessive mutations (-/-). This model implies that as many as 9 or more proteins might participate in the UV repair pathway. An alternative interpretation is that some of the many complementation groups are due to intragenic complementation.

Lambert and Lambert (18) have recently suggested another model in which a smaller number of loci could account for the many complementation groups. Their "recessive co-inheritance" model is based on the idea that the repair-deficient phenotype requires mutation in at least two different loci. This model needs to be critically tested.

The technique of DNA-mediated gene transfer (1,16) has provided, in principle, a way to isolate the individual genes participating in repair and, it is hoped, the encoded proteins. However, to date, the defects in XP cells have not been corrected by the transfer of DNA from normal human cells (26). The reasons for this failure are unclear but could be accounted for if the genes were extremely large, e.g., >100 kbp, or if more than one gene were needed to restore function.

An alternative approach to isolating human repair genes is based on their ability to correct the repair defects in rodent cells. Westerveld and co-workers (45) used a UV-sensitive mutant of CHO cells to isolate a human repair gene that corrects the repair defect in the mutant cells. Their procedure utilized the co-transfer of a dominant selectable marker, the bacterial gpt gene carried on the pSV2-gpt plasmid, in combination with selection for the repair function using mitomycin C (to which the mutant cells are extremely sensitive). The isolated gene, designated ERCC1, was obtained from a cosmid clone library constructed from the co-transformed mutant cells and is approximately 15 kbp in size (J. Hoeijmakers, pers. comm.). This gene can be used to test for its ability to correct the defects in various human and rodent repair mutants. The approach of using rodent-cell mutants to obtain the human repair genes holds considerable promise for being able to perform a detailed molecular genetics analysis of repair functions. In addition to the mutants characterized in our laboratory and summarized below, several other groups have studied rodent-cell mutants that have interesting properties (2,8,9,13,15,27,28,30,32,33,44, 46).

RESULTS

Genetic Properties of the UV-Sensitive CHO Repair Mutants

For several years our laboratory has been characterizing mutant lines of CHO cells that should prove valuable for isolating additional DNA repair genes. Five genetic complementation groups of UV-sensitive mutants were identified, by the endpoint of colony-forming ability after cell fusion and UV irradiation (36,39). Mutants in each of these groups are defective in the incision step of nucleotide excision repair (37). Mutants in Group 2, such as UV4 and UV20 (11,35), correspond (by complementation assay) to the mutant 43-3B, which was used by Westerveld and co-workers to clone the ERCC1 gene (45).

We have shown that the mutants can be divided into two subgroups; mutants in Groups 2 and 4 are extremely hypersensitive to killing by agents that produce DNA interstrand cross-links (11). Recently, we analyzed the UV-sensitive clone UV27-1, which was isolated by Wood and Burki (46) (L. Thompson and E. Salazar, unpubl. observ.), and assigned it to complementation Group 3 (which contains UV24). UV27-1 has similar UV sensitivity to mutants such as UV4 and UV5, but, surprisingly, it has a normal level of resistance to the UV-mimetic agent 4-nitroquinoline 1-oxide (47). UV4 and UV5 showed 8-fold hypersensitivity to this compound (10).

Mutants representing each of the first five UV complementation groups have been fused with human lymphocytes to produce hybrid cells, which segregate human chromosomes. By correlating the presence of individual human chromosomes with the repair-proficient, UV-resistant phenotype, the chromosome(s) that corrects the mutant defect can be identified. In this way, mutant UV20 was found to be complemented by human chromosome 19 (40). Preliminary results with mutants from the other four groups suggest that, with one exception (UV5 in Group 1), the different groups are corrected by human genes on different chromosomes (L. Thompson, M. Siciliano, and A. Carrano, unpubl. observ.). Thus, in most instances, it looks as though the complementation groups will correspond to different genetic loci. The evidence to date suggests that no more than one human chromosome is able to complement each mutant.

Progress Toward Cloning a Human Gene that Corrects Mutant EM9

The mutant line referred to as EM9 is particularly interesting because of its extremely high frequency of SCE (38), which results primarily from the bromodeoxyuridine used in the SCE protocol (25). The high baseline SCE frequency is analogous to that of cells from Bloom's syndrome (BS). However, at the biochemical level EM9 is deficient in some aspect of rejoining DNA strand breaks, a property that has not been found in BS cells. The faulty rejoining does not appear to stem from a defect in DNA ligases (3), AP endonucleases (17), or enzymes of the poly(ADP) ribose system (12).

There exists a human gene that can complement the EM9 defect both in cell hybrids (31) and in the form of purified DNA following DNA-mediated gene transfer using calcium phosphate precipitates (41). The high sensitivity of EM9 cells to incorporated CldUrd (6) has allowed efficient selection for complementation (31,41), which is evidenced by the restoration of normal levels of SCE along with CldUrd resistance (31,41).

In experiments to date, we have made both primary (41) and secondary transformants (Fig. 1) with EM9 using co-transfer of the pSV2-gpt plasmid,

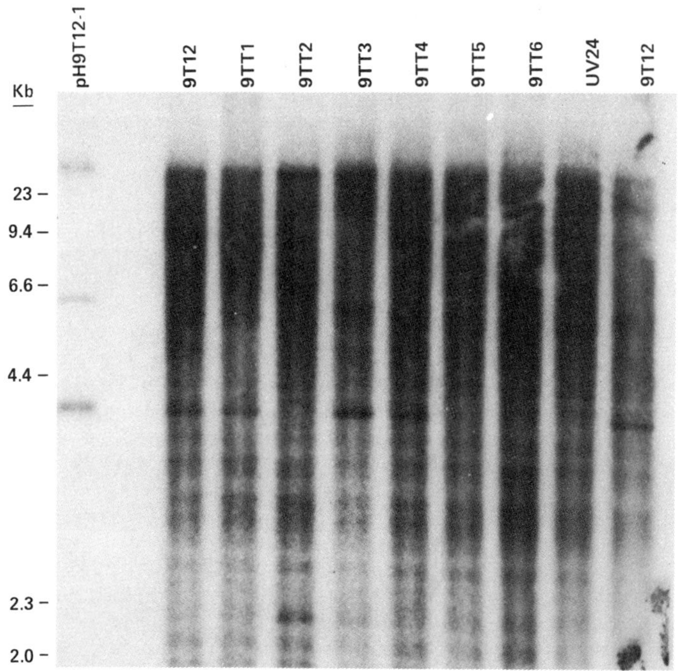

Fig. 1. Autoradiogram of a Southern hybridization to detect human sequences in secondary transformants of EM9. DNA from the primary transformant 9T12 was used in transformation experiments with EM9 to generate the 6 secondary transformants, 9TT1 to 9TT6. DNAs were digested with EcoRI, subjected to electrophoresis in agarose gel, transferred to nitrocellulose, and probed with a nick-translated Alu sequence isolated from a BLUR 8 plasmid. With transformants 9TT1, 9TT3, and 9TT4, a human-specific band at 3.8 kbp is evident, as in 9T12. This band is absent in hamster DNA (UV24). Examination of the intensity of this band in the lane containing cosmid clone pH9T12-1 at the equivalent of 1 copy per diploid genome suggests that the 3.8-kbp sequence is present at 1 copy per cell in lines 9TT1, 9TT3, and 9TT4, and absent in lines 9TT2, 9TT5, and 9TT6. Methods: Prehybridization was done at 65°C in 1.5X SSC for 4 hr, and hybridization was done in 1.5X SSC at 65°C for 19 hr. The probe contained 2.8 x 10^6 cpm ^{32}P in 0.2 μg of DNA. Filters were rinsed in 1.5X SSC, twice for 45 min at room temperature and twice for 20 min at 65°C. Other details of the hybridization were as described (42).

which is selectable in MAXTA medium (see Fig. 2 and Ref. 23). A 3.8-kbp human EcoRI fragment was detected by hybridization and Southern transfer in 6/6 independent primary transformants and 3/6 secondary transformants (Fig. 1, lanes of 9TT1, 9TT3, 9TT4), suggesting that it might be part of the repair gene, or more likely, closely linked. This fragment was isolated in a recombinant cosmid clone (pH9T12-1) (Fig. 1) obtained from a library of the primary transformant 9T12. This clone contained a human insert of about 40 kbp. Although the clone was not active in restoring the repair defect in EM9, it did contain a functional gpt gene. Moreover, using this clone as a probe in northern hybridization to RNA from both human and CHO cells, we could not identify any discrete transcripts except a small RNA [possibly 7SL RNA (43)] containing Alu-related sequences.

DNA from secondary transformants of EM9 was probed with the human Alu family sequence BLUR 8 (5) to obtain the Southern blot shown in Fig. 1. A comparison can be made between the intensity of the 3.8-kbp fragment from the cosmid clone (Fig. 1, lane 1), which is present at the equivalent of 1 copy per diploid genome, and that of each of the six secondary transformants. The results show that the 3.8-kbp sequence appears to be present at the single-copy level in lines 9TT1, 9TT3, and 9TT4. In the other three transformants it is missing. Thus, the sequence does not seem to be part of the repair gene. For the goal of gene cloning, these results suggest that the Alu probe does have adequate sensitivity for detecting human sequences present in transformants. Experiments in progress are aimed at estimating the size of the gene by transfer of repair gene activity in sheared DNA from a secondary transformant.

UV Radiation Selection Procedures for Gene Transfer

Mutants in complementation Groups 2 and 4 readily lend themselves to experiments to isolate repair genes because the repair-proficient phenotype can be selected using mitomycin C at about 10 nM in the medium. However, for mutants from the other groups (Groups 1, 3, and 5), which are not highly sensitive to mitomycin C, we have not been able to identify a chemical that will efficiently select for resistance. Although decarbamoyl mitomycin C was used with some success as a selecting agent to obtain CHO/human hybrids, variations in the effective toxicity of this chemical led us to abandon it in favor of UV radiation. Repeated exposure to UV was successfully used to obtain hybrids of mutants UV5, UV24, and UV135.

For gene transfer, a selection protocol based on exposure to UV radiation was devised as illustrated in Fig. 2. This method worked well with UV135 and should be applicable to the other groups of UV mutants. In these experiments one desires to reduce the survival of the UV-sensitive mutant cells to below 10^{-6}. With genomic DNA the frequency of transfer events for repair genes is approximately 10^{-7} under our conditions (41; unpubl. results with UV20).

Our protocol, similar in approach to that of MacInnes and co-workers (20), aims to kill the sensitive cells incrementally with 3 successive UV doses. The dose of 4 J/m^2 is sufficient to reduce the survival of the mutant cells to about 10^{-2}, while allowing cells that have the wild-type sensitivity to survive with 80-90% efficiency.

After DNA removal, one day is allowed for gpt expression. This seems to be adequate time for full expression (C. Hoy and L. Thompson, unpubl. observ.). An additional 3 days are allowed for expression of the repair

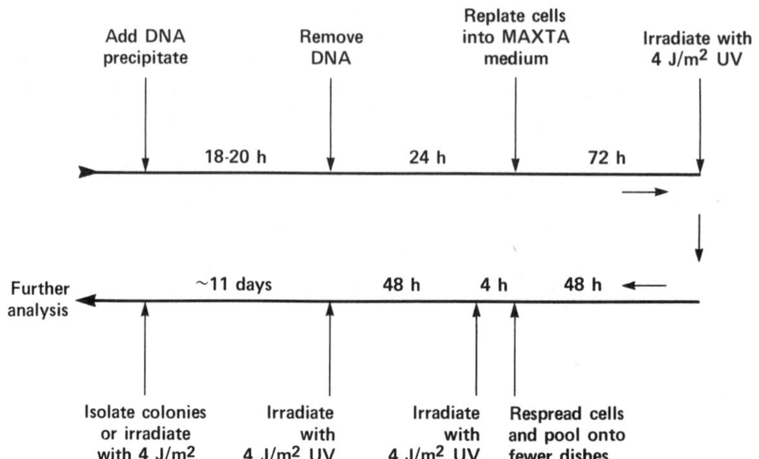

Fig. 2. Outline of UV radiation selection protocol for transferring repair genes into UV-sensitive CHO cells. CHO mutant UV135 was treated with a co-precipitate of genomic DNA + pSV2-gpt DNA (41), and putative transformants were isolated by repeated exposure to UV radiation. The protocol is designed to produce "jackpot" cultures containing many UV-resistant colonies when repair gene transfer occurs. MAXTA medium contains mycophenolic acid (10 μg/ml), adenine (25 μg/ml), xanthine (250 μg/ml), thymidine (10 μg/ml), and amethopterin (0.25 μg/ml).

gene before giving the first dose of UV radiation. (At this time nearly all the gpt-negative cells are dead, and gpt-positive colonies form at a frequency of about 4×10^{-3}.)

During the first 5 days after plating, UV-resistant colonies can multiply to several hundred cells per colony. On day 5, the remaining cells are respread with trypsin and pooled 3:1 for more efficient handling. This step accomplishes two things. First, the cells within truly resistant colonies are distributed into a much larger number of colony-forming units. Second, the dispersal of the foci of partially killed sensitive cells helps to eliminate shielding of live cells at the second irradiation. On day 7 a third dose of UV is given.

With this procedure the UV-sensitive cells are efficiently killed. On a dish there are typically 0, 1, or 2 sensitive colonies from 10^6 initial cells. Dishes that contain transfer events for the relevant repair gene will stand out as "jackpots," containing as many as 200 to 300 colonies. The UV resistance of these colonies can be verified by a fourth irradiation, if desired, either by "in situ" irradiation or by first respreading the cells to allow more uniform exposure.

Transfer of a Hamster Gene that Complements Mutant UV135

Gene transfer into UV135 was performed by the procedure shown in Fig. 2. The source of the wild-type hamster gene was the DNA from a hybrid cell line (24HL5). (This hybrid does not carry the human chromosome that corrects UV135.) Four initial dishes were each treated with 20 μg of genomic DNA + 20 μg of pSV2-gpt DNA. After allowing colony formation and giving a fourth dose of UV, we found that two of the four dishes contained resistant colonies. One dish had 3 large colonies and another had 167, i.e., a "jackpot." The other two dishes contained a total of 3 sensitive colonies.

To further assess whether the two putative transformants (135T1 and 135T2) contain an introduced repair gene and are not simply revertants, a test of phenotype stability was done. Transformants tend to be phenotypically unstable (as compared with revertants). Cells were plated and allowed to form small colonies (5 days old), which were challenged with 4 J/m^2 of UV radiation and examined microscopically 2 days later to determine the percentage of sensitive colonies. The 135T1 culture contained about 20% sensitive cells, while in 135T2 none of 1,600 colonies was sensitive. Therefore, we conclude that 135T1 is a transformant for the repair gene, but the evidence for 135T2 is inconclusive.

In a subsequent experiment, 135T1 DNA was used in a secondary transfer. Of 20 cultures, 9 contained the following numbers of resistant colonies per dish: 1, 2, 4, 5, 11, 33, 141, 154, and 346. Of the other 11 cultures, 6 had zero colonies, and 5 had 1 to 3 sensitive colonies. Thus, it appears that a gene complementing UV135 can be successively transferred. These results confirm and extend previous observations (41).

CONCLUSIONS

In Chinese hamster cell lines there are now a variety of mutants in DNA repair that can be used in a molecular genetics approach toward understanding this complex cellular process. A significant finding to date is that human genes exist that will complement (correct) the defects in each of the six CHO cell mutants we have examined. There is also an example in which a CHO gene(s) will correct the repair defect in transformed XP cells of complementation Group A (14). Together, these results suggest that there are corresponding genetic loci for repair in rodent and human cells and that the repair systems may be very similar.

In addition to allowing the chromosomal mapping of the complementing human repair genes, the rodent-cell mutants provide a means of isolating these genes. It seems likely that a number of human repair genes will soon be obtained, as was done by Westerveld and co-workers with ERCC1. These genes will be valuable probes to test for correction of the defects in mutant human cells from XP, Fanconi's anemia, ataxia telangiectasia, and other genetic diseases where faulty DNA repair/metabolism is associated with the clinical syndrome (29). Once the isolated genes have been placed into expression vectors to obtain the gene products, we will be a significant step further in deciphering the biochemistry of repair in animal cells. The many complementation groups of XP (7), ataxia telangiectasia (24), and Cockayne syndrome (19,34) suggest that the genetics of human-cell repair may be just as complex as it is in lower organisms such as yeast and bacteria.

ACKNOWLEDGEMENTS

This work was performed under the auspices of the U.S. Department of Energy by the Lawrence Livermore National Laboratory under contract W-7405-ENG-48.

REFERENCES

1. Abraham, I. (1985) DNA-mediated gene transfer. In Molecular Cell Genetics, M.M. Gottesman, ed. John Wiley, New York, pp. 181-210.
2. Busch, D.B., J.E. Cleaver, and D.A. Glaser (1980) Large-scale isolation of UV-sensitive clones of CHO cells. Somat. Cell Genet. 6:407-418.
3. Chan, J.Y.H., L.H. Thompson, and F.F. Becker (1984) DNA ligase activities appear normal in the CHO mutant EM9. Mutat. Res. 131:209-214.
4. Cleaver, J.E. (1983) Xeroderma pigmentosum. In The Metabolic Basis of Inherited Disease, 5th ed., J.B. Stanbury, J.B. Wyngaarden, D.S. Fredrickson, J.L. Goldstein, and M.S. Brown, eds. McGraw-Hill, New York, pp. 1227-1248.
5. Deininger, P.L., D.J. Jolly, C.M. Rubin, T. Friedmann, and C.W. Schmid (1981) Base sequence studies of 300 nucleotide renatured repeated human DNA clones. J. Mol. Biol. 151:17-33.
6. Dillehay, L.E., L.H. Thompson, and A.V. Carrano (1984) DNA strand breaks associated with halogenated pyrimidine incorporation. Mutat. Res. 131:129-136.
7. Fischer, E., W. Keijzer, H.W. Thielmann, O. Popanda, E. Bohnert, L. Edler, E.G. Jung, and D. Bootsma (1985) A ninth complementation group in xeroderma pigmentosum. Mutat. Res. 145:217-225.
8. Hama-Inaba, H., N. Hieda-Shiomi, T. Shiomi, and K. Sato (1983) Isolation and characterization of mitomycin-C-sensitive mouse lymphoma cell mutants. Mutat. Res. 108:405-416.
9. Hori, T., T. Shiomi, and K. Sato (1983) Human chromosome 13 compensates a DNA repair defect in UV-sensitive mouse cells by mouse-human cell hybridization. Proc. Natl. Acad. Sci., USA 80:5655-5659.
10. Hoy, C.A., E.P. Salazar, and L.H. Thompson (1984) Rapid detection of DNA-damaging agents using repair-deficient CHO cells. Mutat. Res. 130:321-332.
11. Hoy, C.A., L.H. Thompson, C.A. Mooney, and E.P. Salazar (1985) Defective DNA cross-link removal in Chinese hamster cell mutants hypersensitive to bifunctional alkylating agents. Cancer Res. 45:1737-1743.
12. Ikejima, M., D. Bohannon, D.M. Gill, and L.H. Thompson (1984) Poly-(ADP-ribose) metabolism appears normal in EM9, a mutagen-sensitive mutant of CHO cells. Mutat. Res. 128:213-220.
13. Jeggo, P.A., and L.M. Kemp (1983) X-ray sensitive mutants of Chinese hamster ovary cell line: Isolation and cross-sensitivity to other DNA-damaging agents. Mutat. Res. 112:313-327.
14. Karentz, D., and J.E. Cleaver (1985) Transfer of Chinese hamster DNA repair gene(s) into repair-deficient human cells (xeroderma pigmentosum). Abstract, Thirty-Third Ann. Mtg. Radiat. Res. Soc., Los Angeles, California, p. 92.
15. Kemp, L.M., S.G. Sedgwick, and P.A. Jeggo (1984) X-ray sensitive mutants of Chinese hamster ovary cells defective in double-strand break rejoining. Mutat. Res. 132:189-196.
16. Kucherlapati, R., and A.I. Skoultchi (1984) Introduction of purified genes into mammalian cells. CRC Crit. Rev. Biochem. 16: 349-379.
17. La Belle, M., S. Linn, and L.H. Thompson (1984) Apurinic/apyrimidinic endonuclease activities appear normal in the CHO-cell ethyl methanesulfonate-sensitive mutant, EM9. Mutat. Res. 141:41-44.
18. Lambert, W.C., and M.W. Lambert (1985) Co-recessive inheritance: A model for DNA repair, genetic disease and carcinogenesis. Mutat. Res. 145:227-234.
19. Lehmann, A.R. (1982) Three complementation groups in Cockayne syndrome. Mutat. Res. 106:347-356.

20. MacInnes, M.A., J.M. Bingham, L.H. Thompson, and G.F. Strniste (1984) DNA-mediated co-transfer of excision repair capacity and drug resistance into Chinese hamster ovary mutant cell line UV-135. Mol. Cell. Biol. 4:1152-1158.
21. Maher, V.M., D.J. Dorney, A.L. Mendrala, B. Konze-Thomas, and J.J. McCormick (1979) DNA excision-repair processes in human cells can eliminate the cytotoxic and mutagenic consequences of ultraviolet irradiation. Mutat. Res. 62:311-323.
22. Maher, V.M., and J.J. McCormick (1984) Role of DNA lesions and excision repair in carcinogen-induced mutagenesis and transformation. In Biochemical Basis of Chemical Carcinogenesis, H. Greim, R. Jung, M. Kramer, H. Marquardt, and F. Oesch, eds. Raven Press, New York, pp. 143-159.
23. Mulligan, R.C., and P. Berg (1981) Selection for animal cells that express the Escherichia coli gene coding for xanthine-guanine phosphoribosyltransferase. Proc. Natl. Acad. Sci., USA 78:2072-2076.
24. Murnane, J.P., and R.B. Painter (1982) Complementation of the defects in DNA synthesis in irradiated and unirradiated ataxia telangiectasia cells. Proc. Natl. Acad. Sci., USA 79:1960-1963.
25. Pinkel, D., L.H. Thompson, J.W. Gray, and M. Vanderlaan (1985) Measurement of sister chromatid exchanges at very low BrdUrd substitution levels using a monoclonal antibody. Cancer Res. 45:5795-5798.
26. Royer-Pokora, B., and W.A. Haseltine (1984) Isolation of UV-resistant revertants from a xeroderma pigmentosum complementation group A cell line. Nature 311:390-392.
27. Sato, K., and N. Hieda (1979) Isolation of a mammalian cell mutant sensitive to 4-nitroquinoline-1-oxide. Int. J. Radiat. Biol. 35:83-87.
28. Sato, K., N. Hieda-Shiomi, and H. Hama-Inaba (1983) X-ray sensitive mutant mouse cells with various sensitivities to chemical mutagens. Mutat. Res. 121:281-285.
29. Setlow, R.B. (1978) Repair deficient human disorders and cancer. Nature 271:713-717.
30. Shiomi, T., N. Hieda-Shiomi, and K. Sato (1982) Isolation of UV-sensitive mutants of mouse L5178Y cells by a cell suspension spotting method. Somat. Cell Genet. 8:329-345.
31. Siciliano, M.J., A.V. Carrano, and L.H. Thompson (1985) Assignment of a human DNA repair gene associated with sister chromatid exchange to chromosome 19. Science (Manuscript submitted for publication.)
32. Stamato, T.D., and C.A. Waldren (1977) Isolation of UV-sensitive variants of CHO-K1 by nylon cloth replica plating. Somat. Cell Genet. 3:431-440.
33. Stamato, T.D., R. Weinstein, A. Giaccia, and L. Mackenzie (1983) Isolation of cell cycle-dependent gamma ray-sensitive Chinese hamster ovary cell. Somat. Cell Genet. 9:165-173.
34. Tanaka, K., K. Kawai, Y. Kumahara, M. Ikenaga, and Y. Okada (1981) Genetic complementation groups in Cockayne syndrome. Somat. Cell Genet. 7:445-455.
35. Thompson, L.H., J.S. Rubin, J.E. Cleaver, G.F. Whitmore, and K. Brookman (1980) A screening method for isolating DNA repair-deficient mutants of CHO cells. Somat. Cell Cenet. 6:391-405.
36. Thompson, L.H., D.B. Busch, K. Brookman, C.L. Mooney, and D.A. Glaser (1981) Genetic diversity of UV-sensitive DNA repair mutants of Chinese hamster ovary cells. Proc. Natl. Acad. Sci., USA 78:3734-3737.
37. Thompson, L.H., K.W. Brookman, L.E. Dillehay, C.L. Mooney, and A.V. Carrano (1982) Hypersensitivity to mutation and sister-chromatid-exchange induction in CHO cell mutants defective in incising DNA containing UV lesions. Somat. Cell. Genet. 8:759-773.

38. Thompson, L.H., K.W. Brookman, L.E. Dillehay, A.V. Carrano, J.A. Mazrimas, C.L. Mooney, and J.L. Minkler (1982) A CHO-cell strain having hypersensitivity to mutagens, a defect in DNA strand-break repair, and an extraordinary baseline frequency of sister chromatid exchange. Mutat. Res. 95:427-440.
39. Thompson, L.H., and A.V. Carrano (1983) Analysis of mammalian cell mutagenesis and DNA repair using in vitro selected CHO cell mutants. In Cellular Responses to DNA Damage, E.C. Friedberg and B.R. Bridges, eds. Alan R. Liss, Inc., New York, pp. 125-143.
40. Thompson, L.H., C.L. Mooney, K. Burkhart-Schultz, A.V. Carrano, and M.J. Siciliano (1985) Correction of a nucleotide-excision-repair mutation by human chromosome 19 in hamster-human hybrid cells. Somat. Cell Mol. Genet. 11:87-92.
41. Thompson, L.H., K.W. Brookman, J.L. Minkler, J.C. Fuscoe, K.A. Henning, and A.V. Carrano (1985) DNA-mediated transfer of a human DNA repair gene that controls sister chromatid exchange. Mol. Cell. Biol. 5:881-884.
42. Thompson, L.H., C. Mooney, and K. Brookman (1985) Genetic complementation between UV-sensitive CHO mutants and xeroderma pigmentosum fibroblasts. Mutat. Res. 150:423-429.
43. Ullu, E., and C. Tschudi (1984) Alu sequences are processed 7SL RNA genes. Nature 312:171-172.
44. Waldren, C.A., D. Snead, and T. Stamato (1983) Restoration of normal resistance to killing and of postreplication recovery (PRR) in CHO-UV-1 cells by transformation with hamster or human DNA. In Cellular Responses to DNA Damage, E.C. Friedberg and B.R. Bridges, eds. Alan R. Liss, Inc., New York, pp. 637-646.
45. Westerveld, A., J.H.J. Hoeijmakers, M. van Duin, J. de Wit, H. Odijk, A. Pastink, R.D. Wood, and D. Bootsma (1984) Molecular cloning of a human DNA repair gene. Nature 310:425-429.
46. Wood, R.D., and H.J. Burki (1982) Repair capability and the cellular age response for killing and mutation induction after UV. Mutat. Res. 95:505-514.
47. Zdzienicka, M.Z., and J.W.I.M. Simons (1985) Analysis of repair processes by the determination of the induction of cell killing and mutations in two repair deficient Chinese hamster ovary cell lines. (Manuscript submitted for publication.)

GENETIC AND CHEMICAL FACTORS AFFECTING CHEMICAL MUTAGENESIS IN CULTURED MAMMALIAN CELLS

Y. Kuroda

Laboratory of Phenogenetics
National Institute of Genetics
Mishima, Shizuoka 411, Japan

INTRODUCTION

Cultured mammalian cells are useful test systems for detecting chemical mutagens in our environment. These test systems can operate at relatively moderate initial costs in the short time of about 2 weeks, resulting in savings in costs and time, compared with animal test systems. In addition, the effects of mutagens can be estimated quantitatively in cultured mammalian cell systems. Thus, the dose-rate effect of chemicals can be demonstrated along with the possible existence of repair mechanisms in mammalian cells (22,37).

Recently, it has been found that cultured mammalian cell systems can be used for detecting some carcinogens that cannot be detected for their mutagenicity by microbial test systems, including the Ames test. The mutagenicity of aniline, nitrobenzene, safrole, and o-toluidine was shown in Chinese hamster V79 cells (23,24). In this chapter, the genetic and chemical factors affecting chemical mutagenesis in cultured Chinese hamster V79 cells are described.

MATERIALS AND METHODS

Cell and Culture

The cell line used was the strain of Chinese hamster lung (V79) cells isolated by Ford and Yerganian (11). The cells, maintained in Eagle's Minimum Essential Medium (Nissui Seiyaku Co., Tokyo, Japan) supplemented with 10% fetal bovine serum (Gibco Lab., Grand Island, New York), were held in 60-mm plastic petri dishes (Lux Sci. Corp., Newbury Park, California, No. 5216, 2-mm grid) under a controlled atmosphere of 5% CO_2 and 95% air at 37.5°C. They were mycoplasm-free and gave a colony-forming activity of more than 80% in the above culture conditions. Cells at exponential growth phase in monolayer were dissociated by treatment with 0.25% trypsin (Difco. 1:250) solution, and centrifuged at 1,500 rpm for 5 min. The cells were resuspended in culture medium and used for experiments.

Cytotoxicity Assay

The cytotoxic effect of chemicals was examined by determining the colony-forming activity of cells, as we have shown elsewhere (22,24). Inocula of 1-4 x 10^2 cells were incubated in 60-mm petri dishes in normal medium for 20 hr, washed 2 times with Hanks' solution, treated with chemicals in Hanks' solution for 3 hr at 37.5°C, washed again 2 times with Hanks' solution, and incubated in normal medium for 6 days. The colonies formed were fixed in absolute methyl alcohol and stained with May-Grünwald-Giemsa and Giemsa. The number of colonies containing more than 50 cells was scored under a binocular microscope.

Mutagenicity Assay

Inocula of 2 x 10^5 cells were incubated in normal medium for 20 hr, washed twice in Hanks' solution, and treated with chemicals in Hanks' solution for 3 hr at 37.5°C. These cells were again washed 2 times with Hanks' solution and incubated in normal medium at 37.5°C for expression times of 6 days for 8-azaguanine (8AG)- and 6-thioguanine (6TG)-resistant mutation induction, or 3 days for ouabain (OUA)-resistant mutation induction.

The cells were dissociated by treatment with 0.25% trypsin solution, centrifuged, and inoculated into fresh dishes. Inocula of 2 x 10^5 cells were incubated in medium containing 20 μg/ml 8AG, 5 μg/ml 6TG, or 1 mM OUA for 2 weeks. Colonies that formed were fixed in methyl alcohol and stained with May-Grünwald-Giemsa and Giemsa. The number of colonies containing more than 50 cells was scored by the same procedure as described in the cytotoxicity assay. Inocula of 1-4 x 10^2 dissociated cells were incubated in normal medium for 6 days. The colonies formed were fixed, stained, and scored. The numbers of mutant colonies were corrected by the colony-forming activity of replated cells cultured in normal medium. The induced mutation frequency was calculated from the number of mutant colonies of treated cells reduced by the number of mutant colonies of untreated control cells and divided by the colony-forming activity of replated cells. The induced mutation frequency was expressed as the number of induced mutant colonies per 10^5 survivors.

Chemicals

Chemicals used for inducing mutations were ethyl methanesulfonate (EMS) (Aldrich Chemical Co., Inc., Milwaukee, Wisconsin), methyl methanesulfonate (MMS) (Aldrich), N-methyl-N'-nitro-N-nitrosoguanidine (MNNG) (Aldrich), Trp-P-1 (tryptophan pyrolysis product; a gift from Dr. T. Sugimura, National Cancer Center, Japan), aniline, o-chloroaniline, m-chloroaniline, nitrobenzene, hexachlorobenzene, and chloroform (Wako Pure Chem. Ind., Osaka, Japan). Norharman (Aldrich) was used as a co-mutagen with aniline. Vitamin C (L-ascorbic acid; Wako) was used as an antimutagen.

Chemicals used for selecting mutant colonies were 8AG (Sigma Chemical Co., St. Louis, Missouri), 6TG (Wako Pure Chem. Ind.), and OUA (Aldrich).

RESULTS

Genetic Factors Affecting Chemical Mutagenesis

The frequency of mutations induced by the same chemical is markedly affected by the genetic markers used. In the present experiments, 8AG-,

6TG-, and OUA-resistances ($8AG^r$, $6TG^r$, and OUA^r) were used as markers to detect mutations induced by chemicals.

Figure 1 shows the comparison between $6TG^r$ and OUA^r mutations induced by EMS. The frequency of $6TG^r$ mutations induced by EMS was more than 3 times higher than that of OUA^r mutations induced by EMS at the same concentrations. It was found that the mutation frequency was also affected by different selective agents which permitted the growth of mutant colonies deficient in the same enzyme, hypoxanthine-guanine-phosphoribosyl transferase (HGPRT). Figure 2 indicates that $8AG^r$ mutations were induced more frequently than $6TG^r$ mutations by treatment with EMS at the same concentrations.

Methyl methanesulfonate also induced $6TG^r$ mutations at a higher frequency than OUA^r mutations, which were not induced to a significant frequency at concentrations of less than 100 μg/ml (Fig. 3).

Table 1 shows the induced mutation frequency of $8AG^r$, $6TG^r$, and OUA^r after treatment with almost the same concentrations of MNNG. This indicates that $8AG^r$ mutations were more frequently induced than $6TG^r$ mutations, which were, in turn, more frequently induced than OUA^r mutations. These results indicate that mutations induced by alkylating agents, such as EMS, MMS and MNNG, were more frequent at the HGPRT locus ($8AG^r$ and $6TG^r$) than at the Na^+/K^+ ATPase locus (OUA^r), and that the $8AG^r$ mutations were induced at a higher frequency than the $6TG^r$ mutations at the same HGPRT locus.

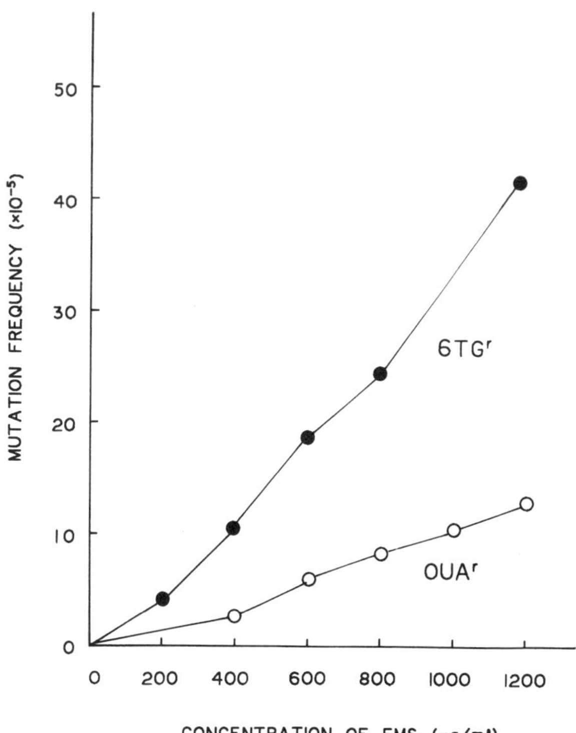

Fig. 1. Comparison of $6TG^r$ and OUA^r mutations induced by EMS in Chinese hamster V79 cells.

It has been found that some amino acid pyrolysis products have mutagenic activity in microorganisms (36,38). Trp-P-1, Trp-P-2, and Glu-P-1 induced $8AG^r$ mutations in human diploid cells and Chinese hamster V79 cells (19,21,22). Figure 4 shows the comparison between frequencies of $8AG^r$ and $6TG^r$ mutations induced by Trp-P-1 with and without S-9 mix. This indicates that $8AG^r$ mutations were induced at a higher frequency than OUA^r mutations, and that the metabolic activation of Trp-P-1 with S-9 mix enhanced the ability of Trp-P-1 to induce $8AG^r$ mutations.

Next, the mutagenic activity of some aromatic compounds on $6TG^r$ and OUA^r mutations in Chinese hamster V79 cells was examined. Aniline is widely used in producing dyestuffs, mordants, intermediates of various compounds, rubber, medicines, and many organic compounds. In our environment, aniline has been detected in the water and substratum of most Japanese rivers, as revealed by the investigations carried out by the Environmental Agency.

It has been reported that aniline was carcinogenic in rats, and produced sarcomas, fibromas, and fibrosarcomas in the spleen and adenomas in the pituitary gland (7). Aniline was, however, negative in the Ames tests (26). It did not induce chromosome aberrations or sister-chromatid exchanges in Chinese hamster Don and CHL cells (1,15). No transformation was induced by aniline in Syrian hamster BHK cells (35).

In the present experiments, aniline induced $8AG^r$ mutations which increased gradually with increased concentrations of aniline (Fig. 5). With the addition of S-9 mix, the mutation frequency induced by aniline

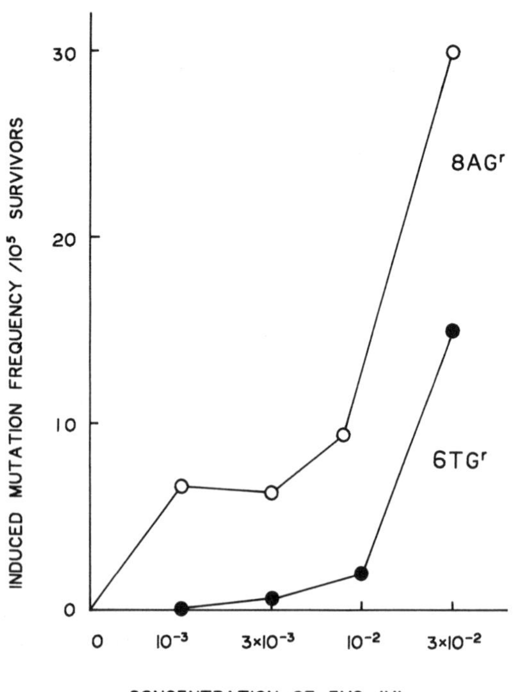

Fig. 2. Comparison of $8AG^r$ and $6TG^r$ mutations induced by EMS in Chinese hamster V79 cells.

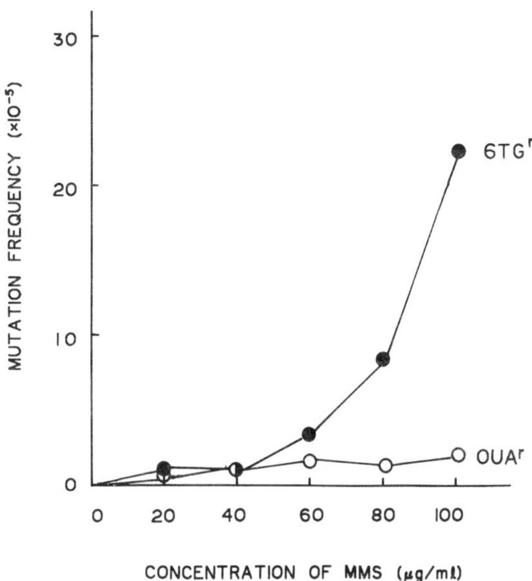

Fig. 3. Comparison of 6TG^r and OUA^r mutations induced by MMS in Chinese hamster V79 cells.

decreased to a low level. OUA^r mutations were not significantly induced by aniline with or without S-9 mix. The 8AG^r mutations induced by aniline were enhanced by the addition of norharman. At as low a concentration as 0.5 μg/ml of aniline, which induced mutations at a low frequency, the addition of norharman enhanced mutations effectively. This enhancing effect of norharman was detected only when cells were treated with norharman simultaneously with aniline. When cells were treated successively with norharman and then with aniline, or pretreated with norharman 4 hr before treatment with aniline, the induced mutation frequency was reduced to one-fourth that in the simultaneous treatment with norharman and aniline.

Halogen derivatives of aniline and o- and p-nitroaniline had no effects on the induction of morphological transformations in Syrian hamster cells (29). In the present experiments, o-, m-, and p-chloroaniline also induced 8AG^r mutations, but did not induce OUA^r mutations (Fig. 6 and 7).

Tab. 1. Comparison of frequencies of 8AG^r, 6TG^r, and OUA^r mutations induced by MNNG in Chinese hamster V79 cells.

Mutagen	Concentration (M)	Number of cells inoculated	Selective drug	Concentration	Number of mutants	Induced mutation frequency per 10 survivors
MNNG	10^{-5}	4×10^4	8AG	20 μg/ml	140	348
MNNG	10^{-5}	4×10^4	6TG	5 μg/ml	100	244
MNNG	3×10^{-5}	5×10^4	OUA	1 mM	60	96

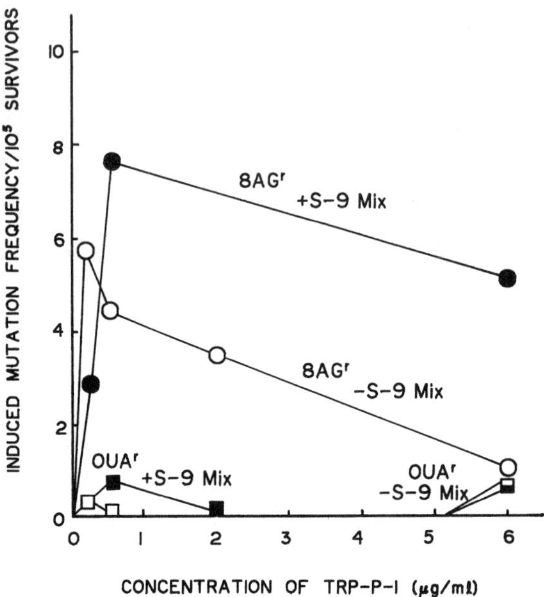

Fig. 4. Comparison of $8AG^r$ and OUA^r mutations induced by Trp-P-1 with and without S-9 mix in Chinese hamster V79 cells.

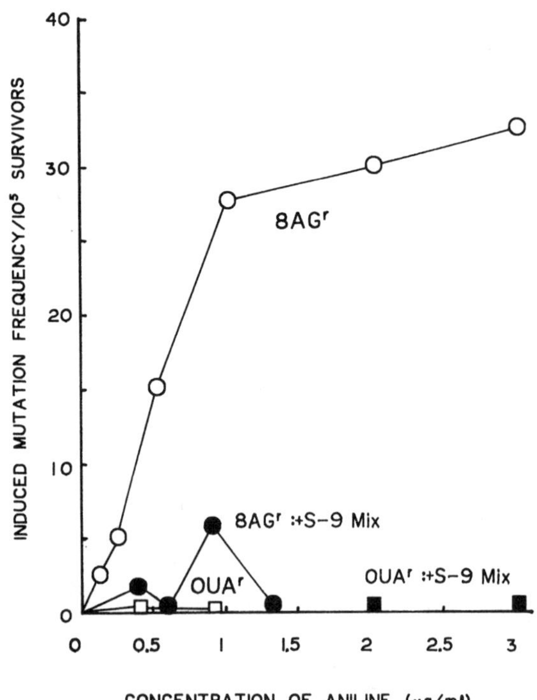

Fig. 5. Comparison of $8AG^r$ and OUA^r mutations induced by aniline with and without S-9 mix in Chinese hamster V79 cells.

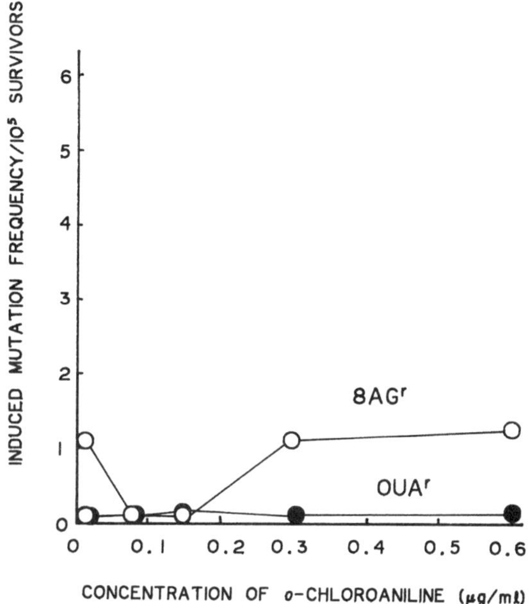

Fig. 6. Comparison of $8AG^r$ and OUA^r mutations induced by o-chloroaniline in Chinese hamster V79 cells.

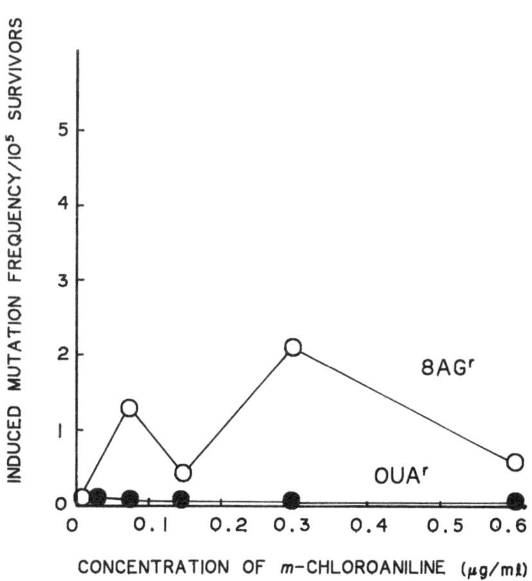

Fig. 7. Comparison of $8AG^r$ and OUA^r mutations induced by m-chloroaniline in Chinese hamster V79 cells.

Nitrobenzene is a compound that was found negative in the Ames *Salmonella* mutagenicity tests (8), and that did not induce transformations in Syrian hamster BHK21 cells (35). This chemical is produced in large amounts and used for dyestuffs, oxidants, mordants, and dust protectors. Nitrobenzene induced $8AG^r$ mutations and enhanced the mutation frequency when S-9 mix was added to the medium (Fig. 8). Nitrobenzene induced OUA^r mutations at low frequencies when it was activated by S-9 mix. Without S-9 mix, no OUA^r mutations were induced.

Hexachlorobenzene is a chemical also used for intermediates of various compounds and rubber additives. It has been reported that hexachlorobenzene did not induce dominant lethal mutations in Wistar rats (18). In the present experiments, hexachlorobenzene induced $8AG^r$ mutations at low frequencies, but did not induce OUA^r mutations (Fig. 9).

Chloroform is a solvent and a cleaning agent, and is used as an anesthetic in medical treatment and as a carminative and flavoring agent in medications. In the present experiments, chloroform also induced $8AG^r$ mutations, but did not induce OUA^r mutations (Fig. 10).

Table 2 summarizes the results of the inducibility of mutations at the HGPRT and the Na^+/K^+ ATPase loci by chemicals tested in the present experiments. These results indicate that all compounds that were positive in Ames tests (EMS, MMS, MNNG, and Trp-P-1) induced $8AG^r$ and/or $6TG^r$ and OUA^r mutations. On the other hand, compounds that were negative in Ames tests (nitrobenzene, aniline, o- and m-chloroaniline, and hexachlorobenzene) induced $8AG^r$ mutations, but did not induce OUA^r mutations.

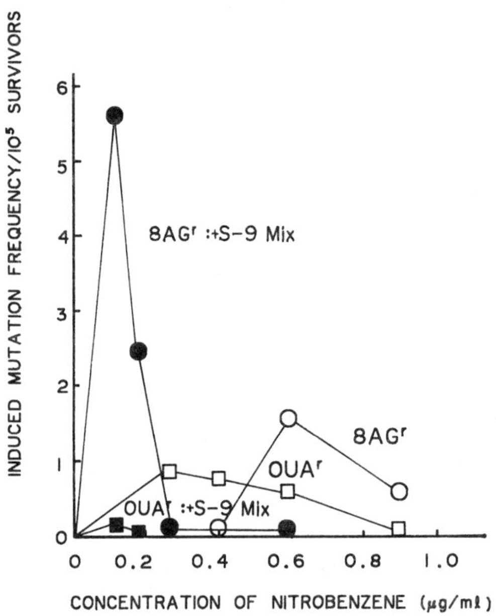

Fig. 8. Comparison of $8AG^r$ and OUA^r mutations induced by nitrobenzene with and without S-9 mix in Chinese hamster V79 cells.

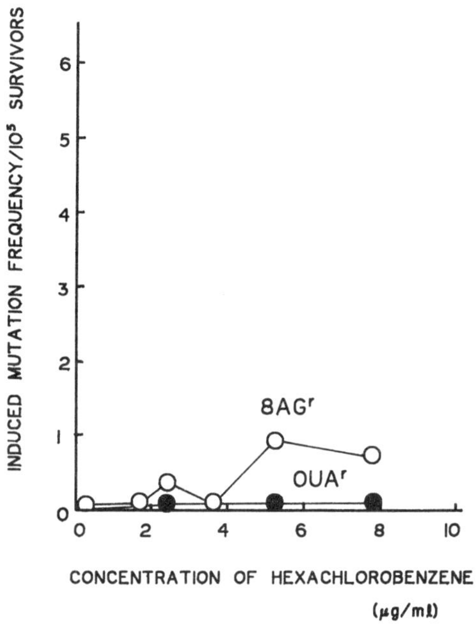

Fig. 9. Comparison of $8AG^r$ and OUA^r mutations induced by hexachlorobenzene in Chinese hamster V79 cells.

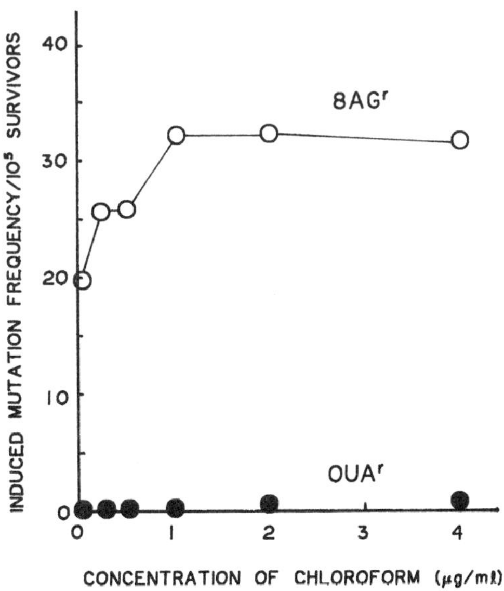

Fig. 10. Comparison of $8AG^r$ and OUA^r mutations induced by chloroform.

Tab. 2. Difference in inducibility of mutations among different genetic markers by various chemicals in Chinese hamster V79 cells.

Mutagen	Ames test	Genetic markers		
		$8AG^r$	$6TG^r$	OUA^r
Ethyl methanesulfonate	+	+	+	+
Methyl methanesulfonate	+	+	+	±
MNNG	+	+	+	+
Trp-P-1	+	+		±
Nitrobenzene	–	+		±
Aniline	–	+		–
o-Chloroaniline		+		–
m-Chloroaniline		+		–
Hexachlorobenzene		+		–
Chloroform		+		–

Chemical Factors Affecting Chemical Mutagenesis

It has been reported that the frequency of mutations induced by chemicals was influenced by various other chemicals. Norharman, as described above, is an example of co-mutagenic activity with aniline. Chemicals added together with mutagens may activate or inactivate the mutagens, resulting in enhancing or lowering the induced mutation frequency.

We found that vitamin C had a marked effect in reducing $6TG^r$ mutations induced by EMS in Chinese hamster V79 cells. Vitamin C is abundantly contained in green tea, some fruits (such as strawberries, pineapples, and oranges), and some vegetables (such as cabbages, spinach, and green peppers). It is effective in preventing scurvy, and its antioxidant activity is useful for detoxification of some drugs and for protection against rancidity in foodstuffs.

Vitamin C alone had no detectable effect on cell survival at concentrations of less than 100 μg/ml. At a concentration of 300 μg/ml, the colony-forming activity of cells was completely inhibited (Tab. 3). On the other hand, EMS had a cytotoxic effect on survival of V79 cells (Fig. 11). In the presence of vitamin C, the cytotoxic effect of EMS decreased markedly.

Vitamin C also had a marked effect in reducing $6TG^r$ mutations induced by EMS. Vitamin C alone had no detectable activity in inducing $6TG^r$ mutations (Tab. 4). EMS had a strong inducing activity of $6TG^r$ mutations in the absence of vitamin C (Fig. 12). This mutagenic activity of EMS was reduced by the addition of 100 μg/ml vitamin C.

Table 5 shows the comparison of survival and induced mutations of cells treated with EMS in the absence and in the presence of vitamin C. The survival of cells treated with EMS was maintained at almost the control level when vitamin C was present at the time of EMS treatment. The frequency of $6TG^r$ mutations induced by EMS was also decreased to about one-fourth by the addition of vitamin C.

Tab. 3. Effect of vitamin C on survival of Chinese hamster V79 cells.

Concentration of vitamin C (μg/ml)	Time of treatment (hr)	Number of cells inoculated	Number of colonies formed	Surviving fraction
0.0	3	100	87	1.00
0.1	3	100	87	1.00
1.0	3	100	87	1.00
5.0	3	100	80	0.92
10.0	3	100	75	0.86
50.0	3	100	74	0.85
100.0	3	100	69	0.79
300.0	3	100	0	0.00

DISCUSSION

The frequency of mutations induced by chemicals was strikingly different among the genetic loci used as markers for selection of mutant colonies. Four chemicals which were positive in the Ames tests induced $8AG^r$ mutations at higher frequencies than OUA^r mutations. Seven other chemicals which were negative or not tested in the Ames tests induced only $8AG^r$ mutations, but did not induce OUA^r mutations.

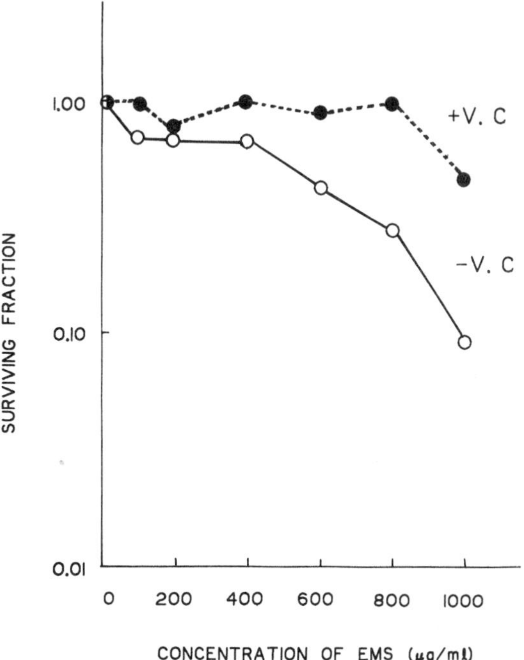

Fig. 11. Effect of vitamin C on the cytotoxic effect of EMS on Chinese hamster V79 cells.

Tab. 4. Effect of vitamin C on the induction of $6TG^r$ mutations in Chinese hamster V79 cells.

Concentration of vitamin C (μg/ml)	Time of treatment (hr)	Cell survival	Number of $6TG^r$ mutants	Induced mutation frequency per 10^5 survivors
0.0	3	0.81	0.2	
0.1	3	0.84	0.4	0.2
1.0	3	1.10	0.0	0.0
5.0	3	1.09	0.0	0.0
10.0	3	0.94	1.0	0.5
50.0	3	0.62	1.0	0.7
100.0	3	1.15	0.2	0.0

The $8AG^r$ and $6TG^r$ mutations are usually mediated by a mutation leading to the loss of or reduction in HGPRT activity. It has been reported, however, that variant clones resistant to high levels of 8AG can often contain significant levels of HGPRT, sometimes at or near the wild-type level (9, 12). On the other hand, $6TG^r$ mutants were completely deficient in HGPRT activity with rare exceptions (12,32). This indicates that $6TG^r$ resistance is better for the HGPRT-deficient marker than $8AG^r$ resistance.

Beaudet et al. (4) found that $8AG^r$ mutants of Chinese hamster cells induced by EMS had cross-reacting materials to HGPRT and the ability to induce reversions to wild-type. This suggests that mutations in the HGPRT enzyme activity may be produced by alterations in the structural gene, such as base changes, frameshifts, and also various lengths of deletions in the DNA coding the HGPRT enzyme molecule.

Resistance to OUA is completely different from resistance to 8AG or 6TG. It is the result of an alteration of the membrane ATPase involved in Na^+/K^+ transport (27). This resistant allele behaves as a co-dominant in somatic hybridization experiments (3). The human analog, Lesch-Nyhan cells, provides a genetic basis for resistance to 8AG or 6TG. For resistance to OUA, however, there is no evidence of a human analog for this mutant.

A comparison of $8AG^r$ and OUA^r mutation induction systems showed that there was broad agreement between the two systems for many mutagens tested (2). The γ-irradiation, however, was nonmutagenic in the OUA^r system, but induced $8AG^r$ mutations. This suggests that OUA^r mutations may be produced by point mutations at the DNA base level, base changes, and frameshifts, but not by deletions or deficiencies. The spontaneous mutation rates to OUA^r (Na^+/K^+ ATPase locus) in mammalian cells in culture have been reported to be 10- to 100-fold lower than those of other loci, such as $8AG^r$ (HGPRT locus) (9,16,17). Under improved conditions, the spontaneous mutation rate to OUA^r was 2- to 10-fold less than that to $8AG^r$ in human fibroblasts (10).

In the present experiments, all chemicals that were positive in the Ames test, alkylating agents, and a tryptophan pyrolysis product induced $6TG^r$ or $8AG^r$ mutations at higher frequencies than those of OUA^r mutations

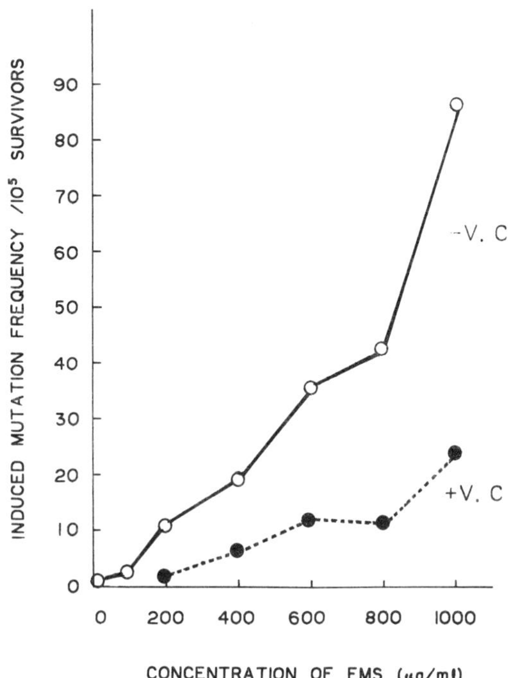

Fig. 12. Effect of vitamin C on $6TG^r$ mutations induced by EMS in Chinese hamster V79 cells.

in Chinese hamster V79 cells. Furthermore, all chemicals that were negative or weak in the Ames test, aniline, and some aromatic compounds induced only $8AG^r$ mutations, but did not induce OUA^r mutations.

The cDNA sequence of the HGPRT gene was identified from the wild-type mouse neuroblastoma cell line with amplified genomic sequences. By using these sequences as probes, it was found that one recombinant clone contained an 1100-bp insert corresponding to 70% of the HGPRT mRNA. The 1.6-kilobase mRNA species exceeded the requirement (700 nucleotides) for encoding the HGPRT subunit; 50% of the HGPRT mRNA was noncoding and the recombinant contained a significant amount of HGPRT-encoding sequence (6).

The difference in the inducibility of mutations among the HGPRT and the Na^+/K^+ ATPase loci may be due to differences in the target sizes of both structural genes and to the nature of mutations produced.

It has been reported that vitamin C inhibited mutations in microbial systems induced by N-nitroso compounds (14) and by carcinogens (30). Ascorbate has also been shown to reduce the DNA-damaging action of several carcinogens (25), and to inhibit carcinogen-induced neoplasia in animals (28,31,33). The morphological transformation induced by 3-methylcholanthrene was also inhibited by the addition of vitamin C in C3H10T1/2 cells (5). No reports appeared to be directly relevant to the inhibition by vitamin C of $6TG^r$ mutations induced by EMS in mammalian cells.

On the other hand, vitamin C is known to induce chromosome aberrations and unscheduled DNA synthesis (34), and was mutagenic in the Ames

Tab. 5. Effects of vitamin C on the cytotoxic and mutagenic activity of EMS in Chinese hamster V79 cells.

Concentration of EMS (μg/ml)	Survival		Mutation (per 10^5 survivors)	
	Without vitamin C	With vitamin C	Without vitamin C	With vitamin C
0	1.00	1.00		
100	0.72	1.29	3.0	2.5
200	0.70	0.99	11.0	1.2
400	0.68	1.39	19.5	7.7
600	0.44	1.26	36.0	12.3
800	0.29	1.29	43.0	12.2
1,000	0.09	0.58	87.5	24.7

Salmonella test (13,34). These deleterious effects of vitamin C were produced by a high dose of vitamin C at more than 10^{-3} M. At concentrations of less than 10^{-3} M, vitamin C had no such effects. Without extra vitamin C in the diet, concentrations in human tissues are reported to range from 2 to 50 μg/100 g or approximately 1-25 x 10^{-4} M. Vitamin C at these low concentrations may play a role in the prevention of mutagenic actions of chemicals taken into the body from our environment.

In the present experiments, 100 μg/ml (5.6 x 10^{-4} M) of vitamin C, while having had no marked effect on cell survival and mutation induction, reduced the cytotoxic effect of EMS to the nontreated control level, and reduced the EMS-induced $6TG^r$ mutation frequency to one-fourth the level of EMS-treated cells without vitamin C.

The mechanism by which vitamin C inhibited MNNG-induced mutagenesis in *Salmonella* TA1530 was through direct reaction of MNNG with vitamin C, which led to the decomposition of MNNG and to its deactivation as a mutagen (14). The inhibitory effects of vitamin C on the cytotoxic action and mutations induced by EMS in the present experiments are an interesting problem in understanding the mechanism of vitamin C.

SUMMARY

The genetic and chemical factors affecting chemical mutagenesis in cultured Chinese hamster V79 cells are described. The frequency of mutations induced by the same chemicals at the same concentrations was markedly affected by the genetic markers used. Some chemicals that were positive in the Ames tests induced $8AG^r$ and $6TG^r$ mutations at higher frequencies than OUA^r mutations. Carcinogens that were negative in the Ames tests induced only $8AG^r$ mutations, but did not induce OUA^r mutations.

As a compound affecting mutations induced by EMS, vitamin C showed marked effects in reducing the cytotoxicity and frequency of $6TG^r$ mutations induced by EMS when cells were treated simultaneously with vitamin C at a concentration of 100 μg/ml.

ACKNOWLEDGEMENTS

This work was supported in part by a Grant-in-Aid for Environmental Science from the Ministry of Education, Science and Culture of Japan. The author wishes to thank Miss Y. Takada for her technical assistance during this work. This is contribution No. 1660 from the National Institute of Genetics, Mishima, Japan.

REFERENCES

1. Abe, S., and M. Sasaki (1977) Chromosome aberrations and sister chromatid exchanges in Chinese hamster cells exposed to various chemicals. J. Natl. Cancer Inst. 58:1635-1641.
2. Arlett, C.F., D. Turnbull, S.A. Harcourt, A.R. Lehman, and C.M. Colella (1975) A comparison of the 8-azaguanine and ouabain-resistance systems for the selection of induced mutant Chinese hamster cells. Mutat. Res. 33:261-278.
3. Baker, R.M., D.M. Brunette, R. Mankovitz, L.H. Thompson, G.F. Whitmore, L. Siminovitch, and J.E. Till (1974) Ouabain-resistant mutants of mouse and hamster cells in culture. Cell 1:9-12.
4. Beaudet, A.L., D.J. Roufa, and C.T. Caskey (1973) Mutations affecting the structure of hypoxanthin:purine phosphoribosyl transferase of cultured Chinese hamster cells. Proc. Natl. Acad. Sci., USA 70:320-324.
5. Benedict, W.F., W.L. Wheatley, and P.A. Jones (1980) Inhibition of chemically induced morphological transformation and reversion of the transformed phenotype by ascorbic acid in C3H/10T1/2 cells. Cancer Res. 40:2796-2801.
6. Brennand, J., A.C. Chinault, D.S. Konecki, D.W. Melton, and C.T. Caskey (1982) Cloned cDNA sequences of the hypoxanthine/guanine phosphoribosyltransferase gene from a mouse neuroblastoma cell line found to have amplified genomic sequences. Proc. Natl. Acad. Sci., USA 79: 1950-1954.
7. Carcinogenesis Testing Program (1978) Report on the Bioassay of Aniline Hydrochloride for Possible Carcinogenicity, U.S. Department of Health, Education and Welfare, National Institutes of Health, DHEW Publication No. 78-1385, pp. 1-53.
8. Chiu, C.W., H.L. Lee, C.Y. Wang, and G.T. Bryan (1978) Mutagenicity of some commercially available nitro compounds of *Salmonella typhimurium*. Mutat. Res. 58:11-12.
9. DeMars, R., and K.R. Held (1972) The spontaneous azaguanine-resistant mutants of diploid human fibroblasts. Humangenetik 16:87-110.
10. Elmore, E., and J.C. Barrett (1982) Measurement of spontaneous mutation rates at the Na^{+}/K^{+} ATPase locus (ouabain resistance) of human fibroblasts using improved growth conditions. Mutat. Res. 97:393-404.
11. Ford, D.K., and G. Yerganian (1958) Observations on the chromosomes of Chinese hamster cells in tissue culture. J. Natl. Cancer Inst. 21: 393-425.
12. Gillin, F.D., D.S. Roufa, A.L. Beaudet, and T.T. Caskey (1972) 8-Azaguanine resistance in mammalian cells. I. Hypoxanthin-guanine phosphoribosyltransferase. Genetics 72:239-252.
13. Guttenplan, J.B. (1977) Inhibition by L-ascorbate of bacterial mutagenesis induced by two *N*-nitroso compounds. Nature 268:368-370.
14. Guttenplan, J.B. (1978) Mechanisms of inhibition by ascorbate of microbial mutagenesis by *N*-nitroso compounds. Cancer Res. 38:2018-2022.
15. Ishidate, Jr., M., and S. Odashima (1977) Chromosome tests with 134 compounds in Chinese hamster cells *in vivo*--A screening for chemical carcinogens. Mutat. Res. 48:337-354.

16. Jacobs, L., and R. DeMars (1978) Quantification of chemical mutagenesis in diploid human fibroblasts: Induction of azaguanine resistant mutants by N-methyl-N'-nitro-N-nitrosoguanidine. Mutat. Res. 53:29-53.

17. Jones, G.E., and P.A. Sargent (1974) Mutants of cultured Chinese hamster cells deficient in adenine phosphoribosyl transferase. Cell 2: 43-54.

18. Khera, K.S. (1975) Teratogenicity and dominant lethal studies on hexachlorobenzene in rats. Food Cosmet. Toxicol. 12:471-477.

19. Kuroda, Y., (1979) Mutagenic activity of tryptophan pyrolysis products on embryonic human diploid cells in culture. Ann. Rep. Natl. Inst. Genet. Japan 29:39.

20. Kuroda, Y. (1980) Dose-rate effects of Trp-P-1 on survival and mutation induction in cultured human diploid cells. Ann. Rep. Natl. Inst. Genet. Japan 30:47-48.

21. Kuroda, Y. (1981) Mutagenic activity of Trp-P-2 and Glu-P-1 on embryonic human diploid cells in culture. Ann. Rep. Natl. Inst. Genet. Japan 31:45-46.

22. Kuroda, Y. (1984) Dose-rate effects of chemicals on mutation induction in mammalian cells in culture. In Problems of Threshold in Chemical Mutagenesis, Y. Tazima, S. Kondo, and Y. Kuroda, eds. Environmental Mutagen Society Japan, Mishima, Shizuoka, pp. 99-108.

23. Kuroda, Y., and M. Asakura (1982) Co-mutagenic activity of non-mutagens in cultured Chinese hamster cells. Ann. Rep. Natl. Inst. Genet. Japan 32:52.

24. Kuroda, Y., A. Yokoiyama, and T. Kada (1985) Assays for the induction of mutations to 6-thioguanine resistance in Chinese hamster V79 cells in culture. In Problems in Mutation Research, Vol. 5, J. Ashby and F.J. de Serres et al., eds. World Health Organization, Elsevier Science Publishers, Amsterdam, Oxford, New York, pp. 537-545.

25. Lo, L.W., and H.F. Stich (1978) The use of short-term tests to measure the preventive action of reducing agents on formation and activation of carcinogenic nitroso compounds. Mutat. Res. 57:57-67.

26. MacCann, J., E. Choi, E. Yamasaki, and B.N. Ames (1975) Detection of carcinogens as mutagens in the Salmonella/microsome test: Assay of 300 chemicals. Proc. Natl. Acad. Sci., USA 72:5135-5139.

27. Mankovitz, R., M. Buchwald, and R.M. Baker (1974) Isolation of ouabain-resistant human diploid fibroblasts. Cell 3:221-226.

28. Mirvish, S.S., A.F. Pelfrene, H. Garcia, and P. Shubik (1976) Effect of sodium ascorbate on tumor induction in rats treated with morpholine and sodium nitrite, and with nitrosomorphiline. Cancer Lett. 2:109-114.

29. Pienta, R.J., J.A. Poiley, and W.B. Lebherz III (1977) Morphological transformation of early passage golden Syrian hamster embryo cells derived from cryopreserved primary cultures as a reliable in vitro bioassay for identifying diverse carcinogens. Intl. J. Cancer 19:642-655.

30. Rosin, M.P., and H.F. Stich (1979) Assessment of the use of the Salmonella mutagenesis assay to determine the influence of antioxidants on carcinogen-induced mutagenesis. Intl. J. Cancer 23:722-727.

31. Schlegel, J.R. (1975) Proposed uses of ascorbic acid in prevention of bladder carcinoma. Ann. N.Y. Acad. Sci. 258:432-437.

32. Sharp, J.D., N.E. Capecchi, and M.R. Capecchi (1973) Altered enzymes in drug-resistant variants of mammalian tissue culture cells. Proc. Natl. Acad. Sci., USA 70:3145-3149.

33. Slaga, T.J., and W.M. Bracken (1977) The effects of antioxidants on skin tumor inhibition and aryl hydrocarbon hydroxylase. Cancer Res. 37:1631-1635.

34. Stich, H.F., J. Karim, J. Koropatnick, and L. Lo (1976) Mutagenic action of ascorbic acid. Nature 26:722-724.
35. Styles, J.A. (1978) Mammalian cell transformation in vitro. Brit. J. Cancer 37:931-936.
36. Sugimura, T., T. Kawachi, M. Nagao, T. Yahagi, Y. Seino, T. Okamoto, K. Shudo, T. Kosugi, K. Tsuji, K. Wakabayashi, T. Iitaka, and A. Itai (1977) Mutagenic principle(s) in tryptophan and phenylalanine pyrolysis products. Proc. Japan Acad. 53:58-61.
37. Sugiura, K., M. Goto, and Y. Kuroda (1978) Dose-rate effects of ethyl methanesulfonate on survival and mutation induction in cultured Chinese hamster cells. Mutat. Res. 51:99-108.
38. Yamamoto, T., K. Tsuji, T. Kosuge, T. Okamoto, K. Shudo, K. Takeda, Y. Iitaka, K. Yamaguchi, Y. Seino, T. Yahagi, A. Nagao, and T. Sugimura (1978) Isolation and structure determination of mutagenic substances in L-glutamic acid pyrolysate. Proc. Japan Acad. 54B:248-250.

MECHANISMS OF CARCINOGENICITY AND ANTICARCINOGENICITY--PART I

MECHANISMS OF CARCINOGENICITY AND ANTICARCINOGENICITY:

ROLE OF DIETARY COMPONENTS

Shahbeg S. Sandhu and Michael D. Waters

Genetic Toxicology Division
Health Effects Environmental Research Laboratory
U.S. Environmental Protection Agency
Research Triangle Park, North Carolina 27711

The role of nutrition in the etiology of human disease, including cancer, has been recognized for a very long time. This knowledge and, perhaps, personal preference are reflected by the choice and balance of edibles included in the daily diet. Even at present, in certain societies where strong emphasis on traditional values has preserved the dietary patterns of ancient times, the daily diet of most people includes a wide variety of legumes, high in protease inhibitors, and selected vegetables, high in β-carotene. With our current state of knowledge, we can appreciate the role of tobacco in causing lung cancer and the role of betel quid in the etiology of oral cancer. It may be of interest to note that in certain societies, the use of tobacco and betel quid was strictly prohibited as early as the sixteenth-century because of the suspicion that these agents may be involved in the etiology of cardiovascular diseases and cancer.

Recent epidemiological studies conducted under diverse cultural conditions have shown marked differences in geographic distribution of specific types of cancer. The higher incidences of cancer have further been associated with higher consumption of certain categories of food and/or their method of preparation (1,3). The difference between the cancer incidence among immigrants with their traditional dietary habits and that of the local population has given further credence to the role of diet in causing cancer.

Our scientific views on the role of nutrition in cancer induction or prevention are based on the studies undertaken by Tannenbaum and his colleagues in the early 1940s. Although their studies were concerned with caloric intake and cancer, today we know that a variety of nutritional components are involved in the initiation or intervention, progression, or regression of cancer. The protective role of certain dietary factors such as vitamin E, carotene, selenium, glutathione, ascorbic acid, uric acid, phenols, and tocopherols is noteworthy in this regard.

Considerable progress has been made in identifying the cellular targets involved in interaction between cells and initiating carcinogens and promoters. It has been evident since the work of Miller and Miller (2) that modification of cellular DNA (leading to mutagenesis) is an important event in initiating carcinogenesis. There are, of course, exceptions to the correlation between mutagenicity and carcinogenicity. Some chemicals show strong mutagenicity when tested in in vitro test systems, but they do not induce cancer in animals. Similarly, certain carcinogens do not cause gene mutations, but induce their effects either through induction of chromosome aberrations, including aneuploidy, or alteration of processes leading to the synthesis or function of DNA, or through other less direct mechanisms.

We are becoming increasingly aware that carcinogenesis is an extremely complex phenomenon mediated through genetic and epigenetic mechanisms. Cancer is an outcome of an interaction between a specific genotype and a specific chemical (or physical) agent. The problem of mechanistic understanding of chemical carcinogenesis becomes even more intricate when we consider the fact that man is seldom exposed to individual chemicals but always to an array of chemical mixtures. As a result, the biological effects of complex chemical mixtures are difficult to comprehend and predict.

Several chapters of this Volume are devoted to the understanding of the mechanisms of carcinogenicity and anticarcinogenicity, as well as the role of diet or hormones in the induction or inhibition of carcinogenesis.

REFERENCES

1. Correa, P., E. Fontham, L.W. Pickle, V. Chen, Y. Lin, and W. Haenszel (1985) Dietary determinants of gastric cancer in south Louisiana inhabitants. J. Natl. Cancer Inst. 75:645-654.
2. Miller, J.A., and E.C. Miller (1969) The metabolic activation of carcinogenic aromatic amines and amides. Prog. Exp. Tumor Res. 11:273-301.
3. Wattenberg, L.W. (1985) Chemoprevention of cancer. Cancer Res. 45:1-8.

REDUCING THE GENOTOXIC DAMAGE IN THE ORAL MUCOSA OF BETEL QUID/TOBACCO CHEWERS

Hans F. Stich

Environmental Carcinogenesis Unit
British Columbia Cancer Research Centre
Vancouver, British Columbia, Canada V5Z 1L3

INTRODUCTION

Many millions around the world chew areca nuts (several Asiatic countries, South Pacific islands), sun-dried (Khaini tobacco; Bihar) or fermented tobacco leaves (snuff/chewing tobacco; Europe, North America), miang leaves (Thailand), khat (Yemen, Somalia), and coca leaves (South America). These plant products are frequently mixed with slaked lime, as in the case of betel quids (India, Taiwan, Papua, New Guinea), Khaini tobacco, nass or naswar (Uzbekistan, Iran, Afghanistan), or coca leaves (Peru). Some of these oral habits, in particular, the chewing of tobacco-containing betel quids, are causally involved in the etiology of oral carcinomas (7).

The abandonment of various chewing habits would appear to be the simplest and most efficient step in reducing the incidence of oral cancer. Unfortunately, this approach may not always be feasible. For example, over thousands of years, areca nut has become an integral part of Hindu religious and cultural heritage. Persons steeped in a traditional lifestyle pattern may resist rapid changes (2).

Purely economic considerations are often hurdles that are difficult to bypass. In many developing countries, tobacco is a cash crop that cannot simply be replaced by another commodity (37). In Taiwan, the planting of areca trees along rice paddies is actively encouraged. The trees do not take away any agriculturally valuable land, and the areca nut provides a relatively cheap source of enjoyment.

Finally, in countries such as Indonesia, where betel quid chewing is losing its attraction, the male population has switched to the more hazardous habit of cigarette smoking (18). Obviously, alternative ways must be sought to reduce the risk for cancer.

In this chapter we describe two different approaches for preventing oral cancers. First, a modification of the betel quid composition could conceivably result in less carcinogenic chewing mixtures. This involves the tracing of the carcinogenic compounds and the identification of those interactions within the saliva of chewers that lead to the enhancement or the reduction of genotoxic and carcinogenic effects.

Second, an elevation of beta-carotene levels in the target tissue could convey an increased resistance to the genotoxic and carcinogenic actions of betel quid/tobacco-related compounds.

MICRONUCLEI IN EXFOLIATED ORAL MUCOSA CELLS OF BETEL QUID CHEWERS

A great variety of different chewing mixtures and patterns can be found in Asiatic and South Pacific countries. By comparing the effects of these chewing habits, it becomes possible to assess the involvement of the various quid ingredients in the formation of micronucleated exfoliated cells (MECs), preneoplastic lesions such as leukoplakia, lichen planus or submucous fibrosis, and carcinomas of the oral cavity. Since it is possible to compare the action of chewing mixtures that differ in only one ingredient, a high degree of resolution power can be achieved. The examples shown in Tab. 1 exemplify such an approach.

A few observations relevant to cancer protection should receive particular attention. First, the chewing of the unripe green areca nut or the green areca nut with betel leaf in the absence of lime does not increase the frequency of MECs in the oral mucosa and does not seem to elevate the incidence of leukoplakias and oral cancer (6).

Second, the addition of slaked lime to the areca nut and betel leaf seems to result in genotoxic activity, as revealed by an increase in MEC frequency in the oral mucosa (e.g., Taiwanese). Whether the use of this chewing mixture leads to a significant elevation in cancer risk remains a debatable issue (7,8). On the one hand, it cannot simply be ignored that in Taiwan, 53.4% of patients with oral carcinomas were betel quid chewers as compared to 0.0% in the matched controls (9). Moreover, the incidence of oral cancer in patients chewing betel nut only was still significantly higher than in patients who only smoked cigarettes and drank alcoholic beverages. Similarly, the incidence of leukoplakias is increased in subjects who chew betel quid without tobacco (3). On the other hand, there is a barely 2-fold increase of oral cancers in areas of heavy betel quid chewing as compared to nonchewing regions. This increase appears to be very small considering that an average of 41 green areca nuts are chewed daily and up to 120 nuts are used daily by taxi, bus, and truck drivers (25).

Third, the incidence of MECs differs significantly between Taiwanese and Khasis (northern hill regions of India) although the types of ingredients are comparable. Fourth, the addition of tobacco (either in the quid or placed separately into the mouth as a plug) leads to an increase in the frequency of MECs.

MICRONUCLEATED CELLS VERSUS LEUKOPLAKIAS AS MARKERS IN CHEMOPREVENTION TRIALS

Markers that are applicable to tissues from which cancer develops and that respond to the administration of chemopreventive agents are urgently needed. The frequency of MECs appears to be a useful indicator in pilot trials designed to test the most efficient treatment protocol before embarking on long-term clinical studies. The morphologically detectable leukoplakia is another marker used as an endpoint in intervention trials. The

Tab. 1. The effect of different chewing mixtures on the frequency of micronucleated exfoliated cells (MECs) of the oral mucosa and the risk for oral cancer.

Chewing ingredients	Population group	Number of individuals	Percent MECs in oral mucosa[a]	Elevated risk for oral cancer[b]
Areca nut	Guamanians	13	0.40 (0.0-0.6)	-
Nut + betel leaf	Guamanians	23	0.27 (0.0-0.8)	-
Nut + betel leaf + lime[c]	Taiwanese	35	1.69 (0.6-3.7)	±
Nut + betel leaf + lime	Khasis (India)	17	4.68 (2.5-7.5)	+
Nut + betel leaf + lime + tobacco	Ifugaos (Philippines)	120	3.94 (0.6-7.8)	+
Nut + betel leaf + lime + tobacco + catechin + spices	Indians (Orissa)	36	6.11 (3.0-11.7)	+++

[a] MECs in nonchewers of various groups averaged between 0.32 and 0.50 (29).
[b] See Ref. 6, 8, and 9.
[c] All the examined betel quid chewers smoked a half to 2 packs of cigarettes (local brand) per day.

question must be raised as to whether both preneoplastic lesions (micronucleated cells and leukoplakias) indicate comparable tissue responses to the action of carcinogens and to the administration of chemopreventive agents. In nass (mainly tobacco, slaked lime, and oil) users of Uzbekistan, leukoplakias had neither a higher nor a lower frequency of MECs as compared to nonaffected oral mucosa (25). Similarly, in betel quid chewers, the frequency of micronucleated oral mucosa cells was linked to the type of chewing mixture and was independent of the presence or absence of leukoplakia.

The two preneoplastic lesions seem to differ in their stability. A relatively high spontaneous regression rate occurs in leukoplakias of Indian betel quid chewers (13). On the other hand, the frequency of MECs varies only slightly in persons who continued to chew betel quid or who used snuff in a regular manner (27,34). It is conceivable and even likely that oral leukoplakias are due, at least in part, to repeated irritation of the mucosa by the caustic action of slaked lime. These preinvasive lesions are virtually absent in the oral mucosa of chewers who do not include lime with the areca nuts and betel leaves (6).

The most convincing evidence pointing to a predominant role of lime in the induction of leukoplakias comes from natives of Papua, New Guinea. They deliver the lime into the mouth by wiping lime sticks against the buccal mucosa. Leukoplakias develop predominantly along the streaky lines made by the lime stick (1). The involvement of slaked lime as an etiological factor of leukoplakias finds support from experiments on hamsters (5). The repeated application of slaked lime to the cheek pouch resulted in hyperplastic lesions and cellular atypia resembling that of leukoplakia in man. Considering that micronuclei in oral mucosa cells reflect genetic damage caused by areca nut- or tobacco-related carcinogens, and that, on the other hand, leukoplakia may result from repeated and prolonged tissue irritation, it should not be surprising to find that both endpoints respond differently to the administration of chemopreventive agents.

PROTECTION BY MODULATING THE QUID COMPOSITION

Considering the results shown in Tab. 1, a change in the composition of the betel quid should comprise a promising approach in cancer prevention. For example, slaked lime is one of the universally used betel quid ingredients that demands closer scrutiny. In vitro tests with lime samples prepared from marine shells (Orissa), snail shells (Philippines), or rock (Meghalaya) did not reveal any clastogenic activity in Chinese hamster ovary (CHO) cells (Tab. 2). In addition, no genotoxic effect was seen when slaked lime was "chewed" and the saliva examined for its capacity to induce chromosome or chromatid aberrations in CHO cells. Thus, any involvement of slaked lime in the etiology of leukoplakias or carcinomas must be more indirect in nature.

In the oral mucosa slaked lime can cause severe irritation, particularly when licked separately from a stick, as is the case in Papua, New Guinea (1), or from the fingers, a common practice among Ifugaos of the Philippines. Such irritation could lead to hyperplasia, thus increasing the number of dividing cells that are the targets of the actual areca nut- or tobacco-related carcinogens.

Another involvement of slaked lime in carcinogenesis could be due to its alkalinity. Phenolics released by areca nuts or betel leaves into the

Tab. 2. The capacity of slaked lime, areca nut, betel leaf, and catechu to induce chromatid breaks/exchanges and micronuclei.

Quid ingredient	Dose	Percent cells with chromatid anomalies	Percent cells with micronuclei
Extracts			
Slaked lime (white, shells)	25 mg/ml	0.2	0.8
Slaked lime (white, rock)	25 mg/ml	0.7	1.0
Slaked lime (white, snails)	25 mg/ml	0.4	1.2
Slaked lime (with catechu)	25 mg/ml	0.0	1.9
Areca nut (green)	50 mg/ml	15.1	5.2
Betel leaf (Taiwan)	3 mg/ml	4.0	4.5
Saliva of chewers			
Areca nut (green)	12.0% saliva*†	13.8	5.6
Betel leaf (Taiwan)	1.5% saliva†	3.5	3.2
Saliva only	12.0% saliva†	0.0	1.6

* The saliva of chewers was diluted with culture medium to a level that did not inhibit cell division.
† These percentages were the first in the dilution series that did not produce a cell-killing effect.

saliva (31) generate genotoxic agents at alkaline pH levels (19) that can be found in the saliva of quid chewers.

Finally, lime could conceivably alter the permeability of cells, leading to a loss of intracellular chemopreventive agents. In this way a retinol or beta-carotene deficiency could be created in an area of the oral mucosa which then becomes more sensitive to the action of carcinogens, resulting in a biological autarchy of a clone of cells, as suggested by De Luca (4). A reduction in the amount of lime to levels that do not shift the pH of the saliva of a betel quid chewer should prevent irritation of the mucosa and the formation of genotoxic radicals generated by the oxidation of phenolics.

The betel leaf is another consistent ingredient of betel quids that merits attention as a possible modifier of carcinogenesis. Experimental studies on rodents have revealed a protective effect of the betel leaf when given concurrently with areca nut extracts, as part of a quid, or in combination with a carcinogen (16,23,24). The nature of the inhibitory compounds in betel leaves is still unknown. Carotenoids could be candidates. Approximately 9,600 IU of beta-carotene per 100 g of betel leaves is not a negligible amount compared to the 8,000 to 12,000 IU per 100 g of carrots or 10,000 IU per 100 g of spinach. Thus, the daily chewing and ingestion of betel leaves could significantly increase the amount of the chemoprotective beta-carotene in chewers.

The exclusion of tobacco from the quid could probably have a major effect on the carcinogenicity of the chewing mixture. Unfortunately, precise information on the contribution of differently prepared tobaccos in the etiology of micronucleated cells, preinvasive lesions, and carcinomas is still lacking (26). The addition of tobacco to the quid resulted in a definite increase in micronucleated buccal mucosa cells in chewers from Guam, and in a barely detectable elevation in Ifugao chewers (Tab. 1). Both these increases in micronucleated cells are relatively small compared to that found in Indian chewers of complex betel quids. Thus, factors other than tobacco itself must be considered when assessing the hazard of a chewing mixture.

The possible number of interactions of all compounds released into the saliva during a quid-chewing period must be staggering. It is therefore not surprising to find both antigenotoxic as well as enhancing effects (31). Against the above-mentioned inhibitory actions of betel leaf, one must place the genotoxic activity of its extracts (20,30) and the supply of transition elements which can greatly enhance the free radical formation and clastogenicity of phenolics (28,31). Similarly, the phenolics of areca nuts are strongly genotoxic (31); but on the other hand, they can inhibit the endogenous formation of nitrosamines (32,33).

PROTECTION BY MODULATING THE DIET

Recently, convincing evidence has been gathered pointing to the anticarcinogenic and/or antigenotoxic action of beta-carotene. Epidemiological results indicate a protective role of beta-carotene-rich vegetables rather than vitamin A-containing food products against snuff-induced oral cancer (41). In animal models, the administration of beta-carotene delayed the onset of invasive carcinomas (12), protected against the carcinogenic effect of ionizing radiation (22), reduced the incidence of cancers in benzo(α)pyrene-injected rodents (21), and inhibited the formation of chromatid bridges and fragments in hepatocytes of rats injected with dimethylnitrosamine (42).

Of particular relevance to our topic is the observation that a vitamin A deficiency enhanced the sensitivity of rats to the carcinogenic effect of an areca nut/calcium hydroxide mixture (36). In human subjects, the twice weekly oral administration of beta-carotene (180 mg/week) for a total of 9-10 weeks resulted in a significant reduction of micronucleated buccal mucosa cells among betel quid chewers (34) or snuff dippers (35). During the treatment period, all individuals continued to use betel quid or snuff, respectively.

Based on the results currently available, one appears to be justified in proposing a beta-carotene-rich diet as part of a program designed to reduce the incidence of oral cancer among betel quid/tobacco chewers and snuff dippers. The question has been raised as to whether beta-carotene actually does occur in the oral mucosa, which is the target of carcinogens released from various chewing mixtures. To gain information on this issue, we estimated beta-carotene levels in exfoliated oral mucosa cells (27). The amount of beta-carotene per 10^6 cells appears to depend on the ingestion of carrots and spinach (Tab. 3). The administration of beta-carotene capsules (total of 360 mg given over a 4-day period) greatly increased the beta-carotene levels in exfoliated oral mucosa cells (Fig. 1). Thus, by manipulating the diet, beta-carotene levels in the target cells for carcinogens can be considerably increased. The observed difference (Tab. 1) in the frequency of micronucleated cells between betel quid chewers in Taiwan and those in the mountains of Luzon or India (Khasis) could conceivably be due to a carotenoid deficiency among the two latter groups.

Tab. 3. The link between beta-carotene levels in exfoliated cells of the oral cavity and the consumption of green/yellow vegetables.

Individuals belonging to the lowest and highest quintile	Vegetable consumption (days/week)	β-Carotene ($ng/10^6$ cells) individuals
Low group	2.1	0, 0, 0, 0, 0, 0, 0.08, 0.09
High group	3.6	1.1, 1.6, 1.7, 1.8, 2.1, 2.1, 2.3, 2.8

OUTLOOK

Although numerous epidemiological and experimental studies point to the protective effect of carotenoids and retinoids against the development of cancer, no countrywide application of chemopreventive agents has yet been undertaken. One of the unresolved questions is the logistics of bringing a chemopreventive regime to a large population group. A daily or weekly administration of vitamin pills and capsules is feasible in a hospital setting, but it is difficult to achieve with rural or urban groups involving thousands or millions of subjects. In addition, the cost of distributing chemopreventive pills to individuals is beyond the resources that can be allocated for primary health care. Finally, our experience with the administration of vitamin A or beta-carotene to betel quid chewers in the Philippines revealed a recurrence of micronucleated oral mucosa cells once the vitamin administration had ceased (29). A comparable experience was reported on retinoid-treated patients with Bowen's disease (10). The skin lesions reappeared within 2 to 3 months following their regression during a

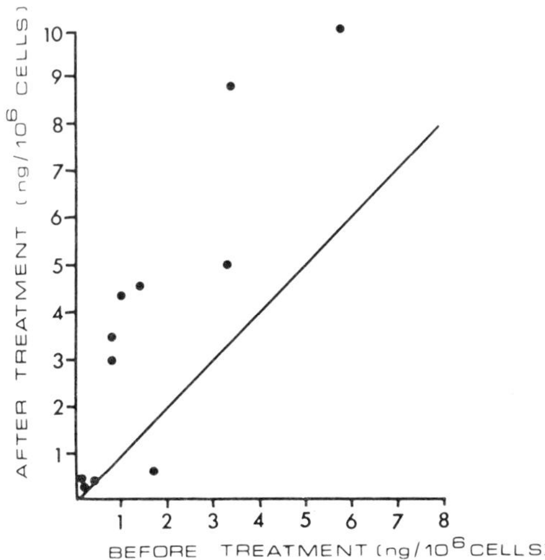

Fig. 1. Effect of beta-carotene administration (360 mg over a 4-day period) on beta-carotene levels in exfoliated oral mucosa cells of 11 volunteers. An increase was detectable in 7 individuals sampled 14 days after the onset of treatment.

Tigason treatment period. Thus, chemopreventive doses must either be administered continuously, or be replaced by lower and more spaced maintenance doses.

The possibility must be explored whether dietary sources that are readily accessible to all members of a high cancer risk group could provide the required maintenance doses. Very little attention is being given to this crucial topic. Its solution could move the idea of chemoprevention from an academic level to a worldwide application. In many Asiatic countries, sweet potatoes would appear to be a promising source of beta-carotene. However, the implementation of even such an apparently simple proposal would require a major effort by various levels of society. The white native varieties of sweet potatoes, which contain only negligible amounts of beta-carotene (0.4 mg/100 g), must be replaced by the new "yellow" carotenoid-rich varieties (e.g., the "Julien" variety contains 24.8 mg/100 g beta-carotene) (38). The daily consumption of sweet potatoes, which is a common event in many Asiatic countries, and, in addition, the inclusion of sweet potato flour in the diet, could readily lead to the ingestion of beta-carotene levels that were found to protect against the induction of micronucleated oral mucosa cells.

However, there appear to be a large number of plants that could provide beta-carotene when correctly included in the diet. Many green vegetable leaves (17) or red palm oil (15) could supply enough beta-carotene to maintain a normal vitamin A and beta-carotene level, thus conveying a protective effect for night blindness and xerophthalmia, and acting as an anticarcinogenic agent.

Changes in the composition of the betel quid would appear to be a second promising approach in cancer prevention. However, tampering with the main ingredients, such as areca nut (which provides the pharmacological effect), betel leaf (which adds a spicy flavor), and tobacco (to which an addiction may have been formed), would be met with considerable resistance.

However, the use of slaked lime could be modified without facing any major opposition. A reduction in the amount of lime would decrease the generation of free radicals from the phenolics released into the saliva. It would also prevent irritation of the mucosa, thus avoiding the increase in the number of target cells. Actually, a trend towards the use of diluted lime has been initiated, since chewers in modern societies dislike the red mouth, gums, and teeth that are due to the oxidation products of polyphenolics released from the areca nut (11).

The areca nut could also be modified to provide a less hazardous betel quid. Unfortunately, our understanding of this particular issue is still incomplete. The saliva of chewers of betel quids collected in Bombay contained areca nut-specific nitrosamines, including N-nitrosoguvacoline and N-nitrosoguvacine (14,39,40), whereas the saliva of green areca nut chewers from Taiwan lacked even traces of nitrosated arecoline and guvacoline (unpubl. data). A systematic study of the composition of the various areca nuts (entire green, unripe nuts with husk, fresh soft nuts, dried nuts, fermented nuts, spiced nuts, etc.) may uncover the pretreatment that leads to the formation of potentially carcinogenic N-nitroso compounds (40).

In summary, one can state that betel quid/tobacco-induced oral carcinomas are among the very few human cancers where our understanding of the etiology has reached a state where practical intervention programs can be designed and implemented.

ACKNOWLEDGEMENTS

These studies were supported by the National Cancer Institute of Canada and by the Natural Sciences and Engineering Research Council of Canada. Dr. H.F. Stich is a Terry Fox Cancer Research scientist of the National Cancer Institute of Canada.

REFERENCES

1. Atkinson, L., I. Chester, F. Smyth, and T. Ten Seldam (1964) Oral cancer in New Guinea: A study in demography and epidemiology. Cancer 17:1289-1298.
2. Burton-Bradley, B.G. (1979) Arecaidinism. Betel chewing in transcultural perspective. Can. J. Psychiat. 24:481-488.
3. Chang, K.-M. (1966) Betel nut chewing and mouth cancer in Taiwan. J. Formosan Med. Assoc. 63:437-444.
4. De Luca, L.M. (1983) Essential function deficiency as the result of tumor promotion and establishment of biological autarchy. J. Natl. Cancer Inst. 70:405-407.
5. Dunham, L.J., C.S. Muir, and J.E. Hamner, III (1966) Epithelial atypia in hamster cheek pouches treated repeatedly with calcium hydroxide. Brit. J. Cancer 20:588-593.
6. Gerry, R.G., S.T. Smith, and M.L. Calton (1952) The oral characteristics of Guamanians, including the effects of betel chewing on the oral tissues. Oral Surg. Med. Pathol. 5:762-1011.
7. Gupta, P.C. (1984) Epidemiologic study of the association between alcohol habits and oral leukoplakia. Commun. Dent. Oral Epidemiol. 12: 47-50.
8. International Agency for Research on Cancer (1985) Tobacco Habits other than Smoking; Betel Quid and Areca Nut Chewing; and Some Related N-Nitroso Compounds, Monographs on the Evaluation of the Carcinogenic Risk of Chemicals to Humans, Vol. 37, International Agency for Research on Cancer, Lyon.
9. Kwan, H.-W. (1976) A statistical study on oral carcinomas in Taiwan with emphasis on the relationship with betel nut chewing: A preliminary report. J. Formosan Med. Assoc. 75:497-505.
10. Lu, Y.C. (1985) Chronic arsenic poisoning in Taiwan (in Chinese). Clin. Dermatol. 27:271-278.
11. Mathew, A.G. (1971) Formation of red colour on chewing areca nut with slaked lime. J. Food Sci. Technol. 8:140-142.
12. Mathews-Roth, M.M. (1982) Antitumor activity of β-carotene, canthaxanthin and phytoene. Oncology 39:33-37.
13. Mehta, F.S., and J.J. Pindborg (1974) Spontaneous regression of oral leukoplakias among Indian villagers in a 5-year follow-up study. Commun. Dent. Oral Epidemiol. 2:80-84.
14. Nair, J., H. Ohshima, M. Friesen, A. Croisy, S.V. Bhide, and H. Bartsch (1985) Tobacco-specific and betel nut-specific N-nitroso compounds: Occurrence in saliva and urine of betel quid chewers and formation in vitro by nitrosation of betel quid. Carcinogenesis 6:295-303.
15. Perissé, J., and W. Polacchi (1980) Geographical distribution and recent changes in world supply of vitamin A. Food and Nutrit. 6:21-27.
16. Ranadive, K.J., S.N. Ranadive, N.M. Shivapurkar, and S.V. Gothoskar (1979) Betel quid chewing and oral cancer: Experimental studies on hamsters. Int. J. Cancer 24:835-843.

17. Rao, A.R. (1984) Modifying influences of betel quid ingredients on B(a)P-induced carcinogenesis in the buccal pouch of hamster. Int. J. Cancer 33:581-586.
18. Reid, A. (1985) From betel-chewing to tobacco-smoking in Indonesia. J. Asian Stud. 44:529-547.
19. Rosin, M.P. (1984) The influence of pH on the convertogenic activity of plant phenolics. Mutat. Res. 135:109-113.
20. Sadasivan, G., G. Rani, and C.K. Kumari (1978) Chromosome-damaging effect of betel leaf. Mutat. Res. 57:183-185.
21. Santamaria, L., A. Bianchi, A. Arnaboldi, L. Andreoni, and P. Bermond (1983) Benzo(α)pyrene carcinogenicity and its prevention by carotenoids. In Modulation and Mediation of Cancer by Vitamins, F.L. Meyskens and K.N. Prasad, eds. Karger, Basel, pp. 81-88.
22. Seifter, E., G. Rettura, J. Padawer, F. Stratford, J. Weinzweig, A.A. Demetriou, and S.M. Levenson (1984) Morbidity and mortality reduction by supplemental vitamin A or β-carotene in CBA mice given total-body γ-radiation. J. Natl. Cancer Inst. 73:1167-1177.
23. Shirname, L.P., M.M. Menon, J. Nair, and S.V. Bhide (1983) Correlation of mutagenicity and tumorigenicity of betel quid and its ingredients. Nutrit. Cancer 6:87-91.
24. Shivapurkar, N.M., S.N. Ranadive, S.V. Gothoskar, S.V. Bhide, and K.J. Ranadive (1980) Tumorigenic effect of aqueous and polyphenolic fractions of betel nut in Swiss strain mice. Ind. J. Exp. Biol. 18:1159-1161.
25. Stich, H.F. (1986) Micronucleated exfoliated cells as markers in chemoprevention trials. J. Natl. Cancer Inst. (in press).
26. Stich, H.F. (1986) The use of micronuclei in tracing the genotoxic damage in the oral mucosa of tobacco users. In New Aspects of Tobacco Carcinogenesis, Banbury Report, Cold Spring Harbor Laboratory, Cold Spring Harbor, New York (in press).
27. Stich, H.F., A.P. Hornby, and B.P. Dunn (1986) Beta-carotene levels in exfoliated mucosa cells of population groups at low and elevated risk for oral cancer. Int. J. Cancer (in press).
28. Stich, H.F., and M.P. Rosin (1984) Naturally occurring phenolics as antimutagenic and anticarcinogenic agents. In Nutritional and Toxicological Aspects of Food Safety, M. Friedman, ed. Plenum Press, New York, pp. 1-29.
29. Stich, H.F., and M.P. Rosin (1985) Towards a more comprehensive evaluation of a genotoxic hazard in man. Mutat. Res. 150:43-50.
30. Stich, H.F., and W. Stich (1982) Chromosome-damaging activity of saliva of betel nut and tobacco chewers. Cancer Lett. 15:193-202.
31. Stich, H.F., B. Bohm, K. Chatterjee, and J. Sailo (1983) The role of saliva-borne mutagens and carcinogens in the etiology of oral and esophageal carcinomas of betel nut and tobacco chewers. In Carcinogens and Mutagens in the Environment, Vol. III, Naturally Occurring Compounds: Epidemiology and Distribution, H.F. Stich, ed. CRC Press, Boca Raton, Florida, pp. 43-58.
32. Stich, H.F., H. Ohshima, B. Pignatelli, J. Michelon, and H. Bartsch (1983) Inhibitory effect of betel nut extracts on endogenous nitrosation in man. J. Natl. Cancer Inst. 70:1047-1050.
33. Stich, H.F., B.P. Dunn, B. Pignatelli, H. Ohshima, and H. Bartsch (1984) Dietary phenolics and betel nut extracts as modifiers of N-nitrosation in rat and man. In N-Nitroso Compounds: Occurrence, Biological Effects and Relevance to Human Cancer, I.K. O'Neill, R.C. von Borstel, C.T. Miller, J. Long, and H. Bartsch, eds. IARC Sci. Publ. No. 57, International Agency for Research on Cancer, Lyon, France, pp. 213-222.

34. Stich, H.F., W. Stich, M.P. Rosin, and M.O. Vallejera (1984) Use of the micronucleus test to monitor the effect of vitamin A, beta-carotene and canthaxanthin on the buccal mucosa of betel nut/tobacco chewers. Int. J. Cancer 34:745-750.
35. Stich, H.F., A.P. Hornby, and B.P. Dunn (1985) A pilot beta-carotene intervention trial on Inuits using smokeless tobacco. Int. J. Cancer 36:321-327.
36. Tanaka, T., H. Mori, M. Fujii, M. Takahashi, and I. Hirono (1983) Carcinogenicity examination of betel quid. II. Effect of vitamin A deficiency on rats fed semi-purified diet containing betel nut and calcium hydroxide. Nutrit. Cancer 4:260-266.
37. Taylor, P. (1984) Smoke Ring. The Politics of Tobacco, The Bodley Head Ltd., London, Sydney, Toronto.
38. Wang, H. (1982) The breeding of sweet potatoes for human consumption. In Sweet Potato, R.L. Villareal and T.D. Griggs, eds. Asian Vegetable Research and Development Center, Tainan, Taiwan, pp. 297-311.
39. Wenke, G., A. Rivenson, K.D. Brunnemann, D. Hoffmann, and S.V. Bhide (1984) A study of betel quid carcinogenesis. II. Formation of N-nitrosamines during betel quid chewing. In N-Nitroso Compounds: Occurrence, Biological Effects and Relevance to Human Cancer, I.K. O'Neill, R.C. von Borstel, C.T. Miller, J. Long, and H. Bartsch, eds. IARC Sci. Publ. No. 57, International Agency for Research on Cancer, Lyon, France, pp. 859-866.
40. Wenke, G., A. Rivenson, and D. Hoffmann (1984) A study of betel quid carcinogenesis. 3. 3-(Methylnitrosamino)propionitrile, a powerful carcinogen in F344 rats. Carcinogenesis 5:1137-1140.
41. Winn, D.M., R.G. Ziegler, L.W. Pickle, G. Gridley, W.J. Blot, and R.N. Hoover (1984) Diet in the etiology of oral and pharyngeal cancer among women from the southern United States. Cancer Res. 44:1216-1222.
42. Yee, C.A., R.H.C. San, and H.F. Stich (1985) The protective effect of beta-carotene on genotoxicity and cytotoxicity of dimethylnitrosamine in vitamin A-deficient rats (abstr.). Annual Meeting of the Genetics Society of Canada, June 24-28, London, Ontario.

THE CONCENTRATION OF BILE ACIDS IN THE FECAL STREAM AS A RISK FACTOR FOR COLON CANCER

W.R. Bruce and R. Bird

Department of Medical Biophysics
Ludwig Institute for Cancer Research
University of Toronto
Toronto, Canada M4Y 1M4

INTRODUCTION

It is usual when considering the origin of human cancers to focus on the initiating carcinogenic factors. For colon cancer we think in terms of fecal and food mutagens--the fecapentaenes (8), the aromatic amines (2), the aromatic hydrocarbons (18), lipid oxidation products such as malonaldehyde (10), and cholesterol oxidation products (17). But of perhaps equal or greater importance are co-carcinogenic and promoting factors that increase the vulnerability of normal cells to the carcinogens in their environment. With colon cancer we think here of the dietary fat and fecal bile acids. The secondary bile acids are known to reduce the integrity of the mucosal barrier (5), to increase the sensitivity of cells to mutagenic (15) and carcinogenic agents (16), to increase the proliferation rate of crypt cells (7), and to act as promoters of experimental colon carcinogenesis (6). Yet despite these factors, an unambiguous role of bile acids in human colon carcinogenesis has not been established.

Seven new lines of evidence suggest, however, that the appropriate risk factor may not have been considered. The biological effects of the bile acids are probably not due to their total mass but to their concentration in solution. This simple distinction may explain the failure of previous studies to correctly assess the importance of bile acids, may explain the origin of these tumors in terms of our food preparation practices, and may lead to simple intervention methods for these common human tumors. These lines of evidence are as follows.

Intrarectal Bile Acids and Oral Calcium in Mice

In our first studies of the bile acids, sodium deoxycholate was administered intrarectally at a concentration approximating the concentration of bile acid in the fecal stream. The degree of epithelial damage was intense and resembled that seen with ethylene dinitrilotetra-acetic acid (EDTA) (4). It was subsequently demonstrated that calcium, as the lactate or carbonate salt, was able to inhibit the damage when it was given orally shortly before the bile salt (19). This result prompted us to define a hypothesis for colon cancer causation which proposed that the bile and fatty acids in the "ionized" form were responsible for the disease and that the oral calcium produced its effect by leading to the formation of insoluble calcium soaps which reduced the concentration of the toxic soluble salts (13).

Colonic Perfusion Studies with Deoxycholic Acid and Calcium in Rats

The effect of the bile acids can be examined in a more defined way with the use of a colon perfusion model (12). Groups of rats were perfused with graded doses of the bile acid, deoxycholic acid, for 80 minutes, and the degree of damage was quantitated by cell morphometry and by measuring the degree of loss of DNA in the perfusion stream. A marked effect of 2 mM of the bile acid was evident. When a 4 mM concentration was used, the damage was extensive. This damage was, however, inhibited if the bile acid was mixed with a calcium salt solution prior to the perfusion (14). Under this condition the concentration of the bile acid in solution was greatly reduced, and the damage reflected the concentration of the solution and not the total mass of the deoxycholate in the perfusion fluids.

Dietary Cholic Acid and Dietary Calcium in Mice

While the perfusion studies allow a detailed analysis of the effects of the bile acids, they are quite removed from the physiological exposure. As a step closer to the reality of dietary regimens, we utilized the oral cholic acid model (1). In this system cholic acid is added at a low concentration to the diet. Such a dietary regimen has been shown to increase the concentration of deoxycholic acid in the fecal stream, to increase the proliferation rate in colonic epithelial cells, and to increase the yield of colon tumors in animals given colon carcinogens (6,7). The effect of dietary calcium on this model system was thus determined.

Laboratory animal diets are usually made up to contain in excess of 0.5% of the weight of the diet as calcium. The effects of synthetic diets in which calcium made up 0.1 to 1.0%, from the levels more closely approximating the levels in human diets to the levels seen commonly in animal chows, were examined. At the low dietary calcium levels the colonic epithelium was extremely sensitive to dietary cholic acid (1). Proliferation was markedly increased at cholic acid levels as low as 0.1%, but this degree of proliferation was reduced to the normal rate when higher levels of calcium were incorporated in the diet mix. These results showed that dietary calcium could have a significant effect on the effects of bile acids in feeding studies.

Oral Boluses of Fat in Mice

Oral boluses of fat provided another method for examining the action of fat and calcium on the colon. In the first experiments, mice were given oral boluses of 0.4 ml of beef tallow (3). Two hours later there was a substantial loss of cells from the surface of the colonic epithelium, 12 hours later there was a wave of mitotic activity at the base of the crypts, while at 24 hours following the bolus there was no evidence of damage produced by the fat. Similar though less pronounced effects were observed following 0.1 and 0.2 ml of beef tallow and after corn oil. These first experiments were carried out with mice on a dietary level of about 1% calcium. When the studies were repeated on animals fed lower levels of dietary calcium, it was evident again that low dietary calcium could augment the effects of the oral fat (Bird, pers. comm.). The damage evident in animals receiving 0.1% calcium was substantially greater than that seen in animals receiving 1% calcium.

Effect of Dietary Fat and Calcium on the Colonic Proliferation Rate in Mice

Our dietary fat intake is in the form of meals that may resemble boluses, but fat is more usually consumed with other components in the diet and not just by itself. Consequently, we began to examine the effect of defined diets that differed in their contents of fat and calcium (Bird, pers. comm.). Although these studies have not been carried out over long periods of time or under many different dietary conditions, it is already clear that increased dietary fat increases the proliferation rate of colonic crypt cells as measured by tritiated thymidine labeling. This result is consistent with a damaging effect of components of the fecal stream for animals on high fat diets. Presumably the components are bile acids in solution.

Effect of Dietary Fat and Calcium on Bile Acids in Solution in the Human Fecal Stream

The effect of dietary fat and calcium on the concentration of bile acids in solution in the human fecal stream can be examined in humans. A preliminary study has been carried out with 20 volunteers. They were randomized into two groups. One group was placed on a low fat, high calcium diet, while the other was put on a diet closer to their usual high fat, low calcium diet. Prior to the study and at the end of a 4-day diet, the volunteers collected fecal samples that were analyzed for their levels of bile acids. The level of bile acids in the low fat group dropped to an average of 0.2 mM, while that in the high fat group increased to 0.5 mM, with a range from 0.15 to 0.8 mM (J. Rafter, pers. comm.). Studies are now being carried out over longer periods of time and with independent control of fat, calcium, and fiber levels.

Dietary Fat and Calcium and Fecal Bile Acids in Solution as Risk Factors in Colon Cancer

If fecal bile acid levels are indeed responsible for colon cancer, the levels of these compounds should vary greatly in different populations corresponding to their risk for the disease. We have tested the feasibility of simultaneously testing a number of such dietary hypotheses (11). More detailed studies in Chinese populations with greatly different colon cancer rates are now in progress.

In a second group of experiments we have begun to compare the concentrations of bile acids in fecal water in patients with colon cancer or colonic polyps, and in patients reaching the gastrointestinal clinic with no evidence of cancer or its precursors. These studies are not completed as yet and the results have not been completely decoded. However, the results do indicate that such patients can have bile acid concentrations that extend well into the millimolar range, concentrations that will have rapid effects on the colonic epithelium of the experimental animals.

CONCLUSION

The hypothesis we have considered here is a refinement of the much older suggestion that bile acid mass in the fecal stream is the risk factor for colon cancer (9). The studies that we have carried out make it clear that the damaging effects of bile acids on the colonic epithelium can be inhibited, and thus that the total mass of these compounds is an inappropriate measure of risk. The risk appears to be more closely related to the concentration of these compounds in solution in the fecal stream. This modified bile hypothesis may allow a more adequate explanation for the great differences in colon cancer rates in different populations throughout the world. It may lead to more easily implemented intervention for colon cancer with combined fat reduction and calcium supplementation. It could explain the origin of colon cancer as a problem with our present food processing in which we are separating fat from other components including calcium in the "natural" food.

REFERENCES

1. Bird, R.P. (1985) Effect of dietary components on the pathobiology of colonic epithelium: Possible relationship with colon tumorigenesis. Lipids (in press).
2. Bird, R.P., and W.R. Bruce (1984) Damaging effect of dietary components to colon epithelial cells in vivo: Effect of mutagenic heterocyclic amines. J. Natl. Cancer Inst. 73:237-240.
3. Bird, R.P., A. Medline, R. Furrer, and W.R. Bruce (1985) Toxicity of orally administered fat to the colonic epithelium of mice. Carcinogenesis 6:1063-1066.
4. Cassidy, M.M., and C.S. Tidball (1967) Cellular mechanism of intestinal permeability alterations produced by chelation depletion. J. Cell Biol. 32:685-698.
5. Chadwick, V.S., T.S. Gaginella, G.L. Carlson, J.C. Debognie, S.F. Phillips, and A.F. Hofmann (1979) Effect of molecular structure on bile acid-induced alterations in absorptive function, permeability and morphology in the perfused rabbit colon. J. Lab. Clin. Med. 94:661-674.
6. Cohen, B.I., R.F. Raicht, E.E. Deschner, M. Takahashi, A.N. Sarwal, and E. Fazzini (1980) Effect of cholic acid feeding on N-methyl-N-nitrosourea induced colon tumors and cell kinetics in rats. J. Natl. Cancer Inst. 64:573-578.
7. Deschner, E.E., B.I. Cohen, and R.F. Raicht (1981) Acute and chronic effect of dietary cholic acid on colonic epithelial cell proliferation. Digestion 21:290-296.
8. Gupta, I., J. Baptista, W.R. Bruce, T.C. Che, R. Furrer, J.S. Gingerich, A.A. Grey, L. Marai, P. Yates, and J.J. Krepinsky (1983) Structures of fecapentaenes, the mutagens of bacterial origin isolated from human feces. Biochemistry 22:241-245.
9. Hill, M.J. (1974) Bacteria and the etiology of colonic cancer. Cancer 34:815-818.
10. Jacobson, E.A., H.L. Newmark, and R. Bird (1983) Increased excretion of malonaldehyde equivalents in the urine after consumption of cooked stored meats. Nutrit. Rpts. Intl. 28:509-517.
11. Jacobson, E.A., H.L. Newmark, G.E. McKeown-Eyssen, K. Hay, E. Bright-See, W.R. Bruce, K. Suzuki, and T. Mitsuoka (1985) Feasibility of assessing diet and other colon cancer risk factors through biochemical markers. (Manuscript submitted for publication.)
12. Mekhjian, H.S., and S.F. Phillips (1975) Perfusion of the canine colon with unconjugated bile acids. Effect on water and electrolyte transport, morphology and bile acid absorption. Gastroenterology 69:380-386.

13. Newmark, H.L., M.J. Wargovich, and W.R. Bruce (1984) Colon cancer and dietary fat, phosphate and calcium: A hypothesis. J. Natl. Cancer Inst. 72:1323-1325.
14. Rafter, J., V.W.S. Eng, R. Furrer, A. Medline, and W.R. Bruce (1985) The mucosal damage produced by deoxycholic acid in the rat colon depends on the amount of bile acid in solution and not on the total bile acid concentration. Gut (in press).
15. Silverman, S.J., and A.W. Andrews (1977) Bile acids: Co-mutagenic activity in the Salmonella-mammalian-microsome mutagenicity test (Brief communication). J. Natl. Cancer Inst. 59:1557-1559.
16. Suzuki, K., and W.R. Bruce (1985) Deoxycholic acid increases the colonic nuclear damage induced by known carcinogens in C57BL/6 mice. (Manuscript submitted for publication.)
17. Suzuki, K., W.R. Bruce, J. Baptista, and J.J. Krepinsky (1985) Characterization of nucleotoxic steroids in human faeces: Do they cause colorectal cancer? (Manuscript submitted for publication.)
18. Wang, C.X., K. Watanabe, J.H. Weisburger, and G.M. Williams (1985) Induction of colon cancer in inbred Syrian hamsters by the intrarectal administration of benzo(α)pyrene, 3-methylcholanthrene and N-methyl-N-nitrosourea. Cancer Lett. 27:309-314.
19. Wargovich, M.J., V.W.S. Eng, H.L. Newmark, and W.R. Bruce (1983) Calcium ameliorates the toxic effect of deoxycholic acid on colonic epithelium. Carcinogenesis 4:1205-1207.

HORMONES AND DIETARY FACTORS CONTROLLING GENE ACTIVATION AND EXPRESSION IN CARCINOGENESIS

Carmia Borek

Radiological Research Laboratory
and Department of Pathology
College of Physicians and Surgeons
Columbia University
New York, New York 10032

INTRODUCTION

The development of a malignant phenotype involves complex interactions among endogenous and environmental factors. Cellular transformation induced by radiation or chemical carcinogens proceeds through a set of cellular pathways initiated by localized changes in DNA and leading to an abnormal expression of altered cellular genes (3,7,18,23,28).

Cell cultures of both rodent and human origin provide powerful tools for defining the various stages in neoplastic development, i.e., initiation and promotion (3,7). They afford the opportunity to identify co-carcinogenic interactions among various agents and to determine at cellular and molecular levels factors that alter the course and frequency of malignant transformation.

INITIATION AND PROMOTION

The initiation of the neoplastic process by physical or chemical agents takes place via irreversible genetic alterations. These may include mutations, gene amplification, oncogene activation, or other genetic rearrangements (7,18,23,28).

Initiation requires DNA metabolism and replication to fix the events as a hereditary property of the cell (13,14). Later events associated with the expression of the transformed phenotype require additional cell divisions (13,14). The onset of the neoplastic process can be triggered by radiation, the most ubiquitous carcinogen, and by a variety of environmental and dietary chemicals that provide a public threat worthy of attention (7).

The kinetics of events initiated by various carcinogens differ. Radiation imparts its oncogenic action within a fraction of a second (7). Its actions can be amplified in a supra-additive fashion by other agents, such as the food-derived tryptophan pyrolysates (11), serving as co-carcinogens. The effects of chemicals depend on their persistence in cells and tissues, on periods of exposure, and on pharmacokinetics associated with their metabolism and destruction. When given to the right target cell and tissue, and when administered in sufficiently high doses, radiation and various chemicals can act alone or in concert as complete carcinogens, producing tumors in vivo or malignant transformations in vitro (Fig. 1).

Over the years, evidence has accumulated both in vivo and in vitro showing that chemicals, which in themselves are noncarcinogenic, can modulate the neoplastic process by acting as tumor promoters (2). These chemicals, such as 12-O-tetradecanoylphorbol-13-acetate (TPA), a phorbol ester derivative (15,29), or teleocidin (16,26), can effectively enhance the rate of malignant transformation following initiation by a carcinogen and modify the expression of neoplastic transformation in a manner that is partly reversible (15,16).

The reversibility of promotion at its early stages clearly offers hope for anticarcinogenic agents that serve as antipromoters and can block the progression of the neoplastic process and its detrimental effects (6).

Oncogenes

Studies in vivo and in vitro (18,23,28) have indicated that the activation of specific transforming genes plays a role in cellular conversion to malignancy. DNA from hamster embryo or mouse C3H/10T1/2 cells transformed in vitro by radiation can impart the neoplastic phenotype to normal cells by means of DNA-mediated gene transfer (transfection) into recipient NIH/3T3 or C3H/10T1/2 cells (18,23). The transforming genes are not of the ras family of genes and they differ in hamster and mouse cells as determined by their sensitivity to restriction enzymes (18).

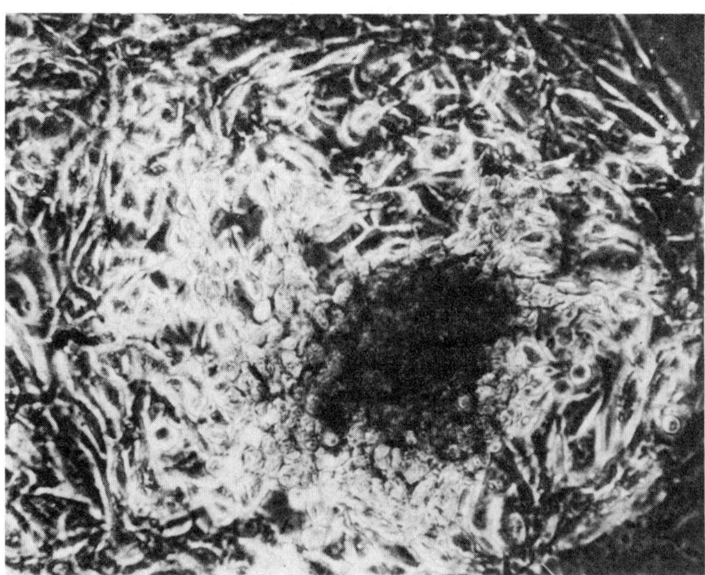

Fig. 1. Human diploid cells transformed in vitro by X-rays. (From Ref. 3.)

Permissive and Protective Factors

While DNA is the target in radiogenic and chemically induced transformation (18,23) and defective DNA repair enhances cell susceptibility to transformation (9), a variety of endogenous and exogenous factors play a determining role in modifying the onset and expression of the neoplastic process (5,7). Physiological factors as well as cell contact may modify the expression of transformation in genetically homogenous, cloned populations of cells (13,14). Cell-cell communication between transformed cells (17) as well as between normal and transformed cells is modified and may lead to an inhibition of growth in the transformed cell populations by contact with their normal counterparts (12). Serum factors also modify cell-cell communication among transformed cells (17) and modulate their "metastasizing" phenotype in vitro (12,17).

Hormones as Permissive and Potentiating Factors in Transformation

Hormones have long been known to exert an important influence on neoplastic transformation in vivo, yet the underlying mechanisms at a cellular level have been hard to define. In vitro, where cells are grown and treated under defined conditions free from homeostatic events, we investigated the role of hormones in transformation (7).

Thyroid Hormones

Our studies, using hamster embryo cells and C3H/10T1/2 mouse cells or NRK rat cells, show that thyroid hormones play a critical role in cellular transformation by radiation (27), chemical carcinogens (21), and tumor viruses (20).

Removal of thyroid hormones from the serum did not modify cell growth or survival, but made the cells refractory to transformation by X-rays (27), benzo(α)pyrene (BP), and N-methyl-N'-nitro-N-nitrosoguanidine (MNNG) (21), and by the Kirsten murine sarcoma virus (20).

When triiodothyronine (T3) was added at physiological levels (10^{12} to 10^{10} M), cell transformation was induced in a T3-dose-dependent manner (20, 21,27) (Fig. 2). The induction of transformation took place within a confined window in time (20,21,27). Maximum transformation was observed when the hormone was added 12 hr prior to exposure to the oncogenic agents, and it had no effect on transformation frequencies when added after exposure to radiation, chemicals, or the virus (20,21,27).

The action of thyroid hormones in transformation appears to be mediated via more than one route. Our studies indicate that T3 may influence the synthesis of a transformation-associated protein that may play a role in the neoplastic process (21). This possibility is supported by the fact that the T3 dose-response relationship for transformation is similar to the T3 dose-dependence for the cellular enzyme Na/K^{+} (21,27).

Our recent data indicate that T3 plays a role in cellular transfection by DNA from radiation-transformed cells (18,23), indicating that the hormone regulates cellular gene expression in transformation (Tab. 1).

An additional mechanism for thyroid hormone regulation of transformation resides in the ability of the hormones to modify the pro-oxidant state of the cells (16). This possibility is underscored by the fact that superoxide dismutase (SOD) suppresses the effects of T3 in potentiating

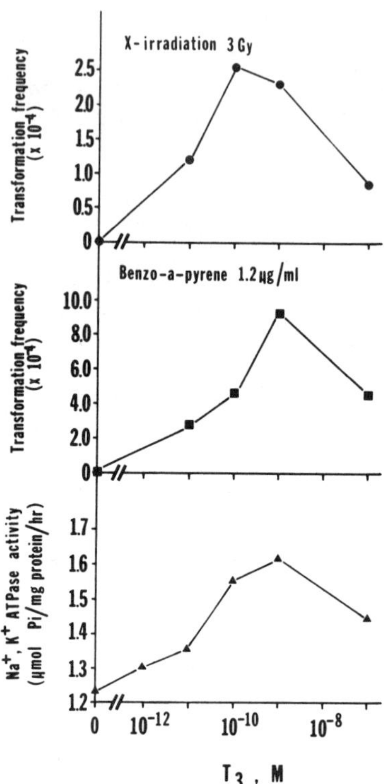

Fig. 2. The effect of varying concentrations of T3 on X-ray-induced (●) and BP-induced (■) transformation and on Na^+, K^- ATPase activity (▲) in stationary-phase C3H/10T1/2 cells. For transformation experiments, cells were pretreated with various doses of T3 one week prior to treatment with BP (1.2 μg/ml) or X-rays (4 Gy) and maintained under the same conditions for the remainder of the experiment. (From Ref. 21.)

radiogenic transformation as well as inhibiting the promoting action of TPA (29) and teleocidin (16,26), tumor-promoting agents that generate free oxygen species (16,31).

Enzymatic and Nutritional Factors as Protective Factors in Transformation

The interaction of cells with radiation, both X-ray and ultraviolet (UV) light, as well as with a variety of chemicals, results in an enhanced generation of free oxygen species and free radical products and in a modified pro-oxidant state (6,30). The result is a loss in the optimal cellular balance between the oxidative challenge, a source of DNA damage, and the inherent mechanisms that protect the cell. These include enzymes (SOD, catalase, peroxidases, transferases) and thiols. Also included are a variety of nutrients that directly or indirectly prevent peroxidation and autoxidation of macromolecules: vitamin A, β-carotene, vitamin C, selenium, and vitamin E (6,10,24).

In recent years, increasing evidence has implicated free radical mechanisms in the initiation and promotion of malignant transformation in vivo

Tab. 1. Thyroid hormone modulation of cellular transforming genes.

		Efficiency of transformation	
	Foci/μg of DNA	Foci/μg of DNA	Foci/μg of DNA
Donor DNA source	Recipient	Thyroid-containing serum	Thyroid-depleted serum
C3H/10T1/2 normal	NIH/3T3	0.00	0.00
C3H/10T1/2 X-ray-transformed	NIH/3T3	0.14	0.00
Hamster embryo normal	NIH/3T3	0.00	0.00
Hamster embryo X-ray-transformed	NIH/3T3	0.15	0.03

and in vitro. Much of the evidence has come from the fact that agents that scavenge free radicals directly or that interfere with the generation of free radical-mediated events inhibit the neoplastic process (1,10,15,24, 32). We have shown in hamster embryo cells that SOD inhibits transformation by radiation and bleomycin and suppresses the promoting action of TPA (15). Catalase had no effect in this cell system, perhaps because of the inherent high level of the enzyme in the cells (15).

Our work on dietary factors in modifying carcinogenesis can be summarized as follows.

Retinoids. We have shown that a variety of retinoids inhibit transformation by radiation, chemicals, and carcinogens (BP and tryptophan pyrolysates) (4,24), and by inhibiting the promoting action of TPA (4). While retinoids have been shown to scavenge free radicals, their action is mediated by modifying cellular adhesion and Na/K^+ ATPase (4,6).

Selenium and vitamin E. Other nutrients important in controlling free radical damage are selenium (10,24), a component of glutathione peroxidase, and vitamin E, a component of the cell membrane (24). Our work in vitro shows that selenium and vitamin E, alone and in additive fashion, inhibit cell transformation by X-rays, BP, and tryptophan pyrolysate (24) (Fig. 3). The underlying mechanism in selenium protection appears to be mediated in part by the enhancement of cellular scavenging systems (10,24) and by a doubling of peroxide breakdown (7,10,24) (Tab. 2). Vitamin E does not modify these enzymes nor does it alter glutathione levels, indicating that its action in inhibiting transformation, while complementary with selenium, proceeds via another mechanism (24).

Bisulfites. Bisulfites serve as food additives and, at low levels, can act as anticarcinogens, inhibiting radiogenic and chemically induced transformation (19) (Fig. 4). Their action is possibly mediated via the inhibition of free radical processes.

Protease inhibitors. Protease inhibitors of high and low molecular weight prevail in many foods. Earlier studies have shown that these agents, including antipain, can inhibit radiogenic transformation in vitro (for review, see Ref. 7).

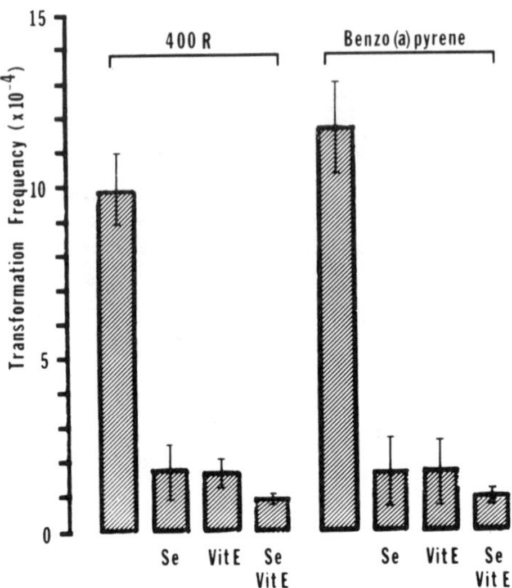

Fig. 3. The inhibitory effect of vitamin E and selenium on oncogenic cell transformation produced by X-rays and benzo(α)pyrene. (Data from Ref. 24.)

Our current experiments indicate that low molecular weight protease inhibitors suppress poly(ADP) ribosylation (8,25). Poly(ADP) ribose synthetase action is triggered by DNA damage induced by radiation and by a variety of alkylating agents (22). The events that follow result in the transfer of the poly(ADP) ribose moieties from nicotinamide-adenine dinucleotide (NAD), their polymerization, and covalent binding to a variety of nuclear proteins, with a resulting alteration in gene expression (8,22,25).

We have found that poly(ADP) ribose plays a critical role in transformation and that its inhibition is associated with suppression of radiation (UV and X-ray) and chemically induced transformation (22). It is thus of interest that dietary components such as low molecular weight protease inhibitors act as anticarcinogens by inhibiting the consequences of DNA damage and by altering gene expression (Fig. 5, Tab. 2).

Tab. 2. Protease inhibitors that suppress poly(ADP) ribosylation and transformation.

Protease inhibitor	Effect on transformation	Effect of heat on poly(ADP) ribosylation
Antipain	Inhibition	Suppression
Leupeptin	Inhibition	Suppression
Soybean	Inhibition	No effect
Bowman-Birk	Inhibition	No effect

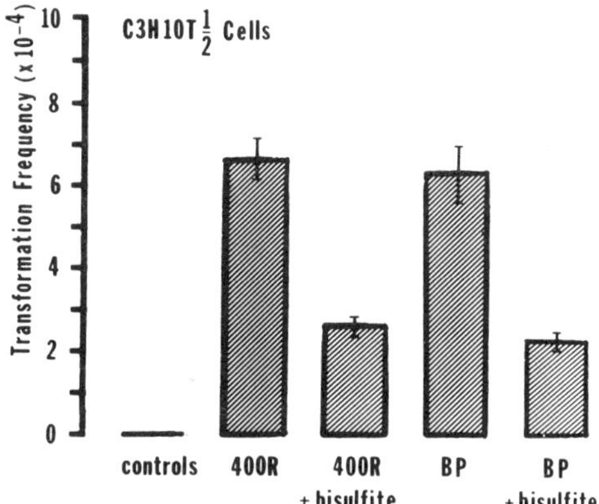

Fig. 4. The inhibitory effect of 0.5 ppm sodium bisulfite on oncogenic cell transformation produced by X-rays or benzo(α)pyrene in C3H/10T1/2 mouse cells. (Data from Ref. 19.)

CONCLUSIONS

There is no complete escape from ubiquitous physical and chemical agents that prevail in our environment. These can induce DNA damage directly or indirectly, resulting in altered gene expression. Some of these processes are mediated via free radical mechanisms, which cause a deterioration in cellular structure and function. Such damage can trigger oncogenic events. Carcinogens, as well as tumor promoters, which produce free radicals in the cell, can therefore challenge cellular integrity by DNA or protein modification, resulting in abnormal expression of cellular genes.

Hormones, such as thyroid hormones, serve as permissive factors for such events. The cell, however, possesses powerful antioxidant mechanisms that, under optimal conditions, monitor oxidant stress and prevent its hazardous consequences.

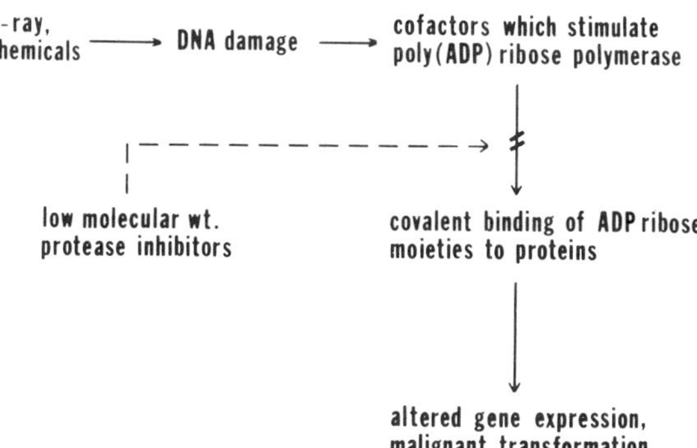

Fig. 5. A scheme describing the possible role of low molecular weight protease inhibitors in poly(ADP) ribosylation and oncogenic cell transformation.

Under enhanced stress during exposure to radiation or chemicals, environmental pollutants, or some dietary carcinogens, one can enhance cellular defenses. This can be done by external enzymatic factors that may directly scavenge free oxygen species, and by nutritional antioxidants that inhibit the cascade of free radical-mediated events (vitamin E) or enhance the cellular capacity to scavenge free radicals and detoxify their products (selenium). Other agents that may suppress transformation are low molecular weight protease inhibitors. By suppressing poly(ADP) ribose polymerization and its covalent modification of cellular protein, these agents modify gene expression and suppress the oncogenic consequences of DNA damage (8,22,25).

ACKNOWLEDGEMENTS

This study was supported by Grant No. CA-12536-11 from the National Cancer Institute, DHHS, and by a contract from the National Foundation for Cancer Research.

REFERENCES

1. Ames, B.N. (1983) Dietary carcinogens and anticarcinogens. Oxygen radical and degenerative diseases. Science 211:1256-1264.
2. Berenblum, I. (1982) Sequential aspects of chemical carcinogenesis: Skin. In Cancer: A Comprehensive Treatise, F.F. Becker, ed. Plenum Press, New York, pp. 451-484.
3. Borek, C. (1980) X ray induced in vitro neoplastic transformation of human diploid cells. Nature 283:776-778.
4. Borek, C. (1982) Vitamins and micronutrients modify carcinogenesis and tumor promotion in vitro. In Molecular Interrelations of Nutrition and Cancer, M.S. Arnott, J. van Eys, and Y.M. Wang, eds. Raven Press, New York, pp. 337-350.
5. Borek, C. (1984) Permissive and protective factors in malignant cells in culture. In The Biochemical Basis of Chemical Carcinogenesis, H. Griem, R. Juna, M. Kraemer, H. Marquardt, and F. Oesch, eds. Raven Press, New York, pp. 175-188.
6. Borek, C. (1985) Free radical, dietary antioxidants and mechanisms in cancer prevention: In vitro studies. In Nutrition and Cancer, Proceedings of the 2nd International Symposium on Modulation and Mediation of Cancer by Vitamins and Micronutrients (Tucson, Arizona), F. Meyskens and K. Prasad, eds. Humana Press, Clifton, New Jersey, pp. 65-82.
7. Borek, C. (1985) The induction and control of radiogenic transformation in vitro: Cellular and molecular mechanisms. Pharm. Ther. 27: 99-142.
8. Borek, C. (1985) The role of nutritional factors in cellular protection against DNA damage, altered gene expression and malignant transformation. In Mechanisms in DNA Damage and Repair, M.G. Simic, L. Grossman, and A.D. Upton, eds. Plenum Press, New York (in press).
9. Borek, C., and A. Andrews (1983) Oncogenic transformation of normal, XP and Bloom syndrome cells by x-rays and ultraviolet irradiation. In Human Carcinogenesis, C.C. Harris and H. Antrup, eds. Academic Press, New York, pp. 519-541.
10. Borek, C., and J.E. Biaglow (1984) Factors controlling cellular peroxide breakdown: Relevance to selenium protection against radiation and chemically induced carcinogenesis. Proc. Am. Assoc. Cancer Res. 25: 125.

11. Borek, C., and A. Ong (1981) The interaction of ionizing radiation and food pyrolysis products in producing oncogenic transformation in vitro. Cancer Lett. 12:61-66.

12. Borek, C., and L. Sachs (1966) The difference in contact inhibition of cell replication between normal cells and cells transformed by different carcinogens. Proc. Natl. Acad. Sci, USA 56:1705-1711.

13. Borek, C., and L. Sachs (1967) Cell susceptibility to transformation by x-irradiation and fixation of the transformed state. Proc. Natl. Acad. Sci., USA 57:1522-1527.

14. Borek, C., and L. Sachs (1968) The number of cell generations required to fix the transformed state in x-ray induced transformation. Proc. Natl. Acad. Sci., USA 59:83-85.

15. Borek, C., and W. Troll (1983) Modifiers of free radicals inhibit in vitro the oncogenic actions of x-rays, bleomycin and the tumor promoter 12-O-tetradecanoylphorbol 13-acetate. Proc. Natl. Acad. Sci., USA 80:5749-5752.

16. Borek, C., J.E. Cleaver, and H. Fujiki (1984) Critical biochemical and regulatory events in malignant transformation and promotion in vitro. In Cellular Interaction by Environmental Tumor Promoters and Relevance to Human Cancer, H. Fujiki et al., eds. Japan Scientific Societies, Tokyo (in press).

17. Borek, C., S. Higashino, and W.R. Loewenstein (1969) Intercellular communication and tissue growth. IV. Conductance of membrane junctions of normal and cancerous cells in culture. J. Memb. Biol. 1:274-293.

18. Borek, C., A. Ong, and H. Mason (1986) Distinctive transforming genes in x ray transformed mammalian cells. (Manuscript submitted for publication.)

19. Borek, C., A. Ong, and H. Mason (1985) Sodium bisulfate protects against radiogenic and chemically induced transformation in hamster embryo and mouse C3H/10T1/2 cells. Toxicol. Indust. Health 1:69-74.

20. Borek, C., A. Ong, and J.S. Rhim (1985) Thyroid hormone modulation of transformation induced by Kirsten Murine Sarcoma virus. Cancer Res. 45:1702-1706.

21. Borek, C., D.L. Guernsey, A. Ong, and I.S. Edelman (1983) Critical role played by thyroid hormone in induction of neoplastic transformation by chemical carcinogens in tissue culture. Proc. Natl. Acad. Sci., USA 80:5749-5752.

22. Borek, C., W.R. Morgan, A. Ong, and J.E. Cleaver (1984) Malignant transformation in vitro is blocked by inhibitors of poly(ADP-ribose) synthesis. Proc. Natl. Acad. Sci., USA 81:243-247.

23. Borek, C., A. Ong, J. Bresser, and D. Gillespie (1984) Transforming activity of DNA of radiation transformed mouse cells and identification of activated oncogenes in the donor cells. Proc. Am. Assoc. Cancer Res. 25:100.

24. Borek, C., A. Ong, L. Donahue, and J.E. Biaglow (1986) Selenium and vitamin E inhibit radiogenic and chemically induced transformation in vitro via different mechanisms of action. Proc. Natl. Acad. Sci., USA (in press).

25. Cleaver, J.E., M. Banda, W. Troll, and C. Borek (1986) Some protease inhibitors are also inhibitors of poly(ADP)ribose. Carcinogenesis (in press).

26. Fujiki, H., M. Mori, M. Nakayasu, M. Terada, and T. Sugimura (1979) A possible naturally occurring tumor promoter, teleocidin B, from streptomyces. Biochem. Biophys. Res. Commun. 90:976-983.

27. Guernsey, D.L., A. Ong, and C. Borek (1980) Modulation of x-ray induced neoplastic transformation in vitro by thyroid hormone. Nature 288:591-592.

28. Guerrero, I., P. Calzada, A. Mayer, and A. Pellicer (1984) A molecular approach to leukomogenesis. Mouse lymphoma contained an activated c-ras oncogene. Proc. Natl. Acad. Sci., USA 81:202-205.
29. Hecker, E. (1971) Isolation and characterization of the cocarcinogenic principles from croton oil. In Methods in Cancer Research, Vol. 6, H. Busch, ed. Academic Press, New York, pp. 439-484.
30. Pryor, H.C., and A.E. Sirica (1980) The role of free radical reactions in biological systems. In Free Radicals in Biology, Vol. 1, W.A. Pryor, ed. Academic Press, New York, pp. 1-49.
31. Troll, W., G. Witz, B. Goldstein, D. Stone, and T. Sugimura (1982) The role of free oxygen radicals in tumor promotion and carcinogenesis. In Carcinogenesis, Vol. 7, Raven Press, New York, pp. 593-597.
32. Zimmerman, R., and P. Cerutti (1984) Active oxygen acts as a promoter of transformation in mouse embryo C3H 10T1/2/C18 fibroblasts. Proc. Natl. Acad. Sci, USA 81:2085-2087.

DIETARY PROMOTERS AND ANTIPROMOTERS

Michael J. Wargovich

Section of Gastrointestinal Oncology and Digestive Diseases
Department of Medical Oncology
University of Texas
M.D. Anderson Hospital and Tumor Institute
Houston, Texas 77030

INTRODUCTION

Changes in human dietary habits can be correlated with changes in the incidence of certain human epithelial cancers such as cancer of the colon and stomach. From this it can be inferred that the diet may harbor chemicals that may cause cancer. It can also be similarly postulated that factors that help to prevent cancer exist in the diet. New testing techniques for mutagenicity and carcinogenicity enable one to identify a considerable number of dietary chemicals. Some of these chemicals influence carcinogenesis by reacting with initiating carcinogens. This presents the exciting possibility that certain cancerous diseases may be prevented by eliminating or binding dietary carcinogens. Although simple in concept, this approach to cancer control is hampered by a lack of carcinogenicity data for dietary carcinogens in humans, and by the logistics of such an endeavor.

It is not now possible, nor is it likely to be in the future, to totally prevent initiation events by environmental carcinogens in human cells. Eluding innate cellular defenses against carcinogenic attack and protection afforded by naturally occurring anticarcinogens, initiated cells are subject to secondary interactions with factors in the diet that promote carcinogenesis and accelerate the progression to a tumor. Little is known about blocking the effects of tumor promoters. However, this is one stage of carcinogenesis at which practical cancer prevention can be achieved. Initiated cells may remain dormant for lengthy periods without untoward effects on the surrounding tissue. Discovery of naturally occurring "antipromoters" may minimize postinitiation consequences and provide an effective means of cancer prevention.

DEFINITIONS: PROMOTERS AND ANTIPROMOTERS

It is difficult to define rigidly the properties of a tumor promoter. One biological feature common to many tumor promoters is the ability to stimulate focal proliferation of initiated cells. Of the possible ways in which initiated cells can be encouraged to proliferate focally, the concept of "differential necrosis with regeneration" as proposed by Farber (12) appears to be the most workable in terms of dietary promoters. In this setting, the dietary promoter induces necrosis in the cellular tissue that includes the initiated cells. Uninitiated cells regenerate following exposure to the necrotoxic agent and eventually return to a normal steady-state as growth restriction and differentiation proceed. Progressive growth of the initiated cells proceeds independently of growth-controlling factors, thus accounting for selective development of cells with neoplastic potential.

The critical question of whether promotional steps in nutritional carcinogenesis can be inhibited has yet to be fully answered. Antipromoters may exert their inhibitory effects in some of the same ways that blockers of initiation work. An antipromoter may chemically bind to the promoting agent, prohibiting further interactions with initiated cells. This is an example of direct inhibition of a tumor promoter.

Another possibility is the disruption of cellular enzymes linked to increased cellular division. In this scenario, a dietary promoter induces enzymes that enhance cell division. Inhibition of these enzymes reduces the biochemical capacity of the population to sustain proliferation in defiance of normal growth-controlling factors. An example is the induction of ornithine decarboxylase (ODC) activity, a rate-limiting enzyme controlling polyamine synthesis. The synthetic inhibitor difluoromethylornithine (DFMO) irreversibly inhibits ODC activity, thus preventing tumor promotion in skin models (41).

For brevity, the discussion of dietary promoters and antipromoters will be limited to organs on which the impact of dietary factors is direct, i.e., the gastrointestinal mucosa.

DIETARY PROMOTERS

Dietary Lipids and Promotion

Because of the difficulty in defining the concept of promotion, dietary studies involving lipids and experimental induction of colon cancer are not uniformly comparable. In many studies, dietary fats varying in degree of saturation and source have been fed to animals throughout the course of the experiment. In the usual protocol, rodents are randomized to receive diets that vary in the amount and source of fat. After a short acclimation period, animals are exposed to an initiating carcinogen and are maintained on the lipid diets until they are killed. A number of studies conducted in this fashion have reported that dietary fat increases colon tumor incidence and tumor burden (Tab. 1). In mechanistic terms, these results implicate dietary lipids as co-carcinogens since their cellular effects cannot be excluded from the time during which the carcinogen was given. This point is critical, since the dietary lipids may have affected the activation and metabolic distribution of the initiating carcinogen.

Experiments in which diets are varied in lipid content in the post-initiation period formally meet the requirements for "promotional" effects. Bull et al. (5) designed their experiment to examine whether dietary beef tallow enhanced colon cancer induction in rats following azoxymethane (AOM) treatment. This sequence induced the greatest frequency of small and large bowel tumors. In another study, Reddy et al. (28) found that diets

Tab. 1. Animal studies examining dietary lipid and colon cancer.

Type of lipid	Duration of lipid diet	Species	Carcinogen	Tumor result/ comment	Reference
15% Vegetable vs 30% tallow	During	Rat	DMH	± Incidence	37
24% Beef fat vs 24% Crisco vs 24% corn oil	During	Rat	DMH	No difference in incidence	22
5% Linoleic vs 5% stearic	During	Rat	AOM	+ Incidence in unsaturated group	38
20% Safflower vs 20% lard	During	Rat	DMH	+ Incidence in both groups	44
5% vs 20% Beef fat	During	Rat	DMH/MAM/MNU	+ Incidence in 20% group	36
5% vs 23% Beef fat	During	Rat	DMAB	+ Incidence in 23% group	29
5% vs 20% Corn oil	During	Rat	DMH	+ Incidence in 20% group	34
5% vs 30% Beef fat	During	Rat	AOM	+ Incidence in 30% group	25
5% vs 30% Beef fat	After	Rat	AOM	+ Incidence in 30% group	5
5% vs 23% Corn, safflower, coconut, and olive oil	After	Rat	AOM	+ Incidence in 23% corn and safflower groups	28

Abbreviations: AOM, azoxymethane; DMAB, 3,2-dimethyl-4-aminobiphenyl; DMH, 1,2-dimethylhydrazine; MAM, methylazoxymethanol; MNU, methylnitrosourea.

enriched with 20% polyunsaturated fat (corn oil or safflower oil) significantly increased colon tumor incidence and tumor frequency compared to a diet enriched with a monounsaturated fat (20% olive oil) when fed to rats in the postcarcinogen period.

Enhancement of Tumorigenesis by Lipid Components: Bile Acids

Several lines of experimental evidence suggest that total fat alone may not account for tumor-promoting effects. Some of these studies are summarized in Tab. 2. In animals as well as humans, high levels of fecal primary and secondary bile acids accompany the ingestion of foods with a high fat content. Both forms of bile acids have subsequently been tested for tumor-promoting activity. Utilizing markers of cellular proliferation, that is, enhanced tritiated thymidine (^{3}HTdR) uptake or ODC activity, several investigators have found that both forms of bile acids are potent inducers of mucosal proliferation (Tab. 2).

In accord with induction of abnormal cellular proliferation, additional colonic studies have revealed that primary and secondary bile acids,

Tab. 2. Promotion of experimental colon cancer by lipid components.

Agent	Carcinogen	Species	Procedure	Result	Reference(s)
Tumors					
Lithocholate	MNNG	Rat	MNNG + LCA i.r.	+ Incidence	30
Cholate DMH		Rat	DMH + Cholate i.g.	± Incidence	8
Deoxycholate	MNNG	Rat	DMH + DCA i.r.	+ Incidence, + multiplicity	35,36
Lithocholate Taurocholate	MNNG	Rat	MNNG + LCA or TCA	+ Incidence	21
Cellular proliferation					
Cholic	MAM	Rat	0.5% Cholic acid in diet	No differences in ^{3}HTdR rates	51
Deoxycholate	None	Mouse	DCA i.r.	+ ^{3}HTdR uptake	49
Oleic acid	None	Mouse	Oleic i.r.	+ ^{3}HTdR uptake	48
Hydroperoxy fatty acids	None	Rat	HPFA i.r.	+ ^{3}HTdR uptake	4

Abbreviations: i.r., intrarectal; i.g., intragastric; DMH, 1,2-dimethylhydrazine.

when given after an initiating carcinogen, increase the incidence and multiplicity of colon tumors. In classical promotion studies by Reddy and colleagues (32,33,36), rat colons treated topically with N-methyl-N'-nitro-N-nitrosoguanidine (MNNG), followed by intrarectal instillations of cholic, lithocholic, or deoxycholic acids, showed increases in the number of colon tumors, although actual incidence of tumors did not vary from the groups given MNNG alone.

The mechanism of action by which bile acids affect colon carcinogenesis is unclear. One attractive possibility relates the physiological effects of bile acids to the epithelial mucosa with subsequent morphological and cell kinetic responses. Chadwick et al. (6) have correlated secretory and permeability changes in colonic mucosa exposed to 6 mM deoxycholate with epitheliolysis, that is, observable mucosal damage in the same colon. Cyclical episodes of bile acid-induced tissue damage might serve as the trigger for enhanced cellular proliferation and provide the environment for selective development of initiated cells.

Enhancement of Tumorigenesis by Lipid Components: Fatty Acids

Of the components of lipid digestion that reach the colon on a daily basis, the majority are unesterfied fatty acids (26). Fatty acids in the ionized form irritate the mucosal layer of the colonic epithelium (15).

Evidence that fatty acids are stimulants for cellular proliferation in the colon has been developed in our laboratory (48) and elsewhere (see Tab. 3). Whether the chemical nature of fatty acid is a factor for enhanced cellular proliferation remains unclear, although our studies and those of Bull et al. (4) implicate unsaturated fatty acids or their oxidation products as efficient inducers of DNA synthesis in rodent colonic mucosa.

Tab. 3. Inhibition studies with food additives butylated hydroxyanisole (BHA) and butylated hydroxytoluene (BHT).

Agent	Procedure	Species	Carcinogen/tumor site inhibited	Reference
BHA	In diet	CF, mice	MAM/colon (+)	46
BHA	In diet	CF, mice	MAM/colon (+)	31
BHA	In diet	Ha-ICR	BPDE/forestomach (+)	47
BHT	In diet	Rats	AAF/bladder (-)	52

These results may link unsaturated components of certain dietary fats with the well-known ability of these lipids to enhance tumor formation (Tab. 1).

Dietary Fiber

Although the bulk of studies performed with dietary fiber indicate that it is an inhibitor of carcinogenesis, a few investigators have studied the opposing view: that is, can the abrasive nature of fiber act as a stimulant for cellular proliferation and thereby enhance tumorigenesis? The biological basis for this question stems from investigations into patterns of colonic cell renewal following fasting and refeeding. Stragand et al. (40) observed enhanced colonic mucosal proliferation in proximal segments of ligated intestine. The portion of the colon proximal to the ligature was subjected to distension from fluid accumulation and exhibited higher ^{3}HTdR incorporation activity than the segment distal to the ligature. Physical stimulus may thus serve as a signal for cellular proliferation.

Of the many factors attributed to fiber, increased fecal bulk is universally accepted. In view of the potential effects of increased fecal bulk, Jacobs (19) measured the cellular growth response of the small and large bowels of rats fed a 20% wheat bran diet. He found that after a 4-week feeding, the thymidine-labeling indices of the colonic mucosa shifted to a profile indicative of increased cellular proliferation, but jejunal rates were unaffected. Subsequently, Jacobs has observed that feeding of a 20% wheat bran diet during and/or after exposure to 1,2-dimethylhydrazine (DMH) increased colonic crypt cellularity (17), and he demonstrated enhanced colon tumorigenesis when a 20% oat bran diet was ingested during administration of DMH (18).

These reports suggest a co-carcinogenic relationship between initiating carcinogens and dietary fiber; however, accelerated tumor formation when fiber is fed <u>after</u> the initiation period remains to be shown.

Food Additives

Food additives are commonly added to the food supply in order to preserve freshness and taste. They aid in preventing spoilage by microorganisms and help to retard autoxidation. Two widely used food additives are the substituted phenols butylated hydroxyanisole (BHA) and butylated hydroxytoluene (BHT). Although approved by the Food and Drug Administration for use in U.S. foodstuffs, BHA and BHT have been extensively tested for carcinogenicity.

Both phenols have been tested for carcinogenicity alone, with conflicting results (16,39). In protocols where BHA or BHT has been co-administered in the diet with carcinogens, inhibition of tumorigenesis was observed (see Tab. 3). The tumors that were inhibited included those of the liver, lung, skin, forestomach, and colon. Williams et al. (52), however, demonstrated that BHT, while inhibiting acetylaminofluorene-induced tumors of the liver, actually enhanced bladder carcinogenesis in the same rats. Deschner and Wattenberg (10) examined the effect of dietary BHA in mice during methylazoxymethanol (MAM) treatment. They found that this phenol reduced the amount of proliferation in the colonic mucosa responding to MAM treatment, thus potentially interfering with the clonal development of mutated cells.

DIETARY ANTIPROMOTERS

Protease Inhibitors

Inhibitors of proteolysis are found commonly in many plant species and are present in considerable amounts in leguminous vegetables (43). In support of epidemiological evidence linking vegetable consumption with negative risk for certain epithelial cancers, protease inhibitors have been tested in the laboratory for the ability to inhibit chemical carcinogenesis. One source of protease inhibitors is the soybean. Diets that include soy protein or the Bowman-Birk protease inhibitor in purified form inhibit spontaneous liver tumors (3) and skin carcinogenesis in mice (42). In one of the few studies involving gastrointestinal organs, the Bowman-Birk inhibitor, when fed at 0.5% of the diet, completely prevented the appearance of DMH-induced colon tumors (53).

Little is known about the mechanism of action of protease inhibitors. Proteases are suspected of being involved as inducers of mitogenesis, i.e., enhanced cellular proliferation. They are also suspected of stimulating superoxide formation in response to tumor promoters such as phorbolmyristate acetate (PMA). Whether protease inhibitors are efficient in retarding or delaying the effects of dietary cancer-promoting agents remains to be investigated. Lipid peroxidation is now considered an important contributing factor in carcinogenesis (1). It would be of interest to observe the action of protease inhibitors as inhibitors of fatty acid peroxidation products and hydroxy fatty acids.

Calcium

Until recently, a working hypothesis for inhibiting the effects of dietary lipids in promotion of colon cancer was not available. Newmark et al. (23) presented a theory in 1984 based upon observations of lipid metabolism. Toxicity of ionized bile and fatty acids to the gastrointestinal epithelium is well known (6). The disruption of colonic cellular architecture in response to these ionized lipids is reminiscent of the effects of the divalent cation chelators ethylene dinitrilotetraacetic acid (EDTA) and ethylene bis(oxyethylenenitrilo) tetracetic acid (EGTA) (24). In fact, the affinity of ionized lipids for calcium is as strong as that for the chelating agents; the end product of this association is the insoluble calcium soap of the lipid acid.

In studies in our laboratory, we have examined whether calcium in the form of calcium salts (carbonate or lactate) given prior to an intrarectal instillation of ionized bile or fatty acids could prevent toxicity to the colonic epithelium and interdict the subsequent proliferative response. We found that the effects of intrarectally instilled deoxycholic acid (Fig. 1) or oleic acid could be modified by calcium pretreatment. Extended studies by Rafter (J. Rafter, unpubl. data) suggest that lipid toxicity to colonic cells can be successfully prevented by acidifying the fecal water and by administering sufficient calcium to balance the lipid input to the colon. Whether calcium salts can prevent lipid-promoted colon carcinogenesis is under current investigation in our laboratory and elsewhere.

Calcium as an antidote to fat, however, proves to be a complicated solution. In humans, availability of calcium is compromised by a number of factors that compete for the ion. These include various phosphates, oxalates, and phytate, which avidly bind to calcium (43). As a potential area for further investigation, calcium (and its interaction with calcium competitors and dietary lipids) presents an excellent opportunity for antipromoter research.

Plant Sterols

Plant sterols are similar in structure to mammalian cholesterol and may comprise about 1% of the human diet (7). One of the most intensively studied plant sterols is β-sitosterol to determine the cause of its ability to decrease cholesterol absorption (20). It has been investigated for inhibition of tumor development. Raicht et al. (27) fed Fischer CD rats a diet supplemented with 0.2% β-sitosterol and/or 0.2% cholic acid, or both. Animals were given methylnitrosourea (MNU) by intrarectal instillation and were maintained on these supplemented diets until the experiment was terminated. Rats fed the cholic acid diet did not differ in tumor incidence from those fed the β-sitosterol diet but did differ in tumor multiplicity. These investigators have speculated that β-sitosterol inserts into colonic cellular membranes, thus offering resistance to the detergent action of the bile acid and prohibiting subsequent cell loss and proliferation. The

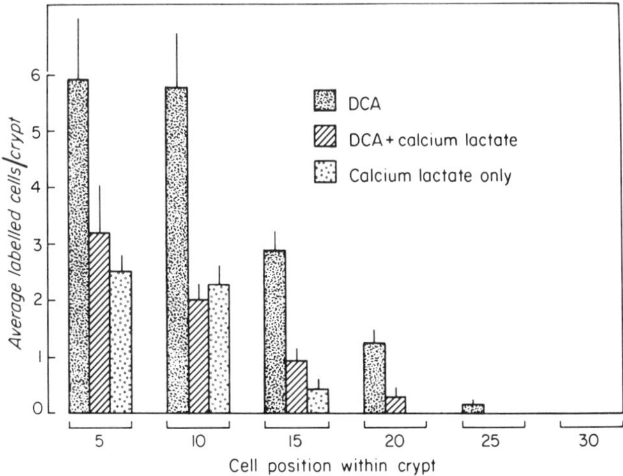

Fig. 1. Patterns of proliferative activity (^{3}HTdR) incorporation in colonic epithelium of mice 24 hr after receiving an intrarectal instillation of deoxycholic acid (DCA) and supplemented with oral calcium lactate. Reprinted from Wargovich et al. (49) with permission from IRL Press.

mechanism of action may be inhibition of the cascade of events by which bile acids induce tissue necrosis and DNA synthesis.

Fiber

It is not surprising that dietary fiber has a multiplicity of physiological effects accounting, in part, for the promoting activity observed in some studies and for the antipromoting effects in others. Studies report diverse results for fiber. Table 4 partly summarizes selected investigations in which fiber components were analyzed for protection against experimentally induced cancer of the colon. In the majority of these experiments, rats were fed the fiber component during the administration of the carcinogen, thus making interpretations of antipromotional effects uncertain. However, two studies indicate a temporal importance for the protective effect of fiber. Bauer et al. (2) found that wheat bran, fed at 20% of the diet, did not decrease DMH-induced incidence or frequency of colon tumors. In their experiment, the bran diet was discontinued after the DMH treatment. Jacobs (18), in a crossover experiment designed to test the potential hazards of a fiber-enriched diet, found that a 20% wheat bran diet fed to rats after DMH treatment reduced the number of benign colon adenomas by 71%, although tumor incidence was unaffected.

Fiber is a complex mixture, and has been subdivided into carbohydrates (cellulose, pectins, gums, guar) and noncarbohydrate (lignin). Pectins, gums, and guars are completely fermented by the human colon, cellulose only partially, and lignin not at all (9). In addition to inducing changes in colonic pH, the degree of digestibility of fiber components varies considerably (45). Thus, not all fiber is available for the protective effects generally attributed to its purported mechanism of protection--the shortening of intestinal transit time and dilution of putative carcinogens. Clearly, the role of fiber in cancer prevention needs to be redefined in terms of antipromoting ability, and experiments designed to test specific fiber components <u>after</u> carcinogen exposure need to be conducted.

Tab. 4. Inhibition of carcinogenesis by dietary fiber.

Fiber component	Carcinogen/organ	Species	Comments	Reference
Wheat bran Citrus fiber	AOM (colon)	Rat	(-) Incidence (-) Multiplicity	32
Corn bran	DMH (colon)	Rat	Early inhibition: None later	14
Alfalfa Pectin Wheat bran	MNU (colon) AOM	Rat	No reduction in incidence	50
Cellulose	DMH (colon)	Rat	(-) Incidence	13
Wheat bran Carrot fiber Pectin	DMH (colon)	Rat	No reduction in incidence for wheat bran; carrot fiber and citrus pectin increased incidence	2
Wheat bran	DMH (colon)	Rat	Crossover study; Wheat bran <u>after</u> DMH = - incidence	18

CONCLUSIONS

The realization that many of the most common forms of cancer are refractory to chemotherapeutic or radiotherapeutic intervention has heightened interest in the causes and prevention of this disease. As part of this refocusing of emphasis, the National Cancer Institute and the American Cancer Society are committed to expansion of research in the field of cancer prevention. Both breast and digestive cancers have a dietary etiology. This is based upon both human and animal studies (11). By understanding how dietary factors relate to the process of carcinogenesis in terms of cause and inhibition, a diet may be prescribed that would aid in the eradication of this cancer epidemic.

In this review, dietary factors that modify carcinogenesis at points subsequent to the initiation event have been explored. Inhibition in the postinitiation phase of carcinogenesis presents a very realistic chance for cancer prevention. Promotional events in carcinogenesis appear to be reversible (12). Removal of the promoting agent may delay carcinogenesis considerably. Antipromoters, which either bind to promoters (thus hindering further cellular interactions) or interfere with critical enzymes that augment cellular proliferation, may prove to be most important in this aspect.

From studies of the carcinogenic process in animals it is clear that animals that are exposed to low doses of carcinogens and fed a diet lacking in promoters develop cancer at a low rate or late in life, if at all. The rate of development of the tumor and tumor frequency are accelerated by dietary promoters. The goal of successful cancer control may be achieved by hindering cancer development during the majority of our life span. The identification of dietary promoters and antipromoters presents an extraordinary opportunity for further research.

ACKNOWLEDGEMENTS

Support from the Elsa U. Pardee Foundation and New Program Development Funds, The University of Texas System Cancer Center, is gratefully acknowledged.

REFERENCES

1. Ames, B.N. (1983) Dietary carcinogens and anticarcinogens. Science 221:1256-1263.
2. Bauer, H.G., N.G. Asp, R. Oste, A. Dahlquist, and P.E. Fredlung (1979) Effect of dietary fiber on the induction of colorectal tumors and fecal B-glucoronidase activity in the rat. Cancer Res. 39:3752-3756.
3. Becker, F.F. (1981) Inhibition of spontaneous hepatocarcinogenesis in C3H/Hen mice by Ed. Pro A, an isolated soy protein. Carcinogenesis 2:1213-1214.
4. Bull, A.W., B.K. Soullier, P.S. Wilson, M.T. Hayden, and N.D. Nigro (1979) Promotion of azoxymethane-induced intestinal cancer by high-fat diet in rats. Cancer Res. 39:956-959.
5. Bull, A.W., N.D. Nigro, W.A. Golembieski, J.D. Crissman, and L.J. Marnett (1984) In vivo stimulation of DNA synthesis and induction of ornithine decarboxylase in rat colon by fatty acid hydroperoxides, auto-oxidation products of unsaturated fatty acids. Cancer Res. 44:4924-4928.

6. Chadwick, V.S., T.S. Gaginella, J.C. Debongie, G.L. Carlson, S.F. Phillips, and A.F. Hofmann (1976) Mucosal epitheliolysis: A mechanism for increased permeability induced by dihydroxy bile acids. Gut 17: 816.
7. Cohen, B.I., and R.F. Raicht (1981) Plant sterols: Protective role in chemical carcinogenesis. In Inhibition of Tumor Induction and Development, M.S. Zedeck and M. Lipkin, eds. Plenum Press, New York, pp. 189-201.
8. Cruse, J.P., M.R. Lewin, and C.G. Clark (1981) The effects of cholic acid and bile salt binding agents on 1,2-dimethylhydrazine-induced colon cancer in the rat. Carcinogenesis 2:439-443.
9. Cummings, J.H. (1982) Polysaccharide fermentation in the human colon. In Colon and Nutrition, H. Kasper and H. Goebell, eds. MTP Press Ltd., Lancaster, pp. 91-102.
10. Deschner, E.E., and L.W. Wattenberg (1982) The proliferative effect of dietary butylated hydroxyanisole on methylazoxymethanol treated colonic mucosa. Cancer Lett. 16:197-202.
11. Doll, R., and R. Peto (1981) The causes of cancer: Quantitative estimates of avoidable risks of cancer in the United States today. J. Natl. Cancer Inst. 66:1191-1308.
12. Farber, E. (1982) Chemical carcinogenesis. Am. J. Pathol. 106:271-291.
13. Freeman, H.J., G.A. Spiller, and Y.S. Kim (1980) A double-blind study on the effects of purified cellulose fiber on 1,2-dimethylhydrazine-induced rat colonic neoplasia. Cancer Res. 38:2912-2917.
14. Freeman, H.J., G.A. Spiller, and Y.S. Kim (1984) Effect of high hemicellulose corn bran on 1,2-dimethylhydrazine induced rat neoplasia. Carcinogenesis 5:261-264.
15. Gaginella, T.S., and S.F. Phillips (1976) Ricinoleic acid (castor oil) alters intestinal surface structure. Mayo Clin. Proc. 51:6-12.
16. Ito, N., S. Fukushima, A. Hagiwara, M. Shibata, and T. Ogiso (1983) Carcinogenicity of butylated hydroxyanisole in F344 rats. J. Natl. Cancer Inst. 70:343-352.
17. Jacobs, L.R. (1983) Effects of dietary fiber on mucosal growth and cell proliferation in the small intestine of the rat: A comparison of oat bran, pectin, and guar with total fiber deprivation. Am. J. Clin. Nutr. 37:954-960.
18. Jacobs, L.R. (1983) Enhancement of rat colon carcinogenesis by wheat bran consumption during the stage of 1,2-dimethylhydrazine administration. Cancer Res. 43:4057-4061.
19. Jacobs, L.R., and F.A. White (1983) Modulation of mucosal cell proliferation in the intestine of rats fed a wheat bran diet. Am. J. Clin. Nutr. 37:945-953.
20. Mattson, F.H., S.M. Grundy, and J.R. Crouse (1982) Optimizing the effect of plant sterols on cholesterol absorption in man. Am. J. Clin. Nutr. 35:697-700.
21. Narisawa, T., N.E. Magadia, J.H. Weisburger, and E.L. Wynder (1974) Promoting effect of bile acids on colon carcinogenesis after intrarectal instillation of N-methyl-N'-nitroso-N-nitrosoguanidine in rats. J. Natl. Cancer Inst. 53:1095-1097.
22. Nauss, K.M., M. Locniskar, and P.M. Newberne (1983) Effect of alterations in the quality and quantity of dietary fat on 1,2-dimethylhydrazine-induced colon carcinogenesis in rats. Cancer Res. 43:4083-4090.
23. Newmark, H.L., M.J. Wargovich, and W.R. Bruce (1984) Colon cancer and dietary fat, phosphate, and calcium: A hypothesis. J. Natl. Cancer Inst. 72:1323-1325.

24. Newmark, H.L., M.J. Wargovich, and W.R. Bruce (1985) Ions and neoplastic development. In Large Bowel Cancer, A.J. Mastromarino and M.G. Brattain, eds. Praeger Scientific, New York, pp. 102-131.
25. Nigro, N.D., D.V. Singh, R.L. Campbell, and M.S. Pak (1975) Effect of dietary beef fat on intestinal tumor formation by azoxymethane in rats. J. Natl. Cancer Inst. 5:429-442.
26. Nordin, B.E., and D.A. Smith (1965) Fat absorption and pancreatic function. In Diagnostic Procedures in Disorders of Calcium Metabolism, Churchill Press, London, pp. 83-97.
27. Raicht, R.F., B.I. Cohen, E.P. Fazzini, A.N. Sarwal, and M. Takahashi (1980) Protective effect of plant sterols against chemically induced colon tumors in rats. Cancer Res. 40:403-405.
28. Reddy, B.S., and Y. Maeura (1984) Tumor promotion by dietary fat in azoxymethane-induced colon carcinogenesis in female F344 rats: Influence of amount and source of fat. J. Natl. Cancer Inst. 72:745-750.
29. Reddy, B.S., and T. Ohmori (1981) Effect of intestinal microflora and dietary fat on 3,2-dimethyl-4-aminobiphenyl-induced colon carcinogenesis in F344 rats. Cancer Res. 41:1363-1367.
30. Reddy, B.S., and K. Watanabe (1979) Effect of cholesterol metabolites and promoting effect of lithocholic acid in colon carcinogenesis in germ-free and conventional F344 rats. Cancer Res. 39:1521-1524.
31. Reddy, B.S., Y. Maeura, and J.H. Weisburger (1983) Effect of various levels of dietary butylated hydroxyanisole on methylazoxymethanol acetate-induced colon carcinogenesis in CFI mice. J. Natl. Cancer Inst. 71:1299-1305.
32. Reddy, B.S., H. Mori, and M. Nicolais (1981) Effect of wheat bran and dehydrated citrus fiber on azoxymethane-induced intestinal carcinogenesis in Fischer 344 rats. J. Natl. Cancer Inst. 66:553-557.
33. Reddy, B.S., K. Watanabe, and J.H. Wesiburger (1977) Effect of high fat diet on colon carcinogenesis in F344 rats treated with 1,2-dimethylhydrazine, methylazoxymethanol acetate and methylnitrosourea. Cancer Res. 37:4156-4159.
34. Reddy, B.S., J.H. Weisburger, and E.L. Wynder (1974) Effect of dietary fat level and dimethylhydrazine on fecal acid and neutral sterol excretion and colon carcinogenesis in rats. J. Natl. Cancer Inst. 52: 507-511.
35. Reddy, B.S., T. Narisawa, J.H. Weisburger, and E.L. Wynder (1976) Promoting effect of sodium deoxycholate on colon adenocarcinoma in germ free rats. J. Natl. Cancer Inst. 56:441-442.
36. Reddy, B.S., K. Watanabe, J.H. Weisburger, and E.L. Wynder (1977) Promoting effect of bile acids in colon carcinogenesis in germ-free and conventional F344 rats. Cancer Res. 37:3238-3242.
37. Rogers, A.E., G. Lenhart, and G. Morrison (1980) Influence of dietary lipotrope and lipid content on aflatoxin B, N-1 fluorenylacetamide, and 1,2-dimethylhydrazine carcinogenesis in rats. Cancer Res. 40: 2801-2807.
38. Sakaguchi, M., Y. Hitamatsu, H. Takada, M. Yamamura, K. Hioki, K. Saito, and M. Yamamoto (1984) Effect of dietary unsaturated and saturated fats on azoxymethane-induced colon carcinogenesis in rats. Cancer Res. 44:1472-1477.
39. Shirai, T., A. Hagiwara, Y. Kurata, M. Shibata, S. Fukushima, and N. Itso (1982) Lack of carcinogenicity of butylated hydroxytoluene on long term administration to B6C3F mice. Food Chem. Toxicol. 20:861-865.
40. Stragand, J.J., and R.F. Hagemann (1977) Effect of lumenal contents on colonic cell replacement. Am. J. Physiol. 233:E208-E211.
41. Takigawa, M., A.K. Verma, R.C. Simsiman, and R.K. Boutwell (1982) Polyamine biosynthesis and skin tumor promotion: Inhibition of 12-O-

tetradecanoylphorbol-13-acetate-promoted mouse skin tumor formation by the irreversible inhibitor of ornithine decarboxylase, α-difluoromethylornithine. Biochem. Biophys. Res. Commun. 105:969-976.

42. Troll, W., and R. Wiesner (1982) Protease inhibitors as anticarcinogens and radioprotectors. In Radioprotectors and Anticarcinogens, O.F. Nygaard and M. Simic, eds. Academic Press, New York, pp. 567-575.

43. Troll, W., S. Belman, S. Wiesner, and C. Shellabarger (1979) Protease action in carcinogenesis. In Biological Function of Proteases, H. Holzer and T. Tschesche, eds. Springer-Verlag, Berlin, pp. 165-170.

44. Trundel, J.L., M.K. Senterman, and R.A. Brown (1983) The fat/fiber antagonism in experimental colon carcinogenesis. Surgery 94:691-696.

45. Van Soest, P.J., and J.B. Robertson (1976) Chemical and physical properties of dietary fibre. In Proceedings of the Miles Symposium, Nova Scotia, Canada, pp. 13-25.

46. Wattenberg, L.W., and V.L. Sparnins (1979) Inhibitory effects of butylated hydroxyanisole on methylazoxymethanol acetate-induced neoplasia of the large intestine and on nicotinamide adenine dinucleotide dependent alcohol dehydrogenase activity in mice. J. Natl. Cancer Inst. 63:219-222.

47. Wattenberg, L.W., D.M. Jerina, L.K. Lam, and H. Yagi (1979) Neoplastic effects of oral administration of (±) trans 7,8-dihydroxybenzol(a)pyrene and their inhibition by butylated hydroxyanisole. J. Natl. Cancer Inst. 62:1103-1106.

48. Wargovich, M.J., V.W.S. Eng, and H.L. Newmark (1984) Calcium modification of the promoting stimulus of fatty acids to colonic epithelium. Cancer Lett. 23:256-261.

49. Wargovich, M.J., V.W.S. Eng, H.L. Newmark, and W.R. Bruce (1983) Calcium ameliorates the toxic effects of deoxycholic acid on colonic epithelium. Carcinogenesis 4:1205-1207.

50. Watanabe, K., B.S. Reddy, J.H. Weisburger, and D. Kritchevsky (1979) Effect of dietary alfalfa, pectin, and wheat bran on azoxymethane or methylnitrosourea induced colon carcinogenesis in F344 rats. J. Natl. Cancer Inst. 63:141-145.

51. Weidema, W.F., E.E. Deschner, B.I. Cohen, and J.J. DeCosse (1985) Acute effects of dietary cholic acid and methylazoxymethanol acetate on colon epithelial cell proliferation; metabolism of bile salts and neutral sterols in conventional and germ free SD rats. J. Natl. Cancer Inst. 74:665-670.

52. Williams, G.M., Y. Maeura, and J.H. Weisburger (1983) Simultaneous inhibition of liver carcinogenicity and enhancement of bladder carcinogenicity of N-2 fluorenylacetamide by butylated hydroxytoluene. Cancer Lett. 19:55-60.

53. Yavelow, J., T.H. Finlay, A.R. Kennedy, and W. Troll (1983) Bowman-Birk soybean protease inhibitor as an anticarcinogen. Cancer Res. 43:2434s-2459s.

MECHANISMS OF CARCINOGENICITY AND ANTICARCINOGENICITY--PART II

A MUTAGEN IS A MUTAGEN, NOT NECESSARILY A CARCINOGEN

Earle R. Nestmann

Mutagenesis Section
Department of National Health and Welfare
Ottawa, Canada K1A 0L2

A largely unspoken but underlying assumption that has been made by authors throughout these proceedings is that processes involved in antimutagenesis are related to anticarcinogenesis, a corollary of the somatic mutation theory of carcinogenesis. It is generally known that most carcinogens are mutagens, but certainly not all chemicals that are mutagenic in short-term in vitro mutagenicity tests are carcinogens. Reasons that contribute to this lack of concordance are discussed later.

The link between mutation and cancer has been a major factor in the dramatic increase of research efforts in mutagenesis, both in mechanistic studies and in more applied aspects (i.e., genetic toxicology). Driven partly by the phrase "carcinogens are mutagens" (1) that was interpreted by many as a promise, and partly by the wish of some geneticists to become involved in what was perceived as more relevant research, this applied subdiscipline grew rapidly. International, national, and regional environmental mutagen societies arose, journals were born, symposia were held, and books of proceedings have appeared.

As the field of genetic toxicology has matured, emphases have changed. The discovery phase of mutagen detection and of new test development, using favorite organisms, has been replaced by an evaluation phase that involves research directed toward understanding test results and rethinking appropriate test strategies.

In keeping with the idea that carcinogens are mutagens, performance of short-term tests has been judged in comparison with results from cancer bioassays. Thus, the sensitivity of a test has been defined as the proportion of carcinogens that are positive, and its specificity as the proportion of noncarcinogens that are negative (3). By these criteria, many short-term tests have been shown to be quite sensitive in their ability to detect carcinogens as mutagens (4). Little attention was paid to test specificity until more recently (5), partly because few compounds were proven noncarcinogens.

In answer to the need for known noncarcinogens that could be used in balanced validation studies, Shelby and Stasiewicz (6) published a list of

70 noncarcinogens. No evidence of induced carcinogenicity by these compounds had been found in four study groups (male and female mice and rats) in cancer bioassays performed by the National Cancer Institute and the National Toxicology Program. Results from four in vitro short-term mutagenicity tests also were summarized by Shelby and Stasiewicz (6) for these 70 noncarcinogens, and many were positive in one or more of the short-term tests. Of course, test performance cannot be evaluated just with the results for these noncarcinogens, out of the context of the much larger databases that are available for these tests. For this reason, Shelby and Stasiewicz (6) did not attempt to rate the tests. Nor did they discuss the reasons why 100% associations are not expected between in vitro mutagenesis and rodent carcinogenesis, although this point generally, and these 70 noncarcinogens specifically, were discussed subsequently by Ashby and Purchase (2).

The words carcinogens and mutagens often are used synonymously, but it is critical to make the distinction. Compounds that induce mutations in vitro may not always be expected to cause cancer in rodents. Chemicals are delivered directly to cells in vitro where they can induce point mutations, chromosome aberrations, or even both types of genetic events. In contrast, many diverse chemical, biological, and physical processes occur in animals. For example, before a mutagenic event (initiation) can occur, a compound must be transported from the site of exposure, metabolized to a reactive form that is not subsequently inactivated, transported to various tissues without decomposition or removal, taken up by target cells, and finally must enter cell nuclei. Such a complex sequence of events is partly responsible for the failure of in vivo short-term mutagenicity tests to detect certain mutagens that are active in vitro. Furthermore, initiation is a preliminary event in the multistage process of carcinogenesis, and many initiating compounds are unable to act also as promoters. Thus, it is clear that in addition to important biological differences between in vitro and in vivo systems, the phenomena of mutation and cancer are divergent endpoints. For all these reasons, evaluation of in vitro test performance on the basis of carcinogenesis alone is inappropriate (7).

REFERENCES

1. Ames, B.N., W.E. Durston, E. Yamasaki, and F.D. Lee (1973) Carcinogens are mutagens: A simple test system combining liver homogenates for activation and bacteria for detection. Proc. Natl. Acad. Sci., USA 70:2281-2285.
2. Ashby, J., and I.F.H. Purchase (1985) Significance of the genotoxic activities observed in vitro for 35 of 70 NTP noncarcinogens. Environ. Mutag. 7:747-758.
3. Cooper, II, J.A., R. Saracci, and P. Cole (1979) Describing the validity of carcinogen screening tests. Brit. J. Cancer 39:87-89.
4. McCann, J., E. Choi, E. Yamasaki, and B.N. Ames (1975) Detection of carcinogens as mutagens in the Salmonella/microsome test. Part 1. Assay of 300 chemicals. Proc. Natl. Acad. Sci., USA 72:5135-5139.
5. Purchase, I.F.H. (1982) ICPEMC Working Paper 2/6, an appraisal of predictive tests for carcinogenicity. Mutat. Res. 99:53-71.
6. Shelby, M.D., and S. Stasiewicz (1984) Chemicals showing no evidence of carcinogenicity in long-term, two-species rodent studies: The need for short-term test data. Environ. Mutag. 6:871-878.
7. Trosko, J.E. (1984) A new paradigm is needed in toxicology evaluation. Environ. Mutag. 6:767-769.

DNA REPAIR AND REPLICATION IN XERODERMA PIGMENTOSUM AND RELATED DISORDERS

James E. Cleaver

Laboratory of Radiobiology
and Environmental Health
University of California
San Francisco, California 94143

ABSTRACT

Xeroderma pigmentosum (XP), ataxia telangiectasia (AT), and Cockayne syndrome (CS) are human diseases that exhibit increased sensitivity to environmental carcinogens [e.g., ultraviolet (UV) light, ionizing radiations, chemicals] because of genetic defects in the patient's capacity to repair and replicate damaged DNA accurately. The major defect in XP is a failure to repair UV damage to DNA; in AT, the failure is in repair or replication of double-strand breaks in DNA; in CS, the failure is in recovery of DNA replication after UV irradiation. Cancer is a major clinical feature of XP and AT, but not of CS. Each disease is complex, with multiple groups defined by complementation in cell-cell hybridization. Overlap is reported between some XP and CS groups. UV-sensitive hamster cell mutants are also known: most of these complement XP groups, and a human gene on chromosome 19 can correct the defects in hamster mutants, but not XP. XP group C is distinct from the other groups in exhibiting a strongly clustered mode of repair, as if only certain regions of the genome can be mended. This mode mainly occurs in confluent group C cells under conditions that permit much greater survival than in exponential growth, and therefore represents a more efficient mode of repair. These diseases all represent important examples of perturbation in the way carcinogen damage in DNA is metabolized, and further research aimed at identifying the kinds of molecular changes involved in the malignancy will be important.

INTRODUCTION

A small number of human hereditary diseases exhibit abnormal cellular responses to DNA-damaging agents, i.e., they are hypersensitive (15). In XP, the reason for the hypersensitivity is known to be a defect in the repair of damaged DNA (10,15). In other hypersensitive diseases there may be more complex abnormalities. In AT there are alterations in both semiconservative replication (65,66) and repair (67,81), especially with respect to agents that make double-strand breaks in DNA (19). In CS there are alterations in the recovery of DNA replication after irradiation by UV light

(45,47,58). Fanconi's anemia exhibits a large increase in sensitivity to DNA cross-linking agents, but the biochemical basis is not known. In addition, defects in repair of O^6-methylguanine, a highly mutagenic lesion produced by endogenous and exogenous alkylating agents, develop readily in cell culture (20,76). This is manifested in the mer^- or mex^- phenotype, but as yet no parallel defects have been identified in vivo.

XERODERMA PIGMENTOSUM: THE CLINICAL CONDITION*

Xeroderma pigmentosum is an autosomal recessive inherited skin disease in which homozygotes show a marked tendency to develop skin cancers from exposure to sunlight (13,15,69). Patients with severe neurological involvement, including microcephaly, mental retardation, and areflexia, have been classified as a distinct form of XP, the de Sanctis-Cacchione syndrome. Some immunological dysfunction also occurs (60). Heterozygotes are generally asymptomatic, although a few cancers and a slight excess of neurological abnormalities have been reported (13,15). Another form of XP with distinct biochemistry is known as the XP variant or pigmented xerodermoid (13,15).

The frequency of cancers in XP patients in sites other than the skin is uncertain because of the difficulty in obtaining enough statistically significant data (41). A retrospective survey of clinical reports of tumors in skin and other organs indicated a 2,000-fold increase in skin (basal and squamous cell carcinoma, and melanoma), ocular, and tongue tumors in comparison to the general population (41). In comparison, there was a 12-fold increase in cancers in other sites, of which the majority involved the brain and the oral cavity. These internal symptoms may reflect exposure to carcinogens at a lower dose rate than that from sunlight.

Cockayne syndrome consists of dwarfism, premature senility, retinal pigment degeneration, optic atrophy, deafness, mental retardation, absence of subcutaneous fat, and sensitivity to sunlight (45,47,58). CS patients do not show the characteristic malignant changes of the skin seen in XP, but at least two patients have now been reported who appear to suffer from a conjunction of XP and CS symptoms (13,69).

XERODERMA PIGMENTOSUM: BIOCHEMICAL ABNORMALITIES

Pyrimidine dimers (cyclobutane and azetidine) and other carcinogen adducts in DNA constitute the major substrates for the nucleotide excision pathway that is important for XP (Fig. 1). These major chemical changes in DNA require extensive reconstruction of the damaged regions (Fig. 1). The initial incision in mammalian cell DNA appears to be endonucleolytic to one or both sides of the damaged sites, and it is this step that is defective in most XP cells (11,13,15). Damage from UV light is approximately random, but repair of moderate to high levels of damage ($>6\ J/m^2$) passes through a stage at which the repair sites are clustered (18), suggesting that repair enzymes are partially processive such that adjacent damaged sites will tend to be repaired sequentially.

---

* Full original references to the clinical features described in this section will be found in the bibliographies of Ref. 13, 15, and 69.

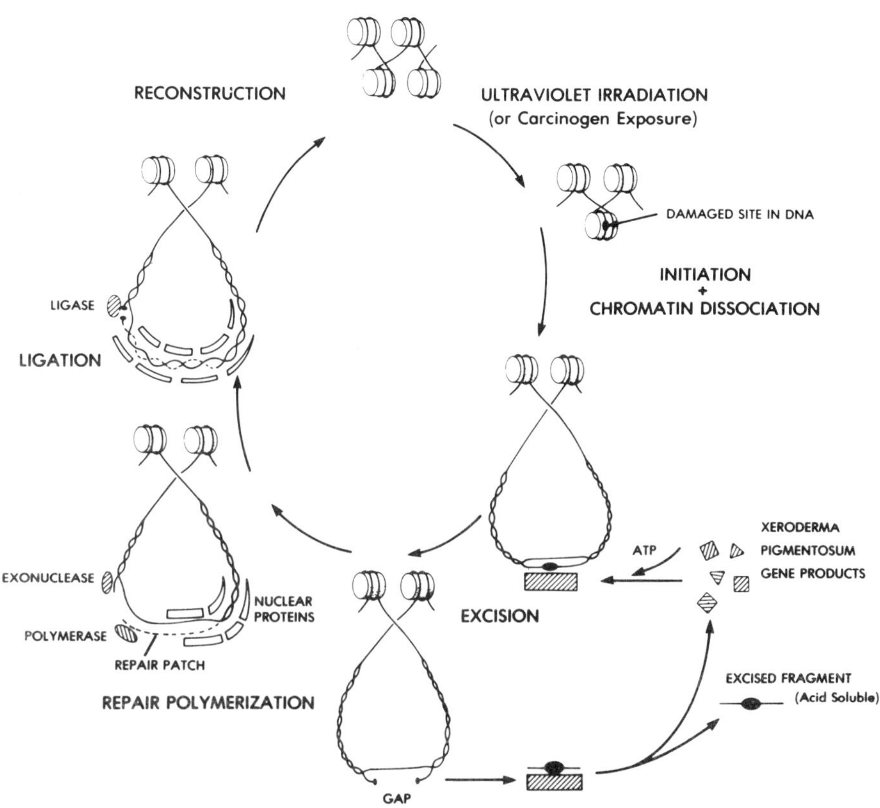

Fig. 1. Excision repair cycle for human chromatin illustrating changes in nucleosomal conformation. Excision involves temporary loosening of chromatin structure in the vicinity of the damaged site, making the region accessible to repair enzymes. The multiple gene products specified by XP complementation groups form an excision complex that removes an oligonucleotide that may initially be acid precipitable before release and further breakdown. As repair replication proceeds, chromatin proteins rapidly become associated with the new patch. Once ligation is complete, the region rapidly acquires native nucleosomal structure with a halftime of 10 to 20 min.

In XP cells, the rate of removal of damaged sites, such as pyrimidine dimers (90) or other chemical adducts (68), is markedly reduced. As a consequence, later stages of repair are reduced to low or negligible levels, including strand breakage (23) and the incorporation of new bases by repair replication (5,10). Cells therefore exhibit a large increase (up to approximately 10-fold) in their sensitivity to killing by UV light (3) (Fig. 2). Although XP appears functionally to be lacking endonucleolytic activity, enzymes have proved difficult to characterize in extracts. Indirect evidence suggests that the enzymatic function may lie in an extremely large aggregate of polypeptides (26,29,84), but their endonucleolytic function on DNA and chromatin remains enigmatic (35,61).

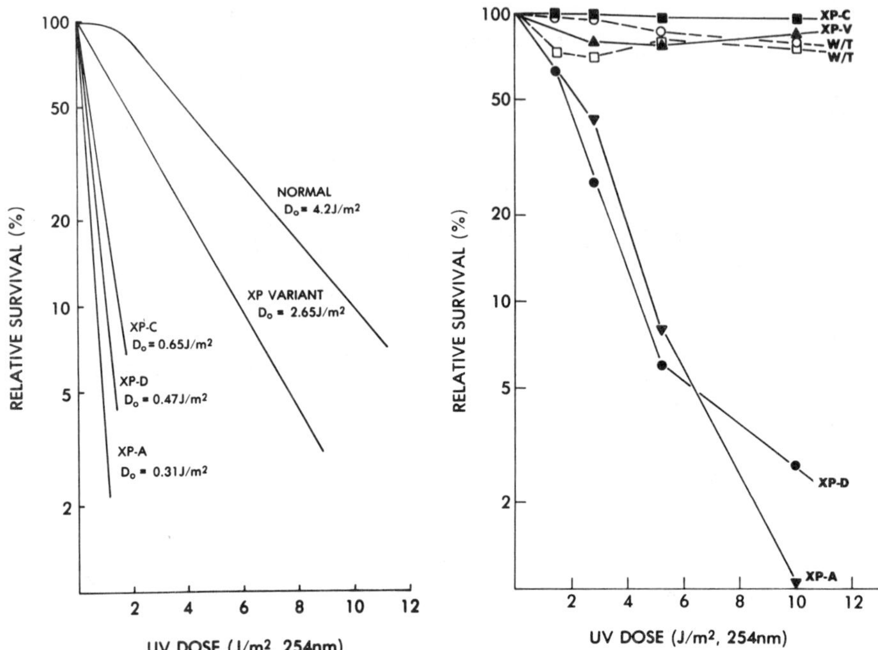

Fig. 2. Relative survival of normal and XP cells irradiated with UV light and assayed either for colony formation (left) or for survival as confluent cultures (right). Colony formation curves represent the mean curves for each group from the table presented by Andrews et al. (23). Confluent survival was measured either by prelabeling cells with [^{14}C]thymidine (0.01 µCi/ml; 54 mCi/mmol) and assaying for ^{14}C activity, or by measuring the capacity of cells to incorporate [^{3}H]hypoxanthine (5 µCi/ml; 10 Ci/mmol) into acid-insoluble RNA and DNA.

Xeroderma pigmentosum cells also show an increased sensitivity to a large variety of chemical carcinogens because of the role of the nucleotide excision repair pathway in repair of bulky adducts in DNA. Certain chemicals raise problems, however. XP cells are sensitive to asbestos (86), antipain (34), and interferon (79), none of which are known to damage DNA (7). UV-sensitive mutants of *Escherichia coli* did not, however, show increased sensitivity to antipain (34), suggesting that XP cells exhibit additional abnormalities unrelated to their repair deficiencies. Similar anomalies have been reported in other diseases. AT fibroblasts are sensitive to tumor promoters under conditions in which DNA damage is not involved (75). Mer^{-} cells, which characteristically are sensitive to alkylating agents, exhibit linked sensitivity to interferon (87).

GENETIC HETEROGENEITY IN XERODERMA PIGMENTOSUM

Genetic heterogeneity in XP cells has been analyzed by somatic cell hybridization, and numerous complementation groups, designated A through I (13,15,69), are known (Tab. 1). The same groups are obtained when chemicals such as 4-nitroquinoline 1-oxide are used as the damaging agent (89). Neurological abnormalities are concentrated in patients in groups A, B, D,

Tab. 1. Complementation groups in xeroderma pigmentosum and UV-sensitive Chinese hamster ovary (CHO) cells.

XP		CHO
A	CNS[1]	I
A - Subgroup	-	II[2]
B (Cockayne)	CNS	III
C	-	IV
D[3]	CNS	V
E	-	VI
F	-	
G	CNS	
H (Cockayne)	CNS	
I	CNS	

[1] Central nervous system.
[2] One CHO complementation group, but no XP groups, is complemented by a cloned gene (ER CC1) from human chromosome 19 (Bootsma, unpubl. observ., 1985).
[3] Group D appears to be linked to trichothiodystrophy, a sulfur-deficient skin and hair disease, on the basis of unrelated patients in the United States and Europe.

and G, although there are a few exceptions (13,15,69). Fusions between groups A and D result in rapid complementation, whereas those involving group C are much slower (27,57). Similar rapid and slow complementation occurs when enucleated cells (cybrids) are used, implying that the gene products can be found in the cytoplasm (38). These studies suggest that the group A gene product is present in excess, turns over rapidly, and promptly reassociates with other products during hybridization, whereas the group C gene product may be synthesized slowly. The repair complex itself appears to be synthesized slowly and is unaffected by inhibition of protein synthesis with cycloheximide (24). Complementation is not prevented by cycloheximide, suggesting that the gene products are not induced (28).

Xeroderma pigmentosum group A cells can be subjected to a selection procedure that produces revertants with normal sensitivity to UV light (70). The production of revertants after treatment with a chemical mutagen (unpubl. observ.) implies that the original gene defect is a point mutation rather than a deletion, although suppression by amplification of some related DNA sequences cannot be excluded. Group A cells also can be complemented by microinjection of a partially purified mRNA of approximately 11S (43), suggesting that this group represents a mutant polypeptide gene product rather than a regulatory gene.

Patients in complementation group C appear to have a milder defect than those in group A, and have no neurological abnormalities. Several observations indicate that these cells have abnormalities in the rates of repair. If these cells are maintained in confluence, for example, they can recover to near normal extents (Fig. 2), whereas cells from groups A and D show little recovery (8,50).

Excision repair sites are strongly clustered in XP group C cells compared to group D and normal cells (36,54,62). We recently have observed that this clustering is confined to confluent group C cells (Fig. 3); random repair occurs in exponentially growing cells (unpubl. observ.). The marked clustering of repair in confluent group C cells therefore correlates with greater survival, and clustering may represent a more efficient way to produce functionally mended genes.

CHROMOSOME LOCATION AND CLONING OF EXCISION-DEFECTIVE XERODERMA PIGMENTOSUM AND RELATED GENES

Identification of the chromosomes carrying the XP genes has been attempted using interspecies hybridization. Unfortunately, although repair-deficient hamster cells have been isolated (82), none of these are known to have precisely the same genetic abnormality as XP (Tab. 1). Most UV-sensitive hamster mutants complement most XP complementation groups, indicating that the genes for repair are different in these two species (77). In hybridization experiments using rodent cells, chromosomes 1 (37,48), 3 (42), and 13 (32) have been identified as important in UV sensitivity.

Experiments designed to clone the genes involved in XP have encountered serious and unexpected technical problems (12,70,80). On the other hand, DNA sequences from human chromosome 19 that complement UV-sensitive hamster cells have been identified and cloned (71,72,85). These sequences do not correct the XP defects and therefore may not correspond to any of the XP genes, but may correspond to a function so vital that homozygote deficiencies are lethal in man.

DNA REPLICATION ABNORMALITIES IN XERODERMA PIGMENTOSUM VARIANT AND OTHER HYPERSENSITIVE DISORDERS

One particular category of XP patients appears to have normal excision repair of UV damage but an abnormality in DNA replication, and has been classified as XP variant (13,15,69). After irradiation, replication forks appear to stop or be interrupted at most sites of damage in the variant (16,17,44,46), whereas normal cells replicate in lengths longer than interdimer distances. The XP variant also shows a delayed recovery of DNA synthesis after irradiation (17) and an extended delay in transition from G_1 to S (33).

The most likely interpretation of the XP variant is that these cells lack gene products whose primary role is in the replication of damaged DNA. In normal cells these gene products would facilitate the rapid passage of replication forks past damaged sites with minimal interruption of daughter strand continuity.

Excision-defective XP, CS, and AT cells also show different abnormal responses to irradiation. In excision-defective cells, greater levels of unrepaired damage interrupt DNA synthesis more than in normal cells. CS cells show an inability to reinitiate replication after UV irradiation. AT cells, in contrast, fail to switch off DNA synthesis after irradiation with X-rays (65,66).

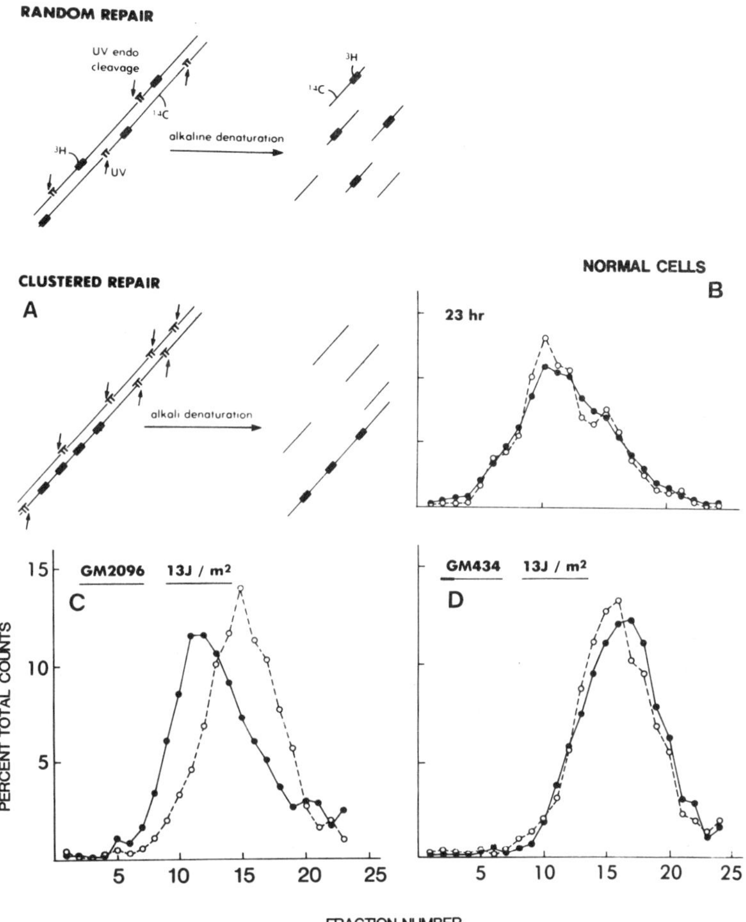

Fig. 3. (**A**) Principle of method used to detect clustering of ^{3}H-labeled excision repair patches in [^{14}C]thymidine uniformly labeled DNA. (TT) pyrimidine dimers; (■) ^{3}H-labeled repair sites. Random repair will give repair patches interspersed among unrepaired lesions, and size distributions of repaired and unrepaired DNA will be similar. Clustered repair will result in larger sizes of DNA containing repaired sites than of DNA between unrepaired sites. Alkaline sucrose gradients of DNA isolated from (**B**) normal (HF1), (**C**) XP group C (GM2096, XP1M1), and (**D**) XP group D (GM434, XP3NE) fibroblasts grown in [^{14}C]thymidine (0.005 μCi/ml; 50 mCi/mmol), irradiated with UV light (13 or 26 J/m^{2}), grown for 24 hr in [^{3}H]thymidine (20 μCi/ml; 80 Ci/mmol) plus hydroxyurea (2 mM), and treated with crude Micrococcus luteus UV endonuclease (45 min, 37°C) before centrifugation. (●) ^{3}H; (○) ^{14}C. A clustered repair mode that produces distributions of ^{3}H-labeled DNA with higher molecular weight than the ^{14}C distributions is evident for only XP group C.

THE ORIGINS OF CANCER IN XERODERMA PIGMENTOSUM: MUTAGENESIS, CHROMOSOME ABERRATIONS, AND TRANSFORMATION

One theory of carcinogenesis invokes an accumulation of genetic damage to some critical region(s) of the genome by exposure to environmental mutagens (2). Damaged regions could become sites for somatic mutation, chromosomal rearrangement, gene amplification, or other alterations in the genome. Evidence bearing on some of these possible mechanisms in carcinogenesis has been obtained from studies of mutagenesis, chromosomal abnormalities, and transformation in XP cells. The life spans of XP patients (41) and of XP cells in culture (14) appear normal (excepting early deaths due to the malignancies), which implies that aging, unlike cancer, is unrelated to the kinds of repair abnormalities seen in XP. Mutation rates after irradiation of XP cells are elevated approximately in proportion to their increase in sensitivity to killing by DNA-damaging agents and their decrease in repair (21,30,31,51).

Although this would suggest that mutagenesis and cell killing are altered in the same quantitative sense by excision repair deficits, other studies point to important differences in the mechanisms of these two processes. In normal fibroblasts, for example, irradiation in early G_1 leaves cells with a long period for repair to occur before DNA synthesis. The consequence of this time delay before replication is a reduction in the mutation rates, as compared to cells irradiated late in G_1, but no change in cell killing (40). The XP variant cells are very sensitive to the induction of mutations by UV light, even though their sensitivity to killing is only slightly increased (52,63). The defect in these cells, therefore, is one that preferentially affects mutagenesis. One hypothesis consistent with these observations would be that XP variant cells may excise cyclobutane dimers correctly but fail to excise the minor but highly mutagenic azetidine photoproducts.

The repair deficiencies in XP have a major effect on the chromosome stability in irradiated cells. Unexposed cells do not, however, exhibit spontaneous chromosome abnormalities or sister chromatid exchanges (SCEs), unlike some other disorders such as Bloom's syndrome, Fanconi's anemia, or AT (25). Aberrations of several types are produced at increased levels in XP cells after UV irradiation (49,56,64,78).

Exposure to UV light causes increased frequencies of SCEs in both excision-defective XP fibroblasts and lymphoblastoid cells, but not in XP variant cells (9,22,83). In contrast, increased SCE frequencies are observed in CS cells, which have a DNA replication abnormality different from that of the XP variant (9,55). The normal frequency of SCEs in XP variant cells implies that SCEs have little relevance to mutation or carcinogenesis in this system. Similarly, elevated SCE frequencies in CS, which does not exhibit increased carcinogenesis, support this view.

Transformation of human cell types in vitro involves experimental systems that are more difficult to use routinely for quantitative purposes than are those for rodent cell types (3,6,53). Also, although transformed cells produced from mouse and hamster cells in vitro can form invasive tumors in vivo in immunosuppressed animals, human transformed cells tend to form benign sarcomas that regress. Transformation in vitro, therefore, mimics some but not all aspects of carcinogenesis in vivo. Normal cells have been transformed by irradiation immediately before DNA synthesis, but lower frequencies are obtained when irradiation occurs earlier in G_1 (53). Several studies have shown enhanced transformation of XP cells in vitro from either short-wavelength (254 nm) (53) or medium-wavelength (∿300 nm) (3) UV light.

Xeroderma pigmentosum implicates mutagenesis and chromosomal abnormalities as mechanisms for an environmentally induced cancer. Gene amplification that can be caused by asymmetric segregation of acentric chromosomal fragments and aberrant replication is also enhanced by DNA damage (59). Many of these mechanisms are associated with activation of oncogenes (39). Point mutations have been identified in the ras series of oncogenes (88). Translocation has been identified for the myc oncogene in lymphoma (1). Amplification of genes associated with epidermal growth control (e.g., growth factors, receptors) could be a relevant molecular event in XP as found for amplified myc-related sequences in some human tumors (73).

Replication of excess unrepaired DNA damage from solar radiation, and aberrant replication of normal levels of DNA damage before time for repair, are mechanisms that may converge to produce similar cellular and clinical consequences. The extreme organ specificity of tumors in XP patients (15, 41,60,69) points to a direct mutagenic effect of sunlight and other carcinogens on the organs at risk. The rapid onset of tumors (41) without the many decades of lag commonly observed for skin cancer in normal individuals suggests that sunlight might also have a strong promoting effect on skin carcinogenesis in XP.

Substantial evidence from microorganisms and mammalian cells supports a theory in which initiating events in carcinogenesis are akin to mutagenic events. We would therefore expect XP tumors to contain mutations or other changes in some of the oncogenes recently identified, especially the ras family of oncogenes (39,88). The codons at 12 and 61 of the ras gene (GGT and CAA, respectively) appear to be the major codons in which mutation can lead to transforming activity in these genes (74,88). Analysis of these codons in XP tumors would be of particular interest because only codon 61 has adjacent thymines (in the complementary strand) subject to UV mutagenesis. Therefore, if mutagenesis arises in ras gene codons from direct damage to the bases within these sequences, we would expect a bias to mutations in codon 61 or participation of flanking bases in adjacent codons.

These considerations suggest that the various cellular parameters studied experimentally, such as DNA damage, repair, mutation, and transformation, are but a few of the complex mechanisms involved in carcinogenesis. Further elucidation of details of the biochemical, cellular, and clinical characteristics of XP should provide a better understanding of one kind of carcinogenesis.

ACKNOWLEDGEMENT

This work was supported by the U.S. Department of Energy (DE-AC03-76-SF01012).

REFERENCES

1. Adams, J.M., S. Gerondakis, E. Webb, L.M. Corcoran, and S. Cory (1983) Cellular myc oncogene is altered by chromosome translocation to an immunoglobulin locus in murine plasmacytomas and is rearranged similarly in human Burkitt lymphomas. Proc. Natl. Acad. Sci., USA 80:1982.

2. Ames, B.N. (1983) Dietary carcinogens and anticarcinogens. Science 221:1256.
3. Andrews, A.D., S.F. Barrett, and J.H. Robbins (1978) Xeroderma pigmentosum neurological abnormalities correlate with colony-forming ability after ultraviolet radiation. Proc. Natl. Acad. Sci., USA 75:1984.
4. Andrews, A.D., and C. Borek (1982) UVB light-induced in vitro transformation of skin fibroblasts from xeroderma pigmentosum and Bloom's syndrome patients (abstract). J. Invest. Dermatol. 78:355.
5. Bootsma, D., M.P. Mulder, F. Pot, and J.A. Cohen (1970) Different inherited levels of DNA repair replication in xeroderma pigmentosum cell strains after exposure to ultraviolet irradiation. Mutat. Res. 9:507.
6. Borek, C. (1980) X-ray-induced in vitro neoplastic transformation of human diploid cells. Nature 283:776.
7. Borek, C., and J.E. Cleaver (1981) Protease inhibitors neither damage DNA nor interfere with DNA repair or replication in human cells. Mutat. Res. 82:373.
8. Chan, G.L., and J.B. Little (1979) Resistance of plateau-phase human normal and xeroderma pigmentosum fibroblasts to the cytotoxic effect of ultraviolet light. Mutat. Res. 63:401.
9. Cheng, W.-S., R.E. Tarone, A.D. Andrews, J.S. Whang-Peng, and J.H. Robbins (1978) Ultraviolet light-induced sister chromatid exchanges in xeroderma pigmentosum and in Cockayne's syndrome lymphocyte cell lines. Cancer Res. 38:1601.
10. Cleaver, J.E. (1968) Defective repair replication of DNA in xeroderma pigmentosum. Nature 218:652.
11. Cleaver, J.E. (1969) Xeroderma pigmentosum: A human disease in which an initial stage of DNA repair is defective. Proc. Natl. Acad. Sci., USA 63:428.
12. Cleaver, J.E. (1983) Workshop summary: DNA repair in normal and repair-defective human cells. In Cellular Responses to DNA Damage, E.C. Friedberg and B.A. Bridges, eds. Alan R. Liss, Inc., New York, p. 327.
13. Cleaver, J.E. (1983) Xeroderma pigmentosum. In Metabolic Basis of Inherited Disease, 5th ed., J.B. Stanbury, J.B. Wyngaarden, D.S. Fredrickson, J.L. Goldstein, and M.S. Brown, eds. McGraw-Hill, New York, p. 1227.
14. Cleaver, J.E. (1984) DNA repair deficiencies and cellular senescence are unrelated in xeroderma pigmentosum cell lines. Mech. Aging Devel. 27:189.
15. Cleaver, J.E. (1986) Xeroderma pigmentosum. In Photomedicine, E. Ben-Hur and I. Rosenthal, eds. CRC Press, Boca Raton, Florida (in press).
16. Cleaver, J.E., R.M. Arutyunyan, T. Sarkisian, W.K. Kaufmann, A.E. Greene, and L. Coriell (1980) Similar defects in DNA repair and replication in the pigmented xerodermoid and the xeroderma pigmentosum variants. Carcinogenesis 1:647.
17. Cleaver, J.E., G.H. Thomas, and S.D. Park (1979) Xeroderma pigmentosum variants have a slow recovery of DNA synthesis after irradiation with ultraviolet light. Biochim. Biophys. Acta 564:122.
18. Cohn, S.M., and M.W. Lieberman (1984) The distribution of DNA excision-repair sites in human diploid fibroblasts following ultraviolet irradiation. J. Biol. Chem. 259:12463.
19. Cornforth, M.N., and J.S. Bedford (1985) On the nature of the defect in cells from individuals with ataxia-telangiectasia. Science 227:1589.
20. Day, III, R.S., C.H.J. Ziolkowski, D.A. Scudiero, S.A. Meyer, and M.R. Mattern (1980) Human tumor cell strains defective in the repair of alkylation damage. Carcinogenesis 1:21.

21. Deluca, J.G., D.A. Kaden, E.A. Komives, and W.G. Thilly (1984) Mutation of xeroderma pigmentosum lymphoblasts by far-ultraviolet light. Mutat. Res. 128:47.
22. de Weerd-Kastelein, E.A., W. Keijzer, G. Rainaldi, and D. Bootsma (1977) Induction of sister chromatid exchanges in xeroderma pigmentosum cells after exposure to ultraviolet light. Mutat. Res. 45:253.
23. Fornace, Jr., A.J., and D.S. Seres (1983) Detection of DNA single-strand breaks during the repair of UV damage in xeroderma pigmentosum cells. Radiat. Res. 93:107.
24. Gautschi, J.R., B.R. Young, and J.E. Cleaver (1973) Repair of damaged DNA in the absence of protein synthesis in mammalian cells. Exp. Cell Res. 76:87.
25. German, J. (1972) Genes which increase chromosomal instability in somatic cells and predispose to cancer. Prog. Med. Genet. 8:61.
26. Giannelli, F., and S.A. Pawsey (1974) DNA repair synthesis in human heterokaryons. II. A test for heterozygosity in xeroderma pigmentosum and some insight into the structure of the defective enzyme. J. Cell Sci. 15:163.
27. Giannelli, F., and S.A. Pawsey (1976) DNA repair synthesis in human heterokaryons. III. The rapid and slow complementing varieties of xeroderma pigmentosum. J. Cell Sci. 20:207.
28. Giannelli, F., P.M. Croll, and S.A. Lewin (1973) DNA repair synthesis in human heterokaryons formed by normal and UV-sensitive fibroblasts. Exp. Cell Res. 78:175.
29. Giannelli, F., S.A. Pawsey, and J.A. Avery (1982) Differences in patterns of complementation of the more common groups of xeroderma pigmentosum: Possible implications. Cell 29:451.
30. Glover, T.W., C.-C. Chang, J.E. Trosko, and S.S.-L. Li (1979) Ultraviolet light induction of diphtheria toxin-resistant mutants in normal and xeroderma pigmentosum human fibroblasts. Proc. Natl. Acad. Sci., USA 76:3982.
31. Grosovsky, A.J., and J.B. Little (1983) Mutagenesis and lethality following S phase irradiation of xeroderma pigmentosum and normal human diploid fibroblasts with ultraviolet light. Carcinogenesis 4:1389.
32. Hori, T.-A., T. Shiomi, and K. Sato (1983) Human chromosome 13 compensates a DNA repair defect in UV-sensitive mouse cells by mouse-human cell hybridization. Proc. Natl. Acad. Sci., USA 80:5655.
33. Imray, P., T. Mangan, A. Saul, and C. Kidson (1983) Effects of ultraviolet irradiation on the cell cycle in normal and UV-sensitive cell lines with reference to the nature of the defect in xeroderma pigmentosum variant. Mutat. Res. 112:301.
34. Ishizaki, K., T. Yagi, and H. Takebe (1980) Cytotoxic effects of protease inhibitors on human cells. 1. High sensitivity of xeroderma pigmentosum cells to antipain. Cancer Lett. 10:199.
35. Kano, Y., and Y. Fujiwara (1983) Defective thymine dimer excision from xeroderma pigmentosum chromatin and its characteristic catalysis by cell-free extracts. Carcinogenesis 4:1419.
36. Karentz, D., and J.E. Cleaver (1986) Excision repair in xeroderma pigmentosum group C but not group D is clustered in a small fraction of the total genome. Mutat. Res. (in press).
37. Keijzer, W., M. Stefanini, A. Westerveld, and D. Bootsma (1984) Mapping of the XPAC gene involved in complementation of the defect in xeroderma pigmentosum group A cells (abstract). Cytogenet. Cell Genet. 37:508.
38. Keijzer, W., A. Verkerk, and D. Bootsma (1982) Phenotypic correction of the defect in xeroderma pigmentosum cells after fusion with isolated cytoplasts. Exp. Cell Res. 140:119.

39. Klein, G., and E. Klein (1984) Oncogene activation and tumor progression. Carcinogenesis 5:429.

40. Konze-Thomas, B., R.M. Hazard, V.M. Maher, and J.J. McCormick (1982) Extent of excision repair before DNA synthesis determines the mutagenic but not the lethal effect of UV radiation. Mutat. Res. 94:421.

41. Kraemer, K.H., M.M. Lee, and J. Scotto (1984) DNA repair protects against cutaneous and internal neoplasia: Evidence from xeroderma pigmentosum. Carcinogenesis 5:511.

42. Lalley, P.A., J.A. Diaz, A.A. Francis, W.C. Dunn, and J.D. Regan (1984) The expression and chromosomal assignments of genes required for repair of UV-induced DNA damage (abstract). Cytogenet. Cell Genet. 37:516.

43. Legerski, R.J., D.B. Brown, C.A. Pederson, and D.L. Robberson (1984) Transient complementation of xeroderma pigmentosum cells by microinjection of poly(A)$^+$ RNA. Proc. Natl. Acad. Sci., USA 81:5676-5689.

44. Lehmann, A.R. (1979) The relationship between pyrimidine dimers and replicating DNA in UV-irradiated human fibroblasts. Nucl. Acids Res. 7:1901.

45. Lehmann, A.R. (1982) Three complementation groups in Cockayne syndrome. Mutat. Res. 106:347.

46. Lehmann, A.R., S. Kirk-Bell, C.F. Arlett, M.C. Paterson, P.H.M. Lohman, E.A. de Weerd-Kastelein, and D. Bootsma (1975) Xeroderma pigmentosum cells with normal levels of excision repair have a defect in DNA synthesis after UV-irradiation. Proc. Natl. Acad. Sci., USA 72:219.

47. Lehmann, A.R., S. Kirk-Bell, and L. Mayne (1979) Abnormal kinetics of DNA synthesis in ultraviolet light-irradiated cells from patients with Cockayne's syndrome. Cancer Res. 39:4237.

48. Lin, P.-F., and F.H. Ruddle (1981) Murine DNA repair gene located on chromosome 4. Nature 289:191.

49. Lo, L.W., and H.F. Stich (1975) DNA damage, DNA repair and chromosome aberrations of xeroderma pigmentosum cells and controls following exposure to nitrosation products of methylguanidine. Mutat. Res. 30:397.

50. Maher, V.W., D.J. Dorney, A.L. Mendrala, B. Konze-Thomas, and J.J. McCormick (1979) DNA excision-repair processes in human cells can eliminate the cytotoxic and mutagenic consequences of ultraviolet irradiation. Mutat. Res. 62:311.

51. Maher, V.M., J.J. McCormick, P.L. Grover, and P. Sims (1977) Effect of DNA repair on the cytotoxicity and mutagenicity of polycyclic hydrocarbon derivatives in normal and xeroderma pigmentosum human fibroblasts. Mutat. Res. 43:117.

52. Maher, V.M., L.M. Ouellette, R.D. Curren, and J.J. McCormick (1976) Frequency of ultraviolet light-induced mutations is higher in xeroderma pigmentosum variant cells than in normal human cells. Nature 261:593.

53. Maher, V.M., L.A. Rowan, K.C. Silinskas, S.A. Kateley, and J.J. McCormick (1982) Frequency of UV-induced neoplastic transformation of diploid human fibroblasts is higher in xeroderma pigmentosum cells than in normal cells. Proc. Natl. Acad. Sci., USA 79:2613.

54. Mansbridge, J.N., and P.C. Hanawalt (1983) Domain-limited repair of DNA in ultraviolet irradiated fibroblasts from xeroderma pigmentosum complementation group C. In Cellular Responses to DNA Damage, E.C. Friedberg and B.A. Bridges, eds. Alan R. Liss, Inc., New York, p. 195.

55. Marshall, R.R., C.F. Arlett, S.A. Harcourt, and B.A. Broughton (1980) Increased sensitivity of cell strains from Cockayne's syndrome to sister-chromatid-exchange induction and cell killing by UV light. Mutat. Res. 69:107.

56. Marshall, R.R., and D. Scott (1976) The relationship between chromosome damage and cell killing in UV-irradiated normal and xeroderma pigmentosum cells. Mutat. Res. 36:397.
57. Matsukuma, S., B. Zelle, W. Keijzer, F. Berends, and D. Bootsma (1981) Different rates of restoration of the repair capacity in complementing xeroderma pigmentosum cells after fusion. Exp. Cell Res. 134:103.
58. Mayne, L.V., and A.R. Lehmann (1982) Failure of RNA synthesis to recover after UV irradiation: An early defect in cells from individuals with Cockayne's syndrome and xeroderma pigmentosum. Cancer Res. 42: 1473.
59. Morgan, W.F., J. Bodycote, M.L. Fero, P.J. Hahn, L.N. Kapp, G.E. Pantelias, and R.B. Painter (1985) A cytogenetic investigation of DNA rereplication after hydroxyurea treatment: Implications for gene amplification. Chromosoma (in press).
60. Morison, W.L., C. Bucana, N. Hashem, M.L. Kripke, J.E. Cleaver, and J.L. German (1985) Impaired immune function in patients with xeroderma pigmentosum. Cancer Res. 45:3929.
61. Mortelmans, K., E.C. Friedberg, H. Slor, G. Thomas, and J.E. Cleaver (1976) Defective thymine dimer excision by cell-free extracts of xeroderma pigmentosum cells. Proc. Natl. Acad. Sci., USA 73:2757.
62. Mullenders, L.H.F., A.C. van Kesteren, C.J.M. Bussmann, A.A. van Zeeland, and A.T. Natarajan (1984) Preferential repair of nuclear matrix associated DNA in xeroderma pigmentosum complementation group C. Mutat. Res. 141:75.
63. Myhr, B.C., D. Turnbull, and J.A. DiPaolo (1979) Ultraviolet mutagenesis of normal and xeroderma pigmentosum variant human fibroblasts. Mutat. Res. 62:341.
64. Natarajan, A.T., E.A.M. Verdegaal-Immerzeel, M.J. Ashwood-Smith, and G.A. Poulton (1981) Chromosomal damage induced by furocoumarins and UVA in hamster and human cells including cells from patients with ataxia telangiectasia and xeroderma pigmentosum. Mutat. Res. 84:113.
65. Painter, R.B. (1981) Radioresistant DNA synthesis: An intrinsic feature of ataxia-telangiectasia. Mutat. Res. 84:183.
66. Painter, R.B., and B.R. Young (1980) Radiosensitivity in ataxia-telangiectasia: A new explanation. Proc. Natl. Acad. Sci., USA 77:7315.
67. Paterson, M.C., B.P. Smith, P.H.M. Lohman, A.K. Anderson, and L. Fishman (1976) Defective excision repair of γ-ray-damaged DNA in human (ataxia-telangiectasia) fibroblasts. Nature 260:444.
68. Regan, J.D., and R.B. Setlow (1974) Two forms of repair in the DNA of human cells damaged by chemical carcinogens and mutagens. Cancer Res. 34:3318.
69. Robbins, J.H., K.H. Kraemer, M.A. Lutzner, B.W. Festoff, and H.G. Coon (1974) Xeroderma pigmentosum: An inherited disease with sun sensitivity, multiple cutaneous neoplasms, and abnormal DNA repair. Ann. Intern. Med. 80:221.
70. Royer-Pokora, B., and W.A. Haseltine (1984) Isolation of UV-resistant revertants from a xeroderma pigmentosum complementation group A cell line. Nature 311:390.
71. Rubin, J.S., A.L. Joyner, A. Bernstein, and G.F. Whitmore (1983) Molecular identification of a human DNA repair gene following DNA-mediated gene transfer. Nature 306:206.
72. Rubin, J.S., V.R. Prideaux, H.F. Willard, A.M. Dulhanty, G.F. Whitmore, and A. Bernstein (1985) Molecular cloning and chromosome localization of DNA sequences associated with a human DNA repair gene. Mol. Cell Biol. 5:398-405.
73. Schwab, M., J. Ellison, M. Busch, W. Rosenau, and H.E. Varmus (1984) Enhanced expression of the human gene N-*myc* consequent to amplifica-

tion of DNA may contribute to malignant progression of neuroblastoma. Proc. Natl. Acad. Sci., USA 81:4940.

74. Seeburg, P.H., W.W. Colby, D.J. Capon, D.V. Goeddel, and A.D. Levinson (1984) Biological properties of human c-Ha-ras 1 genes mutated at codon 12. Nature 312:71.

75. Shiloh, Y., E. Tabor, and Y. Becker (1985) Cells from patients with ataxia-telangiectasia are abnormally sensitive to the cytotoxic effects of a tumor promoter, phorbol-12-myristate-13-acetate. Mutat. Res. 149:283-286.

76. Sklar, R., and B. Strauss (1981) Removal of O^6-methylguanine from DNA of normal and xeroderma pigmentosum-derived lymphoblastoid lines. Nature 289:417.

77. Stefanini, M., W. Keijzer, and D. Bootsma (1985) Interspecies complementation analysis of xeroderma pigmentosum and UV-sensitive Chinese hamster cells. Exp. Cell Res. 161:373.

78. Stich, H.F., W. Stich, and R.H.C. San (1973) Chromosome aberrations in xeroderma pigmentosum cells exposed to the carcinogens, 4-nitroquinoline-1-oxide and N-methyl-N'-nitro-N-nitrosoguanidine. Proc. Soc. Exp. Biol. Med. 142:1141.

79. Suzuki, N., T. Kojima, T. Kuwata, J. Nishimaki, Y. Takakubo, and T. Miki (1984) Cross sensitivity between interferon and uv in human cell strains: IF^r, HEC-1, and CRL1200. Virology 135:20.

80. Takano, T., M. Noda, and T.-A. Tamura (1982) Transfection of cells from a xeroderma pigmentosum patient with normal human DNA confers UV resistance. Nature 296:269.

81. Taylor, A.M.R., J.A. Metcalfe, J.M. Oxford, and D.G. Harnden (1976) Is chromatid-type damage in ataxia-telangiectasia after irradiation at G_0 a consequence of defective repair? Nature 260:441.

82. Thompson, L.H., D.B. Busch, K. Brookman, C.L. Mooney, and D.A. Glaser (1981) Genetic diversity of UV-sensitive DNA repair mutants of Chinese hamster ovary cells. Proc. Natl. Acad. Sci., USA 78:3734.

83. Tohda, H., and A. Oikawa (1983) Differential features of sister-chromatid exchange responses to ultraviolet radiation and caffeine in xeroderma pigmentosum lymphoblastoid cell lines. Mutat. Res. 107:387.

84. Waldstein, E.A., S. Peller, and R.B. Setlow (1979) UV-endonuclease from calf thymus with specificity toward pyrimidine dimers in DNA. Proc. Natl. Acad. Sci., USA 76:3746.

85. Westerveld, A., J.H.J. Hoeijmakers, M. van Duin, J. de Wit, H. Odijk, A. Pastink, R.D. Wood, and D. Bootsma (1984) Molecular cloning of a human DNA repair gene. Nature 310:425.

86. Yang, L.L., R.E. Kouri, and R.D. Curren (1984) Xeroderma pigmentosum fibroblasts are more sensitive to asbestos fibers than are normal human fibroblasts. Carcinogenesis 5:291.

87. Yarosh, D.B., D.A. Scudiero, T. Yagi, and R.S. Day, III (1985) Human tumor strains both unable to repair O^6-methylguanine and hypersensitive to killing by human α and β interferon. Carcinogenesis 6:883-886.

88. Yuasa, Y, S.K. Srivastava, C.Y. Dunn, J.S. Rhim, E.P. Reddy, and S.A. Aaronson (1983) Acquisition of transforming properties by alternative point mutations within c-bas/has human proto-oncogene. Nature 303:775.

89. Zelle, B., and D. Bootsma (1980) Repair of DNA damage after exposure to 4-nitroquinoline-1-oxide in heterokaryons derived from xeroderma pigmentosum cells. Mutat. Res. 70:373.

90. Zelle, B., and P.H.M. Lohman (1979) Repair of UV-endonuclease susceptible sites in the 7 complementation groups of xeroderma pigmentosum A through G. Mutat. Res. 62:363.

ROLE OF INTERCELLULAR COMMUNICATION IN MODIFYING THE CONSEQUENCES OF MUTATIONS IN SOMATIC CELLS

J.E. Trosko and C.C. Chang

Department of Pediatrics and Human Development
Center for Environmental Toxicology
Michigan State University
East Lansing, Michigan 48824

POTENTIAL ROLE OF MUTATIONS IN SOMATIC DISEASES

While there is little question that mutations could play a role in several somatic disease states, it is hard to understand how a mutation in a single cell per se could affect the organism. In the context of these proceedings, prevention of the production of mutations seems to be one approach to the problem of eliminating mutation-dependent diseases. Another approach, which is the aim of this analysis, might be the prevention of the physiological impact of the mutated cell.

Conceptually, to understand mutagenesis for the purpose of possibly preventing its contribution to diseases, one must recognize that, for at least gene mutations in the somatic tissues, there exist three critical barriers to be overcome before the mutation can be produced and expressed: (a) the pre-DNA damage stage; (b) the DNA mutation fixation stage; and (c) the postmutation fixation stage (93).

There are many genetic and environmental/physiological factors which could enhance or prevent DNA damage (88,89,91,92,94,96,98,100,107). Albinism or hyperpigmentation could, for example, either enhance or reduce ultraviolet light induction of DNA damage, thereby modifying the probability of a DNA lesion, such as a pyrimidine dimer, serving as a substrate for a mutation. Genetic levels of metabolizing enzymes or environmentally induced drug detoxifying enzymes could also protect the DNA from damage and the biological effects of such damage (90,109). In addition, in those in vivo situations where a chemical must be metabolized in one cell type (i.e., hepatocyte) in order to induce a mutation in another cell type, it appears that gap-junctional intercellular communication might be necessary for the transfer of the reactive compound (33,70).

Once the DNA has been damaged, the error-free repair of that damage might be modified by processes that could enhance or reduce the probability of mutation fixation. For example, since DNA replication seems necessary for mutation fixation (see Ref. 94), factors that either stimulate or suppress DNA synthesis after DNA damage would enhance or reduce the probability of mutation fixation.

At this level, even without lesions in the DNA template, genetic or physiological/environmental modulation of DNA synthesis could enhance (or reduce) mutation fixation. Modulation of nucleotide pool precursors (38, 52,53), fidelity of DNA polymerases (115), and the kind of DNA polymerases used for replication (30) could also alter the mutation fixation process. The recent finding of Dunn et al. (15) suggests that the hypermutability of the xeroderma pigmentosum (XP) variant might be due to a nucleotide pool imbalance.

Finally, once a mutation is produced in a somatic cell, its nature (i.e., recessive, co-dominant) and its expression (i.e., derepressed or repressed gene) will determine its role in the final consequence of this process. However, for the purpose of this analysis, it is hypothesized that in order for most mutations produced in somatic cells to make an impact on the organism, they must occur in stem or progenitor cells which can be increased in number. If this is true (see below for explanation), then it should be apparent that one of the major rate-limiting steps in the mutation-dependent disease processes in somatic tissues would be at this level--the clonal multiplication of a single cell with an expressed mutated gene critical to the differentiated state of that cell/tissue (95).

INITIATION/PROMOTION CONCEPT IN CHRONIC SOMATIC DISEASES

The terms "initiation" and "promotion" are operational concepts which were introduced to explain distinct observations made during mouse skin tumorigenesis (5). Subsequently, they have been applied to the carcinogenic process in organ systems in other species, including human beings (56,81). In fact, the concept has been extended to another disease state, namely atherogenesis (3,49,93). It will be one objective of this analysis to extend the concept to a much broader range of disease states in somatic tissue to explain how mutations, in genes other than those affecting cancer and atherosclerosis, might affect health in all organs during the aging process (Fig. 1).

Initiation has been defined as an irreversible change in the genome of a cell, which when exposed to promoting conditions can lead to a malignant phenotype. Promotion, on the other hand, is seen as the clonal expansion of initiated cells which enhances the probability of the conversion of one of these cells to malignancy (93,108) (Fig. 2).

Clearly, these are circular definitions, since one process depends on the other. They, of course, refer to the observations made at the biological level. The cellular and molecular mechanisms responsible for the "initiation" and "promotion" phases of carcinogenesis or atherosclerosis are not known. However, those mechanisms that could bring about the "irreversible" change in a cell's genome would be initiators, while those that would induce a clonal amplification of the initiated cell to enhance its probability to become a malignant tumor or an atherosclerotic plaque would be considered promoters. One specific hypothesis which has been offered for the <u>cellular</u> mechanism for initiation is mutagenesis (102,104). Clearly, the <u>molecular</u> mechanisms for mutagenesis (gene and chromosomal) in mammalian cells are not known. Conceptually, those processes that would prevent mutagenesis in this theory would be "anti-initiators."

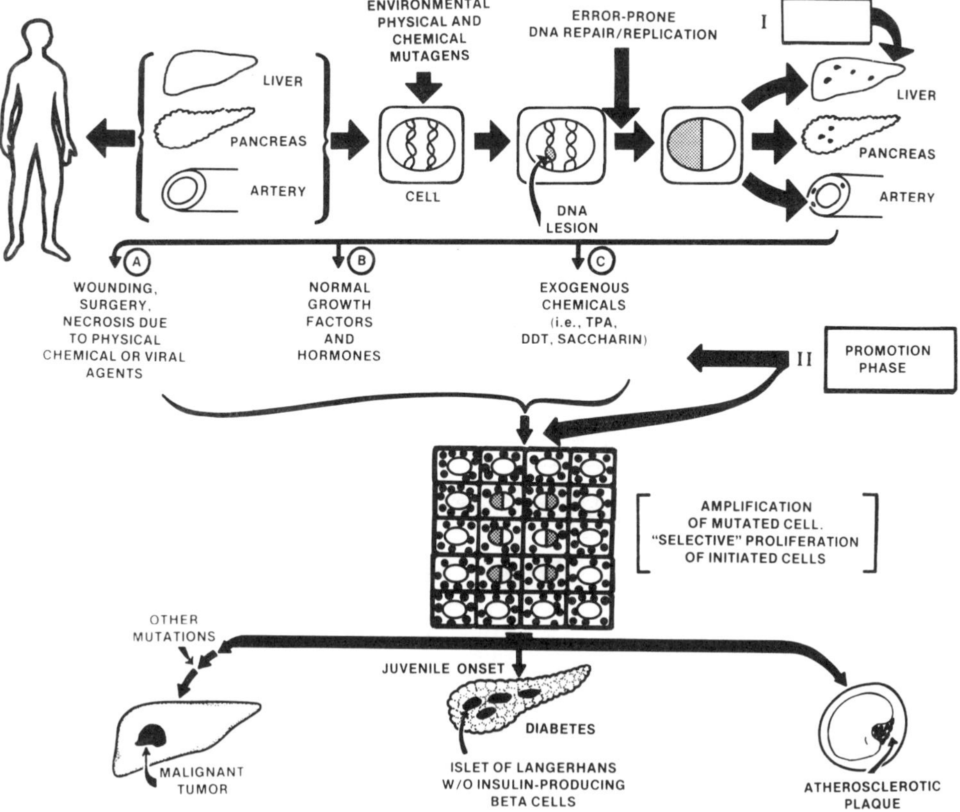

Fig. 1. Initiation and promotion concepts in the genesis of chronic diseases such as carcinogenesis, atherosclerosis, and diabetes. A mutation, depending on the gene and cell type, if clonally amplified can lead to a "critical" mass of dysfunctional tissue in a given organ. Reprinted from Trosko and Chang (100) with permission from Scope, Paris.

A possible cellular mechanism for the amplification phase of promotion would be mitogenesis (93,105). While the molecular mechanisms of mitogenesis are also unknown, the role of inhibited intercellular communication has been postulated to contribute to the clonal expansion of the initiated cell (105,117).

In skin and liver tumorigenesis, the appearance of papillomas and liver nodules (clonal collections of dysfunctional, but benign, cells) occurs after continuous exposure of an initiated organ to physical or chemical conditions that stimulate the selection/growth of the initiated cells. In the case of carcinogenesis, depending on whether additional genomic changes occur during this promotion process, a few of these skin papillomas and liver nodules can be converted to a malignant phenotype (21,28,67,68,76).

In the cancer field, it is usually assumed that if these foci of dysfunctional tissues are benign, they are of no importance because they are not malignant. However, in the context of the overall health of an organism, the appearance of multiple benign foci of dysfunctional tissue in all

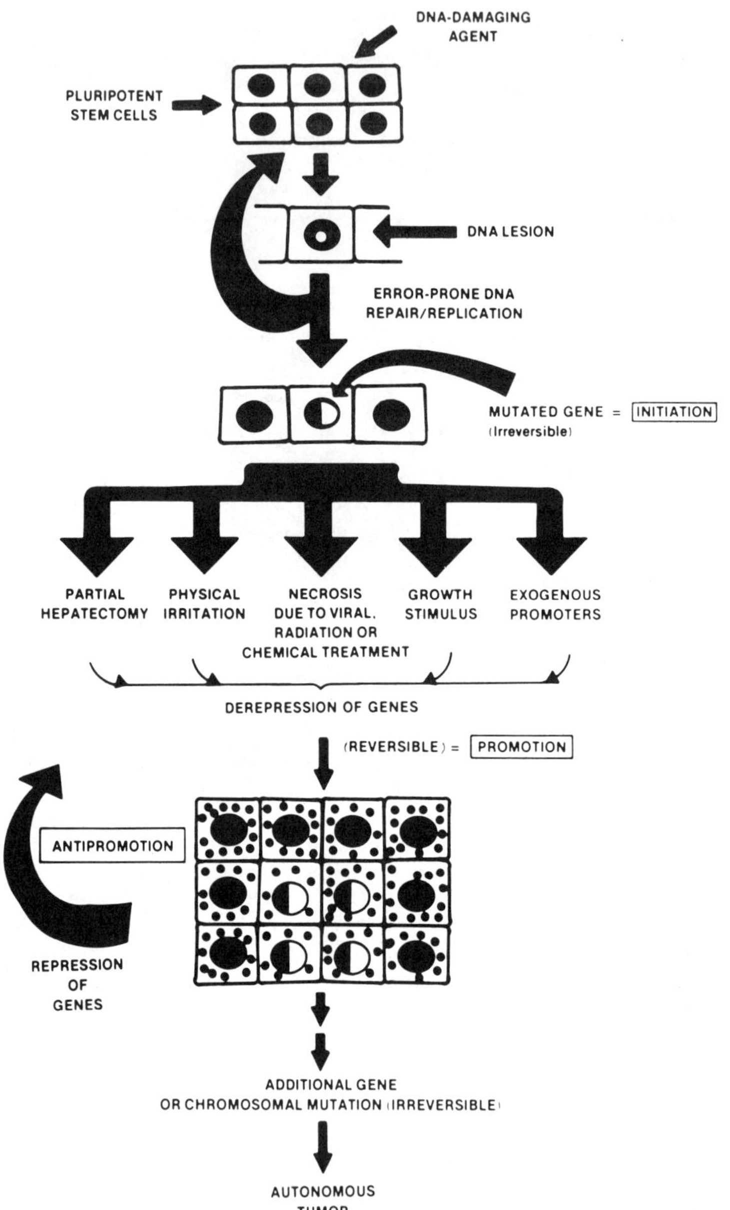

organs, brought about by the initiation/promotion process, could cause physiological disruption in the homeostatic process. In other words, if individual stem/progenitor cells of each organ are initiated but not amplified, the organ/organism will not be affected by a few dysfunctioning cells. On the other hand, if these initiated stem cells are promoted to large masses of benign cells within an organ and are unable to function in the differentiated capacity of that tissue, then the organ/organism will have a problem maintaining its normal physiological role (107).

←

Fig. 2. A diagrammatic heuristic scheme to depict the postulated mechanisms of the initiation and promotion phases of carcinogenesis, diabetes, and atherosclerosis. DNA lesions, induced by physical or chemical mutagens, are substrates that can be fixed if they are not removed in an error-free manner prior to DNA replication. Promotion includes those conditions (i.e., wounding, cytotoxicity, exogenous promoters) in which a pluripotent, but surviving, initiated cell can escape the nonproliferative state. The build-up of initiated cells allows them to "resist" the antimitotic influence of neighboring noninitiated cells. This, together with a second mutation, might allow a given cell to have the autonomous, invasive properties of a malignant cell, or to become a clone of dysfunctional cells in a specialized tissue. Reprinted from Trosko and Chang (93) with permission from Medical Hypothesis, New York.

POSSIBLE ROLE OF MUTAGENESIS IN INITIATION OF CHRONIC SOMATIC DISEASES

Based on our current understanding, maintenance of the integrity of the genetic information in both germ and somatic cells is of significant importance for the prevention of a variety of genetic and somatic diseases. Consequences of gene and chromosomal mutations in the germ cells have been attributed to embryo and fetal lethality, as well as to congenital defects and metabolic disorders (37,51,61,62). Mutations in the somatic tissue have been postulated to play roles in carcinogenesis (7,91,102), atherosclerosis (3,93), cataract formation (27), and in chronic diseases, in general (46), as well as in the aging process (102,103).

The molecular mechanisms responsible for the various kinds of gene and chromosomal mutations, such as base substitutions, gene deletions, gene duplication and amplification, chromosomal deletions, chromosomal rearrangements, nondisjunctions, etc., are not only unknown in mammalian systems, but must be quite different; thus, it would not be possible at present to analyze those factors that might prevent or enhance the actual generation of mutations since we lack detailed information regarding the mechanisms of each type of mutation. In addition, at least for mammalian (including human) mutagenesis, the techniques to measure a given type of mutation are not yet perfected to allow one to unequivocally interpret the results (87, 111). This is because, in the absence of direct analysis of changes in the DNA at the molecular level, the presence of a mutation is based on changes in phenotype.

For example, resistance to 6-thioguanine signifies a possible mutation in the gene for hypoxanthine guanine phosphoribosyltransferase (HG-PRT). Resistance to bromodeoxyuridine signifies a mutation in the gene for thymidine kinase (TK). Ouabain resistance is interpreted to indicate a mutation in the Na^+/K^+ ATPase. However, various factors, which could modify these phenotypes without mutating these genes, could interfere with the interpretation (41,42). Stable epigenetic mechanisms which could repress the HG-PRT or TK genes might be misinterpreted as a mutational process. Physiological factors, such as metabolic cooperation or modulation in membrane chemistry, could alter the frequency of these phenotypes, thereby creating an incorrect assessment of mutation frequencies.

At other levels, there appear two basic mechanisms leading to mutations at the gene level in eukaryotic systems, namely, error-prone DNA replication and error-prone DNA repair (Fig. 3). In the former, it is implied that during the replication of undamaged DNA templates, intrinsic thermodynamic factors could contribute to a low, but finite, amount of mispairing errors, and endogenous factors could modulate the fidelity of DNA polymerases (14), and hence the production of mutations.

The demonstration of the existence of excision DNA repair enzymes in human cells (69) and of their absence in the cells of the cancer-prone xeroderma pigmentosum syndrome (10), with the subsequent correlation with hypermutability of the xeroderma pigmentosum cells (25,48,60), led to the speculation that nonrepaired or misrepaired DNA lesions could lead to gene mutations.

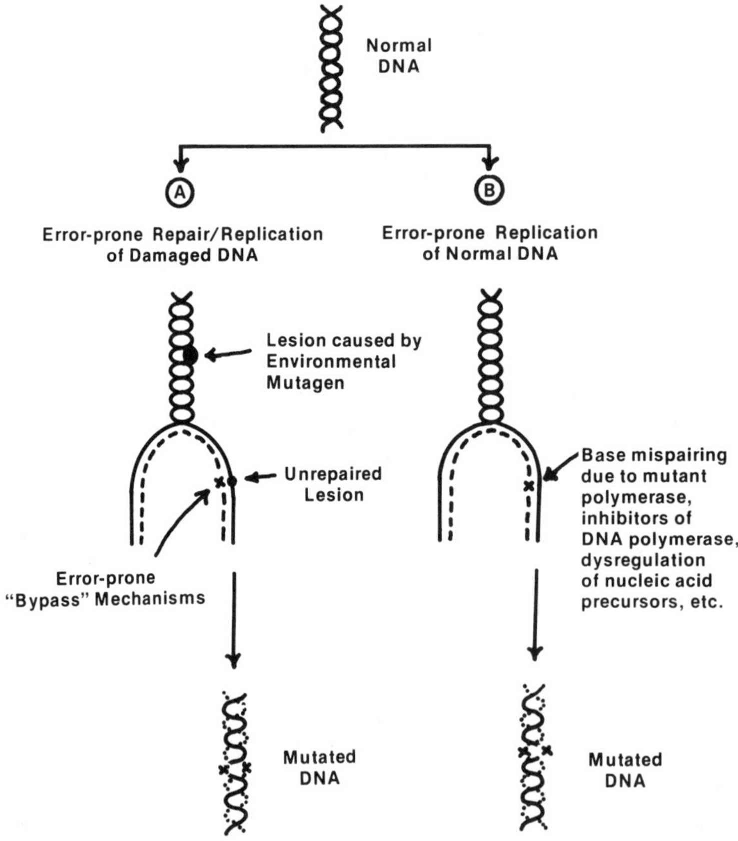

Fig. 3. Diagram illustrating two theoretical mechanisms of mutagenesis. One involves the lesions in the template DNA that act as "substrates" for "error-prone" DNA repair mechanisms. The other demonstrates that mutations can be produced by error-prone replication of newly synthesized DNA off of normal DNA templates.

How either "spontaneous" (74) or induced gene mutations (41) are produced in mammalian cells is not known. Evidence for the presence or absence of induced error-prone repair mechanisms is conflicting (29,73; however, see Ref. 1 and Ref. 20,84,85). The roles of α-DNA polymerase and β-DNA polymerase in both error-free (or error-prone) DNA replication and repair are also unclear (12,30,42,57,79,82).

Since gene and chromosomal amplification, conceptually a mutational process, has been speculated to play a role in carcinogenesis (63,77), some clues to this process have been offered (30). In brief, if α-DNA polymerase is inhibited by either DNA lesions in the template or by extra-template factors such as depleted nucleotide pools or inhibitors of α-DNA polymerase, the β-DNA polymerase seems to take over. It appears that under these conditions, in at least Chinese hamster V79 cells, β-DNA polymerase can substitute for α-DNA polymerase. However, it appears that β-DNA polymerase is unable to recognize signals which tell it that a round of semiconservative replication has been completed and that it can start the process over. Endoreduplication of chromosomes has been observed under these conditions. One could speculate that on the gene level, this same process might occur, leading to a mechanism of gene amplification.

ROLE OF INTERCELLULAR COMMUNICATION IN REGULATING PROLIFERATION AND DIFFERENTIATION OF EUKARYOTIC CELLS

In mammalian, including human, organisms, the homeostatic control of hundreds of billions of cells, both stem/progenitor and those of the differentiated state, is accomplished by several interdependent forms of intercellular communication, namely, communication between cells of one type over a distance to another and between continuous cells linked by a membranous structure, called the gap junction (105) (Fig. 4).

Gap-junctional communication is mediated by a dynamic protein structure on membranes of cells which facilitates the transfer of ions and low molecular weight molecules (45) (Fig. 5). It seems to be found in most cells in all tissues of all metazoans at some time during the lifetime of the cell (19). It seems that during the initial stages after fertilization, the cells of the early embryo are without gap junctions; but as differentiation and development proceed, selective coupling of cells via gap junctions occurs (43). Developmental (i.e., early blastocyst cells with repressed gap junction genes) (43), endogenous (hormone regulation of gap junctions) (40), exogenous (environmental chemicals) (110), and mutational (cancer cells lacking the regulation of gap junction function) (47) factors seem to modulate the structure/function of these gap junctions. The correlation of gap junctions with contact inhibition and reduction of cell proliferation (6,13,24,54,83), development and differentiation (4,23,39,55,65, 78,80) suggests that they play a vital role in the control of normal cell proliferation and differentiation.

As a logical extension of these observations, it would seem that there might be no or very selective gap-junctional communication between stem/-progenitor cells of all tissues. The lack of gap-junctional communication between the stem/progenitor cells and their terminally differentiated daughter cells would allow physiological independence from each other. If nongap-junctional communication existed between the stem/progenitor cells and their differentiated lineage, a homeostatic control would, however, provide a means to keep these cells in balance (11,72,97).

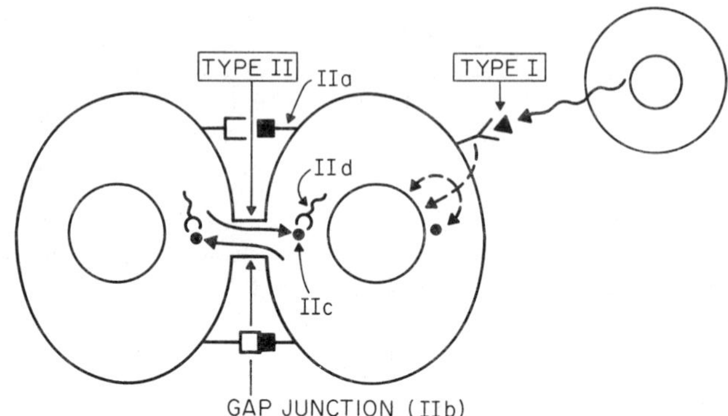

TYPE I — COMMUNICATION BETWEEN DIFFERENT CELL TYPES VIA GROWTH FACTORS, HORMONES, ETC.

TYPE II — GAP JUNCTIONAL COMMUNICATION

IIa = CELL RECOGNITION AND/OR ADHESION MOLECULES
IIb = FUNCTIONAL GAP JUNCTIONS
IIc = MOLECULAR REGULATORY SIGNALS TRANSPORTED VIA JUNCTIONS
IId = MOLECULAR RECEPTOR FOR REGULATORY MOLECULAR SIGNAL

Fig. 4. Heuristic diagram illustrating two general forms of intercellular communication. One involves the production and transmission of "signal" molecules over a distance through extracellular space to a target tissue. The other involves the transfer of signal molecules via permeable intercellular junctions between coupled cells. Clearly, the one form could influence the other form of intercellular communication if the transmitted signal stimulates molecules of the target cell that modulates the function of gap junctions.

INHIBITION OF GAP-JUNCTIONAL COMMUNICATION AND THE CLONAL EXPANSION OF CELLS

Since the original postulation and demonstration of the existence of gap-junctional intercellular communication (36,71), the idea that it is generally absent in most malignant cells (44; see Ref. 19 and 39 for discussions of exceptions to this generalization) has stimulated much research. The fact that cancer has been described as a "disease of differentiation" (64), or as "oncogeny is blocked or partially blocked ontogeny" (66), links gap-junctional communication to cancer, since the gap junction has been linked to the normal control of proliferation and differentiation.

In addition, there have been recent demonstrations that many chemicals, known to be tumor promoters of various species' organ systems, inhibit intercellular communication in vitro (32,110) and in vivo (34,35). Conceptually, other types of tumor promotion (i.e., wounding, abrasion, nonspecific cell killing, and growth factor hormone-induced hyperplasia) could also, by releasing initiated cells from contact inhibition, cause the clonal expansion of initiated cells (Fig. 1).

Techniques such as electrocoupling (17), dye transfer (16), autoradiography or scintillation counting of radioactive nucleotide transfer (12, 58), metabolic cooperation between genetically proficient and deficient

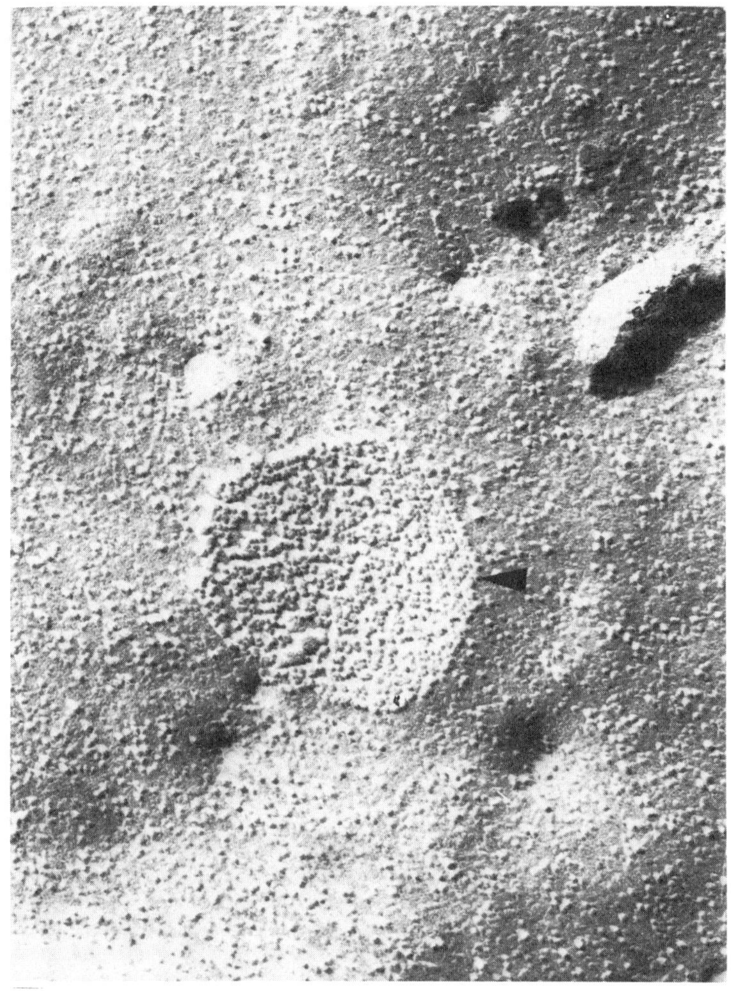

Fig. 5. A large gap junction found on the membrane of a Chinese hamster V79 cell (X99,000). Reprinted from Yancey et al. (116) with permission from Academic Press, Orlando, Florida.

cells (15), photobleaching (112), and freeze fracture (116) have shown that tumor promoters can inhibit gap-junctional intercellular communication.

Another piece of evidence that links gap junctions with carcinogenesis is the well-known correlation between teratogenesis and carcinogenesis (8, 99). Many chemicals that are "carcinogenic" are also teratogenic. Since the normal regulation of cell proliferation and differentiation is very critical during early organogenesis (22), it has been hypothesized that inhibition of intercellular communication during critical phases of development could lead to teratogenesis (106) (Fig. 6). The fact that some known tumor promoters are also teratogens (31) and that antibodies to gap-junctional proteins interfere with normal development (113) is consistent with this idea.

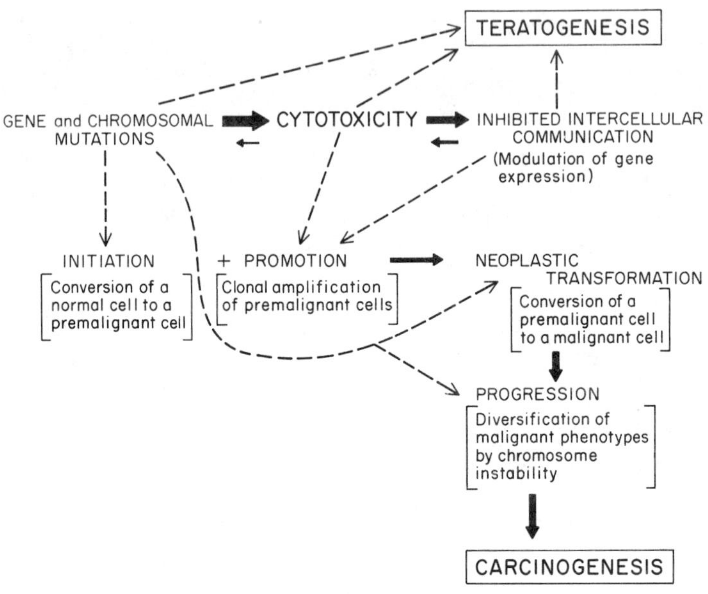

Lastly, the role of cellular oncogenes has been speculated to play a vital role in normal cell functions because they seem to be highly evolutionarily conserved. By definition, they are involved in some aspect of the complex carcinogenic process. Recently, several oncogenes have been implicated in normal embryonic development (18,26,59,75,86,101,114). Since the gap junction structure/function might be viewed as an important and highly evolutionarily conserved function needed to regulate differentiation in metazoans, genes controlling the gap junction structure/function might be viewed as cellular "oncogenes."

It has now been demonstrated that several oncogenes (i.e., src, mos, ras) seem to interfere with gap-junctional communication (1,2,9). The fact that several other oncogenes (i.e., Erb-B and sis) seem to code for growth factor or growth factor receptors (50) indicates, again, that these oncogenes might play a role in the proliferation of initiated cells.

IMPLICATIONS OF THE INHIBITION OF INTERCELLULAR COMMUNICATION HYPOTHESIS FOR THE IMPACT OF SOMATIC MUTATIONS ON HUMAN DISEASES

If the hypothesis that gap-junctional communication regulates cell proliferation via the phenomenon of "contact inhibition" is correct, there are several direct predictions and implications for both the formation and expression of mutations in somatic tissues. The first is that only potential proliferable cells (stem or progenitor cells) will be relevant at the somatic level. Also, any proliferation condition could enhance the probability of a mutation fixation, either by "spontaneous" errors of replication or by the error-prone replication/repair of spontaneous or induced lesions in the DNA. If the mutation is, per se, nonlethal, and if the cell is not terminally differentiated (i.e., can act as a stem or progenitor cell), then, depending upon whether it remains quiescent (suppressed by its surrounding normal neighbors) or it is stimulated to proliferate, the

←

Fig. 6. A heuristic scheme to classify chemicals on the basis of three biological endpoints: mutagenicity, cytotoxicity, and inhibition of intercellular communication. The arrows between the biological endpoints are designated to mean that chemicals can have multiple biological consequences in the direction indicated (i.e., mutagens can kill cells, and the death of cells can cause the modulation of gene expression in some surviving cells). In addition, some chemicals that can inhibit intercellular communication at noncytotoxic levels could, at much higher concentrations, kill cells, possibly also inducing some chromosomal mutations at these high doses. The scheme further depicts how mutagenesis, cell death, and disrupted intercellular communication can play a role in teratogenesis and carcinogenesis. An early conceptus or fetus that is exposed to a mutagen at a low dose could conceivably have a critical gene mutated in a stem cell, leading to a congenital defect. At high cytotoxic doses of a mutagen, fetal toxicity would be expected. Exposure to nongenotoxic, but cytotoxic, chemicals could also, conceivably, lead to congenital defects or fetal toxicity, depending on the type and amount of cells killed by the exposure. Chemicals that inhibit intercellular communication could disrupt regulation of proliferation and differentiation of tissues if given at a critical period of development. The consequence would lead either to congenital defects in specific organs or to embryo or fetal toxicity, depending on the time, amount, distribution, metabolism, or receptors for the chemical. These same chemicals and actions of chemicals can influence carcinogens. A low dose of a mutagen could initiate or mutate a stem cell of a given tissue. If that initiated cell, which is held "in check" by surrounding and communicating normal cells, is allowed to escape the "metabolic cooperation" of the communicating normal cells, then it can conceivably clonally expand itself. During that clonal expansion, additional genetic changes could occur which "stabilize" the neoplastic phenotype which resists the suppressive influence of normal tissue. Cytotoxic doses of the mutagen itself could act as an "indirect" promoter by allowing surviving, but initiated, cells to repopulate the dead tissue (i.e., compensatory hyperplasia), and therefore it would be considered a "complete" carcinogen. Alternatively, removal of cells by nongenotoxic chemicals, surgery, or normal cell death could act as a "promoter" to the surviving initiated stem cell. Lastly, exposure of initiated tissue to noncytotoxic, intercellular communication, inhibitors could cause promotion of the initiated cell, resulting in cancer. Reprinted from Trosko and Chang (99) with permission from Plenum Press, New York.

mutation might or might not have a physiological impact on the tissue/organ in which it occurs.

Clearly, exposure of mitotically active cells to mutagens enhances the probability of creating potentially clonable mutated cells. Likewise, stimulation of previously mutagen-exposed cells could cause clonable expansion of quiescent mutated cells. Since there are periods of time during normal development when some tissues go through peaks of hyperplasia, and since there are certain tissues throughout life that are constantly going through cell division, suppression of this mitotic activity, if possible, might reduce mutation fixation, but it would cause other undesirable consequences. Only by minimizing chronic hyperplasia in those tissues that are not normally dependent on constant cell renewal/repair could one minimize the clonal amplification of mutated cells, and therefore the probability of accruing additional detrimental genetic alterations which might lead to malignancy.

However, even without accruing those changes leading to a malignant phenotype, the clonal amplification of a mutated cell (depending on the gene mutated, the nature of the mutation, and the tissue in which the expressed mutation is found) could have a physiological impact by forming a mass of nonfunctioning cells in a critical organ. In other words, promotion is a rate-limiting step in the ultimate physiological expression of a mutation in a somatic cell. Therefore, potential exposure to agents that might prevent the clonal expansion, or the avoidance of those agents that cause the amplification of cells known to have been exposed to mutagens, might minimize the contribution of somatic mutations to human health. Because of the lack of knowledge of how this might be done without causing harm to nonaffected tissues and by only suppressing the affected tissues, it appears that this hypothesis, if correct, has little practical significance.

ACKNOWLEDGEMENTS

Research on which this chapter was based was supported by a grant from the National Cancer Institute (CA 21104) and a gift from Dow Chemical, U.S.A.

REFERENCES

1. Atkinson, M., and J. Sheridan (1984) Decreased junctional permeability in cells transformed by 3 different viral oncogenes: A quantitative video analysis. J. Cell Biol. 99:401a.
2. Azarnia, R., and W.R. Loewenstein (1984) Intracellular communication and the control of growth: X. Alteration of junctional permeability by the src gene. A study with temperature-sensitive mutant Rous sarcoma virus. J. Membr. Biol. 82:191-205.
3. Benditt, E.P. (1977) The origin of atherosclerosis. Sci. Am. 236:74-85.
4. Bennet, M.V.L., D.C. Spray, and A.L. Harris (1981) Electric coupling in development. Am. Zool. 21:413-427.
5. Berenblum, I., and P. Shubik (1947) A new quantitative approach to the study of the stages of chemical carcinogenesis. Brit. J. Cancer 1:383-391.
6. Borek, C., and L. Sachs (1966) The difference in contact inhibition of cell replication between normal cells and cells transformed by different carcinogens. Proc. Natl. Acad. Sci., USA 56:1705-1711.
7. Boveri, T. (1914) Zur Frage der Entstehung Maligner Tumoren, Gustav Fisher, Pub. Jena, Federal Republic of Germany.
8. Caruso, R., and J.E. Trosko (1985) Inhibited intercellular communication as a mechanistic link between teratogenesis and carcinogenesis. CRC Crit. Rev. Toxicol. 16:157-183.
9. Chang, C.C., J.E. Trosko, H.-J. Kung, D. Bombick, and F. Matsumura (1985) Potential role of the src gene product in inhibition of gap-junctional communication in NIH/3T3 cells. Proc. Natl. Acad. Sci., USA 82:5360-5364.

10. Cleaver, J.E. (1969) A human disease in which an initial stage of DNA repair is defective. Proc. Natl. Acad. Sci., USA 63:428-435.
11. Cline, M.J., and D.W. Golde (1979) Cellular interactions in haematopoiesis. Nature 277:177-181.
12. Collins, A. (1983) DNA repair in ultraviolet-irradiated HeLa cells is disrupted by aphidicolin: The inhibition of repair need not imply the absence of repair synthesis. Biochim. Biophys. Acta 741:341-347.
13. Corsaro, C.M., and B.R. Migeon (1977) Comparison of contact-mediated communication in normal and transformed human cells in culture. Proc. Natl. Acad. Sci., USA 74:4476-4480.
14. Davidson, J.S., I.M. Baumgarten, and E.H. Harley (1985) Effect of 12-O-tetradecanoylphorbol-13-acetate and retinoids on intercellular junctional communication measured with a citrulline incorporation assay. Carcinogenesis 6:645-650.
15. Dunn, W.C., J.D. Regan, and R.D. Snyder (1985) Elevation of dCTP pools in xeroderma pigmentosum variant human fibroblasts alters the effects of DNA repair arrest by arabinoturanosyl cytosine. Cell Biol. Toxicol. 1:75-86.
16. Enomoto, T., and H. Yamasaki (1984) Lack of intercellular communication between chemically transformed and surrounding nontransformed BALB/c 3T3 cells. Cancer Res. 44:5200-5203.
17. Enomoto, T., Y. Sasaki, Y. Kanno, and H. Yamasaki (1981) Tumor promoters cause a rapid and reversible inhibition of the formation and maintenance of electrical cell coupling in culture. Proc. Natl. Acad. Sci., USA 78:5628-5632.
18. Filmus, J., and R.N. Buick (1985) Relationship of c-myc expression to differentiation and proliferation of HL-60 cells. Cancer Res. 45:822-825.
19. Finbow, M.E., and S.B. Yancey (1981) The roles of intercellular junctions. In Biochemistry of Cellular Regulation, Vol. IV, M.J. Clemens, ed. CRC Press, Boca Raton, Florida, pp. 215-249.
20. Frosina, G., S. Bonatti, and A. Abbondandolo (1984) Negative evidence for an adaptive response to lethal and mutagenic effects of alkylating agents in V79 Chinese hamster cells. Mutat. Res. 129:243-250.
21. Fry, R.J.M., R.D. Ley, D. Grube, and E. Staffeldt (1982) Studies on the multistage nature of radiation. In Carcinogenesis, Vol. 7, E. Hecker, N.E. Fusenig, W. Kunz, F. Marks, and H.W. Thielmann, eds. Raven Press, New York, pp. 155-165.
22. Gilula, N.B. (1980) Cell to cell communication and development. In Mediator of Developmental Process, S. Subtelny and N.K. Wessells, eds. Academic Press, New York, pp. 23-41.
23. Gilula, N.B., D.W. Fawcell, and A. Aoki (1976) The sertoli cells occluding junctions and gap junctions in mature and developing mammalian testes. Devel. Biol. 51:142-168.
24. Gilula, N.B., O.R. Reeves, and A. Steinbach (1972) Metabolic coupling, ionic coupling and cell contacts. Nature 235:262-265.
25. Glover, T.W., C.C. Chang, J.E. Trosko, and S.S.L. Li (1979) Ultraviolet light induction of diphtheria toxin-resistant mutants in normal and xeroderma pigmentosum human fibroblasts. Proc. Natl. Acad. Sci., USA 76:3982-3986.
26. Grosso, L.E., and H.C. Pitot (1985) Transcriptional regulation of c-myc during chemically-induced differentiation of HL-60 cultures. Cancer Res. 45:847-850.
27. Hartman, P.E. (1983) Mutagens: Some possible health impacts beyond carcinogenesis. Environ. Mutag. 5:139-152.
28. Hennings, H., R. Shores, M.L. Wenk, E.F. Spangler, R. Tarone, and S.H. Yuspa (1983) Malignant conversion of mouse skin tumors is

increased by tumor initiators and unaffected by tumor promoters. Nature 304: 67-69.

29. Herrlich, P., U. Mallick, H. Ponta, and H.J. Rahmsdorf (1984) Genetic changes in mammalian cells reminiscent of an SOS response. Hum. Genet. 67:360-368.
30. Huang, Y., C.C. Chang, and J.E. Trosko (1983) Aphidicolin-induced endoreduplication in Chinese hamster cells. Cancer Res. 43:1361-1364.
31. Jone, C.M., L.E. Parker, J.E. Trosko, M.L. Netzloff, and C.C. Chang (1985) Inhibition of metabolic cooperation by the anticonvulsants diphenylhydantoin and phenobarbital. Terat. Carcinog. Mutag. (in press).
32. Jone, C., J.E. Trosko, C.F. Aylsworth, L. Parker, and C.C. Chang (1985) Further characterization of the in vitro assay for inhibitors of metabolic cooperation in the Chinese hamster V79 cell line. Carcinogenesis 6:361-366.
33. Jongen, W.M.F., B.C. Hakkert, and M.L.M. van de Poll (1985) Inhibitory effects of the phorbol ester TPA and cigarette smoke condensate on the mutagenicity of benzo(α)pyrene in a cocultivation system. Mutat. Res. (in press).
34. Kalimi, G.H., and S.M. Sirsat (1984) Phorbol ester tumor promoter affects the mouse epidermal gap junctions. Cancer Lett. 22:343-350.
35. Kalimi, G.H., and S.M. Sirsat (1984) The relevance of gap junctions to stage I tumor promotion in mouse epidermis. Carcinogenesis 5: 1671-1677.
36. Kanno, Y., and W.R. Loewenstein (1964) Low-resistance coupling between gland cells. Some observations on intercellular contact membranes and intercellular space. Nature 201:194-195.
37. Knudson, Jr., A.G. (1979) Our load of mutations and its burden of disease. Am. J. Hum. Genet. 31:401-413.
38. Kunz, B.A. (1982) Genetic effects of deoxyribonucleotide pool imbalances. Environ. Mutag. 4:695-725.
39. Larsen, W.J. (1983) Biological implications of gap junction structure, distribution and composition: A review. Tissue and Cell 9: 373-394.
40. Larsen, W.J., A.N. Tung, and C. Polking (1981) Response of granulosa cell gap junctions to human chorionic gonadotropin at ovulation. Biol. Reprod. 25:1119-1134.
41. Lemontt, J.F., and W.M. Generoso, eds. (1982) Molecular and Cellular Mechanisms of Mutagenesis, Plenum Press, New York.
42. Liu, P., C.C. Chang, and J.E. Trosko (1984) Evidence of mutagenic repair in V79 cell mutant with aphidicolin-resistant DNA polymerase α. Somat. Cell Molec. Genet. 10:235-245.
43. Lo, C.W., and N.B. Gilula (1979) Gap junctional communication in the postimplantation mouse embryo. Cell 18:411-422.
44. Loewenstein, W.R. (1979) Junctional intercellular communication and the control of growth. Biochim. Biophys. Acta Cancer Rev. 560:1-65.
45. Loewenstein, W.R. (1981) Junctional intercellular communication: The cell to cell membrane channel. Physiol. Rev. 61:829-913.
46. Lower, G.M., and M.S. Kanarek (1982) The mutation theory of chronic, noninfectious disease: Relevance to epidemiologic theory. J. Epidemiol. 115:803-817.
47. MacDonald, C. (1982) Genetic complementation in hybrid cells derived from two metabolic cooperation defective mammalian cell lines. Exp. Cell Res. 138:303-310.
48. Maher, V.M., and J.J. McCormick (1980) Comparison of the mutagenic effect of ultraviolet radiation and chemical in normal and DNA repair-deficient human cells in culture. In Chemical Mutagens, Vol. 6,

F.J. de Serres and A. Hollaender, eds. Plenum Press, New York, pp. 309-329.

49. Majesky, M.W., M.A. Reidy, E.P. Benditt, and M.R. Juchau (1985) Focal smooth muscle proliferation in the aortic intima produced by an initiation-promotion sequence. Proc. Natl. Acad. Sci., USA 82:3450-3454.
50. Marx, J.L. (1984) What do oncogenes do? Science 223:673-676.
51. Matsunaga, E. (1982) Perspectives in mutation epidemiology. Mutat. Res. 99:95-128.
52. Meuth, M. (1981) Role of deoxynucleoside triphosphate pools in the cytotoxic and mutagenic effects of DNA alkylating agents. Som. Cell Genet. 7:89-102.
53. Meuth, M. (1984) The genetic consequences of nucleotide precursor pool imbalance in mammalian cells. Mutat. Res. 126:107-112.
54. Meyer, R.A., and J. Overton (1983) Changes in intercellular junctions. Devel. Biol. 99:172-180.
55. Midgley, Jr., A.R. (1983) Ovarian communication and control. In Factors Regulating Ovarian Function, G.S. Greenwald and P.F. Terranova, eds. Raven Press, New York, pp. 1-10.
56. Moolgavkar, S.H., N.E. Day, and R.G. Stevens (1980) Two-stage model for carcinogenesis: Epidemiology of breast cancer in females. J. Natl. Cancer Inst. 65:559-569.
57. Mosbaugh, D.W., and S. Linn (1984) Gap-filling DNA synthesis by HeLa DNA polymerase α in an in vitro base excision DNA repair scheme. J. Biol. Chem. 259:10247-10251.
58. Mosser, D.D., and N.C. Bols (1982) The effect of phorbols on metabolic cooperation between human fibroblasts. Carcinogenesis 3:1207-1212.
59. Muller, R., and E.F. Wagner (1984) Differentiation of F9 teratocarcinoma stem cells after transfer of c-fos protooncogenes. Nature 311: 433-437.
60. Myhr, B.C., D. Turnbull, and J.A. DiPaolo (1979) Ultraviolet mutagenesis of normal and xeroderma pigmentosum variant human fibroblast. Mutat. Res. 62:341-353.
61. Neel, J.V. (1978) Mutation and disease in man. Can. J. Genet. Cytol. 20:295-306.
62. Newcombe, H.B. (1978) Problems of assessing the genetic impact of mutagens on man. Can. J. Genet. Cytol. 20:459-470.
63. Pall, M.L. (1981) Gene amplification model of carcinogenesis. Proc. Natl. Acad. Sci., USA 78:2465-2468.
64. Pierce, G.B. (1974) Neoplasms, differentiation and mutations. Am. J. Path. 77:103-118.
65. Pitts, J.D. (1980) The role of junctional communication in animal tissues. In Vitro 16:1043-1048.
66. Potter, V.R. (1978) Phenotypic diversity in experimental hepatomas: The concept of partially blocked ontogeny. Brit. J. Cancer 38:1-23.
67. Potter, V.R. (1981) A new protocol and its rationale for the study of initiation and promotion of carcinogenesis in rat liver. Carcinogenesis 2:1375-1379.
68. Reddy, A.L., and P.J. Fialkow (1983) Papillomas induced by initiation-promotion differ from those induced by carcinogen alone. Nature 304: 69-71.
69. Regan, J., J.E. Trosko, and W.L. Carrier (1968) Evidence for excision of ultraviolet induced pyrimidine dimers from the DNA of human cells in vitro. Biophys. J. 8:319-325.
70. Reiners, J.J., and S. Herlick (1985) Cell-cell proximity is a parameter affecting the transfer of mutagenic polycyclic aromatic hydrocarbons between cells. Carcinogenesis 6:793-795.

71. Revel, J.-P., and M.J. Karnovsky (1967) Hexagonal array of subunits in intercellular junctions of the mouse heart and liver. J. Cell Biol. 33:C7.
72. Roth, J., D. LeRoith, J. Shiloach, and C. Rubinovitz (1983) Intercellular communication: An attempt at a unifying hypothesis. Clinical Res. 31:354-363.
73. Sarasin, A., F. Bourre, and A. Benoit (1982) Error-prone replication of ultraviolet-irradiated Simian virus 40 in carcinogen-treated monkey kidney cells. Biochimie 64:815-821.
74. Sargentini, N.J., and K.C. Smith (1985) Spontaneous mutagenesis: The roles of DNA repair, replication and recombination. Mutat. Res. 154: 1-27.
75. Schartl, M., and A. Barnekow (1984) Differential expression of the cellular src gene during vertebrate development. Devel. Biol. 105: 415-422.
76. Scherer, E., A.W. Feringa, and P. Emmelot (1984) Initiation-promotion-initiation. Induction of neoplastic foci within islands of precancerous liver cells in the rat. In Models, Mechanisms and Etiology of Tumor Promotion, M. Borzonyi, N.E. Day, K. Lapis, and H. Yamasaki, eds. IARC Scientific Publ., No. 56, Lyon, pp. 57-66.
77. Schimke, R.T. (1984) Gene amplification, drug resistance and cancer. Cancer Res. 44:1735-1742.
78. Schultz, R.M. (1985) Roles of cell to cell communication in development. Biol. Reprod. 32:17-42.
79. Seki, S., H. Nobuo, and T. Oda (1984) Differential sensitivity to aphidicolin of replicative DNA synthesis and ultraviolet-induced unscheduled DNA synthesis in vivo in mammalian cells. Acta Med. Okayama 38:227-237.
80. Sheridan, J.D. (1976) Cell coupling and cell communication during embryogenesis. In The Cell Surface in Animal Embryogenesis and Development, G. Poste and G.L. Nicolson, eds. Elsevier/North-Holland Biomedical Press, New York, pp. 409-447.
81. Slaga, T.J. (1979) Mechanisms of Tumor Promotion, Vol. IV, CRC Press, Boca Raton, Florida.
82. Smith, C.A., and D.S. Ikumoto (1984) Nature of DNA repair synthesis resistant to inhibitors of polymerase α in human cells. Biochemistry 43:1383-1391.
83. Stamatoglou, S.C., and C.J. Marshall (1978) Ruthenium red-positive surface layer, extracellular filamentous material and intercellular junctions in hybrids between tumour and normal cells: Abundant gap junctions correlate with density-dependent inhibition of growth. J. Cell Sci. 31:323-329.
84. Takimoto, K. (1983) Lack of enhanced mutation of UV- and gamma-irradiated herpes virus in UV-irradiated CV-1 monkey cells. Mutat. Res. 121:159-166.
85. Taylor, W.D., L.E. Bockstahler, J. Montos, M.A. Babick, and C.D. Lytle (1982) Further evidence that ultraviolet radiation-enhanced reactivation of simian virus 40 in monkey kidney cells is not accompanied by mutagenesis. Mutat. Res. 105:291-298.
86. Thiele, C.J., C.P. Reynolds, and M.A. Israel (1985) Decreased expression of N-myc precedes retinoic acid-induced morphological differentiation of human neuroblastoma. Nature 313:404-406.
87. Trosko, J.E. (1984) A new paradigm is needed in toxicology evaluation. Environ. Mutag. 6:767-769.
88. Trosko, J.E., and C.C. Chang (1976) Role of DNA repair in mutation and cancer production. In Aging, Carcinogenesis and Radiation Biology, K.C. Smith, ed. Plenum Press, New York, pp. 399-444.

89. Trosko, J.E., and C.C. Chang (1978) Environmental carcinogenesis: An integrative model. Quart. Rev. Biol. 53:115-141.
90. Trosko, J.E., and C.C. Chang (1978) Genes, pollutants and human diseases. Quart. Rev. Biophys. 11:503-527.
91. Trosko, J.E., and C.C. Chang (1978) The role of mutagenesis in carcinogenesis. In Photochemical and Photobiological Reviews, K.C. Smith, ed. Plenum Press, New York, pp. 135-162.
92. Trosko, J.E., and C.C. Chang (1979) Chemical carcinogenesis as a consequence of alterations in the structure and function of DNA. In Chemical Carcinogens and DNA, Vol. II, P.L. Grover, ed. CRC Press, Boca Raton, Florida, pp. 181-195.
93. Trosko, J.E., and C.C. Chang (1980) An integrative hypothesis linking cancer, diabetes and atherosclerosis: The role of mutations and epigenetic changes. Med. Hypoth. 6:455-468.
94. Trosko, J.E., and C.C. Chang (1981) The role of radiation and chemicals in the induction of mutations and epigenetic changes during carcinogenesis. In Advances in Radiation Biology, Vol. 9, J. Lett and H. Adler, eds. Academic Press, New York, pp. 1-36.
95. Trosko, J.E., and C.C. Chang (1983) Potential role of intercellular communication in the rate-limiting step in carcinogenesis. J. Am. Coll. Toxicol. 2:5-22.
96. Trosko, J.E., and C.C. Chang (1983) The role of DNA repair capacity and somatic mutations in carcinogenesis and aging. In Handbook of Diseases of Aging, Van Nostrand-Reinhold, New York, pp. 252-296.
97. Trosko, J.E., and C.C. Chang (1984) Adaptive and nonadaptive consequences of chemical inhibition of intercellular communication. Pharmacol. Rev. 36:137s-144s.
98. Trosko, J.E., and C.C. Chang (1984) Error-prone DNA repair and replication in relation to malignant transformation. Transplant. Proc. 16:363-365.
99. Trosko, J.E., and C.C. Chang (1984) A possible mechanistic link between teratogenesis and carcinogenesis: Inhibited intercellular communication. In Mutation, Cancer and Malformation, E.H.Y. Chu and W.M. Generoso, eds. Plenum Press, New York, pp. 529-547.
100. Trosko, J.E., and C.C. Chang (1985) Implications of genotoxic and nongenotoxic mechanisms in carcinogenesis to risk assessment. In Methods for Estimating Risk of Chemical Injury: Human and Non-Human Biota and Ecosystems, V.B. Vouk, G.C. Butler, D.G. Hoel, and D.B. Peakall, eds. Scope, Paris, pp. 181-200.
101. Trosko, J.E., and C.C. Chang (1985) Oncogene and chemical inhibition of gap-junctional intercellular communication: Implications for teratogenesis and carcinogenesis. In Fourth International Conference on Environmental Mutagens, C. Ramel, ed. Alan R. Liss, Inc., New York (in press).
102. Trosko, J.E., and E.H.Y. Chu (1975) The role of DNA repair and somatic mutation in carcinogenesis. In Advances in Cancer Research, Vol. 21, G. Klain, S. Weinhouse, and A. Haddow, eds. Academic Press, New York, pp. 391-425.
103. Trosko, J.E., and R.W. Hart (1976) DNA mutation frequencies in mammals. Interdiscipl. Topics Geront. 9:168-197.
104. Trosko, J.E., C.C. Chang, and T. Glover (1977) Analysis of experimental evidence of the relation between mutagenesis and carcinogenesis. Role of DNA repair in carcinogenesis. In Mecanismes d'Alteration et de Reparation du DNA: Relations avec la Mutagenese et la Cancerogenese Chimique, P. Daudel, ed. Centre National de la Recherche Scientifique, Paris, pp. 353-388.

105. Trosko, J.E., C.C. Chang, and A. Medcalf (1983) Mechanisms of tumor promotion: Potential role of intercellular communication. Cancer Invest. 1:511-526.

106. Trosko, J.E., C.C. Chang, and M. Netzloff (1982) The role of inhibited cell-cell communication in teratogenesis. Terat. Carcinog. Mutag. 2:31-45.

107. Trosko, J.E., C.C. Chang, and M.H. Wade (1984) Mammalian mutagenesis. In Genetics: New Frontiers, Vol. 1, V.L. Chopra, B.C. Joshi, R.P. Sharma, and H.C. Bansal, eds. Oxford and IBH Publishing Co., New Delhi, pp. 329-346.

108. Trosko, J.E., C. Jone, and C.C. Chang (1983) The role of tumor promoters on phenotypic alterations affecting intercellular communication and tumorigenesis. Ann. N.Y. Acad. Sci. 407:316-324.

109. Trosko, J.E., V.M. Riccardi, C.C. Chang, S.T. Warren, and M. Wade (1985) Genetic predispositions to initiation or promotion phases in human carcinogenesis. In Biomarkers, Genetics and Cancer, H.T. Lynch and H.A. Guirgis, eds. Van Nostrand-Reinhold, New York (in press).

110. Trosko, J.E., L.P. Yotti, S.T. Warren, G. Tsushimoto, and C.C. Chang (1982) Inhibition of cell-cell communication by tumor promoters. In Carcinogenesis, Vol. 7, E. Hecker, N.E. Fusenig, W. Kunz, F. Marks, and H.W. Thielmann, eds. Raven Press, New York, pp. 565-585.

111. Trosko, J.E., C.C. Chang, S.T. Warren, P.K. Liu, M.H. Wade, and G. Tsushimoto (1981) The use of mammalian cell mutants to study the mechanisms of carcinogenesis. In Mutation, Promotion and Transformation In Vitro, N. Iani, T. Kuroki, M. Yameda, and C. Heidelberger, eds. Gann Monograph on Cancer Research, No. 27, Japan Scientific Society Press, Tokyo, pp. 159-179.

112. Wade, M., J.E. Trosko, and M. Schindler (ms. in prep.).

113. Warner, A.E., S.C. Guthrie, and N.B. Gilula (1984) Antibodies to gap-junctional protein selectively disrupt junctional communication in the early amphibian embryo. Nature 311:127-131.

114. Watanabe, T., E. Sariban, T. Mitchell, and D. Kufe (1985) Human c-myc and N-ras expression during induction of HC-60 cellular differentiation. Biochem. Biophys. Res. Commun. 126:998-1005.

115. Weymouth, L.A., and L.A. Loeb (1978) Mutagenesis during in vitro DNA synthesis. Proc. Natl. Acad. Sci., USA 75:1294-1298.

116. Yancey, S.B., J.E. Edens, J.E. Trosko, C.C. Chang, and J.-P. Revel (1982) Decreased incidence of gap junctions between Chinese hamster V79 cells upon exposure to the tumor promoter TPA. Exp. Cell Res. 139:329-340.

117. Yotti, L.P., C.C. Chang, and J.E. Trosko (1979) Elimination of metabolic cooperation in Chinese hamster cells by a tumor promoter. Science 206:1089-1091.

MECHANISMS OF CARCINOGENICITY AND ANTICARCINOGENICITY--PART III

GENETIC ASPECTS OF CANCER EPIDEMIOLOGY

F.H. Sobels

Department of Radiation Genetics and Chemical Mutagenesis
University of Leiden, Sylvius Laboratories
Leiden, The Netherlands

INTRODUCTION

In this chapter, I will first briefly review to what extent various genetic factors predispose to cancer; second, I will discuss the possible significance in terms of cancer proneness of heterozygotes for ataxia telangiectasia; and third, I want to emphasize the importance of follow-up studies of individuals with an increased level of chromosome aberrations after exposure to mutagens.

The idea that genomic changes may play a causative role in the etiology of cancer goes back to Boveri in 1914. This notion received further support by the findings of Muller, Stadler, Auerbach, and Oelkers that physical and chemical agents known to act as carcinogens were capable of producing mutations and other changes in the genetic material, and, more recently, from the work of Ames and others demonstrating the impressive overlap between carcinogens and mutagens. The discovery of oncogenes and their mechanisms of activation provided final evidence that, indeed, base change mutations or chromosomal rearrangements may lead to cancer. All these changes in the genetic material are obviously of considerable importance in tumorigenesis. What can we say now about genetics and cancer epidemiology?

In a recent paper, Mulvihill (8) points out that the genetics of any disorder, including cancer, can be discussed under the following headings: chromosomes, single genes and polygenes, or multifactorial inheritance.

CANCER CYTOGENETICS

First with regard to chromosomal effects, it appears that leukemia and lymphoma provide at present the best known association of cytogenetic defects with cancer. This phenomenon may well be an artifact due to the facility of repeated blood and bone marrow sampling (8). Among these conditions, the most predominant type of abnormality is an acquired translocation, which is seen only in the malignant tissue and which does not seem to result in loss of genetic material. As an example, the Philadelphia chromosome involved in chronic myeloid leukemia is the product of a reciprocal translocation usually between chromosomes 9 and 22, and this translocation is probably associated with the transfer and activation of the oncogene c-abl, which is normally located on chromosome 9. This cytogenetic defect is not associated with congenital anomalies or a constitutionally abnormal karyotype. Another example is the translocation involving chromosomes 8 and 14 in Burkitt lymphoma. As a result of the translocation, the *myc* gene of chromosome 8 is transferred to chromosome 14 to a site within the Ig_H region. The *myc* gene has now come under the influence of the active Ig^H domain and is expressed at an elevated level, leading to its oncogenic activity.

In contrast, cytogenetic abnormalities of solid tumors, especially embryonal neoplasms, are: (a) associated with loss of genetic material; (b) present in every body cell; and (c) associated with birth defects, often in the same organ system that eventually becomes malignant. Wilms tumor presents an example of a constitutional cytogenetic defect; in this case a deletion of the short arm of chromosome 11, resulting in various congenital malformations, such as absence of the iris, growth and mental retardation, and a predisposition to cancer of the kidney.

SINGLE GENE TRAITS PREDISPOSING TO CANCER

Out of a total of 2,300 traits listed in McKusick's fourth edition (7), 200 are associated with neoplasia. Thus, altogether 9% of all known genes confer predisposition to cancer. About 50 dominantly inherited tumors have been recognized. Retinoblastoma and polyposis of the colon are examples of this class. Though, what by classical genetic analysis may appear as dominant transmission, may well be recessive as in the case of retinoblastoma involving deletion of the normal allele of the oncogene involved (4). These inherited tumors probably represent only a small proportion of the total cancers. Those that can be recognized appear to be among the more uncommon tumors; inherited tumors manifest a tendency to have an earlier age of onset.

In this context, the findings obtained by Nomura (9) on the induction of dominant mutations leading to formation of tumors in offspring of radiated mice are of interest. The basic observation is that X-irradiation of ICR mice leads to dominant mutations with low penetrance of 40% resulting in tumors. In the lung, only 1 or 2 tumor nodules were observed, although all lung cells carried mutations. However, when progeny from X-irradiated parents were postnatally exposed to urethane, a considerable proportion of mice had a large number of tumor nodules. Increased penetrance of tumor mutations is ascribed to a promoting effect of urethane in initiated cells already carrying mutations that can potentially lead to tumor formation. For the induction of germline mutation leading to tumors, the frequency per rad per gamete for postmeiotic stages (spermatogonia and oocytes) is 10 times higher than those recorded for dominant skeletal mutations, and 100 times higher than those leading to specific locus mutations.

FAMILIAL CANCER

Apart from cytogenetic disorders and single gene traits, cancer may run in families. Since at least 1 in 4 people will get cancer in a lifetime, every family may have some affected relatives by chance. Mulvihill

(8) defines an excess frequency of cancer in a family if malignancy occurs, for example, in multiple generations, in a high percentage of siblings, at an unusual age, in an atypical sex ratio, or in association with other cancers or genetic disorders. The identification of genetic factors of determinants of familial cancer is sometimes facilitated by laboratory markers with pathogenetic significance in normal cells. Classical single-cell gene traits that show abnormal response to radiation in vivo and in vitro are ataxia telangiectasia (AT) or xeroderma pigmentosum (XP).

The frequencies with which genes predisposing to cancer occur in the population may not be quite high enough to give rise to cancer families, yet may significantly affect cancer incidence (3). One example may be the gene that predisposes smokers to lung cancer and that is associated with the metabolism of debrisoquine (1). Extensive metabolizers of debrisoquine were found more frequently among lung cancer patients, who had been smoking, whereas smoking controls were characterized by poor metabolism. Thus, debrisoquine 4-hydroxylation may represent a marker to assist in assessing individual risk of smokers.

ATAXIA TELANGIECTASIA-HETEROZYGOTES AND CANCER

A well-known example of a genetic condition predisposing to cancer is AT. Homozygotes for AT are deficient in immune response, show progressive neurological degeneration and defects in organ maturation, and are prone to develop cancer, particularly in cells of the immune system. Their peripheral lymphocytes show spontaneous chromosome aberrations and are characterized by hypersensitivity to the chromosome-breaking and lethal effects of ionizing radiation and chemicals, such as bleomycin, adriamycin, and a tumor promoter (TPA) that is supposed to produce free radicals [Shilo, cited by Bridges et al. (3)]. This suggests that AT cells have a defect in the processing or repair of DNA exposed to free radicals.

Homozygotes occur with a low frequency of about 1 per 10^5. Ataxia telangiectasia heterozygotes are, however, occurring at a much higher frequency, in the order of 1-2% of the population (3). A study by Swift (12), which suggested that AT heterozygotes likewise may exhibit increased proneness to develop cancer, elicited a great deal of interest. Characterization of AT heterozygotes formed the topic of a recent workshop organized by the International Agency for Research on Cancer (IARC) and the International Commission for Protection against Environmental Mutagens and Carcinogens (ICPEMC) (3). The results of a second, more broadly based study by Swift appear to be consistent with those of his first study in showing that AT heterozygotes have indeed a greater predisposition to develop cancer (see Ref. 3). Data are now available from 8 different laboratories where AT heterozygote cells have been examined for their in vitro sensitivity to the chromosome-breaking and lethal effects induced by ionizing radiation or neocarzinostatin. These results are consistently showing greater radiosensitivity in AT heterozygotes than in homozygous wild-type cells.

The above studies may serve to illustrate an important point put forward by Mulvihill (8) that progress in understanding cancer genetics is likely to be best advanced by bringing etiologic observations made by clinicians and epidemiologists to the attention of laboratory scientists who have the tools to dissect mechanisms in cells and chromosomes.

FOLLOW-UP ON INDIVIDUALS WITH CYTOGENETIC DAMAGE IN LYMPHOCYTES

In this context, I would like to make a suggestion on how efforts in monitoring cytogenetic damage in peripheral blood lymphocytes could be employed to yield data relevant to future health risks of the exposed populations. To date, the only method to detect genetic damage in humans consists of measuring chromosome aberrations or mutations to thioguanine resistance in peripheral blood lymphocytes. An increase in, say, the frequency of chromosome aberrations tells us that cells in the exposed individual have experienced damage to DNA, but we do not know what this means in terms of pathogenic effects. It would seem that a great deal more could be learned if careful follow-up studies on individuals responding with cytogenetic or mutational damage were taken into consideration.

First of all, a data bank should be established to register whether exposed subjects show an increased tendency of developing malignancy at a later age. Furthermore, one should consider the possibility that subjects characterized by increased response to carcinogens or mutagens reflect a cancer-prone section of the population. Consequently, it would be appropriate to determine their capacity to repair DNA damage, for which simple tests are now available. Another parameter that could be considered is measurement of their capacity to metabolize carcinogens by employing simple test substances, as outlined by Vesell (13), or a substance like debrisoquine. Populations of cured cancer patients who underwent cytostatic treatment (2,6), heavy smokers (11), alcoholics (10), and individuals exposed to ionizing radiation (5) might well qualify for such investigations. As a rule, the number of responding individuals in such monitoring surveys is small. International coordination of efforts using standardized protocols by the various investigators would increase the statistical evaluation and significance of the data.

ACKNOWLEDGEMENT

I would like to thank my colleague Professor K. Sankaranarayan for critically reading this manuscript.

REFERENCES

1. Ayesh, R., J.R. Idle, J.C. Ritchie, M.J. Crothers, and M.R. Hetzel (1984) Metabolic oxidation phenotypes as markers for susceptibility to lung cancer. Nature 312:169-170.
2. Bohr, V., and L. Kϕber (1985) DNA repair in lymphocytes from patients with secondary leukemia as measured by strand joining and unscheduled DNA synthesis. Mutat. Res. 146:219-225.
3. Bridges, B.A., G. Lenoir, and L. Tomatis (1985) Workshop on ataxia-telangiectasia heterozygotes and cancer. Cancer Res. Vol. 45 (in press).
4. Cavenee, W.K., T.P. Dryja, R.A. Phillips, W.F. Benedict, R. Godbout, L. Gallie, A.L. Murphree, L.C. Strong, and R.L. White (1983) Expression of recessive alleles by chromosomal mechanisms in retinoblastoma. Nature 305:779-781.
5. Evans, H.J. (1984) Genetic damage and cancer. In Genes and Cancer, J.M. Bishop, J.D. Rowley, and M. Greaves, eds. Alan R. Liss, Inc., New York, pp. 3-18.

6. Gebhard, E. (1984) Chromosomal aberrations In lymphocytes of patients under chemotherapy. In Mutations in Man, G. Obe, ed. Springer-Verlag, Berlin, pp. 198-222.
7. McKusick, V.A. (1975) Mendelian Inheritance in Man: Catalogs of Autosomal Dominant, Autosomal Recessive, and X-Linked Phenotypes, 4th edition, The Johns Hopkins University Press, Baltimore, London, 837 pp.
8. Mulvihill, J.J. (1984) Clinical ecogenetics and human cancer. In Genes and Cancer, J.M. Bishop, J.D. Rowley, and M. Greaves, eds. Alan R. Liss, Inc., New York, pp. 19-36.
9. Nomura, T. (1984) Quantitative studies on mutagenesis, teratogenesis and carcinogenesis in mice. In Problems of Threshold in Chemical Mutagenesis, Y. Tazima, S. Kondo, and Y. Kuroda, eds. The Environmental Mutagen Society of Japan, Mishima, Shizuoka, pp. 27-34.
10. Obe, G., D. Göbel, H. Engeln, J. Herka, and A.T. Natarajan (1980) Chromosomal aberrations in peripheral lymphocytes of alcoholics. Mutat. Res. 73:377-386.
11. Obe, G., W.D. Heller, and H.J. Vogt (1984) Mutagenic activity of cigarette smoke. In Mutations in Man, G. Obe, ed. Springer-Verlag, Berlin, pp. 223-246.
12. Swift, M. (1982) Disease predisposition of ataxia-telangiectasia heterozygotes. In Ataxia-Telangiectasia a Cellular and Molecular Link between Cancer, Neuropathology and Immune Deficiency, B.A. Bridges and D.G. Harnden, eds. John Wiley and Sons, Inc., Chichester, England, pp. 355-361.
13. Vesell, E.S. (1980) Genetic and environmental factors affecting the metabolism of carcinogens in short term assays for detection of environmental carcinogens. In Molecular and Cellular Aspects of Carcinogen Screening Tests, R. Montesano, H. Bartsch, and L. Tomatis, eds. IARC Scientific Publication No. 27, International Agency for Research on Cancer, Lyon, pp. 23-40.

PROTECTIVE EFFECTS OF BETA-CAROTENE AGAINST PSORALEN PHOTOTOXICITY: RELEVANCE TO PROTECTION AGAINST CARCINOGENESIS

Andrija Kornhauser, Wayne Wamer, and Albert Giles, Jr.

Division of Toxicology
Food and Drug Administration
Washington, D.C. 20204

INTRODUCTION

Improvement in prevention and therapy of various types of cancers represents the major challenge in modern medicine. Surgery, chemotherapy, and radiotherapy result in temporary or sometimes complete remission only in a certain percentage of cancer patients. Immunotherapy, although promising, is still in an experimental stage. Increasingly, emphasis has therefore focused on the prevention of cancer formation. One of the ways to achieve this goal is to design and administer agents that modulate the process of carcinogenesis.

In recent years, there has been an upsurge of interest in various biological antioxidants, including carotenoids, retinoids, ascorbic acid, and vitamin E, as potential anticancer agents (1,39). The protective potential of one such antioxidant, β-carotene, is receiving particular attention.

Beta-carotene is a naturally occurring pigment present in most of the green and yellow vegetables common in our diets. Its primary function in all photosynthetic organisms is to protect the cell against its own endogenous photosensitizer, chlorophyll. The protective function of the β-carotene molecule is expressed in a wide variety of living systems, including single cells, higher plants, animals, and humans. Table 1 lists the carotenoid content of some selected fruits and vegetables. In addition, β-carotene is used both as a food color additive and as a drug for management of the photosensitivity exhibited by persons having the genetic disorder erythropoietic protoporphyria (EPP) (32).

Recently, the protective role of dietary β-carotene against chemical- and light-induced carcinogenesis has become an area of vigorous investigation in both epidemiological (1,39,54) and experimental studies (27). Experimental studies of β-carotene effects have been limited by a lack of appropriate animal models, i.e., models in which dietary β-carotene is accumulated and in which the toxicological endpoint of interest is easily observed.

Tab. 1. Carotenoid content of various fruits and vegetables.

Fruit or vegetable	Fresh weight (mg/kg)
Apricots	0.9–5.4
Asparagus	22.0
Broccoli	6.0–46.0
Carrots	54.0
Cherries	5.0–11.0
Cling peaches	27.0
Collards	36.0
Cranberries	5.8
Japanese persimmons	54.0
Lettuce	8.0–68.0
Navel orange pulp	23.0
Parsley	32.0–110.0
Pears	0.3–1.2
Prunes	21.0
Pumpkin	59.0
Spinach	26.0–76.0
Strawberry	0.6–1.5
Tomato	51.0
Turnip greens	46.0–55.0
Yams	4.3

Note: Data compiled from Ref. 4.

Several species are known to accumulate dietary β-carotene; humans are some of the best accumulators (8). It has been reported that rodents do not accumulate dietary β-carotene (8). We (53a) and others, however, have found that mice (30) and rats (this chapter) given a β-carotene-fortified diet for extended periods accumulate β-carotene to a limited degree. In this chapter we present kinetic data of β-carotene profiles in serum and in skin, establishing the rat as a useful animal model for studying the protective effects of dietary β-carotene.

RELEVANCE OF PHOTOTOXICITY STUDIES

A phototoxic reaction can be defined as a nonimmunological light-induced tissue response with a basic clinical appearance that resembles a sunburn. In the last decade, phototoxic reactions caused by drugs, cosmetics, and a large number of environmental chemicals have become a major health problem (13). Chronic phototoxic exposure can lead to neoplastic skin changes, including basal and squamous cell carcinomas, and, to a certain extent, malignant melanomas. Nonmelanoma cancer of the skin is the most common malignant neoplasia in the white population of the United States (45). The incidence rate of skin cancer is increasing steadily, and the disease represents a major health and economic problem in the modern industrialized world (45).

Tab. 2. Some natural sources of psoralens.

Source	Total psoralens (mg/kg)
Celery	1.3[a]
Diseased celery	98.7[a]
Parsnips	40.0[b]
Bergamot oil	3,000.0[c]

[a] See Ref. 44.
[b] See Ref. 17.
[c] See Ref. 56.

Psoralens are a group of naturally occurring (Tab. 2) and synthetic substances. The photosensitizing ability of psoralens in a variety of biological systems is well established. These biological effects have generally been attributed to the ability of psoralens to form both mono- and diadducts to DNA (6,35). On the molecular level, the following facts are known:

(a) Psoralens intercalate into DNA--that is, slip in between adjacent base pairs--by forming molecular complexes involving weak chemical interactions ("dark reaction").

(b) On ultraviolet A (UVA) light (320-400 nm) irradiation of such a system, in vitro or in vivo, covalent bond formation between a pyrimidine base and the furocoumarin molecule takes place (C_4 cycloaddition). Because of their structure, psoralens in this reaction can react either at their 3,4 double bond or at their corresponding 4',5' site, yielding monoadducts.

(c) On absorption of an additional photon, a further chemical reaction, yielding a "cross-linked DNA," may take place. Thus psoralens can behave as photoreactive bifunctional agents, one molecule reacting with 2 pyrimidines in opposite strands of DNA. The structures of psoralen mono- and diadducts with thymine are shown in Fig. 1. Figure 2 schematically shows DNA cross-linked by a psoralen molecule. The result is a cross-linked DNA in which the individual strands cannot be separated by standard denaturation conditions. Both types of lesions, the monofunctional adduct and the cross-linked product, can be repaired in vivo (2,38).

The covalent addition of psoralens to DNA has been evoked to explain the major biological effects of psoralens. These include mutation and lethality in prokaryotic and eukaryotic systems, inhibition of DNA synthesis, sister chromatid exchange, and carcinogenesis. Skin photosensitization is one of the most widely studied properties of some psoralens. Several clinical types of photodermatoses occur when skin comes into contact with plants or vegetable products and is later exposed to sunlight (35). Much less is known about potential adverse cutaneous effects after chronic ingestion of foods that contain psoralens (20). Psoralens are also used in photochemotherapy for management of such disorders as vitiligo, psoriasis, and mycosis fungoides (14,36). This application involves oral

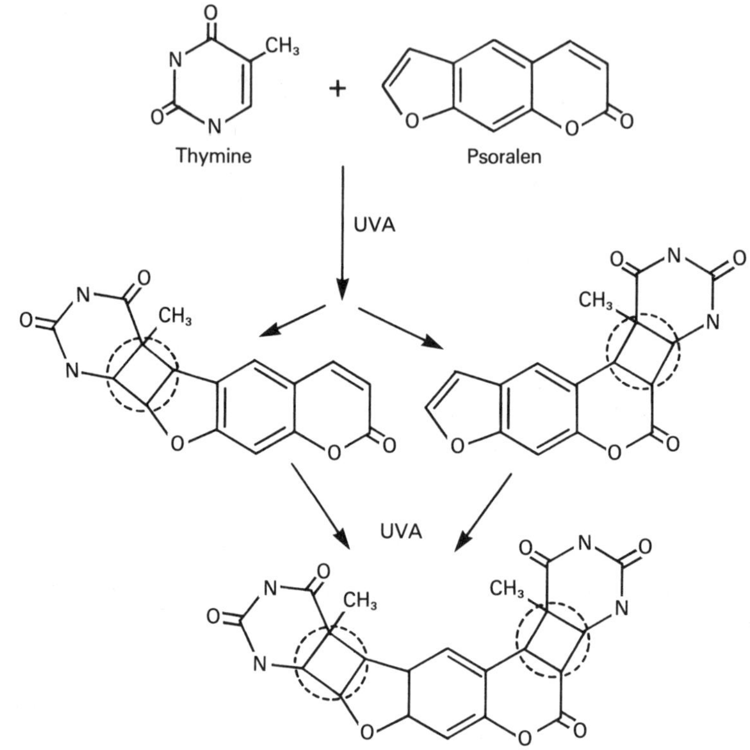

Fig. 1. Photoaddition products of psoralen with thymine after UV irradiation in vivo.

administration of the chosen psoralen, typically 8-methoxypsoralen (8-MOP), followed by irradiation with UVA (PUVA = psoralens + UVA). 5-Methoxypsoralen (5-MOP), whose topical phototoxicity is comparable to that of 8-MOP, shows significantly reduced phototoxicity when given orally (18,19).

In the past, psoralen photoreactions were described as proceeding under strictly anaerobic conditions. More recently, it has been suggested that psoralen-induced phototoxicity in animals and man resulting in erythema may be produced via a different mechanism. It has been reported that psoralens are effective photosensitizers for the formation of singlet oxygen ($^1O_2$) (9,40).

Several investigators have found that psoralens are effective sensitizers for the photo-oxidation of a range of molecular (12,16,23,40,41,52, 53,56) and cellular (24,47) targets. In addition, photosensitized $^1O_2$ production has been proposed as contributing to the observed in vitro photomutagenicity of psoralens (10). More recently, an apparent correlation between the ability of psoralens to photosensitize $^1O_2$ formation in vitro and to produce erythema has been reported (11,37). Other investigators, however, have observed that skin photosensitization is related to psoralen photobinding to DNA rather than to $^1O_2$ production (51).

To date, the importance of $^1O_2$ photo-oxidative mechanisms in psoralen-induced photosensitization of skin has not been established in vivo. Potapenko et al. (42) have shown that the antioxidant vitamin, α-tocopherol,

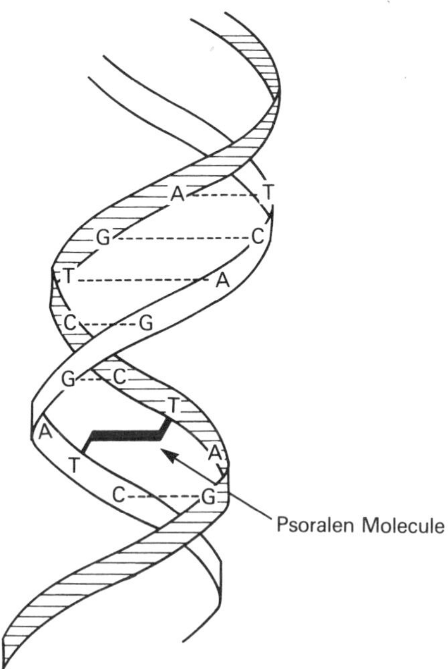

Fig. 2. Schematic representation of DNA cross-linked by a psoralen molecule.

when applied topically, reduces erythema induced by 8-MOP in both human and rat skin. This protective effect was interpreted as suggesting $^1O_2$ or free radical involvement in 8-MOP phototoxicity.

In this chapter we describe an animal model developed to investigate the role that $^1O_2$ or other activated oxygen species involved in photo-oxidation play in 8-MOP-induced phototoxicity in vivo. For this purpose β-carotene, a well-established quencher of photo-oxidation (21), was fed to groups of rats. The protective effect of β-carotene against both PUVA- and UVB-induced erythema was then determined in this animal model.

Since it has been shown, both in animal models and in clinical studies, that chronic psoralen phototoxicity can lead to carcinogenesis, the observed protective effect may have implications for protection against carcinogenesis. It is our belief that dietary (not topical) β-carotene, a naturally occurring antioxidant with extremely low toxicity, both in experimental animals and in man, has potential as an anticancer agent. A few facts that support this hypothesis are presented below.

EXPERIMENTAL APPROACH

A total of 105 albino female rats of the Osborne-Mendel strain were assigned to 7 groups of 15 each. The groups were fed daily up to 18 weeks as follows: 3 groups received Purina rodent chow (control diet), 3 groups received the 1% β-carotene-fortified diet, and 1 group received the placebo diet. The 1% β-carotene-fortified diet was prepared from β-carotene beadlets (Hoffman-LaRoche, Inc., Nutley, New Jersey) and Purina chow, as described previously (15). Each batch of β-carotene diet was analyzed for β-carotene content and was found to be stable for up to 1 month at room temperature. The placebo diet was similarly prepared with placebo beadlets. Placebo beadlets were identical in composition to the β-carotene beadlets but lacked β-carotene.

To determine the extent to which the rats on a 1% β-carotene diet accumulated β-carotene in both serum and skin, 3 animals from a group receiving the 1% β-carotene-fortified diet were sacrificed at each of the selected time intervals (i.e., 3, 4, 5, 6, and 8 weeks). Serum and skin samples were analyzed by high-performance liquid chromatography (HPLC) (15). Successful quantitation of serum, and particularly skin levels, required the use of HPLC methods, both to increase sensitivity and to eliminate background interferences common in the alternative spectrophotometric assay.

After a 14-week feeding period, animals from the β-carotene, placebo, and control diet regimens were chosen for comparison of erythema produced by PUVA treatment. PUVA-treated rats received a 20 mg/kg oral dose of 8-MOP (P. Elder Co., Bryan, Ohio, 0.2% wt/vol in corn oil) by gavage. A fourth group of rats on the control diet served as a solvent control by receiving an equivalent dose of corn oil only. Before irradiation, the backs of all animals were clipped. Two hours after dosing, the animals were immobilized on a wooden board and singly irradiated with 5 J/cm^2 UVA. The light source used in this study was a 1 kilowatt Hg-Xe Arc Lamp (Hanovia) (15). Suitable filters were used to isolate UVA or UVB spectral regions. The spectral irradiance in each portion of the UV spectrum was determined using a Gamma Scientific Spectral DH spectroradiometer system. In addition, routine light measurements were made before and during each experiment with an International Light IL 700 research radiometer (15).

Phototoxicity reactions (erythema) were observed at appropriate times (24, 48, 72, and 96 hr after irradiation) and graded as 0, tr, ±, +, ++, +++, or ++++. For calculating fractional response, ratings of 0, tr, and ± were taken as negative and the remaining ratings were considered positive. The ratio of the number of positive responses to the total number of animals in each group is presented as the fractional response. An average intensity was calculated by assigning numerical values 0, 1, 2, 3, 4, 5, and 6 to ratings 0, tr, ±, +, ++, +++, and ++++, respectively. These numerical values were then averaged for each group. After grading, the animals were sacrificed, and serum and skin analyzed for β-carotene.

For the UVB experiment, a group of animals on the control and β-carotene-fortified diets (18 weeks) were clipped and immobilized as previously described. The animals were singly irradiated with 0.7 J/cm^2 UVB, graded for erythema at 24, 48, and 72 hr after irradiation, and then sacrificed. Serum and skin samples were analyzed as above.

RESULTS

The results obtained in this study indicate that the rat is capable of accumulating β-carotene from a dietary source to a limited degree. Figure 3 presents the results of the serum and skin analyses for rats sacrificed at selected times within the feeding period. These data indicate that rats continue to accumulate β-carotene in serum, and particularly in skin, throughout the feeding period.

Table 3 presents the fractional response for groups of animals receiving PUVA treatment. Animals receiving the β-carotene-fortified diet exhib-

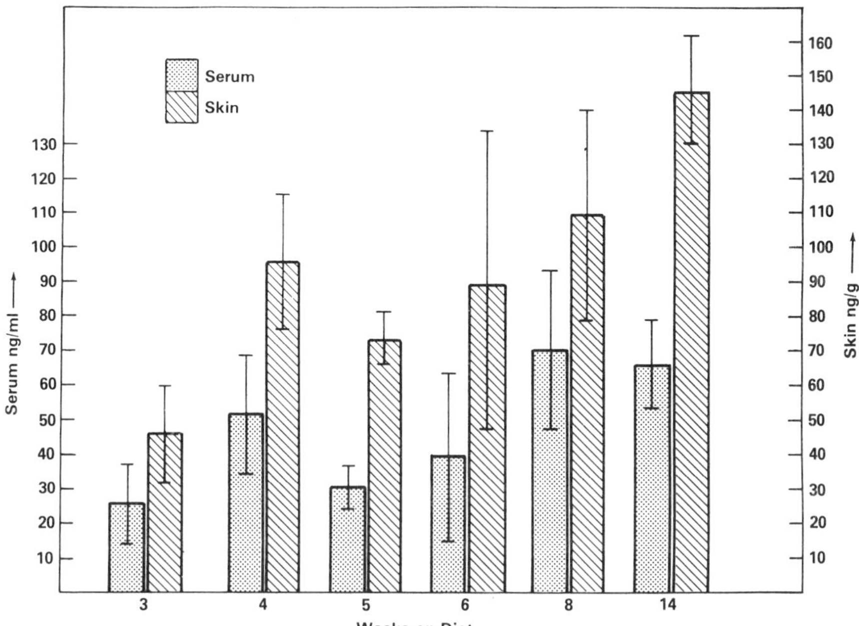

Fig. 3. Beta-carotene serum and skin levels were measured for the 15 animals that were on a 1% β-carotene-fortified diet for 14 weeks. The average with the standard error of the mean is given. Averages from 3 to 8 weeks are from groups containing 3 animals each.

it reduced fractional responses relative to the other PUVA-treated groups. In addition, the average erythema intensity induced by PUVA treatment was significantly reduced for the group fed the β-carotene diet (Tab. 4). We have found that the degree of protection observed correlates well with the level of β-carotene, particularly β-carotene skin levels. Figure 4 shows the relationship between β-carotene levels and phototoxicity (erythema at 72 hr after irradiation) for each of the 15 PUVA-treated animals that received the β-carotene-fortified diet. Those animals having high β-carotene levels (particularly skin levels) showed decreased erythema scores. Furthermore, there is an apparent correlation between β-carotene serum and skin levels in this animal model.

Beta-carotene showed no protective effect against UVB-induced erythema, although β-carotene skin levels of UVB-irradiated animals receiving a β-carotene-fortified diet were similar to those of PUVA-treated animals (135 ± 22 ng/g and 145 ± 16 ng/g, respectively).

DISCUSSION

The role of β-carotene as a protective agent against photosensitization has been established in a variety of biological systems (29). However, to our knowledge, this is the first reported demonstration of dietary β-carotene protection against psoralen-induced phototoxicity in vivo. The results presented here clearly demonstrate that dietary β-carotene provides protection against PUVA-induced erythema in the rat.

Tab. 3. Fractional response (FR) after 8-methoxypsoralen (8-MOP) (20 mg/kg) and UVA (5 J/cm^2) treatment.

Group	FR[a] (72 hr)	FR[a] (96 hr)
Control diet PUVA-treated	14/14	14/14
Placebo diet (14 weeks) PUVA-treated	11/15 $P < 0.06$[b]	14/15 NS*
β-Carotene diet (14 weeks) PUVA-treated	6/15 $P < 0.001$	8/15 $P < 0.004$
Control diet UVA-treated	0/15 $P < 0.001$	0/15 $P < 0.001$

[a] The FR is calculated by counting the number of animals having a + response or greater, then dividing by the number of animals in the group.
[b] All P values given are one-tailed comparisons with the control diet, PUVA-treated.
* NS = Not significant.

Since the role of oxygen radicals and related activated oxygen species in mutagenesis and carcinogenesis represents one of the interests of this conference, a discussion of the mechanisms of β-carotene protection is warranted.

Several mechanisms are possible for the observed β-carotene protection against PUVA-induced phototoxicity. In general, both triplet sensitizer and $^1O_2$ can be quenched by β-carotene in energy transfer reactions, thereby generating the ground state of the sensitizer and $^3O_2$, respectively. In addition, radical species produced in type I photoreactions can also be quenched by carotenoids. It is well known that β-carotene is an effective quencher of these reactive intermediates (5,21). All three mechanisms could, therefore, serve as protective reactions against PUVA-induced phototoxicity.

Figure 5 is a simplified scheme of several proposed mechanisms for psoralen-induced erythema. Possible formation of DNA adducts from the singlet excited state of psoralen (49) and formation of $^1O_2$ from triplet state psoralen, intercalated or bound as the 4',5'-monoadduct (9), have been omitted for clarity. Figure 5 also shows three points at which β-carotene may affect psoralen-photosensitized intermediates and photoproducts.

As previously mentioned, a correlation between skin photosensitizing potential and ability to form $^1O_2$ for many psoralen derivatives has been reported (37). Therefore, one mechanism for the observed protection may involve quenching of PUVA-produced activated oxygen species (e.g., $^1O_2$ and/or free radicals) by β-carotene (Fig. 5, point C).

The quenching of activated oxygen species would explain the photoprotective effect of orally administered β-carotene in treating EPP in patients as well as in ameliorating phototoxicity in animal models of EPP

Tab. 4. Average intensity (AI) after 8-methoxypsoralen (8-MOP) (20 mg/kg) and UVA (5 J/cm^2) treatment.

Group	AI[a] (72 hr)	AI[a] (96 hr)
Control diet PUVA-treated	4.4 ±0.3	5.8 ±0.1
Placebo diet (14 weeks) PUVA-treated	3.5 ±0.4 $P < 0.10$	5.5 ±0.3 NS*
β-Carotene diet (14 weeks) PUVA-treated	1.4 ±0.4 $P < 0.005$	2.9 ±0.7 $P < 0.005$
Control diet UVA-treated	0.0 ±0.0 $P < 0.005$	0.0 ±0.0 $P < 0.005$

[a] The AI is calculated by assigning the numerical equivalent to each animal score in a group. The assigned numerical equivalents are added and then divided by the number of animals in the group. Entries are AI ± standard error of the mean.

* NS = Not significant.

(26,32). In EPP, the photosensitizing compound, protoporphyrin, is a photodynamic agent and generator of $^1O_2$ (50). The involvement of $^1O_2$ in PUVA-induced erythema, however, has not yet been conclusively proven.

Indeed, it has been pointed out that both the production of $^1O_2$ and the formation of the mono- and, particularly, the diadduct proceed via a common intermediate, the psoralen triplet state (11). Thus psoralens that undergo efficient intersystem crossing should readily photosensitize the formation of $^1O_2$ and photoreact with DNA, unless low DNA binding or steric constraints predominate. It is understandable that correlations have been reported between $^1O_2$ formation and erythema production (37) as well as between DNA photobinding and erythema production (51). However, the causal relationship between these two photoproducts ($^1O_2$ or DNA adducts) and erythema production is still under active investigation.

Another possible mechanism for the observed β-carotene protective effect is direct quenching of triplet states. Direct quenching of the 8-MOP triplet state at points A or B in Fig. 5 could result in reduction of both $^1O_2$ and DNA photoadduct formation, although not necessarily with equal efficiency. It has been demonstrated spectroscopically that β-carotene in benzene quenches the triplet states of furocoumarins, including psoralen, 5-MOP, 8-MOP, and angelicin (3). Significant quenching of the 8-MOP triplet state would result in reduction of 8-MOP cross-linked DNA.

The photochemical intermediates quenched by β-carotene may depend in large part on the availability of β-carotene within the cell. It is likely that β-carotene, because of its hydrophobic nature, localizes in cellular membranes. This has been shown experimentally in bacteria (33). Similar data for intracellular β-carotene distribution in skin are not available.

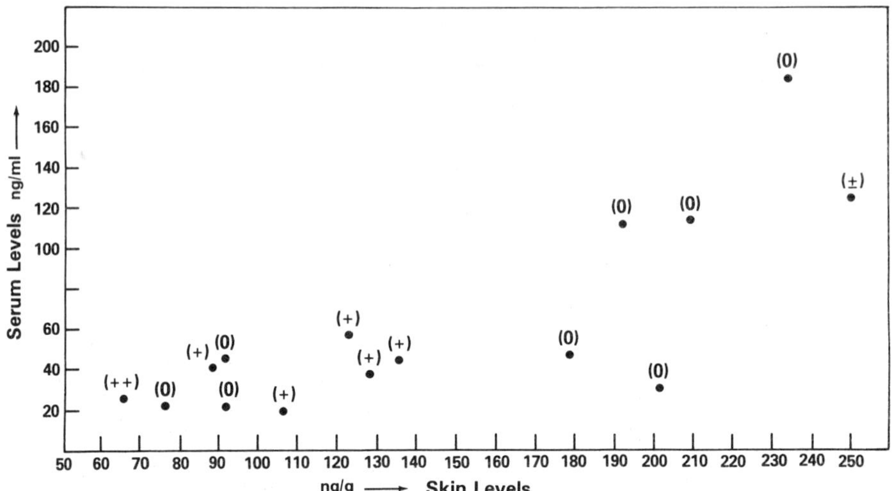

Fig. 4. Relationship between phototoxicity grades at 72 hr and β-carotene skin and serum levels. Each point represents data derived from one animal.

We are therefore unable to predict the availability of β-carotene in membranes, cytoplasm, or the nucleus in viable skin or epidermal cells.

As noted by other investigators (22,43), a simple filter protective effect caused by β-carotene (in the UVB range) in the skin is not probable. Indeed, we estimate that the highest observed β-carotene skin levels in this study would result in less than 0.5% attenuation of UVA light on passage through the skin.

In summary, this report represents an in vivo demonstration of a protective effect of dietary β-carotene against PUVA-produced phototoxicity. This animal model may also be useful in further studies to better elucidate the role of psoralen photoadducts to DNA and/or photodynamic oxidation in PUVA-induced erythema. In addition, the animal model developed here may be applicable in future studies of dietary β-carotene protection against carcinogenesis. The mechanisms discussed here should also be relevant for explaining β-carotene protection by quenching reactive oxygen species generated in vivo as a consequence of normal or pathologic metabolic pathways.

OBSERVATIONS SUGGESTING ANTICARCINOGENIC EFFECTS OF BETA-CAROTENE

The following observations are based on studies performed in various systems:

In Vitro

Som et al. (48) have recently suggested that β-carotene may inhibit cellular transformation induced by 7,12-dimethylbenz(α)anthracene (DMBA). They showed that β-carotene is capable of inhibiting DMBA-induced transformation of epithelial cells in an organ culture of whole mammary glands from BALB/c mice. No conversion of β-carotene to vitamin A was detectable in

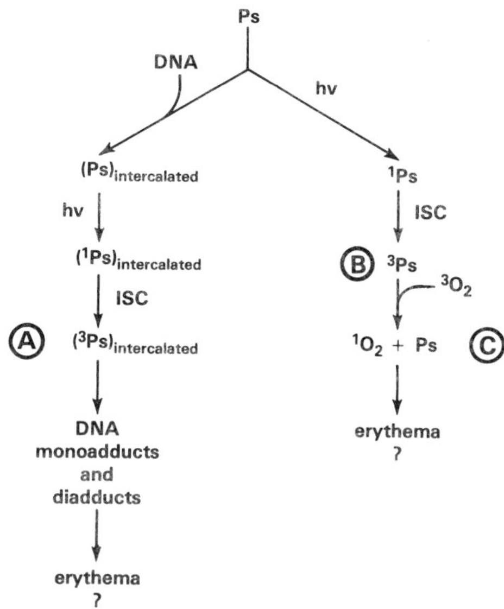

Fig. 5. Scheme showing proposed mechanism for PUVA-induced erythema and points (A, B, and C) at which β-carotene may quench photoproducts. Ps, ground state psoralen; ^{1}Ps, excited singlet state; ^{3}Ps, triplet state; ISC, intersystem crossing.

this system, implicating β-carotene itself as the protective agent. The authors suggest that the carotenoid pigment inhibits both the initiation and promotional stages in this system.

Animal Studies

The inhibitory effect of several carotenoid pigments against UV light and chemically induced neoplasia of the skin has been studied in the mouse (27,28). These interesting preliminary results are compiled and summarized in Tab. 5.

Tab. 5. Carotenoid protection against carcinogenesis induced by UVB or 7,12-dimethylbenz(α)anthracene (DMBA) in the mouse.

Carotenoid	Protection against UVB carcinogenesis	Protection against DMBA carcinogenesis	Provitamin A activity
1. β-Carotene	+	+	+
2. Canthaxanthin	+	-	-
3. Phytoene	+	-	-
4. Crocetin	-	±	-

Note: Compiled from Ref. 27 and 28.

The detailed mechanism of carotenoid protective function in this system is unknown at this time. It has been proposed that β-carotene, canthaxanthin, and phytoene inhibit UVB-induced tumors through quenching of free radicals and activated forms of oxygen that have been implicated in UVB-induced carcinogenesis. It is suggested that crocetin is less effective in inhibiting UVB-induced neoplasia because of two structural features. First, crocetin has fewer conjugated double bonds than β-carotene or canthaxanthin, which makes it a less effective free-radical scavenger (31). Second, lipophilic carotenoids, such as β-carotene, have been found to be associated with cellular membranes in several organisms (33,34). Crocetin, a water-soluble carotenoid, may show a different pattern of intracellular localization.

Another important finding extracted from these studies is the fact that provitamin A activity, which is lacking in canthaxanthin, phytoene and crocetin, appears to be needed for inhibition of chemically induced skin carcinogenesis. Thus different mechanisms must be operating in the inhibition of UV-induced skin tumors, as compared to chemically induced tumors. We must point out, however, that histologically these tumors are indistinguishable, regardless of their etiology. It is interesting to note that a reduced potency of canthaxanthin (as compared to β-carotene) in protection against some forms of human oral carcinogenesis was reported (Stich, this Volume). An additional study on antitumor action of β-carotene in mice inoculated with an oncogenic virus has also been reported (46).

Epidemiological Studies

A number of reports have been published in recent years which indicate that human cancer risks are inversely correlated with blood retinol and also with blood carotene levels resulting from dietary sources (39,54). To confirm these findings, the compound of interest might be administered prophylactically to a large number of people at high risk to develop the particular kind of cancer in question, along with an appropriate control group.

The National Cancer Institute is presently supporting a few such studies in which carotenoids are administered to a population at high risk for developing certain kinds of cancer. Some studies are centered on the prevention of nonmelanoma skin cancers; one such study to be carried out in Tanzania will involve the administration of carotenoids to albinos, who are at very high risk for developing these tumors. Another study will involve administration of β-carotene to a group of subjects who have already developed at least one skin tumor. The other studies deal with the possible prevention of colon and lung cancers by β-carotene in selected populations. A study of special interest involves a large group of healthy physicians and is aimed at seeing whether the administration of β-carotene and aspirin, on alternate days, will reduce the incidence of all types of cancer and also the mortality from cardiovascular disease.

CONCLUSIONS AND FUTURE RECOMMENDATIONS

One of the contributions that stimulated epidemiological as well as experimental studies of carotenoids as potential anticancer agents was the work of Peto et al. (39). The authors were aware of the fact that the protective effect could have been elicited by an artifact due merely to an association of β-carotene with some other protective dietary component. There is additional reported evidence to suggest that the consumption of certain carotene-rich vegetables is associated with a reduction in the incidence of cancer at several sites in humans (7). Accumulated data strongly suggest that β-carotene is responsible for the beneficial effect, although a number of other components present in these vegetables might modify carcinogenesis in laboratory animals and humans. The majority of epidemiological studies performed to date suggest an anticarcinogenic effect of carotenoids; a few studies, however, have failed to confirm this (55). Further epidemiological studies would be useful in clarifying the role of β-carotene.

The final results of ongoing epidemiological studies are still many years away. In the meantime, more animal work should be pursued in order to evaluate the role of carotenoids in cancer prevention. It is surprising that, in view of increased interest in the role of dietary antioxidants in general, only a very few animal experiments using β-carotene have been performed. This fact can be explained partly by the lack of adequate protocols. Carotenoids are generally unstable compounds with poor bioavailability; therefore it is difficult to administer them in pure chemical form and results are difficult to reproduce. To avoid these difficulties and achieve a reasonable accumulation, carotenoids have to be administered in a diet in special formulations over a prolonged time period.

The mechanism of protection by β-carotene is still not proven. Carotenoids are among the most efficient substances known to quench activated oxygen species and trap certain free radicals. These species seem to play a role in both initiation and promotion of carcinogenesis. Furthermore, the mechanism of β-carotene protection could also be based on its known provitamin A activity. We again point out that testing β-carotene in lieu of retinoids has the advantage of avoiding the risk of hypervitaminosis A, and hypercarotenosis (as far as we know) is harmless in both rodents and man. The tolerable daily intake of β-carotene for an adult is estimated to be 350 mg per day (39).

Another line of evidence should be pointed out in this context. We have recently become aware that oxygen-derived free radicals play an important role in various pathologic processes of postischemic tissue injury (25). There is preliminary evidence that antioxidant vitamins may protect against this type of tissue damage (1). The potency of carotenoids and other antioxidants in modifying these processes, which result in conditions including heart disease and cerebral ischemia, should be established. A number of reports have been communicated recently on the effects of carotenoids in modification of the immune response (2a,7a,46a) and the aging

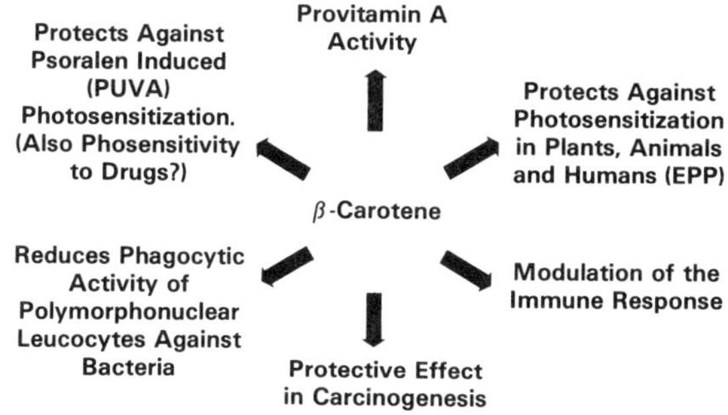

Fig. 6. Selected biological effects of β-carotene.

process (7a). (Although these are interesting hypotheses, a detailed discussion exceeds the scope of this work. The reader is encouraged to consult the original publications for further information.) A few selected biological properties of β-carotene are summarized in Fig. 6.

The last decade has greatly increased our knowledge and awareness of both carcinogens and anticarcinogens in our environment and particularly in our diets. Based on the present scientific and public interest in this field, the next decade promises to furnish solid evidence of the role of biological antioxidants and also of their mechanism of action in modifying carcinogenesis and some other degenerative diseases in man.

REFERENCES

1. Ames, B.N. (1983) Dietary carcinogens and anticarcinogens. Science 221:1256-1264.
2. Baden, H.P., J.M. Parrington, J.D.A. Delhanty, and M.A. Pathak (1972) DNA synthesis in normal and xeroderma pigmentosum fibroblasts following treatment with 8-methoxypsoralen and long wave ultraviolet light. Biochim. Biophys. Acta 262:247-255.

2a. Bendich, A., and S.S. Shapiro (1984) Effect of dietary β-carotene on lymphocyte responses to mutagens. Fed. Proc. 41:2936.

3. Bensasson, R.V., E.J. Land, and C. Salet (1978) Triplet excited state of furocoumarins: Reaction with nucleic acid bases and amino acids. Photochem. Photobiol. 27:273-280.
4. Borenstein, B., and R.H. Bunnel (1966) Carotenoids: Properties, occurrence and utilization in foods. In Advances in Food Research, Vol. 15, C.O. Chichester, E.M. Mrak, and G.F. Stewart, eds. Academic Press, Inc., New York.
5. Burton, G.W., and K.U. Ingold (1984) β-Carotene: An unusual type of lipid antioxidant. Science 224:569-573.
6. Cole, R.S. (1970) Light-induced cross-linking of DNA in the presence of a furocoumarin (psoralen). Studies with phage λ, Escherichia coli, and mouse leukemia cells. Biochim. Biophys. Acta 217:30-39.
7. Committee on Diet, Nutrition and Cancer; National Research Council (1982) Diet, Nutrition, and Cancer, National Academy Press, Washington, D.C.

7a. Cutler, R.G. (1985) Antioxidants and longevity of mammalian species. In Molecular Biology of Aging, A.D. Woodhead, A.D. Blackett, and A. Hollaender, eds. Plenum Press, New York, pp. 15-73.

8. Deuel, H.J. (1957) The Lipids, Their Chemistry and Biochemistry, Vol. 3, Interscience Publishers, Inc., New York.
9. De Mol, N.J., G.M.J. Beijersbergen van Henegouwen, and B. Van Beele (1981) Singlet oxygen formation by sensitization of furocoumarins complexed with, or bound covalently to DNA. Photochem. Photobiol. 34: 661-666.
10. De Mol, N.J., G.M.J. Beijersbergen van Henegouwen, G.R. Mohn, B.W. Glickman, and P.M. van Kleef (1981) On the involvement of singlet oxygen in mutation induced by 8-methoxypsoralen and UVA irradiation in Escherichia coli K-12. Mutat. Res. 82:23-30.
11. De Mol, N.J., and G.M.J. Beijersbergen van Henegouwen (1981) Relation between some photobiological properties of furocoumarins and their extent of singlet oxygen production. Photochem. Photobiol. 33:815-819.
12. De Mol, N.J., and G.M.J. Beijersbergen van Henegouwen (1979) Formation of singlet molecular oxygen by 8-methoxypsoralen. Photochem. Photobiol. 30:331-335.

13. Fitzpatrick, T.B., M.A. Pathak, L.C. Harber, M. Seiji, and A. Kukita (1974) An introduction to the problem of normal and abnormal responses of man's skin to solar radiation. In Sunlight and Man, M.A. Pathak, L.C. Harber, M. Seiji, and A. Kukita, eds. University of Tokyo Press, Tokyo, pp. 3-14.
14. Gilchrest, B., J. Parrish, L. Tannenbaum, H. Haynes, and T. Fitzpatrick (1976) Oral methoxsalen photochemotherapy of mycosis fungoides. Cancer (Philadelphia) 38:683-689.
15. Giles, Jr., A., W. Wamer, and A. Kornhauser (1985) In vivo protective effect of β-carotene against psoralen phototoxicity. Photochem. Photobiol. 41:661-666.
16. Granger, M., F. Toulme, and C. Helene (1982) Photodynamic inhibition of Escherichia coli DNA polymerase I by 8-methoxypsoralen plus near ultraviolet irradiation. Photochem. Photobiol. 36:175-180.
17. Ivie, G.W., D.L. Holt, and M.C. Ivey (1981) Natural toxicants in human foods: Psoralens in raw and cooked parsnip root. Science 213:909-910.
18. Kornhauser, A., W.G. Wamer, and A.L. Giles, Jr. (1982) Psoralen phototoxicity: Correlation with serum and epidermal 8-methoxypsoralen and 5-methoxypsoralen in the guinea pig. Science 217:733-735.
19. Kornhauser, A., W.G. Wamer, and A.L. Giles, Jr. (1982) Difference in topical and systemic reactivity of psoralens: Determination of epidermal and serum levels. In Photobiologic, Toxicologic, and Pharmacologic Aspects of Psoralens (NCI Monograph 66), M.A. Pathak and J.K. Dunnick, eds. National Cancer Institute, pp. 97-101.
20. Kornhauser, A. (1980) Molecular aspects of phototoxicity. Ann. N.Y. Acad. Sci. 346:398-414.
21. Krinsky, N.I., and S.M. Deneke (1982) Interaction of oxygen and oxyradicals with carotenoids. J. Natl. Cancer Inst. 69:205-210.
22. Lamola, A.A., and W.E. Blumberg (1976) The effectiveness of β-carotene and phytoene as systemic sunscreens. In Proceedings of the 4th Annual Meeting of the American Society for Photobiology, Denver, Colorado, Abstract No. FAM-C4, p. 109.
23. Lerman, S., J. Megaw, and I. Willis (1980) The photoreactions of 8-methoxypsoralen with tryptophan and lens proteins. Photochem. Photobiol. 31:235-242.
24. Luo, R.F. (1978) Photobiological action of 8-methoxypsoralen to the respiratory response and lipid peroxide formation of mitochondria. Kurume Med. Assoc. J. 41:723-735.
25. McCord, J.M. (1985) Oxygen-derived free radicals in postischemic tissue injury. New Engl. J. Med. 312:159-163.
26. Mathews-Roth, M.M. (1984) Prophyrin photosensitization and carotenoid protection in mice: In vitro and in vivo studies. Photochem. Photobiol. 40:63-67.
27. Mathews-Roth, M.M. (1982) Antitumor activity of β-carotene and canthaxanthin. Oncology 39:33-37.
28. Mathews-Roth, M.M. (1982) Effect of crocetin on experimental skin tumors in hairless mice. Oncology 39:362-364.
29. Mathews-Roth, M.M. (1981) Medical application and uses of carotenoids. In Carotenoid Chemistry and Biochemistry, G. Britton and T.W. Goodwin, eds. Pergamon Press, New York, pp. 297-307.
30. Mathews-Roth, M.M., D. Hummel, and C. Crean (1977) The carotenoid content of various organs of animals administered large amounts of beta-carotene. Nutr. Rep. Int. 16:419-423.
31. Mathews-Roth, M.M., T. Wilson, E. Fujimori, and N.I. Krinsky (1974) Carotenoid chromophore length and protection against photosensitization. Photochem. Photobiol. 19:217-222.

32. Mathews-Roth, M.M., M.A. Pathak, T.B. Fitzpatrick, L.C. Harber, and E.H. Kass (1974) β-Carotene as an oral photoprotective agent in erythropoietic protoporphyria. J. Am. Med. Assoc. 228:1004-1008.
33. Mathews, M. (1963) Studies on the localization, function and formation of the carotenoid pigments of a strain of Mycobacterium marinum. Photochem. Photobiol. 2:1-8.
34. Mathews, M., and W.R. Sistrom (1959) Intracellular location of carotenoid pigments and some respiratory enzymes in S. lutea. J. Bacteriol. 78:778-787.
35. Musajo, L., G. Rodighiero, G. Caporale, F. Dall'acqua, S. Marciani, F. Bordin, F. Bacchichetti, and R. Bevilacqua (1974) Photoreactions between skin-photosensitizing furocoumarins and nucleic acids. In Sunlight and Man, M.A. Pathak. L.C. Harber, M. Seiji, and A. Kukita, eds. University of Tokyo Press, Tokyo, pp. 369-387.
36. Parrish, J., T. Fitzpatrick, L. Tannenbaum, and M. Pathak (1974) Photochemotherapy of psoriasis with oral methoxypsoralen and long-wave ultraviolet light. New Engl. J. Med. 291:1207-1222.
37. Pathak, M.A., and P.C. Joshi (1984) Production of active oxygen species ($^1O_2$ and O_2^-) by psoralens and ultraviolet radiation (320-400 nm). Biochim. Biophys. Acta 798:115-126.
38. Pathak, M.A., and D.M. Kramer (1969) Photosensitization of skin in vivo by furocoumarins (psoralens). Biochim. Biophys. Acta 195:197-206.
39. Peto, R., R. Doll, J.D. Buckley, and M.B. Sporn (1981) Can dietary β-carotene materially reduce human cancer rates? Nature 291:201-208.
40. Poppe, W., and L.I. Grossweiner (1975) Photodynamic sensitization by 8-methoxypsoralen via the singlet oxygen mechanism. Photochem. Photobiol. 22:217-219.
41. Potapenko, A.Ya., M.V. Moshnin, A.A. Krasnovsky, Jr., and V.L. Sukhorukov (1982) Dark oxidation of unsaturated lipids by the photoxidized 8-methoxypsoralen. Z. Naturforsch. 37c:70-74.
42. Potapenko, A.Ya., G.A. Abiev, and F. Pliquett (1980) Inhibition of erythema of the skin photosensitized with 8-methoxypsoralen by α-tocopherol. Bull. Exp. Biol. Med., USSR 89:611-615.
43. Santamaria, L., L. Bianchi, R. Pizzala, A. Bianchi, and P. Bermond (1984) Antimutagenic action of two carotenoids on mutagenicity induced by 8-methoxypsoralen and long UV irradiation in Salmonella typhimurium TA 102. In Proceedings of the 9th International Congress on Photobiology and 12th Annual Meeting of the American Society for Photobiology, Philadelphia, Pennsylvania, Abstract No. MPM-C4, p. 22S.
44. Scheel, L.D., V.B. Perone, R.L. Larkin, and R.E. Kupel (1963) The isolation and characterization of two phototoxic furanocoumarins (psoralens) from diseased celery. Biochemistry 2:1127-1131.
45. Scotto, J., T.R. Fears, and J.F. Fraumeni, Jr. (1981) Incidence of Nonmelanoma Skin Cancer in the United States, National Institutes of Health Publication No. 82-2433.
46. Seifter, E., G. Rettura, and J. Padawer (1982) Moloney murine sarcoma virus tumors in CBA/J mice: Chemopreventive and chemotherapeutic action of supplemental β-carotene. J. Natl. Cancer Inst. 68:835-840.
46a. Shapiro, S.S., and A. Bendich (1985) Effect of dietary carotenoids on lymphocyte responses to mutagens. Fed. Proc. 45:775.
47. Singh, H., and J.A. Vadasz (1978) Singlet oxygen: A major reactive species in the furocoumarin photosensitized inactivation of E. coli ribosomes. Photochem. Photobiol. 28:539-545.
48. Som, S., M. Chatterjee, and M.R. Banerjee (1984) β-Carotene inhibition of 7,12-dimethylbenz(a)anthracene-induced transformation of murine mammary cells in vitro. Carcinogenesis 5:937-940.

49. Song, P.S., M.L. Harter, T.A. Moore, and W.C. Herndon (1971) Luminescence spectra and photocycloaddition of the excited coumarins to DNA bases. Photochem. Photobiol. 14:521-530.
50. Spikes, J.D. (1975) Porphyrins and related compounds as photodynamic sensitizers. Ann. N.Y. Acad. Sci. 244:496-508.
51. Vedaldi, D., F. Dall'acqua, A. Gennaro, and G. Rodighiero (1983) Photosensitized effects of furocoumarins: The possible role of singlet oxygen. Z. Naturforsch. 38c:866-869.
52. Veronese, F.M., O. Schiavon, R. Bevilacqua, R. Bordin, and G. Rodighiero (1982) Photoinactivation of enzymes by linear and angular furocoumarins. Photochem. Photobiol. 36:25-30.
53. Veronese, F.M., O. Schiavon, R. Bevilacqua, and G. Rodighiero (1979) Drug-protein interaction: 8-Methoxypsoralen as photosensitizer of enzymes and amino acids. Z. Naturforsch. 34c:392-396.
53a. Wamer, W., A. Giles, and A Kornhauser (1985) Accumulation of dietary β-carotene in the rat. Nutr. Rep. Int. 32:295-301.
54. Willett, W.C., and B. MacMahon (1984) Diet and cancer--An overview. New Engl. J. Med. 310:633-638.
55. Willett, W., B. Polk, B. Underwood, M. Stampfer, S. Pressel, B. Rosner, J. Taylor, K. Schneider, and C. Hames (1984) Relation of serum vitamins A and E and carotenoids to the risk of cancer. New Engl. J. Med. 310:430-434.
56. Zaynoun, S.T., B.E. Johnson, and W. Frain-Bell (1977) A study of oil of bergamot and its importance as a phototoxic agent. Brit. J. Dermatol. 96:475-482.

ANTICARCINOGENIC AND OTHER PROTECTIVE EFFECTS OF DITHIOLTHIONES

Ernest Bueding, Sherry Ansher,* and Patrick Dolan

Department of Immunology and Infectious Disease
School of Hygiene and Public Health
Johns Hopkins University
Baltimore, Maryland 21205

In this chapter we discuss a class of chemicals that protect against carcinogenesis, radiation, and injury by a variety of toxic agents. These compounds are dithiolthiones (Fig. 1). They are characterized by a pentacyclic ring, consisting of two adjacent sulfurs with a third sulfur attached to a carbon, and an aromatic ring attached to another carbon. As found in collaboration with Dr. Wattenberg, a single dose of a dithiolthione called oltipraz protects against the carcinogenic effects of benzo(α)pyrene (BP) in mice. When oltipraz was administered 24 hr before BP, the number of malignancies in the lung was reduced from 23.4 to 4.6, and the number in the forestomach from 5.1 to 2.8. This protection lasted for at least two days because, if oltipraz was administered 48 hr (instead of 24 hr) before BP, there was a similar reduction of tumors in both the lung and the forestomach.

The anticarcinogenic effects of oltipraz are not limited to BP, but extend also to another potent carcinogen, aflatoxin B. Shortly after the intraperitoneal administration of radioactively labeled aflatoxin B to rats, the presence of DNA-aflatoxin B adducts is demonstrable in the liver and the kidney of these animals. According to recent findings by Kensler et al. (12), the concentration of these adducts in rats maintained on a diet containing 0.1% oltipraz is reduced by 76% in the liver and by 64% in the kidney. Furthermore, the concentration of aflatoxin B is reduced significantly in these tissues. Some antioxidants [tert. butyl-hydroxyanisole (BHA) and ethoxyquin] have similar effects, but considerably higher concentrations are required. Also, oltipraz is less toxic, as well as more potent, than BHA or ethoxyquin.

In a preliminary study with Dr. Wattenberg, the protective activity of oltipraz against two other carcinogens was studied. Again, after only a

---

* Present address: Section on Enzymes and Cellular Metabolism, National Institute of Arthritis, Diabetes, Digestive Disorders and Kidney Diseases, National Institutes of Health, Bethesda, Maryland 20205.

single dose of oltipraz, there were significant reductions in numbers of carcinomas in the lung and forestomach induced by uracil mustard and diethylnitrosamine. It is likely that greater protection against these two carcinogens can be obtained after optimal dosage schedules and other conditions of administration have been developed.

The anticarcinogenic effects of oltipraz, as well as those of phenolic antioxidants, are associated with an enhancement of biochemical defense mechanisms. Several years ago it was found in our laboratory that administration of some antioxidants such as BHA results in an increase in glutathione levels in a variety of tissues (4). Dithiolthiones have the same effects. On a molar basis, oltipraz and other dithiolthiones are more potent than BHA, except on the upper jejunal mucosa; on the other hand, the dithiolthione 129L is far more effective in raising glutathione levels in the upper jejunal mucosa than is BHA.

In some instances, but by no means all, differences in the effects of various dithiolthiones are quite pronounced. In the case of 129L, the marked response of the upper jejunal mucosa is contrasted by a much lower increase in hepatic glutathione levels produced by this compound. A chemoprotective effect of increased glutathione levels could be brought about by direct scavenging or trapping of electrophilic carcinogens or their metabolites by this nucleophilic tripeptide. Glutathione is involved also in the removal of hydroperoxides serving as a substrate for glutathione peroxidase.

In addition, glutathione is a substrate for a number of enzymes catalyzing the inactivation and subsequent excretion of toxic compounds such as carcinogens. These enzymes are the glutathione transferases (10). The conjugation of electrophilic compounds such as carcinogens with glutathione catalyzed by these enzymes, or their binding of these xenobiotics, can provide a powerful means for protection against such compounds. The products of glutathione transferases are thioethers, and these are metabolized further to water soluble acetylated conjugates with cysteine, called mercapturic acids, that are excreted in the urine. For the transferase activities, chlorodinitrobenzene (CDNB) and dichloronitrobenzene (DCNB) were used as substrates.

As shown by Benson et al. (6), administration of BHA in the diet results in marked increases in the activities of glutathione transferases. Only a single oral dose of either oltipraz or of other dithiolthiones is sufficient to produce marked increases in the activities of glutathione transferases in liver, lung, upper jejunal mucosa, and kidney (3). In the liver, depending on the dithiolthione used, 4- to 10-fold increases in enzyme activities were observed. In the lung these effects were much smaller, with intermediate elevations of enzymatic activities in the other two tissues. Pearson et al. (15) have shown that the elevated levels of one of the glutathione transferase activities induced by BHA, as well as by oltipraz, are mediated by a 20- to 30-fold increase in messenger RNA coding for a major glutathione transferase.

Another enzyme whose activities are markedly enhanced by frequent administrations of BHA (5) and by one or, at the most, two doses of dithiolthiones, is quinone reductase. This enzyme catalyzes the reduction of quinones, some of which are formed by the oxidative metabolism of phenolic compounds including some carcinogens. This reduction facilitates their detoxication by subsequent conjugation. Among the dithiolthiones tested, 129L proved to be the most active. Administration of a single dose of this

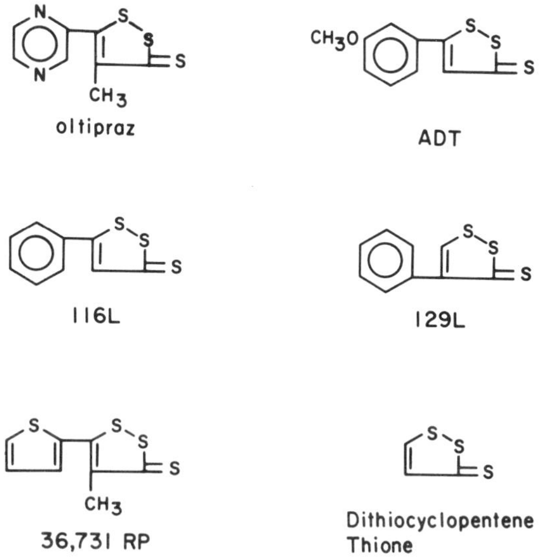

oltipraz

ADT

116L

129L

36,731 RP

Dithiocyclopentene Thione

Fig. 1. Structures of some dithiolthiones.

dithiolthione resulted in a 9-fold increase in activity of quinone reductase in the liver and almost a 7-fold increase in the upper intestinal mucosa.

Elevated glutathione levels could, at least in part, be accounted for by increased glutathione reductase activity. Administration of dithiolthiones consistently gave rise to elevations in glutathione reductase activities. However, they were less pronounced than the increases observed with glutathione transferase and with quinone reductase activities.

The obligatory substrate used for the enzymatic reduction of oxidized glutathione is NADPH. Reduction of nicotinamide adenine dinucleotide phosphate (NADP) can be brought about by the activities of glucose-6-phosphate and of 6-phosphogluconate dehydrogenases. The activities of both these enzymes are increased by the administration of dithiolthiones. In this manner more substrate is provided for the generation of reduced glutathione.

It appears that the nature of the substituent aromatic ring is not critical to the biochemical properties of a dithiolthione, because no marked differences were detected whether a pyrazine, a phenol, or a thiophene ring was attached to the dithiolthione (for structures, see Fig. 1). As a matter of fact, even the unsubstituted dithiolthione had considerable activity in increasing the levels of glutathione and of detoxication enzymes. However, it was far more toxic. Therefore, the major function of the aromatic side chain seems to be to reduce the toxicity of dithiolthiones.

It is noteworthy that biochemical effects of dithiolthiones on mouse and rat tissues are not observable in tumors. Administration of even high doses of dithiolthiones did not increase glutathione levels or protective enzyme activities in murine mammary adenoma transplants. With one dithiolthione, 116L, there was even a slight, but significant, decrease; nor did the protective enzymes of this tumor respond to dithiolthiones administered to the host. Conversely, the same enzymes, glutathione transferase, glutathione reductase, quinone reductase, and 6-phosphogluconate dehydrogenase, in the liver and the lung of the tumor-bearing animals exhibited the usual dithiolthione-induced increases in their activities (3). Therefore, the protective effects of these compounds on mammalian tissues are likely to be selective and should not alter the effectiveness of the treatment of malignancies by anticancer drugs or X-rays.

The dithiolthione-induced increases in glutathione levels and in protective enzyme activities suggested that pretreatment with them might accelerate the detoxication of xenobiotics generating electrophilic metabolites. Two model compounds were chosen to study this problem: carbon tetrachloride and acetaminophen. Pretreatment with dithiolthiones significantly reduced the mortality of mice that had been challenged with either carbon tetrachloride or acetaminophen (2). This protective effect was demonstrable also when the hepatotoxic action of acetaminophen was monitored, with the aid of liver function tests, by determination of serum transaminases or sorbitol dehydrogenase activities. Administration of carbon tetrachloride or of acetaminophen markedly depleted the glutathione stores of the liver, but this was at least partially reversed by treatment with dithiolthiones. Therefore, it is likely that this reversal can increase the interaction of electrophilic metabolites of carbon tetrachloride and of acetaminophen with glutathione. This could account for the chemoprotective effects of dithiolthiones. In addition, or alternatively, the toxic metabolites may be conjugated with glutathione because of increased levels of glutathione transferase activities induced by dithiolthiones. Either or both of these effects can account for the chemopreventive actions of dithiolthiones.

Based mainly on epidemiological evidence, consumption of cruciferous vegetables, such as cabbage and cauliflower, is associated with a reduction in the incidences of cancer in man (8,14). Twenty-eight years ago, two Czech investigators (11) reported that cruciferous vegetables contain dithiolthiones. Accordingly, the presence of one or of several dithiolthiones in these vegetables can account, at least in part, for the inhibition of carcinogenesis brought about by these foods, although other vegetable constituents such as indoles and isothiocyanates may also play a role.

If there is a relationship between cancer prevention and protective biochemical effects, feeding of cabbage should produce biochemical actions similar to those induced by dithiolthiones. Examination of this problem is complicated by the presence in commercial rodent food of antioxidants such as BHA and ethoxyquin. Therefore, the levels of tissue glutathione and of protective enzyme activities were compared in mice fed a commercial rodent food (Purina) with those in mice on a semisynthetic (AIN-US Biochemicals) antioxidant- free diet. Regardless of the tissues analyzed and of the enzymatic activities measured, both glutathione levels and enzymatic activities from animals fed the commercial Purina diet were higher than those from mice on the semisynthetic diet. Therefore, the elimination of food additives supplied with commercial rodent diets can provide a better background for evaluating the biochemical effects of putative protective agents.

When lyophilized cabbage was fed to mice on a semisynthetic diet for 28 days, the glutathione levels of the tissues became elevated. There was a progressive increase with increasing cabbage concentration in the diet. Similarly, there were dose-related increases in the activities of protective enzymes, such as glutathione transferase, glutathione reductase, quinone reductase, and phosphogluconate dehydrogenase. Boyd et al. (7) have reported that feeding of cabbage or of cauliflower reduced aflatoxin B-induced carcinogenesis in rats. Therefore, both the anticarcinogenic and the biochemical effects of dithiolthiones are mimicked by the presence of cabbage in the diet.

On the other hand, the anticarcinogenic effects of vegetables have been ascribed to the presence of beta-carotene; however, feeding of beta-carotene to mice on a semisynthetic diet, either equivalent to its concentration in a diet containing 20% cabbage or 4 times more, failed to produce biochemical changes induced by cabbage or by dithiolthiones.

In order to mimic the mode of ingestion of dithiolthiones by man consuming cruciferous vegetables, mice were fed low doses of dithiolthiones by including them in the diet for 10 days. Originally, a concentration of 0.05% was fed. The effects of such a diet on glutathione levels and some enzyme activities were compared with those of a single dose administered orally. These low dietary concentrations of dithiolthiones produced significant elevations of glutathione levels and induced some enzyme activities. The elevations were approximately equal to those induced by the single dose of oltipraz found to produce protection against carcinogenic effects.

Presently, we are testing the effects of an even lower concentration, that is, 0.02%. Preliminary data suggest that this lower concentration is still effective. Once the minimum dietary concentration producing the biochemical effects has been established, animals on such a daily diet will be exposed to carcinogens, radiation, and some toxic compounds to determine whether or not any protective effects are detectable.

The association of protective with biochemical effects of dithiolthiones raises the question whether a given biochemical event is causally related to a protective effect. This is recognizable when, in a given tissue, the administration of a dithiolthione is responsible for only a single biochemical event. So far, this has been observed only rarely. In one instance, there was a close correlation between reduction in DNA-aflatoxin B adducts by oltipraz or by phenolic antioxidants, and an induction of a major glutathione transferase activity (12). On the other hand, no such correlation was detectable with other biochemical effects, such as increases in glutathione levels, or in the activities of other enzymes. Therefore, in the case of aflatoxin B, conjugation of the carcinogen and/or of its metabolite(s) with glutathione appears to account for the major chemoprotective effect of oltipraz.

In another instance, a different correlation is observable. Whole body irradiation leads to depletion of hematopoietic spleen cells. Under given conditions of dosage and timing, administration of oltipraz inhibits markedly the depletion of hematopoietic splenic stem cells. While under these conditions no increased protective enzyme activities are detectable, there is a close correlation between increased spleen colony formation and elevated glutathione levels. Therefore, in this case the protective effect of oltipraz may be related directly to increased glutathione concentrations, rather than to activation of an enzymatic reaction.

In yet a third instance, the marked protective and biochemical actions of dithiolthiones in some tissues are contrasted by the failure of oltipraz to exert any biochemical effects on the pancreas. This is consistent with a complete lack of protection by this dithiolthione against azaserine-induced pancreatic tumors (Roebuck et al., unpubl. data). Further studies are required to elucidate the biochemical basis of most chemoprotective and radioprotective effects of dithiolthiones.

The evidence available so far indicates that dithiolthiones are endowed with low mammalian toxicity. Two of these compounds have been administered already to a fairly large number of human subjects: oltipraz for the treatment of schistosomiasis (9), and anetholdithiolthione (ADT) in France and Canada to counteract the inhibition of salivation exerted by antidepressant drugs (13,16).

Dithiolthiones have potential for inhibiting the development of secondary tumors after treatment of malignancies by chemotherapy or radiation, but many more studies are needed before the precise mechanisms of their protective actions can be fully understood, and before the properties of these compounds can be used safely for the protection of man against injury by chemical and physical agents.

ACKNOWLEDGEMENTS

These studies were supported in part by grants from Biothrust and the American Cancer Society (Sig 3).

REFERENCES

1. Ali, H.-A., M.A. Homeida, S.M. Sulaiman, and S.M. Bennett (1984) Diet control blood levels of oltipraz in male subjects. J. Antimicrob. Chemotherapy 13:465.
2. Ansher, S., P. Dolan, and E. Bueding (1983) Chemoprotective effects of two dithiolthiones and of butylhydroxy-anisole against carbon tetrachloride and acetaminophen toxicity. Hepatology 3:932.
3. Ansher, S., P. Dolan, and E. Bueding (1985) Biochemical effects of dithiolthiones. Food Chem. Toxicol. (in press).
4. Batzinger, R., S. Ou, and E. Bueding (1978) Antimutagenic effects of 2(3)-tert-butyl-4-hydroxyanisole and of antimicrobial agents. Cancer Res. 38:4478.
5. Benson, A., M. Hunkeler, and P. Talalay (1980) Increase of NAD(P)H:-quinone reductase by dietary antioxidants: Possible role in protection against carcinogenesis and toxicity. Proc. Natl. Acad. Sci., USA 77:5216.
6. Benson, A., R. Batzinger, S. Ou, E. Bueding, Y.-N. Cha, and P. Talalay (1978) Elevation of hepatic glutathione-S-transferase activities and protection against mutagenic metabolites of benzo(α)pyrene by dietary antioxidants. Cancer Res. 38:4486.
7. Boyd, J.N., J.G. Babish, and G.S. Stoewesand (1982) Modification by beet and cabbage diets of aflatoxin B1 induced rat plasma alphafoetoprotein elevation, hepatic tumorigenesis and mutagenicity of urine. Food Chem. Toxicol. 20:47.
8. Colditz, G.A., L.G. Branch, R.J. Lipnick, W.C. Wilson, B. Rosner, B.M. Posner, and C.H. Hennekens (1985) Increased green and yellow vegetable intake and lowered cancer deaths in an elderly population. Am. J. Clin. Nutr. 41:32.
9. Gentilini, M., M. Brucker, G. Danis, G. Niel, and G. Charmot (1979) Premiers essays therapeutiques chez l'homme de l'antibilharzien 35 972. R.P. Bull. Soc. Pathol. Exot. 72:466.

10. Jakoby, W. (1977) The glutathione S-transferases: A group of multi-functional detoxification proteins. Adv. Enzymol. 46:383.
11. Jirousek, L., and J. Starka (1958) Uber das vorkommen von trithionen (1,2-dithiacyclopent-4-en-3-thione) in Brassicapflanzen. Naturwiss. 45:386.
12. Kensler, T.W., P.A. Egner, M.A. Trush, E. Bueding, and M.D. Groopman (1985) Modification of aflatoxin B1 binding to DNA in vivo in rats fed phenolic antioxidants and a dithiolthione. Carcinogenesis 6:759.
13. Lelord, G., C. Mercat, and J. Fuseiller (1969) Interet du trithiopara-methoxy phenylpropene (sulfarlem) dans la prevention des complications salivaires des traitments psychotropes. Gaz. Med. France 76:2257.
14. National Academy of Sciences (1982) Diet, Nutrition and Cancer, National Academy Press, Washington, D.C.
15. Pearson, W.R., J.J. Windle, J.F. Morrow, A.M. Benson, and P. Talalay (1983) Increased synthesis of glutathione-S-transferase in response to anticarcinogenic antioxidants. Cloning of messenger RNA. J. Biol. Chem. 258:2050.
16. Schiano, P.J. (1985) Bouches seches: Epidemiologie et efficacite du sulfarlem S25 multicentrique nationale portant sur 526 cas. Tribune Med. 127:61.

ANTIOXIDANTS AS ANTITUMOR PROMOTERS

Walter J. Kozumbo and Peter A. Cerutti

Department of Carcinogenesis
Swiss Institute for Experimental Cancer Research
1066 Epalinges s, Lausanne, Switzerland

INTRODUCTION

Phorbol 12-myristate 13-acetate (PMA) is a potent mouse skin tumor promoter which has been researched most intensively over the years in an attempt to understand the biochemical mechanisms involved in tumor promotion. PMA stimulates a multitude of cellular and biochemical reactions (82). Three of these, induction of ornithine decarboxylase (ODC) activity, hyperplasia and inflammation, correlate well with tumor promotion in mouse skin, and each is considered to be a necessary but insufficient event in this process (22,27,51,58,64,70,83). The release of metabolites of arachidonic acid (AA) is also thought to be important in view of the inhibitory effects of nonsteroidal anti-inflammatory drugs on PMA-induced ODC activity and tumor promotion (19,48).

For several years now evidence has accumulated indicating that promoters may act at least in part via the induction of a cellular prooxidant state (9,35). Characterized by increases in intra- or extracellular levels of active oxygen (AO) species (O_2^-, HO•, $^1O_2^*$, and H_2O_2), radicals and organic peroxides and their degradation products, this prooxidant state would effectively enhance the rate and number of tumors that are formed in tissue. Because of difficulties in measuring in vivo intracellular concentrations of these species, much of the current evidence linking prooxidant states with tumor promotion stems from investigations where antioxidants have antagonized either promotion itself or biochemical and biological events elicited by tumor promoters. Thus, effects that are induced by promoters and inhibited by antioxidants potentially serve to delineate mechanisms of action through which prooxidants mediate their tumor-promoting activities. Some of these activities and their possible relationships to tumor promotion are discussed in this chapter.

PROOXIDANT STATES ASSOCIATED WITH TUMOR PROMOTION

Antioxidants Inhibit Tumor Promotion in Mouse Skin

Slaga and co-workers (68) have shown that, in mouse epidermis, PMA and anthralin, a nonphorbol-ester tumor promoter, cause a rapid and sustained decrease in the activities of superoxide dismutase (SOD) and catalase. These antioxidant enzymes are in part responsible for preventing the accumulation of toxic intracellular levels of superoxide anion (O_2^-) and hydrogen peroxide (H_2O_2), respectively. Also, in mouse skin, Goldstein et al. (25) presented evidence that PMA, in fact, does elevate levels of H_2O_2. Kensler and co-workers (36) postulated that if tumor promotion in mouse epidermis is related to increases in intracellular superoxide anions (as a result of PMA-inhibited SOD activity), then the resupplementation of superoxide dismutating activity to such cells could conceivably modulate tumor growth. Indeed, the exogenous addition to mouse skin of a lipophilic biomimetic SOD, copper (II) (3,5-diisopropylsalicylic acid)$_2$ (CuDIPS), inhibited PMA-induced ODC activity and, as shown in Tab. 1, tumor formation as well. In experiments by Friedman and Cerutti (21), scavengers of AO (SOD, catalase, and mannitol) displayed similar effects by antagonizing PMA-induced ODC activity in mouse mammary tumor cells.

It has been well established that O_2^- can give rise to species of higher reactivity, such as perhydroxyl ($HO_2\cdot$) and hydroxyl ($HO\cdot$) radicals, which in turn can initiate lipid-peroxidizing free-radical chain reactions (54, 75). Such prooxidant stress can alter the cellular redox state and reduce levels of intracellular glutathione (GSH), leading to both subtle and gross forms of cytopathology (67). This kind of action may also bear some relationship to tumor promotion (59). In fact, a recent report showed that PMA causes a decrease in ratios of reduced to oxidized GSH in mouse epidermal

Tab. 1. Effects of CuDIPS on tumor promotion in mouse skin.

Group	Treatment[a] Modifier (μmol)	Treatment[a] PMA (nmol)	Number: Mice alive at 20 weeks	Number: Papillomas per mouse at 20 weeks	Number: Mice with papillomas at 20 weeks
1	CuDIPS (0.5)	PMA (4)	18	6.9[b]	17
2	CuDIPS (2.0)	PMA (4)	18	0.9[b]	6[c]
3	Diethyl ether	PMA (4)	18	12.3	16
4	Cupric acetate (2.0)	PMA (4)	17	11.3	16
5	DIPS (4.0)	PMA (4)	17	7.5	13
6	Acetone and H_2O	PMA (4)	17	9.4	15
7	Diethyl ether	Acetone	18	0.0	0
8	CuDIPS (2.0)	Acetone	18	0.0	0

[a] Shaven female CD-1 mice at 18 per group were initiated with 0.2 μmol of 7,12-dimethylbenz(α)anthracene dissolved in 0.2 ml acetone. After 10 days, mice received topical applications of PMA (4.0 nmol) twice weekly for 20 weeks. Mice were also treated with either 0.5 or 2.0 μmol of CuDIPS, 4.0 μmol of 3,5-diisopropylsalicylic acid (DIPS), or 2.0 μmol of cupric acetate 30 min before each tumor-promoting agent application. Control mice received the appropriate vehicle only. The number and incidence of papillomas were recorded weekly.

[b] Differs from control ($P < 0.05$ for group 1 and $P < 0.01$ for group 2) by Duncan's new multiple-range test on data transformed by $\sqrt{x+1}$ to stabilize the variance. Krushkal-Wallis test followed by Dunn's multiple comparison procedure gave essentially the same results.

[c] Differs from control ($P < 0.01$) by χ^2 test.

Note: Data reproduced from Ref. 36 with permission of Science, Vol. 221, 1983.

cells and that this decrease is prevented by the antioxidants GSH, cysteine, and α-tocopherol. In the same system these antioxidants inhibit PMA-induced ODC activity as well as tumor growth (55). Furthermore, the free-radical scavengers butylated hydroxyanisole (BHA) and butylated hydroxytoluene (BHT), which are well-known inhibitors of lipid peroxidation, also inhibit PMA-induced tumor promotion in mouse skin (65).

In a study (Tab. 2) which approximates the relative antitumor-promoting activities of 8 BHA analogs by measuring their capacity to inhibit PMA-induced ODC activity in mouse skin, it was found that, with the exception of methyl-BHA, both the antioxidant potential (as assessed by inhibition of lipid peroxidation) and the lipophilic nature of each analog correlate favorably with its relative antitumor-promoting activity (41,42). (It should be pointed out that the anomalous inhibition of ODC activity by the nonantioxidant analog methyl-BHA is possibly explained by its in vivo O-demethylation to BHA.) Such a correlative analysis indicates that important determinants of the antitumor-promoting activity of a compound would include not only the antioxidant character of that compound, but also its lipophilicity and potential for being metabolized to active or inactive forms.

A direct demonstration of the involvement of free radicals and/or peroxides in tumor promotion occurred when benzoyl peroxide, lauroyl peroxide, and several other radical-free generating compounds, including H_2O_2 (albeit to a weak extent), were found to promote tumor growth in mouse skin (39,65, 66). Not surprisingly, benzoyl peroxide-induced tumor promotion was also antagonized by BHA and BHT (65). Taken together, these studies suggest

Tab. 2. Inhibition of PMA-induced ODC activity and ascorbate-initiated lipid peroxidation by butylated hydroxyanisole (BHA) analogs.

BHA Analog	ID_{50} (μmol)[a] PMA-induced ODC activity in mouse skin	IC_{50} (μM) Nonenzymatic lipid peroxidation	LogP Octanol:water partition coefficient
t-Butylhydroquinone	2	0.1	2.1
3-t-Butyl-4-hydroxyanisole	5	0.2	1.9
2-t-Butyl-hydroxyanisole	7	0.2	1.9
2-t-Butylphenol	15	15.0	1.5
p-Hydroquinone	50	28.0	0.2
4-Hydroxyanisole	50	47.0	0.6
Phenol	nonactive[b]	200.0	0.8
2-t-Butyl-1,4-dimethoxybenzene (methyl-BHA)	18	nonactive	1.7

[a] Inhibition of PMA-induced ODC activity by BHA analogs was determined in the following manner. Various analogs were topically applied to shaven mice 30 min prior to PMA (17 nmol) treatment. At the time of peak ODC activity (7 hr post-PMA application), mice were sacrificed and ODC activity was determined as described by Kozumbo et al. (41). Inhibitors of lipid peroxidation (as measured by malondialdehyde formation) in mouse microsomes were determined as described by Kozumbo et al. (42).

[b] No activity at 50 μM or 50 μmol.

Note: Data reproduced from Ref. 41 and 42 with permission of <u>Cancer Research</u>, Vol. 43, 1983, and <u>Chem. Biol. Interact.</u>, Vol. 40, 1985.

that PMA causes an increase in intracellular AO and free radicals which are chemically reduced and inactivated by antioxidants or, alternatively, react with other biological molecules that mediate processes related to tumor promotion.

Antioxidants Inhibit In Vitro Tumor-Promoting Activities

A number of in vitro experiments further attest to the antitumor-promoting action of antioxidants. For example, Colburn and co-workers (50) established an in vitro model system for late-stage tumor promotion that uses JB-6 mouse epidermal cells which transform to anchorage-independent growth in agar upon PMA administration. In this system, PMA-induced transformation is potently inhibited by SOD and CuDIPS and, to a lesser but significant degree, by other antioxidants such as n-propylgallate, tannic acid, and BHA. On the other hand, catalase and GSH peroxidase are ineffective as inhibitors.

Other experiments by Borek and Troll (6), using primary cultures of hamster embryo cells, demonstrated that transformation resulting from X-ray and bleomycin treatments is also inhibited by SOD but not by catalase. Moreover, the PMA-induced enhancement of transformation is completely eliminated by SOD, even when added together with PMA at 24 hr postirradiation. The authors contend that the inhibitory action of SOD during later stages of in vitro transformation is remarkably similar to the inhibition by antioxidants of PMA-induced promotion in mouse skin.

A direct demonstration of the promoting activity of AO was given by Zimmermann and Cerutti (85). They showed that O_2^-, which is generated enzymatically by xanthine/xanthine oxidase, can essentially substitute for PMA and promote the morphological transformation of initiated cells C3H10T1/2. This effect is also inhibited by SOD.

These in vitro results strengthen the argument that AO plays a significant role in promoting tumor growth and, furthermore, stress the importance of O_2^- in the promotional process. The antitumor-promoting effects of antioxidants such as BHA, n-propylgallate, α-tocopherol, and GSH suggest the involvement in tumor promotion of peroxides or of other free radicals which are possibly derived in secondary reactions from O_2^-.

CLASTOGENIC AND PARACRINE ACTIONS OF PMA IN TUMOR PROMOTION

Paracrine Actions of PMA-Stimulated Inflammatory Cells in Tumor Promotion

There are multiple mechanisms by which xenobiotics can create a prooxidant state in epithelial cells themselves (9,35). Alternatively, they may be exposed to AO, organic hydroperoxides and radicals which were generated in the promoter-treated tissue by infiltrated leukocytes. Such a model was set forth initially by Goldstein, Troll, and co-workers, and later by others (4,15,23,79) who argued that persistent oxidative damage to epidermal DNA, via the intermediacy of PMA-stimulated release of AO by inflammatory cells, plays a role in the promotional process. Several interesting observations support such a possibility.

(a) Since all mouse skin tumor promoters are inflammatory agents, inflammation has been recognized as a necessary, although insufficient, event in the process of tumor promotion (64).

(b) All phorbol ester tumor promoters are inflammatory agents which cause phagocytic leukocytes to release AO in a respiratory burst whose magnitude is generally found to be proportional to the tumor-promoting activity of the corresponding phorbol ester (23).

(c) Inhibitors of the respiratory burst, such as retinoic acid, protease inhibitors, BHA, BHT, CuDIPS, and others (23,24,34,37,42), are all antitumor promoters (36,56,65,76,77).

(d) In cells co-cultured with PMA-stimulated leukocytes, AO causes cytotoxicity, mutagenicity, and DNA damage (11,44,80,81).

Leukocytes treated with PMA cause various forms of DNA damage, such as single strand breaks in co-cultured epidermal cells (11), ring-saturated thymines in fibroblasts (44), and sister chromatid exchanges (SCEs) in hamster ovarian cells (78). The suppression of this genetic damage by the antioxidants catalase and/or SOD suggests that AO mediates the damage. Work by Cerutti, Emerit, and co-workers (14,15) has shown that PMA-treated monocytes release diffusible lipophilic clastogenic factors (CFs) whose formation is also inhibited by SOD and catalase (15,17). This finding offered the interesting possibility that genetic damage may result from the formation of secondary oxidative products rather than directly from AO itself.

Under conditions of CF formation, human monocytes were found to release O_2^-, H_2O_2, thromboxane B_2, prostaglandin $F_2\alpha$ ($PGF_2\alpha$), prostaglandin E_2 (PGE_2), 12-hydroxy-5,8,10-heptadecatrienoic acid (HHT), hydroxy- and hydroperoxy-derivatives of AA, free AA, and small amounts of mono- and diacylglycerols and phospholipids (40). It is believed that in the presence of O_2^- and H_2O_2 this mixture of lipids undergoes free-radical chain reactions, initiating autoxidation in the target cell and ultimately inducing chromosomal damage. The fact that SOD inhibits CF formation (15,17) further suggests that superoxide-reduced iron, reacting with H_2O_2 (7) in a superoxide-driven Fenton reaction to form HO•, plays a critical role in this process. Lymphocytes, which normally produce neither active oxygen nor CFs, will form CFs on prolonged exposure to active oxygen from an exogenous enzymatic source. This AO-induced formation of CFs correlates with the appearance of cellular lipid peroxides and is inhibited by SOD, catalase, GSH peroxidase, and scavengers of HO• (16).

The hydroxy-AA metabolites released by PMA-treated monocytes (Fig. 1) were derived from their corresponding hydroperoxy-precursors, i.e., 5-, 11-, and 15-HETE from 5-, 11-, and 15-HPETE; thromboxane B_2 and PGE_2 from the corresponding peroxide intermediate prostaglandin G_2 (PGG_2). In vivo, these hydroperoxy-metabolites are thought to be sufficiently stable to exert their effects on nearby cells (46) and perhaps to mediate DNA damage. Using alkaline elution analysis, we in fact recently demonstrated that a variety of hydroperoxy-derivatives of AA and AA itself cause DNA strand breaks in mouse fibroblasts (52). The relative order of potency was determined to be hydroperoxy-AA > hydroxy-AA > AA. The damage depended on the presence of calcium and, unlike CF-mediated clastogenicity, the damage was not inhibited by antioxidants. These observations suggest an indirect and unique mode of action which may not, in a simple way, be related to a free-radical mechanism. (The relationship between prooxidant-induced DNA damage and tumor promotion is discussed in a following section.)

Active oxygen and AA metabolites from monocytes may also exert non-clastogenic tumor-promoting effects on neighboring cells. For example, because of the essential roles reportedly played by PGE_2, PGF_2, and various lipoxygenase products in mouse skin tumor promotion (19,22,48), it is of interest that PMA-stimulated monocytes not only release these AA metabolites themselves, but also produce free radical-generating peroxides, such as 15-HPETE, PGG_2, and H_2O_2, which could trigger the further formation of prostaglandins in neighboring target cells. Lands and co-workers (29) have demonstrated that peroxides are required in order to initiate and maintain the catalytic activity of purified fatty-acid cyclooxygenase. In their system, 15-HPETE and PGG_2, as well as the tumor-promoting compounds H_2O_2, t-butylhydroperoxide, and cumene peroxide (39,65,66), can each activate the enzymatic catalysis of AA to prostaglandins (28,29). Apparently this activation is not just limited to enzymes in the purified form, since cultured cells readily respond to H_2O_2 and t-butylhydroperoxide with the synthesis of prostaglandins E_2 and $F_2\alpha$ (1,73).

Furthermore, the inhibition of this peroxide-induced enzymatic activation by various antioxidants, including the antitumor-promoting compounds BHT (28) and GSH (10), suggests both a peroxide-dependent free-radical mechanism of catalysis (28,29) and a potential site of action for antitumor-promoting antioxidants. The involvement of inflammatory cells in creating such a "peroxide tone" and in activating enzymes of the AA cascade was originally proposed by Lands and co-workers (28) and could represent part of an overall prooxidant spectrum of action in the promotional process.

In view of the above discussion, it may be relevant to note that tumor promotion in initiated mouse skin is caused and enhanced by endoperoxide analogs (U-46619 and U-44069) and $PGF_2\alpha$ (19,45), respectively, and that ODC activity and DNA synthesis in the colonic mucosa of rats are induced by both hydroxy- and hydroperoxy-derivatives of AA (8). It is not known whether the DNA-damaging effects of these organic peroxides are related to the induction of ODC activity and DNA synthesis or to the stimulation of tumor growth. Nevertheless, with chronic application of PMA to mouse skin, it appears likely that AO and other diffusible oxidation products originating from leukocytes could exert their actions on neighboring epidermal cells. Supporting a probable role for inflammation in tumor promotion is the interesting observation that the skin of an athymic nude mouse fails to demonstrate an inflammatory or hyperplastic response to PMA and, at the same time, is resistant to the PMA promotional protocol unless grafted with a normal thymus (30).

Direct Prooxidant Action of PMA on Target Cells

Although PMA may facilitate tumor growth indirectly through its effects on leukocytes, it also has direct prooxidant effects on target cells themselves. For example, SOD suppresses the in vitro PMA-induced formation of chromosomal aberrations in pure lymphocytes (essentially free of phagocytic leukocytes) (17), of SCEs in hamster ovary cells (49), and of transformation to anchorage-independent growth in JB-6 cells (50). Thus, effects of PMA apparently can be mediated by the production of O_2^- from cells other than phagocytic leukocytes. If SOD were not capable of penetrating cells, then one could reasonably conclude that its scavenging and, therefore, its anticlastogenic and antipromoting activities were working at the cell surface. Because of evidence indicating that native bovine Cu-SOD can both penetrate fairly rapidly (15 min) to intracellular cytosol and associate with the membranous cellular fraction (47), one can exclude neither the lipophilic nor the aqueous intracellular compartments as possible sites of action and/or generation of O_2^-. However, in the event that O_2^- acts at the level of the plasma membrane, it does not necessarily follow that the cell

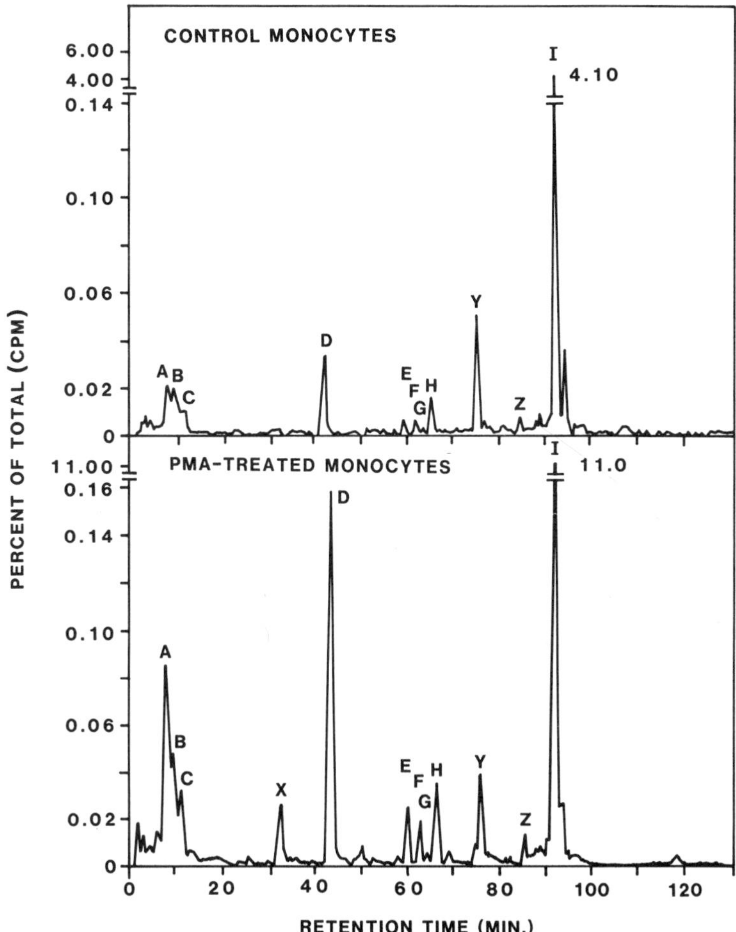

Fig. 1. Representative HPLC profiles of arachidonic acid (AA) metabolites released from control and PMA-treated human monocytes. Human monocytes (essentially platelet-free) were radiolabeled with [^{3}H]AA, washed and incubated in serum-free medium in the absence or presence of PMA (30 ng/ml) for 1 hr at 37°C. Ethyl acetate extractions were evaporated and resuspended in 0.35 ml methanol for HPLC analysis modified slightly from that described by Eling et al. (13). AA metabolites were identified by ^{14}C internal standards or by the retention times of nonradiolabeled standards absorbing light at 214 nm. The amount of AA metabolites is expressed as percent of total cpms contained in cells plus culture medium. Identity of peaks: A, thromboxane B_2; B, $PGF_{2}\alpha$; C, PGE_2; D, HHT; E, 15-HETE; F, 11-HETE; G, 12-HETE; H, 5-HETE; I, AA; X, Y, and Z, unidentified.

surface exclusively generates superoxide in response to PMA. The intracellular production of univalently reduced oxygen by mitochondria and endoplasmic reticulum (9,35), for example, coupled with the PMA-induced decrease in SOD and catalase activities (68), could allow AO to reach the cell surface.

Fischer and Adams (20) recently presented indirect evidence that PMA-treated mouse epidermal cells do indeed produce AO as well as hydroperoxy-AA metabolites. In view of such evidence, one may consider the pro-oxidant actions of PMA in mouse skin to be attributed to the effects of phorbol esters on epidermal cells themselves. However, since the actual phenomenon of tumor promotion essentially occurs only in situ, and since monocyte/macrophage products possess tumor-promoting activities, promotion may result both from the direct interaction of promoter with target cell as well as from the indirect paracrine and clastogenic effects of inflammatory cell products.

Prooxidant-Induced DNA Damage and Tumor Promotion

To date, the role of DNA damage by prooxidant promoters in the process of tumor promotion remains unclear. There is some suggestion that the formation of relatively large numbers of random DNA breaks could be involved in modulating gene expression and in altering cellular growth and differentiation (18,21,33,38,72). For example, it was found that the induced differentiation of Friend erythroleukemic cells by agents such as ultraviolet light, butyrate, and dimethyl sulfoxide is associated with DNA single-strand breaks (57), and, furthermore, that cells that are refractory to dimethyl sulfoxide-induced differentiation are also resistant to DNA breakage (74). If modulation of cellular development can result from DNA damage by various agents, then it follows that the DNA breaks produced directly or indirectly by prooxidant compounds with tumor-promoting activity [such as PMA, anthralin, benzoyl peroxide, hydrogen peroxide, hydroperoxy AA, and sodium metaperiodate (3,26,52,69)] may likewise serve to regulate the growth state of initiated cells.

Studies by Yuspa and co-workers (26) have shown that PMA does indeed produce DNA strand breaks in cultured mouse epidermal cells, but that the damage was only specific for a differentiating subpopulation. The nondifferentiating population, representing the actual target cells in tumor promotion, was resistant to strand breaks (26), but not to the proliferating effects of PMA (82). The authors argue that strand breakage was the effect rather than the cause of PMA-induced differentiation and, furthermore, suggest that shifts in the differentiating and proliferating subpopulations rather than DNA breaks are more closely aligned with the tumor-promoting activity of PMA.

The failure of SOD and catalase in this system to inhibit either DNA breaks or terminal differentiation most likely dissociates these direct PMA-induced effects in epidermal cells from the antitumor-promoting action of antioxidants. This dissociation further implies the existence of other prooxidant tumor-promoting pathways that can be modulated by antioxidants but which are unrelated to the clastogenic action of PMA on differentiating cells. That the proliferating subpopulation is indeed sensitive to the DNA-breaking effects of a tumor-promoting peroxide (benzoyl peroxide) but not to those of PMA, may indicate that still other peroxides, including those derived from PMA-stimulated inflammatory cells, could also cause clastogenicity in the proliferating subset. In this case, however, since

PMA alone can stimulate proliferation (82), the leukocyte-mediated DNA damage would be expected to fulfill some additional (and as yet undefined) requirement of tumor promotion.

Could it be that in vivo damage to the DNA of proliferating cells is also of importance in promoting tumor growth, and that the inhibitory action of antioxidants is to suppress the formation or action of the leukocyte mediators of this damage? Or, on the other hand, are the antioxidants responsible for suppressing in this proliferating population some nonclastogenic, but vital, component of the promotional process? The recent observation by Singh et al. (60) that a prooxidant state mediates the induction of poly(ADP)-ribosylation of nuclear proteins in response to PMA, as discussed in the following section, addresses this latter question and provides a model for PMA action at the genomic level.

ANTIOXIDANT-INHIBITABLE POLY(ADP)-RIBOSYLATION OF NUCLEAR PROTEINS BY PMA

The proliferating and differentiating states of cells are affected by prooxidant promoters such as PMA (26,82), and these states require the expression of specific genes. Therefore, it is of central interest to determine the ways in which gene expression can be modulated by PMA and by AO in general.

The post-translational modification of proteins represents one possible mechanism for the modulation of gene expression. PMA has already been shown to cause the phosphorylation of many proteins, including the receptors for epidermal growth factor (31) and insulin (71) and the chromatin proteins H2B and H4 (53).

Another type of post-translational protein modification, namely poly-(ADP)-ribosylation, is of special interest because it specifically involves chromosomal proteins and, therefore, could provide the cell with a mechanism for modulating gene expression via changes in chromatin conformation. There is evidence to implicate poly(ADP) ribose in DNA repair (12,32), cell differentiation, proliferation (2,18,33,38,72), and malignant transformation (5,43). Since the alteration of the cellular redox state is related both to poly(ADP)-ribosylation and to tumor promotion by PMA, we determined the effect of PMA on the poly(ADP)-ribosylation of nuclear proteins.

It was found that PMA induces the poly(ADP)-ribosylation of nuclear proteins in human and mouse fibroblasts (C3H10T1/2) (Fig. 2) as well as in human monocytes (61). In mouse fibroblasts the nuclear proteins serving as acceptors included the core histones H2B, A24, H3d, and smaller amounts of H2A/H3 and H1 (62). PMA also induces the poly(ADP)-ribosylation of numerous medium and high molecular weight nonhistone proteins. According to western blot analysis with antibody directed against ADP ribosyl transferase, the major acceptors are ADP ribosyl transferase itself and its proteolytic fragments of 20-25, 45, and 72-95 kDa (63). That the response, which is optimal at 3 hr in mouse cells, is suppressed by inhibitors of RNA and protein synthesis indicates a requirement for macromolecular synthesis (60).

The potent inhibition of the response by the antioxidants SOD, catalase, GSH peroxidase and BHT, as shown in Tab. 3, underscores an oxidative mechanism of action in the mediation of this nuclear effect. The failure to detect DNA strand breakage (with either alkaline elution or alkaline

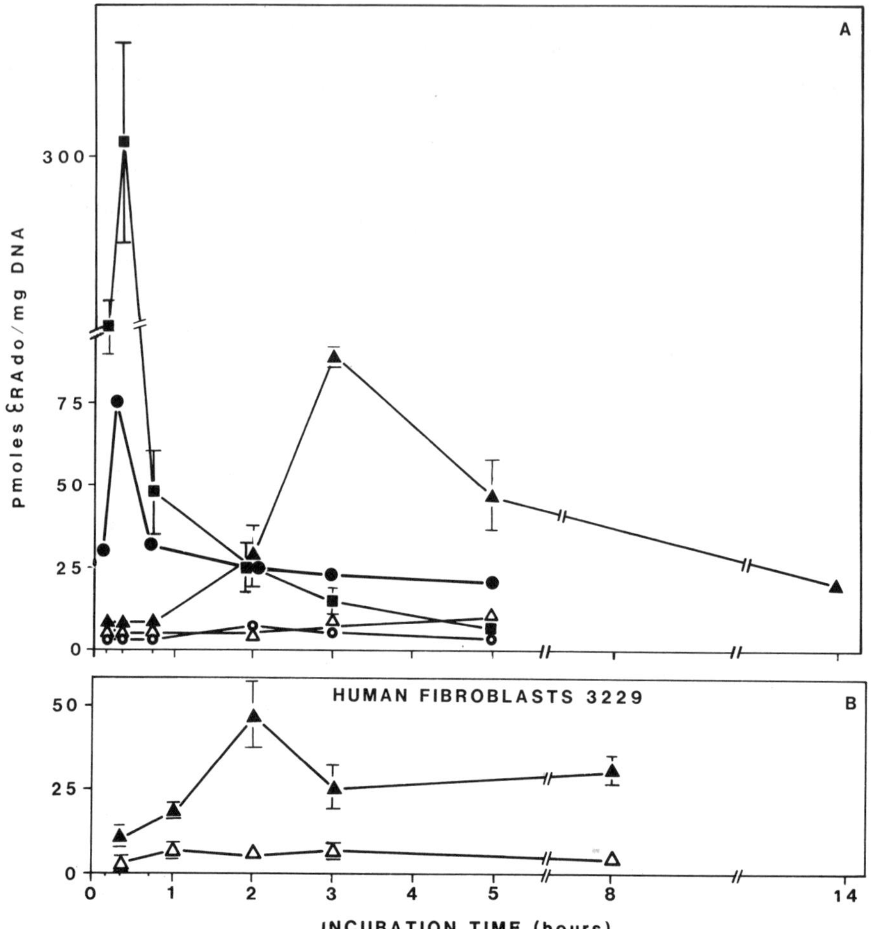

Fig. 2. Kinetics of poly(ADP) ribosyl accumulation in fibroblasts treated with PMA or MNNG. Intracellular content of poly(ADP) ribose following treatment of fibroblasts with PMA (25 ng/ml) and MNNG (136 μM). Each value is a mean of 2 separate experiments, each performed in duplicate. (A) Mouse embryo fibroblasts C3H 10T1/2; (B) human fibroblasts 3229. (Δ) Control; (▲) 25 ng/ml PMA; (o) control; (■) 20 μg/ml MNNG; (•) 5 μg/ml MNNG. Data reproduced from Ref. 60.

unwinding assays) indicates that this response was unrelated to PMA clastogenic activity. Since NAD is the substrate of poly(ADP) ribosyl transferase, the levels of intracellular oxidized pyridine nucleotides usually diminish when poly(ADP) ribosyl activity is stimulated by alkylating agents such as N-methyl-N'-nitro-N-nitrosoguanidine (MNNG). In contrast, PMA failed to elicit any change in NAD concentrations. In fact, in comparison to the response induced by PMA, an alkylating agent such as MNNG is also faster acting, does not require de novo macromolecular synthesis, produces detectable DNA breaks, and is not inhibited by antioxidants (60). These differences reflect a mode of action by PMA that is distinct from alkylating agents and that involves a prooxidant state and probably gene

Tab. 3. Effects of antioxidants on the accumulation of poly(ADP) ribose in PMA- and MNNG-treated C3H10T1/2 cells.

Antioxidants	PMA	MNNG	Poly(ADP) ribose	% Inhibition
-	-	-	9.2 ± 5.0	-
-	+	-	79.0 ± 3.9	0
SOD	+	-	16.0 ± 4.8	90
CAT	+	-	20.7 ± 5.4	83
GSH-P	+	-	17.0 ± 2.2	89
BHT	+	-	25.7 ± 11.9	76
Heated CAT	+	-	82.0 ± 9.8	0
Heated GSH-P	+	-	55.1 ± 4.1	34
Heated BSA	+	-	75.9 ± 12.5	4
-	-	-	9.5 ± 3.1	-
-	-	+	315.2 ± 31.7	0
SOD	-	+	287.9 ± 55.0	9
CAT	-	+	315.6 ± 66.3	0
GSH-P	-	+	273.4 ± 66.2	13

Note: Mouse C3H10T1/2 cells were preincubated for 30 min with the above antioxidants SOD (25 μg/ml), catalase (CAT) (50 μg/ml), GSH-P (0.3 μU/ml), or BHT (10 μM) prior to the addition of PMA (25 ng/ml) or MNNG (20 μg/ml). Poly(ADP) ribose levels (εRAdo pmol/mg DNA) were measured as described by Singh et al. (60) after 3 hr of PMA treatment or 20 min of MNNG treatment. BSA = bovine serum albumin. Mean values with standard deviations of 2 experiments carried out in duplicate are given. Data reproduced from Ref. 60 with permission EMBO J., Vol. 4, 1985.

activation. This antioxidant-inhibitable post-translational modification of chromosomal proteins suggests a mechanism for the modulation of chromatin structure and gene expression that may be related to the prooxidant tumor-promoting action of PMA.

ACKNOWLEDGEMENTS

The original work reported in this chapter was supported by grants from the Swiss National Science Foundation and from the Swiss Association of Cigarette Manufacturers.

REFERENCES

1. Ager, A., and J.L. Jordon (1984) Differential effects of hydrogen peroxide on indices of endothelial cell function. J. Exp. Med. 159:592-603.
2. Althaus, F., S.D. Lawrence, Y.Z. He, G.L. Sattler, Y. Tsukda, and H.C. Pitot (1982) Effects of altered (ADP-ribose) metabolism on expression of fetal functions by adult lymphocytes. Nature 300:366-368.

3. Birnboim, H.C. (1983) Importance of DNA strand-break damage in tumor promotion. In Radioprotectors and Anticarcinogens, O.F. Nygaard and M.G. Simic, eds. Academic Press, New York, pp. 539-556.
4. Birnboim, H.C. (1982) DNA strand breakage in human leukocytes exposed to a tumor promoter, phorbol myristate acetate. Science 215:1247-1249.
5. Borek, C., W.F. Morgan, A. Ong, and J.E. Cleaver (1984) Inhibition of malignant transformation in vitro by inhibitors of poly (ADP-ribose) synthesis. Proc. Natl. Acad. Sci., USA 81:243-247.
6. Borek, C., and W. Troll (1983) Modifiers of free radicals inhibit in vitro the oncogenic actions of x-rays, bleomycin, and the tumor promoter 12-O-tetradecanoylphorbol-13-acetate. Proc. Natl. Acad. Sci., USA 80:1304-1307.
7. Bradley, M., and L. Erickson (1981) Comparison of the effects of hydrogen peroxide and X-irradiation on toxicity, mutation and DNA damage repair in mammalian cells. Biochim. Biophys. Acta 654:135-141.
8. Bull, A.W., N.D. Nigro, W.A. Golembieski, J.D. Crissman, and L.J. Marnett (1984) In vivo stimulation of DNA synthesis and induction of ornithine decarboxylase in rat colon by fatty acid hydroperoxides, autoxidation products of unsaturated fatty acids. Cancer Res. 44:4924-4928.
9. Cerutti, P.A. (1985) Prooxidant states and tumor promotion. Science 227:375-381.
10. Cook, H.W., and W.E.M. Lands (1976) Mechanism for suppression of cellular biosynthesis of prostaglandins. Nature 260:630-632.
11. Dutton, D.R., and G.T. Bowden (1985) Indirect induction of a clastogenic effect in epidermal cells by a tumor promoter. Carcinogenesis 6(9):1279-1284.
12. Durkacz, B.W., O. Omidiji, D.A. Gray, and S. Shall (1980) (ADP-ribose)n participates in DNA excision repair. Nature 283:593-596.
13. Eling, T., B. Tainer, A. Ally, and R. Warnock (1982) Separation of arachidonic acid metabolites by high pressure liquid chromatography. In Methods of Enzymology, Vol. 86, W.E.M. Lands and W.L. Smith, eds. Academic Press, New York, pp. 511-517.
14. Emerit, I., and P.A. Cerutti (1982) Tumor promoter phorbol 12-myristate 13-acetate induces a clastogenic factor in human lymphocytes. Proc. Natl. Acad. Sci., USA 79:7509-7513.
15. Emerit, I., and P. Cerutti (1983) Clastogenic action of tumor promoter phorbol-12-myristate-13-acetate in mixed human leukocyte cultures. Carcinogenesis 4:1313-1316.
16. Emerit, I., S.H. Khan, and P.A. Cerutti (1985) Treatment of lymphocyte cultures with a hypoxanthine-xanthine oxidase system induces the formation of transferable clastogenic material. J. Free Radicals in Biol. and Med. 1(1):51-58.
17. Emerit, I., and P. Cerutti (unpublished observations).
18. Farzaneh, F., R. Zalin, D. Brell, and S. Shall (1982) DNA strand breaks and ADP-ribosyl transferase activation during cell differentiation. Nature 300:362-366.
19. Fischer, S.M. (1985) Arachidonic acid metabolism and tumor promotion. In Arachidonic Acid Metabolism and Tumor Promotion, S.M. Fischer and T.J. Slaga, eds. Martinus Nijhoff Publishing, pp. 22-47.
20. Fischer, S.M., and L.M. Adams (1985) Suppression of tumor promoter-induced chemiluminescence in mouse epidermal cells by several inhibitors of arachidonic acid metabolism. Cancer Res. 45:3130-3136.
21. Friedman, J., and P. Cerutti (1983) The induction of ornithine decarboxylase by phorbol 12-myristate 13-acetate or by serum is inhibited by antioxidants. Carcinogenesis 4:1425-1427.

22. Furstenberger, G., and F. Marks (1985) Prostaglandins, epidermal hyperplasia and skin tumor promotion. In Arachidonic Acid Metabolism and Tumor Promotion, S.M. Fischer and T.J. Slaga, eds. Martinus Nijhoff Publishing, pp. 50-72.
23. Goldstein, B.D., G. Witz, M. Amoruso, D.S. Stone, and W. Troll (1981) Stimulation of human polymorphonuclear leukocyte superoxide anion radical production by tumor promoters. Cancer Lett. 11:257-262.
24. Goldstein, B.D., G. Witz, M. Amoruso, and W. Troll (1979) Protease inhibitors antagonize the activation of polymorphonuclear leukocyte oxygen consumption. Biochem. Biophys. Res. Commun. 88:854-860.
25. Goldstein, B., G. Witz, J. Zimmerman, and C. Gee (1983) Free radicals and reactive oxygen species in tumor promotion. In Oxy Radicals and their Scavenger Systems, Vol. II, Cellular and Medical Aspects, R.A. Greenwald and G. Cohen, eds. Elsevier, New York, pp. 321-325.
26. Hartley, J.A., N.W. Gibson, L.A. Zwelling, and S.H. Yuspa (1985) Strand breaks in mouse epidermal DNA are associated with phorbol ester induced differentiation. Proc. Am. Assoc. Cancer Res. 26:42.
27. Hecker, E. (1978) Structure-activity relationships in diterpene esters irritant and cocarcinogenic to mouse skin. In Carcinogenesis--A Comprehensive Survey, Vol. II, Mechanisms of Tumor Promotion and Cocarcinogenesis, T.J. Slaga, A. Sivak, and R.K. Boutwell, eds. Raven Press, New York, pp. 11-48.
28. Hemler, M.E., H.W. Cook, and W.E.M. Lands (1979) Prostaglandin biosynthesis can be triggered by lipid peroxides. Arch. Biochem. Biophys. 193(2):340-345.
29. Hemler, M.E., and W.E.M. Lands (1980) Evidence for a peroxide-initiated free-radical mechanism of prostaglandin biosynthesis. J. Biol. Chem. 255(13):6253-6261.
30. Holland, J.M., E.H. Perkinds, and L.C. Gipson (1977) Resistance of germ free athymic nude mice to two-stage epidermal carcinogenesis. Proc. Am. Assoc. Cancer Res. 18:10.
31. Iwashita, S., and C. Fox (1984) Epidermal growth factor and potent phorbol tumor promoters induce epidermal growth factor receptor phosphorylation in a similar but distinctively different manner in human epidermoid carcinoma A431 cells. J. Biol. Chem. 259:2559-2567.
32. Jacobson, E.L., K.M. Antol, H. Juarez-Salinas, and M.K. Jacobson (1983) Poly (ADP-ribose) metabolism in ultraviolet-irradiated human fibroblasts. J. Biol. Chem. 258:103-107.
33. Johnstone, A.P., and G.T. Williams (1982) Role of DNA breaks and ADP-ribosyl transferase activity in eukaryotic differentiation demonstrated in human lymphocytes. Nature 300:368-370.
34. Kensler, T.W. (personal communication).
35. Kensler, T.W., and M.A. Trush (1984) Role of oxygen radicals in tumor promotion. Environ. Mutag. 6:593-616.
36. Kensler, T.W., D.M. Bush, and W.J. Kozumbo (1983) Inhibition of tumor promotion by a biomimetic superoxide dismutase. Science 221:75-77.
37. Kensler, T.W., and M.A. Trush (1983) Inhibition of oxygen radical metabolism in phorbol ester-activated polymorphonuclear leukocytes by antitumor-promoting copper complex with superoxide dismutase-mimetic activity. Biochem. Pharmacol. 32(22):3485-3487.
38. Kidwell, W.R., and M.G. Mage (1976) Changes in poly (adenosine diphosphate ribose) and poly (adenosine diphosphate) polymerase in synchronous Hela cells. Eur. J. Biochem. 15:1213-1217.
39. Klein-Szanto, A.J.P., and T.J. Slaga (1982) Effects of peroxides on rodent skin: Epidermal hyperplasia and tumor promotion. Invest. Derma. 79:30-34.
40. Kozumbo, W.J., D. Muhlematter, A. Jorg, I. Emerit, and P.A. Cerutti (1985) Release of membrane lipids from human monocytes stimulated by

the tumor promoter phorbol-12-myristate-13-acetate (submitted for publ.).
41. Kozumbo, W.J., J.L. Seed, and T.W. Kensler (1983) Inhibition by 2(3)-tert-butyl-4-hydroxyanisole and other antioxidants of epidermal ornithine decarboxylase activity induced by 12-O-tetradecanoylphorbol-13-acetate. Cancer Res. 43:2555-2559.
42. Kozumbo, W.J., M.A. Trush, and T.W. Kensler (1985) Are free radicals involved in tumor promotion? Chem. Biol. Interact. 40:199-209.
43. Kun, E., E. Kirsten, G.E. Milo, P. Kurian, and H.L. Kumari (1983) Cell cycle-dependent intervention by benzamide of carcinogen-induced neoplastic transformation and in vitro poly ADP-ribosylation of nuclear proteins in human fibroblasts. Proc. Natl. Acad. Sci., USA 80:7219-7223.
44. Lewis, J.G., and D.O. Adams (1985) Induction of 5,6-ring-saturated thymine bases in NIH-3T3 cells by phorbol ester-stimulated macrophages: Role of reactive oxygen intermediates. Cancer Res. 45:1270-1275.
45. Lupulescu, A. (1984) Tumorigenic potential of endoperoxide analogs. Experientia 40:209-211.
46. Marnett, L.J. (personal communication).
47. Michelson, A.M., K. Puget, P. Durosay, and A. Rosselet (1980) Penetration of erythrocytes by superoxide dismutase. In Biological and Clinical Aspects of Superoxide and Superoxide Dismutase, W.H. Bannister and J.V. Bannister, eds. Elsevier, New York, pp. 348-366.
48. Nakadate, T., S. Yammamoto, M. Ishii, and R. Kato (1982) Inhibition of 12-O-tetradecanoylphorbol-13-acetate-induced epidermal ornithine decarboxylase activity by lipooxygenase inhibitors: Possible role of product(s) of lipoxygenase pathway. Carcinogenesis 3:1411-1414.
49. Nagasawa, H., and J.B. Little (1981) Factors influencing the induction of sister chromatid exchanges in mammalian cells by 12-O-tetradecanoyl-phorbol-13-acetate. Carcinogenesis 2:601-607.
50. Nakamura, Y., N.H. Colburn, and T.D. Gindhart (1985) Role of reactive oxygen in tumor promotion: Implication of superoxide anion in promotion of neoplastic transformation in JB-6 cells by TPA. Carcinogenesis 6:229-235.
51. O'Brien, T.G. (1976) The induction of ornithine decarboxylase as an early, possibly obligatory, event in mouse skin carcinogenesis. Cancer Res. 36:2644-2653.
52. Ochi, T., and P.A. Cerutti (unpublished observations).
53. Patskan, G., and C. St. Baxter (1985) Specific stimulation by phorbol esters of the phosphorylation of histones H2B and H4 in murine lymphocytes. Cancer Res. 45:667-672.
54. Pederson, T.C., and S.D. Aust (1973) The role of superoxide and singlet oxygen in lipid peroxidation promoted by xanthine oxidase. Biochem. Biophys. Res. Commun. 52:1071-1078.
55. Perchellet, J.P., M.D. Owen, T.D. Posey, D.K. Orten, and B.A. Schneider (1982) Inhibitory effects of glutathione level-raising agents and D-tocopherol on ornithine decarboxylase induction and mouse skin tumor promotion by 12-O-tetradecanoylphorbol-13-acetate. Carcinogenesis 6:567-573.
56. Rossman, T.G., and W. Troll (1980) Protease inhibitors in carcinogenesis, possible sites of action. In Modifiers of Chemical Carcinogenesis, Vol. 5, T.J. Slaga, ed. Raven Press, New York, pp. 127-143.
57. Scher, W., and C. Friend (1978) Breakage of DNA and alterations in folded genomes by inducers of differentiation in Friend erythroleukemia cells. Cancer Res. 38:841-849.
58. Schmidt, R., W. Adolf, A. Marston, H. Roser, B. Sorg, H. Fujiki, T. Sugimura, R.E. Moore, and E. Hecker (1983) Inhibition of specific

binding of ^{3}H phorbol-12,13-dipropionate to an epidermal fraction by certain irritants and irritant promoters of mouse skin. Carcinogenesis 4:77-81.

59. Shamberger, R.J. (1972) Increase of peroxidation in carcinogenesis. J. Natl. Cancer Inst. 48(5):1491-1497.
60. Singh, N., G. Poirier, and P. Cerutti (1985) Tumor promoter phorbol-12-myristate-13-acetate induces poly (ADP)-ribosylation in fibroblasts. EMBO J. 4:1491-1494.
61. Singh, N., G. Poirier, and P. Cerutti (1985) Tumor promoter phorbol-12-myristate-13-acetate induces poly ADP-ribosylation in human monocytes. Biochem. Biophys. Res. Commun. 126:1208-1214.
62. Singh, N., and P.A. Cerutti (1985) Poly ADP-ribosylation of histones in tumor promoter phorbol-12-myristate-13-acetate treated mouse embryo fibroblasts C3H10T1/2 (submitted for publ.).
63. Singh, N., Y. Leduc, G. Poirier, and P. Cerutti (1985) Non-histone chromosomal protein acceptors for poly ADP-ribose in phorbol-12-myristate-13-acetate-treated mouse embryo fibroblasts C3H10T1/2. Carcinogenesis (in press).
64. Slaga, T.J., S.M. Fischer, A. Viaje, D.L. Berry, W.M. Bracken, S. Le Clerc, and D.R. Miller (1978) Inhibition of tumor promotion by anti-inflammatory agents: An approach to the biochemical mechanism of promotion. In Carcinogenesis, Vol. 2, Mechanisms of Tumor Promotion and Cocarcinogenesis, T.J. Slaga, A. Sivak, and R.K. Boutwell, eds. Raven Press, New York, pp. 173-195.
65. Slaga, T.J., V. Solanki, and M. Logani (1983) Studies on the mechanisms of action of antitumor promoting agents: Suggestive evidence for the involvement of free radicals in promotion. In Radioprotectors and Anticarcinogens, O.F. Nygaard and M.G. Simic, eds. Academic Press, New York, pp. 471-485.
66. Slaga, T.J., A.J.P. Klein-Szanto, L.L. Triplett, L.P. Yotti, and J.E. Trosko (1981) Skin tumor-promoting activity of benzoyl peroxide, a widely used free radical-generating compound. Science 213:1023-1025.
67. Smith, M.T., H. Thor, S.A. Jewell, G. Bellomo, M.S. Sandy, and S. Orrenius (1984) Free radical-induced changes In the surface morphology of isolated hepatocytes. In Free Radicals in Molecular Biology, Aging and Disease, D. Armstrong, R.S. Sohal, R.G. Cutler, and T.F. Slater, eds. Raven Press, New York, pp. 103-118.
68. Solanki, V., R.S. Rana, and T.J. Slaga (1981) Diminution of mouse epidermal superoxide dismutase and catalase activities by tumor promoters. Carcinogenesis 2:1141-1146.
69. Srinivas, L., and N.H. Colburn (1984) Preferential oxidation of cell surface sialic acid by periodate leads to promotion of transformation in JB-6 cells. Carcinogenesis 5:515-519.
70. Suganuma, M., H. Fujiki, T. Tomoko, C. Cheuk, R.E. Moore, and T. Sugimura (1984) Estimation of tumor-promoting activity and structure-function relationships of aplysiatoxins. Carcinogenesis 5:315-318.
71. Takayma, S., M. White, V. Lauris, and C. Kuhn (1984) Phorbol esters modulate insulin receptor phosphorylation and insulin action in cultured hepatoma cells. Proc. Natl. Acad. Sci., USA 81:7797-7801.
72. Tanigawa, Y., A. Kitamura, M. Kawamura, and M. Shimoyama (1978) Effect of poly (ADP-ribose) formation on DNA synthesis in chick-embryo-liver nuclei. Eur. J. Biochem. 92:261-269.
73. Taylor, L., M.J. Menconi, and P. Polgar (1983) The participation of hydroperoxides and oxygen radicals in the control of prostaglandin synthesis. J. Biol. Chem. 258(11):6855-6857.
74. Terada, M., U. Nudel, E. Fibach, R.A. Rifkind, and P.A. Marks (1978) Changes in DNA associated with induction of erythroid differentiation

by dimethyl sulfoxide in murine erythroleukemia cells. Cancer Res. 38:835-840.

75. Thomas, M.J., S. Mehl, and W.A. Pryor (1978) The role of the superoxide anion in the xanthine oxidase-induced autoxidation of linoleic acid. Biochem. Biophys. Res. Commun. 83:927-932.

76. Verma, A.K., C.L. Astendel, and R.K. Boutwell (1980) Inhibition by prostaglandin synthesis inhibitors of the induction of epidermal ornithine decarboxylase activity, the accumulation of prostaglandins, and tumor promotion caused by 12-O-tetradecanoylphorbol-13-acetate. Cancer Res. 40:308-315.

77. Verma, A.K., T.J. Slaga, P.W. Wertz, G.C. Mueller, and R.K. Boutwell (1980) Inhibition of skin tumor promotion by retinoic acid and its metabolites 5,6-epoxyretinoic acid. Cancer Res. 40:2367-2371.

78. Weitberg, A.B., S.A. Weitzman, E.P. Clark, and T.P. Stossel (1985) Effects of antioxidants on oxidant-induced sister chromatid exchange formation. J. Clin. Invest. 75:1835-1841.

79. Weitberg, A.B., S.A. Weitzman, M. Destrempes, S.A. Latt, and T.P. Stossel (1983) Stimulated human phagocytes produce cytogenetic changes in cultured mammalian cells. New Eng. J. Med. 308:26-30.

80. Weiss, S.J., A.F. Lobuglio, and H.B. Kessler (1980) Oxidative mechanisms of monocyte-mediated cytotoxicity. Proc. Natl. Acad. Sci., USA 77:584-587.

81. Weitzman, S.A., and T.P. Stossel (1982) Effects of oxygen radical scavengers and antioxidants on phagocyte-induced mutagenesis. J. Imm. 128:2770-2772.

82. Yuspa, S.H. (1984) Mechanisms of tumor promotion. In Tumor Promotion and Carcinogenesis In Vitro ("Mechanisms of Tumor Promotion," Vol. 3), T.J. Slaga, ed. CRC Press, Boca Raton, Florida, pp. 1-11.

83. Yuspa, S.H., U. Lichti, T. Ben, E. Patterson, and H. Hennings (1976) Phorbol-esters stimulate DNA synthesis and ornithine decarboxylase activity in mouse epidermal cell cultures. Nature 262:402-404.

84. Yuspa, S.H., T. Ben, H. Hennings, and U. Lichti (1982) Divergent responses in epidermal basal cells exposed to the tumor promoter 12-O-tetradecanoylphorbol-13-acetate. Cancer Res. 42:2344-2349.

85. Zimmermann, R., and P. Cerutti (1984) Active oxygen acts as a promoter of transformation in mouse embryo C3H10T1/2C18 fibroblasts. Proc. Natl. Acad. Sci., USA 81:2085-2087.

FUTURE DIRECTIONS FOR RESEARCH IN ANTIMUTAGENESIS AND ANTICARCINOGENESIS

INTRODUCTION: FUTURE DIRECTIONS FOR RESEARCH IN ANTIMUTAGENESIS AND ANTICARCINOGENESIS

Elizabeth K. Weisburger

Division of Cancer Etiology
National Cancer Institute
Bethesda, Maryland 20205

The initial experiment demonstrating an anticarcinogenic effect, but under drastic conditions, was reported by Berenblum in 1929. Sulfur mustard treatment decreased the incidence of skin tumors in mice resulting from subsequent application of coal tar [Berenblum, I. (1929) J. Path. Bact. 32:425-434]. A biochemical explanation has not been provided for this phenomenon, but the mustard may have cross-linked the cellular DNA so that further attachment by the polycyclic aromatic hydrocarbons in the tar was prevented.

Further research in carcinogenesis disclosed that there were many situations where the effects of carcinogens were inhibited. These included: (a) administration of a structurally similar noncarcinogen, implying that some site on a cellular receptor was involved; (b) inhibition of metabolic synthesis of toxic (or reactive) intermediates; (c) induction of detoxifying enzymes; (d) hormonal means, either by ablation of an endocrine organ or by administering a drug which led to atrophy of an endocrine organ; and (e) inhibiting the synthesis of cellular constituents. Several of these points have been discussed in these proceedings.

Reports on the anticarcinogenic and antimutagenic activity of natural constituents in ordinary foodstuffs have been presented here. Broccoli, brussels sprouts, and similar vegetables have a protective effect against various carcinogens. Plant flavones or similar substances have even protected against the liver carcinogen, aflatoxin B_1. There are even natural antimutagens against the potent mutagens formed during pyrolysis of proteins, a beneficial action.

We have also learned of various sophisticated techniques to study the genetic effects of carcinogens and mutagens. Conversely, knowledge in this area will eventually facilitate means to inhibit or prevent the action of such compounds. We now look forward to coming trends in research on anticarcinogenic and antimutagenic agents, the subject of the following chapters.

INHIBITORS OF MUTAGENESIS AND THEIR RELEVANCE TO CARCINOGENESIS*

Claes Ramel

Wallenberg Laboratory
University of Stockholm
S106 91 Stockholm, Sweden

The International Commission for the Protection against Environmental Mutagens and Carcinogens (ICPEMC) is an international forum, founded in 1977, with the aim to provide an unbiased scientific basis for the evaluation of the genetic and carcinogenic effects of exposure to chemicals. The ICPEMC has acted by means of different subgroups to which experts in different fields have been selected. One such subgroup has dealt with inhibitors of mutagenesis and their relevance to carcinogenesis (21).

An essential background for the work of this expert subgroup has been the relationship between mutagenicity and carcinogenicity. The somatic mutation theory of cancer induction has received particularly strong support during the last few years through the analyses of oncogenes, which have provided powerful support for the notion that mutagenic alterations of DNA and chromosomes are intimately involved with the carcinogenic process.

Mutations, as a step in the sequence of events leading to cancer, have been connected primarily with initiation of cancer. However, promotion may also involve effects at the DNA level through radical formation, as suggested by Cerutti (6), or through gene amplification, as suggested by Varshavsky (28) and Hayashi et al. (11a).

The fact that mutagenic events seem to be of critical importance in carcinogenicity is of relevance when interpreting the background of human cancer. It has been estimated that at least 75% of cancer cases in humans are caused by environmental factors (9), and it can be assumed that mutagenic factors are of importance in this context. It is, however, unlikely that exposure to mutagenic agents is the only matter of significance in this connection. Exposure to inhibitors of mutagenesis may be equally significant. These are of widespread occurrence in the human environment, comprising a part of virtually every human diet. In connection with a growing interest in chemoprevention of cancer (i.e., using chemical

* An overview of a report by an expert group to the International Commission for the Protection against Environmental Mutagens and Carcinogens.

compounds to prevent the occurrence of cancer), inhibitors of mutation and neoplasia are being identified at an accelerating rate. It is apparent that several mechanisms exist to protect organisms against the mutagenic effects of a wide variety of natural, as well as synthetic, chemical mutagens. Inhibitors of mutagenicity and carcinogenicity can be classified according to the stage during which they exert their protective effects. Two principal stages can be recognized--one implying action extracellularly and one, intracellularly. The following classification of inhibitors of mutagenesis and carcinogenesis has been adopted. The first two groups constitute the first-stage chemicals, those acting outside the cells, while the other groups of chemicals act within the cells.

CLASSIFICATION OF INHIBITORS OF MUTAGENESIS AND CARCINOGENESIS

Inhibitors of Formation of Mutagens

In this group of inhibitors, attention has been focused primarily on the formation of genotoxic nitroso compounds from precursor amines or amides and nitrite. Ascorbic acid constitutes an efficient inhibitor of that process, both in vitro and in vivo (18). Mirvish (18) also proposed that ascorbic acid may prevent the induction of gastric tumors by nitroso carcinogens in high-risk populations. Alpha-tocopherol, tannic acid, phenols, and chlorogenic acid also have the property of preventing nitrosation (19a,26,27).

As is often the case, inhibitors of mutagenicity can themselves have mutagenic and carcinogenic side effects in other combinations and under other circumstances. This applies to phenols and chlorogenic acid, according to Stich et al. (26).

Inactivators of Promutagens and Mutagens

Examples of inactivators of mutagens include various vegetable juices, which act as inhibitors of mutagenic pyrolysis products (13). Proteins have been isolated from vegetables that are active in this way. It also has been shown that tryptophan and glutamic acid pyrolysis products can be inactivated by an irreversible adsorption to vegetable fibers, which may contribute to the anticarcinogenic effect of fiber-rich diets (14).

Blocking Agents

These agents prevent mutagens from reaching or reacting with the genetic material in at least three different ways.

Inhibitors of activation of promutagens. One mechanism is the inhibition of the conversion of promutagens and procarcinogens to their ultimate reactive form. This applies to disulfiram, which inhibits colon neoplasia by 1,2-dimethylhydrazine (29). Disulfiram also has been shown to alter the organotrophy of tumors caused by nitrosamines, presumably as a result of an alteration of the metabolism of these compounds (24). Disulfiram also seems to increase the excretion of glucuronide metabolites from 2-acetylaminofluorene (2-AAF) and N-hydroxy 2-AAF (10). A reverse effect of disulfiram has, however, also been reported with 1,2-dichlorobromide, which became much more carcinogenic with disulfiram (31).

Inducers of detoxifying enzymes. A second mechanism of blocking agents is based on the induction of detoxifying enzymes. Of particular interest are compounds with a selective effect on certain enzymes, particularly those involved in the conjugation of xenobiotic agents. The prototype of such a blocking agent is butylated hydroxyanisole (BHA), which acts by increasing the activity of glutathione S-transferase and UDP-glucuronyl transferase, and by increasing synthesis of glutathione (3). In addition, BHA also enhances the activity of epoxide hydrolase and quinone reductase (i.e., the most efficient detoxifying mechanisms) (3). But BHA does not increase microsomal monooxygenase activity, although it seems to modify it (16). Wattenberg has proposed that BHA induces the enzymes via a receptor, which evidently is different from the TCDD receptor (29). BHA and related compounds are highly effective in detoxifying a wide variety of mutagens and carcinogens. This group of inhibitors of mutagenic and carcinogenic compounds may contain possible candidates for practical use in the chemoprevention of cancer. Wattenberg (30) has pointed out that the increased level of glutathione activity may be used as an assay to trace blocking agents in natural products.

Although BHA and similar compounds have promising properties from the point of view of anticarcinogenesis, these compounds may also have an opposite effect under certain circumstances. It is, for instance, known that 1,2-dichloroethane and the corresponding bromocompound are activated to highly potent mutagens by conjugation to glutathione (22), and it can be suspected that an increase of glutathione-S-transferase and glutathione by addition of BHA may cause an increased mutagenicity, although there are no experimental data yet to support this possibility. Also, butylated hydroxytoluene, which is related to BHA and has similar properties, increases the neoplastic response to urethane and 2-AAF if it is given after the carcinogen (20).

Also, inducers of the microsomal monooxygenase system with P-450 have turned out to act as inhibitors of mutagenic and carcinogenic compounds. Man is exposed to such inducers through vegetables containing inducing agents. Particularly rich in inducers are cruciferous plants, such as cabbage, and it is of interest that in a case-controlled study, Graham et al. (11) found a negative correlation between the consumption of cabbage and the occurrence of colon cancer. Although there is a connection between induction of the mixed-function oxygenase system and protection against chemical carcinogenesis, the detoxification system is indeed complex and is, of course, also involved in activation of procarcinogens. Therefore, the reverse effect of inducers of P-450 can also, a priori, be suspected to occur. A correlation between the induction of P-448 (AHH) and the induction of lung cancer has been found in mice with 3-methylcholanthrene (15a). The complexity is furthermore illustrated by the fact that inhibitors of the mixed-function oxygenase system can function both as enhancers and as inhibitors of carcinogenesis.

<u>Agents reacting directly with electrophilic metabolites</u>. A third mechanism of blocking agents is by direct reactions with electrophiles. The carcinogenic metabolites of benzo(α)pyrene provide an example. In this case, the dihydrodiol epoxide is inactivated by binding to riboflavin-5-phosphate and by binding to a phenolic compound found in various plants, ellagic acid (32,33).

Scavengers of Radicals

Although these agents could logically be included under blocking agents, we have put them in a special category, since the study of the generation of radicals and their role in mutagenesis and carcinogenesis has developed into a specialized field of high current interest. Oxygen stress and radical generation from chemicals in our environment may constitute an important hazard.

It has been emphasized by Ames (2) that many of the natural mutagens in our food may act through the generation of oxygen radicals. The generation of oxygen radicals may also contribute to the carcinogenic effect of a high-fat intake. In vitro studies on Salmonella by Ames and his group have also revealed that well-known carcinogens like benzo(α)pyrene may act largely through radical mechanisms and not only by the formation of electrophilic metabolites (8).

Since the formation of oxygen radicals is a natural process, organisms have developed elaborate endogenous and exogenous defense mechanisms. Consequently, the effects of radicals are not only a question of radical formation, but also of inhibition or enhancement of the defense system. Many agents acting in either way have been recognized. It should be stressed that natural antioxidants like tocopherol, ascorbic acid, and β-carotene can, on the basis of experimental and epidemiological data, be assumed to be important inhibitors of cancer, presumably to a large extent through protection against radicals. The correlation between low selenium intake and cancer has been interpreted to involve the role of selenium in the defense against radicals through glutathione peroxidase, which requires selenium.

Suppressing Agents

These agents prevent neoplastic expression of initiated cells. They act during carcinogenesis, subsequent to the initial attack of genotoxic compounds. Suppressing agents include several mechanisms, probably involving epigenetic effects in most cases. Some suppressing effects presumably are connected with inhibition of the arachidonic acid cascade. Indomethacin, for example, acts as an antipromoter, preventing membrane effects that lead to radical formation (7). The best studied suppressing compounds are vitamin A and its synthetic analogs, retinoids. They have been shown to suppress many carcinogen-induced neoplasias (19,23), but the inhibition is reversible and therefore presumably of an epigenetic nature. Vitamin A and retinoids, however, also modify the mutagenicity of indirect mutagens. Both inhibition and enhancement of mutagenicity have been reported (4,5).

Agents that Affect DNA Repair

Kada and Kanematsu (12) have reported that cobaltous chloride inhibits mutagens both in vitro and in vivo. In vitro experiments on Escherichia coli suggest that cobaltous chloride acts by the promotion of error-free repair by means of recA (25), which may explain the wide range of antimutagenic effects of cobaltous chloride. Cobaltous ions are essential constituents of the antimutagenic principle found in human or animal placental tissue (15). Data on protease inhibition of mutagenesis, such as by antipain and elastatinal, suggest that these substances may also act at the level of DNA repair (17).

The occurrence of inhibitors of mutagenesis. The ICPEMC report discusses the occurrence of natural inhibitors in plants from both a population point of view and from the point of view of edible plants. Alekperov (1) has found, in natural populations of oats and other plants in the USSR, that different ecotypes have different contents of vitamin E and vitamin C, and the contents of vitamins were negatively correlated with the spontaneous and induced mutation rates. He has reported that extreme habitats (high altitudes, for instance) tend to correlate with an increased content of vitamins E and C.

In edible plants a steadily increasing number of inhibitors of mutagenesis and carcinogenesis have been found, particularly through the work of Kada and his group. The presence of these compounds may be at least a contributing factor to the low cancer incidence reported for vegetarians.

PRACTICAL APPLICATIONS

In conclusion, when it comes to the practical application of agents that inhibit mutagenesis and carcinogenesis, there are two principal hypothetical possibilities. It is clear that different foods (in particular, vegetables) contain many protections against mutagenicity and carcinogenicity, and a manipulation of the intake of foods and beverages on the basis of both epidemiological and experimental evidence may lead to protection against genotoxic compounds. Of interest in this context is the use of a fiber-rich diet for chemoprevention of cancer.

The other possibility for chemoprevention is the use of pure chemicals that have been identified as protective agents. The obvious problem is the fact that a compound, which in most cases functions to prevent cancer, may under certain conditions do exactly the opposite, as has been repeatedly observed. Few, if any, chemicals can meet the requirement of enough knowledge of their biological effects to be used deliberately on people. Possible exceptions are vitamins A, C, and E, and also selenium in areas with low selenium intake. Improved understanding of antimutagenic and anticarcinogenic properties of food components may eventually lead to general dietary recommendations designed to improve protection of the population.

REFERENCES

1. Alekperov, U. (1982) Antimutagens and the problem of controlling the action of environmental mutagens. In Environmental Mutagens and Carcinogens, T. Sugimura, S. Kondo, and H. Takebe, eds. Alan R. Liss, Inc., New York, pp. 361-368.
2. Ames, B.N. (1983) Dietary carcinogens and anticarcinogens. Science 221:1256-1263.
3. Benson, A., Y. Cha, E. Bueding, H. Heine, and P. Talalay (1979) Elevation of extrahepatic glutathione S-transferase and epoxide hydratase activities by 2(3)tert-butyl-4-hydroxyanisole. Cancer Res. 39:2971-2977.
4. Busk, L., and U.G. Ahlborg (1982) Retinol (vitamin A) as a modifier of 2-aminofluorene and 2-acetylaminofluorene mutagenesis in the Salmonella/microsome assay. Arch. Toxicol. 49:169-174.
5. Busk, L., U.G. Ahlborg, and L. Albanus (1982) Inhibition of pyrolysate mutagenicity of retinol (vitamin A). Food Chem. Toxicol. 20:535-538.
6. Cerutti, P.A. (1985) Prooxidant states and tumor promotion. Science 227:375-381.
7. Cerutti, P.A., J. Emerit, and P. Amstad (1983) Membrane mediated chromosome damage. In Genes and Proteins in Oncogenesis, I.B. Weinstein and F. Vogel, eds. Academic Press, New York, pp. 55-67.
8. Chesis, P.L., D.E. Levin, M.T. Smith, L. Ernster, and B.N. Ames (1984) Mutagenicity of quinones: Pathways of metabolic activation and detoxification. Proc. Natl. Acad. Sci., USA 81:1696-1700.

9. Doll, R., and R. Peto (1981) The causes of cancer: Quantitative estimates of avoidable risks of cancer in the United States today. J. Natl. Cancer Inst. 66:1192-1265.
10. El-Torkey, N.M., and E.K. Weisburger (1983) Influence of disulfiram on mutagenicity of the urinary metabolites of 2-acetylaminofluorene and N-hydroxy-2-acetylaminofluorene in the Salmonella test system. Mutat. Res. 118:1-6.
11. Graham, S., H. Dagai, M. Swanson, A. Mittleman, and G. Wilkinson (1978) Diet in the epidemiology of cancer of the colon and rectum. J. Natl. Cancer Inst. 61:709-714.
11a. Hayashi, K., H. Fujiki, and T. Sugimura (1983) Effects of tumor promoters on the frequency of metallothionein. I. Gene amplification in cells exposed to cadmium. Cancer Res. 43:5433-5436.
12. Kada, T., and N. Kanematsu (1978) Reduction of N-methyl-N'-nitrosoguanidine induced mutations by cobalt chloride in Escherichia coli. Proc. Japan Acad. 54:234-237.
13. Kada, T., K. Morita, and T. Inoue (1978) Anti-mutagenic action of vegetable factor(s) on the mutagenic principle of tryptophan pyrolysate. Mutat. Res. 53:351-353.
14. Kada, T., M. Kato, and S. Kiriyama (1983) Adsorption of mutagens by vegetable fibres. Proceedings of the 12th Annual Meeting of the Environmental Mutagen Society of Japan, Tokushima, p. 74.
15. Komura, H., H. Minakata, H. Mochizuki, and T. Kada (1983) Antimutagenicity of human placenta extract: Chemical studies. In Carcinogens and Mutagens in the Environment, Vol. II, H.F. Stich, ed. CRC Press, Inc., Boca Raton, Florida, pp. 85-99.
15a. Kouri, R.E., L.H. Billups, T.H. Rude, C.E. Whitmire, B. Sass, and C.J. Henry (1980) Correlation of inducibility of aryl hydrocarbon hydroxylase with susceptibility to 3-methylcholanthrene-induced lung cancers. Cancer Lett. 9:277-284.
16. Lam, L.K., and L.W. Wattenberg (1977) Effects of butylated hydroxyanisole on the metabolism of benzo(α)pyrene by mouse liver microsomes. J. Natl. Cancer Inst. 58:413-417.
17. Meyn, M.S., T. Rossman, and W. Troll (1977) A protease inhibitor blocks SOS functions in Escherichia coli: Antipain prevents lambda repressor inactivation, ultraviolet mutagenesis and filamentous growth. Proc. Natl. Acad. Sci., USA 74:1152-1156.
18. Mirvish, S. (1981) Ascorbic acid inhibition of N-nitroso compound formation in chemical, food and biological systems. In Inhibition of Tumor Induction and Development, M.S. Zedeck and M. Lipken, eds. Plenum Press, New York, pp. 101-126.
19. Moon, R.C., D.L. McCormick, and R.G. Mehta (1983) Inhibition of carcinogenesis by retinoids. Cancer Res. 43(Suppl.):2469-2475.
19a. Newmark, H., and W. Mergens (1981) Alpha-tocopherol (vitamin E) and its relationship to tumor induction and development. In Inhibition of Tumor Induction and Development, M.S. Zedeck and M. Lipken, eds. Plenum Press, New York, pp. 127-168.
20. Peraino, C., R.J. Fry, E. Staffeldt, and J.P. Christopher (1977) Enhancing effects of phenobarbitone and butylated hydroxytoluene on 2-acetylaminofluorene-induced hepatic tumorigenesis in the rat. Food Cosmet. Toxicol. 15:93-96.
21. Ramel, C., U. Alekperov, B.N. Ames, T. Kada, and L.W. Wattenberg (1986) Inhibitors of mutagenesis and their relevance to carcinogenesis. Report by an ICPEMC expert group on antimutagens and desmutagens. Mutat. Res. and Zoologisches Zentralblatt (in press).
22. Rannug, U., A. Sundvall, and C. Ramel (1978) The mutagenic effect of 1,2-dichloroethane on Salmonella typhimurium. I. Activation through conjugation with glutathione in vitro. Chem. Biol. Interact. 20:1-16.

23. Saffiotti, V., R. Montesano, A.R. Sellakumar, and S.A. Borg (1967) Experimental cancer of the lung: Inhibition by vitamin A of the induction of tracheobronchial squamous metaplasia and squamous cell tumor. Cancer 20:857-864.
24. Schmähl, D., F.W. Krüger, M. Habs, and B. Diehl (1976) Influence of disulfiram on the organotropy of the carcinogenic effect of dimethylnitrosamine and diethylnitrosamine in rats. Z. Krebsforschung 85:271-276.
25. Shibata, T., T. Ohtani, P.K. Chang, and T. Ando (1982) Role of superhelicity in homologous pairing of DNA molecules promoted by Escherichia coli recA protein. J. Biol. Chem. 257:370-376.
26. Stich, H.F., P.K.L. Chan, and M.P. Rosin (1982) Inhibitory effects of phenolics, teas and saliva on the formation of mutagenic nitrosation products of salted fish. Int. J. Cancer 30:719-724.
27. Stich, H.F., C. Wu, and W. Powrie (1982) Enhancement and suppression of genotoxicity of food by naturally occurring components in these products. In Environmental Mutagens and Carcinogens, T. Sugimura, S. Kondo, and H. Takebe, eds. Alan R. Liss, Inc., New York, pp. 348-353.
28. Varshavsky, A. (1981) Phorbol ester dramatically increases incidence of methotrexate-resistant mouse cells. Cell 25:561-572.
29. Wattenberg, L.W. (1980) Inhibitors of chemical carcinogens. J. Env. Pathol. Toxicol. 3:35-52.
30. Wattenberg, L.W. (1983) Inhibition of neoplasia by minor dietary constituents. Cancer Res. 43(Suppl.):2448-2453.
31. Wong, L.C.K., J.M. Winstone, C.B. Hong, and H. Plotnick (1982) Carcinogenicity and toxicity of 1,2-dibromoethane in the rat. Toxicol. Appl. Pharmacol. 63:155-165.
32. Wood, A.W., M.-T. Huang, R.L. Chang, H.L. Newmark, R.L. Lehr, H. Yagi, J.M. Sayer, D.M. Jerina, and A.H. Conney (1982) Inhibition of the mutagenicity of bay-region diol epoxides of polycyclic aromatic hydrocarbons by naturally occurring plant phenols: Exceptional activity of ellagic acid. Proc. Natl. Acad. Sci., USA 79:5513-5517.
33. Wood, A.W., J.M. Sayer, H.L. Newmark, M. Yagi, D.P. Michaud, D.M. Jerina, and A.C. Conney (1982) Mechanism of the inhibition of mutagenicity of a benzo(α)pyrene 7,8-diol 9,10-epoxide by riboflavin 5'-phosphate. Proc. Natl. Acad. Sci., USA 79:5122-5126.

POPULATION CONSEQUENCES OF MUTAGENESIS AND ANTIMUTAGENESIS

James F. Crow

Department of Genetics
University of Wisconsin
Madison, Wisconsin 53706

ABSTRACT

Although the progress in basic understanding of mutagenesis and in techniques for precise measurement of mutation rates in test systems has been enormous, there has been very little progress in applying this information to estimates of germline mutation in humans, and even less in translating such estimates into quantitative assessments of the impact on future generations. This doesn't mean that new information about the mutation process, and antimutagens in particular, is not useful. Lowering the human mutation rate would be good, even if we can't say how good.

Some simple population kinetics of a change of mutation are discussed, and it is shown that future environmental changes can be ignored if we assume that the impact of a disease on human welfare is changed by the environment in the same proportion as its effect on fitness.

Since the human mutation rate appears to be much higher in males than in females, it would be especially important to find ways of reducing the male rate. The role of transposable elements in determining human spontaneous mutation rates is unknown, but unless data from experimental organisms are grossly misleading, this role may be substantial.

It is sometimes argued that such responses as error-prone repair systems may be an evolutionary strategy to allow the population to try a larger repertoire of mutations in times of environmental change. They may also be a survival strategy. I suggest that, although such an evolutionary strategy may possibly be adopted in asexual organisms with a very high reproductive rate, it is very unlikely in Mendelian species with limited reproduction such as most higher animals. The amount of existing variability in a large population is so great relative to that which arises in a few generations by mutation that segregation and recombination of existing alleles would appear to be a better way of coping with changing environment.

As the human age of reproduction has increased in the recent evolutionary past, it is possible that the compensatory adjustment of mutation rates has not been fast enough to keep up. Perhaps evolution of mutation rates is more determined by selection to reduce somatic mutation than by selection to reduce germinal mutation. Regardless of the answer to the question of the optimum mutation rate for long-time evolution, in my view, the optimum mutation rate from the standpoint of human welfare for the foreseeable future is zero.

INTRODUCTION

To one who views from the sidelines the cellular and molecular aspects of mutation research, the progress in this area is nothing short of astounding. When I was a graduate student in the 1930s we knew, thanks to the work of Muller (17), that radiation was a mutagen, but it was the only one known. The high activation energy of the mutation process, assumed because of the large temperature coefficient, was often seen as an impediment to finding chemical mutagens. In the 1940s Charlotte Auerbach's (2) discovery of chemical mutagenesis was greeted with great interest, but not with much concern for health effects, since the mutagens of this period were so toxic that there was no expectation of any appreciable population exposure.

In the early 1950s I was very impressed by the chemostat experiments of Novick and Szilard (21), which measured changes in mutation rates more precisely than any previous method. Furthermore, some of the substances found to be mutagenic, such as caffeine, were everyday compounds. I recall asking Novick if this system might be used as a way to assess environmental chemicals. He thought that chemostats were too complicated and difficult for routine use; a person clever enough to make a chemostat work would want to study more interesting problems. More important from the standpoint of the purposes of this conference, Novick and Szilard found that some common nucleosides were antimutagens and suggested that this might be the beginning of a chemical understanding of mutagenesis.

The 1960s brought the discovery of very highly mutagenic compounds, such as ethyl methanesulfonate, that are not overtly toxic. Such compounds might very well exist among the enormous number to which the population is or might be exposed. Largely due to the efforts of Matthew Meselson, the Genetics Study Section of the National Institutes of Health was asked to look into this question and make recommendations. A conference was called and it fell to me, as Chairman of the Study Section, to draft a report which was revised and approved by the Section as a whole and published (10). There were four recommendations, as follows:

(1) That an up-to-date register of mutagenic chemicals be kept and the information made generally available.

(2) That tests for mutagenicity be a routine part, such as toxicity tests now are, of the testing of chemicals that are used as food or drugs, or to which large numbers of persons may be exposed for other reasons.

(3) That research to develop more sensitive and cheaper assays of mutagenicity and chromosome breakage, or assays based on organisms more closely related to man, be encouraged and supported.

(4) That the feasibility of genetic monitoring of the human population for chromosome breakage and increased genetic disease be explored.

It is remarkable that all four recommendations were quickly put into effect. The most significant recommendation for the purposes of this conference is the third. The various systems that Bruce Ames has developed are paradigmatic of what was hoped for. These and many others are summarized in a recent National Academy of Sciences report (19).

As the title of this chapter implies, I shall deal with inherited, not somatic, aspects of mutagenesis and shall confine myself to questions of human welfare and evolution.

As I said, there has been enormous progress both in phenomenology and in fundamental knowledge of the mutation process and what influences it. Yet the gap between this knowledge and any realistic, quantitative assessment of the human risk remains enormous, and the possibility of filling it as elusive as ever. The bottom line is germ line mutagenesis and the translation of this into effects on human welfare. We still have no reliable way to go from DNA damage, however precisely measured, to human well-being *n* generations from now. This doesn't mean that the new information is not useful. Lowering the human mutation rate is good, even if we don't know how good.

THE POPULATION KINETICS OF A CHANGED MUTATION RATE

Any population geneticist can draw curves relating future incidence of disease to a mutation rate change, but only if he knows the relevant parameters, which he usually doesn't. One can nevertheless sketch the general pattern. I have done this several times for mutation rate increase, and I could almost use the same graphs upside down for a mutation rate decrease. And, just as radiation geneticists discuss the doubling dose of radiation, we could speak of the halving dose of antimutagens.

First I'll consider the generations following a temporary reduction of the mutation rate. Suppose that for one generation the mutation rate is reduced by half. In the next generation the number of diseases caused by dominant mutations that, because of sterility or prereproductive death, reduce the fitness to zero will be reduced by half. The effect of this reduction will persist only one generation. Diseases due to mutations causing a smaller reduction in fitness will be reduced by lesser amounts, but the reduction will persist in the population for a correspondingly longer time. Figure 1 gives the qualitative picture--an immediate, temporary reduction of disease incidence or mutation impact (measured in whatever units one chooses), followed by an approach to the original state at an ever-slower rate. The return to the original state will be very slow for polygenic and recessive conditions. Of course, with an improved environment the conditions will change; so what is actually approached is the state that would exist if the temporary reduction had not occurred.

A reduction in mutation rate is likely to last more than one generation. Figure 2 shows what will happen if the reduction is permanent. The genetic disease incidence or mutation impact will decrease somewhat in the first generation and then, at an ever decreasing rate, approach a new value that is half what it would have been at the original mutation rate. In the early generations the reduction is due mainly to a few severe diseases. In later generations the milder mutants, whose individual impact is small but whose total impact may be large because of their large number, make a larger contribution.

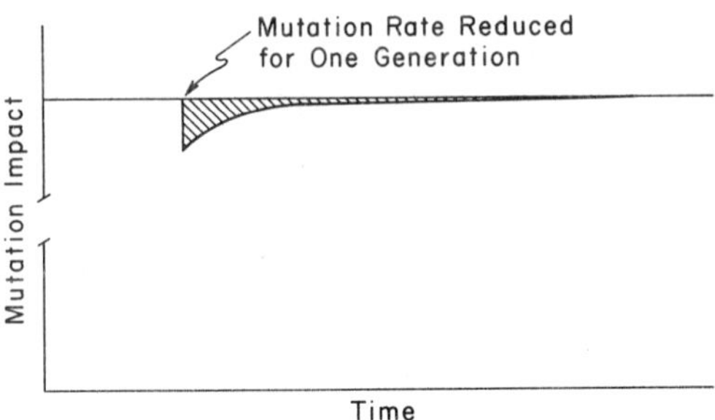

Fig. 1. The effect of a single generation of mutation-rate reduction on the mutational impact in the population. The shaded area represents a net gain. The population size is assumed to be constant, or else the impact is measured per capita. No environmental improvement is assumed; if there is improvement, the horizontal line should slant downward, and would not be expected to be straight.

There is abundant evidence in Drosophila (7) and supporting evidence from humans (9) that completely recessive mutants are very unusual. The typical mutant has some effect, almost always deleterious, in the heterozygous condition. Because there are so many more heterozygotes than homozygotes for a rare allele, most such alleles are eliminated from the population by the individually slight but collectively large amount of selection against heterozygotes. Thus the frequency of recessive alleles is determined mainly by their heterozygous effects. Their impact on the population fitness is also mainly through their heterozygous effects. A change in mutation rate would have almost no effect on the incidence of diseases caused by homozygous recessive alleles; changing heterozygous selective forces would be much more important.

I would now like to treat the problem more formally and quantitatively. Suppose there is an allele that, in the heterozygous state, causes a certain disease or impairment. Assume that this disease reduces survival and fertility (i.e., fitness) by a factor s. The number of persons affected by this disease before it is eliminated from the population is its *persistence*, p. Then the expected number of persons affected by a single mutation is equal to:

$$p = 1 + (1-s) + (1-s)^2 + (1-s)^3 + \dots = 1/s$$

assuming that the population size does not change. Not surprisingly, the number of individuals affected by a mutation before it is eliminated is equal to the reciprocal of s, the probability of its elimination per generation. This simple result is also true when stochastic variability is permitted (15). A simple adjustment can take changing population size into account.

It is more realistic to assume that s is not constant, but is changing (probably decreasing) because of environmental and medical improvements. Let s_i be the decrease in fitness caused by the mutant allele i generations after its occurrence. Then the persistence formula becomes:

$$p = 1 + (1-s_1) + (1-s_1)(1-s_2) + (1-s_1)(1-s_2)(1-s_3) + \dots$$

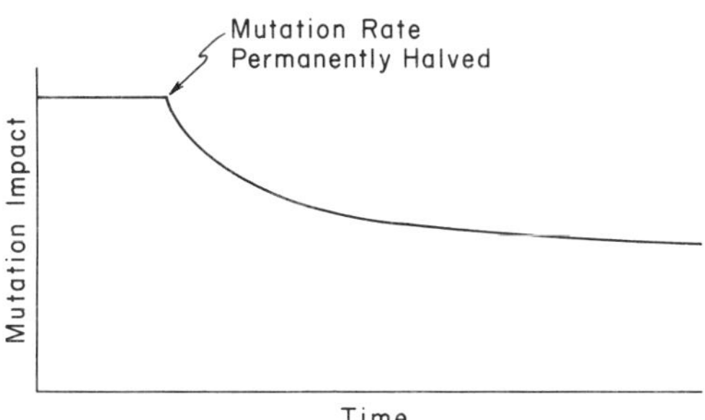

Fig. 2. The reduced mutational impact in the generations following a permanent halving of the mutation rate.

We could, in principle, go through this kind of calculation for each gene. But we know the relevant parameters, even approximately, for only a few genes, and the possibility of being able to do any realistic calculation of this sort for the totality of mutational damage remains remote.

Now let us ask, not for the future incidence of disease following a mutation, but for the effect on fitness. Assume a population of $\underline{N}$ individuals and a mutation rate increment $\Delta\mu$. I want to calculate the decrease in fitness resulting from this. A small decrement in the mutation rate caused by antimutagenesis will cause an equal increase in fitness.

The decrease in fitness, summed over successive generations, is given by:

$$\underline{L} = 2\underline{N}\Delta\mu[\underline{s}_1 + (1-\underline{s}_1)\underline{s}_2 + (1-\underline{s}_1)(1-\underline{s}_2)\underline{s}_3 + \ldots]$$

Now let $\underline{x}_i = 1 - \underline{s}_i$. This leads to:

$$\underline{L} = 2\underline{N}\Delta\mu[(1-\underline{x}_1) + (\underline{x}_2-\underline{x}_1\underline{x}_2) + (\underline{x}_1\underline{x}_2-\underline{x}_1\underline{x}_2\underline{x}_3) + (\underline{x}_1\underline{x}_2\underline{x}_3-\underline{x}_1\underline{x}_2\underline{x}_3\underline{x}_4) \ldots] = 2\underline{N}\Delta\mu$$

We have arrived at the Haldane principle (12), which states that the impact of mutation on fitness, or the number of "genetic deaths" in Muller's (18) terminology, is simply twice the total mutation rate. The equation shows that this principle remains correct even if the effect of the mutation changes erratically in the future.

I have written this as if the population size remained constant. On a per person basis, which is in many ways preferable, the $\underline{N}$ disappears and the expression is independent of population size.

I don't want to assume that decreased fitness is the appropriate way to measure the impact of mutation on human welfare. I would therefore like to introduce a quantity $\underline{C}$ as a factor to convert reduced fitness into social impact. The impact may be measured in years of productive life lost, loss of earning power, loss to society, or possibly in units of debility and suffering. This quantity would differ from disease to disease, but it seems reasonable to assume as a first approximation that $\underline{C}$ stays constant through time for a given disease, i.e., that improved environment decreases the social cost of the disease in proportion to the reduction of fitness. Then the social impact of a mutation rate increment $\Delta\mu$ for a specific disease (or perhaps class of diseases) is:

$$I = 2\Delta\mu\underline{C}$$

Until much more is known, we can make such a calculation for only a few diseases. If the number of diseases under consideration becomes large or includes common ones, we may have to begin to take interactions into account; but that is far in the future. A more realistic and more general approach to measuring the mutational impact was given by Crow and Denniston (8).

MUTATION RATES IN THE TWO SEXES

Curiously, we know more about the relative spontaneous mutation rates in males and females for humans than for any other mammal, including the mouse. There are several reasons to think that the male rate is higher in humans.

One reason is the larger number of cell divisions between zygote and sperm than that between zygote and egg. Another reason is the finding of a higher average age at reproduction in the fathers of persons showing a new autosomal dominant mutation, or in the maternal grandfathers for X-linked recessives. There is no corresponding increase in mother's (or grandmother's) age, except in so far as father's and mother's ages are correlated (9,26). It is interesting that paternal age as a criterion for new mutations was pointed out, probably for the first time, by Wilhelm Weinberg. Ironically, he is best known for the Hardy-Weinberg formula, although he made a number of far more original and profound discoveries.

A third line of evidence comes from recessive X-linked diseases. Here we can be more quantitative. If the male and female mutation rates are the same, one-third of affected males are new mutants while two-thirds come from heterozygous mothers. If mutations were to occur only in males, the mutant genes would be transmitted to daughters and, through them, to grandsons. Hence, to whatever extent the male mutation rate is higher than the female rate, there will be an excess of affected males who inherit the gene from carrier mothers, not as new mutations from homozygous normal mothers. The data for hemophilia and Lesch Nyhan syndrome suggest that the male rate is about 10 times the female rate, although there is a large standard error (9,25). If the mutation rate could be reduced by half, it would be more important that this happen in males than in females.

The data at present are confined largely to two X-linked loci with some supporting data from paternal age for autosomal mutants. It should be possible in the very near future by molecular methods, using restriction site polymorphisms or direct sequencing, to identify the parental origin of new mutants. Then our information will be much more precise.

Another aspect of human mutation rates is that not all dominant mutations show these effects. Specifically, X-linked Duchenne muscular dystrophy has a very high mutation rate, no evidence for rate differences in the two sexes, and no maternal grandfather age effect. This suggests that for this disease there is a different mechanism, possibly related to meiotic recombination, that does not depend on the number of cell divisions (27). Autosomal mutations that appear not to be associated with a paternal age effect may be similar.

Another important area for chemical antimutagenesis research is transposons. It is still quite uncertain what role, if any, transposition plays in causing the human mutational burden. The ubiquity of such phenomena in other species (23), plus evidences of transpositional changes having taken place in the past, makes it seem very unlikely that the human species is exempt.

The alcohol dehydrogenase locus in Drosophila has many transposon-induced duplications outside the coding and other conserved regions, but none inside the coding region (1). This is strong evidence that those that occur in essential regions are harmful and are eliminated by natural selection. Although transposition is harmful, whether it is an important component of the human burden is impossible to determine at present. The rate of progress in this area is very rapid, and we should know a great deal more soon.

It would be good to know whether there are antitransposogens. They may well be found, for it is known that the rate of transposition can be controlled by extrachromosomal cellular influences. The M-P system in Drosophila shows some of the possibilities for the study of such systems (11). There can be an extremely high rate of transposition, but this is under experimental control. Matings of M female by P male (where M and P are arbitrary classifications) produce a high rate of transposition with all its undesirable side effects, while reciprocal matings lead to a very low rate. Furthermore, the effects are confined to the germ cells, yet both natural and artificially created mutations have lost the ability to distinguish between germinal and somatic transposition. The molecular analysis of this system has been nothing short of spectacular, as has its use as an experimental tool (22).

From the standpoint of human welfare, we need to know both the extent to which such elements impact on human health and also what kinds of chemicals might change the impact--downward we hope.

SURVIVAL VERSUS EVOLUTIONARY STRATEGIES

There has been considerable discussion in this conference about SOS, error-prone repair systems, heat-shock responses, "genomic stress," and the like. H. Echols and M. Lambert have both suggested a possible evolutionary role for increasing the mutation rate in times of stress or rapidly changing environmental conditions. Similar suggestions have been made frequently in the history of genetics (e.g., Ref. 20 and 24).

It is not easy to tell whether such systems represent a survival or an evolutionary strategy. An organism with the threat of a high rate of chromosome breakage and cell death may opt for a "dirty" means of repair as the only one rapid enough for survival. On the other hand, it has often been argued that an enhanced mutation rate may help the organism adapt to whatever environmental change brought on the stress. Of course, the existence of exquisitely precise repair mechanisms argues that under most circumstances a low rate of mutation is advantageous.

I believe that the reasonableness of a strategy of enhanced mutation to meet an environmental change depends strongly on the life cycle strategy. In *Escherichia coli*, evolution is mainly clonal and periodic selection is an important factor. Cox, Gibson, and Chao (3,5) found that bacteria adapted more rapidly to chemostat life if they had an increased mutation rate; at least the more mutable strain often increased at the expense of the less mutable. So, it appears that under some circumstances an increased mutation rate can be helpful to the species. I would suggest that, if such a situation ever arises in nature, it will be in organisms with a very high rate of reproduction, either single-celled species or organisms like oysters and mushrooms where one individual can produce enormous numbers of progeny. Sacrificing even a large number of progeny because of bad mutations in order to get one that would do better in the new environment might then be a reasonable strategy.

But what about species with very little excess reproductive capacity, such as elephants and ourselves? It's hard to think of a chemostat experiment with elephants, but we can ask whether in some higher organisms the evolution rate depends on the mutation rate.

It is well known that the Mendelian mechanism is very effective in conserving variance. It is also known that the amount of selectable variance in any large population exceeds enormously that which arises in a single generation of mutation. For example, in Drosophila the amount of variance for bristle number or isozymes is some three orders of magnitude higher than that which arises by mutation in a single generation (4,16); for viability genes, the ratio of standing variance to that from new mutations is about two orders of magnitude (7). In fact, this ratio is simply the average persistence of a mutant allele. Furthermore, this standing variability is pretested, the worst variants already having been discarded by natural selection.

So I believe that in higher organisms, especially those with limited reproductive capacity, the species can better keep up with a changing environment by reshuffling existing alleles than by creating new ones by mutation.

Another line of evidence for this view comes from the fact that selection within inbred lines is so ineffective. One example is DDT resistance in Drosophila, where resistance develops rapidly with selection in genetically heterogeneous strains but hardly at all in inbred lines (6).

I conclude that in higher organisms, and mammals in particular, any SOS-like or heat-shock response is much more likely to be a survival strategy than an evolutionary one. But in _E. coli_ it just might be an evolutionary strategy; there the question remains open.

There is one more point to add to this discussion. One might think that organisms, like many invertebrates, that can afford to sacrifice 99 percent or more of their progeny in order to evolve more rapidly in a changing environment would have evolved more rapidly than species that could not survive such a loss. The facts, however, are otherwise. At least by morphological criteria, mammals have evolved much more rapidly than invertebrates, and slowly reproducing primates, including ourselves, more rapidly than other vertebrates. This argues against a high mutation rate, high reproductive rate, low parental care life pattern as an evolutionary strategy.

I conclude that it is very unlikely that the mutation rate limits the capacity to respond to changes in the environment, especially when the species has a low rate of reproduction, when inheritance is Mendelian, and when the population is not inbred. Of course the mutation rate cannot be zero, but it is not obvious what the optimum rate would be.

IS THE HUMAN MUTATION RATE TOO HIGH OR TOO LOW?

Our main interest in this conference is the human mutation rate. It has been known for a long time that mutation rates are adjusted to the life cycle. Human mutation rates, crudely known as they are, are clearly more comparable to those of mouse and Drosophila if they are compared on a per generation basis rather than on an absolute time basis (13). A striking illustration is as follows. The Drosophila spontaneous X-linked recessive lethal rate is 1-2 per 1,000 X chromosomes per generation. There are about 1,000 Drosophila generations (10-12 days) per human reproductive generation (30 years). If we had the same absolute rate of mutation, we would have one or two new X-linked lethals <u>every</u> generation and would have long been extinct for this reason alone.

Clearly, mutation rates are adjusted to the life cycle. But how precise is this adjustment? In a sexual species there is only a very loose coupling between genes determining the mutation rate and genes whose rate is being governed (14). Suppose there is a mutation that lowers the mutation rate. An individual with such a mutation will have more surviving offspring in the next few generations, because it will transmit fewer deleterious mutants. But a particular mutant-free survivor is unlikely to be the one that inherits the rate-reducing gene. Selection for reducing the mutation rate appears to be a slow, inefficient process. It is entirely possible that, as the human generation length has increased in the recent past, the adjustment of the mutation rate has not kept up. If so, our rate is higher than optimum. So it may well be that, even as an evolutionary strategy, our mutation rate is too high.

As has been abundantly brought out at this conference, there are many editing and post-replication repair processes that collectively produce a system with a very high fidelity of replication. As the life cycle becomes longer and the number of cell divisions per generation greater, can additional mechanisms evolve fast enough to keep the mutation rate somewhere near optimum (if there is such a thing)? Perhaps the failure to do this sets an upper limit on the life cycle. Many years ago it was suggested that one might look for albino seedlings from sequoia and other very long-lived trees. As far as I know, this either was not done or no interesting results were found. Molecular methods would be much more revealing.

If natural selection for lowering the mutation rate is as slow and inefficient as I suspect, are there other mechanisms? One possibility is that reduction of the mutation rate to fit the generation length has come from selection to reduce the somatic mutation rate. Somatic mutations may damage crucial cells or lead to malignancies. Germinal and somatic rates are to a large extent responsive to the same influences, so that selection to lower the somatic rate would also lower the germinal rate. Since selection on somatic rates acts directly on the survival and fertility of the individual in which the mutations occur, selection for reduction of the somatic mutation rate is much more direct than selection for reducing the germinal rate (9). Richard Shaw has independently arrived at the same conclusion (R. Shaw, pers. comm.).

These speculations about evolutionarily optimum mutation rates may or may not be correct, but it seems to me that from the standpoint of human welfare in the foreseeable future the human rate is clearly too high. If our mutation rate were to drop to zero, the only overt effect would be the disappearance of some of our most severe dominant diseases. The store of selectable variance would not be noticeably reduced for hundreds of generations unless humans were to institute some system of inbreeding. Even quite intense selection is not likely to cause significant variance reduction. If, in the remote future, we need new mutations, we certainly know how to produce them.

Better sanitation, vaccination, and other environmental improvements have greatly reduced the effectiveness of natural selection. What happens if this trend goes so far as to insure that each individual survives and produces the same number of progeny? Then, except for chance fluctuations, the gene frequencies would not change. One thing would continue, however-- the accumulation of mutations. Perhaps we shall be able to find ways of lowering the mutation rate enough to compensate for the relaxation of natural selection.

I would opt for zero human mutation rate if this were possible. As often, Haldane said it best. I have often quoted him and will do so once more. He noted that if we consider the enormous genetic variation in physique, in intelligence, in personality, in various talents, and in character in the human population and remember the possibilities that could arise by recombination among the causative genes, we surely have all the genotypes that even the most optimistic eugenicist could desire. Furthermore, there are environmental ways of achieving some of the same ends. But, as Haldane said, if we want to become a race of angels we shall have to have new mutations, for the present gene pool includes neither genes for wings nor for the requisite moral perfection.

A SOCIO-POLITICAL CONCERN

From the results presented at this conference, it is quite likely that ways will be found to reduce substantially the average mutation rate in the human population. We could hope that some simple dietary change, for example, might have this effect. It would very likely also have a similar cancer-reducing effect.

It is also possible that such a compound would have some easily discovered deleterious side effect. However small this might be, it would almost certainly prevent the use of the substance. The saving of 1,000 statistical lives in the near future by cancer reduction, or in the longer future by mutation reduction, would not be directly observed. Two directly observed lives lost would probably have more political impact. This is only an extreme example of the political difficulty of creating a large aggregate benefit to unknown individuals when this is accompanied by much less, but clearly identified, damage to a few.

I hope that in our increasingly litigious and individual-oriented society some political way can be found such that a nearly harmless and effective antimutagen and/or anticarcinogen would have some chance of being used.

ACKNOWLEDGEMENTS

This chapter is paper No. 2856 from the Laboratory of Genetics, University of Wisconsin, Madison.

REFERENCES

1. Aquadro, C.F., S.F. Deese, M.M. Bland, C.H. Langley, and C.C. Laurie-Ahlberg (1986) Molecular population genetics of the alcohol dehydrogenase gene region of *Drosophila melanogaster*. *Genetics* (in press).
2. Auerbach, C., and J.M. Robson (1946) Chemical production of mutations. *Nature* 157:302.
3. Chao, L., and E.C. Cox (1983) Competition between high and low mutating strains of *Escherichia coli*. *Evolution* 37:123-134.
4. Clayton, G.A., and A. Robertson (1955) Mutation and quantitative variation. *Am. Natur.* 89:151-158.
5. Cox, E.C., and T.C. Gibson (1974) Selection for high mutation rates in chemostats. *Genetics* 77:169-184.
6. Crow, J.F. (1957) Genetics of insect resistance to chemicals. *Ann. Rev. Entom.* 2:227-246.
7. Crow, J.F. (1979) Minor viability mutants in *Drosophila*. *Genetics* 92s:165-172.
8. Crow, J.F., and C. Denniston (1981) The mutation component of genetic damage. *Science* 212:888-893.
9. Crow, J.F., and C. Denniston (1985) Mutation in human populations. *Adv. Hum. Genet.* 15:59-123.
10. Crow, J.F., and members of the NIH Genetics Study Section (1968) Chemical risk to future generations. *Scientist and Citizen* June-July, pp. 113-117.
11. Engels, W.R. (1983) The P family of transposable elements in *Drosophila*. *Ann. Rev. Genet.* 17:315-344.
12. Haldane, J.B.S. (1937) The effect of variation on fitness. *Am. Natur.* 71:337-349.
13. Haldane, J.B.S. (1949) The rate of mutation of human genes. *Hereditas* 1949(Suppl.):267-273.
14. Leigh, E. (1973) The evolution of mutation rates. *Genetics* 73s:1-18.
15. Li, W.H., and M. Nei (1972) Total number of individuals affected by a single deleterious mutation in a finite population. *Am. J. Hum. Genet.* 24:667-679.
16. Mukai, T., H. Tachida, and M. Ichinose (1980) Selection for viability at loci controlling protein polymorphisms in *Drosophila melanogaster* is very weak at most. *Proc. Natl. Acad. Sci., USA* 77:4857-4860.
17. Muller, H.J. (1927) Artificial transmutation of the gene. *Science* 66: 84-87.
18. Muller, H.J. (1950) Our load of mutations. *Am. J. Hum. Genet.* 2:111-176.
19. National Academy of Sciences (1983) *Identifying and Estimating the Genetic Impact of Chemical Mutagens*, National Academy Press, Washington, D.C.
20. Neel, J.V. (1942) A study of a case of high mutation rate in *Drosophila melanogaster*. *Genetics* 27:519-536.
21. Novick, A., and L. Szilard (1950) Experiments with the chemostat on spontaneous mutations of bacteria. *Proc. Natl. Acad. Sci., USA* 36: 708-719.
22. Rubin, G.M., and A.C. Spradling (1982) Genetic transformation of *Drosophila* with transposable element vectors. *Science* 218:348-353.
23. Shapiro, J.A. (1983) *Mobile Genetic Elements*, Academic Press, New York.
24. Thompson, J.N., and R.C. Woodruff (1978) Mutator genes--Pacemakers of evolution. *Nature* 274:317-321.
25. Vogel, F. (1977) A probable sex difference in some mutation rates. *Am. J. Hum. Genet.* 29:312-319.

26. Vogel, F., and A.G. Motulsky (1979) Human Genetics, Springer-Verlag, Berlin.
27. Winter, R.M., and M.E. Pembry (1982) Does unequal crossing over contribute to the mutation rate in Duchenne muscular dystrophy? Am. J. Med. Genet. 12:437-441.

A SUMMARY--AND A LOOK AHEAD

John Cairns

Department of Cancer Biology
Harvard School of Public Health
Boston, Massachusetts 02115

As several chapters have pointed out, the principle behind much of cancer research and the principle underlying this Volume rests on the observation that each of the major forms of cancer is common in some parts of the world yet rare in others. Furthermore, it is plain from studies of migrants that these differences reflect the operation of environmental factors. Although there are genetic factors that affect cancer rates, these are not all-important: despite our genes, each of us could significantly reduce our risk of getting cancer if we really knew how it is that the Mormons, the Seventh Day Adventists, and the Protestant clergymen of the United States manage to drop their risk by almost a factor of two. That is the target of many investigations and is part of the stimulus for holding a conference on antimutagenesis and anticarcinogenesis.

Before reviewing the substance of this conference, I would like to pursue a little further some thoughts on the epidemiology of cancer. Recent studies of the trends in cancer mortality in the U.S. show that, leaving aside lung cancer which is a special case, the general trend is downwards. Of course there are exceptions (e.g., malignant melanoma and multiple myeloma), but overall our world seems to be becoming a less carcinogenic place than it used to be. And I suspect that this is just the beginning, because the trend is most conspicuous in the young. When diseases have disappeared in the past (e.g., the decline of tuberculosis in the U.S. in the twentieth century), they have started their disappearing act first in the young. The fact that cancer is doing just this suggests to me that it, too, may be drifting away. Perhaps the affluent West _is_ doing something right.

One more thought comes to me when I consider the epidemiological and demographical underpinnings of this Volume. The National Cancer Institute seems to have put its main effort into developing treatments for cancer, apparently in the mistaken belief that diseases are conquered by treatment. All historic precedents would argue that this is the wrong choice. So far, all of our major conquests have, in each case, been by some form of prevention--either calculated acts of prevention (e.g., immunization against smallpox or the addition of fluoride to water) or prevention by accident (e.g., the prevention of enteric diseases in the nineteenth century that resulted from separating the water supply and the sewage system). Historic precedents do not, however, weigh heavily in the New World; it was, after all, Henry Ford who said that history is bunk. So I shall have to have recourse to an economic argument, to which I'm sure he would have been more amenable. People who advocate high technology treatments for cancer forget that there has been no real issue of cost yet because the treatments do not, on the whole, work. Imagine what would happen if, say, the treatment for childhood leukemia (which is very successful) could be applied equally successfully to all other kinds of lethal cancer. We know that to treat a child with leukemia costs about $70,000. If you are going to do that for 600,000 people every year (which would be the United States' quota), that comes to about 50 billion dollars a year--which is almost 20 percent of the defense budget. To me, that is a reductio ad absurdum. I think we have to face it that what we are after is prevention--in short, anticarcinogenesis.

I think it is almost impossible to discuss negative forces (antimutagens and anticarcinogens) without discussing them in the context of their positive counterparts. You cannot see anything as negative without having something positive for it to work on. So I am going to start by considering the positive things.

It seems to me that our models for mutagenesis and carcinogenesis went through a period of maximum simplicity about 15 or 20 years ago. In the case of carcinogenesis, we start with Rous' study of skin carcinogenesis in the ear of a rabbit. When he stopped applying coal tar he observed that most of the tumors which had formed during the prolonged period of application of the tar would disappear, although they could be made to reappear months later if the exposure to tar was resumed. He therefore proposed that carcinogenesis consists of two processes: (a) initiation, which is nothing more than the production of variant cells capable of giving rise to a cancer, and (b) promotion, which is an essential, second step that allows the initiated cells to form large clones (tumors) and sets going the kind of turmoil within the epithelium that allows progression to ever-increasing levels of malignancy. Such was his picture of carcinogenesis, and it is sufficiently plausible to have been with us in scarcely modified form ever since. But it does have its unsatisfactory features, because it has until recently been quite unclear what are the normal functions of the genes that have to be mutated, and it is even less clear what has to happen during the prolonged period of promotion. Some people, however, are still happy with the conventional picture of promotion as being simply the stage during which mutated genes are given the chance to express themselves.

Because the molecular biology of multicellular systems is so inaccessible to most forms of investigation, attention turned instead to an analysis of the chemistry of carcinogens. Boyland had suggested in the 1930s that most carcinogens require metabolic activation before they can function as the proximate causes of cancer, and he was, I think, the first to suggest that such proximal carcinogens interact with DNA. Out of these ideas there has arisen a veritable industry centered on various short-term tests for mutagenicity, since mutation is the manifestation of DNA damage and the mutagenicity of any compound turns out to be a good predictor of its carcinogenicity.

In the last few years, however, there are signs of dissatisfaction, at least with the practicalities of this discipline. Too many chemicals prove to have measurable levels of mutagenicity and carcinogenicity. Too many substances, with which we come into close contact, have levels of activity that could be interpreted as posing significant risk. It seems hardly compatible with the facts of cancer epidemiology that the risk of cancer could be, for each of us, a linear function of the sum of these risk factors. In something of a volte face, Ames has now raised the possibility that the most important characteristic of each potential carcinogen is not its capacity for producing mutations but simply its direct toxicity. If this proves correct, our picture for the process of carcinogenesis may have to undergo considerable revision.

The experimentalists continue to generate a diversity of results that cry out to be assembled by someone into a coherent model of carcinogenesis. On the one hand, it is possible to show that in certain circumstances every cell exposed to a carcinogen can become initiated (i.e., become capable of yielding tumorigenic progeny), an observation that would seem hardly compatible with the notion that initiation is nothing more than mutagenesis. On the other hand, several groups have now shown that certain forms of chemical carcinogenesis in intact animals are accompanied by mutation in the ras oncogene, apparently as the result of a straightforward direct (but rare) interaction between the proximal carcinogen and the cellular oncogene of the target cell. Which of these two dramatically opposed scenarios more closely resembles the real-life human predicament remains to be determined. We should, however, remember that they both are the result of extremely high levels of exposure, and so neither of them need necessarily have much relation to the real world. After all, the study of the response of bacteria to high levels of potent mutagens has not told us much about spontaneous mutability or about the classes of genetic change that make up most spontaneous forward mutations.

Our understanding of the whole process of carcinogenesis is therefore much less secure now than it seemed to be some 10 or 15 years ago. Interestingly, this retreat from certainty in the field of cancer research mirrors what happened rather earlier for those who study the molecular biology of mutagenesis. In the early days of molecular biology, mutants were thought to result simply from random errors of replication, plus the forced errors that resulted from the incorporation of base analogs or the formation of the equivalent of base analogs as the result of DNA damage. We now see this as a vast oversimplification. Cells have a multitude of pathways for repair, some constitutive and some inducible, some error-prone and some virtually error-free, and some programmed to resist change while others apparently are ready to generate variability under certain particular conditions of stress. Typical of all segments and facets of the biosphere, processes such as mutation and DNA repair are not simply the end result of the laws of physics and chemistry but are elaborated in a completely unpredictable, idiosyncratic way that reflects many millions of years of evolution. If we find that something happens in a biological system one way rather than another, we should always be prepared to consider the possibility that this is because it, as it were, has been chosen to occur this way. We should, in short, think of every biological phenomenon as potentially the manifestation of the operation (or failure) of some program that has been scrutinized by evolution.

To give just one example, it has been suggested that the best nucleotide to put opposite an apurinic site in a DNA template is deoxyadenosine because the commonest noncoding lesion in DNA of organisms inhabiting the surface of this earth is the UV-induced thymine dimer, and this is why apurinic sites tend to lead to the incorporation of adenine and perhaps why some noncoding lesions stimulate certain sequence changes and not others.

This way of looking at biological phenomena is not very fashionable, because it is seen as being dangerously teleological. But I do think it can often save a lot of time. DNA repair was discovered really by accident, when Ruth Hill chanced upon some variants of Escherichia coli that were abnormally sensitive to radiation. But the discovery could, I think, have been made sooner if it had been everyone's reflex to treat each biological phenomenon (even mutation) as at least potentially the end result of some highly evolved program. This is a strategy that, I think, can be profitably carried across to cancer research.

The biosphere can be seen as a system that behaves rather like Maxwell's demon, lowering entropy within its own confines by raising entropy outside; it is creating a tiny corner of immense order by pushing out disorder. The laws of physics are observed but they are, as it were, tampered with and perverted to the greatest possible extent. We have learned to believe that this trick, this demon, arose by chance and was then elaborated and perfected by the inexorable force of natural selection which is the fundamental law of biology. Yet we now see that, with the invention of multicellular systems, the biosphere is evading even its own laws, because it seems to be a characteristic of such systems that they are, on the whole, not subject to natural selection.

For example, although cancers commonly become resistant to chemotherapy, it is significant that the cells in the normally proliferating tissues of the body seem never to become resistant. The cells within a cancer enjoy all the benefits of natural selection and that is why each cancer tends to progress towards increasing malignancy, but the normal cells of the body (apart, perhaps, from those of the immune system and the nervous system) are not allowed to compete with each other. In part, therefore, the study of cancer is the study of the results of introducing natural selection into systems of cells that are normally protected against it.

This is a facet of carcinogenesis that has been rather neglected, certainly compared to the more conventional investigation of mutagenesis. In the last few years, however, we have started to learn something about the network of restraints and controls that operates in the various tissues of the body. In particular, we find more and more evidence showing that cells communicate with each other and determine each other's behavior. It should, for example, have been no surprise to find that normal cells are often closely coupled to their neighbors by gap junctions whereas cancer cells (and, interestingly, also cells undergoing promotion, as Trosko pointed out) tend to be isolated from the cells around them. One could argue, therefore, that the nearest basic science to cancer research is actually developmental biology rather than molecular biology. This has, for some time, been my belief. Though I would not wish to force it on anyone, I take this occasion to offer it up for your contemplation.

For example, I think we may find that peculiarities of developmental lineage and programs of stem cell renewal account for the windows of vulnerability that are found in certain forms of carcinogenesis. Thus, there is only a short period in the life of a rat when it can easily be given mammary cancer; similarly, human breast cancer incidence is simply proportional to the interval between puberty and first completed pregnancy, suggesting that here too there is a limited period of susceptibility. What does this say about risk? Should I have been especially worrying about my daughter's diet when she was in high school? This is surely a question of developmental biology rather than intermediary metabolism or molecular biology.

Having said all that and having dealt, perhaps rather incoherently, with mutagenesis and carcinogenesis (the positive subjects of this Volume), I must now grapple with their negative counterparts. It is about 15 years since the beginnings of mutagenicity testing. It is now an industry in its own right. Open the pages of a journal like Mutation Research and you find the hordes at work--testing, testing, testing. Does much usable truth come out of this? I do not know. But plainly the tests are providing an ecological niche for many people. I mention this because these proceedings have made me worry that the study of anticarcinogens may prove to be an even larger ecological niche. After all, if there are N ways of testing mutagens, will there not be N^2 ways of testing antimutagens?

I therefore find myself concerned with the value of antimutagenicity tests. Like some brilliant metaphor, that through passing into common usage loses its wild originality, the Ames test is now becoming underestimated. It was set up in such a way that minor variations in procedure would have the least chance of influencing the outcome. Its ruggedness lay in the experimental protocol. Now, this is all forgotten, and the "antimutagenists" happily inject their test substances into the Ames test, at one point or another, and expect the final number of mutants to reflect in some way the antimutagenicity of each substance. Surely this is nonsense. Anything that prevents multiplication (even putting the test bacteria for a while into minimal medium) will lower the number of mutants, as the result of what has been called mutation frequency decline (MFD), and it would be a mistake to accept the results of a test so ill-controlled as some of the tests discussed in this Volume.

Having said that, I have to add that my doubts do not extend to studies of anticarcinogenicity. The great hope, for many epidemiologists, is that the rate-determining factors for most cancers--at least those in the industrialized affluent nations of the world--will turn out to be our exposures to protective agents rather than to carcinogens (i.e., to negative factors rather than positive ones). That could make cancer--like rickets, goiter, or dental caries--readily avoidable by quite minor supplements to our diet.

If, instead, the important factors are the carcinogens, we may be in for a difficult time, because it is not going to be easy to get the inhabitants of the affluent West to alter their habits. The nineteenth-century reformers could successfully attack the positive, infectious causes of disease because the reforms they instigated made life more pleasant for the city-dweller. The twentieth- and twenty-first-century reformers may have a more difficult task when they tackle the diseases that are associated with affluence. If, however, cancer rates are determined by a deficiency of anticarcinogens, then our goal of preventing most forms of cancer may actually be attainable. If this Volume encourages work along these lines, it will have served its purpose.

POSTER ABSTRACTS

POSTER ABSTRACTS--TABLE OF CONTENTS

(Poster Abstracts Listed in Alphabetical Order by Presenting Author.)

ANTAGONISTIC EFFECTS OF SOME PLANT COMPOUNDS ON THE ACTIVITY OF FOOD MUTAGENS

A.J. Alldrick, I.R. Rowland, T.M. Coutts, and J. Flynn

British Industrial Biological Research Association
Carshalton, Surrey SM5 4DS, United Kingdom

The human diet contains a number of genotoxic compounds and, in addition, substances that may exert antimutagenic effects. We have studied the effect of 4 plant polyphenols (caffeic, chlorogenic, ferulic, and ellagic acids) and 3 plant flavonoids (morin, myricetin, and quercitin) on the mutagenic activity of a number of cooked-food mutagens (IQ, MeIQ, MeIQx, Trp-P-1, and Trp-P-2) using S. typhimurium TA98, and a Swiss-mouse hepatic S-9 activating system. None of the plant compounds were, in themselves, mutagenic at the concentrations used. Over the range 0-100 μM, none of the plant polyphenols had an effect on any of the mutagens studied. However, inclusion of myricetin or quercitin (range 0-2.0 μM) in the assay resulted in a dose-dependent reduction in mutagenicity irrespective of which mutagen was used (e.g., at a concentration of 2 μM, both myricetin and quercitin reduced MeIQ mutagenicity by approximately 75%). These observations contrast with morin, which was not as effective (at 2 μM, morin reduced MeIQ mutagenesis by approximately 50%). These results suggest that certain plant-derived compounds may have a role in ameliorating the genotoxic effects of these food mutagens in vivo.

(We thank the U.K. Ministry of Agriculture, Fisheries and Food for financial support.)

EFFECT OF DIETARY LIPID ON THE ACTIVATION OF FOOD MUTAGENS BY RAT-HEPATIC S-9 FRACTIONS

A.J. Alldrick, I.R. Rowland, T.M. Coutts, and J. Flynn

British Industrial Biological Research Association
Carshalton, Surrey SM5 4DS, United Kingdom

We have investigated the effect of altered dietary lipid intake of rats on the ability of hepatic S-9 fractions derived from them to activate mutagens known to occur in food. Groups of three male and three female Ola:Sprague-Dawley rats (7 weeks old) were maintained for 28 days on a basal, purified, low-fat (1%) diet or on a diet containing 35% safflower oil, olive oil, coconut oil, or beef dripping. All diets had similar energy/protein ratios. Individual S-9 fractions were prepared from each rat's liver. The ability of activating systems (S-9 mixes) containing 1.64 mg protein/ml to convert the cooked food mutagens IQ, MeIQ, MeIQx, and the nitrosamines N-dimethyl-nitrosamine (NDMA) and N-nitrosopyrrolidine (NPYR) to bacterial mutagens using S. typhimurium was investigated. Sex-related differences in the ability of S-9 mixes derived from rats on the low-fat diet to activate MeIQ and NPYR were observed. Feeding male rats high-fat diets frequently resulted in a reduced ability of S-9 mixes derived from them to activate NDMA and NPYR (between 20 and 40%). However, S-9 mixes derived from rats on the high-fat diet, in general, converted IQ, MeIQ, and MeIQx to mutagens more effectively than S-9 mixes from rats on the low-fat diet (2- to 6-fold). The extent to which this increase was observed depended both on the source of fat and the sex of the rat. This study demonstrates that variations in both the quantity and type of fat included in the diet can modify the ability of hepatic enzymes to activate mutagens commonly found in food.

(We thank the U.K. Ministry of Agriculture, Fisheries and Food for financial support.)

INHIBITING EFFECT OF SOME BIOGENIC AMINES ON THE ACTIVATION OF FOOD MUTAGENS

A.J. Alldrick, I.R. Rowland, T.M. Coutts, and J. Flynn

British Industrial Biological Research Association
Carshalton, Surrey SM5 4DS, United Kingdom

The generation of mutagens frequently accompanies the cooking of proteinaceous foods. We have investigated the ability of the biogenic amines histamine, tryptamine, and tyramine to modify the metabolic conversion of IQ, MeIQ, Trp-P-1, and Trp-P-2 to bacterial mutagens using S. typhimurium TA98. Assays were performed using microsomal, cytosolic, and S-9 activating systems, prepared from the pooled liver homogenates obtained from either control or Aroclor-treated Ola:Sprague-Dawley rats. Inclusion of histamine (150 μM) in the assay failed to have any effect on the mutagenic activity of any of the mutagens irrespective of the source of activating system. Addition of tryptamine (20 μM) resulted in inhibition of mutagenesis by IQ (by 40%), MeIQ (by 75%), and Trp-P-2 (by 25%) but not by Trp-P-1. Tyramine (100 μM) inhibited only IQ (by 50%) and MeIQ (by 17%) mutagenesis. In contrast, the mutagenic activity of all four mutagens using S-9 and/or microsomes from Aroclor-treated rats was enhanced by 30% to 90% in the presence of tryptamine. Since tyramine and tryptamine are found in many foods and beverages, our findings may have important implications for the genotoxic effects of these mutagens in the intact animal.

(We thank the U.K. Ministry of Agriculture, Fisheries and Food for financial support.)

AN INDUCIBLE REPAIR PATHWAY IN *COPRINUS CINEREUS* COMMON TO INACTIVATION EFFECTS OF UVA AND THE CHALLENGING DOSE OF HYDROGEN PEROXIDE

John D. Amirkhanian

Natural History Museum of Los Angeles County
Los Angeles, California 90007

Oidia from the morphological mutant *den*-2 (Amirkhanian and Cowan, 1985) served as an ideal eukaryotic system in the present study. Pre-exposure of the growing oidia in liquid medium with a sublethal dose of H_2O_2 (50 μM) induced an adaptive response, which improved the protective effect of *rad*-1 (hyper-rec mutation) strongly and the wild-strain RAD-1 to a lesser extent from the challenging dose of H_2O_2 (5-10 mM) and UVA (315-365 nm) radiation but not from UVC (254 nm) radiation. The *rad*-1 confers sensitivity to a number of DNA-damaging agents (UVC, MMS, mono- and polyfunctional alkylating agents) and has enhancing effects on spontaneous meiotic and UV-induced mitotic recombination frequencies (Amirkhanian, 1979). Consistent with this observation are the findings that UVA-radiation generates superoxide anions and reactive oxygen species (Peak et al., 1981, 1985; Tuveson, 1980; Tyrrell, 1985) that are capable of producing single-strand breaks (SSBs). It is possible that the portion of reactive anions not destroyed by the inducible "protective mechanism" (Foote, 1976) may lead to accumulation of SSBs in DNA (Nakayama et al., 1985) that have causal significance in carcinogenesis. These SSBs are probably repaired through recombination repair (Howard-Flanders et al., 1984) in the hyper-rec mutant but partially so in the wild strain. In conclusion, DNA lesions other than cyclobutane pyrimidine dimers and most probably SSBs are the main lethal effects of UVA and H_2O_2. This substantiates findings that UVA-induced DNA damage differs from UVC effects in eukaryotes (Sammartano and Tuveson, 1983, 1985).

IN VITRO FORMATION OF MUTAGENIC N-NITROSO COMPOUNDS DERIVED FROM ANTIPARASITARY DRUGS

Alba M. Arriaga, J.J. Espinoza, and C. Cortinas de Nava

Instituto de Investigaciones Biomedicas UNAM
Apartado Postal 70-228 C.P., 04510 Mexico, D.F.

The mutagenic activity of some antiparasitary drugs on *S. typhimurium* has been reported [Cortinas de Nava et al. (1983) *Mutat. Res.* 117:79-91]. Some antiparasitary drugs giving negative results contain aminated groups on their chemical structure that can easily react with sodium nitrite. This work shows that the in vitro reaction of sodium nitrite with piperazine, mebendazole, chloroquine, dehydrohemetine, and pyrantel pamoate produces N-nitroso compounds. These compounds were all able to induce base change mutations in *S. typhimurium*. No N-nitroso compounds were obtained from drugs having groups with tertiary amine-like behavior or quaternary ammonium salts, such as bephenium hydroxinaphtoate and iodo-chloro hydroxiquinoleine.

INDOLE-3-CARBINOL AND BETA-NAPHTHOFLAVONE, INHIBITORS OF AFLATOXIN B_1 HEPATOCARCINOGENESIS, CAN ALSO ACT AS PROMOTERS

G.S. Bailey, J.D. Hendricks, D.W. Goeger, and J. Nixon

Oregon State University
Corvallis, Oregon 97331

The effects of indole-3-carbinol (I3C) and β-naphthoflavone (BNF) on initiation and postinitiation phases of aflatoxin B_1 (AFB_1) hepatocarcinogenesis were examined in rainbow trout. Dietary treatment with I3C (1,000 or 2,000 ppm) or BNF (500 ppm) 6 weeks prior to, and 2 weeks along with, AFB_1 exposure (20 ppb) reproducibly inhibited tumor response in rainbow trout. However, populations initiated with AFB_1 as embryos, and then treated as fingerlings with dietary I3C or BNF, reproducibly showed significantly enhanced tumor response compared to AFB_1-only controls. Mechanisms of anti-initiation of I3C and BNF were examined in vitro and in vivo. Dietary pretreatment by either compound greatly reduced formation of AFB_1-DNA adducts in isolated hepatocytes, and in vivo in liver nuclear DNA. Neither compound influenced adduct stability (repair) in vivo. BNF strongly enhanced cytochrome P-448 and associated mixed function oxidase (MFO) activities in trout liver, and greatly stimulated conversion of AFB_1 to the less carcinogenic aflatoxin M_1 (AFM_1). I3C was not an MFO inducer in trout, and only slightly enhanced AFM_1 production. Both compounds profoundly altered AFB_1 pharmacokinetic distribution in vivo, including enhanced bile elimination of AFB_1 as glucuronides of aflatoxin M_1. Neither compound enhanced production of AFB_1 glutathione conjugated in this species. Thus, stimulation of glutathione detoxication is not a requisite for BNF or I3C anticarcinogenesis in all species. These results also demonstrate that inhibitors that act as anti-initiators may achieve the end result of reduced DNA adduct formation through quite different mechanisms. The mechanisms of promotion by these compounds are not presently understood.

(Supported by National Institutes of Health grants ES 00210, ES 00092, and CA 34732.)

ANTIMUTAGENICITY OF CHLOROPHYLLIN AGAINST TEN COMPLEX MIXTURES IN TA98 OF *SALMONELLA TYPHIMURIUM*

Tong-man Ong,[1] Wen-Zong Whong,[1] John Stewart,[1] and Herman E. Brockman[2]

[1]National Institute for Occupational Safety and Health
Morgantown, West Virginia

[2]Department of Biological Sciences
Illinois State University, Normal, Illinois

Chlorophyllin has been shown by others to be responsible for most of the antimutagenic activity of certain vegetable extracts. Nevertheless, the antimutagenicity of chlorophyllin has not been studied extensively. Because humans are exposed mainly to complex environmental mixtures rather than single chemicals, we have studied the effect of chlorophyllin on complex mixtures known from previous experiments to be mutagenic in strain TA98 of *S. typhimurium*. The complex mixtures were extracted and/or concentrated with organic solvents. A mutagenic concentration of each extract was combined with various concentrations of chlorophyllin in the *Salmonella*/mammalian microsome plate incorporation test. Chlorophyllin (1.25 mg per plate) caused total or close to total inhibition of the mutagenicity of extracts of airborne particles, cigarette smoke, fried shredded pork, and fried beef. Chlorophyllin also inhibited strongly (92-98%) the mutagenic activity of extracts of diesel emission particles, coal dust, red grape juice, and red wine, but concentrations of chlorophyllin greater than 1.25 mg per plate were required. The mutagenic activity of extracts of tobacco snuff and chewing tobacco was inhibited 75-80% by 10 mg of chlorophyllin per plate. Reconstruction and control experiments showed that chlorophyllin was not toxic to Salmonella at the concentrations used in the antimutagenicity experiments. Heating a chlorophyllin solution for 10 min at 100°C did not affect its antimutagenic activity. We conclude that chlorophyllin is an effective antimutagen against these dietary and environmental complex mixtures in TA98 of *S. typhimurium*.

INTERACTION OF RETINOIDS WITH 9,10-DIMETHYL-1,2-BENZ(α)ANTHRACENE IN THE UDS ASSAY AND IN THE CHO/HPGRT MUTATION ASSAY

J.D. Budroe, H.M. Schol, J.G. Shaddock, and D.A. Casciano

National Center for Toxicological Research
Jefferson, Arkansas 72079; and
University of Arkansas for Medical Sciences
Little Rock, Arkansas 72205

The potential ability of retinol and retinoic acid to interfere with the genotoxic activity of 9,10-dimethyl-1,2-benz(α)anthracene (DMBA) was studied in the primary rat hepatocyte/unscheduled DNA synthesis (UDS) assay and in the Chinese hamster ovary/hypoxanthine-guanine phosphoribosyl transferase locus (CHO/HGPRT) mutation assay. UDS was evaluated using an autoradiographic detection method in hepatocytes obtained from Sprague-Dawley rats via an in situ collagenase perfusion technique. Resistance to 6-thioguanine (6-TG) was the criterion used for determination of mutation frequencies in treated CHO cells. Metabolic activation of test chemicals was accomplished using 1-day Aroclor 1254-induced Sprague-Dawley rat S-9 where appropriate. Retinol and retinoic acid both significantly inhibited 2.5 μg/ml DMBA-induced UDS in a dose-dependent manner at concentrations of 10 μM or greater. Retinoic acid was also observed to inhibit 5 μg/ml DMBA-induced UDS at concentrations of 2 μM or greater. In the absence of an exogenous metabolic activation system, both retinol and retinoic acid failed to induce mutations in CHO cells at the HGPRT locus at concentrations up to 50 μM. Retinol elicited a dose-dependent inhibition of 6-TG resistant mutants induced by 2.5 μg/ml DMBA with S-9 activation at concentrations of 5 μM or greater. Preliminary results also indicate that retinoic acid causes a dose-dependent inhibition of CHO mutants induced by DMBA with S-9 activation.

GENOTOXICITY OF A FLAVONOID: METABOLIC CONDITIONS INVOLVED IN THE RESPONSE IN IN VITRO AND IN VIVO ASSAYS

T. Chaveca, A. Laires, J. Rueff, H. Borba, M. Gomes, and M. Halpern

Faculty of Medical Sciences
Campo de Santana 130
P-1198 Lisbon Codex, Portugal

About 2,000 individual flavonoids, mostly glycosides, are described in plants. Glycosides of flavonols (namely, of quercetin) are found in the edible portions of the majority of food plants [Knudsen (1982) Prog. Clin. Biol. Res. 109:315]. Quercetin displays a well-documented genotoxic activity though it seems to be devoid of carcinogenic activity. The lack of carcinogenicity might be due to the fact that quercetin seems to act as an antipromoter [Hoyoku et al. (1983) Cancer Lett. 21:1; Kato et al. (1983) Carcinogenesis 4:1301]. The metabolic fate of quercetin, however, might also help to explain why this genotoxin is not carcinogenic. We have compared the genotoxicity of quercetin in 3 different assay systems under different metabolic conditions in vitro and in an in vivo assay. While in the Ames assay the mutagenicity of quercetin is increased by a factor of 3.0 in the presence of Aroclor-induced rat liver enzymes (S9) and the respective cytosolic fraction (S100), an opposite response is obtained when the induction of SOS responses in *E. coli* is assessed using the SOS Chromotest. Similarly, the induction of sister chromatid exchanges in human lymphocytes is lower when testing is carried out in the presence of liver enzymes. The results obtained seem to suggest that the putative genotoxic metabolites of quercetin are not the same for different genetic endpoints considered. With our negative results in the micronucleus test in mice, which confirm those obtained by others, we tentatively suggest that quercetin might be detoxified in vivo, possibly by cytosolic enzymes, which could help to explain the lack of carcinogenicity.

EFFECT OF BUTYLATED HYDROXYTOLUENE ON PROCARCINOGEN MUTAGENESIS IN HEPATOCYTE/SALMONELLA ASSAYS

J.E. Davies and J.K. Chipman

Department of Biochemistry
University of Birmingham
Birmingham, United Kingdom

Butylated hydroxytoluene (BHT) can protect against the carcinogenicity of 2-acetylaminofluorene (2AAF) and 7,12-dimethylbenz(α)anthracene (DMBA). The effect of BHT on the mutagenicity of these procarcinogens has been studied in Salmonella assays using hepatocytes as activation systems to model in vivo metabolism. Using hepatocytes from uninduced rats, the bacterial mutagenicity of DMBA (337 nmole/plate) was inhibited by BHT in a dose-dependent manner (90% inhibition at 300 nmole BHT/plate). Under these conditions, however, metabolic activation by Salmonella appeared to override the hepatocyte contribution. At 40 nmole DMBA/plate, Aroclor-induced hepatocytes were used to provide metabolic activation, and subtoxic levels of BHT afforded inhibition of up to 15% under these conditions. The mutagenicity of 2-AAF (100 nmole/plate) was also inhibited by BHT in uninduced hepatocyte/Salmonella assays. At 100 nmole/plate, BHT gave approximately 40% inhibition with both rat and human hepatocytes. The reduction in the number of revertant colonies was not due to BHT-mediated toxicity to bacteria in any experiment.

In contrast to these antimutagenic effects, when hepatocytes were prepared from rats pretreated with BHT in the diet (0.5% for 10 days), the mutagenicity of 2-AAF was significantly increased. With 250 nmole 2-AAF/plate, a 72% increase in mutagenicity was observed when compared with assays employing hepatocytes from untreated rats. The possible role of BHT interactions with reactive metabolites and with hepatocyte enzymes is being investigated.

(We acknowledge the financial support of the Medical Research Council.)

INHIBITION OF MUTAGENICITY OF N-METHYL-N-NITROSOUREA BY ELLAGIC ACID

Rakesh Dixit and Barry Gold

Eppley Institute for Research in Cancer
University of Nebraska Medical Center
Omaha, Nebraska 68105

Ellagic acid (EA), a plant phenol present in a variety of soft fruits and vegetables, has been shown to possess antimutagenic and anticarcinogenic properties against bay region diol epoxide of polycyclic aromatic hydrocarbons. It is suggested that EA forms an adduct with diol epoxide of benzo(α)pyrene and thus prevents its binding to DNA. To better understand the mechanism of reactivity and inhibition properties of EA, we studied the effect of EA on mutagenicity and DNA alkylation of carcinogenic N-nitroso compounds, including N-methyl-N-nitrosourea (MNU) and N-methyl-N'-nitro-N-nitrosoguanidine (MNNG). MNU and MNNG are direct-acting mutagens requiring no metabolic activation. MNU showed a linear dose response between the concentration range of 50 to 400 nmole in an Ames/Salmonella mammalian mutagenicity test. EA at concentrations of 100, 250, 500, and 1,000 nmole inhibited the mutagenicity of MNU (400 nmole) by 3, 13, 45, and 60%, respectively. MNNG produced a nonlinear dose response in mutagenicity between the concentrations of 0.5 to 4 nmole. EA showed no appreciable inhibition of MNNG mutagenicity. Inhibition of DNA alkylation by MNU and MNNG by EA was studied by preincubating 50 to 200 nmole of EA with 200 nmole of [^{3}H]-MNU or [^{3}H]-MNNG for 10 min at 37°C, followed by incubation of polymer deoxyguanosine:deoxycytosine(poly dG:dC) (1 unit) overnight. EA caused no inhibitory effect on MNNG alkylation of poly dG:dC. Experiments on the effect of EA on alkylation of DNA and formation of nucleoside adducts by MNU are in progress, and results will be discussed with reference to MNU and MNNG mutagenicity and EA inhibition.

ANTIMUTAGENIC ACTION OF TUMOR PROMOTERS AND CO-MUTAGENIC ACTION OF CO-CARCINOGENS IN YEAST AND MICE

Rudolf Fahrig

Fraunhofer-Institut für Toxikologie und Aerosolforschung
Hannover, Federal Republic of Germany

In experiments with yeast strain MP1, co-carcinogens were found to be co-mutagenic and antirecombinogenic, tumor promoters to be co-recombinogenic and antimutagenic. Substances that were co-carcinogens as well as tumor promoters had an intermediary effect. Bile acids showed all three possible ways of action supporting the hypothesis that mutagenesis is the mechanism by which chemicals induce malignancy, and that co-carcinogens modify the process by enhancement of mutagenicity, whereas tumor promoters effect carcinogenesis by increase of the spontaneous frequency of recombination. These results were confirmed in the mammalian spot test, by in vivo treatment of mice with the following: (a) the co-carcinogens catechol and hydroquinone, (b) the tumor promoter and co-carcinogen TPA, and (c) the tumor promoter limonene. Carcinogen-induced recombination due to mitotic crossing over and that due to gene mutations were reduced and enhanced, respectively.

REGULATION OF REPAIR OF O^6-METHYLGUANINE IN DNA OF MAMMALIAN CELLS

W.C. Dunn,[1] R.S. Foote,[2] P.A. Lalley,[3] R.E. Hand, Jr.,[1] and S. Mitra[1]

[1]Biology Division
[2]University of Tennessee-Oak Ridge Graduate School of Biomedical Sciences
Oak Ridge National Laboratory
Oak Ridge, Tennessee 37831

[3]Institute for Medical Research
Bennington, Vermont 05201

O^6-Methylguanine (m^6Gua), produced in DNA by methylating mutagens and carcinogens, is demethylated in situ by stoichiometric reaction with O^6-methylguanine-DNA methyltransferase (MGMT). The level of MGMT in Mex$^+$ mammalian cells varies over a range of approximately 1×10^4 to 2×10^5 molecules per cell, depending on cell line, and is undetectable in Mex$^-$ cells. Because m^6Gua is reportedly not repaired in mouse C3H 10T1/2 cells during S phase, we have investigated whether the MGMT level is cell cycle-dependent in these cells. Progression of the cell cycle in parasynchronized cultures was monitored by flow cytometry, and the cells were analyzed at various stages for in vivo disappearance of m^6Gua (introduced by treatment with

N-methyl-N'-nitro-N-nitrosoguanidine), as well as for MGMT activity by in vitro assay. Decreased levels of MGMT corresponded to reduced in vivo repair between the G_1-S border and the onset of G_2 phase. The magnitude of the decrease in MGMT was proportional to the maximum fraction of cells in S phase. Preliminary results of MGMT assays performed on S phase cells obtained by cell sorting indicate that these cells contain 10-20% of the maximum level of MGMT.

Fusion of Mex^+ human cells (WI38) with Mex^- mouse cells (RAG) resulted in only Mex^- hybrids, suggesting that the Mex^- phenotype is dominant. Hybrids of RAG cells and Mex^- hamster cells (E36) were also Mex^-.

(Research sponsored by the Office of Health and Environmental Research, U.S. Department of Energy, under contract DE-AC05-84OR21400 with the Martin Marietta Energy Systems, Inc., and by the National Cancer Institute under Grant No. CA-31721.)

A SENSITIVE GENETIC ASSAY FOR CYTOSINE-TARGETED MUTATIONAL EVENTS: STUDIES ON CYTOSINE DEAMINATION

L.A. Frederico,[1,2] B. Ramsay Shaw,[1] and T.A. Kunkel[2]

[1]Department of Chemistry
Duke University
Durham, North Carolina 27706

[2]National Institute of Environmental Health Sciences
Research Triangle Park, North Carolina 27709

The inherent instability of chemical bonds in DNA provides a substantial source of spontaneous and induced mutations. Lindahl has calculated [Biochemistry (1974) 13:3405-3410] that the deamination of cytosine to produce uracil presents a major challenge to cells at physiologically relevant temperatures from an extrapolation of biochemical experiments performed at high temperature. Our interest in cytosine deamination has led us to develop a sensitive genetic assay to monitor this event. The system is based on reversion of a mutant CCC proline codon in the lac Zα gene carried in bacteriophage M13mp2. The mutant produces a colorless plaque phenotype which can be reverted to wild-type dark blue color by C → T transitions at either of the first two cytosine residues in the codon. Transfection of treated DNA into an *E. coli* strain that is not capable of removing uracil produces a C → T transition for each deamination event at the two observable positions. This sensitive assay has permitted a determination of the activation energy for cytosine deamination as well as the rate constant for both double- and single-stranded DNA at physiologically relevant temperatures. The use of additional cytosine-containing codons permits the sequence specificity of cytosine deamination to be determined. In addition to its utility for deamination studies, this cytosine-target reversion system should be applicable to other interesting questions, including the mutational specificity of DNA-damaging agents and fidelity studies with purified proteins.

SUPPRESSION OF MUTAGENIC ACTIVITY OF AMINES OR CARBONYL COMPOUNDS BY THEIR MUTUAL INTERACTION

H. Furukawa, K. Kawai, and N. Okado

Meijo University
Yagoto-urayama 15, Tempaku-cho, Nagoya 468, Japan

An equimolar mixture of aflatoxin B_1 and spermine was allowed to stand for a few days, then the mutagenic activity of aflatoxin B_1 on S. typhimurium TA100 with S-9 mix was suppressed by spermine. The mutagenic activities of citreviridin and aflatoxin G_1 on TA100 without S-9 mix were also suppressed by addition of spermine. The mutagenic activity of citreviridin on TA100 without S-9 mix was suppressed by addition of hematoporphyrin. The mutagenic activity of an equimolar mixture of aflatoxin G_2 and Trp-P-2 on TA100 without S-9 mix was weaker than the sum of their mutagenic activities. The mutagenic activity of an equimolar mixture of methylglyoxal and Trp-P-2 on TA98 with S-9 mix was weaker than the sum of their mutagenic activities. The mutagenic activity of IQ on TA100 with S-9 mix was suppressed by addition of L-ascorbic acid. Studies with NMR, IR, and UV spectra showed hydrogen bond, charge transfer complex, or reaction products formation between carbonyl compounds and amines, suggesting a possible mechanism for these suppressions. But the mixing of carbonyl compounds and amines does not always result in an observed suppression of mutagenic activity. In the case of bilirubin as amine and aflatoxin G_2, sterigmatocystin or psoralen as carbonyl compounds, an enhancement of the mutagenic activity of carbonyl mutagens by adding the bilirubin was observed. Mutagenic activities of amines and/or carbonyl compounds were modulated by their contact with each other. We therefore suggest that such suppression and enhancement of mutagenic activity be termed "contact modulation."

BLOCKAGE OF DNA CHAIN ELONGATION AND THE USE OF ALTERNATE SITES OF REPLICON INITIATION IN CHINESE HAMSTER OVARY CELLS AFTER EXPOSURE TO UV LIGHT

T. Daniel Griffiths and Su Y. Ling

Department of Biological Sciences
Northern Illinois University
DeKalb, Illinois 60115

Wild-type Chinese hamster ovary cells (AA8) and 5 excision-deficient and UV-sensitive clones derived from the AA8 line (UV-4, UV-5, UV-20, UV-24, and UV-41) were exposed to UV light and then analyzed for their ability to incorporate [^{3}H]thymidine and to initiate as well as elongate replicon-sized DNA fragments. For exposures of 4.0 J/m^2 or higher, the wild-type cells recovered normal rates of thymidine incorporation within a few hours, while none of the excision-deficient lines exhibited complete recovery. For fluences below 4.0 J/m^2, all but the UV-5 line exhibited at least some recovery.

The ability to elongate DNA chains appeared to correlate with the thymidine incorporation data, with the UV-5 line exhibiting the strongest blockage of DNA chain elongation, the AA8 line exhibiting the least blockage, and the UV-20 line exhibiting an intermediate response. All cell lines exhibited a decrease in the distance between replication origins after exposure to 2.5 or 5.0 J/m^2 UV. This decrease was evident within the first 25 min after exposure and persisted for 2.5-3.5 hr in the AA8 line. The excision-deficient cell lines appeared generally to exhibit this decrease for a longer time interval after UV exposure. These data support models which propose that exposure to UV results in the use of alternate sites for initiation of replication.

(This work supported by U.S. Public Health Service Grant CA 32579.)

INTRAGENOMIC DNA REPAIR HETEROGENEITY IN MAMMALIAN CELLS

P.C. Hanawalt, V.A. Bohr, I. Mellon,
D.S. Okumoto, and C.A. Smith

Department of Biological Sciences
Stanford University
Stanford, California 94305

We have developed methodology for studying DNA repair in defined genes in mammalian cells. Restricted genomic DNA from UV-irradiated cells is treated or not treated not treated with the pyrimidine dimer-specific T4 endonuclease V, electrophoresed under denaturing conditions, transferred to nitrocellulose paper, and hybridized to specific ^{32}P-labeled probes. The proportion of fragments free of endonuclease-sensitive sites is determined from the difference in amount of probe hybridized at the position of full length fragments for T4 endonuclease-treated and untreated samples. In Chinese hamster ovary (CHO) cells, 70% of the dimers are removed from a 14-kb restriction fragment within the dihydrofolate reductase (DHFR) gene in 24 hr, while little repair is detectable in a 20-kb fragment 30 kb upstream from the gene and only 15% repair is seen in the genome overall. Preferential repair of vital genes such as DHFR may account for the high UV resistance of CHO cells in spite of low overall repair levels [Bohr et al. (1985) Cell 40:359-369]. Repair in the DHFR gene in normal human cells reflects the high level characteristic of these cells overall. However, in cells from xeroderma pigmentosum, group C (XPC), exhibiting UV sensitivity and only 10-20% repair of dimers in the genome overall, repair in the DHFR gene is deficient. It appears that the previously noted heterogeneity in repair in XPC may be nonselective for active genes and therefore of little value for cellular survival. Thus, repairability of damage in mammalian genomes may depend not only upon the type of lesion, but also upon the functional state of its genomic location. We may need to understand the "fine structure" of DNA repair in order to interpret effects of DNA damage upon particular biological endpoints, such as survival, mutagenesis, and transformation.

(Supported by grants from the National Institutes of Health and the American Cancer Society.)

ANTICARCINOGENIC EFFECTS OF GREEN TEA EXTRACT

Yukihiko Hara,[1] Hajime Asai,[2] and Kozo Nakamura[3]

[1]Food Research Laboratories,
Mitsui Norin Co., Ltd.
Fujieda City, 426 Japan

[2]Maruko Pharmaceutical Co., Ltd.
Kasugai City, 486 Japan

[3]National Cancer Center
Tsukiji, Tokyo, 104 Japan

It was reported that green tea extract has antimutagenic effects on spontaneous reverse mutation and MNNG-induced mutation in bacteria (Kada et al.). Based on the above findings, we isolated Tea Tannin Mixture (TTM) and its main component, (-)-epigallocatechin gallate (EGCg), from Japanese green tea and studied their anticarcinogenic effects in rats and mice. Tumors used were Ehrlich-, Sarcoma 180-, and 20MC-induced tumors. The effects were evaluated by the suppression of the growth of solid tumors and by the prolongation of life in ascitic tumors. Compounds were administered orally, subcutaneously, or intraperitoneally every day or every other day. Both TTM and EGCg showed suppressing effects on solid tumor growth when they were administered intraperitoneally. Particularly, TTM showed a marked effect when administered orally before implanting tumors. In ascitic tumors, no life prolongation effects were observed in any case. The facts that TTM, which has an antimutagenic effect, inhibits the growth of the tumors and that the effect is strengthened by giving it orally before implanting tumors are interesting and worth further study.

POTENTIATION OF THE ACTION OF DTIC BY INHIBITORS OF POLY(ADP) RIBOSE POLYMERASE AND INTERACTIONS WITH O^6-METHYLGUANINE-DNA METHYLTRANSFERASE

J.M. Lunn,[1] A.L. Harris,[1] P.M. Brown,[1]
C. Pierpoint,[2] and B.T. Golding[2]

[1]Cancer Research Unit
[2]Department of Organic Chemistry
University of Newcastle upon Tyne
Newcastle upon Tyne NE1 4LP, United Kingdom

The drug 5-(3,3-dimethyl-1-triazeno)imidazole-4-carboxamide (DTIC) is metabolized by humans to the related monomethyl compound (MTIC), whose subsequent breakdown gives rise to short-lived reactive species that methylate cellular DNA. DTIC is the most active chemotherapeutic agent in treatment of malignant melanoma. To obviate the need for this activation step, MTIC has been synthesized and its effects on A549 human lung adenocarcinoma cells (mer^+) examined. Exposure to MTIC caused inhibition of cell proliferation quantitatively similar to that produced by the methylating agent methyl nitrosourea (MNU). DNA strand breaks, detected by an alkaline unwinding technique, increased linearly with increasing concentrations of MTIC. Cellular NAD levels dropped after exposure to MTIC concentrations in excess of 1 mM. Inclusion of 3-acetamidobenzamide (3-AAB), an inhibitor of poly(ADP) ribose synthetase, in the growth medium enhanced MTIC cytotoxicity 4-fold, and higher levels of DNA strand breaks persisted in its presence, showing that 3-AAB can overcome resistance associated with the mer^+ phenotype. Methylation of DNA in vitro by MTIC produced O^6-methylguanine (O^6-MeGua). Since the sensitivity of cells to MTIC has been shown to be dependent on the Mer phenotype, cells are being treated with O^6-MeGua to deplete them of O^6-MeGua-DNA methyltransferase in order to sensitize them to MTIC. MTIC is the only clinically used methylating agent, and these studies suggest ways to potentiate its use against drug-resistant tumors.

THE ROLE OF FREE RADICALS IN GAMMA-RAY-INDUCED DNA LESIONS

P. Hentosh and Richard J. Reynolds

School of Public Health
Harvard University
Boston, Massachusetts 02115

Studies of free radicals have gained impetus from accumulating evidence implicating them as an intermediate step in the production of DNA damage. Moreover, it has been well documented that free radicals are involved in radiation-induced carcinogenesis. Such documentation implies that radical-scavenging agents may be effective anticarcinogens and antimutagens. We have examined at a molecular level the effects of radical scavengers on the induction of DNA damage by ^{60}Co gamma-rays. Using irradiated human lymphoblasts and dilute solutions of purified hamster cell DNA, two classes of DNA lesions were examined by velocity sedimentation: strand breaks and DNA base damage recognized and converted to single strand breaks by enzymatic activities present in extracts of _Micrococcus luteus_. In irradiated lymphoblasts, strand breakage exhibited the expected sensitivity to oxygen and OH• scavengers, but no difference in the yields under N_2 or N_2O irradiation conditions was observed. In contrast to strand break induction, enzyme sensitive sites (ENSSs) had no oxygen enhancement ratio, but decreased when OH• scavengers were present during cellular irradiation under O_2, N_2, or N_2O. In dilute solutions of purified DNA, radiation-induced strand breaks displayed little quantitative change under either O_2 or N_2, but increased appreciably under N_2O. Hypoxic irradiation of purified DNA resulted in a 2-fold increase in ENSSs compared to the yield detected under O_2. These sites were shown to decrease in number when DNA was irradiated in the presence of several OH• or aqueous electron scavengers. Our results suggest a role for both OH radicals and aqueous electrons in the formation of base damage recognized by _M. luteus_ gamma endonuclease, and provide a strong foundation for further studies on the antimutagenic effects of radical scavengers.

(Supported by the National Institutes of Health grant CA 09078. Present address for P.H.: Eleanor Roosevelt Institute for Cancer Research, Denver, Colorado 80262.)

GENETIC COMPLEMENTATION OF CHINESE HAMSTER OVARY DNA REPAIR MUTANTS WITH HUMAN DNA

Ian D. Hickson, Craig N. Robson, and Adrian L. Harris

Department of Clinical Oncology
University of Newcastle upon Tyne
Royal Victoria Infirmary
Newcastle upon Tyne NE1 4LP, United Kingdom

Seven DNA repair mutants of CHO-K1 cells have been isolated on the basis of sensitivity to the anticancer drugs mitomycin C and bleomycin. One mutant, designated MMC-2, exhibits 6-fold hypersensitivity to mitomycin-C and 10-fold hypersensitivity to UV light. Five of the other mutants show approximately this level of sensitivity to mitomycin C but are resistant to UV light. One of these mutants (BLM-2) is also hypersensitive to bleomycin. One additional mutant (BLM-1) exhibits sensitivity to only bleomycin and intercalating drugs. Cell fusion experiments indicate that the mutants represent at least 6 different complementation groups. A human gene bank of approximately 300,000 recombinants has been constructed in the cosmid pNNL (*Nucl. Acids Res.* 10:6715). This DNA has been introduced into the mutants by polycation transfection (*Mol. Cell. Biol.* 4:1172) and repair-competent transfectants selected in MAXT medium (mycophenolic acid, adenine, xanthine, thymidine) containing mitomycin C or bleomycin. In 3 cases, transfectants have been recovered that exhibit essentially wild-type resistance to DNA-damaging agents. Southern hybridization analysis of genomic DNA isolated from the transfectants reveals the presence of both the *gpt* gene and human DNA sequences (using the Alu sequence as probe). We are currently attempting to recover the transfected DNA by cosmid rescue.

INTERACTIONS BETWEEN YEAST REPAIR FUNCTIONS IN MITOTIC RECOMBINATION

Merl F. Hoekstra and R.E. Malone

Department of Microbiology
Loyola University
Maywood, Illinois

The *rem1* mutations of *S. cerevisiae* are mitotic-specific, semidominant, and confer a hyperrec/hypermutable phenotype. Unlike *rad* mutations exhibiting some of the same properties, *rem1* strains are as resistant as wild-type to UV and MMS. We have examined interactions between *rem1* and various *rad* mutations to determine the cause of the phenotype. The interactions are: (a) *rem1* is inviable with *rad50* and *rad52*; (b) *rem1 rad6* strains show *rem1* levels of recombination and mutation; (c) *rem1* with *rad1* or *rad4* shows reduced recombination levels and decreased mutation rates; and (d) *rad1* and *4* restore viability to *rem1 rad50* and *rem1 rad52* double mutants.

Our interpretation is that reml strains contain lesions resulting in double-strand breaks, thereby requiring the RAD50 recombination-repair group for viability. In order for the lesion to become a double-strand break, the recognition and action of the RAD3 excision-repair group are required. Mutationally inactivating excision repair reduces the hyperrec effect and rescues reml rad50 and reml rad52 strains by blocking double-strand break formation. One possibility for the lesion in reml strains is an increase in base mismatches. Excision tracts on both strands could lead to a double-strand break requiring RAD50 and RAD52.

We have cloned REM1. The clone restores normal levels of mitotic recombination when on a high copy episomal plasmid. It also complements reml rad50 inviability by allowing reml ts rad50 strains to survive at 35°C. The clone also complements a UV sensitivity mutation. We currently are cloning by gap-rescue procedures reml alleles in order to determine the nature of the mutations responsible for the phenotype. We feel that the REM1 gene product may have more than one function--an involvement in repairing UV damage as well as affecting mitotic recombination. An obvious possibility is that the REM1 gene codes for a repair function capable of recognizing helix distortions and thus might play a role in DNA replication fidelity.

NUCLEOSIDE ANALOGS AS TOOLS IN THE INVESTIGATION OF THE MECHANISMS OF MISMATCH REPAIR IN *ESCHERICHIA COLI*

Steven G. Wood, Josef Jiricny, Aiko Ubasawa, and Doris Martin

Friedrich Miescher-Institut
CH-4002 Basel, Switzerland

A number of analogs of deoxyadenosine and deoxyguanosine were synthesized and incorporated into synthetic hexadecanucleotides. These were then used as carriers for the stable integration of these lesions into predetermined sites within the α-peptide gene of M13mp9 in such a way that the nucleoside analog was positioned opposite a noncomplementary pyrimidine in the template strand. Following transfection into *E. coli* strains JM101 and BMH 71-18 mutS, the efficiency of repair of these substrates from mismatches was scored as the number of plaques carrying an unaltered phenotype [β-gal(+) or β-gal(-)]. A/C and G/T mismatches were found to be repaired with high efficiency. Inosine/thymidine mismatch was even more effectively removed from duplex DNA, presumably by the action of hypoxanthine DNA glycosylase. The other analogs, which consisted mainly of modifications of the 5-membered ring of the purine residue, were, in general, poorly repaired from mismatches of the purine/pyrimidine type.

In a second set of experiments an attempt was made at reproducing the repair reaction in vitro by using total or partially purified cell extracts from a variety of *E. coli* mutants with mismatch-containing synthetic oligonucleotide duplexes. Our preliminary results indicate that the cell extracts from the mismatch repair-deficient strains BMH 71-18 mutL and mutS are unable to incise these substrates. Reduced mismatch repair activity was also observed with cell extracts from *E. coli* uvrB and C mutants.

A NEW MODEL SYSTEM FOR MAMMALIAN MECHANISMS OF ANTICARCINOGENESIS

H. Joenje, J.J.P. Gille, and P. van der Valk

Institute of Human Genetics, Free University
Amsterdam, The Netherlands

Active oxygen metabolites are considered as the main causative agents in spontaneous and induced carcinogenesis [Ames (1983) Science 221:1256; Cerutti (1985) Science 227:375], so that antioxygenic defenses may play a role in anticarcinogenesis. To study such defenses we have used Chinese hamster ovary (CHO) and HeLa cells to select for oxygen-tolerant substrains by chronic adaptation to stepwise increased partial pressures of O_2. The tolerant substrains obtained were capable of proliferating at 600-720 mm Hg O_2, an O_2 level that readily kills normal tissue culture cells. In CHO cells, increased O_2 tolerance was associated with increased activity levels of superoxide dismutase, catalase, and glutathione peroxidase, which is consistent with their proposed important role in the defense against O_2 toxicity. The mechanism of oxygen tolerance in the HeLa substrain, which did not exhibit increased protective enzyme levels, has not been elucidated yet. Since hyperoxia is known to be highly genotoxic to eukaryotic cells, background levels of chromosomal aberrations (CAs) and sister chromatid exchanges (SCEs) were determined in the hyperoxia-adapted cells. CA and SCE levels were substantially increased in both types of tolerant substrains. When O_2-tolerant CHO cells were shifted back to normoxic conditions, the SCE level went back to normal, whereas the elevated CA level persisted for many cell generations. This study shows (a) that HeLa cells can survive under hyperoxia without increased levels of the well-known "antioxidant" enzymes; (b) that cells growing under hyperoxic stress are genetically unstable; and (c) that chromosomal instability in O_2-adapted CHO cells is a secondary consequence ("after-effect") of adaptation to hyperoxic stress. [References: P. van der Valk et al. (1985) Cell Tissue Res. 239:61; H. Joenje et al. (1985) Lab. Invest. 52:420.]

ANTIMUTAGENIC ACTION OF SPENT MEDIA FROM HUMAN FIBROBLAST CULTURES ON MUTATIONS INDUCED IN *ESCHERICHIA COLI* BY ULTRAVIOLET LIGHT

K. Hunter and R.C. Johnson

Medical University of South Carolina
Charleston, South Carolina 29425

Previous studies in our laboratory demonstrated some asbestos antitoxic properties of spent media from neonatal fibroblasts. In this study, we tested the possible antimutagenic action of spent media from human fibroblast cultures on mutations induced in *E. coli* AB1157 by UV light. Bacteria were scored for survival and mutation to rifampicin resistance after exposure to UV light. Bacteria were either incubated in spent media after UV light exposure or untreated. Spent media were prepared from nearly confluent normal neonatal or adult fibroblast cells. The spent media may be slightly toxic to the bacterial cells but demonstrated mutation frequencies well within the range of spontaneous mutation frequencies. Incubation of bacteria with spent media after exposure to UV light reduced the expected mutation frequency by at least 1 log. Donor age appears to affect the result only slightly. The antimutagenic substance is nondialyzable and heat-sensitive; reduction of γ-irradiation-induced mutation was also observed.

EFFECT OF ANTICANCER COMPOUNDS ON CELLULAR AND SIMIAN VIRUS 40 DNA SYNTHESIS IN VIVO AND IN VITRO

Mira Jung, Muriel K. Smith, and Robert T. Su

Department of Microbiology, University of Kansas
Lawrence, Kansas 66045

Many anticancer drugs are intercalative compounds which inhibit DNA synthesis in cells. The mechanisms of several anticancer drugs (ellipticine, adriamycin, mitoxantrone) on cellular and simian virus 40 (SV40) chromosome synthesis were examined in African green monkey kidney (CV-1) cells. All drugs specifically inhibited initiation of viral and cellular replicons. Using SV40 as a model, elongation and termination of chromosome synthesis were normal in the presence of drugs. The damaged viral chromosomes caused by the interaction of drug and DNA molecules in vivo and in vitro showed a preferential repair synthesis in the nucleosome-free region. Increasing amounts of supercoiled viral DNA could be synthesized in reaction mixtures in the presence of different concentrations of drugs. Most of the ^{32}P radioactivity was found in a region containing the origin of replication. Supercoiled plasmid DNA with or without the insertion of the origin of SV40 was also synthesized when added into a reaction mixture in the presence of drug. To test for the possible competition among different drugs on the target site of viral chromosomes, the recovery of viral DNA synthesis after the removal of drugs was examined. Mitoxantrone appeared to intercalate more firmly on viral chromosome than ellipticine and adriamycin. The inhibition of replicon initiation might contribute to the antineoplastic activity of these drugs.

SPONTANEOUS MUTATION RATES IN CELLS WITH CHROMOSOMAL REARRANGEMENTS

Debra A. Kaden and Ruth Sager

Dana-Farber Cancer Institute
Boston, Massachusetts

The frequent occurrence of chromosomal rearrangements in tumor cells has suggested that some event leading to genetic instability is involved in the neoplastic process. One such event, the mobilization of transposable genetic elements, would also be expected to cause an elevated level of spontaneous mutation. To study this possibility, we have examined the rate of spontaneous mutation at the hgprt locus in the diploid Chinese hamster embryo fibroblast cell lines CHEF/18 and CHEF/16 as well as in their descendant tumor-derived lines which show varying levels of chromosome rearrangement. For each cell line, the conditions for recovering hgprt⁻ mutants were optimized and mutation rates were determined by Luria/Delbrück fluctuation analysis. Our results indicate that the spontaneous mutation rate in tumor-derived cells increases in proportion to the extent of rearrangement in the cell lines.

(This work was supported by National Institutes of Health grant CA 24828. We gratefully acknowledge Lee Bardwell for helpful discussion.)

STRUCTURE-ACTIVITY RELATIONSHIP AND DESIGN OF ANTIMUTAGENS AGAINST THE UV-INDUCED MUTATION OF ESCHERICHIA COLI

K. Kakinuma,[1] J. Koike,[1] K. Ishibashi,[2] W. Takahashi,[2] H. Takei,[2] and T. Kada[3]

[1]Laboratory of Chem. Natural Products
[2]Department of Life Chemistry
Tokyo Institute of Technology, Midori, Yokohama 227, Japan

[3]National Institute of Genetics, Mishima, Shizuoka 411, Japan

The structure-activity relationship of antimutagens was studied by designing and synthesizing analogs and derivatives of 2(5H)-furanone based on the structures of natural antimutagens isolated so far by us as well as by others, and by assaying them (with a method of Kada and Mochizuki) by observing UV-induced reversion of *E. coli* WP2 B/r *uvr trp E*. It was clearly demonstrated that, at least in this assay system, an α,β-unsaturated carbonyl structure is required for the antimutagenic activity. Further modification of 3(2H)-furanones resulted in developing a highly potent antimutagen. Reactivity of these synthetic antimutagens with alkanethiol was also studied, and it was suggested that the antimutagenic effects may be due to alteration of a protein(s) involved in the DNA repair system by modification of the thiol groups.

INHIBITION OF MULTISTAGE TUMOR PROMOTION BY BIOMIMETIC SUPEROXIDE DISMUTASE: ROLE OF REACTIVE OXYGEN

T.W. Kensler, P.A. Egner, and B.G. Taffe

School of Hygiene/Public Health, Johns Hopkins University
Baltimore, Maryland

To probe the role of reactive oxygen species in carcinogenesis, we have examined the effects of selective scavengers of the proximate oxygen radical O_2^- (superoxide anion) on tumor promotion in mouse skin. The principal scavenger of interest, copper (II) (3,5-diisopropylsalicyclase)$_2$ (CuDIPS), is a low molecular weight, lipophilic copper coordination complex that catalytically dismutates O_2^- at a rate comparable to native CuZn SOD. We have previously observed that CuDIPS is a potent inhibitor of tumor promotion in a standard initiation-promotion protocol. Tumor promotion can be divided into 2 stages, conversion (I) and propagation (II), through the use of distinct chemical agents. In the present studies, female CD-1 mice were initiated with a single dose of dimethylbenz(α)anthracene followed by 4 twice-weekly applications of 10 nmol 12-O-tetradecanoylphorbol-13-acetate (TPA) for stage-I promotion. Stage-II promotion was completed by twice-weekly applications of 10 nmol 12-O-retinoylphorbol-13-acetate (RPA), an incomplete promoter, for 18 weeks. CuDIPS (4 μmol) was applied 30 min prior to the phorbol esters. Application of CuDIPS during stage I was without effect on tumor incidence and produced only 40% reduction in tumor multiplicity. By contrast, treatment during stage II inhibited tumor incidence and multiplicity by 50% and 80%, respectively. CuDIPS treatment during both stages I and II produced comparable effects. The induction of ornithine decarboxylase (ODC) activity is a biochemical marker for promoters with stage-II activity. ODC induction in mouse epidermis by either TPA or RPA can be blocked by CuDIPS (IC_{50} = 1 μmol). CuDIPS also blocks the induction of anchorage independence by TPA or RPA in JB6 mouse epidermal cells. This in vitro system is essentially a correlate of stage-II promotion. Collectively, these findings demonstrate that reactive oxygen species are involved in stage-II tumor promotion.

[Supported by grants from the American Cancer Society (Maryland Division and SIG-3) and the National Institutes of Health (CA-36380).]

MECHANISM OF INHIBITION OF AFLATOXIN B_1-INDUCED HEPATOCARCINOGENESIS BY ANTIOXIDANTS

T.W. Kensler,[1] P.A. Egner,[1] N.E. Davidson,[2] B.D. Roebuck,[3] A. Pikul,[4] and J.D. Groopman[4]

[1]School of Hygiene/Public Health
Johns Hopkins University
Baltimore, Maryland

[2]Uniformed Services University of the Health Sciences
Bethesda, Maryland

[3]Dartmouth Medical School
Hanover, New Hampshire

[4]Boston University School of Public Health
Boston, Massachusetts

Antioxidants such as ethoxyquin (EQ) exert chemoprotective effects against tumor induction by several carcinogens including aflatoxin B_1 (AFB_1). The inhibitory effect of EQ on AFB_1 carcinogenesis appears related to induction of detoxication enzymes. Rats maintained on a purified diet supplemented with 0.4% EQ for 7 days showed a 5-fold increase in hepatic cytosolic glutathione transferase (GST) specific activities and elevated GST messenger RNA levels, reflecting increased synthesis of enzyme in response to EQ. Correspondingly, biliary elimination of AFB-glutathione conjugate was increased several fold in animals on the EQ diet shortly following p.o. administration of 250 μg AFB_1/kg.

In other studies, rats were placed on the EQ diet and dosed 1 week later with 250 μg AFB_1/kg, Mon-Fri for 2 weeks. EQ was removed from the diet at week 4. EQ produced a dramatic reduction in the binding of AFB_1 to hepatic DNA: 7-fold initially and 3-fold at the end of the dosing period. Although binding was detectable at 3 and 4 months post-dosing, no effect of EQ was observed, suggesting that these persistent adducts (formamino derivatives of AFB_1-N^7-guanine) are not of primary relevance to AFB_1 carcinogenesis.

Livers were also analyzed at 4 months for putative preneoplastic lesions by the appearance of AFB_1-induced gamma glutamyl transpeptidase (GGT)-positive foci. EQ treatment reduced the number and volume of GGT foci by >95%. Dithiothiones, antioxidants found in cruciferous vegetables, produce a similar spectrum of effects. Thus, induction of enzymes by antioxidants important to AFB_1 detoxication, such as GST, leads to enhanced carcinogen elimination, reduction in AFB_1-DNA adduction and subsequent expression of preneoplastic lesions, and ultimately neoplasia.

[Supported by grants from the American Cancer Society (BC-477 and SIG-3), the American Institute for Cancer Research, BRSG, and Biothrust.]

ASCORBIC ACID (VITAMIN C) DECREASES CYCLOPHOSPHAMIDE-INDUCED SISTER CHROMATID EXCHANGES IN MICE

G. Krishna,[1] J. Nath,[2] and T. Ong[1]

[1]National Institute for Occupational Safety and Health
Morgantown, West Virginia

[2]Plant and Soil Sciences
West Virginia University
Morgantown, West Virginia

Ascorbic acid is known to act as an antimutagen and anticarcinogen under different experimental conditions in several test systems. However, there is no report of its effect on carcinogen-induced chromosomal damage in in vivo situations. The present study was performed to determine its effect on sister chromatid exchanges (SCEs) induced by cyclophosphamide, a known carcinogen and a mutagen.

Mice were exposed to cyclophosphamide, 40 mg/kg, and also to different concentrations of ascorbic acid for 8 hr with appropriate controls. In vivo and in vivo/in vitro (exposure of animals to experimental chemicals followed by culturing of cells) SCE analyses in mouse bone marrow and spleen cells were performed according to the established procedures.

The results indicated that ascorbic acid per se did not cause SCEs in mice. It, however, caused a dose-related decrease in the cyclophosphamide-induced SCE level at all concentrations tested in both bone marrow and spleen cells in vivo. At the highest concentration, 6.67 g/kg, approximately 75% SCE inhibition was noted in both bone marrow and spleen cells in mice. Likewise, under in vivo/in vitro conditions, ascorbic acid caused a dose-related decrease in cyclophosphamide-induced SCEs. At the concentration of 3.33 g/kg, 61% and 41% SCE inhibition was observed in mouse bone marrow and spleen cells, respectively. In addition, ascorbic acid did not cause cell cycle delay under these experimental conditions. Thus, ascorbic acid acts as an anti-SCE agent in both in vivo and in vivo/in vitro studies in mice.

REC-ASSAY IS AVAILABLE FOR SCREENING ANTIMUTAGENS AS WELL AS MUTAGENS

Koichi Kuroda[1] and Young S. Yoo[2]

[1]Osaka City Institute of Public Health
and Environmental Sciences
8-34 Tojo-cho, Tennoji-ku, Osaka 543, Japan

[2]Department of Public Health and Preventive Medicine
Osaka University Medical School
Osaka 545, Japan

The DNA-damaging activity of 33 synthetic flavoring agents widely used in foodstuffs was tested by the spore Rec-assay (Hirano et al., 1982) with _B. subtilis_ strains M45(rec^-) and H17(rec^+). Rec-assay positive agents were then tested for mutagenicity and antimutagenicity (Ohta et al., 1983) using _E. coli_ WP2 _uvrA_(trp^-).

Rec-assay: 23 flavoring agents inhibited growth of the organisms; 18 agents were positive (D=M45-H17 $\geq$ 4 mm) and 5 agents were negative (D < 4 mm). Mutagenicity test: Out of the Rec-assay-positive agents, ethyl acetoacetate (EA) was mutagenic without S-9 mix at the range of 200-1600 μg/plate. The mutation frequency of _trp_$^+$ revertants induced by EA at a dose of 1600 μg/plate was 4.2×10^{-7}. Antimutagenicity test: Out of the Rec-assay-positive agents (except EA), γ-undecalactone (UL), benzyl alcohol (BA), methyl anthranilate (MA), and cinnamic aldehyde (CA) reduced the number of _trp_$^+$ revertants induced by furylfuramide (AF-2) without any decrease of cell viability. Their RD_{50} values (a dose for reducing revertants to 50%) were 73, 1035, 267, and 17 μg/ml, respectively. Furthermore, UL and MA suppressed N-methyl-N'-nitro-N-nitrosoguanidine (MNNG)-induced mutagenesis, but BA and CA did not significantly.

Four antimutagenic agents detected in this study were assumed to disturb the process of mutagenesis, such as DNA replication and/or repair induced by DNA damage, and to suppress the mutagenicity of AF-2 or MNNG as a result. The Rec-assay, a method for detecting DNA damage activity, may be able to screen not only mutagens but also antimutagens which are due to DNA damage.

SEPARATION OF ESTROGENICITY FROM CARCINOGENICITY IN 2-FLUOROESTRADIOL

Joachim G. Liehr, George M. Stancel, Lynn P. Chorich, George R. Bousfield, and Aysegul Ari-Ulubelen

The University of Texas Medical School
Houston, Texas 77225

Estrogens have been associated with tumors in humans and in rodents. For instance, in male Syrian hamsters, estradiol or other synthetic or natural estrogens induce renal carcinoma with high incidence in 6-8 months. In an attempt to clarify the mechanism of estrogen-induced carcinogenesis, the biological activity of 2-fluoroestradiol was compared with that of estradiol, because of the reduced conversion of 2-fluoroestradiol to catechol estrogen metabolites. 2-Fluoroestradiol was administered to male Syrian hamsters by s.c. implantation at 3 times the dose (60 mg) of estradiol (20 mg), which was the positive control in this experiment. After 7 months, 75% of the estradiol-treated animals had renal carcinoma, while kidney tumors could not be detected in 2-fluoroestradiol-treated hamsters. The reduced tumor incidence was not due to a lack of absorption from the pellet; it was not due to a lack of estrogenic activity of the fluorinated steroid. The pituitary LH concentrations in the test animals closely matched those measured in estradiol-treated hamsters. Also, the decrease in testes weights as a result of steroid treatment was comparable in the test and control animals. The estrogenic potency of 2-fluoroestradiol was further illustrated by the uterine wet weight increase in immature female rats. The uterine growth was shown to be accompanied by increases in DNA and protein syntheses comparable to those found in estradiol-treated animals. 2-Fluoroestradiol also stimulated growth of H-301 cells in vivo. These cells are estrogen-dependent for growth and are derived from the primary hamster kidney tumor. The results indicate that hormonal activity and carcinogenicity of estrogens are separable properties.

[This work was supported by the National Institutes of Health (CA 27539 and HD 08615).]

INSERTIONAL MUTAGENESIS IN BACILLUS SUBTILIS: ISOLATION AND INITIAL CHARACTERIZATION OF DAMAGE-INDUCIBLE AND COMPETENCE-INDUCIBLE OPERON FUSIONS

P.E. Love, K.A. Gillespie, and R.E. Yasbin

University of Rochester, Rochester, New York 14642

Operon fusions, which were inducible for synthesis of β-galactosidase in the presence of mitomycin C (MC), UV radiation, and ethyl methanesulfonate (EMS), were generated in B. subtilis utilizing the Tn917-lacZ transposable element [P. Youngman et al. (1985) In Molecular Biology of Microbial Differentiation, J. Hoch and P. Setlow, eds. ASM Publ., Washington, D.C., pp. 47-54]. Seventeen damage-inducible (Din) insertions have been isolated to date. Fifteen of these are as resistant to UV, MC, and EMS as the wild-type. Two of the Din insertions were extremely sensitive to UV and MC and appear to have generated defects in a uvr transcriptional unit. All 17 Din::Tn917-lacZ fusion strains appear to be competence-proficient and are transformed for auxotrophic markers at or near wild-type levels. Additionally, we have found that the competent subpopulation of these Din mutants expresses β-galactosidase at much higher levels than the noncompetent subpopulation. Several of the Din promoter regions have already been subcloned into a retrieval vector. Putative promoters were then inserted into sites upstream from the cat-86 structural gene in plasmid pPL603. CAT synthesis in these plasmids was under the control of the SOB regulon [P.E. Love and R.E. Yasbin (1984) J. Bacteriol. 160:910-920]. Based on data obtained from these and previous experiments, it has become apparent that B. subtilis, like E. coli, possesses an inducible SOS-like (SOB) system, and that specific chromosomal loci become transcriptionally active in the presence of certain DNA-damaging agents. Furthermore, in B. subtilis competence and SOB induction appear to be closely linked events since (a) SOB phenomena are expressed during competence, and (b) several operons (one of which appears to be involved in DNA repair) have been found to be both damage-inducible (Din) and competence-inducible (Cin).

UV- AND EMS-INDUCED MUTAGENESIS IN BACILLUS SUBTILIS DEPENDS UPON A FUNCTIONAL SOB SYSTEM

K. Miehl-Lester and R.E. Yasbin

University of Rochester, Rochester, New York 14642

Bacillus subtilis possesses a set of pleiotrophic phenomena that are coordinately induced in response to DNA damage or the development of competency (1-3). This inducible system is known to consist of filamentous growth, prophage induction, W-reactivation, specific protein induction, and Din gene induction (1-5). This inducible system has been termed the SOB system of B. subtilis to denote the similarities to, yet differences between, the well-characterized SOS system of E. coli (2,6). SOB induction has an absolute dependency upon the $recE^+$ gene product and a partial dependency upon the $recA^+$, $recB^+$, and $recG^+$ gene products of B. subtilis (2,3). This current research communication reports that error-prone repair is also an SOB response. Furthermore, the ability of ethyl methanesulfonate (EMS) to induce mutations in B. subtilis is also dependent upon a functional SOB system. The evidence for these conclusions is as follows. Using the reversion of 3 auxotrophic mutations as the endpoint for induced mutagenesis, isogenic repair-deficient B. subtilis strains were tested for their ability to be mutated by UV or EMS. The recG13-containing strain exhibited only 40% of the level of UV-induced revertants that was observed in a repair-proficient B. subtilis strain. However, no induced mutagenesis was observed following EMS treatment of this same strain. Both UV and EMS treatment failed to induce any revertants in the recE4-containing strain. This dependency upon a functional recE4 gene product implies that error-prone repair is a component of the SOB system. In addition, EMS fails to induce any mutagenesis via a direct pathway in B. subtilis, thus differing significantly from what has been observed with the bacteriophage T_4 and E. coli but resembling what has been observed with yeast.

[References: (1) R.E. Yasbin (1977) Mol. Gen. Genet. 153:211; (2) P.E. Love and R.E. Yasbin (1984) J. Bacteriol. 160:910; (3) P.E. Love et al. (1985) Proc. Natl. Acad. Sci., USA (in press); (4) W.M. deVos and G. Venema (1983) Mol. Gen. Genet. 190:56; (5) C.M. Lovett, Jr. and J.W. Roberts (1985) J. Biol. Chem. 260:3305; (6) E.M. Witkin (1976) Bacteriol. Rev. 40:869.]

LACK OF VITAMIN E INHIBITION OF NITROSAMINE FORMATION FROM NO_2-DERIVED NITROSATING AGENT IN MOUSE SKIN

Sidney S. Mirvish and Kevin Laughlin

Eppley Institute for Research in Cancer
University of Nebraska Medical Center
Omaha, Nebraska 68105

α-Tocopherol (TOC) inhibits nitrosamine formation from N_2O_4 (NO_2 dimer) in CH_2Cl_2 [S. Mirvish (1981) In Cancer 1980, J.H. Burchenal, ed., Grune and Stratton, 1:557]. NO_2 reacts with skin lipids to form NO_2-derived nitrosating agent (NSA) (a component of which is cholesterol-3-β-nitrite; ms. in prep.), which reacts with amines to form nitrosamines [S. Mirvish (1983) Cancer Res. 43:2550]. We now report on the ability of TOC to inhibit NSA-amine reactions. Mice were exposed for 4 hr to 50 ppm NO_2, left 20 hr, and killed. Skin lipids containing NSA were extracted with CH_2Cl_2 and stored at -15°C. Samples (54 mg) were mixed with 25 mg morpholine or N-methylaniline in 75 ml CH_2Cl_2 ± 50 mg TOC, refluxed 30 min in a Kuderna-Danish apparatus, concentrated to 1.5 ml, left 20 hr, and analyzed by gas chromatography-Thermal Energy Analysis. We obtained 4.6 ± 0.6 and 2.5 ± 0.4 nmol N-nitrosomorpholine (NMOR), and 7.4 ± 0.1 nmol N-nitrosomethylaniline (NMA) (in 3 exps., each in duplicate) in absence of TOC; and 4.0 ± 0.1 and 3.2 ± 0.3 nmol NMOR, and 5.8 ± 0.4 nmol NMA in presence of TOC, with inhibitions of 13%, -28%, and 22%, respectively. Thus, TOC might inhibit skin NSA formation from NO_2, but does not inhibit nitrosamine formation after NO_2 exposure. It appears that NSA does not react with amines via liberation of free NO_2 or nitrite.

(Support: National Institutes of Health grant R01-CA-32192.)

RECOMBINATION AND MUTATION CATALYZED BY MAMMALIAN CELL EXTRACTS

Beth A. Montelone and Walter A. Scott

Department of Biochemistry
University of Miami School of Medicine, Miami, Florida

Crude nuclear extracts capable of catalyzing homologous recombination have been prepared from mouse and human cells grown in tissue culture.

Recombination has been assayed using a 411-bp tandem duplication constructed within the *lac i* gene of *E. coli* cloned into pBR322. Removal of the duplication results in a change from blue to white colony color on indicator X-gal plates after the plasmids have been introduced into *lac* i^- *E. coli*. Loss of the duplication is also measured directly by hybridization to a restriction fragment shorter by 411 bp than the starting material. Analysis of plasmids contained within blue and white halves of sectored colonies revealed that reversion of the duplication can occur by intermolecular recombination. Forward mutation has also been detected using the *lac i* plasmid. This is shown by the appearance of blue colonies on indicator medium following introduction of plasmids into *lac* i^- *E. coli*. Treatment of cells with carcinogens prior to extract preparation resulted in enhanced recombinational and mutational activities of the extracts. This suggests that inducible recombinogenic and/or mutagenic activities exist in mammalian cells.

(Supported by National Institutes of Health grant CA 35244 and National Institutes of Health Postdoctoral Fellowship 5 F32 CA 07524.)

STUDIES ON NATURAL DESMUTAGENS: ESPECIALLY, A DESMUTAGEN FROM BURDOCK (*ALCTIUM LAPPA* LINNE)

K. Morita[1] and T. Kada[2]

[1]Biochemistry Laboratory, Kanebo Ltd.
3-28 Kotobuki 5-chome, Odawara, Kanagawa 250, Japan

[2]Department of Induced Mutation
National Institute of Genetics
Mishima, Shizuoka 411, Japan

We have been carrying out detection and identification of vegetable desmutagens which work on pyrolysate mutagen derived from amino acids such as Trp-P-1 and Trp-P-2. In this way, hemoproteins with peroxidase and NADPH-oxidase properties were isolated from cabbage and broccoli. A new type of desmutagen was isolated from burdock. This factor reduced the mutagenicity of mutagens that are active without metabolic activation, such as 2-nitro-1,4-DAB and 4-nitro-1,2-DAB, as well as mutagens such as Trp-P-1 and Trp-P-2 requiring S-9 for metabolic activation. The partially purified principle had a molecular weight higher than 300,000 and showed the characteristics of a polyanionic substance.

Moreover, studies were carried out to elucidate the chemical nature of this desmutagen. The present findings suggest that this factor is not protein. The GC-MS analyses of its pyrolysates and reductive degradation products showed that this desmutagen might be a lignin-like compound having 10% sugar. From a C13-NMR spectrum and some other analyses, this desmutagen probably possesses carboxyl groups which play an important role in its desmutagenic activity. This factor irreversibly diminishes the activity of the mutagens, such as 2-nitro-1,4-DAB or Trp-P-2, presumably through adsorption.

COUMARIN INHIBITS MICRONUCLEI FORMATION IN MALE ICR MICE BY BENZO(α)PYRENE AND ALTERS THE PATTERN OF URINARY METABOLITES

Debra L. Morris, V.M.S. Ramanujam, and M.S. Legator

Department of Preventive Medicine and Community Health
University of Texas Medical Branch, Galveston, Texas 77550

We have demonstrated that pretreatment with coumarin inhibits micronucleus formation in male ICR mice treated with a single injection of benzo(α)pyrene (BP) in olive oil at 150 mg/kg. Male ICR mice were treated daily by oral gavage for one week with coumarin at 130 mg/kg (with one day of no treatment at midweek). There was no significant decrease in micronuclei formation from coumarin-treated animals when bone marrows were harvested at 24 hr after BP injection; but there was a significant decrease in micronuclei in animals pretreated with coumarin at the harvests 36 and 48 hr after the BP injection (a 27% and 52% decrease, respectively). Urines were collected for 24 hr after BP treatment and analyzed by HPLC for some metabolites of BP after incubation with β-glucuronidase. Both 3-OH-BP and 9-OH-BP showed approximately a 50% decrease in the urine of animals pretreated with coumarin compared to those who received only BP. Other metabolites also were decreased in coumarin-pretreated animals. Unmetabolized BP was increased by almost 5X in the urine of animals pretreated with coumarin. In a similar experiment with a week's pretreatment by coumarin at 65 mg/kg daily (except midweek), there was also a reduction in urinary metabolites of BP as well as a reduction in micronuclei. However, in a preliminary experiment with female ICR mice pretreated with coumarin at 65 mg/kg, we showed no difference in micronuclei formation at 12, 24, 36, and 48 hr after BP injection when coumarin-pretreated animals were compared to those gavaged with oil alone. There were little differences in BP metabolism between coumarin-pretreated mice and oil controls. We are currently extending our studies to include a time course analysis of BP metabolism in mice pretreated with coumarin or oil including glucuronides, sulfates, and glutathione conjugates. We are also more closely examining our preliminary observation of sex differences in this experimental protocol.

(Supported by U.S. Environmental Protection Agency R810920.)

REPAIR OF TRANSFECTED GENES IN CHINESE HAMSTER OVARY CELLS

Rodney S. Nairn, Gerald M. Adair, Moon-shong Tang, and Ronald M. Humphrey

University of Texas System Cancer Center
Smithville, Texas 78957

We have introduced cloned genetic sequences into both DNA repair-proficient and repair-deficient Chinese hamster ovary (CHO) cell lines by calcium-phosphate transfection. Unlike the case for human cells, we do not observe stimulation of genetic transformation by UV damage in transfected sequences. DNA adducts caused by NA-AAF treatment of transfecting pSV2gpt, but not adducts caused by N-OH AF or UV light, markedly affected the transformation frequencies in repair-deficient gene transfer recipients. These results are consistent with adduct removal data for repair-proficient and repair-deficient CHO cells. Other transfected genes such as herpes tk and hamster aprt showed different responses than pSV2gpt in the cell lines we analyzed. Co-transformation experiments were designed to determine the mechanism of integration of unrepaired transfecting sequences. Our results suggest that, while mammalian repair deficiencies can be reflected in transfection assays with cloned genes, other genetic and physiological factors may also be important in the generation of stable transformants in DNA repair-deficient cells.

(Supported by National Cancer Institute grants CA 36361 and CA 04484.)

SUPEROXIDE DISMUTASE INHIBITS PROMOTION OF NEOPLASTIC TRANSFORMATION BY TPA IN JB6 CELLS: ROLE OF REACTIVE OXYGEN IN TUMOR PROMOTION

Y. Nakamura,[1] E. Ohki,[1] I. Tomita,[1]
T.D. Gindhart,[2] and N.H. Colburn[2]

[1]Shizuoka College of Pharmaceutical Sciences
2-2-1 Oshika, Shizuoka 422, Japan

[2]Cell Biology Section
LVC, National Cancer Institute, Frederick, Maryland 21701

Concerning the role of reactive oxygen (RO) in tumor promotion, it is proposed that superoxide anion is a required mediator from the effect of a battery of RO scavengers on promotion of transformation by 12-O-tetra-decanoylphorbol-13-acetate (TPA) in JB6 mouse epidermal cells [Nakamura et al. (1985) Carcinogenesis 5:229-235]. We investigated the time of superoxide dismutase (SOD) addition required for the inhibition of soft agar colony induction by TPA and measured RO relevant biological response, such as the activities of SOD, glutathione peroxidase (GSH-Px), catalase and thiobarbituric acid reacting substances (TBA-RS) during TPA exposure in various JB6 and related clones [sensitive (p^+) or resistant (p^-) clones to induction of neoplastic transformation, and transformants (Tx)]. In experiments of inhibition of TPA-induced soft agar colony induction, delay of the addition of SOD by 2 hr or more after exposure of JB6 cells to TPA results in partial or complete loss of promotion inhibitory activity. Endogenous SOD levels were found to distinguish consistently the p^+(C141: low) from the p^-(C130: high) cells; but when these cell lines were exposed to TPA, their SOD activities changed in the same manner, reaching the minimum at 2 hr after TPA-exposure, followed by backing to base line activities, gradually. GSH-Px activity showed the same fluctuation pattern, but catalase and TBA-RS activities showed a few more complicated changes which are different from those of SOD and GSH-Px in p^+ and p^- clones. No change except catalase activity was observed in Tx. These results indicate that the activity of superoxide anion, accompanied by diminution of SOD activity, is critical during the first hour after the phorbol ester interacts with its cell-surface receptor (C-kinase).

SCREENING OF ANTIMUTAGENIC ACTIVITIES OF HERBAL EXTRACTS

T. Nunoshiba,[1] A. Ohtsuka,[1] T. Sotani,[1]
S. Saigusa,[1] H. Yoshimasa,[2] and H. Nishioka[1]

[1]Doshisha University, Kyoto 602, Japan

[2]Noevir Laboratory, Shiga 527, Japan

Activities of approximately 50 herbal extracts, which have been used as "old wives'" remedies for (i) inactivation of 4NQO mutagenicity, and (ii) suppression of UV mutagenesis, were screened by using bacterial assay systems. Herbal extracts samples were obtained by extraction with ethanol and/or distilled water. For mutation studies, the reversion to tryptophan prototroph and the forward mutation to a low concentration (5 μg/ml) of streptomycin resistance (SM^r-5) in WP2uvrA/pKM101, which has been developed by us [Sci. Eng. Rev., Doshisha Univ. (1982) 22:4] for a sensitive assay to detect carcinogen in the environment as mutagen, were examined. For experiment (i), the extract sample, the bacterial suspension (approximately 2×10^8 cells/ml), and 4NQO (0.3 μg/plate) were plated in the two different types of media for the reversion and the forward mutation. For experiment (ii), the UV-irradiated (0.75 J/m^2) bacteria were plated in the similar media containing the extract sample. It was observed that the extracts of Panax ginseng, horsetail, marigold, rosemary, broad-leaved lime, etc., inactivate 50-80% of 4NQO mutagenicity, and that those of Panax ginseng, horsetail, aloe, chamomile, etc., suppress 50-70% of UV mutagenesis. It is interesting that Panax ginseng and horsetail indicated the antimutagenic activities for both experiments (i) and (ii). A similar effect was also observed in riboflavin and chlorophyll Cu-Na salt. From these results, the mechanism of these antimutagenic effects will be discussed.

ANTIMUTAGENIC EFFECTS OF FLAVORINGS ON MUTAGENESIS INDUCED BY CHEMICAL MUTAGENS IN BACTERIA

T. Ohta,[1] M. Watanabe,[1] K. Watanabe,[1]
Y. Shirasu,[1] and T. Kada[2]

[1]Institute of Environmental Toxicology
Kodaira, Tokyo 187, Japan

[2]National Institute of Genetics, Mishima, Shizuoka 411, Japan

Twenty-five flavorings were tested for their antimutagenic potential on mutagenesis induced by several kinds of chemical mutagens in E. coli and S. typhimurium. Bacterial cells were treated with each mutagen in phosphate buffer for 15-60 min and then washed to remove the mutagen. Mutagenized cells were spread on semienriched minimal (SEM) agar plates containing each flavoring at various concentrations and on control plates with no flavoring. Cellular viability was determined on the same SEM plates.

Colonies of revertants and viable cells were counted after 3 days incubation at 37°C. Anisaldehyde, ethyl vanillin, and vanillin showed marked antimutagenic effects on mutagenesis induced by 4NQO, AF-2, Captan, or methylglyoxal in E. coli WP2s. However, they were not effective against mutations provoked by Trp-P-2 or IQ in S. typhimurium TA98. Despite the decrease in the number of mutants, a remarkable increase was observed in the survival of mutagen-treated WP2s cells after exposure to these flavorings. We assume that these compounds may act as bioantimutagens by enhancing an error-free recombinational repair system, because this reactivation of survival was strictly dependent on the recA gene function but not on the lexA and uvrA gene functions.

CHEMICAL STUDIES OF ANTIMUTAGENS OF MICROBIAL ORIGIN

T. Osawa,[1] M. Namiki,[1] S. Udaka,[1] and T. Kada[2]

[1]Nagoya University, Nagoya 464, Japan

[2]National Institute of Genetics, Mishima, Shizuoka 411, Japan

Recently, much attention has been focused on natural antimutagens which could have significant roles in lowering cancer incidence. We have started our project to isolate antimutagens from microbial origins, in particular, from Streptomyces spp. Screening of antimutagens from 293 different strains of Streptomyces has been carried out, monitoring antimutagenic effects on mutagenesis induced by MNNG or UV-irradiation in E. coli MP-1 and on spontaneous mutability in B. subtilis NIG 1125. Strong antimutagenic activity was observed in the metabolites of 2 strains of Streptomyces, Streptomyces griseus var rhodochrous AJ 9029 (AM-1) and an unidentified strain (Z-24). In order to isolate the antimutagens, large-scale incubation of the Z-24 strain (250 l) was carried out. Pure active substance was isolated from alcoholic extracts of the metabolite by combination of different types of chromatographic methods, including HPLC. From the data obtained by gel filtration and amino acid analysis, the active substance is believed to be the first peptide-type antimutagen. Chemical analysis and mechanisms of antimutagenesis will be discussed.

ETHIONINE HYPOMETHYLATING ACTIVITY, CELL CYCLE BLOCK AND SISTER CHROMATID EXCHANGE INDUCTION IN MAMMALIAN CELLS IN VITRO

Paolo Perticone and Ruggero Ricordy

Centro di Genetica Evoluzionistica del C.N.R.
c/o Dipartimento di Genetica e Biologia Molecolare
Università "La Sapienza", 00185 Roma, Italy

Ethionine, the ethyl analog of the essential amino acid methionine, is a well-known carcinogenic, but not mutagenic, agent with a DNA hypomethylating activity exerted by interfering with O^6-methylguanine methyl-transferases. Recent data show that the agent possesses the ability to block reversibly human lymphocytes and Chinese hamster ovary cells in the G_1 phase of the cell cycle. Besides, a clear ability to raise spontaneous and MMC-induced sister chromatid exchange (SCE) rates has been demonstrated.

Here we present a review of ethionine cytogenetic activity in different cell systems, in some of which (Friend erythroleukemia cells) this agent is able to induce differentiation. An attempt to correlate all these activities with the effect on DNA methylation is made.

(Supported by Progetto Finalizzato C.N.R. "Oncologia.")

ANTIMUTAGENIC EFFECTS OF 5-AZACYTIDINE IN BACTERIA

Denis M. Podger

Division of Molecular Biology
CSIRO
North Ryde, NSW 2113, Australia

The mutagenic effects of treatment of log-phase cells of *S. typhimurium* frameshift indicator strain *hisC3076* with 9-aminoacridine (9-AA) or ICR191 are dramatically reduced by pretreatment or concurrent treatment with the base analog, 5-azacytidine (5-azaC). Treatment of cells with 5-azaC immediately after the mutagenic exposure has little effect. The antimutagenic effect occurs rapidly, within a few seconds of exposure to 5-azaC. At the doses used, cell survival is unaltered by 9-AA and 5-azaC treatment. The antimutagenic effect of 5-azaC was not observed when log-phase cells were mutagenized with UV light or nitrosoguanidine. When a comparison was made between the mutagenic effects of UV light on stationary-phase and log-phase cells, no difference was observed in the relative mutability of either phase. In contrast, 9-AA was found to be mutagenic only to log-phase cultures. These results suggest that 9-AA exerts mutagenic effects at the replication fork, and that the antimutagenic effects of 5-azaC are replication-dependent.

Naladixic acid treatment of log-phase cells has little effect on the yield of 9-AA-induced revertants. This result suggests that the antimutagenic effect of 5-azaC is not due to the inhibition of DNA replication.

Induction of *recA*-dependent SOS responses by 5-azaC does not appear to have a role in reducing the yield of frameshift mutations, since the antimutagenic effects of 5-azaC were found to be similar in both wild-type and *recA* derivatives of *hisC3076*. Furthermore, 5-azaC pretreatment of cells has no effect on the levels of induction in the SOS chromotest by ICR191. This result indicates that 5-azaC does not interfere with the entry of the mutagen into the cell, and that the antimutagenic effects occur during protein synthesis.

MECHANISMS OF BENZO(α)PYRENE-DETOXICATION IN MAMMALIAN CELLS

L. Recio and A.W. Hsie

Graduate Center for Toxicology
University of Kentucky
Lexington, Kentucky 40536

Health and Safety Research Division
Oak Ridge National Laboratory
Oak Ridge, Tennessee 37831

The biotransformation of benzo(α)pyrene (BP) by the mixed-function oxidases results in substrates for further detoxication by UDP-glucuronyltransferases and glutathione-S-transferases. We have been investigating the biological roles of these enzymes in the Chinese hamster ovary cells/hypoxanthine-guanine phosphoribosyl transferase (CHO/HGPRT) gene mutational assay, supplemented with a biotransformation system prepared from rat liver homogenate, S-9 mix, and the appropriate enzyme co-factors. We found that addition of uridine diphosphate α-D-glucuronic acid to S-9 mix results in a reduction of BP-induced cytotoxicity without affecting mutagenicity. This is likely due to an elimination of cytotoxic phenols without affecting the formation and conversion of the proximate mutagen BP-7,8-diol-induced cytotoxicity; GSH lowered the mutagenic response of BP-7,8-diol in a concentration-dependent (2.1-4.2 μM) manner. The addition of glutathione-S-transferases to S-9 mix with GSH resulted in a decrease of BP-7,8-diol-induced cytotoxicity and mutagenicity at concentrations of BP-7,8-diol (7.0-14.0 μM) not affected by GSH alone. Our results demonstrate that UDP-glucuronyltransferases and glutathione-S-transferases play a role in modulating the cytotoxicity and mutagenicity of BP.

(Sponsored by the U.S. Environmental Protection Agency and the Oak Ridge National Laboratory under contract with Martin Marietta Energy Systems, Inc. L.R. is an ORAU Graduate Fellow and A.W.H. is an EPA Distinguished Visiting Scientist.)

INHIBITION OF AFLATOXIN MUTAGENICITY BY CHLOROPHYLLIN IN *NEUROSPORA CRASSA*

Edward Robins

Department of Biological Sciences
Illinois State University
Normal, Illinois

Epidemiological studies have shown that a reduction in the incidence of certain cancers is associated with the consumption of raw vegetables. This observation led to the testing of various vegetable extracts for their inhibition of the mutagenicity of certain carcinogens. The extracts of dark green vegetables were found to have the highest inhibitory activity, and chlorophyll was identified as the major factor responsible for the antimutagenic activity. Chlorophyllin, a commercial derivative of chlorophyll, was also found to have antimutagenic activity. I have studied the effect of sodium-copper chlorophyllin (SCC) on the mutagenicity of aflatoxin B_1 (AFB_1) in growing cultures of *N. crassa* in a reverse-mutation assay. AFB_1 was chosen because it is a human carcinogen, is activated metabolically in vivo and in vitro by man and *N. crassa*, and is a potent mutagen in many short-term tests for genetic toxicity. Initially, a dose-response experiment was done to determine a concentration of AFB_1 that would give a high reversion frequency in strain N23, a presumptive base-pair substitution *ad-3* mutant of *N. crassa*. Slants containing various concentrations of AFB_1 were inoculated with conidia from N23. The cultures were incubated for 7 days, after which mycelial growth and conidiation were complete. Conidia from these cultures were harvested, washed, and counted. Twenty million conidia were added to 2 ml of overlay medium and poured onto plates of minimal medium. A concentration of AFB_1 of 10 µg/ml, which produced ∿60 revertants per plate, was chosen to test chlorophyllin as an antimutagen. N23 was grown on slants containing 10 µg of AFB_1 per ml in the presence of various concentrations of SCC. SCC reduced the reversion frequency ∿50% and ∿90% at 150 and 1,500 µM SCC, respectively. SCC was nontoxic to *N. crassa* at all the concentrations used.

ANTIMUTAGENIC PROPERTY OF THE UNSAPONIFIABLE PORTION OF RICE BRAN OIL

C. Rukmini and Kalpagam Polasa

National Institute of Nutrition
Hyderabad, India

In recent years, considerable numbers of dietary constituents have been shown to inhibit carcinogenesis. Food contains a large number of non-nutritive components. These non-nutritive components of foods have been shown to have some beneficial effects, such as hypocholesterolemic, antilipidemic, antidiabetic, antihypertensive, anticarcinogenic, and antimutagenic activities. An attempt has been made to screen various Indian foods for these beneficial properties.

Rice-bran oil is a new source of oil which is undergoing rigorous toxicological evaluation to clear it for human edibility. Rice-bran oil is shown to have hypocholesterolemic activity. The unsaponifiable portion of rice-bran oil contains phytosterols, triterpene alcohols, and waxes. The triterpene alcohols were identified as cycloartenol and 24-methylene cycloartenol. This unsaponifiable fraction elevates the glutathione-S-transferase activity in vitro. When this fraction was tested with *S. typhimurium* strains TA98 and TA100, with and without metabolic activation, it did not alter the number of revertants significantly from the control group. The experimental details and the significance of these results will be discussed in detail.

(The authors express their grateful acknowledgements to Dr. B.S. Narasinga Rao, the Director, National Institute of Nutrition, for his keen interest in this work.)

INHIBITION BY CAFFEINE AND OTHER PHOSPHODIESTERASE INHIBITORS OF THE PROMOTIONAL EFFECT OF TPA ON TRANSFORMATION OF SYRIAN HAMSTER EMBRYO CELLS

Tore Sanner and Edgar Rivedal

Laboratory for Environmental and Occupational Cancer
Institute for Cancer Research
Montebello, N-0310 Oslo 3, Norway

Phorbol ester tumor promoters enhance the induction of morphological transformation of mammalian cells. The purpose of the present investigation was to study the effect of modulators of the cellular cAMP level on the promotional effect of TPA. The phosphodiesterase inhibitors caffeine, theophylline, and aminophylline were found to inhibit induction of morphologically transformed Syrian hamster embryo cell colonies by sequential exposure to benzo(α)pyrene and the tumor promoter TPA. Almost complete inhibition of cell transformation was observed when 50 μg/ml theophylline or aminophylline, or 200 μg/ml caffeine was present together with the tumor promoter. The compounds had no effect on the transformation frequency when present together with the initiator, benzo(α)pyrene, in the first exposure period. Substances that stimulate the adenylate cyclase activity and the addition of exogenous dibutyryl-cAMP had a similar inhibitory effect.

(Supported by the Norwegian Cancer Society.)

DESMUTAGENIC EFFECT OF HUMIC ACID

T. Sato, Y. Ose, and H. Nagase

Gifu Pharmaceutical University
5-6-1, Mitahora-higashi
Gifu City, Gifu, Japan

Various mutagens have been discovered in a water environment. River water contains humic substances, and there is a possibility that these humic substances may influence the mutagenicity of mutagens in a water environment. We examined the inhibitory effect of humic acid on the mutagenicity of various mutagens with the Ames method. Humic acid inhibited the mutagenicity of 3,4-benzo(α)pyrene, 3-aminoanthracene (+ S-9 mix), 1-nitropyrene and 3-nitrofluorene (- S-9 mix), but not the mutagenicity of 4NQO, AF-2, and MNNG (- S-9 mix). Humic acid had no mutagenicity itself and did not inhibit spontaneous mutation. The inhibitory effect was due to desmutagenic activity that acts directly on the mutagens before the mutagens act on cells, and not to an antimutagen that blocks the process, changing normal cells to mutants. The desmutagenic effect did not decrease with the heat treatment (120°C, 15 min). The humic acid was fractionated by molecular weight, and the desmutagenic effect increased with an increase of molecular weight. The desmutagenic effect of the molecular weight fraction above 300,000 decreased by centrifugation. Ferulic acid, protochatechuic acid and resorcinol, which are the components of humic acid, had no desmutagenic effect. Lignin, however, had a desmutagenic effect. Fulvic acid and activated sludge extract inhibited the mutagenicity of benzo(α)pyrene a little.

PROTECTIVE EFFECTS OF LONGER WAVELENGTHS ON UVB-INDUCED MUTAGENESIS AND LETHALITY

Christopher P. Selby,[1,2] John Calkins,[1,2] and Harry G. Enoch[1,3]

[1]Graduate Center for Toxicology
[2]Department of Radiation Medicine
University of Kentucky
Lexington, Kentucky

[3]Kentucky Energy Cabinet

We have conducted UVB action spectroscopy for mutagenesis and survival of Ames' *S. typhimurium* strain TA98 (uvrB, PKM101) using both monochromatic radiation from a dye laser and broader band-width radiation emitted from FS-20 sunlamps. A series of different filters having successively longer transmission cutoffs, together with the sunlamp source, provided different band-widths having successively longer wavelength components. When filters having successively longer wavelength cutoffs were interposed between the sunlamp source and bacterial populations, decreased mutagenesis and lethality were observed.

The "broad-band" action spectra were constructed based upon differences in mutational/lethal responses, energy transmitted, and wavelengths transmitted through successive cutoff filters. The two sets of action spectra differed; the "broad-band" spectra showed roughly a 100-fold reduced effectiveness for both mutagenic and lethal effects of UVB.

These results suggest a large protective effect of the background UVA and/or visible radiations which were present during the broad-spectrum irradiations and which are also present in solar radiation. Additional experiments suggest that this protective effect is neither photoreactivation nor dependent on dose rate.

(This work was supported in part by grants from the National Institutes of Health (RR-1620) and the Kentucky Energy Cabinet, and cooperative agreement #CR810294 from the U.S. Environmental Protection Agency. Although the research described in this article has been funded wholly or in part by the U.S. Environmental Protection Agency under assistance agreement number CR810294 to John Calkins, it has not been subjected to the Agency's required peer and administrative review and, therefore, does not necessarily reflect the view of the Agency, and no official endorsement should be inferred.)

ANTIMUTAGENIC EFFECTS OF PLANT COMPONENTS ON UV-INDUCED MUTAGENESIS

K. Shimoi,[1] Y. Nakamura,[1] I. Tomita,[1] and T. Kada[2]

[1]Shizuoka College of Pharmaceutical Sciences
2-2-1 Oshika, Shizuoka 422, Japan

[2]National Institute of Genetics
1111 Yata, Mishima, Shizuoka 411, Japan

Bioantimutagenic activities of various plant components on UV-induced mutagenesis were studied using E. coli B/r strains with different DNA repair capacities. Tannic acid, gallic acid, (-)-epigallocatechin, (-)-epigallocatechin gallate, and (-)-epicatechin gallate reduced the mutation induction about 70% in the strain of WP2, whereas chlorogenic acid, caffeic acid, quercetin, caffeine, and D-(+)-catechin showed no significant effect. Trihydroxy phenolic structure, therefore, seems to be essential for the bioantimutagenic activity. The compounds with a trihydroxy phenolic structure were also effective to mid and near UV-induced mutagenesis, and a 60-70% decrease of mutation induction was observed in the strain of WP2. It is assumed from the following findings that these trihydroxy phenolic compounds show bioantimutagenic activities by affecting the DNA excision repair system: (a) they did not suppress N-methyl-N'-nitro-N-nitrosoguanidine-induced mutagenesis; (b) this bioantimutagenic effect was not seen in the DNA excision repair-deficient strains $WP2_s$ and ZA159; and (c) inhibition of the expression of a revertant phenotype and the delay of the first cell division after UV irradiation were not observed in the presence of tannic acid [Shimoi et al. (1985) Mutat. Res. 149:17-23]. Their effects in mammalian systems are now under investigation.

LIPID METABOLISM IN MOUSE EPIDIDYMAL ADIPOSE TISSUE DURING THE CONTROLLED GROWTH OF EHRLICH ASCITES CARCINOMA CELLS

Safwat Shoukry,[1] Zeinab Zaki,[1] Fahmi T. Ali,[1] Samia S. Rizk,[2] and Amina Medhat[1]

[1]Department of Biochemistry, Faculty of Science
Ain Shams University, Cairo, A.R.E.

[2]National Research Center, Dokki

An i.p. dose (100 mg/kg) of 7-oxabicyclo(2.2.1)-5-heptene-2,3-dicarboxylic anhydride (OHD) curtailed the in vivo growth of Ehrlich ascites carcinoma (EAC) cells by inhibiting the de novo synthesis of DNA, RNA, and proteins. This effect was optimal during the first 24 hr, after which tumor growth returned slowly to its normal level. OHD had no effect on lipid synthesis in EAC cells. Therefore, OHD was used to slow down the tumor growth and consequently the turnover of lipids, which allowed us to study the relationship between host and tumor lipids. In normally growing EAC cells there was continuous mobilization of lipids from adipose tissue, and the ratio of neutral lipids (NL) to phospholipids (PL) was 2:1. After OHD injection there was a significant increase in adipose tissue lipogenesis; however, the NL:PL was significantly altered (1:3) indicating that the metabolic system regulating the synthesis of NL and PL lost its integrity to meet the requirement of the recovered EAC cells. The fatty acid (FA) composition of EAC cells indicated that values of saturated/unsaturated FAs decreased with the development of the tumor cells. Nevertheless, the ratio of monenoic/polyenoic FAs was 1.7-fold higher than tumor cells controlled with OHD. In adipose tissue, ratios of saturated/unsaturated FAs in normally growing EAC cells and OHD-controlled ones were 7.4 and 9.0, respectively. These results indicate that part of the tumor unsaturated FAs could be derived from adipose tissue lipids. In animals bearing tumors, the FA analysis of the adipose tissue indicated the presence of an unidentified peak (RT 344 sec on 20% PEGA column) which appeared after $C_{16:1}$ and before C_{17}. The concentration of this substance is under investigation.

THE IDEAL ANTIMUTAGEN MAY BE A QUENCHER OF EXCITED STATES

Kendric C. Smith

Department of Radiology
Stanford University School of Medicine
Stanford, California 94305

Much of spontaneous mutagenesis in *E. coli* is due to the error-prone repair of DNA damage; such damage is presumably produced in DNA as a natural consequence of normal metabolism. From knowledge of the genetic control of the DNA repair processes that affect spontaneous mutagenesis, it has been concluded that this spontaneous DNA damage is "UV-like" [N.J. Sargentini and K.C. Smith (1981) *Carcinogenesis* 2:863-872]. This conclusion is based in part upon the fact that a deficiency in error-free excision repair, which enhances the spontaneous mutation rate of cells, also renders cells more mutable by UV irradiation, but has little effect on X radiation mutagenesis. Similarly, mutations that affect DNA repair functions and reduce UV radiation mutagenesis (e.g., *uvrD*, *recB*) have a similar effect on spontaneous mutagenesis [reviewed in N.J. Sargentini and K.C. Smith (1985) *Mutat. Res.* 154:1-27].

A model exists, using horseradish peroxidase and several different substrates, for an enzymatic reaction that requires oxygen and produces excited-state molecules (i.e., "UV-like") that can damage DNA [G. Cilento (1980) *Photochem. Photobiol. Rev.* 5:199-228] and proteins [E.I. Rivas et al. (1984) *Photochem. Photobiol.* 40:565-568]. Furthermore, metabolizing cells exhibit low levels of chemiluminescence [E. Cadenas (1984) *Photochem. Photobiol.* 40:823-830; T.I. Quickenden et al. (1985) *Photochem. Photobiol.* 41:611-615], which is diagnostic for the presence of excited-state molecules. Therefore, the ideal antimutagen may be a quencher of excited states, since such a molecule should reduce or prevent the metabolic production of "UV-like" damage in DNA.

EFFECTS OF SODIUM SELENITE ON DNA AND DNA METABOLIC PROCESSES IN HUMAN DIPLOID FIBROBLASTS

Ronald D. Snyder

Stauffer Chemical Company
Farmington, Connecticut 06032

A number of mechanisms have been proposed for the anticarcinogenic activity of selenium. These include protection against carcinogen-DNA adduct formation by alteration of cellular metabolizing systems or by free radical scavenging due to the antioxidant nature of selenium compounds, inhibition of tumor growth through inhibition of DNA synthesis, and enhancement of cellular DNA repair systems. However, under certain conditions, selenium compounds can themselves induce genotoxic effects, including DNA damage and repair and sister chromatid exchanges. These factors could contribute to the reported carcinogenic activity of selenium compounds in some systems. In an attempt to better understand the nature of selenium interactions in intact cells, a series of experiments was conducted on the effects of sodium selenite on DNA and DNA metabolic processes in human foreskin fibroblasts. It is shown that selenite is a potent inducer of DNA single-strand breaks, but only in the presence of reduced glutathione or an as yet unidentified serum component. This damage is dose-dependent, comprised of true strand breaks rather than alkaline-labile sites, and apparently does not arise via free radical formation. Selenite is further shown to inhibit cellular proliferation at doses above 500 μM. In addition, our studies demonstrate that selenite does not enhance (or inhibit) the repair of MMS-, UV-, or bleomycin-induced damage in human fibroblasts. These results are discussed in terms of the prevailing models for selenium anticarcinogenesis.

MECHANISMS OF ANTIMUTAGENESIS PREDICTED BY A SIMULATION OF BACTERIAL MUTAGENESIS

Jonathan A. Sopher and R.L. Bernstein

Department of Biological Sciences
San Francisco State University
San Francisco, California 94132

Five distinct antimutagenic effects emerge from a theoretical treatment of the mutagenesis dose-response curve. Apparent antimutagens each fall into one of the five categories, depending upon the particular compartment and particular parameter affected.

The model of bacterial mutagenesis centers on induced error-prone SOS repair as the major source of mutation. Mutagens benzo(α)pyrene and ultraviolet light are explicitly treated. The simulated dose-response curve extends into the multi-hit high-dose region. Output of the model compares well with data for *S. typhimurium* strain TA100 (which has enhanced SOS repair and deficient excision repair). When each of the 16 parameters of the model is altered in the direction of antimutagenesis, the dose-response curve changes in a characteristic way. Antimutagenesis appears as (a) a lower specific rate of mutagenesis in the linear "one-hit" region; (b) a lower peak value of mutant yield; (c) a shift of the dose-response curve to higher dose values (to the right); or (d) certain combinations of these changes. For instance, a higher K_m for metabolic activation of benzo(α)pyrene shifts the simulated curve rightwards, which implies that an antimutagenic substance that produces such a shift when tested experimentally may reduce mutagen activation.

The model generates curves for dose-response, survival, mutation frequency, SOS repair, and metabolic activation. Values for premutagenic lesions per genome and mutation frequency agree with values in the literature. The model also predicts values for fixed mutations per cell and mutagen-specific mutability of sites. Because the simulation is continuous rather than discrete, the interesting region of very low-dose mutagenesis may be examined.

PROGRESS IN CLONING HUMAN GENES THAT AFFECT LETHALITY AND MUTABILITY

C. Waldren,[1,2] P. Hentosh,[1] D. Snead,[1] and S. Fixman[1]

[1]The Eleanor Roosevelt Institute for Cancer Research
[2]Department of Radiology
University of Colorado Health Sciences Center
Denver, Colorado 80262

In bacteria, and probably in mammalian cells, mutation depends on DNA repair. Since mutation underlies cancer induction, illumination of mechanisms of repair and mutagenesis is necessary to an understanding of antimutagenesis and anticarcinogenesis. Relationships between repair and mutation in bacteria have been elucidated through the study of repair-deficient mutants, an approach emulated with naturally occurring human cell mutants and, more recently, by induction and isolation in culture of repair mutants.

We have produced and begun to characterize a mutant of Chinese hamster ovary cells (CHO-UV-1) that is hypersensitive to killing by many agents, including alkylating drugs and ultraviolet radiation. UV-1 has wild-type levels of excision repair, dimer exchange, sister chromosome exchange, and O^6 methyl guanine transferase activity, but it is defective in postreplication recovery (PRR). This latter defect and its hypomutability have led us to hypothesize that UV-1 is deficient in error-prone repair which, in wild-type cells, increases mutagenesis. The abnormal phenotype of UV-1 can be corrected by cell-cell hybridization with normal human fibroblasts or by transfection with human genomic DNA. Analysis of reduced hybrids should allow mapping the PRR phenotype to a specific human chromosome(s). Southern blots of several primary transformants using ^{32}P-labeled human genomic

DNA as a probe reveal a small number of bands of human DNA, some of which should represent the gene of interest. The restriction enzyme SmaI, which produces DNA fragments of average size 30 kilobases, does not destroy the transforming activity of digested DNA, indicating that the active gene(s) is of a size suitable for cloning and characterization.

(ERICR contribution #613. Supported by GM34041 and National Institutes of Health grant NIH HD07197.)

THE A_L ASSAY SENSITIVELY DETECTS GENE MUTATIONS

C. Waldren,[1,2] M. Sognier,[1,3] and J. Bedford[3]

[1]Eleanor Roosevelt Institute for Cancer Research
[2]Department of Radiology
University of Colorado Health Sciences Center
School of Medicine
Denver, Colorado 80262

[3]Department of Radiology and Radiation Biology
Colorado State University
Fort Collins, Colorado

Identification of antimutagens and anticarcinogens and illumination of their mechanisms of action will require development of assay systems that sensitively detect the different kinds of mutagenic events that initiate cancer. We have described a method, the A_L system, that provides superior sensitivity for analysis of a spectrum of gene mutations, including single gene mutation, deletion, and chromosome nondisjunction.

The A_L system employs a human Chinese hamster ovary cell hybrid (CHO-A_L) that contains the standard array of CHO chromosomes plus a single human chromosome, number 11. Chromosome 11 provides genetic information required for expression of a set of surface antigens (a_1, a_2, a_3, ...) that map to opposite arms of the chromosome. Mutagen-induced loss of these specific antigen genes is detected by the growth of mutant colonies in the presence of complement plus specific antibodies under conditions that kill wild-type cells. Analysis of the relative fractions of the different patterns of marker loss provides a measure of the kinds of mutations that have taken place. This complement-mediated cytotoxicity assay has permitted analysis of X-ray mutation at a sensitivity 200 times greater than other cell systems. It also detects as a mutagen agents like chrysotile (asbestos) that may escape detection in other systems. The A_L methodology has now been adapted for analysis of promutagens like dimethyl benzanthracene that require metabolic activation. Improvements in the A_L assay using flow cytometry, immunochemistry, and recombinant DNA technology promise greater resolving power and illumination of mechanisms of mutation.

(ERICR contribution #614. Supported by MOD 15-10, CA36447, CA09236, and ES03746.)

CNICIN, A SESQUITERPENE LACTONE DESMUTAGEN ISOLATED FROM SPOTTED KNAPWEED (CENTAUREA MACULOSA LAM.)

Guylyn R. Warren,[1] Rick G. Kelsey,[2] and Rhoda E.R. Craig[3]

[1]Montana State University
Bozeman, Montana

[2]University of Montana
Missoula, Montana

[3]Kalamazoo College
Kalamazoo, Michigan

The sesquiterpene lactone Cnicin was originally isolated and described in Belgium in 1973. It was noted to be a potent antibacterial compound and is suspected to be the phytoallelic chemical responsible for the competitive abilities of spotted knapweed. We compared Cnicin to a number of active antitumor agents of the same chemical class and found it to be more active than any of them on a DNA repair differential inhibition screening assay. Cnicin gives a two-well differential inhibition on a recAuvrA double mutant as compared to a wild-type strain of E. coli. The toxicity of this compound made mutagenesis testing difficult in the standard plate incorporation assay, so we switched to a fluctuation test in microtiter using WP2uvrA- in a trp reversion system. Cnicin is very weakly mutagenic at low concentrations (<1 μg/ml). At higher concentrations, up to 8 μg/ml, Cnicin reduces and finally totally eliminates spontaneous mutations. There is no selection against trp+ mutations. Concomitant dosages of Cnicin and known mutagens NTG and Dexon at levels sufficient to induce 70% positive wells were tested, and 6 μg/ml Cnicin reduced mutagenesis by NTG and Dexon from 70% to 0%. Mutagenesis interactions with ICR191 are presently being tested. Fluctuation tests with Salmonella strain TA102 are in progress to determine the effects of Cnicin on the three classes of revertants identifiable by analogs in that strain.

SELENIUM PREVENTS THE GROWTH STIMULATORY EFFECTS OF CADMIUM ON BENIGN HUMAN PROSTATIC EPITHELIUM

M.M. Webber

Division of Urology
University of Colorado Health Sciences Center
Denver, Colorado 80262

Association of cadmium (Cd) with a variety of animal tumors has been established. Tumors of the muscle and connective tissue arise at the site of injection. Systemic administration of Cd compounds is associated with interstitial cell tumors of the testis. Increased risk of prostate cancer in men, as a result of occupational and environmental exposure to Cd, has been reported. It is not known whether Cd acts as an initiator or as a promoter in these situations. Our preliminary studies show that Cd, at levels as low as 10^{-9} M, stimulates the growth of human prostatic epithelium in vitro. This may be one mechanism by which Cd alters cell behavior. Concentrations higher than 10^{-6} M are toxic. Selenium (Se), an essential trace element for mammals, is being investigated for its role in cancer chemoprevention. It reduces the incidence of spontaneous mammary tumors in susceptible strains of mice and inhibits tumor induction in animals by chemical carcinogens. At concentrations between 10^{-12} M and 10^{-7} M, Se has no stimulatory or inhibitory effects on the growth of benign human prostatic epithelium in vitro. Higher concentrations inhibit growth, and it is cytotoxic at 10^{-5} M level. Of particular significance is the finding that Se, when present at the 10^{-8} M level in culture medium, inhibits the growth-stimulatory effects of Cd on prostatic epithelium. These results suggest that Se may be useful in counteracting the effects of cadmium and may offer possibilities for investigations on the protective effects of selenium in cadmium-related carcinogenesis.

(Supported by Department of Health and Human Services/National Cancer Institute Grant Nos. CA33169 and CA34308.)

ROSTER OF ORGANIZING COMMITTEE, CHAIRMEN, SPEAKERS, AND PARTICIPANTS

Algaier, Joe, University of Kansas, Lawrence, KS 66045
Alldrick, A.J., British Ind. Biological Research, Carshalton, Surrey, England SM5 4DS
Ames, Bruce N., University of California, Berkeley, CA 94720
Amirkhanian, John D., Los Angeles County Museum of Natural History, Los Angeles, CA 90007
Amirmozafari, Nour, University of Kansas, Lawrence, KS 66045
Arriaga-Alba, Myriam, Inst. de Invest. Biomed. UNAM, San Miguel Chap., Mexico, D.F. 11850
Awadalla, Safwat S., Ain Shams University, Cairo, Egypt 706546
Ayres, Deborah J., Philip Morris, Richmond, VA 23234

Bailey, George S., Oregon State University, Corvallis, OR 97331
Baker, Raymond M., Roswell Park Memorial Institute, Buffalo, NY 14263
Balbinder, Elias, University of Colorado, Denver, CO 80262
Bartsch, Helmut, International Agency for Research on Cancer, Lyon Cedex 08, France 69372
Beije, Brita, Wallenberg Laboratory, Stockholm University, Stockholm S-10691, Sweden
Benedict, William F., Children's Hospital, Los Angeles, CA 90027
Bird, Ramjama, Ludwig Institute for Cancer, Toronto, Ontario, Canada M4Y 1M4
Borek, Carmia, College of Physicians and Surgeons, Columbia University, New York, NY 10032
Bradford, Lawrence, University of Kansas, Lawrence, KS 66045
Breidt, Fred, University of Kansas, Lawrence, KS 66045
Brockman, Herman E., Illinois State University, Congerville, IL 61729
Bronzetti, Giorgio, Istituto di Mutagenesi e Differenziamento CNR, 56100 Pisa, Italy
Bruce, W. Robert, Ludwig Institute for Cancer Research, Toronto, Ontario, Canada M4Y 1M4
Budroe, John D., National Center for Toxicological Research, Jefferson, AR 72079
Bueding, Ernest, The Johns Hopkins University, Baltimore, MD 21205.
Burns, Phil, York University, Toronto, Ontario, Canada M3J 1P3

Cairns, John, Harvard University, Boston, MA 02115
Causse, Catherine, L'Oreal, 93600 Aulnay-sous-Bois, Paris, France 75001
Cerutti, Peter A., Swiss Institute for Experimental Cancer Research, Lausanne, Switzerland
Chaveca, Maria T., University of Lisbon, Lisbon, Portugal
Cheng, C.C., University of Kansas Medical Center, Kansas City, KS 66103
Chipman, James Kevin, University of Birmingham, Birmingham, England B15 2TT

Christoffersen, R.E., The Upjohn Company, Kalamazoo, MI 49001
Clarke, Colin H., University of East Anglia, Norwich, England
Cleaver, James E., University of California, San Francisco, CA 94143
Corno, Demetrio, A.I.I.P.A., Lombardia, Italy 20040
Crow, James F., University of Wisconsin, Madison, WI 53706

Demple, Bruce, Harvard University, Cambridge, MA 02138
Dixit, Rakesh, Eppley Cancer Institute, University of Nebraska, Omaha, NE 68105
Drake, Steve, University of Kansas, Lawrence, KS 66045
Dunn, William C., Oak Ridge National Laboratory, Oak Ridge, TN 37831

Ebner, Kurt, University of Kansas Medical Center, Kansas City, KS 66103
Echols, Harrison, University of California, Berkeley, CA 94720
Eisenstark, A., University of Missouri, Columbia, MO 65211
Epler, James L., Oak Ridge National Laboratory, Oak Ridge, TN 37831
Ershadi, Mandana, University of Kansas, Lawrence, KS 66045
Evenchik, Zigmund, Israel Institute for Biological Research, Ness-Ziona, Israel

Fahrig, Rudolf, Fraunhofer-Institut für Toxikologie, D-3000 Hannover 61, Federal Republic of Germany
Faiman, Morris, University of Kansas, Lawrence, KS 66045
Fix, Douglas, York University, Toronto, Ontario, Canada M3J 1P3
Foote, Robert S., University of Tennessee, Oak Ridge National Laboratory, Oak Ridge, TN 37830
Frederico, Lisa, Duke University, Durham, NC 27706
Friedberg, Errol C., Stanford University Medical Center, Stanford, CA 94305
Friesen, Benjamin S., University of Kansas, Lawrence, KS 66045
Furukawa, Hideyuki, Meijo University, Tempaku-ku, 468 Nagoya, Japan

Ghosh, Anil K., University of Kansas, Lawrence, KS 66045
Glickman, Barry, York University, Toronto, Ontario, Canada M3J 1P3
Gollapudi, Sitaraghav, University of Kansas, Lawrence, KS 66045
Grace, Allison, University of Kansas, Lawrence, KS 66045
Graham, Harold N., Thomas J. Lipton, Inc., Englewood Cliffs, NJ 07632
Gregoire, Kalopissis, Ste. L'Oreal-Centre Eugene Schueler, Clichy, France 92117
Griffiths, T. Daniel, Northern Illinois University, De Kalb, IL 60115
Grigg, Geoffrey W., Division of Molecular Biology, CSIRO, Sidney, NSW, Australia 2113
Gronke, Robert, University of Kansas, Lawrence, KS 66045
Gross, Clark L., U.S. Army Medical Research Institute of Chemical Defense, Aberdeen Proving Ground, MD 21010
Gupta, Ranjan, Illinois State University, Normal, IL 61761

Hafner, Louise, LaTrobe University, Ivanhoe, Bundoora, Victoria, Australia 3079
Hallagin, Janet, University of Kansas, Lawrence, KS 66045
Hanawalt, Philip C., Stanford University, Stanford, CA 94305
Hara, Yukihiko, Mitsui Norin Company Ltd., Fujieda City 426, Japan
Hardin, Steven E., University of Kansas, Lawrence, KS 66045
Harris, Adrian L., Royal Victoria Infirmary, University of Newcastle upon Tyne, Newcastle upon Tyne, United Kingdom NE14LP
Hartman, Philip, The Johns Hopkins University, Baltimore, MD 21218
Haynes, Robert H., York University, Toronto, Ontario, Canada M3J 1P3

Hentosh, Patricia, Eleanor Roosevelt Institute for Cancer Research, Denver, CO 80262
Hersh, Robert, University of Kansas, Lawrence, KS 66045
Hickson, Ian David, Royal Victoria Infirmary, University of Newcastle upon Tyne, Newcastle upon Tyne, United Kingdom NE14LP
Higuchi, Takeru, University of Kansas, Lawrence, KS 66045
Hirano, Koichi, Frederick Cancer Research Center, National Cancer Institute, NIH, Frederick, MD 21701
Hoekstra, Merl, Loyola University, Maywood, IL 60153
Hollaender, Alexander, Council for Research Planning in Biological Sciences, Inc., Washington, D.C. 20036-2077
Honeyman, Allen, University of Kansas, Lawrence, KS 66045
Hudson, Michael C., University of Kansas, Lawrence, KS 66045
Hughes, Thomas J., Research Triangle Institute, Research Triangle Park, NC 27709

Inoue, Tadashi, National Institute of Genetics, Mishima, 411 Shizuoka-ken, Japan

Jaberaboansari, Armin, University of Kansas, Lawrence, KS 66045
Jae-ho, Kim, University of Kansas, Lawrence, KS 66045
Jagannath, D.R., Litton Bionetics, Inc., Kensington, MD 20895
Jensen, David, Fels Research Institute, Temple Medical School, Philadelphia, PA 19140
Jiricny, Josef, Friedrich Miescher Institut, CH-4002, Basel, Switzerland
Joenje, Hans, Free University, Amsterdam, The Netherlands
Johnson, Robert C., Medical University of South Carolina, Charleston, SC 29425
Jung, Mira, University of Kansas, Lawrence, KS 66045

Kada, Tsuneo, National Institute of Genetics, Mishima, 411 Shizuoka-ken, Japan
Kaden, Debra A., Dana Farber Cancer Institute, Boston, MA 02115
Kakinuma, Kataumi, Tokyo Institute of Technology, Yokohama, Kanagawa 227, Japan
Katz, D. Sue, Oak Ridge National Laboratory, Oak Ridge, TN 37831
Kensler, Thomas W., The Johns Hopkins University, Baltimore, MD 21205
Ketterer, Brian, Courtald Institute of Biochemistry, The Middlesex Hospital, London, England W1P 7PN
Kitos, Paul A., University of Kansas, Lawrence, KS 66045
Kitos, Theresa, University of Kansas, Lawrence, KS 66045
Kornhauser, Andrija, Food and Drug Administration, Washington, D.C. 20204
Krishna, Gopala, National Institute for Occupational Safety and Health, Morgantown, WV 26505
Kuo, Simon, University of Kansas, Lawrence, KS 66045
Kuroda, Yukiaki, National Institute of Genetics, Mishima, 411 Shizuoka-ken, Japan
Kuroda, Koichi, Osaka City Institute of Public Health and Environmental Sciences, 543 Osaka, Japan

Lambert, Michael E., Cold Spring Harbor Laboratory, Cold Spring Harbor, NY 11724
LaRossa, Robert A., E.I. duPont de Nemours and Company, Wilmington, DE 19898
Liehr, Joachim G., University of Texas Medical School, Houston, TX 77225
Lockhart, Michael L., N.E. Missouri State University, Kirksville, MO 63501
Loeb, Lawrence, University of Washington School of Medicine, Seattle, WA 98195

Long, Marilyn, University of Kansas, Lawrence, KS 66045
Loprieno, Nicola, Universita di Pisa, 56100 Pisa, Italy
Love, Paul, University of Rochester School of Medicine, Rochester, NY 14642

MacPhee, Donald G., LaTrobe University, Bundoora, Victoria, Australia 3083
Marshall, Milton V., Southwest Foundation for Biomedical Research, San Antonio, TX 78284
Massoud, Aly, Ain Shams Faculty of Medicine, Abbassia, Cairo, Egypt
Matney, Tom, University of Texas Health Science Center, Houston, TX 77225
Matthews, Edwin J., Hazelton Biotechnology Corporation, Kensington, MD 20895
Miehl-Lester, Rosemarie, University of Rochester, Rochester, NY 14642
Miller, Daniel G., Strang Clinic, Preventive Medicine Institute, New York, NY 10016
Mirvish, Sidney, Eppley Cancer Institute, University of Nebraska, Omaha, NE 68105
Mitscher, Les, University of Kansas, Lawrence, KS 66045
Mohn, Georges, State University of Leiden, 2300 Leiden, The Netherlands
Monteione, Beth A., University of Miami School of Medicine, Miami, FL 33101
Morgenstern, Kurt, University of Kansas, Lawrence, KS 66045
Morita, Kazuyoshi, Kanebo, Ltd., Odawara-shi, Kanagawa-ken, Japan
Morris, Debra L., The University of Texas Medical Branch, Galveston, TX 77550
Muller, Ewald W., Technical University, D-1600 Darmstadt, Federal Republic of Germany
Murray, Joseph S., University of Kansas, Lawrence, KS 66045
Myhr, Brian, Hazelton Biotechnology Corporation, Kensington, MD 20895

Nagao, Minako, National Cancer Center for Research Institute, Tokyo 104, Japan
Nairn, Rodney S., University of Texas System Cancer Center, Smithville, TX 78957
Nakamura, Yoshiyuki, Shizuoka College of Pharmacology, 422 Shizuoka-ken, Japan
Namiki, Mitsuo, Nagoya University, 464 Nagoya, Japan
Nestmann, Earle, Environmental Health Centre, Tunney's Pasture, Ottawa, Ontario, Canada K1A 0L2
Nishioka, Hajime, Doshisha University, 602 Kyoto, Japan
Nunoshiba, Tatsuo, Doshisha University, 602 Kyoto, Japan

Ohta, Toshihiro, Institute of Environmental Toxicology, 187 Kodaira, Tokyo, Japan
Ong, Tong-man, National Institute for Occupational Safety and Health, Morgantown, WV 26505
Osawa, Toshihiko, Chicusa, 464 Nagoya, Japan
Oskouian, Babak, University of Kansas, Lawrence, KS 66045

Pangier, Mlle., L'Oreal, 92117 Clichy, Paris, France
Patel, Rajesh D., University of Kansas, Lawrence, KS 66045
Pellicer, Angel, New York University Medical Center, New York, NY 10016
Perticone, Paolo, Centro di Genetica Evoluzionistica del CNR, Università "La Sapienza," 00185 Roma, Italy
Picking, William, University of Kansas, Lawrence, KS 66045
Podger, Denis M., Division of Molecular Biology, CSIRO, Sidney, NSW, Australia 2113

Poth, Albrecht, Technical University, D-6100 Darmstadt, Germany
Pryor, William A., Louisiana State University, Baton Rouge, LA 70803

Raeissi, Shami, University of Kansas, Lawrence, KS 66045
Ramel, Claes, Wallenberg Laboratory, University of Stockholm, S-10691 Stockholm, Sweden
Ramsay-Shaw, Barbara, Duke University, Durham, NC 27706
Rayford, Walter, University of Kansas, Lawrence, KS 66045
Recio, Leslie, Oak Ridge National Laboratory, Oak Ridge, TN 37830
Regan, James D., Martin Marietta Energy Systems, Oak Ridge, TN 37831
Ricordy, Ruggero, Centro di Genetica Evoluzionistica del CNR, Università "La Sapienza," 00185 Roma, Italy
Robins, Edward, Illinois State University, Normal, IL 61761
Rosin, Miriam, British Columbia Cancer Research Centre, Vancouver, British Columbia, Canada V5Z 1L3

Sandhu, Shahbeg S., U.S. Environmental Protection Agency, Research Triangle Park, NC 27709
Sanner, Tore, Institute for Cancer Research, N-0310 Oslo 3, Norway
Sato, Takahiko, Gifu Pharmaceutical University, Gifu City, 502 Gifu, Japan
Sekiguchi, Mutsuo, Kyushu University, 812 Fukuoka, Japan
Selby, Christopher, University of Kentucky Medical Center, Lexington, KY 40536
Shahin, Majdi M., L'Oreal, 93600 Aulnay/Bois, Paris, France
Shane, Barbara S., Louisiana State University, Baton Rouge, LA 70803
Shankel, Carol, Helen Foresman Spencer Museum of Art, University of Kansas, Lawrence, KS 66045
Shankel, Delbert M., University of Kansas, Lawrence, KS 66045
Shimoi, Kayoko, Shizuoka College of Pharmacology, Shizuoka-shi, 422 Shizuoka-ken, Japan
Shirasu, Yasuhiko, Institute of Environmental Toxicology, Kodaira, 187 Tokyo, Japan
Simic, Michael G., National Bureau of Standards, Gaithersburg, MD 20899
Singer-Ansher, Sherry, The Johns Hopkins University, Baltimore, MD 21205
Smith, C.G., Revlon Health Care Group, Tuckahoe, NY 10707
Smith, Kendric C., Stanford University School of Medicine, Stanford, CA 94305
Snyder, Brad, University of Kansas, Lawrence, KS 66045
Snyder, Ronald D., Stauffer Chemical Company, Farmington, CT 06032
Sobels, Fritz H., University of Leiden, Leiden, The Netherlands
Sonnenblick, B.P., Rutgers University-Newark, Millburn, NJ 07041
Sopher, Jonathan A., San Francisco State University, San Francisco, CA 94121
Stewart, George, University of Kansas, Lawrence, KS 66045
Stich, Hans F., British Columbia Cancer Research Centre, Vancouver, B.C., Canada
Stone, Kent, University of Kansas, Lawrence, KS 66045
Strobel, Henry W., University of Texas Health Science Center, Houston, TX 77225
Su, Michael, Duke University Medical Center, Durham, NC 27710

Tan, Wai-Yuan, Memphis State University, Memphis, TN 38152
Tennant, Kevin D., University of Kansas, Lawrence, KS 66045
Thomas, Jr., Charles A., Helicon Foundation, San Diego, CA 92109
Thomas, Susan M., Princeton University, Princeton, NJ 08544
Thompson, L.H., Lawrence Livermore National Laboratory, Livermore, CA 94550
Tomita, Isao, Shizuoka College of Pharmacy, 111 Mishima, Japan

Trosko, James E., Michigan State University, East Lansing, MI 48824

Voepel, Kai, University of Kansas, Lawrence, KS 66045
von Borstel, R.C., University of Alberta, Edmonton, Alberta, Canada T6G 2E9
Wagner, Trish, University of Kansas, Lawrence, KS 66045
Waitley, Jennifer D., University of Kansas, Lawrence, KS 66045
Wakabayashi, Kazuhiko, University of Yamanashi Medical School, 409 38 Yamanashi, Japan
Waldren, Charles, Eleanor Roosevelt Institute for Cancer Research, Denver, CO 80262
Walker, G.C., Massachusetts Institute of Technology, Cambridge, MA 02139
Wall, Monroe E., Research Triangle Institute, Research Triangle Park, NC 27709
Wallen, C. Anne, University of Kansas, Lawrence, KS 66045
Wang, Hong, University of Kansas, Lawrence, KS 66045
Wargovich, Michael J., M.D. Anderson Hospital, Houston, TX 77030
Waritz, Richard S., Hercules, Inc., Wilmington, DE 19894
Warren, Guylyn R., Montana State University, Bozeman, MT 59717
Wattenberg, Lee, University of Minnesota, Minneapolis, MN 55455
Webber, Mukta M., University of Colorado Health Sciences Center, Denver, CO 80262
Weisburger, Elizabeth, National Cancer Institute, NIH, Bethesda, MD 20205
Wheeler, Kenneth, University of Kansas, Lawrence, KS 66045
Wilson, Claire M., Council for Research Planning in Biological Sciences, Inc., Washington, D.C. 20036-2077
Wuenscher, Michael, University of Kansas, Lawrence, KS 66045

Yasbin, Ronald E., University of Rochester School of Medicine, Rochester, NY 14642
Yatagai, F., York University, Toronto, Ontario, Canada M3J 1P3
Yoo, Young S., Osaka City University, 545 Osaka, Japan
Yue-Wei, Lee, Research Triangle Institute, Research Triangle Park, NC 27709

Zennie, Thomas M., Purdue University, West Lafayette, IN 47967

INDEX